STUDIES OF MIND AND BRAIN

BOSTON STUDIES IN THE PHILOSOPHY OF SCIENCE

EDITED BY ROBERT S. COHEN AND MARX W. WARTOFSKY

VOLUME 70

STEPHEN GROSSBERG
Department of Mathematics, Boston University

STUDIES OF MIND AND BRAIN

Neural Principles of Learning, Perception, Development, Cognition, and Motor Control

D. REIDEL PUBLISHING COMPANY
DORDRECHT : HOLLAND / BOSTON : U.S.A.
LONDON : ENGLAND

Library of Congress Cataloging in Publication Data

Grossberg, Stephen, 1939–
Studies of mind and brain.

(Boston studies in the philosophy of science ; v. 70)
Bibliography: p.
Includes index.
1. Learning–Physiological aspects–Collected works. 2. Mind and body–Collected works. 3. Neural circuitry–Adaptation–Collected works. 4. Memory–Collected works. 5. Neuropsychology–Collected works. I. Title. II. Series.
Q174.B67 vol. 70 [QP408] 501s [153] 81-23453
ISBN 90-277-1359-6 AACR2
ISBN 90-277-1360-X (pbk.)

Published by D. Reidel Publishing Company,
P.O. Box 17, 3300 AA Dordrecht, Holland.

Sold and distributed in the U.S.A. and Canada
by Kluwer Boston Inc.,
190 Old Derby Street, Hingham, MA 02043, U.S.A.

In all other countries, sold and distributed
by Kluwer Academic Publishers Group,
P.O. Box 322, 3300 AH Dordrecht, Holland.

D. Reidel Publishing Company is a member of the Kluwer Group.

Printed in The Netherlands

To Gail,

My Parents,

and My Friends.

TABLE OF CONTENTS

EDITORIAL PREFACE

Throughout the history of philosophy, the project of a naturalistic epistemology - of a theory of knowledge based upon a scientific account of the natural processes of perception and cognition, and of learning - occupied such major thinkers as Aristotle, Descartes, Hume, Reid, Peirce, and recently philosophers and scientists from Helmholtz and Mach to Piaget, Popper and Gibson. The question of how knowledge is acquired is two-sided. On the one hand, there is the epistemological questions *par excellence*: what is truth? by what criteria, or under what conditions, are cognitive claims warranted? On the other hand, there is the question of how the human organism, with its structure of sense perception, language and thought, can acquire veridical knowledge of this world?

With the advent of evolutionary theory in biology, human perceptual and cognitive activity came to be seen in its relation to the more general acquisitions of animal learning or animal intelligence, from which it was believed to have evolved. Attention to the comparative anatomy and physiology of the nervous systems of different species focussed on both the gross structure of behavior as an interaction between organism and environment, and on the fine structure of the neural response subtleties of the sense modalities, and on the cross-modal and higher integrative functions of the brain. In the modern period, naturalistic theories of knowledge therefore have been framed in terms of both biological and psychological description, and have aspired to mathematical formulation in the image of the natural sciences.

Stephen Grossberg's studies, gathered in this volume, lie at the intersection of psychology, neurophysiology, and mathematics. The problem he sets for himself, however, is deeply philosophical and methodological: is a mathematical model of a dynamic, evolving, adaptive system possible? Can such a mathematical model adequately account for such psychological phenomena as arousal, attention, memory, or more generally learning, perception, cognition? Grossberg approaches this not as a formal problem but as a concrete research task. He posits the two major constraints: neural anatomy and function of the brain, and operations in real time. Given these spatial or topological, and temporal constraints, and basing his analysis on the mass of experimental data from current research in psychology and physiology, Grossberg proposes and develops a non-linear mathematics as a model for specific functions of mind and brain. He finds the classic approach to the mathematical modelling of mind and brain systematically inadequate. This inadequacy, he holds, arises from the attempt to describe adaptive systems in the mathematical language of a physics developed to describe "stationary", i.e. non-adaptive and non-evolving systems. In place of this linear mathematics, Grossberg develops his non-linear approach. His method is at once imaginative, rigorous, and philosophically significant: it is the thought experiment. It is here that the richness of his interdisciplinary mastery, and the power of his methods, constructions and proofs, reveal themselves. The method is what C. S. Peirce characterized as the method of abduction, or of hypothetical inference in theory construction: given the output of the system as a psychological phenomenon (e.g.learning, perception, cognition) and interpreting such activities in an evolutionary context, as adaptive behavior with respect to complex and changing patterns of the environment, how can the known structures and properties of neural networks account for the known behavior or features of neural and psychological activity given by the experimental data?

Thus Grossberg deals with such general problems as "how does the brain build a cognitive code?", and such specific ones as, "how does an on-center off-surround anatomy of networks of nerve cells lead to such characteristics of the neural processing as contour enhancement in vision or short-term memory?"

Grossberg's papers in this volume seem to us to make a major contribution to the theoretical formulation of problems in the study of mind and brain, and to their mathematical and empirical solution.

Boston University
Center for the Philosophy
and History of Science
February 1982

ROBERT S. COHEN
MARX W. WARTOFSKY

ACKNOWLEDGEMENTS

The author, the editor, and the publisher are grateful to the following persons and institutions for permission to reprint the papers included in this volume:

"How Does a Brain Build a Cognitive Code?". (First published in *Psychological Review* **87** (1980), 1–51.)

"Some Physiological and Biochemical Consequences of Psychological Postulates". (First published in *Proceedings of the National Academy of Sciences* **60** (1968), 758–765.)

"Classical and Instrumental Learning by Neural Networks". (First published in *Progress in Theoretical Biology*, Vol. 3, Academic Press, New York, 1974, pp. 51–141.)

"Pattern Learning by Functional-Differential Neural Networks with Arbitrary Path Weights". (First published in K. Schmitt (ed.), *Delay and Functional Differential Equations and their Applications*, Academic Press, 1972, pp. 121–160.)

"A Neural Theory of Punishment and Avoidance. II: Quantitative Theory". (First published in *Mathematical Biosciences* **15** (1972), 253–285.)

"A Neural Model of Attention, Reinforcement and Discrimination Learning". (First published in *International Review of Neurobiology* **18** (1975), 263–327.)

"Neural Expectation: Cerebellar and Retinal Analogs of Cells Fired by Learnable or Unlearned Pattern Classes". (First published in *Kybernetik* **10** (1972), 49–57.)

"Contour Enhancement, Short Term Memory, and Constancies in Reverberating Neural Networks". (First published in *Studies in Applied Mathematics* **LII** (1973), 213–257.)

"Biological Competition: Decision Rules, Pattern Formation, and Oscillations". (First published in *Proceedings of the National Academy of Sciences* **77** (1980), 2338–2342.)

"Competition, Decision, and Consensus". (First published in *Journal of Mathematical Analysis and Applications* **66** (1978), 470–493.)

"Behavioral Contrast in Short Term Memory: Serial Binary Memory Models or Parallel Continuous Memory Models?". (First published in *Journal of Mathematical Psychology* **17** (1978), 199–219.)

"Adaptive Pattern Classification and Universal Recoding. I: Parallel Development and Coding of Neural Feature Detectors". (First published in *Biological Cybernetics* 23 (1976), 121–134.)

"A Theory of Human Memory: Self-Organization and Performance of Sensory-Motor Codes, Maps, and Plans". (First published in *Progress in Theoretical Biology*, Vol. 5, Academic Press, 1978, pp. 233–374.)

INTRODUCTION

How is psychology different from physics? What new philosophical and scientific ideas will explicate this difference? Why were the inspiring interdisciplinary successes of Helmholtz, Maxwell, and Mach a century ago followed by a divergence of psychological and physical theory rather than a synthesis? Why has physics rapidly deepened and broadened its theoretical understanding of the world during this century, while psychology has spawned controversy after controversy, as well as dark antitheoretical prejudices?

My scientific work on problems related to mind and brain began in 1958 when I was an undergraduate, too young and enthusiastic to know about, let alone to worry about, these issues. After twenty years of scientific inquiry, answers are emerging which clarify some of the philosophical and scientific questions as well as the sociological ones. The answers suggest the following observations.

The difference between psychology and physics centers in the words evolution and self-organization. Classical physical theory focusses on a stationary world and the transitions between known physical states. Studies of mind and brain focus on a nonstationary world in which new organismic states are continually being synthesized to form a better adaptive relationship with the environment. These new states can thereupon be maintained in a stable fashion to form a substrate for the synthesis of yet more complex states in a continuing evolutionary progression. Perhaps no better example of this evolutionary process exists than language learning, which is one of the defining characteristics of human civilization.

Whereas physics has gradually fashioned a measurement theory for a stationary world, psychology needs to discover an evolutionary measurement theory, or universal developmental code. Whereas physics has been well served by linear mathematics, the evolutionary psychological processes (development, learning, perception, cognition) depend on nonlinear mathematics. Since the time of Helmholtz, Maxwell, and Mach, nineteenth century linear mathematics has stood ready to express and analyse the intuitive insights of physicists interested in electromagnetic theory, relativity, and quantum theory. Students of mind cannot turn to a well-developed body of appropriate mathematics with which to express their deepest intuitions. New nonlinear mathematics must be found that is tailored to these ideas.

Scientific revolutions wherein both physical intuitions and mathematical concepts need to be developed side-by-side are especially complex and confusing, but they also offer special intellectual rewards. In the present instance, understanding self-organizing systems is a necessary step towards understanding life itself, both in its individual and collective forms.

Brain studies play a central role in this pursuit for more than the egocentric reason that brains are the crucibles of all human experience. The brain is a universal measurement device acting on the quantum level. Data from all of our senses, – even a few light quanta! – are synthesized by our minds into a common dynamical coin that supports a unitary experience, rather than a series of dislocated experiential fragments. This universality property is the scientific reason, I believe, that brain studies are starting to play a role as central to evolutionary studies as black body radiation played in the development of quantum theory. This universality property clarifies the usefulness of brain theory laws towards explaining a growing body of data about living systems other than brains.

We find ourselves today in a paradoxical and disturbing situation. After physicists abandoned the study of mind, psychological experimentalists were left with an inappropriate world view for understanding each other's data. Personal experimental replication became a major source of security in an atmosphere of conceptual solipsism. Experimentalists dug into paradigms that were sufficiently narrow to maintain the replication criterion. Experimental approaches to mind hereby shattered into a heap of mutually suspicious fiefdoms, and mind theorists became persona non grata. This tendency has been exascerbated by short-sighted governmental policies that deny adequate funding of both the experimental body and the theoretical mind of our discipline. The same governmental policies encourage the search for easy and fast scientific fame. The nature of the crisis and the opportunity facing the brain sciences suggests that a long-range dialog between data and theory should be fostered instead.

Such a dialog plays a central role in the progress of my scientific work. My method of studying adaptive systems starts by identifying a fundamental environmental constraint, or problem, to which a species must adapt in order to survive. The solution to this problem takes the form of a principle of behavioral organization. The behavioral principle is translated into its minimal realization as a mathematical law. Minimality plays the role of an Occam's razor, or a principle of atrophy due to disuse, in the theory. I shall soon say how the theory overcomes the possibility that the prior evolutionary history of a system prevents the minimal solution from occurring. These mathematical laws have always possessed a vivid interpretation as neural networks. The formal mathematical language hereby bridges the gap between macroscopic psychology and microscopic physiology, much as a mathematical bridge exists between thermodynamics and statistical mechanics.

The reader might well ask: "Why have not all behavioral theories generated neurological insights?" An important part of the answer is this: All the principles in my theory describe how the organism solves the environmental problem in real-time. The theory is not merely formal or probabilistic. It attempts to describe the unfolding of individual behavior through time. This demand for individual real-time laws, simple though it seems, places strong constraints on the form that the solution can take.

Having expressed the behavioral principle in mathematical form, one is now confronted by a nonlinear mathematical system, and one must classify the possibilities inherent in this system. Unaided physical intuition has, time and again, proved unequal to this task. This is because the interactive, or collective, properties of the system control its interesting behavioral properties. The human mind does not easily grasp nonlinear interactions between billions of cells without mathematical tools. A rigorous mathematical method is needed to reveal the implications of the behavioral principle. Among the most comforting and rewarding facts of my life has been that mathematical methods could be invented for the understanding of behavioral principles. These mathematical methods effect a great conceptual simplification by structuring and predicting a large body of complex psychophysiological data as manifestations of a simple behavioral principle. If nothing else, this procedure confronts us with unexpected consequences of our present empirical beliefs, and provides a rigorous and transparent conceptual superstructure with whose aid new concepts can more effectively be fashioned.

It would be hard for me to overemphasize the importance of mathematics in these conceptual advances, although I was myself at first unsure of the need for a rigorous attack, as opposed to an intuitive attack. On many occasions, mathematical work has revealed a totally unexpected property, moreover a property so fundamental that it forced a whole series of new intuitive insights. On other occasions, by being able to recognize a general principle at work in several ostensibly unrelated bodies of data, I could regroup the data in terms of underlying principles, rather than in terms of experimental techniques. Each experimental technique can probe only certain aspects of a principle, but by pooling the results from several techniques that are used in seemingly distinct, but mechanistically related, situations, one can understand the underlying mechanisms much better than one could have by relying only on the techniques applicable in one situation.

The use of thought experiments to derive adaptive behavioral principles from environmental pressures, and the reorganization of data in terms of principles rather than experimental procedures, provide a powerful theoretical method for understanding brain and behavior. This method can detect information that eludes experimental techniques for several reasons: It shows how many system components work together; it compresses into a unified description environmental pressures that act over long, or at least nonsimultaneous times; and most importantly, it explicates design constraints that are needed to adapt in a real-time setting.

The mathematical classification theory approaches the question of minimality by admitting that several principles can simultaneously constrain the adaptive design of a given neural structure. The classification theory expresses its ambivalence towards minimality by suggesting species-specific variations on the same organizational theme which have adapted to principles other than the one under study.

Another important task of a classification theory is to clarify what a behavioral principle cannot achieve. In every case, a sharper understanding of a principle's limitations has suggested which other principles, which solve different environmental problems, are also at work in a given situation. Then the theoretical cycle begins again, and leads us in an evolutionary progression to a small set of adaptive principles and mechanisms capable of organizing and predicting a large variety of psychological and physiological data.

As I mentioned above, the collective or interactive properties of the mathematical laws subserve the adaptive behavioral properties that solve these environmental problems. In this sense, my theory is a 'field' theory which attempts to discover the conceptual level, and the functional transformations acting on this level, that drive particular aspects of the adaptive or evolutionary process. The evolutionary method also 'embeds' the properties of one principle into the properties of several principles acting together. For these reasons, the name *embedding field theory* still seems to be a convenient rubric for the method after the twenty-three years since its inception.

The ensuing papers are loosely grouped according to organizational principles and publication dates. The prefaces that introduce each paper sketch some of the issues, whether about nonequilibrium physical theory, language learning, mental illness, epistemology, or new engineering horizons, that in my mind stand above the scientific results as signposts for further scientific work and philosophical inquiry. The papers in this volume were published between 1968 and 1980. I spent most of the decade between the theory's inception and the first appearance of these articles acquiring the interdisciplinary skills that I knew would be needed. The foundations of the theory were laid while I was an undergraduate at Dartmouth College from 1957 to 1961. The theory continued to expand while I pursued graduate studies at Stanford University until 1964. Then I transferred to the Rockefeller University to write my Ph.D. thesis on this subject. A long monograph marked the first stage of my thesis writing. This experience was torrential and liberating after six years of silent but rapid accumulation of results. My Rockefeller professors generously funded the distribution of this 1964 monograph to one hundred leading laboratories in the U.S. and abroad. The monograph contained many of the physical laws and results which later appeared in papers of 1967–1969, but the theory still lacked a precise mathematical method for analyzing the nonlinear dynamics whereby arbitrarily many cells can learn. I found such a mathematical apparatus while I was a student at Rockefeller and it was the subject of my Ph.D. thesis. To my own surprise, this mathematical theory greatly amplified my physical intuition, and carried me through the first complete cycle of the evolutionary method. The prefaces to the papers sketch the several cycles that the theory has undergone since that time. Because of space limitations, some of the articles that developed a given theoretical cycle and forced the next cycle have been omitted. The prefaces indicate how both enclosed and omitted articles contributed to each cycle.

CHAPTER 1

HOW DOES A BRAIN BUILD A COGNITIVE CODE?

PREFACE

This article provides a self-contained introduction to my work from a recent perspective. A thought experiment is offered which analyses how a system as a whole can correct errors of hypothesis testing in a fluctuating environment when none of the system's components, taken in isolation, even knows that an error has occurred. This theme is of general philosophical interest: How can intelligence or knowledge be ascribed to a system as a whole but not to its parts? How can an organism's adaptive mechanisms be stable enough to resist environmental fluctuations which do not alter its behavioral success, but plastic enough to rapidly change in response to environmental demands that do alter its behavioral success? To answer such questions, we must identify the functional level on which a system's behavioral success is defined.

The article suggests that the functional unit of perception and cognition is a state of resonant activity within the system as a whole. Only the resonant state enters consciousness. Only the resonant state can drive adaptive changes in system structure, such as learned changes. The resonant state is therefore called an *adaptive resonance.* Adaptive resonance arises when feedforward (bottom-up) and feedback (top-down) computations within the system are consonant. The feedback computations correspond to our intuitive notion of expectancies. Feedback expectancies help to stabilize the code against errosive effects of irrelevant environmental fluctuations.

The adaptive resonance concept sheds new light on epistemological problems such as those which Heidegger considered. Is the Johnny I see today the same Johnny that I saw yesterday? Usually not. The resonant state constitutes the recognition act, but it also subverts itself by altering its own defining parameters. Tomorrow's resonance need not be the same as today's, yet certain invariant properties of the resonance remain unchanged, such as being able to say: Here's Johnny!

How Does a Brain Build a Cognitive Code?*

This article indicates how competition between afferent data and learned feedback expectancies can stabilize a developing code by buffering committed populations of detectors against continual erosion by new environmental demands. The gating phenomena that result lead to dynamically maintained critical periods, and to attentional phenomena such as overshadowing in the adult. The functional unit of cognitive coding is suggested to be an adaptive resonance, or amplification and prolongation of neural activity, that occurs when afferent data and efferent expectancies reach consensus through a matching process. The resonant state embodies the perceptual event, or attentional focus, and its amplified and sustained activities are capable of driving slow changes of long-term memory. Mismatch between afferent data and efferent expectancies yields a global suppression of activity and triggers a reset of short-term memory, as well as rapid parallel search and hypothesis testing for uncommitted cells. These mechanisms help to explain and predict, as manifestations of the unified theme of stable code development, positive and negative aftereffects, the McCollough effect, spatial frequency adaptation, monocular rivalry, binocular rivalry and hysteresis, pattern completion, and Gestalt switching; analgesia, partial reinforcement acquisition effect, conditioned reinforcers, underaroused versus overaroused depression; the contingent negative variation, P300, and ponto-geniculo-occipital waves; olfactory coding, corticogeniculate feedback, matching of proprioceptive and terminal motor maps, and cerebral dominance. The psychophysiological mechanisms that unify these effects are inherently nonlinear and parallel and are inequivalent to the computer, probabilistic, and linear models currently in use.

How do internal representations of the environment develop through experience? How do these representations achieve an impressive measure of global self-consistency and stability despite the inability of individual nerve cells to discern the behavioral meaning of the representations? How are coding errors corrected, or adaptations to a changing environment effected, if individual nerve cells do not know that these errors or changes have occurred? This article describes how limitations in the types of information available to individual cells can be overcome when the cells act together in suitably designed feedback schemes. The designs that emerge have a natural neural interpretation, and enable us to explain and predict a large variety of psychological and physiological data as manifestations of mechanisms that have evolved

---

* This work was supported in part by the National Science Foundation (NSF MCS 77-02958).

Requests for reprints should be sent to Stephen Grossberg, Department of Mathematics, Boston University, Boston, Massachusetts 02215.

to build stable internal representations of a changing environment. In particular, various phenomena that might appear idiosyncratic or counterintuitive when studied in isolation seem plausible and even inevitable when studied as a part of a design for stable coding.

Some of the themes that will arise in our discussion have a long history in psychology. To achieve an exposition of reasonable length, the article is built around a thought experiment that shows us in simple stages how cells can act together to achieve the stable self-organization of evironmentally sensitive codes. If nothing else, the thought experiment is an efficient expository device for sketching how organizational principles, mechanisms, and data are related from the viewpoint of code development, using a minimum of technical preliminaries. On a deeper level, the thought experiment provides hints for a future theory about the types of developmental events that can generate the neural structures in which the codes are formed. It does this by correlating the types of environmental pressures to which the developmental mechanisms are sensitive with the types of neural structures that have evolved to cope with these pressures. References to previous theories and data have been chosen to clarify the thought experiment, to contrast its results with alternative viewpoints, to highlight areas in which more experimentation can sharpen or disconfirm the theory, or to refer to more complete expositions that should be consulted for a thorough understanding of particular results. The thought experiment and its consequences do not, however, depend on these references, and the reader will surely know many other references that can be used to confront and interpret the thought experiment.

1. A Historical Watershed

Some of the themes that will arise were already adumbrated in the work of Helmholtz during the last half of the 19th century (Boring, 1950; Koenigsberger, 1906). Unfortunately, the conceptual and mathematical tools needed to cast these themes as rigorous science were not available until recently. This fact helped to precipitously terminate the productive interdisciplinary activity between physics and psychology that had existed until Helmholtz's time, as illustrated by the perceptual contributions of Mach and Maxwell (Boring, 1950; L. Campbell & Garnett, 1882; Ratliff, 1965) in addition to those of Helmholtz (1866, 1962); to create a schism between psychology and physics that has persisted to the present day; and to unleash a century of controversy and antitheoretical dogma within psychology that led Hilgard and Bower (1975) to write the following first sentence in their excellent review of *Theories of Learning:* "Psychology seems to be constantly in a state of ferment and change, if not of turmoil and revolution" (p. 2).

One illustrative type of psychological data that Helmholtz studied concerned color perception. Newton had noted that white light at a point in space is composed of light of all visible wavelengths in approximately equal measure. Helmholtz realized, however, that the light we perceive to be white tends to be the average color of a whole scene (Beck, 1972). Thus perception at each point is nonlocal; it is due to a psychological process that averages data from many points to define the perceived color at each point. Moreover this averaging process must be nonlinear, since it is more concerned with relative than absolute light intensities. Unfortunately, most of the mathematical tools that were available to Helmholtz were local and linear.

There is a good evolutionary reason why the light that is perceived to be white tends to be the average color of a scene. We rarely see objects in perfectly white light. Thus our eyes need the ability to average away spurious coloration due to colored light sources, so that we can see the "real" colors of the objects themselves. In other words, we tend to see the "reflectances" of objects, or the relative amounts of light of each wavelength that they reflect, not the total amount of light reaching us from each point. This observation is still a topic of theoretical interest and is the starting point of the modern theory of lightness (Cornsweet, 1970; Grossberg, 1972a; Land 1977).

A more fundamental difficulty faced Helmholtz when he considered the objects of perception. Helmholtz was aware that cognitive factors can dramatically influence our perceptions and that these factors can evolve or be learned through experience. He referred to all such factors as *unconscious inferences*, and developed his belief that a raw sensory datum, or *perzeption*, is modified by previous experience via a learned imaginal increment, or *vorstellung*, before it becomes a true perception, or *anschauung* (Boring, 1950). In more modern terms, sensory data activate a feedback process whereby a learned template, or expectancy, deforms the sensory data until a consensus is reached between what the data "are" and what we "expect" them to be. Only then do we "perceive" anything.

The struggle between raw data and learned expectations also has an evolutionary rationale. If perceptual and cognitive codes are defined by representations that are spread across many cells, with no single cell knowing the behavioral meaning of the code, then some buffering mechanism is needed to prevent previously established codes from being eroded by the flux of experience. It will be shown below how feedback expectancies establish such a buffer.

Unfortunately, Helmholtz was unable to theoretically represent the nonstationary, or evolutionary, process whereby the expectancy is learned, the feedback process whereby it is read out, or the competitive scheme whereby the afferent data and efferent expectancy struggle to achieve consensus. Helmholtz's conceptual and mathematical tools were linear, local, and stationary.

Section 4 begins to illustrate how nonlinear, nonlocal, and nonstationary concepts can be derived as principles of organization for adapting to a fluctuating environment. The presentation is nontechnical, but it will become apparent as we proceed that without a rigorous mathematical theory as a basis, the heuristic summary would have been impossible, since some of the properties that we will need are not intuitively obvious consequences of their underlying principles, and were derived by mathematical analysis. Furthermore, it will emerge that several design principles for adapting to different aspects of the environment operate together in the same structure. One of the facts that we must face about evolutionary systems is that their simple organizational principles can imply extraordinarily subtle properties. Indeed, part of the dilemma that many students of mind now face is not that they do not know enough facts on which to base a theory, but rather they do not know which facts are principles and which are epiphenomena, and how to derive the multitudinous çonsequences that occur when a few principles act together. A rigorous theory is indispensable for drawing such conclusions.

The next two sections summarize some familiar experiments whose properties will reappear from a deeper perspective in the thought experiment. These experiments are included to further review one of the themes that Helmholtz confronted, and to prepare the reader for the results of the thought experiment. The sections can be skipped on a first reading.

2. Overshadowing: A Multicomponent Adult Phenomenon With Developmental Implications

Psychological data are often hard to analyze because many processes are going on simultaneously in a given experiment. This point is illustrated below in a classical conditioning paradigm that will be clarified by the theoretical development. Classical conditioning is considered by many to be the most passive type of learning and to be hopelessly inadequate as a basis for cognitive studies. The overshadowing phenomenon illustrates the fact that even classical conditioning is often only one component of a multicomponent process in which attention, expectation, and other "higher order" feedback processes play an important role (Kamin, 1969; Trabasso & Bower, 1968; Wagner, 1969).

Consider the four experiments depicted in Figure 1. Experiment 1 summarizes the simplest form of classical conditioning. An unconditioned stimulus (UCS), such as shock, elicits an unconditioned response (UCR), such as fear, and autonomic signs of fear. The conditioned stimulus (CS), such as a briefly ringing bell, does not initially elicit fear, but after preceding the UCS by a suitable interval on sufficiently many conditioning trials, the CS does elicit a conditioned response (CR) that closely resembles the UCR. In this way, persistently pairing an indifferent cue with a

I: $CS - UCS$

$CS \rightarrow CR$

II: $(CS_1 + CS_2) - UCS$

$CS_i \rightarrow CR, \; i=1,2$

III: $CS_1 - UCS$

$(CS_1 + CS_2) - UCS$

$CS_2 \not\rightarrow CR$

IV: $CS_1 - UCS_1$

$(CS_1 + CS_2) - UCS_2$

$CS_2 \rightarrow CR_{12}$

Figure 1. Four experiments illustrate overshadowing. (Experiment I summarizes the standard classical conditioning paradigm: conditioned stimulus–unconditioned stimulus [CS–UCS] pairing enables the CS to elicit a conditioned response (CR). Experiment II shows that joint pairing of two CSs with the UCS can enable each CS separately to elicit a CR. Experiment III shows that prior CS_1–UCS pairing can block later conditioning of CS_2 to the CR. Experiment IV shows that CS_2 can be conditioned if its UCS differs from the one used to condition CS_1. The CR that CS_2 elicits depends on the relationship between both UCSs, hence the notation CR_{12}.)

significant cue can impart some of the effects of the significant cue to the indifferent cue.

In Experiment 2, two CSs, CS_1 and CS_2, occur simultaneously before the UCS on a succession of conditioning trials; for example, a ringing bell and a flashing light both precede shock. It is typical in vivo for many cues to occur simultaneously, or in parallel, and the experimental question is, Is each cue separately conditioned to the fear reaction or is just the entire cue combination conditioned? If the cues are equally salient to the organism and are in other ways matched, then the answer is yes. If either cue CS_1 or CS_2 is presented separately after the conditioning trials, then it can elicit the CR.

Experiment 3 modifies Experiment 2 by performing the conditioning part of Experiment 1 on CS_1 before performing Experiment 2 on CS_1 and CS_2. In other words, first condition CS_1 until it can elicit the CR. Then present CS_1 and CS_2 simultaneously on many trials using the same UCS as was used to condition CS_1. Despite the results of Experiment 2, the CS_2 does not elicit the CR if it is presented after conditioning trials. Somehow prior pairing of CS_1 to the CR "blocks" conditioning of CS_2 to the CR.

The meaning of Experiment 3 is clarified by Experiment 4, which is the same as Experiment 3, with one exception. The UCS that follows CS_1 is not the same UCS that follows the stimulus pair CS_1 and CS_2 taken together. Denote the first UCS by UCS_1 and the second UCS by UCS_2. Suppose, for example, that UCS_1 and UCS_2 are different shock levels. Does CS_2 elicit a CR in this situation? The answer is yes if the two shock levels are sufficiently different. If the shock UCS_2 exceeds UCS_1 by a sufficient amount, then CS_2 elicits fear, or a negative reaction. If, however, the shock level UCS_1 exceeds UCS_2 by a sufficient amount, then CS_2 elicits relief, or a positive reaction.

How can the difference between Experiments 3 and 4 be summarized? In Experiment 3, CS_2 is an irrelevant or uninformative cue, since adding it to CS_1 does not change the expected consequence UCS. In Experiment 4, by contrast, CS_2 is informative because it predicts a change in the UCS. If the change is for the worse, then CS_2 eventually elicits a negative reaction (Bloomfield, 1969). If the change is for the better, then CS_2 eventually elicits a positive reaction (Denny, 1970).

Thus many learners are minimal adaptive predictors. If a given set of cues is followed by expected consequences, then all other cues are treated as irrelevant, as is CS_2 in Experiment 3. Each of us can define a given object using different sets of cues without ever realizing that our private sets are different, so long as the situations in which each of us uses the object always yield expected consequences. By contrast, if unexpected consequences occur, as in Experiment 4, then we somehow enlarge the set of relevant cues to include cues that were erroneously disregarded.

Several important qualitative conclusions can be drawn from these remarks. First, what is conditioned depends on our expectations, and these in turn help to regulate the cues to which we pay attention. Second, cues are conditioned, and indeed codes that interrelate these cues are built up, only if we pay attention to these cues because of their potential informativeness. Third, the mismatch between expected consequences and real events occurs only after attention has been focused on certain cues that thereupon generate the expectancy. Somehow this mismatch "feeds backwards in time" to amplify cues that have previously been overshadowed but that must have contained relevant information that we have erroneously ignored. Fourth, whenever we are faced with unexpected consequences, we do not know which cues have erroneously been ignored. The feedback process must be capable of amplifying all of the cues that are still being stored, albeit in a suppressed state. In other words, the feedback process is nonspecific. Finally, the nonspecific feedback process that is elicited by unexpected events competes with the specific consummatory channels that have focused our attention on the wrong set of cues. This competition between specific and nonspecific mechanisms helps us to reorganize our attentional focus until expected consequences are once again achieved.

This brief discussion reveals several basic processes working together in the overshadowing paradigm:

(a) classical conditioning, (b) attention, (c) learned expectancies, (d) matching between expectancies and sensory data, and (e) a nonspecific system that is activated by unexpected or novel events and competes with the specific consummatory system that focuses attention on prescribed cues.

Thus even classical conditioning is not a passive process when it occurs in realistic behavioral situations. Furthermore, its understanding requires the analysis of such teleological concepts as expectancy and attention. Helmholtz's doctrine of unconscious inference is readily called to mind.

Attention is to many individuals a holistic, if not unscientific, concept that does not mesh well with recent technological advances, say in microelectrode recording from individual nerve cells. Perhaps for this reason the fact that attentional variables can significantly influence what codes will be learned seems to have been ignored by some neurophysiologists who study the development of the visual cortex. For example, Stryker and Sherk (1975) were unable to replicate the Blakemore and Cooper (1970) study of visual code development in kittens. In the Blakemore and Cooper study, kittens were raised in a cylindrical chamber whose walls were painted with vertical black and white bars. The visual cortices of the kittens were reported to possess abnormally small numbers of horizontally tuned feature detectors. Hirsch and Spinelli (1970) performed experiments that did replicate in later experiments. In their experiments, the cats wore goggles, one lens with vertical stripes and the other with horizontal stripes. The corresponding visual cortices were reported to possess abnormally small numbers of feature detectors that were tuned to the orthogonal orientation. The entire controversy focused on such technical details as possible sampling errors due to Blakemore and Cooper's method of placing their electrodes. It is obvious, however, that the two experimental paradigms are attentionally inequivalent. Even perfect experimental technique would not necessarily imply similar experimental results.

3. Parallel Processing and the Persistence of Learned Meanings

The fact that classical conditioning, and for that matter any form of code development or learning, cannot be divorced from feedback processes that are related to attention is also made clear by the example illustrated by Figure 2. In Figure 2a, two classical conditioning experiments are depicted, one in which stimulus S_2 is the UCS for response R_2 and S_1 is its CS, and one in which S_1 is the UCS for R_1 and S_2 is its CS. What would happen if each cue S_1 and S_2 is conditioned to its own response R_1 or R_2, respectively, before a classical conditioning experiment occurs in which S_1 and S_2 are alternately scanned? This is the typical situation in real life, when we scan many cues in parallel, or intermittently, and many of these cues already have their own associations. If classical conditioning were a

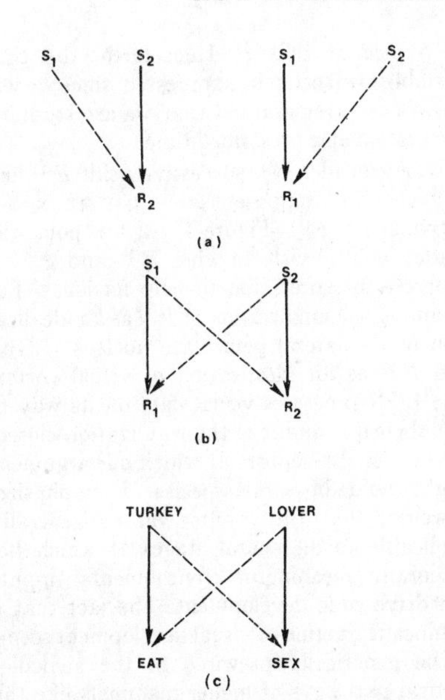

Figure 2. Classical conditioning cannot be a passive feed-forward process during real behavior. (In (a), S_1 acts as a conditioned stimulus (CS) for S_2, whereas S_2 acts as a CS for S_1. In (b), parallel processing of S_1 and S_2, each previously conditioned to responses R_1 and R_2, would yield cross-conditioning. In (c), some of the disastrous consequences of cross-conditioning are illustrated.)

passive feed-forward process, then cross-conditioning from S_1 to R_2 and from S_2 to R_1 would rapidly occur, as in Figure 2b.

However, this is absurd, as the particular example in Figure 2c vividly illustrates. Figure 2c schematizes the situation that would occur due to having a turkey dinner with one's lover. One alternately looks at lover and turkey, with lover associated with sexual responses (among others!) and turkey associated with eating responses. Why do we not come away from dinner wanting to eat our lover and to have sex with turkeys? Somehow the persistence of learned meanings can endure despite the fact that cues that are processed in parallel often generate incompatible responses. This is not always true, however, since if we, for example, consistently use a turkey as a discriminative cue for shock, or even sex, then turkeys might well become associated with fear or sexual arousal. Figure 2 depicts a situation in which the free reorganization of attention, rather than a forced pairing of a CS with a UCS, maintains the learned persistence of meanings. Grossberg (1975) developed a thought experiment in which overcoming the environmentally imposed dilemma of Figure 2 leads to attentional mechanisms that imply the overshadowing phenomena in Figure 1.

Before leaving the subject of overshadowing, we might ask why this adult attentional phenomenon is related to the development of sensory and cognitive codes, even in infants. This article argues that feedback is necessary to stabilize the development of behaviorally meaningful codes in a rich input environment. The feedback processes include attentional mechanisms, and the stabilization of developing codes leads to gating phenomena, or the emergence of critical periods, that are dynamically maintained by the feedback processes.

From this perspective, the structure of an environmentally adaptive tissue is a dynamic scheme whose parameters change very slowly only because of the nature of its maintaining feedback. Death itself is a dramatic example of how seemingly persistent structures can rapidly disintegrate when maintaining feedback is disturbed. When the development of a structure is driven by a particular type of experience, one of the structure's maintaining factors is that variety of experience. A subtle feature of such a developing structure is its ability to selectively amplify those experiences that tend to maintain its structure. Next I

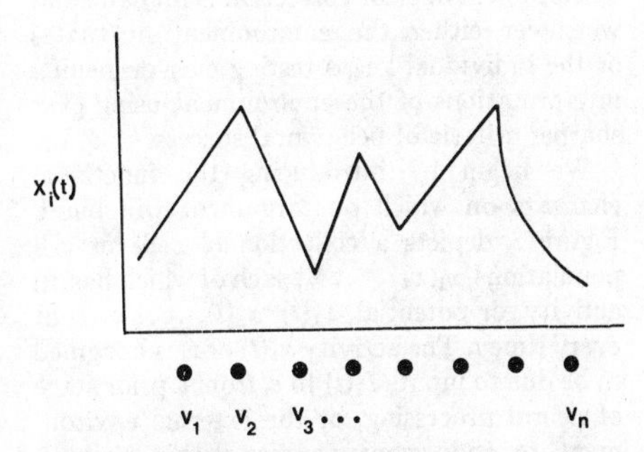

Figure 3. Each cell (or cell population) v_i possess an activity or potential $x_i(t)$, at every time t, $i = 1, 2, \ldots, n$. (The vector $(x_1(t), x_3(t), \ldots, x_n(t))$ of all these activities is a spatial pattern of activity.)

will discuss how feedback expectancies help to accomplish this end.

4. A Thought Experiment: The Need for Learned Feedback Expectancies

We now start to build a framework in which to discuss environmentally driven and behaviorally meaningful code development. Wherever possible, mathematical details will be suppressed, and the *minimal* structure capable of achieving our ends will be defined. This procedure will clarify what mathematical problems have to be solved, what their relationship is to each other, and what types of thematic variations on the minimal structures can be anticipated in different species and different neural locations in the same individual.

The central theoretical theme will be, How can a coding error be corrected if no individual cell knows that one has occurred? The importance of this issue becomes clear when we realize that erroneous cues can accidentally be incorporated into a code when our interactions with the environment are simple and will only become evident when our environmental expectations become more demanding. Even if our code perfectly matched a given environment, we would certainly make errors as the environment itself fluctuates. Furthermore, we never have an absolute criterion of whether our understanding of a fixed environment is faulty, or the environment that we thought we understood is no longer the same. The problem of error correction is fundamental whenever either the environment fluctuates or the individual keeps testing ever-deepening interpretations of the environment using ever sharper criteria of behavioral success.

We begin by introducing the functional elements on which our argument will build. Figure 3 depicts a collection of cells or cell populations, $v_1, v_2, \ldots, v_n$, each of which has an activity, or potential, $x_1(t), x_2(t), \ldots, x_n(t)$ at every time t. The activity $x_i(t)$ or v_i is imagined to be due to inputs $I_i(t)$ to v_i from a prior stage of neural processing, or the external environment, or endogenous sources within v_i itself. At every time t, these activities form a *pattern* $x(t) = (x_1(t), x_2(t), \ldots, x_n(t))$ across the cells $v_1, v_2, \ldots, v_n$, to which we will refer collectively as a *field* of cells F. Henceforth, the time variable t will often be suppressed, since we will always take for granted that we are studying the system at a prescribed time.

Now consider two successive fields $F^{(1)}$ and $F^{(2)}$ of cells. Suppose that a pattern $x^{(1)}$ is active across $F^{(1)}$ (Figure 4). At this point the reader might wish to give $F^{(1)}$ and $F^{(2)}$ a concrete interpretation to help fix ideas. For example, one might think of $F^{(1)}$ as an idealization of the lateral geniculate nucleus (LGN) and $F^{(2)}$ as an idealization of visual cortex. The LGN processes visual data on its way to visual cortex, and it is the way station closest to the visual receptors at which our argument might hold in some species. I emphasize, however, that the results will be generally applicable to all neural stages at which behaviorally meaningful environmental inputs can drive code development. The fact that a significant fraction of visual development seems to be genetically prewired in the geniculocortical pathways of higher mammals like the monkey (Hubel & Wiesel, 1977) will not weaken the general conclusions that we will reach, and in fact various predictions and recent data about LGN, among other structures, will emerge from the analysis.

Suppose that the signal-carrying pathways from $F^{(1)}$ to $F^{(2)}$ act to filter the pattern $x^{(1)}$, and that due to prior developmental experience, this filter "codes" pattern $x^{(1)}$ by eliciting pattern $x^{(2)}$ across $F^{(2)}$. Knowing the detailed structure of this code is unnecessary to make our argument. However, we must be able to show how signal pathways can act as a filter that can be tuned by experience. This is done in Appendix A.

Suppose after the system learns to code $x^{(1)}$ by $x^{(2)}$ that another pattern is presented to $F^{(1)}$ and is erroneously coded at $F^{(2)}$ by $x^{(2)}$. To describe this situation conveniently, I introduce some subscripts. Denote $x^{(1)}$ and $x^{(2)}$ by $x_1^{(1)}$ and $x_1^{(2)}$, respectively, and denote the erroneously coded pattern at $F^{(1)}$ by $x_2^{(1)}$. In Figure 5 we draw the pattern $x_2^{(2)}$ that codes $x_2^{(1)}$ to equal $x_1^{(2)}$. Equality is meant to imply functional equivalence rather than actual identity. We now ask the central question, How can this coding error be corrected if no individual cell knows that an error has occurred?

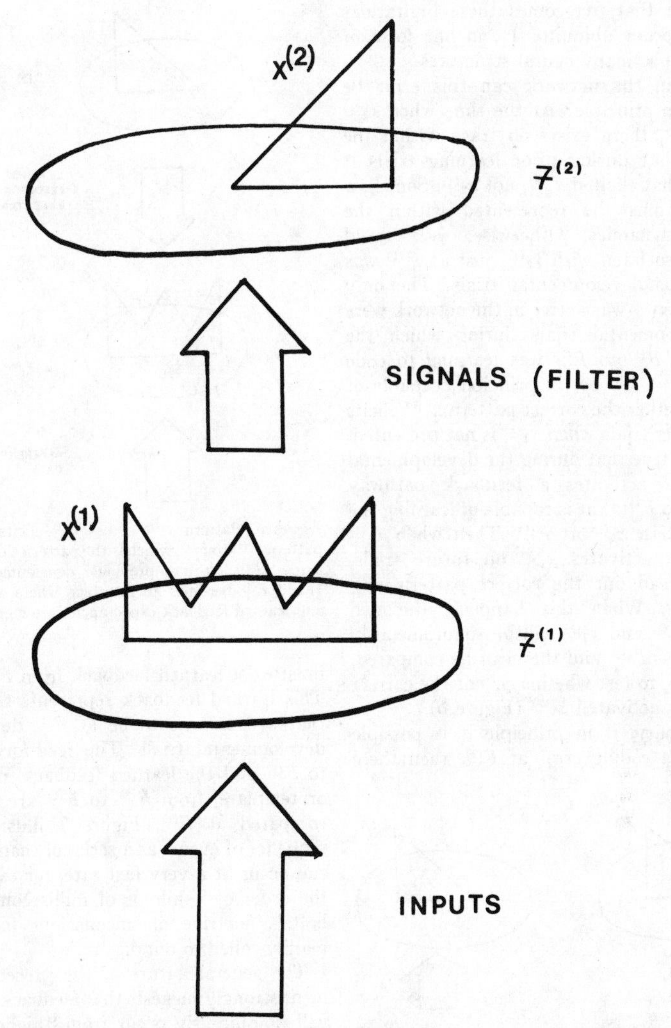

Figure 4. The activity pattern $x^{(1)}$ across $F^{(1)}$ is filtered to elicit a pattern $x^{(2)}$ across $F^{(2)}$.

Our first robust conclusion is now apparent: Whatever the mechanism is that corrects this error, it cannot exist within $F^{(2)}$, since by definition $x_1^{(2)}$ and $x_2^{(2)}$ are functionally equivalent. In principle, $F^{(2)}$ does not have the ability to distinguish the fact that $x_1^{(1)}$, and not $x_2^{(1)}$, should elicit $x_1^{(2)}$, since so far as $F^{(2)}$ knows, $x_1^{(1)}$ *is* active at $F^{(1)}$ rather than $x_2^{(1)}$.

It is important to realize that this argument is independent of coding details. It is based only on the type of information that $F^{(2)}$ cannot, in principle, possess. Much of our argument will be based on similar limitations in the types of information that particular processing stages can, in principle, possess. The robustness of this argument suggests why the design that overcomes these limitations seems to occur ubiquitously, in one form or another, in so many neural structures.

Where in the network can this error be detected in principle? At the time when $x_2^{(1)}$ elicits $x_1^{(2)}$, there exists no trace within the network that during prior learning trials it was $x_1^{(1)}$ that elicited $x_1^{(2)}$, not $x_2^{(1)}$. Somehow this fact must be represented within the network dynamics. Otherwise, $x_1^{(2)}$ could become associated with $x_2^{(1)}$, just as $x_1^{(1)}$ was on previous developmental trials. The only times that $x_1^{(1)}$ was active in the network were the developmental trials during which the filter from $F^{(1)}$ to $F^{(2)}$ was learning to code $x_1^{(1)}$ by $x_1^{(2)}$. To be, in principle, capable of testing whether the correct pattern $x_1^{(1)}$ elicits $x_1^{(2)}$ on later trials when $x_1^{(1)}$ is not presented, it must be true that during the developmental trials, $x_1^{(2)}$ activates a feedback pathway from $F^{(2)}$ to $F^{(1)}$ that is capable of learning the active pattern $x_1^{(1)}$ at $F^{(1)}$. Then when $x_2^{(1)}$ erroneously activates $x_1^{(2)}$ on future trials, $x_1^{(2)}$ can read out the correct pattern $x_1^{(1)}$ across $F^{(1)}$. When this happens, the two patterns $x_1^{(1)}$ and $x_2^{(1)}$ will be simultaneously active across $F^{(1)}$, and they can be compared, or matched, to test whether or not the correct pattern has activated $x_1^{(2)}$ (Figure 6).

In summary, if in principle it is possible to correct a coding error at $F^{(2)}$, then there must exist learned feedback from $F^{(2)}$ to $F^{(1)}$. This learned feedback represents the pattern that $x_1^{(2)}$ *expects* to be at $F^{(1)}$ due to prior developmental trials. The feed-forward data to $F^{(1)}$ and the learned feedback expectancy, or template, from $F^{(2)}$ to $F^{(1)}$ are thereupon compared at $F^{(1)}$. Figure 7 illustrates this sequence of events as a series of snapshots that can occur at a very fast rate, for example, on the order of hundreds of milliseconds. Helmholtz's doctrine of unconscious inference is readily called to mind.

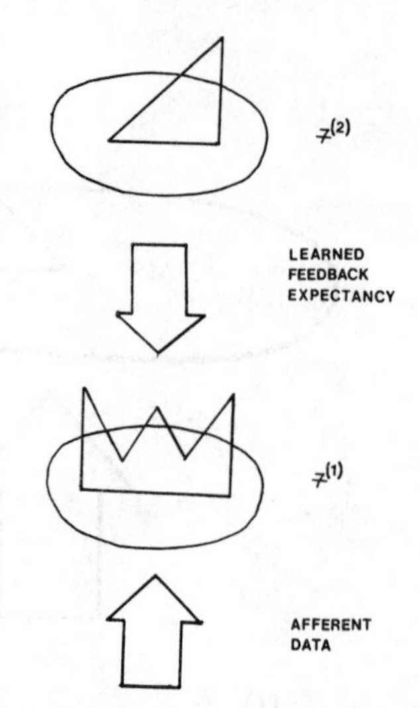

Figure 6. Pattern $x_1^{(2)}$ across $F^{(2)}$ elicits a feedback pattern $x_1^{(1)}$ to $F^{(1)}$, which is the pattern that it sampled across $F^{(1)}$ during previous developmental trials. (Field $F^{(1)}$ becomes an interface where afferent data and learned feedback expectancies are compared.)

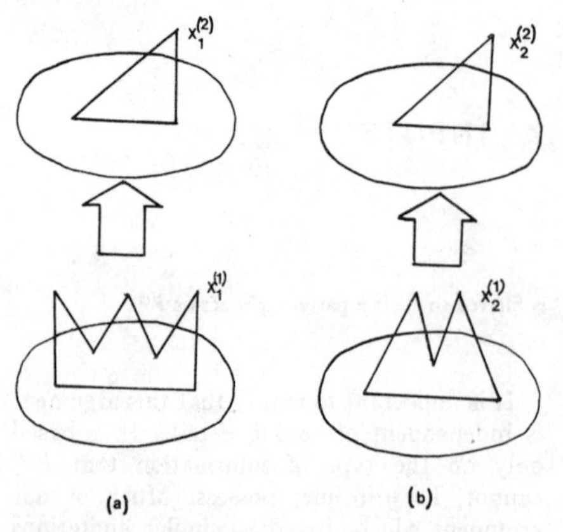

Figure 5. In (a), pattern $x_1^{(1)}$ at $F^{(1)}$ elicits the correct pattern $x_1^{(2)}$ across $F^{(2)}$. In (b), pattern $x_2^{(1)}$ elicits the incorrect pattern $x_2^{(2)}$, which is functionally equivalent to $x_1^{(2)}$ across $F^{(2)}$.

The general nature of the preceding argument strongly suggests that feedback pathways will ubiquitously occur from "higher" neural centers to the relay stations that excite them. In fact, reciprocal thalamocortical connections seem to exist in all thalamo–neocortical systems (Macchi & Rinvik, 1976; Tsumoto, Creutzfeldt, & Legéndy, 1978).

At this point, we also recognize two more design problems for mathematics. The first problem is, How do feedback pathways from $F^{(2)}$ learn a pattern of activity across $F^{(1)}$? (See Appendix B for a summary of this mechanism.)

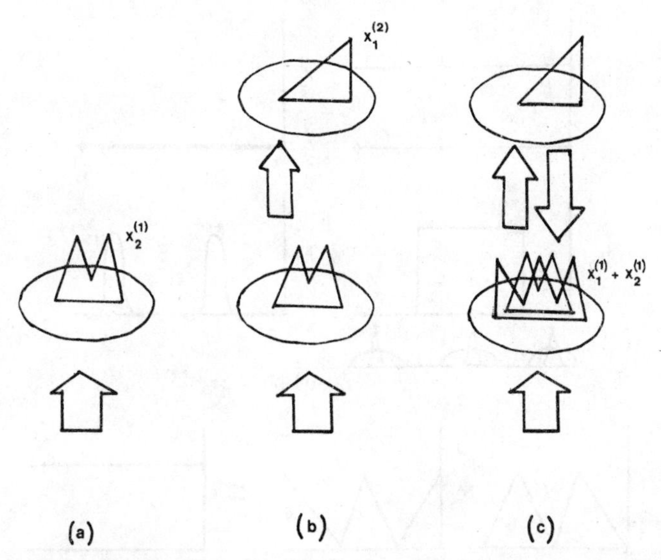

Figure 7. The stages (a), (b), and (c) schematize the rapid sequence of events whereby afferent data is filtered and activates a feedback expectancy that is matched against itself.

5. Noise Suppression, Pattern Matching, and Spatial Frequency Detection

The second design problem that we must face is this: Somehow the mismatch between the patterns $x_1^{(1)}$ and $x_2^{(1)}$ must rapidly shut off activity across $F^{(1)}$. Otherwise, $x_1^{(2)}$ would learn to code $x_2^{(1)}$ much as $x_1^{(2)}$ learned to code $x_1^{(1)}$ on the preceding developmental trials. Pattern $x_1^{(2)}$ must also be rapidly shut off if only to prevent behavioral consequences of $x_1^{(2)}$ from being triggered by further network processing. Moreover, $x_1^{(2)}$ must be shut off in such a fashion that $x_2^{(1)}$ can thereupon be coded by a more suitable pattern across $F^{(2)}$.

The only basis on which these changes can occur is the mismatch of $x_1^{(1)}$ and $x_2^{(1)}$ across $F^{(1)}$. We must therefore ask, How does the mismatch of patterns across a field $F^{(1)}$ of cells inhibit activity across $F^{(1)}$? The mathematical details are summarized in Appendix C. Here, however, it is useful to make the important distinction between mechanisms that develop due to evolutionary pressures and properties that are merely consequences of these mechanisms. One might well worry that the design of a mismatch mechanism is a rather sophisticated evolutionary task. We now indicate that such a mechanism is a consequence of a more basic property, namely noise suppression, and that noise suppression is itself a variation of a basic evolutionary principle. Moreover, other useful properties follow from noise suppression, such as spatial frequency detection and edge enhancement.

The environmental problem out of which the noise suppression property emerges is the *noise-saturation dilemma*. This dilemma has been discussed in detail elsewhere (e.g., Grossberg, 1977, 1978d). The dilemma confronts all noisy cellular systems that process input patterns, as in Figure 3. If the inputs are too small, they can get lost in the noise. If the inputs are amplified to avoid the noisy range, they can saturate all the cells by activating all of their excitable sites, and thereby reduce to zero the cells' sensitivity to differences in the input intensities. Appendix C reviews how competitive interactions among the cells automatically retune their sensitivity to overcome the saturation problem. In a neural context, the competitive interactions are said to be shunting interactions, and they are carried by an on-center off-surround anatomy. The retuning of sensitivity is due to automatic gain control by the inhibitory

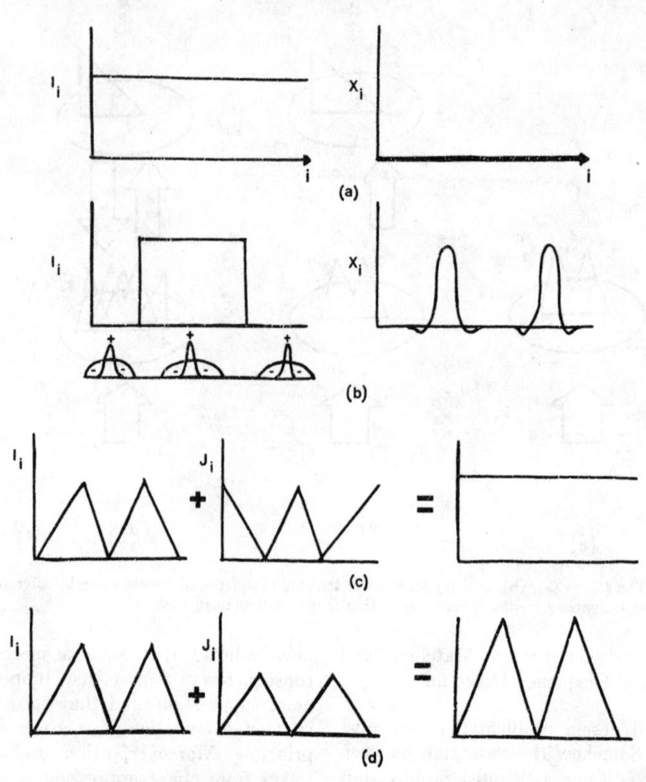

Figure 8. In (a), noise suppression converts a uniform input pattern into a zero activity pattern. In (b), a rectangular input pattern elicits differential activity at its edges because the cells within its interior and beyond its boundary perceive uniform fields. (This is a special case of spatial frequency detection.) In (c), two mismatched patterns add to generate an approximately uniform total input pattern, which will be suppressed by the mechanism of (a). In (d), two matched patterns add to yield a total input pattern that can elicit more vigorous activation than either input pattern taken separately.

off-surround signals. This fundamental property does not exist in additive models of lateral inhibition, such as the Hartline–Ratliff model (Ratliff, 1965). Appendix C shows how the automatic gain control mechanism can inhibit a uniform pattern of inputs, no matter how intense the inputs are. This is the property of noise suppression that we seek.

Figure 8a depicts this noise suppression property. A uniform pattern does not distinguish any cell from any other cell. For example, when the cells are feature detectors of one kind or another, a uniform input pattern contains no information that can distinguish one feature from any other feature. Noise suppression eliminates this irrelevant activity and allows the network to focus on informative discriminations.

Once noise suppression is guaranteed, several consequences automatically follow. For example, Figure 8b shows that such a network responds to the edges of a rectangular input, or to spatial gradients in more general input patterns. This is because cells whose inhibitory surrounds fall outside the rectangle perceive a uniform field, and cells with inhibitory surrounds that are near the center of the rectangle also perceive a uniform field. Both types of cells suppress their inputs. Only cells near the edges of the rectangle do not perceive a uniform pattern. Consequently only the edges of the rectangle elicit large activation.

This argument tacitly supposes that the lateral inhibitory interactions affecting each cell have a prescribed spatial extent, and that the width of the rectangle exceeds this spatial scale. More generally, spatial gradients in an input pattern are matched against the spatial scale of each cell's excitatory and inhibitory interactions. Only those spatial gradients in the input pattern that are nonuniform with respect to the cell's interaction scales generate large activities. By varying the inhibitory scales across cells, one can tune different cells to respond to different spatial frequencies. Thus spatial frequency detectors are a natural consequence of noise suppression properties within cells having a prescribed inhibitory scale. Since all networks in which shunting inhibition occurs have such scales, the existence of spatial frequency detectors should come as no surprise and does not imply that neural networks are Fourier analyzers in the spatial domain (Robson, 1976). Indeed, Fourier analyzers are linear mechanisms. By contrast, shunting networks that are capable of short-term memory contain feedback pathways, and all such networks must be nonlinear to be stable (Grossberg, 1973, 1978d).

Finally, Figure 8c and 8d indicate how a noise suppression mechanism can accomplish pattern matching. Figure 8c supposes that two mismatched patterns feed into $F^{(1)}$, where they add before coupling into the shunting dynamics. Because of the mismatch, the peaks of I_i fill in the troughs of J_i. The total input pattern is approximately uniform and is consequently quenched as noise. By contrast, in Figure 8d, the two patterns match. Their peaks and troughs mutually reinforce each other, so the resultant activities can be amplified beyond the effect of just one pattern. In summary, mismatched input patterns quench activity, whereas matched patterns amplify activity across a field $F^{(1)}$ that is capable of noise suppression.

A subsidiary mathematical question is now evident: How uniform must a pattern be for it to be suppressed? Part of the answer is determined by the choice of structural parameters, such as the strength and spatial distribution of lateral inhibitory coefficients (Appendix C). However, the field $F^{(1)}$ can also be dynamically tuned, or sensitized, by fluctuations in the level of nonspecific arousal that perturbs it through time. An arousal increment can, for example, act by inhibiting the inhibitory interneurons of the network (Ellias & Grossberg, 1975; Grossberg, 1973, 1978e; Grossberg & Levine, 1975). Such a tuning mechanism can simultaneously alter the spatial frequency properties of the network by multiplicatively strengthening or weakening the inhibitory interactions of the cells (Barlow & Levick, 1969a, 1969b). Such mechanisms will arise in a natural fashion as our argument continues.

6. Triggering of Nonspecific Arousal by Unexpected Events

Having suppressed $x_2^{(1)}$ at $F^{(1)}$ due to mismatch with the feedback expectancy $x_1^{(1)}$, we must now use this suppression to inhibit $x_1^{(2)}$ at $F^{(2)}$, since the mismatch at $F^{(1)}$ is the only mechanism in the network that can, in principle, distinguish that an error has occurred at $F^{(2)}$. Moreover, until $x_1^{(2)}$ is quenched, it will continue to read out the template $x_1^{(1)}$ to $F^{(1)}$, which will prevent $x_1^{(2)}$ from eliciting a new signal to $F^{(2)}$.

We were led to the mismatch mechanism at $F^{(1)}$ by noting that $F^{(2)}$ could not discriminate whether an error had occurred. Now we note that $F^{(1)}$s information is also limited At $F^{(1)}$ it cannot be discerned *which* pattern across $F^{(2)}$ caused the mismatch at $F^{(1)}$. It could have been any pattern whatsoever. All $F^{(1)}$ knows is that a mismatch has occurred. Whatever pattern across $F^{(2)}$ caused the mismatch must be inhibited. Consequently, a mismatch at $F^{(1)}$ must have a *nonspecific* effect on all of $F^{(2)}$, since any of the cells in $F^{(2)}$ might be one of the cells that must be inhibited.

We are therefore led to the following questions: How does mismatch and subsequent quenching of activity across $F^{(1)}$ elicit a nonspecific signal (arousal!) to $F^{(2)}$? Where does the activity that drives this nonspecific arousal pulse come from?

Before answering these questions, we should realize that we have been led to a familiar conclusion: Unexpected or novel events are arousing. (To forcefully remind yourself of this basic fact, test a friend's reaction by unexpectedly slamming your hand on a table).

Now we will consider how such arousal is initiated and how it contributes to attentional processing.

Where does the activity that drives the arousal come from, and why is it released when quenching of activity at $F^{(1)}$ occurs? There are two possible answers to the first part of the question, but only one of them survives closer inspection. The activity is either endogenous (internally and persistently generated) or the activity is elicited by the sensory input. If the activity were endogenous, then arousal would occur whenever $F^{(1)}$ was inactive, whether this inactivity was due to active quenching by mismatched feedback from $F^{(2)}$ or to the absence of sensory inputs. This leads to the unpleasant conclusion that $F^{(2)}$ would be tonically flooded with arousal whenever nothing interesting was happening at $F^{(1)}$ or $F^{(2)}$. Therefore, sensory inputs to $F^{(1)}$ bifurcate before they reach $F^{(1)}$. One pathway is *specific:* It delivers information about the sensory event $F^{(1)}$. The other pathway is *nonspecific:* It activates the arousal mechanism that is capable of nonspecifically influencing $F^{(2)}$. The idea that cues have both informative (specific) and arousal (nonspecific) functions has been empirically known at least since the work of Moruzzi and Magoun on the reticular formation (Hebb, 1955; Moruzzi & Magoun, 1949).

Given that the sensory inputs to $F^{(1)}$ also activate an arousal pathway, what prevents this pathway from being activated except when activity at $F^{(1)}$ is quenched? The answer is now clear: Activity at $F^{(1)}$ inhibits the arousal pathway, and quenching of this activity disinhibits the arousal pathway. Figure 9 schematizes the (very rapid) sequence of events to which we have been led. First, a sensory event elicits a pattern $x_2^{(1)}$ across $F^{(1)}$ as it begins to activate the arousal pathway $\mathcal{A}$. This activation at $\mathcal{A}$ is inhibited by activity from $F^{(1)}$. Simultaneously, pattern $x_2^{(1)}$ activates pathways to $F^{(2)}$ that act as a filter that erroneously activates $x_1^{(2)}$. Pattern $x_1^{(2)}$ reads out the learned feedback expectancy $x_1^{(1)}$ to $F^{(1)}$. Mismatch of $x_1^{(1)}$ and $x_2^{(1)}$ at $F^{(1)}$ quenches activity across $F^{(1)}$. The inhibitory signal from $F^{(1)}$ to $\mathcal{A}$ is also quenched, and the arousal pathway is disinhibited. A nonspecific arousal pulse is hereby unleashed on $F^{(2)}$.

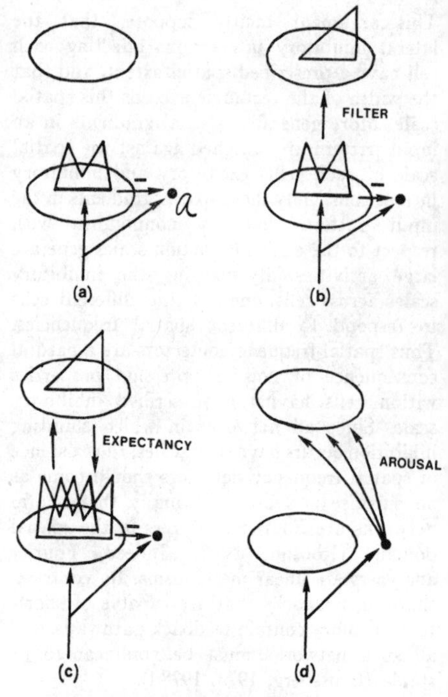

Figure 9. In (a), afferent data elicit activity across $F^{(1)}$ and an input to the arousal source $\mathcal{A}$ that is inhibited by $F^{(1)}$. In (b), the pattern at $F^{(1)}$ maintains inhibition of $\mathcal{A}$ as it is filtered and activates $F^{(2)}$. In (c), the feedback expectancy from $F^{(2)}$ is matched against the pattern at $F^{(1)}$. In d, mismatch attenuates activity across $F^{(1)}$ and thereby disinhibits $\mathcal{A}$, which releases a nonspecific arousal signal to $F^{(2)}$.

7. Parallel Hypothesis Testing in Real Time: The Probabilistic Logic of Complementary Categories

The next design problem is now clearly before us: How does the increment in nonspecific arousal differentially shut off the active cells in $F^{(2)}$? The active cells are the cells that elicited the feedback expectancy to $F^{(1)}$, and since mismatch occurred at $F^{(1)}$, these cells must have been erroneously activated. Consequently, they should be shut off. Furthermore, inactive cells at $F^{(2)}$ should not be inhibited, because these cells must be available for possible coding of $x_2^{(1)}$ during the next time interval. Thus a differential suppression of cells is required: The cells that are most active when arousal occurs should be most

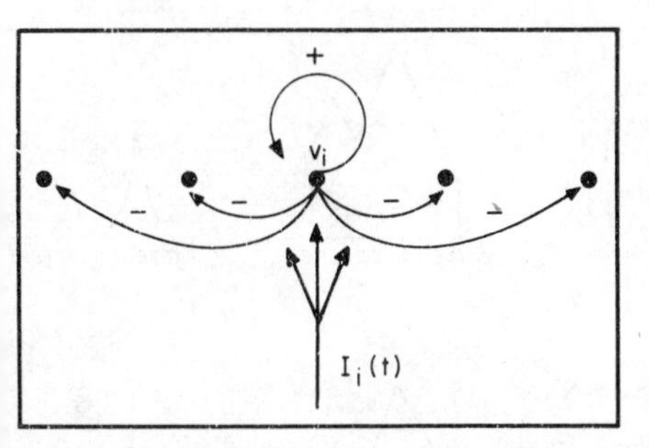

Figure 10. A recurrent shunting on-center off-surround network is capable of contrast-enhancing its input pattern, normalizing its total activity, and storing the contrast-enhanced pattern in short-term memory (STM). (If its feedback signals are properly chosen—e.g., sigmoid, or S-shaped signals—then a quenching threshold exists that defines the activity level below which activity is treated as noise and quenched, and above which activity is contrast enhanced and stored in STM.)

inhibited. This property realizes a kind of probabilistic logic in real time. If activating cell v_i in $F^{(2)}$ to a given degree leads to a certain degree of error or mismatch at $F^{(1)}$, then cell v_i should be inhibited to a degree that is commensurate both with its prior activation and with the size of the arousal increment, or the amount of error. If saying "yes" at v_i leads to error, then change the "yes" to "no," and do it in a graded fashion across the field $F^{(2)}$. Since cells that were only minimally active could have contributed only a small effect to the feedback expectancy, their inhibition will consequently be less, and they can contribute more to the correct coding of $x_2^{(1)}$ during the next time interval.

The arousal-initiated inhibition of cells across $F^{(2)}$ must be enduring as well as selective. Otherwise, as soon as $x_1^{(2)}$ is inhibited, the feedback expectancy, $x_1^{(1)}$, would be shut off, and $x_2^{(1)}$ would be free to reinstate $x_1^{(2)}$ across $F^{(2)}$ once again. The error would perseverate, and the network would be locked into an uncorrectable error. The inhibited cells must therefore stay inhibited long enough for $x_2^{(1)}$ to activate a different pattern across $F^{(2)}$ during the next time interval. The inhibition is therefore slowly varying compared to the time scale of filtering, feedback expectancy, and mismatch.

Once this selective and enduring inhibition is accomplished, the network has a capability for rapid hypothesis testing. By enduringly and selectively inhibiting $x_1^{(2)}$, the network "renormalizes" or "conditionalizes" the field $F^{(2)}$ to respond differently to pattern $x_2^{(1)}$ during the next time interval. If the next pattern elicited by $x_2^{(1)}$ across $F^{(2)}$ also creates a mismatch at $F^{(1)}$, then it will be suppressed, and $F^{(2)}$ will be renormalized again. In this fashion, a sequence of rapid pattern reverberations between $F^{(1)}$ and $F^{(2)}$ can successively conditionalize $F^{(2)}$ until either a match occurs or a set of uncommitted cells is found with which $x_2^{(1)}$ can build a learned filter from $F^{(1)}$ to $F^{(2)}$, and a learned expectancy from $F^{(2)}$ to $F^{(1)}$.

8. The Parallel Dynamics of Recurrent Competitive Networks: Contrast Enhancement, Normalization, Quenching Threshold, Tuning

At this point one can justifiably wonder how $x_2^{(1)}$ elicits a supraliminal pattern across $F^{(2)}$ after $x_1^{(2)}$ is inhibited? If $x_1^{(2)}$ is the pattern that $x_2^{(1)}$ originally excites, and $x_1^{(2)}$ is inhibited, then won't the next pattern elicited by $x_2^{(1)}$ across $F^{(2)}$ have very small activity? In other words, why was the second pattern not also active when $x_1^{(2)}$ was active?

It would have been if the anatomy within $F^{(2)}$ contained only feedforward, or nonrecurrent pathways. Thus we are forced to conclude that the anatomy within $F^{(2)}$ contains feedback, or recurrent pathways. Since all cellular systems face the noise-saturation dilemma, these pathways are distributed in a competitive geometry, or an on-center off-surround anatomy (Figure 10). Mathematical analysis demonstrates that the normalization property holds when recurrent pathways are distributed in a competitive geometry. When these competitive networks are designed to overcome noise amplification and saturation, they enjoy several properties that we need (Appendix D). First, they are capable of contrast-enhancing small differences in initial pattern activities into large and easily discriminable differences that are thereupon stored in short-term memory (STM; see

Figure 11). This property is necessary to build up the codes for $F^{(1)}$ patterns at $F^{(2)}$. Before an $F^{(1)}$ pattern is coded by $F^{(2)}$, it might elicit an almost uniform activity pattern across $F^{(2)}$. The recurrent dynamics within $F^{(2)}$ quickly contrast enhances and stores the contrast-enhanced pattern in STM, where it can be sampled and stored in long-term memory (LTM) by the pathways from $F^{(1)}$ to $F^{(2)}$. When the next occurrence of the same pattern at $F^{(1)}$ occurs, these pathways therefore elicit a more differentiated pattern across $F^{(2)}$, which is again contrast enhanced and stored in STM. The feedback enhancement between STM and LTM continues until the two processes equilibrate other things being equal.

Another property of such a network is its tendency to conserve, or adapt, the total activity that it stores in STM. This is the normalization property that we seek. If certain cells in the network are prevented from sharing the STM activity, say due to arousal-initiated inhibition, then the total activity is renormalized by being distributed to the other cells. Thus after $x_1^{(2)}$ is inhibited across $F^{(2)}$, the network will respond to the signals due to $x_2^{(1)}$ by differentially amplifying them in a way that tends to preserve the total STM activity across $F^{(2)}$. This new STM pattern will inherit much of the STM activity that $x_1^{(2)}$ had before it was suppressed, but the new STM pattern across $F^{(2)}$ will be a quite different pattern than $x_1^{(2)}$, since it is built from $F^{(1)}$ signals that previously fared poorly in the competition for STM activity. The normalization property manifests itself in a large class of psychological data, notably data about behavioral contrast and ratio scales in choice behavior (Grossberg, 1975, 1978a).

These recurrent networks also possess a *quenching threshold* (QT), which is a parameter whose size determines what activities will be suppressed, or quenched, and what activities will be stored in STM (Grossberg, 1973). Activities in populations that start below the QT will be suppressed; activities that exceed the QT will be contrast enhanced and stored in STM. Thus the QT is the cutoff point that defines noise in a recurrent network. All networks that possess a QT can be tuned; that is, by varying the QT, the criterion of which data shall be stored in STM and which data shall be quenched can be altered through time. Several parameters work together to determine QT size, notably the strength of recurrent lateral inhibitory pathways within the network. For example, if a nonspecific arousal pulse multiplicatively inhibits or shunts the inhibitory interneurons of a recurrent network, then its QT will momentarily decrease—the network's inhibitory "gates" will open—to facilitate STM storage.

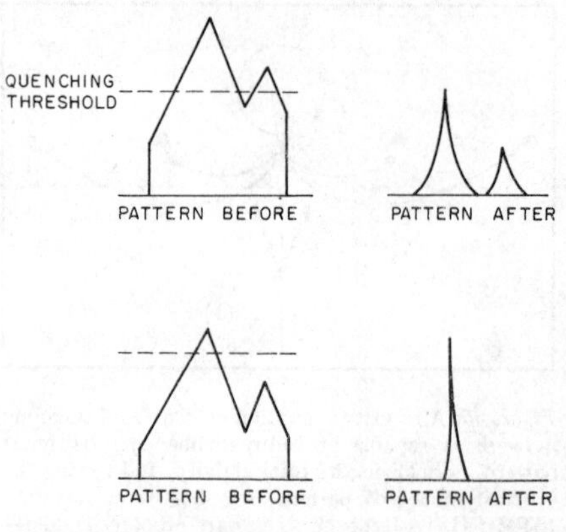

Figure 11. If the quenching threshold is variable, for example, due to shunting signals that nonspecifically control the size of the network's inhibitory feedback signals, then the network's sensitivity can be tuned to alter the ease with which inputs are stored in short-term memory.

The normalization property also helps us to understand the relevance of probabilistic models of hypothesis testing to cognitive processing. Normalization plays the role of summing all the probabilities to equal 1. Shunting, or multiplicative network dynamics, plays the role of multiplying the probabilities of independent events. However, probabilistic concepts only approximately describe some aspects of shunting competitive dynamics. A most serious difference is that although the network's hypothesis testing mechanism might produce a serial sequence of renormalizations in time, these operations are performed by parallel, rather than serial mechanisms. Serial mechanisms of hypothesis testing are not equivalent to the parallel theory.

More generally, serial behavioral properties do not imply that the control processes that subserve them are also serial. In particular, various serial, notably computer, models of memory and cognitive processing have been shown to be fundamentally inequivalent to parallel neural interactions. This inequivalence is noted for the Atkinson and Shiffrin (1968, 1971) theory of free recall in Grossberg (1978a) and for the Schneider and Shiffrin (1976) theory of automatic versus controlled visual information processing in Grossberg (1978e, Section 61). Different predictions of the two types of theory are also described in these articles, and some data that are inexplicable by the serial theories are explained using parallel properties, such as normalization, in a basic way.

9. Antagonistic Rebound Within On-Cell Off-Cell Dipoles

We are now faced with a subtle design problem: How can a nonspecific event, such as arousal, have specific consequences of any kind, let alone generate an exquisitely graded, enduring, and selective suppression of active cells? Here again mathematical analysis was absolutely essential, since the theory could not progress beyond this step had not the answer already been derived (to my own surprise) during work on reinforcement mechanisms (Grossberg, 1972b, 1972c, 1975). In this work, the mechanism helped explain such nontrivial effects as learned helplessness, vicious circle behavior, superconditioning, overshadowing, asymptotically nonchalant avoidance, and peak shift with behavioral contrast when it was joined with suitable conditioning and cognitive mechanisms that were all derived as evolutionary solutions to prescribed environmental pressures. These results extend such popular learning theories as those of Irwin (1971), Kamin (1969), Rescorla and Wagner (1972), and Seligman, Maier, and Solomon (1971) by explicating mechanisms, conceptual distinctions, and predictions in a psychophysiological framework that are invisible to descriptive theories. One might wish to know what reinforcement mechanisms have to do with the development of cognitive codes. The answer is that the property in question occurs whenever optimally designed chemical transducers, or transmitters, occur in competing network channels, or dipoles, whether these channels arise in reinforcement mechanisms, attentional mechanisms (Grossberg, 1975), developmental mechanisms (Grossberg, 1976b), or mechanisms of motor control (Grossberg, 1978e). The property is a robust consequence of a ubiquitous neural design principle, and it guarantees a type of rapid hypothesis testing and error correction wherever this principle is used.

First let us consider some familiar behavioral facts that help to motivate the mechanism. Suppose that I wish to press a lever in response to the offset of a light. If light offset simply turned off the cells that code for light being on, then there would exist no cells whose activity could selectively elicit the lever-press response after the light was turned off. Clearly, offset of the light not only turns off the cells that are turned on by the light, but it also selectively turns on cells that will transiently be active after the light is shut off. The activity of these "off"-cells—namely the cells that are turned on by light offset—can then activate the motor commands leading to the lever press. Let us call the transient activation of the off-cell by cue offset *antagonistic rebound*.

Antagonistic rebound also occurs in a variety of other behavioral situations. For example, shock can unconditionally elicit the emotion of fear and various autonomic consequences of fear (Dunham, 1971; Estes, 1969; Estes & Skinner, 1941). Offset of shock is (other things equal) capable of eliciting relief or a complementary emotional reaction (Denny, 1970; Masterson, 1970; McAllister & McAllister, 1970). In a similar fashion I suggest that when motor command cells are organized in agonist–antagonist pairs, offset of the agonist input can elicit a rebound in the antagonist command cell that acts to rapidly brake the motion in the muscles controlled by the agonist command cell.

When such on-cell off-cell interactions are modeled, one finds examples akin to Figure 12. In Figure 12a, a nonspecific, or adaptation level, input I is delivered equally to both channels, whereas a test input J is delivered to the on-cell channel. These inputs create signals S_1 and S_2 in both channels, and the signals are multiplicatively gated by slowly varying chemical transmitters z_1 and z_2, respectively. The gated signals S_1z_1 and S_2z_2 thereupon compete and yield the on-cell off-cell responses that are depicted in Figure 12a. Appendix E describes the details that are needed for a better understanding, but the main idea behind antagonistic rebound is easy to describe. Consider Figure 12a. Here the transmitters z_1 and z_2 are depleted by being released at rates proportional to S_1z_1 and S_2z_2, respectively. More depletion of z_1 than z_2 occurs if the signal S_1 exceeds S_2. While the test input J is on, the on-channel receives a larger input than the off-channel, since its total input is J plus the nonspecific input I, whereas the off-cell channel only receives the input I. Consequently, $S_1 > S_2$, so that depletion of transmitter leads to the inequality $z_1 < z_2$. Despite this fact, one can prove that the gated signals satisfy the inequality $S_1z_1 > S_2z_2$. Consequently, the on-channel receives a larger gated signal than the off-channel, so that after competition takes place, there is a net on-reaction.

What happens when the test input is shut off? Both channels receive only the equal nonspecific input I. The signals S_1 and S_2 rapidly equalize until $S_1 = S_2$. However, the transmitters are more slowly varying in time so that the inequality $z_1 < z_2$ continues to hold. The gated signals therefore satisfy $S_1z_1 < S_2z_2$. Now the off-channel receives a larger signal. After competition takes place there is an antagonistic rebound in response to offset of the test input.

Why is the rebound transient in time? The equal signals S_1 and S_2 continue to drive the depletion of the transmitters z_1 and z_2. Gradually the amounts of z_1 and z_2 also equalize so that S_1z_1 and S_2z_2 gradually equalize. As the gated signals equalize, the competition shuts off both the on-channel and the off-channel. These facts are summarized in Table 1.

Table 1
Antagonistic Rebound at Offset of Phasic Input

Test input J is on	Right after offset of J	After dipole equilibrates to offset of J
$I + J > I$	$I = I$	$I = I$
$x_1 > x_2$	$x_1 = x_2$	$x_1 = x_2$
$S_1 > S_2$	$S_1 = S_2$	$S_1 = S_2$
$z_1 < z_2$	$z_1 < z_2$	$z_1 = z_2$
$S_1z_1 > S_2z_2$	$S_1z_1 < S_2z_2$	$S_1z_1 = S_2z_2$
$x_3 > x_4$	$x_3 < x_4$	$x_3 = x_4$
$x_5 > 0 = x_6$	$x_5 = 0 < x_6$	$x_5 = 0 = x_6$

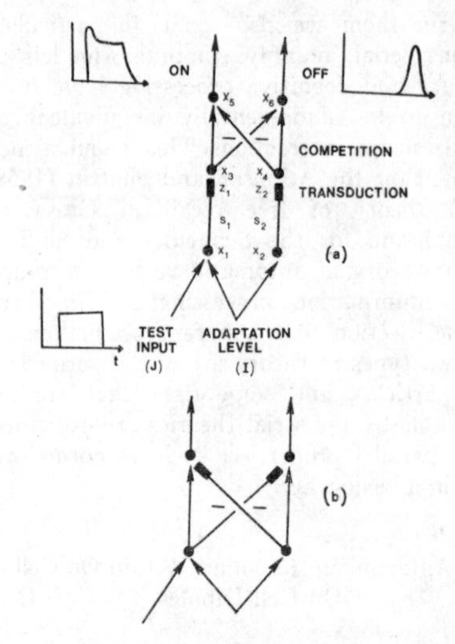

Figure 12. Two examples of on-cell off-cell dipoles. (In (a), the test input J and adaptation level input I add in the on-channel. The adaptation level input perturbs the off-channel. Each input is gated by a slowly varying excitatory transmitter [square synapses]. Then the channels compete before eliciting a net on-response or off-response. In (b), the slowly varying transmitters are inhibitory, and participate in the competition process.)

10. Analgesia, Escape, Partial Reward, and Underaroused Versus Overaroused Depression

In Figure 12a, the two transmitters are excitatory and generate gated signals before competition occurs. Similar effects occur in Figure 12b in which the transmitters are inhibitory and act both as gates and as competing channels. There exist many variations on this theme in vivo. For example, by analyzing more complex learning situations, in particular, experiments on secondary conditioning phenomena, or on transfer between instrumental and classical conditioning, one can show that feedback pathways must exist within the channels that subserve incentive motivation. These feedback channels lead to meaningful comparisons with psychophysiological data when they are interpreted as a formal analogue of the medial forebrain bundle (Grossberg, 1972c, 1975).

Even the feed-forward networks already have surprising and important properties, however. For example, consider a network in which the on-channel supplies negative incentive motivation ("fear") and the off-channel supplies positive incentive motivation ("relief") in a conditioning paradigm. Choose shock reduction as the experimental manipulation. Let shock excite the on-channel, and suppose that the size of the positive rebound after shock terminates is monotonically related to the rewarding effect of the manipulation. Then one can derive a quantitative formula for rebound size (Grossberg, 1972c) that orders infinitely many possible experiments in terms of how rewarding they will be. In particular, reducing J units of shock to $J/2$ units is less rewarding than reducing $J/2$ units of shock to 0 units, despite the fact that shock reduction equals $J/2$ units in both cases. This analgesic effect is due to intracellular adaptation of the chemical transmitters. Analogous data have been reported by Campbell (1968); B. Campbell and Kraeling (1953); Gardner, Licklider, and Weisz (1961); and Myers (1969). Moreover, it is predicted that three indices should all covary as a function of the reticular formation arousal level, which is interpreted to be a source of nonspecific input to the incentive motivational dipoles. These indices are (a) the rewarding effect due to switching J units to $J/2$ units of shock, (b) the ability of an animal to learn to escape from presentation of a discrete fearful cue, and (c) the relative advantage of partial reward over continuous reward (Grossberg, 1972c).

One also finds that two types of depressed emotional affect exist in the dipole: an underaroused syndrome and an overaroused syndrome. These syndromes are manifestations of the dramatic changes in the net incentive motivation that occur when the arousal level is parametrically changed (Grossberg, 1972c). The two syndromes are the endpoints in an inverted U of net incentive as a function of arousal level. At underaroused levels, the behavioral threshold is abnormally high, but the system is hyperactive after this threshold is exceeded. At overaroused levels, the behavioral threshold is abnormally low, but the system is so hypoactive that little net incentive is ever generated. Parkinson's patients and certain hyperactive children seem to exhibit the underaroused syndrome (Fuxe & Ungerstedt, 1970; Ladisich, Volbehr, & Matussek, 1970; Ricklan, 1973), which is paradoxical because behavioral threshold is inversely related to suprathreshold reactivity. Such underaroused individuals can be brought "down" behaviorally by a drug that acts as an "up"; that is, it raises the adaptation level to the normal range. In Parkinson's patients, this up is L-dopa, and in certain hyperactive children, it is amphetamine.

A general question now presents itself: Do *all* neural dipoles share these properties whether they occur in motivational, sensory, or motor representations? This question is considered for the case of cortical red–green dipole responses to white light in Section 12.

11. Arousal Elicits Antagonistic Rebound: Surprise and Counterconditioning

A surprising feature of the on-cell off-cell dipole is its reaction to rapid temporal fluctuations in arousal, or adaptation level. This reaction allows us to answer the following question posed in Section 9: How can a nonspecific event, such as arousal, selectively suppress active on-cells? Appendix E shows that arousal fluctuations can reset the dipole, despite the fact that they generate equal inputs to the on-cell and off-cell channels. In particular, a sudden increment in arousal can, by itself, cause an antagonistic rebound in the relative activities of the dipole. Moreover, the size of the arousal increment that is needed to cause rebound can be independent of the size of the test input that is driving the on-channel. When this occurs, an arousal increment that is sufficiently large to rebound any dipole will be large enough to rebound all dipoles in a field. In other words, if the mismatch is "wrong" enough to trigger a large arousal increment, then all the errors will be simultaneously corrected. This cannot, in principle, happen in a serial processor. Moreover, the size of the rebound is an increasing

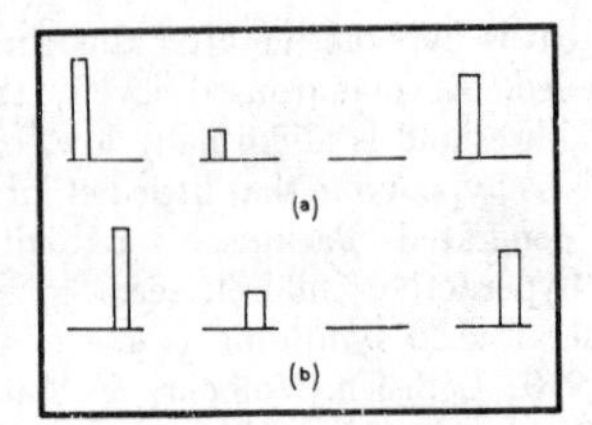

Figure 13. A rebound from on-cell activation to off-cell activation can be elicited by a rapid increment in the arousal or adaptation level of the dipole. (The size of the rebound is determined by the size of the on-cell activation. In (a) are depicted the on-responses of four cells. In (b) are depicted possible rebounds by their off-cells in response to a nonspecific increment.)

function of the size of the on-cell test input (Figure 13). Thus the amount of antagonistic rebound is precisely matched to the amount of on-cell activation that is to be inhibited. Finally, in previously inactive dipoles no rebound occurs, but the arousal increment can sensitize the dipole to future signals by changing by equal amounts the gain, or temporal averaging rate, of the on-cell and off-cell. In summary, the on-cell off-cell dipole is superbly designed to selectively reset $F^{(2)}$, and to do so in an enduring fashion because of the slow fluctuation rate of the transmitter gates.

In a reinforcement context, the rebound due to arousal shows how surprising or unexpected events can reverse net incentive motivation and thereby drive counterconditioning of a behavior's motivational support (Grossberg, 1972b, 1972c). Once the rebound capabilities of surprising events are recognized, one must evaluate with caution such general claims as "the surprising omission of . . . shock . . . can hardly act as a reinforcing event to produce excitatory conditioning" (Dickinson, Hall, & Mackintosh, 1976, p. 321).

The above mechanisms indicate how dynamical critical periods might be laid down by learned feedback expectancies. These expectancies modulate an arousal mechanism that buffers already coded populations by shutting them off so rapidly in response to erroneous STM coding that LTM recording is impossible. In other words, the mechanism helps to stabilize the LTM code against continual erosion by environmental fluctuations.

The thought experiments from which these conclusions follow are purely abstract. One experiment describes how limitations in the types of information available to individual cells can be overcome when the cells act together in suitably designed feedback schemes. Another experiment describes a solution to the noise-saturation dilemma, and yet another experiment describes how to design a chemical transducer and how dipoles formed when such transducers compete in parallel channels can achieve antagonistic rebound. As the thought experiments proceed, however, the resultant network designs take on increasingly neural interpretations. To test the theory by psychophysiological experiments, these empirical connections must be made more explicit. The next three sections discuss three of the major design features in more detail to suggest that some psychophysiological designs are examples of our abstract designs, and to explain and predict some psychophysiological phenomena using formal properties of the abstract designs as a guide. These examples are hardly exhaustive, but they will perhaps be sufficient to enable the reader to continue making new connections. Further details are in the articles of Grossberg (1972b, 1972c, 1975, 1976b, 1978e). The next three sections can be skipped on a first reading if the reader wishes to immediately study Section 15 to find out what happens when the patterns at $F^{(1)}$ and $F^{(2)}$ mutually reinforce each other.

12. Dipole Fields: Positive and Negative Aftereffects, Spatial Frequency Adaptation, Rivalry, and the McCollough Effect

Section 8 noted that $F^{(2)}$ possesses a recurrent on-center off-surround anatomy that is capable of normalizing its total STM activity within its functional channels. Section 9 showed that the cells in this recurrent anatomy are the on-cells of on-cell off-cell dipoles. I therefore conclude that $F^{(2)}$ consists of a field of on-cell off-cell dipoles such that the on-cells interact within a recurrent on-center off-surround anatomy and the off-cells also interact within a recurrent on-center off-surround anatomy. Denote by $F_+^{(2)}$ the recurrent subfield of on-cells, and by $F_-^{(2)}$ the recurrent subfield of off-cells (Figure 14). The

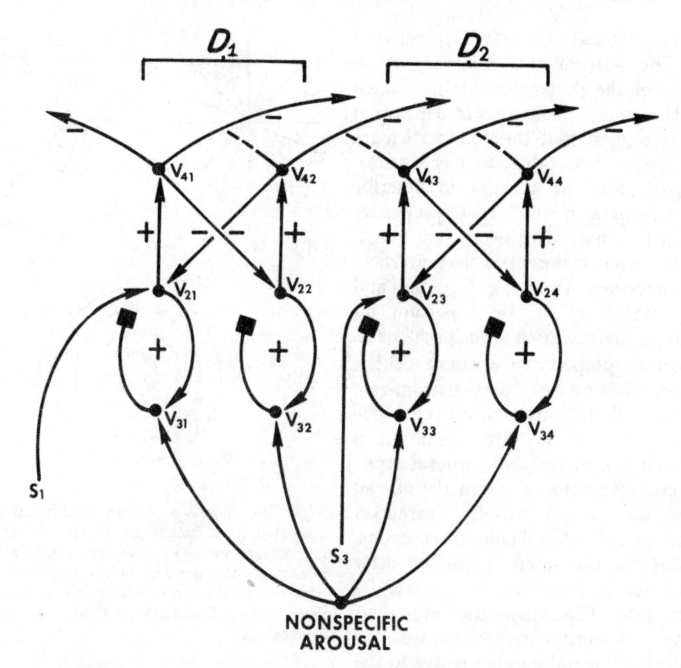

Figure 14. A possible anatomy of two dipoles (D_1 and D_2) is depicted, embedded in recurrent subfields of on-cells and off-cells. (The nonspecific arousal signal is gated by transmitters in the pathways $v_{3i} \rightarrow v_{2i}$, $i = 1, 2, \ldots$. The transmitter gates are depicted by square synapses. The arousal level hereby determines an overall level of transmitter adaptation across the dipole field. The signal S_1 turns on the cell v_{21}, which inhibits its off-cell v_{22} via the inhibitory interneuron v_{41}. Simultaneously, the on-cell v_{21} begins to differentially deplete its transmitter gate via the feedback pathway $v_{21} \rightarrow v_{31} \rightarrow v_{21}$. The interneurons v_{4i}, $i = 1, 2, \ldots$, also activate the recurrent interactions among on-cells and among off-cells that normalize their respective subfields.)

existence of neural, in particular, cortical on-cells and off-cells, and the joining together of nerve cells in on-center off-surround anatomies are familiar neural facts. Moreover, these facts have often been used to explain psychophysiological data (Carterette & Friedman, 1975; Cornsweet, 1970). The present treatment is novel in several respects, however. That a dipole field is a major tool to reset an error and to search for a correct code is, to the best of my knowledge, a new insight. Moreover, the way in which arousal fluctuations interact with slowly varying, competing transmitter gates to cause rebound or a shift in adaptation level, and the way in which shunting interactions define a quenching threshold, normalize field activity, and regulate contrast enhancement also seem to be new insights.

There exists a basic difference between the recurrent inhibition within a subfield and the dipole inhibition between on-cells and their off-cells. Dipole inhibition creates a balance between mutually exclusive categories or features. Intrafield inhibition normalizes and tunes its subfield. For example, suppose that the on-cells in a given field respond to white bars of prescribed orientation on a black field, and their corresponding off-cells respond to black bars of similar orientation on a white field. A continuous shift in the position of a white bar can induce a continuous shift of activity within the on-field, but at each position there can exist either a white bar on a black field or a black bar on a white field, but not both. Next are summarized some of the phenomena that are due to continuous changes within subfields and complementary changes when dipole rebounds cause a flip between subfields. The goal of this summary is to clarify some of the properties through which dipole fields manifest themselves in perceptual data, and to suggest that these properties are manifestations of code stabilizing mechanisms. The summary will not attempt to describe the global schemata in which these properties are embedded during a live perceptual event, although the article makes clear that interfield signaling processes, such as filtering and expectancy matching, will be important ingredients in the classification of such schemata.

An important property of a dipole field is this: If a test input excites a particular on-cell, then the on-cell inhibits its off-cell. The inhibited off-cell can, in turn, disinhibit a nearby off-cell due to the tonic arousal input and the recurrent anatomy within the off-cell field. The disinhibited off-cell thereupon inhibits its on-cell via dipole interactions. Suppose that the test input is shut off after it has been on long enough to deplete its transmitter gate. (To make this argument quantitative, we must carefully control the duration of experimental inputs relative to the transmitter depletion rate.) Then antagonistic rebound within its dipole can turn on its off-cell, which inhibits the nearby off-cell, whose on-cell is hereby disinhibited and responds by rebounding onward. Negative aftereffects are hereby generated. For example, suppose that the on-cells are orientationally selective such that nearby orientations recurrently excite each other, whereas more distinct orientations inhibit each other (Figure 15a). Then persistent inspection of a field with radial symmetry (Figure 15a) can elicit an aftereffect with circular symmetry (Figure 15c), as MacKay (1957) has reported.

In Section 5 I noted that the noise suppression properties of shunting lateral inhibition also imply spatial frequency properties. Consequently, dipole fields whose subfield inhibition is of shunting type are capable of spatial frequency adaptation. A grating with a sinusoidal luminance profile of prescribed spatial frequency will excite a band of cell types whose inhibitory fields permit maximal excitation by the input. If the input stays on for awhile, the activated transmitter gates will be differentially depleted. Test inputs

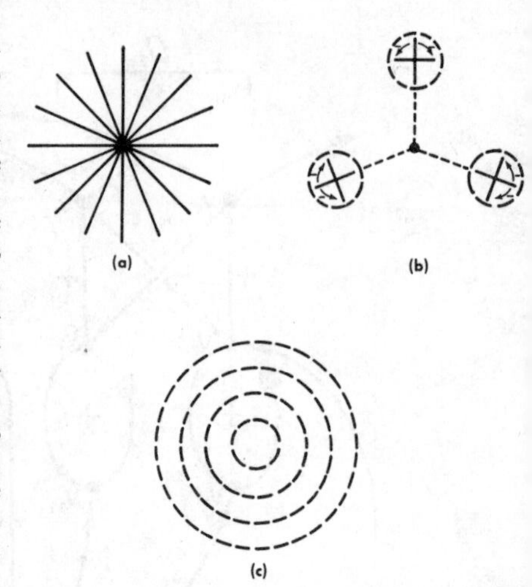

Figure 15. In (a), a pattern with radial symmetry is inspected for a long time. In (b), the net inhibitory interactions among mutually perpendicular orientations at each position are depicted. In (c), offset of the radial pattern elicits antagonistic rebounds across the field that differentially activate the perpendicular orientations.

with similar spatial frequencies share some of these gated pathways, so the overall sensitivity of response to these inputs will be less (Grossberg, 1976b). This view of spatial frequency adaptation contrasts with the view developed by Wilson (1975) that spatial frequency adaptation is due to classical conditioning of an inhibitory transmitter. It is often assumed that a slowly varying effect implies a conditioned change. The alternative notions that "fatigue" and antagonistic activity can yield perceived changes are also very old (see Brown, 1965, for a review).

The present theory refines the latter view by noting how slowly varying changes can follow from dipole adaptation without any conditioning taking place. In particular, even if the adaptational differences decay until they are very similar, contrast enhancement due to fast recurrent competitive interactions can bootstrap these differences into the perceivable range. An interaction between slow transmitters and fast recurrent interactions can hereby create behavioral effects that are much more enduring than the transmitter decay rate would suggest. This suggestion is made again later for the McCollough effect. The Wilson model differs from the present theory in its STM properties as well as in its description of slow adaptation effects. Wilson used the Wilson–Cowan equations to describe fast intercellular interactions. Among other differences, these equations do not incorporate automatic gain control by lateral inhibitory signals (Grossberg 1973). Consequently, the Wilson–Cowan equations cannot retune their sensitivity in response to shifts in background input intensity, a difficulty that also occurs in all additive models of lateral inhibition.

Pattern-contingent colored aftereffects can also be generated in a dipole field. Suppose that a prescribed field of feature detectors is color coded. Let the on-cells be maximally turned on by red light and the off-cells be maximally turned on by green light for definiteness. Then white light will excite both on-cells and off-cells; that is, white light acts like an adaptation level in this situation. Suppose that a red input whose features are extracted by the field is turned on long enough to substantially deplete its transmitter. What happens if a white input replaces the red input on test trials? The depletion caused by the prior red input now causes the white adaptation level to generate a larger gated signal to the green channel, so a green pattern-contingent aftereffect will be generated.

How enduring will this aftereffect be? Here we must recall that the anatomies of $F_+^{(2)}$ and $F_-^{(2)}$ are recurrent, and that one property of such recurrent anatomies is their ability to contrast enhance small differences in net input into large differences that can then be stored in STM (Section 8). Thus, even if the large initial differences in transmitter depletion within the on-cell off-cell dipoles decay steadily to small differences, the recurrent anatomy can contrast enhance these small differences into a perceptually visible aftereffect when the white test pattern is presented. For this to happen, however, the feature field must be protected from new inputs that can disrupt the pattern of small differences until the test trial occurs. Sleep can hereby prolong the apparent duration of the aftereffect. These properties are familiar ones in the McCollough effect (MacKay & MacKay, 1975; McCollough, 1965).

Various authors have suggested that the long duration of the McCollough effect implicates classical conditioning mechanisms. Montalvo (1976) presented a particularly ingenious application of this idea. This approach seems to trade-off one paradox for another, since the classical conditioning must produce a negative aftereffect during test trials, rather than the positive effect that was experienced during learning trials. Unless one can isolate a large class of phenomena in which classical conditioning reverses the effect on test trials, this explanation is hard to understand from the viewpoint of basic neural design. The present theory points out that slowly varying transmitter gates supplemented by rapid contrast enhancement and STM storage in a recurrent anatomy can also generate long-term effects whose duration is much longer than the transmitter decay rate would suggest. Such long-term effects must unambiguously be ruled out before classical conditioning is invoked as a unitary explanation.

Dipole field structure also helps to explain monocular rivalry (Rauschecker, Campbell, & Atkinson, 1973), whereby two superimposed gratings with the same sinusoidal luminance profile, one vertically oriented and one horizontally oriented, and each illuminated by white light or by different (say complementary) colors, are seen to alternate through time. The tendency toward rivalry can be explained by the recurrent inhibition across orientationally tuned on-cells and across orientationally tuned off-cells; the vertical on-cells tend to inhibit the horizontal on-cells, and conversely. The tendency to alternate can be explained by the fact that persistent STM reverberation of the active vertical on-cells tends to deplete their transmitter gates, thereby weakening their reverberation and providing a relative advantage to the inhibited, and therefore relatively undepleted, horizontal on-cells. When the vertical on-cell depletion reaches a critical value, the horizontal on-cells are sufficiently disinhibited to allow the recurrent dynamics to contrast enhance the horizontally coded inputs into STM. The horizontally coded on-cells thereupon reverberate in STM until the cycle repeats itself. Thus the main effect can be ascribed to combined effects of slow transmitter depletion, recurrent inhibition across orientations, and the contrast-enhancing capabilities of the recurrent network, even if there are no changes in gaze.

Of particular interest is the fact that the alternation rate depends on the color of the gratings. Two white and black gratings, or two monochromatic gratings, alternate up to three times slower than gratings that are illuminated by complementary colors. This can be discussed in terms of the rebound behavior that occurs between subfields that are orientationally coded and whose dipoles code for complementary colors. When two white and black gratings of sufficient contrast are used, the white inputs can excite both on-cells and off-cells of the color-coded dipoles, thereby inhibiting them. It is therefore assumed that apart from altering their gain, intense black and white gratings cause net excitation primarily in feature fields whose on-cells respond unselectively to light-on and whose off-cells respond to light-off. In such a feature field, the horizontal and vertical white bars excite the same subfield, and the horizontal and vertical black bars excite the complementary subfield. Each subfield tends to adapt or conserve its total STM activity (within its functional channels!) so that there exists a tendency for the horizontal and vertical inputs to compete for STM activity, and to thereby decrease the transmitter depletion rate in active cells.

By contrast, consider what happens in a color-coded dipole field in response to two gratings that use the field's complementary colors, say red-vertical and green-horizontal. Here the red-vertical bars deplete only the red field, and the green-horizontal bars deplete only the green field. There is no direct inhibition within a given subfield between horizontal and vertical orientations. Thus, other things equal, greater STM activation of red-verticals or green-horizontals is possible than in the black–white case because less intrafield competition for STM activity occurs. Greater STM activation implies faster transmitter depletion and faster alternation rates. If this explanation is correct, then it is a special case of a more general phenomenon; namely, that the frequency of perceptual oscillations can be pattern contingent due to the intrafield normalization property.

Other aftereffects provide more direct evidence for the existence of slowing varying transmitter gates. In particular, the effects of changing background illumination, or the *secondary field*, on aftereffects are remarkably similar to the effects of changing arousal level on the rebound. If a secondary field is turned on during the observation of a positive afterimage in darkness, then a rapid transition to a negative afterimage can be generated (Brown, 1965, p. 483; Helmholtz, 1866, 1924). If the secondary field is then turned off, the afterimage can revert in appearance to that of the stage when the secondary field was first turned on. In a dipole, an increase of adaptation level tends to rebound the relative dipole activities. If the arousal level is then decreased, the slowly varying transmitter levels can still be close to their original values, so that the original relative dipole activities are rapidly restored. The higher the luminance of the secondary field, the shorter the afterimage latency, and the more rapidly the afterimage is extinguished (Juhasz, 1920). In a dipole, a higher adaptation level more rapidly equalizes the amounts of transmitter in the two dipole channels by depleting them both at a faster, more uniform rate. When approximately equal levels of transmitter are achieved, the inhibitory interneurons between the dipole's populations kill any relative advantage of one population over the other. The duration of an afterimage increases with an increase in primary stimulus luminance (Brown, 1965, p. 493). In a dipole, increasing the intensity of an input to one population increases the rebound at the other population when the input terminates, much as termination of a more intense shock causes greater relief, other things being equal (Grossberg, 1972c, 1976b).

The preceding considerations lead to some experimental predictions. Some of these concern red and green cortical dipoles. For example, suppose that a red stimulus has activated a red-cell long enough to substantially deplete the transmitter. Does an increment in white light cause a green-cell rebound? Does a decrement in red light from J units to $J/2$ units cause a smaller rebound when white light is on than a decrement from $J/2$ units to 0 units? Is there an inverted U in dipole responsiveness as a function of the arousal level or the intensity of white light? Does the relative rebound size increase as a function of arousal level size for intermediate levels of arousal? In other words, are visual dipoles designed the same way as motivational dipoles?

Another set of predictions concerns the McCollough effect. For example, how does the McCollough effect depend on the intensity of white light during test trials? A more intense white light should yield an initially larger aftereffect unless the white is so intense that overarousal occurs. Moreover, more intense white should equalize the relative transmitter stores more rapidly than less intense white. This suggests an experiment in which a double test is made. The first test uses prolonged inspection of white bars whose intensity differs across subjects. Before the second test is made, some visual experience should occur to blot out whatever small differences in transmitter storage might still exist after the bright white bars are examined. Then a second test with white bars is given. Subjects who saw less intense bars on the first test should perceive a larger aftereffect.

An experiment concerning spatial frequency adaptation is also suggested. This experiment is analogous to the experiment on aftereffects due to changes in the secondary field. Speaking generally, if spatial frequency adaptation and certain other aftereffects are all due to dipole depletion, albeit in different fields of feature detectors, then they should undergo similar transformations in response to analogous experimental manipulations, other things being equal, notably the persistence with which each feature field is disrupted by uncontrolled inputs. Suppose that when a series of vertical sinusoids drifts horizontally across the visual field, those on-cells and off-cells whose recurrent inhibitory signals collide with visually induced inputs will have their activities suppressed. Consider the on-cells and off-cells that can be activated by the prescribed spatial frequency. What happens as the contrast of the visual pattern is parametrically increased across subjects?

This is a delicate question because more than one dipole field in the coding hierarchy can be activated by such an input. Let us consider what would happen if only one dipole field is activated. In the limit of absolute black and very bright white verticals, both the on-cells and the off-cells would be almost equally excited on the average, albeit at different times, as the light and dark verticals drift over their receptive fields. Neither on-cell nor off-cell would gain a large *relative* advantage, but both would have their transmitter stores significantly depleted by the persistence of the horizontally drifting input. Hence, significant spatial frequency adaptation would occur, but not due to large relative imbalances in the dipoles. What happens as the contrast between the white and black verticals is decreased? Then other things being equal, the off-cells will be depleted more than the on-cells. Hence, a greater relative depletion within the dipoles can be induced at smaller contrast levels than at larger contrast levels. How can this conclusion be tested? Consider two groups of subjects. Let Group 1 be adapted and tested using high contrast gratings. Let Group 2 be adapted on a lower contrast grating and tested using the same higher contrast grating used to test Group 1. The net on-responses at a black–white interface as the test grating slowly drifts across the visual field should be greater in Group 2 than in Group 1. Can such differentially enhanced boundaries between the trailing edge of black and the leading edge of white be perceived? If the answer is yes, then one can properly claim that the effect is a functional analogue within the visual system of the partial reinforcement acquisition effect in the motivational system (Grossberg, 1975).

13. Reset Wave: Reaction Time, P300, and Contingent Negative Variation

The nonspecific arousal that is triggered by unexpected events (or mismatch) selectively and enduringly inhibits active population across $F^{(2)}$. In vivo, do there exist broadly distributed inhibitory waves that are triggered by unexpected events? In average evoked potential experiments, one often finds such a wave, namely the P300 (Rohrbaugh, Donchin, & Eriksen, 1974; Squires, Wickens, Squires, & Donchin, 1976). The theory's relationship to P300 is discussed in Grossberg (1978e), in which the following properties of P300 are shown to be analogous to properties of the resetting wave: Reaction time is an increasing function of P300 size (Squires et al., 1976); P300 is not the same average evoked potential as the contingent negative variation (CNV) (Donchin, Tueting, Ritter, Kutas, & Heffley, 1975; cf. Section 16); P300 can be elicited in the absence of motor activity (Donchin, Gerbrandt, Leifer, & Tucker, 1972); resetting the STM codes of longer sequences of events can take longer than resetting the STM codes of shorter sequences of events, and due to the relationship between reaction time and P300 size, longer sequences will elicit larger P300s (Remington, 1969; Squires et al., 1976). Moreover, Chapman, McCrary, and Chapman (1978) showed that in a number- and letter-comparison task, there existed an evoked potential component with a poststimulus peak at about 250 msec that is related to the storage of cue-related information in STM. This latency fits well with the idea that STM storage occurs if the feedback expectancy does not create a mismatch. The extra 50 or so msec needed to generate a P300 would also be necessary in the network to trigger the reset wave if a mismatch does occur.

If the P300 is indeed a reset wave of the type that the thought experiment describes, then several types of experiments can be undertaken to test this hypothesis. On the anatomical side, Where does the expectancy matching take place? What pathways subserve the arousal? On the physiological side, Do dipole rebounds cause the inhibition? On the psychophysiological side (e.g., average evoked potential experiments), Is there a more direct experimental paradigm for testing whether P300 directly inhibits STM? In particular, Can a succession of P300s be reliably triggered when information is disconfirmed in successive stages? On a deeper functional level, Does the P300 act to buffer committed cells against continual recording by the flux of experience? If P300 is inhibited, can previously committed cells be recoded? In other words, when we consider cognitive coding, does a chemical switch contribute to code stability, or is code stability entirely dependent on buffering by dynamic reset mechanisms?

As was noted in Section 4, feedback expectancies that trigger STM reset mechanisms should occur in many thalamocortical systems, so that there should exist different reset waves corresponding to each functionally distinct system. In Grossberg (1978e), the preceding scheme is generalized to a variety of examples in which competition occurs between attentional, or consummatory, pathways and novelty, or orienting, pathways. A matching process goes on within the attentional system and computes such information as follows: Are the sensory cues the ones that are expected? Do the proprioceptive motor cues match the terminal motor map that is guiding the limb? If the answer is yes, then goal-oriented arousal systems are activated to support the matching process and its consequences, such as posture. If the answer is no, then complementary arousal systems are activated that support rapid reset and orienting reactions aimed at acquiring new information with which to correct the error. Given that the P300 helps to reset sensory STM in response to unexpected events, does there exist a complementary wave that occurs along with expected events? The CNV would appear to be such a wave (Cohen, 1969), since it is associated with an animal's expectancy, decision (Walter, 1964), motivation (Cant & Bickford, 1967; Irwin, Rebert, McAdam, & Knott, 1966), volition (McAdam, Irwin, Rebert, & Knott, 1966), preparatory set (Low, Borda, Frost, & Kellaway, 1966), and arousal (McAdam, 1969).

If the P300 and the CNV are indeed complementary waves, then experiments should be undertaken to determine the neural loci at which the generators of these waves compete. For example, Section 16 suggests that the hippocampus provides output that contributes to the CNV. Does expectancy mismatch occur within the hippocampus, or in a cell nucleus that activates hippocampus, and thereby release a P300 by disinhibiting its generator?

Having noted the existence of reset and attentional waves that are triggered by sensory events, it is natural to ask whether there exist analogous waves that are triggered by motor events? To answer this question, the next section considers how eye movements can modulate the LGN's sensitivity to afferent visual signals and the related questions of whether the LGN has a dipole field organization and whether feedback from visual cortex to LGN can selectively attenuate or amplify afferent visual signals. This discussion leads to a reinterpretation of LGN data and to some predictions. These predictions concern the possible existence of a reset motor wave and the timing of certain developmental events relative to the end of the critical period for plasticity in the primary visual cortices.

14. Template Matching and Reset: PGO Wave, Geniculate Dipoles, and Corticogeniculate Feedback

An example of an "attentional" motor wave seems to be the ponto-geniculo-occipital (PGO) wave whose effects on the LGN are admirably reviewed by Singer (1977). Singer (1977) distinguished at least two types of inhibitory interneurons in his discussion of LGN dynamics:

> There apparently are two inhibitory mechanisms with two different functions. One is based on intrinsic interneurons and presumably conveys the retinotopically organized and highly selective inhibitory interactions between adjacent retinocortical channels. . . . This inhibition seems to be mainly of the feed-forward type. . . . The second inhibitory pathway is exclusively of the recurrent type and is relayed via cells in nucleus reticularis thalami. . . . This extrinsic inhibitory loop is probably involved in more global modifications of LGN excitability as they occur during changes in the animal's state of alertness and during orienting responses associated with eye movements. (p. 394)

Singer noted that mesencephalic reticular formation (MRF) stimulation leads to field potentials in the LGN and the visual cortex that closely resemble PGO waves. LGN transmission is facilitated during PGO waves and during the analogous negative field potential that occurs after MRF stimulation.

One mechanism of MRF facilitation is inhibition of the cells in the nucleus reticularis thalami, which are recurrent inhibitory interneutrons between LGN relay cells. From a theoretical viewpoint, this type of disinhibition would be expected to have nonspecific effects like decreasing the quenching threshold of an entire recurrent subfield of cells, and thereby facilitating transmission of signals through these cells (Grossberg, 1973; Grossberg & Levine, 1975). Such an effect seems to occur in LGN. Since MRF stimulation can completely suppress inhibitory postsynaptic potentials elicited from optic nerve or optic radiation, Singer (1977) concluded

> that the intrinsic inhibitory pathways also get inactivated. However, it cannot yet be decided whether the inhibitory interneurons in the main laminae are also subject to direct reticular inhibition as is the case for cells in nucleus reticularis thalami. (p. 409)

Singer went on to suggest that corticogeniculate feedback could partially accomplish the intrinsic cell inhibition.

For present purposes, the main point is Singer's (1977) functional interpretation of the MRF-induced LGN disinhibition. He claimed

> that the brief phase of disinhibition serves to reset the thalamic relay each time the point of fixation is changed. . . . To assure a bias-free initial processing of the pattern viewed after a saccade . . . inhibitory gradients ought to be erased before the eyes come to rest on the new fixation point . . . the concomitant disinhibition occurs only towards the end of the saccade right before the eyes come to rest. (p. 411)

Singer's remarks can be mechanistically interpreted as follows: As the proprioceptive coordinates of the eye muscles approach the terminal motor coordinates that control the saccade, the two sets of coordinates match, a PGO wave is initiated, it disinhibits LGN relay cells, and prepares the LGN to transmit retinal signals to the visual cortex. If the PGO wave is indeed elicited by a matching process between the terminal motor map and proprioceptive coordinates of the eye muscles, then this matching process should be capable of exciting cells that inhibit the LGN interneurons within the nucleus reticularis thalami. In what neural structure does this matching process take place? One component of this structure might already have been discovered by Tsumoto and Suzuki (1976), who report a pathway from the frontal eye fields to the perigeniculate nucleus in which are found the LGN inhibitory interneurons. Electrical stimulation of the frontal eye fields inhibits the perigeniculate cells and facilitates LGN transmission.

Singer (1977) claimed that the PGO wave resets the LGN so that it can respond to retinal signals without bias. However, nonspecifically reducing the quenching threshold is not the type of selective reset that I have discussed earlier. Indeed, Singer's discussion of LGN dynamics emphasizes the wiping away of all inhibitory gradients as a reset mechanism. But what if excitatory activities already exist in the LGN when this happens? Why do these activities not get amplified and thereupon maximally bias LGN activity in response to the next retinal input volley? I suggest that the LGN reset that is due to the nucleus thalami reticularis occurs while the eye is moving and the extrinsic inhibitory interneurons are active. This extrinsic inhibitory feedback resets the LGN by generating a high quenching threshold and thereby wiping out the LGN's excitatory patterns. As the eye comes to rest at its intended position, I suggest that matching occurs between the terminal and proprioceptive motor maps of the eye muscles, thereby activating the attentional system, in particular the PGO wave, which sensitizes the LGN to retinal and cortical signals.

Even if the preceding interpretation of Singer's argument is correct, it discusses a nonspecific effect on the QT and the sensitivity of visual pattern processing, but not the selective reset that aims at reorganizing attention in response to an error, or other unexpected event.

Is there a wave that is functionally complementary to the PGO wave, that can precede it, and that drives a selective reset of LGN dynamics in response to unexpected events? If such a wave does exist, it would be functionally analogous to the P300. In this regard, Singer (1977) parenthetically mentions the work of Foote, Manciewicz, and Mordes (1974) to explain the inhibition of LGN transmission that sometimes occurs shortly after MRF stimulation but before the facilitatory phase. Foote et al. suggest that this inhibitory pathway is due to serotonergic fibers originating in the dorsal raphe nucleus. Are these fibers the pathway over which selective reset can occur?

For a selective reset wave to exist, it must operate on on-cell off-cell dipoles. Do such dipoles exist in the LGN? Much of the data discussed by Singer was collected in the cat LGN. Singer (1977) reports here

> that reciprocal inhibitory connections exist between adjacent neurons driven by the same eye that have the same receptive field center characteristics; i.e., between on-center cells and between off-center cells, respectively. (p. 390)

These interneurons are analogous to the intrafield lateral inhibition that was postulated within $F_+^{(2)}$ and $F_-^{(2)}$, but which we now recognize as a prerequisite for total activity adaptation and quenching threshold tuning in any recurrent network. In addition, there exist "reciprocal inhibitory interactions between neurons with antagonistic field center characteristics—that is, between on- and off-center units with spatially overlapping receptive fields" (Singer, 1977, p. 390). These cells would appear to form dipoles. If they are dipoles of the type discussed, then the arousal system that triggers their rebounds will feed into them—from the dorsal raphe nucleus—and activating this arousal system will rebound their relative activities.

These hypotheses should be easier to test in the monkey than the cat, because Schiller and Malpeli (1978) have reported that of the four parvocellular layers in the monkey, the two layers committed to the left eye are subdivided into an on-cell layer and an off-cell layer, and the two layers committed to the right eye are also subdivided into an on-cell layer and an off-cell layer. Do dipole interactions occur between the on-cell and off-cell layers of each eye representation? Does a suitable arousal increment rebound the relative activities of these dipoles? If so, we will have found an elegant functional reason for the existence of this structure in the monkey: Each eye has its own dipole field to carry out its selective reset modes. We will also have found an elegant reason for the existence of intrinsic and extrinsic inhibitory systems: Attentional reduction of the quenching threshold is functionally distinct from, and even complementary to, selective reset.

Another important point of Singer's (1977) article concerns the role of corticogeniculate feedback.

> In a highly selective way the cortex permits transmission of binocular information that can be fused and evaluated in terms of disparity depth cues while it leaves it to the intrinsic LGN circuits to cancel transmission of signals that give rise to disturbing double images. (p. 398)

In other words, the corticogeniculate feedback acts as a template that selectively enhances the type of data that the cortex is capable of coding in a globally self-consistent way.

In summary, the LGN seems to enjoy a dipole field structure whose sensitivity to afferent sensory signals is modulated both by corticogeniculate feedback, which acts like a sensory expectancy-matching mechanism, and by MRF arousal, which lowers the LGN QT in response to proprioceptive-terminal map matching within the eye movement system.

If we interpret the geniculocortical relay as an example of our thought experiment, then several experimental predictions arise. These predictions are made with caution, since a significant part of visual development seems to be genetically prewired in the geniculocortical pathways of higher mammals (Hubel & Wiesel, 1977). It is still not clear, however, to what extent corticogeniculate feedback does help to terminate the visual critical period in these animals. Nor is it clear whether the same neural design that is used in some species, or in individual neural relays, to terminate a critical period using feedback is also used in others wherein a chemical switch or other prewired mechanisms are appended. The predictions flow from the observation that if the geniculocortical system is an example of the thought experiment, *albeit* vestigially, then its reset and search mechanisms must develop before the end of the visual critical period. In particular, if lateral inhibition within the LGN is used to help match cortical and retinal data, then these inhibitory connections must develop before the end of the critical period. The dipole field structure of the cortex must also develop before the end of the critical period. Moreover, mismatch within the LGN system should disinhibit an arousal system capable of rebounding the cortical dipoles. There exists a catecholamine arousal system to neocortex, among other structures (Fuxe, Hökfelt, & Ungerstedt, 1970; Ungerstedt, 1971; Jacobowitz, 1973; Lindvall & Björklund, 1974; Stein, 1974). Is this the arousal system being sought? Does it develop before the end of the critical period? Is this arousal system capable of driving antagonistic rebound in cortical dipoles? Is a catecholamine transmitter always used in arousal systems that drive antagonistic rebound, for example, the catecholamine system originating in the dorsal raphe nucleus that was described by Foote et al. (1974)? Finally, is there a structural similarity between all pairs of attentional and selective reset waves, as between CNV and P300, or PGO and its hypothesized complementary reset wave?

15. Adaptive Resonance, Code Stability, and Attention

The preceding sections discuss some of the network events that occur when feedforward data mismatch feedback expectancies. What happens if an approximate match occurs? Then the activity patterns at $F^{(1)}$ and $F^{(2)}$ elicit interfield signals that mutually reinforce each other, and activities at both levels are amplified and locked into STM. Because the STM activities can now persist much longer in time than the passive decay rates of individual cells, the slowly varying feed-forward filters and feedback expectancies have sufficient time to sample the STM patterns and store them in LTM. I call this dynamical state an *adaptive resonance*. The resonant state provides a global interpretation of the afferent data, or a context-dependent code, that explicates in neural terms the idea that the network is paying attention to the data. The resonance idea suggests that many individual neural events, such as cell potentials and axonal signals, are behaviorally irrelevant until they are bound together by resonant feedback. Of special importance is the observation that unless resonance occurs, no coding in LTM can take place. This observation clarifies from a mechanistic viewpoint the psychological fact that a relationship exists between paying attention to an event and coding it in LTM (Craik & Lockhart, 1972; Craik & Tulving, 1975).

The resonant state provides a context-dependent code due to several factors acting together. For example, the pattern of expectancy feedback can alter, through a matching process, the activity that a given feature detector would have experienced if only afferent signals were operative. Similarly, competitive interactions within a subfield can rapidly alter the net input pattern before storing it in STM. Thus when an activity pattern at a field $F^{(1)}$ is projected by interfield signaling to a field $F^{(2)}$, feedback from $F^{(3)}$ to $F^{(2)}$ can deform this pattern before it is further reorganized by competition within $F^{(2)}$. Because the resonant state provides a context-dependent code whose resultant patterns in STM and LTM depend on all active components of the system, it is impossible to determine the code from the measurements taken by any single microelectrode, no matter how precise its calibration. I claim that adaptive resonances are the functional units of cognitive coding, and that classification of the resonances that occur in prescribed situations is a central problem for cognitive psychology. The structural substrates of these cognitive units are nonlinear feedback modules involving whole fields of cells rather than individual nerve cells.

The technical details needed to rigorously build up the resonance idea are derived in Grossberg (1976a, 1976b), in which a summary of related coding models is also given, and further developed in Grossberg (1978e). In particular, Grossberg (1976a) points out that a coding theory that depends on a feedforward anatomy with any fixed number of cells is faced with a crippling dilemma: Either a chemical switch turns off code development at a prescribed time, but then the code will be behaviorally meaningless with a high likelihood, or the code is unstable through time whenever the number of patterns in the environment significantly exceeds the number of coding cells. It is also proved that a developing code can be stable in a sparse input environment, but this does not address the typical situation in vivo, where a continuous visual flow, and therefore a nondenumerable series of visual patterns, must be dealt with. Computer models of code development missed these basic points because they typically used small numbers of inputs and small numbers of coding cells.

Once the main point was vividly made, one could see that feedback was essential to stabilize a developing code in a rich input environment, and that the types of feedback that were needed resembled attentional mechanisms that had previously been derived from different considerations, namely classical and instrumental conditioning postulates (Grossberg, 1975). The examples in Sections 2 and 3 illustrate these attentional phenomena. Two pleasing conclusions were thereby drawn: Adult mechanisms, in this case attentional mechanisms, are often continuations along a developmental continuum of infantile mechanisms, in this case code development mechanisms; and the rather mysterious rubrics of "paying attention" and "expectancy" could be attached to the more substantial theme of "code stability and consistency," and the establishment of dynamically maintained critical periods.

Anderson, Silverstein, Ritz, & Jones (1977) also recognized the importance of feedback in defining the functional units of neural network. Their model differs, however, from the present theory in several notable respects. The recurrent STM interactions in their model are defined by linear feedback signals. Grossberg (1973, 1978d) shows how linear feedback signals among cells that are capable of saturating create unphysical instabilities such as noise amplification and compression of an input pattern. Furthermore, known neural nonlinearities, such as sigmoid signals between cells, overcome these instabilities and contrast enhance the input pattern (Appendix D). The LTM interactions in the Anderson et al. model are described by summing up a large number of mutually orthogonal LTM vectors

$$z = \sum_{k=1}^{n} z_k$$

to form the total LTM trace across a field $F^{(1)}$ of cells. When a signal pattern S from $F^{(1)}$ to $F^{(2)}$ is gated by the total LTM trace, as in

$$S \cdot z = \sum_{k=1}^{n} S \cdot z_k$$

(see Appendix A), it might be perpendicular to all but one of the increments, say z_1. Consequently, the net signal is $S \cdot z_1$. This concept gets into difficulty because in vivo the total LTM trace z must be composed of small quantities (e.g., transmitter concentrations). Each of the summands z_k must therefore be a very small quantity unless n is also very small, but then the theory is powerless. If each z_k is very small, then the net signal depends on gating by a sum of very small quantities. This creates an unstable situation. Furthermore, the LTM trace z_{ij} from cell v_i to cell v_j in the

Anderson et al. model is assumed to equal the LTM trace z_{ji} from v_j to v_i. This symmetry assumption is too restrictive for our purposes, since we do not want the filter from $F^{(1)}$ to $F^{(2)}$ to necessarily equal the expectancy from $F^{(2)}$ to $F^{(1)}$; this would limit the tendency to achieve greater abstractness of feature extraction in a hierarchy of fields. A more serious problem for the LTM symmetry assumption is its implication that the signal from every cell be proportional to its STM trace. This follows because the growth rate of LTM trace z_{ij} is proportional to the product of signal S_{ij} from v_i and v_j times STM trace x_j of v_j. To achieve $z_{ij} \equiv z_{ji}$, it is necessary for $S_{ij}x_j \equiv S_{ji}x_i$, which is possible if $S_{ij} = \alpha x_i$ and $S_{ji} = \alpha x_j$. In particular, the recurrent signals must be linear functions of the STM traces, and the usual instabilities that recurrent linear signals generate among cells will be generated.

The next three sections summarize a few resonant schemes and predictions pertaining thereto that suggest the scope of the resonance phenomenon. Other resonances, notably the olfactory resonance that is described by the distinguished work of Freeman (1975), are discussed more completely in Grossberg (1976b, 1978e). Freeman discovered a resonant phenomenon by performing parallel electrode experiments on the cat prepyriform cortex. When the cat smells an expected scent, its cortical potentials are amplified until a synchronized oscillation of activity is elicited across the cortical tissue. The oscillation organizes the cortical activity into a temporal sequence of spatial patterns. The spatial patterns of activity across cortical cells carry the olfactory code. By contrast, when the cat smells an unexpected scent, then the cortical activity is markedly suppressed. Freeman traces the differences in cortical activity after expected versus unexpected scents to gain changes within the cortical tissue. Appendix C shows how a matching mechanism in a shunting network simultaneously changes gains as it amplifies or attenuates network activity. Freeman also notes a tendency for the most active populations to phase-lead less active populations. This also occurs automatically in a shunting network due to the correlation between gain and asymptote. Because the cortex oscillates, Freeman models his data using second-order differential equations whose coefficients are changed by expectations in a manner that is descriptively stated, but not dynamically explained, by his model. I suggest that the oscillations are caused by feedback between cells that obey first-order differential equations whose gains are changed by signals coupled to the shunting mechanism.

Grossberg (1978b) claimed that adaptive resonances also occur in nonneural tissues, where they are suggested to be a basic design principle in a universal developmental code. Syncytium formation during sea urchin gastrulation is identified as a possible adaptive resonance phenomenon—in particular, the law whereby pseudopods from the mesenchymal cells adhere to ectodermal cells to form a syncytium has the same form as the law for an LTM trace—and some predictions are made to test this hypothesis.

16. Are Conditioned Reinforcer Pathways and Conditioned Incentive Motivational Pathways Reciprocal Pathways in an Adaptive Resonance?

Grossberg (1975) has described a psychophysiological theory of attention in which an adaptive resonance occurs. This resonance helps to explain why the dilemma of cross-conditioning that is depicted in Figure 2 does not routinely occur. Figure 16 idealizes this resonance. Speaking intuitively, the internal representations of external cues elicit signals that are, before conditioning takes place, distributed nonspecifically across the various internal drive representations. During conditioning trials, the pattern of reinforcement and drive levels strengthens the LTM traces within certain of these signal pathways and weakens the LTM traces within other pathways. These conditioned changes in the signal pathways endow the external cues with conditioned reinforcer properties. On recall trials, these conditioned signals combine with internal drive inputs to determine whether or not feedback signals will be elicited. The feedback signals play the role of incentive motivation in the network. Incentive motivation is released in a given feedback pathway only if the momentary balance of conditioned reinforcer signals plus drive inputs compete

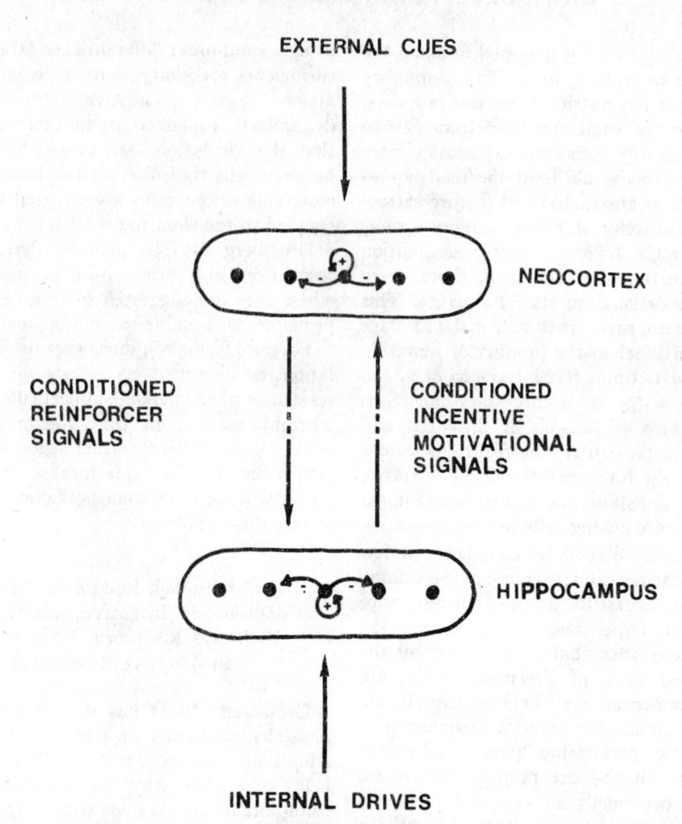

Figure 16. An adaptive resonance between neocortex and hippocampus is suggested to occur when external cues are compatible with internal needs. (The conditionable feedback pathways in this module subserve conditioned reinforcer properties and conditioned incentive motivational properties.)

favorably against these factors within the other feedback pathways.

Before conditioning occurs, each of the incentive motivational channels nonspecifically projects to the external cue representations. As in the case of the conditioned reinforcers, the incentive motivational channels are conditionable, and their LTM traces can be strengthened when their signals are large and contiguous to active external cue representations. Thus after conditioning occurs, an internal drive representation can deliver incentive motivational signals preferentially to those external cue representations with which it was previously associated. In this way, activating a given external cue representation can sensitize an ensemble of motivationally related external cue representations via incentive motivational feedback. A type of subliminal psychological set is hereby formed. Since the external cue representations compete among themselves for storage in STM, the conditioned incentive motivational feedback abets the storage of compatible cues and tends to overshadow the storage of incompatible cues. Thus, during alternate scanning of incompatible cues, attentional switching between resonances compatible with one cue class and then the other class buffers each class against indiscriminate cross-conditioning with incompatible cues. Various data and related theories about reinforcement and attention are analyzed in the light of these concepts in the articles of Grossberg (1972b,

1972c, 1975). Herein I suggest a neural substrate for this resonance and a psychophysiological experiment to test its existence.

In Figure 16, $F^{(1)}$ and $F^{(2)}$ both possess recurrent on-center off-surround interactions, and both the $F^{(1)} \to F^{(2)}$ and $F^{(2)} \to F^{(1)}$ pathways are conditionable. Region $F^{(1)}$ contains external cue representations, and region $F^{(2)}$ contains internal drive representations. When this network is embedded into a more complete system of interactions, an interpretation of $F^{(1)}$ as neocortex and of $F^{(2)}$ as hippocampus is suggested. Given this interpretation, the conditioned reinforcer pathways $F^{(1)} \to F^{(2)}$ should have a final common pathway at hippocampal pyramidal cells, and their LTM traces should be sensitive to the balance of drives and reinforcements through time. Relevant data have been collected by Berger and Thompson (1977), who describe neural plasticity at the hippocampal pyramids during classical conditioning of the rabbit nictitating membrane response.

The conditioned incentive motivational pathways $F^{(2)} \to F^{(1)}$ should have a final common pathway at neocortical pyramidal cells, and their LTM traces should be sensitive to the balance of motivation and cue saliency through time. The CNV is a conditionable neocortical potential shift that has been associated with an animal's motivational state (Cant & Bickford, 1967; Irwin et al., 1966), and Walter (1964) has hypothesized that the CNV shifts the average baseline of the cortex by depolarizing the apical dendrites of its pyramidal cells. If the conditioned incentive motivational feedback is indeed realized by the CNV and if adaptive resonances between conditioned reinforcers and conditioned incentives do exist, then there should exist neural feedback loops between neocortex and hippocampus such that while conditioned reinforcer properties are being established with the hippocampal pyramid cells as a final common pathway, simultaneously conditioned incentive properties are being conditioned with the apical dendrites of neocortical pyramid cells as a final common pathway. Experiments to test this prediction would require either simultaneous measurement from electrodes in the neocortical and hippocampal loci of the resonant circuit, or correlation of electrode measurements in the hippocampus simultaneous with CNV measurements.

17. Pattern Completion, Hysteresis, and Gestalt Switching

Consider what happens to an adaptive resonance as its afferent data are slowly and continuously deformed through time, say from the letter O to the letter D. By "slowly" I mean slowly relative to the rate with which resonant feedback can be exchanged. Recall that feedback from $F^{(2)}$ to $F^{(1)}$ can deform what "is" perceived into what "is expected to be" perceived. Otherwise expressed, the feedback is a prototype, or higher order Gestalt, that can deform and even complete activity patterns across lower order feature detectors. For example, suppose that a sensory event is coded by an activity pattern across the feature detectors of a field $F^{(1)}$. The $F^{(1)}$ pattern is then coded by certain populations in $F^{(2)}$. If the sensory event has never before been experienced, then the $F^{(2)}$ populations that are chosen are those whose codes most nearly match the sensory event because the pattern at $F^{(1)}$ is projected onto $F^{(2)}$ by the positional gradients in the $F^{(1)} \to F^{(2)}$ pathways (Appendix A). If no approximate match is possible, then mismatch at $F^{(1)}$ will trigger a reset wave that selectively inhibits $F^{(2)}$ and elicits a search routine. If an approximate match is possible, however, then the feedback signals from $F^{(2)}$ to $F^{(1)}$ will elicit the template of the sensory events that are optimally coded by the $F^{(2)}$ pattern. These feedback signals rapidly deform the $F^{(1)}$ pattern until this STM pattern is a mixture of feedforward codes and feedback templates. Otherwise expressed, $F^{(2)}$ tries to complete the $F^{(1)}$ pattern using the prototype, or template, that its active populations release.

In Grossberg (1978e, Section 40), another completion mechanism is also suggested, namely a normative drift. This mechanism generalizes the line neutralization phenomenon that was described by Gibson (1937). In suitably designed feature fields, STM activity at a particular coding cell can spontaneously drift toward the "highest order" coding cell in its vicinity, due either to the existence of more cell sites, or to larger and spatially more broadly distributed feedback signals, at the highest order cells. After STM activity drifts to its local norm, the highest order cell can thereupon release its feedback template. It was shown in Levine and Grossberg (1976) that such drifts are a type of lateral masking due to the recurrent interactions within the feature field. I suggest that many Gestalt-like pattern completions are manifestations of intrafield competitive transformations, such as normative drifts, and the deformation by feedback expectancies of lower order STM patterns. Such global dynamical transformations transcend the capabilities of classical pattern discrimination models (e.g., Duda & Hart, 1973).

Two important manifestations of the completion property are hysteresis and Gestalt switching. For example, once an STM resonance is established in response to the letter O, the resonance resists changing its codes when small changes in the sensory event occur—this is hysteresis. Hysteresis occurs because the active $F^{(2)} \rightarrow F^{(1)}$ template keeps trying to deform the shifting $F^{(1)}$ STM pattern back to one that will continue to code the $F^{(2)}$ populations that originally elicited this template.

If, however, the sensory event changes so much that the mismatch of test and template patterns becomes too great, then the arousal-and-reset mechanism is triggered. This event inhibits the old code at $F^{(2)}$ and forces a search for a distinct code. A dramatic switch between global percepts can hereby be effected. The global nature of the switch is due not only to the rapid suppression of the previously active $F^{(2)}$ code but also to the fact that $F^{(2)}$ contains populations that can synthesize data from many feature detectors in $F^{(1)}$, and the feedback templates of these populations can reorganize large segments of the $F^{(1)}$ field. I suggest that an analogous two-stage process of hysteresis and reset is operative in various visual illusions, such as Necker's cube (Graham, 1965). When ambiguous figures are presented, these mechanisms can elicit spontaneous switches of perceptual interpretation due either to shifts of gaze or to the input-induced cyclic rates of transmitter depletion that can occur even if the gaze remains relatively fixed (Section 12).

If Gestalt switching is a two-stage process, then at the moment of switching, a reset wave should occur. Does a P300 occur at the moment of perceived switching? If not, can this paradigm be used to discover what average evoked potential, if any, parallels activation of the reset mechanism?

18. Binocular Resonance and Rivalry

My final example indicates that adaptive resonances need not be hierarchically organized, and points to a class of resonances of particular importance. Hysteresis can occur between two reciprocally connected fields even if they are not hierarchically organized, since the individual cells do not know whether they are in a hierarchy. For example, suppose that each eye activates a field of monocularly coded feature detectors. Suppose that each monocular field is endowed with a recurrent on-center off-surround anatomy, indeed with a recurrent dipole field of on-cell and off-cells. Let the on-cells in each monocular field be capable of exciting corresponding on-cells in the other monocular field. In other words, signals from a given monocular field act as a template for the other monocular field. It does not matter what features are coded by these detectors to draw the following conclusion. Once a resonance is established between the two monocular fields, hysteresis will prevent small and slow changes in the input patterns from changing the coded activity. Julesz (1971) introduced a field of physical dipoles to model the binocular hysteresis that he and Fender described (Fender & Julesz, 1967). Resonance between two recurrent on-center off-surround anatomies undergoing shunting dynamics provides a neural model of the phenomenon. Such a binocular resonance will generate properties of binocular rivalry, since competition within each subfield of the recurrent networks will inhibit feature detectors that do not participate in the resonance.

The construction of monocular representations whose binocular resonances code globally self-consistent invariants of stereopsis is presently being undertaken. Although this construction is not yet complete, some observations can be made in broad strokes to guide the reader who is interested in pursuing the elucidation of perceptual and motor resonances.

Before the two eyes can fixate on the scenes that will drive binocular development, there must already exist enough prewired visual feature detectors to direct the eye movement system to lock in the fixation process. Thus the existence of prewired visual feature fields does not argue against the need for visual tuning by experience. Such tuning seems necessary to achieve accurate stereopsis in the face of significant variations in bodily parameters due to individual differences and growth (Daniels & Pettigrew, 1976). An effort should be made to correlate individual and species differences in the motor mechanisms that are used to accumulate visual data and to act on the visual environment with corresponding differences in prewired sensory feature detectors and the ultimate feature fields that can be synthesized (Arbib, 1972; Creutzfeldt & Nothdurft, 1978).

Even if feature development can continue in the absence of visual experience, this does not imply that visual experience does not alter visual development. Just as imprinting can be driven by endogenous drive sources that are later supplanted by environmentally reactive drive sources (Sluckin, 1964), an effort should be made to test whether visual development is driven by endogenous arousal sources before these sources are supplanted by visual experience, in particular by visually reactive arousal sources.

Binocular visual resonances seem to be a special case of bilateral resonances that are due to the bilateral organization of the body, for example, binaural auditory resonances. As in the case of binocular corticogeniculate feedback (Singer, 1977), bilateral interactions at each of several anatomical stages help to select the activity patterns that elicit and are modulated by hierarchical signals. The hierarchical signals are supplemented by environmental feedback signals to complete the sensorimotor loops that control the circular reactions of a developing individual (Piaget, 1963). An effort should be made to correlate the structures that emit the environmental signals with those that receive them, for example, the algebraic properties of motor speech commands with the corresponding properties of auditory feedback patterns (Grossberg, 1978e).

19. Symmetry and Symmetry Breaking in Sensory and Motor Systems

An important theme in the design of adaptive resonances will be the analysis of their symmetry and symmetry-breaking properties. This theme is unavoidable when sensory resonances are studied side by side with their motor counterparts, as Section 18 suggests. For example, the system schematized in Figure 9 shows a manifest asymmetry in the construction of its arousal and pattern analysis components. However, this system forms only one part of a larger system that enjoys a much more symmetric structure in which two subsystems compete, namely an attentional and an orienting subsystem (Lynn, 1966). The component in Figure 9 is part of the attentional system, which also includes incentive motivational and CNV components (Grossberg, 1975). This subsystem focuses attention on cues that are expected to generate prescribed consequences of behavior. It can overshadow irrelevant cues, as in Section 2, by selectively amplifying certain patterns at the expense of others. The complementary orienting system is also capable of selectively amplifying patterns, but these are not the patterns that code for sensory or cognitive events. They are, rather, the motor maps that are capable of directing the subject toward sources of unexpected environmental events.

The dichotomous but interdependent nature of these subsystems is illustrated by the existence of X-cells and Y-cells in mammalian retinas and by the neural pathways that these cells excite. The X-cells project primarily to the LGN, where their signals are processed as visual data, whereas the Y-cells have axons that bifurcate to send branches both to the LGN and the superior colliculus (Fukuda & Stone, 1974; Robson, 1976), which has been identified as an area in which a visuomotor map for eye movements is elaborated (Wurtz & Goldberg, 1972). The competitive nature of these two subsystems is illustrated by considering how different our motor reactions can be when a loud sound to the left is unexpected versus when it is a learned discriminative cue for rapid button pushing that will be highly rewarded if it is sufficiently rapid. In the former case, our eyes and head rotate rapidly to the left. In the latter case, rotation can be inhibited and supplanted by a rapid button push.

Competition between attentional and orienting subsystems may clarify certain paradoxes about mental illness. As just summarized, the attentional system focuses attention on cues that are expected to generate prescribed consequences of behavior and can thereby overshadow irrelevant cues. The competing system is triggered by unexpected events (novelty) and allows the network to redefine the set of relevant cues to avoid unexpected consequences. Overarousal of either subsystem can yield attentional deficits (Grossberg, 1972c; Grossberg & Pepe, 1970, 1971), but the exact nature of the deficit and its proper treatment depends on the particular subsystem that is overaroused. For example, a schizophreniclike syndrome of reduced attentional span and contextual collapse can be elicited by overarousal of the incentive-motivational system, but would not necessarily be cured by a depressant that acted differentially on the novelty (reticular formation) system. In fact, depressing the wrong arousal system can cause a paradoxical deterioration of a syndrome by disinhibiting the hyperactive competing arousal system that caused the syndrome. Complicating the situation further is the inverted-U in responsiveness that can be caused by parametrically exciting either of the arousal systems separately (Section 10).

Alternation between attentional and orienting reactions seems also to occur, and in a cyclic fashion, within the motor system during the performance of a familiar sequence of skilled movements. Grossberg (1978e, Sections 48–54) used a thought experiment concerning the information available to a behaving infant to derive a minimal network for the learning of circular reactions. A central mechanism in this network is the matching or mismatching of a terminal motor map, or where the end organ expects to go, and a proprioceptive motor map, or where the end organ now is. Proprioceptive-terminal map matching is the analogue within motor systems of expectancy matching in sensory systems (Tanji & Evarts, 1976). Proprioceptive-terminal map matching means that the end organ has reached the location where it expects to be. I suggest that such matching is capable of eliciting signals that not only support the motor postures and perceptual sensitivity needed to pay attention—reflected in the PGO wave—but also release from STM the next motor command in a goal-directed motor sequence. The new motor command instates a new terminal motor map that mismatches the current proprioceptive map, thereby inhibiting the attentional arousal and releasing the new orienting reaction. Thus the matching process seems to cyclically sow the seeds of its own destruction, at least until the entire motor plan is executed. An effort should be made to test whether proprioceptive-terminal matching does indeed elicit signals that reset motor commands in a goal-directed motor plan.

The minimal dimension of the symmetry that is needed to design bilateral hierarchical resonances between competing subsystems is a 16-fold symmetry, since each subsystem contains at least two levels capable of matching their patterns, and each level contains a pair of dipole fields to compute a bilateral resonance.

Despite the greater symmetry that manifests itself by studying competing subsystems side by side, it is inevitable that neural system design will exhibit substantial symmetry breaking. In addition to the asymmetry between excitatory and inhibitory configurations that supports neural development and evolution (e.g., on-center off-surround anatomy), such environmental asymmetries as between light versus dark and between up versus down must be reflected in the neural machinery that has adapted to them. Some insights concerning this neural machinery are suggested in terms of the preceding discussion. For example, if certain off-cells are tonically on in darkness, and if offset of a light triggers a transient output signal from the corresponding off-cell, then why does the tonic activity of this off-cell in the dark not drive a tonic output signal? If the off-field is normalized, then when all the off-cells are on in the dark, none of them is sufficiently active to exceed the output threshold, which is chosen higher than the quenching threshold. After a light is turned off, a particular off-cell's activity is differentially rebounded for a short time during which its activity exceeds the output threshold. Tonic activity and transient outputs are hereby reconciled. This example illustrates the importance of carefully tuning the relative levels of overall network activity and output threshold.

By contrast, suppose that the output threshold is lowered by disinhibiting the output cells' axon hillocks, or that the overall network activity is enhanced by lowering its quenching threshold—perhaps as in the nucleus reticularis thalami. Then the off-field can deliver tonic output signals to its target cells. If, for example, the target cells control the contraction of muscles, then the tonic muscle signals can maintain a posture that resists the effects of gravity, for example, standing. In this situation, periodic phasic inputs to the on-cells, whether due to external sources or to feedback signals from the off-cells, can cause an oscillatory motor reaction during every cycle of which agonist contraction is followed by an antagonist rebound, for example, walking. Thus, differential tuning of output threshold and normalized activity can convert transient off-cell output signals, as in phasic sensory responses, into tonic off-cell output signals that either balance a persistent asymmetry in environmental influences, as in standing, or energize rhythmic output bursts, as in walking.

20. Cerebral Dominance: The Anatomy of Temporal Versus Spatial Encoding

A more profound type of symmetry breaking occurs between the attentional and orienting subsystems, due to the different nature of cognitive and motor data, and within the attentional subsystem itself, due to the different processing of data about space and time. A pattern of activity across a field of populations at a given time is inherently ambiguous. Does the pattern code a single event in time, such as the features in a visual scene, or does it code the order information in a series of events? Because of this fundamental ambiguity, it is suggested in Grossberg (1978e) that different STM reset mechanisms are needed to reset spatial versus temporal data. The spatial reset mechanism is a matching mechanism such as I have just discussed. The temporal reset mechanism is derived from a study of free recall and serial learning. The output signal from a population in a temporal processor is suggested to activate a self-destructive inhibitory feedback signal. This feedback inhibition prevents perseverative performance of the same item, and conditionalizes the order information among the populations that remain active, with the most active population performed first, since its reaction time for generating an output signal is smallest. The readout of order information from a field of active populations is suggested to be accomplished by either a nonspecific decrease in all the output thresholds or a nonspecific amplification of the total STM activity in the field. Again the relative size of these two levels is a crucial parameter in determining network performance. Thus, the readout of sensory order information is suggested to be mechanistically analogous to the activation of a sequential motor program. By this scheme, a list of items can be performed in a perfect serial ordering despite the fact that all the mechanisms in the network are parallel mechanisms. Serial properties do not imply serial processes.

I suggest that the cortical microanatomy that subserves spatial versus temporal processing will be found to exhibit these different STM reset mechanisms. Consequently, to unambiguously decode temporal versus spatial data, somehow the populations that code the different types of data must be spatially segregated so that they can be endowed with their disparate STM reset mechanisms. The ambiguous meaning of spatial patterns hereby suggests the need to spatially segregate the processing of sequential, including language-like, codes from codes concerning themselves with spatial integration. This dichotomy might be one reason for the emergence of cerebral dominance (Gazzaniga, 1970, chap. 8), despite the fact that a typical speech act can include both spatial and temporal coding elements, and thus requires cyclic resetting of both types of codes. Visual and auditory processing are sensory prototypes of higher codes that emphasize spatial and temporal processing, respectively. Since visual and auditory representations are bilateral, the trend toward segregation of spatial versus temporal processing in separate hemispheres can be viewed as a symmetry-breaking operation with a drift of visuallike processing into the non-dominant hemisphere and auditorylike processing into the dominant hemisphere. The symmetry between bilateral resonances in these regions should be correspondingly broken, leading to a generalized avalanche or command structure between the two hemispheres to coordinate the temporal unfolding of spatial representations. An effort should be made to test whether the cortical microanatomy in spatial versus temporal processors exhibits traces of different reset mechanisms in the anatomy of inhibitory feedback interneurons.

21. Conclusion: How to Understand Evolutionary Data?

The thought experiment in this article illustrates a general method for discovering the mechanisms behind psychological data. Many psychological phenomena are facets of the evolutionary process—variously called chunking, unitization, or automation—whereby behavioral fragments are grouped into new control units that become the fragments of still higher behavioral units in a continuing process of hierarchical organization and command synthesis. By its very nature, this evolutionary process hides the mechanistic substrate on which it is built, so that we can behave in a world of percepts, feelings, and plans rather than of cells, signals, and transmitters. Because our brains are these evolutionary devices, we have immediate introspective evidence about basic psychological processes, and can consensually define concepts like reward, punishment, frustration, expectation, memory, and plan even without a scientific understanding of their mechanistic substrates. To represent these consensual concepts in our scientific work by processes that mirror their introspective properties is, however, a fundamental mistake. Then the consensual impression of events blinds us to their functional representation.

For example, language processes whose properties seem discrete and serial are often realized by continuous and parallel control processes (Grossberg, 1978a, 1978e). The two types of representation are not fundamentally equivalent and generate different predictions. Similarly, behavioral properties that seem linear are often controlled by nonlinear processes (Grossberg, 1978d). Again the two types of description are fundamentally not equivalent. When a theory is erroneously built on consensual properties, it soon meets data that it finds paradoxical. Then the theory either collapses or is decorated with a succession of ad hoc hypotheses. Theoretical epicycles soon crowd the scientific landscape, and theory gets a bad name even though we cannot live without it.

An alternative procedure is to respect the wisdom of evolution by trying to imitate it. To do this, at each stage of theory construction, prescribed environmental pressures are identified that force adaptive designs on the behaving brain. Most of us know these pressures; they are familiar precisely because they are among the constraints to which we have successfully adapted. Thus the theory is grounded on a firm basis. By contrast with the consensual method, these pressures are properties of the environment rather than of our behavior. The thought experiments show how these environmental constraints generate explicit minimal mechanisms for coping with them. Such experiments include information that eludes experimental techniques for several reasons. For example, they show how many system components work together, and they compress into a unified description environmental pressures that act over long, or at least nonsimultaneous, times. Most importantly, the thought experiments explicate design constraints that are needed to adapt in a real-time setting. These real-time constraints are often the most crucial ones, and they are invisible to descriptive or purely formal theories.

Once the minimal mechanisms that realize several environmental pressures are constructed, mathematical analysis shows how they work together to generate data and predictions whose complexity and subtlety transcend the apparent simplicity of the environmental pressures, as well as unaided intuition. This procedure defines new conceptual categories into which to divide the data, and also points to important environmental pressures that have been overlooked, by clearly delineating what the mechanisms can and cannot do. In this way, a small number of principles and mechanisms is organized in an evolutionary progression, and large bodies of data are hierarchically grouped as manifestations of these principles.

In the present article a thought experiment shows how limitations in the types of information available to individual cells can be overcome when the cells act together in suitably designed feedback schemes. The explication of these schemes in a rigorous setting (see the appendices) forces us to study a series of general design problems whose complete solution includes many examples that go beyond the thought experiment; for example, competitive systems (their decision schemes, self-tuning, adaptation, fast pattern transformations, and STM), nonstationary prediction systems (their filtering, pattern learning, and LTM), dipole systems (their transduction and rebound properties), and resonant systems (their hysteresis, deformation, and reset properties). This thought experiment is just one in a series that has helped to unravel psychological mechanisms and to generate as yet untested predictions.

An early thought experiment used the simplest classical conditioning postulates, interpreted in real time (see Grossberg, 1974, for a review), to derive explicit neural networks. When, for example, these networks are exposed serially to long lists, a variety of serial learning properties automatically occur, such as bowing, skewing, anchoring, primacy dominating recency, anticipatory and perseverative generalization gradients, and response oscillation (Grossberg, 1969b; Grossberg & Pepe, 1970, 1971). In addition, mathematical analysis unexpectedly showed how overarousal can cause an attentional deficit with reduced attentional span and collapsed contextual constraints. This overaroused syndrome includes a change toward less skewing of the bowed error curve and toward recency dominating primacy. These formal properties have not yet been empirically tested.

Using these results on classical conditioning, another thought experiment about classical conditioning became necessary. The time intervals between CS and UCS presentations on successive learning trials are not always the same. In a real-time theory, this trivial fact creates a severe synchronization problem whose solution unexpectedly led to explicit mechanisms of instrumental conditioning (Grossberg, 1971a, 1972b, 1972c). Many insights about instrumental mechanisms and their relationship to Pavlovian mechanisms were hereby derived. One of them is especially pertinent to this article. A dipole mechanism was forced on the theory to control net incentive motivation through time. Mathematical analysis of the dipole revealed several unexpected properties (Sections 10 and 11) including the ability of arousal, and hence of unexpected events, to adapt or rebound the dipole. The detailed understanding of dipole dynamics helped to clarify many novelty-related phenomena, such as learned helplessness, superconditioning, and vicious circle behavior. It also forced on the theory the realization that cognitive events, via expectancy matching, can directly influence reinforcement, via the dipole. In summary, a simple environmental pressure concerning a real-time synchronization problem in classical conditioning was solved by mechanisms of instrumental conditioning and led to a role for cognitive processing in the direct evaluation of reinforcement.

With these results in hand, a thought experiment about feature fields came into view. The parallel activation of many cells by external cues can easily destroy decision rules that regulate the balance of net incentive through time. The minimal solution of this difficulty is to impose a normalization property at the processing stages where cues are stored in STM (Grossberg, 1972c). This normalization property had already been noticed as a property by which competitive shunting networks solve the saturation problem (Grossberg, 1970). These results from reinforcement theory made it clear that further progress concerning feature extraction and related perceptual phenomena required a frontal attack on the mathematics of competitive systems. The early results in this direction (Grossberg, 1973) eventually led to many surprising properties, the most general being that every competitive system induces a decision scheme that can be used to predict its behavior through time (Grossberg, 1978c). For present purposes, the normalization and quenching threshold properties are particularly important, since they show how arousal can tune STM, and thereby help to control what cues are overshadowed vs. what cues are processed. Another role for cognitive events, again acting on arousal via expectancies, was hereby discerned.

Once the normalization and quenching threshold properties were discovered, a thought experiment was suggested that joins together facets of perceptual and motivational processing: How can cues with incompatible motivational consequences be processed in parallel without causing chaotic cross-conditioning (Figure 2)? This thought experiment showed how incentive motivational feedback can influence STM storage to yield stable self-consistent coding and, as side benefits, explanations of attentional data such as overshadowing and discrimination-learning data such as peak shift and behavioral contrast (Grossberg, 1975). Several other theoretical stages then followed as the attentional phenomena were recognized to be special cases of the resonance idea. It became possible to build a theory of stable code development (Grossberg, 1976a, 1976b), which, in turn, suggested a psychophysiological foundation for cognitive theory (Grossberg, 1978e), one of whose facets is heuristically summarized by the present thought experiment.

The evolutionary procedure thus embodies a program of real-time theory construction in psychological studies that underscores the need to understand the collective properties of hierarchically organized nonlinear neural networks. Because the rigorous analysis of such networks is well under way, we can anticipate an emergent resonance between experimental psychology and psychophysiological theory during our generation.

References

Anderson, J. A., Silverstein, J. W., Ritz, S. A., & Jones, R. S. Distinctive features, categorical perception, and probability learning: Some applications of a neural model. *Psychological Review*, 1977, *84*, 413–451.

Arbib, M. A. *The metaphorical brain*. New York: Wiley, 1972.

Atkinson, R. C., & Shiffrin, R. M. Human memory: A proposed system and its control processes. In K. W. Spence & J. T. Spence (Eds.), *Advances in the psychology of learning and motivation research and theory* (Vol. 2). New York: Academic Press, 1968.

Atkinson, R. C., & Shiffrin, R. M. The control of short-term memory. *Scientific American*, 1971, *225*, 82–90.

Barlow, H. B., & Levick, W. R. Changes in the maintained discharge with adaptation level in the cat retina. *Journal of Physiology*, 1969, *202*, 699–718. (a)

Barlow, H. B., & Levick, W. R. Three factors limiting the reliable detection of light by retinal ganglion cells of the cat. *Journal of Physiology*, 1969, *200*, 1–24. (b)

Beck, J. *Surface color perception*. Ithaca, N.Y.: Cornell University Press, 1972.

Berger, T. W., & Thompson, R. F. Limbic system interrelations: Functional division among hippocampal–septal connections. *Science*, 1977, *197*, 587–589.

Blakemore, C., & Cooper, G. F. Development of the brain depends on the visual environment. *Nature*, 1970, *228*, 477–478.

Bloomfield, T. M. Behavioral contrast and the peak shift. In R. M. Gilbert & N. S. Sutherland (Eds.), *Animal discrimination learning*. New York: Academic Press, 1969.

Boring, E. G. *A history of experimental psychology* (2nd ed.). New York: Appleton-Century-Crofts, 1950.

Brown, J. L. Afterimages. In C. H. Graham (Ed.), *Vision and visual perception*. New York: Wiley, 1965.

Campbell, B. A. Interaction of aversive stimuli: Summation or inhibition? *Journal of Experimental Psychology*, 1968, *78*, 181–190.

Campbell, B. A., & Kraeling, D. Response strength as a function of drive level and amount of drive reduction. *Journal of Experimental Psychology*, 1953, *45*, 97–101.

Campbell, F. W., & Howell, E. R. Monocular alternation: A method for the investigation of pattern vision. *Journal of Physiology*, 1972, *225*, 19–21.

Campbell, L., & Garnett, W. *The life of James Clerk Maxwell*. London: Macmillan, 1882.

Cant, B. R., & Bickford, R. G. The effect of motivation on the contingent negative variation (CNV). *Electroencephalography and Clinical Neurophysiology*, 1967, *23*, 594.

Carterette, E. C., & Friedman, M. P. (Eds.). *Handbook of perception: Seeing* (Vol. 5). New York: Academic Press, 1975.

Chapman, R. M., McCrary, J. W., & Chapman, J. A. Short-term memory: The "storage" component of human brain response predicts recall. *Science*, 1978, *202*, 1211–1214.

Cohen, J. Very slow brain potentials relating to expectancy: The CNV. In E. Donchin & D. B. Lindsley (Eds.), *Average evoked potentials*. Washington, D.C.: National Aeronautics and Space Administration, 1969.

Cornsweet, T. N. *Visual perception*. New York: Academic Press, 1970.

Craik, F. I. M. & Lockhart, R. S. Levels of processing: A framework for memory research. *Journal of Verbal Learning and Verbal Behavior*, 1972, *11*, 671–684.

Craik, F. I. M., & Tulving, E. Depth of processing and the retention of words in episodic memory. *Journal of Experimental Psychology: General*, 1975, *104*, 268–294.

Creutzfeldt, O. D., & Northdurft, H. C. Representation of complex visual stimuli in the brain. *Naturwissenschaften*, 1978, *65*, 307–318.

Daniels, J. D., & Pettigrew, J. D. Development of neuronal responses in the visual system of cats. In G. Gottlieb (Ed.), *Neural and behavioral specificity* (Vol. 3). New York: Academic Press, 1976.

Denny, M. R. Relaxation theory and experiments. In F. R. Brush (Ed.), *Aversive conditioning and learning*. Academic Press: New York, 1970.

Dickinson, A., Hall, G., & Mackintosh, N. J. Surprise and the attenuation of blocking. *Journal of Experimental Psychology: Animal Behavior Processes*, 1976, *4*, 313–322.

Donchin, E., Gerbrandt, L. A., Leifer, L., & Tucker, L. Is the contingent negative variation contingent on a motor response? *Psychophysiology*, 1972, *9*, 178–188.

Donchin, E., Tueting, P., Ritter, W., Kutas, M., & Heffley, E. *Electroencephalography and Clinical Neurophysiology*, 1975, *38*, 1–13.

Duda, R. O., & Hart, P. E. *Pattern classification and scene analysis*. New York: Wiley, 1973.

Dunham, P. J. Punishment: Method and theory. *Psychological Review*, 1971, *78*, 58–70.

Ellias, S. A., & Grossberg, S. Pattern formation, contrast control, and oscillations in the short term memory of shunting on-center off-surround networks. *Biological Cybernetics*, 1975, *20*, 69–98.

Estes, W. K. Outline of a theory of punishment. In B. A. Campbell & R. M. Church (Eds.), *Punishment and aversive behavior*. New York: Appleton-Century-Crofts, 1969.

Estes, W. K., & Skinner, B. F. Some quantitative properties of anxiety. *Journal of Experimental Psychology*, 1941, *29*, 390–400.

Fender, D., & Julesz, B. Extension of Panum's fusional area in binocularly stabilized vision. *Journal of the Optical Society of America*, 1967, *57*, 819–830.

Foote, W. E., Manciewicz, R. J., & Mordes, J. P. Effect of midbrain raphe and lateral mesencephalic stimulation on spontaneous and evoked activity in the lateral geniculate of the cat. *Experimental Brain Research*, 1974, *19*, 124–130.

Freeman, W. J. *Mass action in the nervous system*. New York: Academic Press, 1975.

Fukuda, Y., & Stone, J. Retinal distribution and central projections of X-, Y-, and W-cells of the cat's retina. *Journal of Neurophysiology*, 1974, *37*, 749–772.

Fuxe, K., Hökfelt, T., & Ungerstedt, U. Morphological and functional aspects of central monoamine neurons. *International Review of Neurobiology*, 1970, *13*, 93–126.

Fuxe, K., & Ungerstedt, U. Histochemical, biochemical, and functional studies on central monoamine neurons after acute and chronic amphetamine administration. In E. Costa & S. Garattini (Eds.), *Amphetamines and related compounds*. New York: Raven Press, 1970.

Gardner, W. J., Licklider, J. C. R., & Weisz, A. Z. Suppression of pain by sound. *Science*, 1961, *132*, 32–33.

Gazzaniga, M. S. *The bisected brain*. New York: Appleton-Century-Crofts, 1970.

Gibson, J. J. Adaptation with negative aftereffect. *Psychological Review*, 1937, *44*, 222–244.

Graham, C. H. Visual form perception. In C. H. Graham (Ed.), *Vision and visual perception*. New York: Wiley, 1965.

Grossberg, S. *The theory of embedding fields with applications to psychology and neurophysiology*. New York: Rockefeller Institute for Medical Research, 1964.

Grossberg, S. On learning and energy–entropy dependence in recurrent and nonrecurrent signed networks. *Journal of Statistical Physics*, 1969, *1*, 319–350. (a)

Grossberg, S. On the serial learning of lists. *Mathematical Biosciences*, 1969, *4*, 201–253. (b)

Grossberg, S. On the production and release of chemical transmitters and related topics in cellular control. *Journal of Theoretical Biology*, 1969, *22*, 325–364. (c)

Grossberg, S. Neural pattern discrimination. *Journal of Theoretical Biology*, 1970, *27*, 291–337.

Grossberg, S. On the dynamics of operant conditioning. *Journal of Theoretical Biology*, 1971, *33*, 225–255. (a)

Grossberg, S. Pavlovian pattern learning by nonlinear neural networks. *Proceedings of the National Academy of Sciences*, 1971, *68*, 828–831. (b)

Grossberg, S. Neural expectation: Cerebellar and retinal analogs of cells fired by learnable or unlearned pattern classes. *Kybernetik*, 1972, *10*, 49–57. (a)

Grossberg, S. A neural theory of punishment and avoidance. I. Qualitative theory. *Mathematical Biosciences*, 1972, *15*, 39–67. (b)

Grossberg, S. A neural theory of punishment and avoidance, II. Quantitative theory. *Mathematical Biosciences*, 1972, *15*, 253–285. (c)

Grossberg, S. Pattern learning by functional-differential neural networks with arbitrary path weights. In K. Schmitt (Ed.), *Delay and functional-differential equations and their applications*. New York: Academic Press, 1972. (d)

Grossberg, S. Contour enhancement, short-term memory, and constancies in reverberating neural networks. *Studies in Applied Mathematics*, 1973, *52*, 217–257.

Grossberg, S. Classical and instrumental learning by neural networks. In R. Rosen & F. Snell (Eds.), *Progress in theoretical biology* (Vol. 3). New York: Academic Press, 1974.

Grossberg, S. A neural model of attention, reinforcement, and discrimination learning. *International Review of Neurobiology*, 1975, *18*, 263–327.

Grossberg, S. Adaptive pattern classification and universal recoding, I: Parallel development and coding of neural feature detectors. *Biological Cybernetics*, 1976, *23*, 121–134. (a)

Grossberg, S. Adaptive pattern classification and universal recording, II: Feedback, expectation, olfaction, and illusions. *Biological Cybernetics*, 1976, *23*, 187–202. (b)

Grossberg, S. Pattern formation by the global limits of a nonlinear competitive interaction in n dimensions. *Journal of Mathematical Biology*, 1977, *4*, 237–256.

Grossberg, S. Behavioral contrast in short-term memory: Serial binary memory models or parallel continuous memory models? *Journal of Mathematical Psychology*, 1978, *17*, 199–219. (a)

Grossberg, S. Communication, memory, and development. In R. Rosen & F. Snell (Eds.), *Progress in theoretical biology* (Vol. 5). New York: Academic Press, 1978. (b)

Grossberg, S. Decisions, patterns, and oscillations in the dynamics of competitive systems with applications to Volterra-Lotka systems. *Journal of Theoretical Biology*, 1978, *73*, 101–130. (c)

Grossberg, S. Do all neural models really look alike? A comment on Anderson, Silverstein, Ritz, and Jones. *Psychological Review*, 1978, *85*, 592–596. (d)

Grossberg, S. A theory of human memory: Self-organization and performance of sensory-motor codes, maps, and plans. In R. Rosen & F. Snell (Eds.), *Progress in theoretical biology* (Vol. 5). New York: Academic Press, 1978. (e)

Grossberg, S., & Levine, D. S. Some developmental and attentional biases in the contrast enhancement and short term memory of recurrent neural networks. *Journal of Theoretical Biology*, 1975, *53*, 341–380.

Grossberg, S., and Pepe, J. Schizophrenia: Possible dependence of associational span, bowing, and primacy vs. recency on spiking threshold. *Behavioral Science*, 1970, *15*, 359–362.

Grossberg, S., & Pepe, J. Spiking threshold and overarousal effects on serial learning. *Journal of Statistical Physics*, 1971, *3*, 95–125.

Hebb, D. O. Drives and the CNS (conceptual nervous system). *Psychological Review*, 1955, *62*, 243–254.

Helmholtz, H. von. *Handbuch der physiologischen optik* (1st. ed.). Hamburg, Leipzig: Voss, 1866.

Helmholtz, H. von. *Physiological optics* (Vol. 2) (J. P. C. Southall, Ed.). New York: Dover, 1962.

Hilgard, E. R., & Bower, G. H. *Theories of learning* (4th ed.). Englewood Cliffs, N.J.: Prentice-Hall, 1975.

Hirsch, H. V. B., & Spinelli, D. N. Visual experience modifies distribution of horizontally and vertically oriented receptive fields in cats. *Science*, 1970, *168*, 869–871.

Hubel, D. H., & Wiesel, T. N. Functional architecture of macaque monkey visual cortex. *Proceedings of the Royal Society of London* (B), 1977, *198*, 1–59.

Irwin, F. W. *Intentional behavior and motivation: A cognitive theory*. Philadelphia, Pa.: Lippincott, 1971.

Irwin, D. A., Rebert, C. S., McAdam, D. W., & Knott, J. R. Slow potential change (CNV) in the human EEG as a function of motivational variables. *Electroencephalography and Clinical Neurophysiology*, 1966, *21*, 412–413.

Jacobowitz, D. M. Effects of 6-hydroxydopa. In E. Usdin & H. S. Snyder (Eds.), *Frontiers in catecholamine research*. New York: Pergamon Press, 1973.

Juhasz, A. Über die komplementärge-färbten nachbilder. *Zeitschrift fur Psychologie*, 1920, *51*, 233–263.

Julesz, B. *Foundations of cyclopean perception*. Chicago: University of Chicago Press, 1971.

Kamin, L. J. Predictability, surprise, attention, and conditioning. In B. A. Campbell & R. M. Church (Ed.), *Punishment and aversive behavior*. New York: Appleton-Century-Crofts, 1969.

Koenigsberger, L. *Hermann von Helmholtz*. (F. A. Welby, trans.). Oxford, England: Clarendon, 1906.

Ladisich, W., Volbehr, H., & Matussek, N. Paradoxical effect of amphetamine on hyperactive states in correlation with catecholamine metabolism in brain. In E. Costa & S. Garattini (Eds.), *Amphetamines and related compounds*. New York: Raven Press, 1970.

Land, E. H. The retinex theory of color vision. *Scientific American*, 1977, *237*, 108–128.

Levine, D. S., & Grossberg, S. Visual illusions in neural networks: Line neutralization, tilt aftereffect, and angle expansion. *Journal of Theoretical Biology*, 1976, *61*, 477–504.

Lindvall, O., & Björklund, A. The organization of the ascending catecholamine neuron systems in the rat brain as revealed by the glyoxylic acid fluorescence method. *Acta Physiologia Scandinavia Supplement*, 1974, *412*, 1–48.

Low, M. D., Borda, R. P., Frost, J. D., & Kellaway, P. Surface negative slow potential shift associated with conditioning in man. *Neurology*, 1966, *16*, 711–782.

Lynn, R. *Attention, arousal, and the orientation reaction*. New York: Pergamon Press, 1966.

Macchi, G., & Rinvik, E. Thalamo-telencephalic circuits: A neuroanatomical survey. In A. Rémond (Ed.), *Handbook of electroencephalography and clinical neurophysiology* (Vol. 2, Pt. A). Amsterdam: Elsevier, 1976.

MacKay, D. M. Moving visual images produced by regular stationary patterns. *Nature*, 1957, *180*, 849–850.

MacKay, D. M., & MacKay, V. What causes decay of pattern-contingent chromatic aftereffects? *Vision Research*, 1975, *15*, 462–464.

Masterson, F. A. Is termination of a warning signal an effective reward for the rat? *Journal of Comparative and Physiological Psychology*, 1970, *72*, 471–475.

McAdam, D. W. Increases in CNS excitability during negative cortical slow potentials in man. *Electroencephalography and Clinical Neurophysiology*, 1969, *26*, 216–219.

McAdam, D. W., Irwin, D. A., Rebert, C. S., & Knott, J. R. Conative control of the contingent negative variation. *Electroencephalography and Clinical Neurophysiology*, 1966, *21*, 194–195.

McAllister, W. R., & McAllister, D. E. Behavioral measurement of conditioned fear. In F. R. Brush (Ed.), *Aversive conditioning and learning*. New York: Academic Press, 1970.

McCollough, C. Color adaptation of edge-detectors in the human visual system. *Science*, 1965, *149*, 1115–1116.

Montalvo, F. S. A neural network model of the McCollough effect. *Biological Cybernetics*, 1976, *25*, 49–56.

Moruzzi, G., & Magoun, H. W. Brain stem reticular formation and activation of the EEG. *Electroencephalography and Clinical Neurophysiology*, 1949, *1*, 455–473.

Myers, A. K. Effects of continuous loud noise during instrumental shock-escape conditioning. *Journal of*

Comparative and Physiological Psychology, 1969, *68*, 617–622.

Piaget, J. *The origins of intelligence in children*. New York: Norton, 1963.

Ratliff, F. *Mach bands: Quantitative studies of neural networks in the retina*. San Francisco: Holden-Day, 1965.

Rauschecker, J. P. J., Campbell, F. W., & Atkinson, J. Colour opponent neurones in the human visual system. *Nature*, 1973, *245*, 42–45.

Remington, R. J. Analysis of sequential effects in choice reaction times. *Journal of Experimental Psychology*, 1969, *2*, 250–257.

Rescorla, R. A., & Wagner, A. R. A theory of Pavlovian conditioning: Variations in the effectiveness of reinforcement and nonreinforcement. In A. Black & W. F. Prokasy (Eds.), *Classical conditioning II*. New York: Appleton-Century-Crofts, 1972.

Ricklan, M. *L-dopa and parkinsonism: A psychological assessment*. Springfield, Ill.: Charles C Thomas, 1973.

Robson, J. G. Receptive fields: Neural representation of the spatial and intensive attributes of the visual image. In E. C. Carterette & M. P. Friedman (Eds.), *Handbook of perception* (Vol. 5). New York: Academic Press, 1976.

Rohrbaugh, J., Donchin, E., & Eriksen, C. Decision making and the P300 component of the cortical evoked response. *Perception & Psychophysics*, 1974, *15*, 368–374.

Schiller, P. H., & Malpeli, J. G. Functional specificity of lateral geniculate nucleus laminae of the rhesus monkey. *Journal of Neurophysiology*, 1978, *41*, 788–797.

Schneider, W., & Shiffrin, R. M. Automatic and controlled information processing in vision. In D. LaBarge & S. J. Samuels (Eds.), *Basic processes in reading: Perception and comprehension*. Hillsdale, N.J.: Erlbaum, 1976.

Seligman, M. E. P., Maier, S. F., & Solomon, R. L. Unpredictable and uncontrollable aversive events. In F. R. Brush (Ed.), *Aversive conditioning and learning*. New York: Academic Press, 1971.

Singer, W. Control of thalamic transmission by corticofugal and ascending reticular pathways in the visual system. *Physiological Review*, 1977, *57*, 386–420.

Sluckin, W. *Imprinting and early learning*. London: Methuen, 1964.

Squires, K., Wickens, C., Squires, N., & Donchin, E. The effect of stimulus sequence on the waveform of the cortical event-related potential. *Science*, 1976, *193*, 1142–1146.

Stein, L. Norepinephrine reward pathways: Role in self-stimulation, memory consolidation, and schizophrenia. In J. K. Cole & T. B. Sonderegger (Eds.), *Nebraska Symposium on Motivation* (Vol. 22). Lincoln: University of Nebraska Press, 1974.

Stryker, M., & Sherk, H. Modification of cortical orientation selectivity in the cat by restricted visual experience: A reexamination. *Science*, 1975, *190*, 904–905.

Tanji, J., & Evarts, E. V. Anticipatory activity of motor cortex neurons in relation to direction of an intended movement. *Journal of Neurophysiology*, 1976, *39*, 1062–1068.

Thomas, G. B., Jr. *Calculus and analytic geometry*. Reading, Mass.: Addison-Wesley, 1968.

Trabasso, T., & Bower, G. H. *Attention in learning: Theory and research*. New York: Wiley, 1968.

Tsumoto, T., Creutzfeldt, O. D., & Legéndy, C. R. Functional organization of the corticofugal system from visual cortex to lateral geniculate nucleus in the cat. *Experimental Brain Research*, 1978, *32*, 345–364.

Tsumoto, T., & Suzuki, D. A. Effects of frontal eye field stimulation upon activities of the lateral geniculate body of the cat. *Experimental Brain Research*, 1976, *25*, 291–306.

Ungerstedt, U. Stereotoxic mapping of the monoamine pathways in the rat brain. *Acta Physiologica Scandinavia*, 1971, *82* (Supplement 367), 1–48.

Wagner, A. R. Frustrative nonreward: A variety of punishment. In B. A. Campbell & R. M. Church (Eds.), *Punishment and aversive behavior*. New York: Appleton-Century-Crofts, 1969.

Walter, W. G. Slow potential waves in the human brain associated with expectancy, attention, and decision. *Arch. Psychiat. Nervenkr.*, 1964, *206*, 309–322.

Wilson, H. A synaptic model for spatial frequency adaptation. *Journal of Theoretical Biology*, 1975, *50*, 327–352.

Wurtz, R. H., & Goldberg, M. E. The role of the superior colliculus in visually evoked eye movement. In J. Dichgans & E. Bizzi (Eds.), *Cerebral control of eye movement and motion perception*. Basel, Switzerland: Karger, 1972.

Appendix A

This section summarizes some of the mechanisms whereby an activity pattern across $F^{(1)}$ elicits signals to $F^{(2)}$ that filter the pattern. The filtered signal pattern is then rapidly transformed by recurrent competitive interactions within $F^{(2)}$ before the resultant pattern is stored in STM. The STM pattern endures long enough to alter the interfield path strengths that define the filter. This is an LTM change. Then the process repeats itself until STM and LTM equilibrate.

Filter

Denote the cells of $F^{(1)}$ by $v_i^{(1)}$, $i = 1, 2, \ldots, n$, and the cells of $F^{(2)}$ by $v_j^{(2)}$, $j = 1, 2, \ldots, N$. Let the activity of $v_i^{(1)}$ at time t be $x_i^{(1)}(t)$. Suppose that the activity $x_i^{(1)}(t)$ elicits a signal $S_i = S_i(x_i^{(1)}(t))$ in the pathways from $v_i^{(1)}$ to $F^{(2)}$. Let the net signal from $v_i^{(1)}$ to $v_j^{(2)}$ be $S_i z_{ij}$, where z_{ij} provides a measure of the efficiency of the pathway e_{ij} from $v_i^{(1)}$ to $v_j^{(2)}$. In other words, z_{ij} *gates* signal S_i on its way to $v_j^{(2)}$. Then the *total* signal from $F^{(1)}$ to $v_j^{(2)}$ is

$$T_j = \sum_{i=1}^{n} S_i z_{ij}.$$

This equation for T_j has an informative geometrical interpretation in terms of the vectors $S = (S_1, S_2, \ldots, S_n)$ of signals and $z_j = (z_{1j}, z_{2j}, \ldots, z_{nj})$ of path strengths from $F^{(1)}$ to $v_j^{(2)}$. Namely, T_j is the dot product, or inner product, of S and z_j, which is written $T_j = S \cdot z_j$ (Thomas, 1968). The dot product can be evaluated in terms of the vector lengths

$$\|S\| = \sqrt{\sum_{i=1}^{n} S_i^2}$$

and

$$\|z_j\| = \sqrt{\sum_{i=1}^{n} z_{ij}^2},$$

and the cosine of the angle between S and z_j by the formula $T_j = \|S\|\,\|z_j\| \cos(S, z_j)$. In particular, if all $\|z_j\|$ are equal, then the cell $v_j^{(2)}$ in $F^{(2)}$ that receives the largest signal is the cell whose $\cos(S, z_j)$ is maximal. The cosine can be increased by choosing the coefficients of z_j more proportional, or parallel, to S, and can be decreased by choosing the coefficients more perpendicular, or orthogonal, to S. Thus each z_j *filters* S by producing a net signal T_j whose size depends on how parallel z_j is to S. Otherwise expressed, z_j *projects* S onto $v_j^{(2)}$. The pattern $T = (T_1, T_2, \ldots, T_N)$ of inputs to $F^{(2)}$ represents pattern S by projecting S onto all the cells $F^{(2)}$ with relative input sizes that depend on the choice of all the vectors $z_1, z_2, \ldots, z_N$.

Contrast Enhancement

When S is first presented to $F^{(1)}$, the pattern T of inputs that it elicits across $F^{(2)}$ might be approximately uniform. Recurrent on-center off-surround interactions within $F^{(2)}$ rapidly contrast enhance this input pattern in order to produce a sharper pattern of STM activities across $F^{(2)}$ (Grossberg, 1976a, 1976b). I illustrate this concept with the simplest case: Suppose that $F^{(2)}$ can choose the cell whose initial input is maximal for storage in STM. Denoting the activity of $v_j^{(2)}$ by $x_j^{(2)}$, this law says that

$$\begin{aligned} x_j^{(2)} &= 1 \text{ if } T_j > \max\{\epsilon, T_k, k \neq j\} \\ x_j^{(2)} &= 0 \text{ if } T_j < \max\{\epsilon, T_k, k \neq j\}. \end{aligned} \tag{A1}$$

The coefficient ϵ designates a quenching threshold that must be exceeded before any STM storage is possible. Suppose for definiteness that $T_1 > \max\{\epsilon, T_k, k \neq 1\}$. Then the activity of $v_1^{(2)}$ is rapidly contrast enhanced and stored in STM, whereas all other activities across $F^{(2)}$ are suppressed.

Coding

The path strengths z_{ij} are LTM traces that can slowly adapt to the signal pattern S from $F^{(1)}$ and the STM pattern across $F^{(2)}$. In the simplest case, z_{ij} changes only if $x_j^{(2)} > 0$. Then

$$\frac{d}{dt} z_{ij} = (-z_{ij} + S_i) x_j^{(2)}. \tag{A2}$$

For example, if $x_1^{(2)} = 1$ and all $x_j^{(2)} = 0$, $j \neq 1$, then this LTM law causes the signal $T_1 = S \cdot z_1$ to be maximized as S is practiced by making z_1 become parallel to S. In this way, presentation of S at $F^{(1)}$ can induce a differentiated STM pattern across $F^{(2)}$ by changing the LTM vectors $z_1, z_2, \ldots, z_N$.

Grossberg's (1976a, 1976b, 1978b, 1978e) articles describe these mechanisms in greater detail. They also show how to generalize the mechanisms to include more complex STM and LTM interactions. Despite these generalizations, the mechanisms are there shown to be unstable in a complex input environment. A precise understanding of this difficulty forced the use of learned feedback expectancies.

Appendix B

This section indicates how a single cell population in $F^{(2)}$ can learn a spectial pattern of activity across $F^{(1)}$. Analogous arguments then show how many simultaneously active cell populations across $F^{(2)}$ can learn a spatial pattern across $F^{(1)}$, albeit not necessarily the same spatial pattern that would have excited a single cell in $F^{(2)}$ by interfield signaling from $F^{(1)}$ to $F^{(2)}$.

Associative Learning

Our laws for associative learning appeared in a monograph by Grossberg (1964), and were mathematically analyzed in a series of articles, leading to a universal theorem of associative learning in Grossberg (1969a, 1971b, 1972d). The universal theorem proves that if these associative learning laws were invented at a prescribed time during the evolutionary process, then they could be used to guarantee unbiased associative learning in essentially any later evolutionary specialization. That is, the laws are capable of learning arbitrary spatial patterns in arbitrarily many, simultaneously active sampling channels that are activated by arbitrary continuous data preprocessing in an essentially arbitrary anatomy. Learning of arbitrary space-time patterns is also guaranteed given modest requirements on the temporal regularity of stimulus sampling. (See Grossberg, 1974, for a review.) Herein I summarize the fact that the unit of LTM is a *spatial pattern*. This is done by considering the *minimal* anatomy that is capable of classical conditioning.

STM and LTM Laws That Factor Pattern from Activity

Let presentation of a CS create an input $I_0(t)$ that activates the cell population v_0. Let the UCS create an input pattern $(I_1(t), I_2(t), \ldots, I_n(t))$ that activates the cell populations $v_1, v_2, \ldots, v_n$, whose outputs elicit the UCR. Let the STM trace of v_i be $x_i(t)$, j = 0, 1, ..., n, and let the LTM trace of the axon pathway e_{0i} from v_0 to v_i be $z_{0i}(t)$, i = 1, 2, ..., n (Figure A1). Suppose that the STM and LTM traces obey the laws

$$\frac{d}{dt}x_0 = -A_0x_0 + I_0(t) \quad \text{(A3)}$$

$$\frac{d}{dt}x_i = -Ax_i + Bz_{0i} + I_i(t), \quad \text{(A4)}$$

and

$$\frac{d}{dt}z_{0i} = -Cz_{0i} + Dx_i, \quad \text{(A5)}$$

$i = 1, 2, \ldots, n$. The terms A_0 and A are STM decay rates. The term C is the LTM decay rate. The terms B and D are signals from v_0 along all the pathways e_{0i}, $i = 1, 2, \ldots, n$; for example, $B(t) = f(x_0(t - \tau))$, where $f(w)$ is a sigmoid function of w. The LTM trace z_{0i} is computed at the interface of the synaptic knob S_{0i} (at the end of e_{0i}) and the postsynaptic cell v_i—that is, at the synaptic knob and/or postsynaptic membrane —where it can gate the signals B on their way to v_i, as in term Bz_{0i} of Equation A4, and simultaneously time-average (term $-Cz_{0i}$) the product of signals D and postsynaptic STM trace x_i (term Dx_i), in A5. In particular, A2 is a special case of A5.

A *spatial pattern* is a UCS whose *relative* activities remain fixed, even though their absolute activities can fluctuate through time,

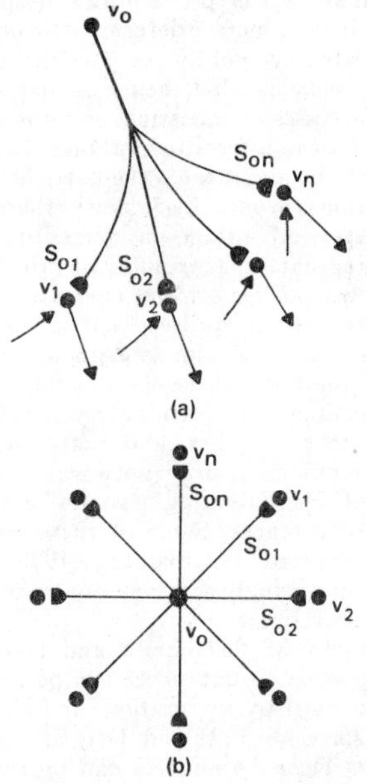

Figure A1. In (a) the conditioned stimulus (CS)-activated population v_0 samples the unconditioned stimulus (UCS)-activated populations $v_1, v_2, \ldots, v_n$; in (b) the *outstar* is the minimal network capable of classical conditioning.

namely, $I_i(t) = \theta_i I(t)$, $i = 1, 2, \ldots, n$, where θ_i is the fixed relative activity and $I(t)$ is the total UCS activity. The convention

$$\sum_{i=1}^{n} \theta_i = 1$$

guarantees the normalization

$$I(t) = \sum_{i=1}^{n} I_i(t).$$

The relative values $\theta = (\theta_1, \theta_2, \ldots, \theta_n)$ are like generalized "reflectances" that carry the information in the UCS pattern, whereas $I(t)$ provides the UCS activity that drives system changes in response to θ. It is shown below how this system, which I call an *outstar* (Figure A1), can factorize pattern information θ from information about total activity $I(t)$. This property has many important implications. For example, θ is a probability distribution, since each $\theta_i \geq 0$ and

$$\sum_{k=1}^{n} \theta_k = 1.$$

The system learns probabilities despite the fact that it can generate deterministic behavior. There exists a type of "wave-particle" dualism in these systems that helps to explain the partial successes of statistical learning models, and provides an interesting vantage point from which to think about the wave-particle dualism of quantum theory. Also, since there is no evolutionary advantage in perceptually discriminating data that cannot, in principle, be learned, we can expect the neural perceptual apparatus also to process spatial patterns. The brightness and hue constancies of vision illustrate this fact. These observations clarified how perceptual and learning mechanisms are matched to each other, and suggested study of the minimal neural networks that are capable of discriminating a spatial pattern θ; that is, reflectances. Some of these networks were constructed in Grossberg (1970, 1972a) and, not surprisingly, have an anatomy that is remarkably retinal.

System A3–A5 *factorizes* θ and $I(t)$ in the following sense. Equation A3 can be explicitly solved for $x_0(t)$ by integration, and the result used to solve for $B(t)$ and $D(t)$ as functions of time t. Then A4 and A5 can be rewritten in terms of the relative STM traces

$$X_i = x_i\left(\sum_{k=1}^{n} x_k\right)^{-1}$$

and relative LTM traces

$$Z_i = z_{0i}\left(\sum_{k=1}^{n} z_{0k}\right)^{-1}$$

as follows:

$$\frac{d}{dt}X_i = E(Z_i - X_i) + F(\theta_i - X_i) \quad \text{(A6)}$$

and

$$\frac{d}{dt}Z_i = G(X_i - Z_i). \quad \text{(A7)}$$

The coefficients E, F, and G depend only on $I(t)$, on the total STM activity

$$x = \sum_{k=1}^{n} x_k,$$

and on the total LTM activity

$$z = \sum_{k=1}^{n} z_{0k}.$$

By A4 and A5,

$$\frac{d}{dt}x = -Ax + Bz + I \quad \text{(A8)}$$

and

$$\frac{d}{dt}z = -Cz + Dx. \quad \text{(A9)}$$

Equations A8 and A9 are independent of θ; they depend only on the total activity $I(t)$. These equations *decouple* total activity data (I, x, z) from pattern data (θ, X, Z), where $X = (X_1, X_2, \ldots, X_n)$ and $Z = (Z_1, Z_2, \ldots, Z_n)$ are also probability distributions. The total activity data influence the pattern data only via the coefficients E, F, and G, which are always nonnegative. No matter how wildly the CS input $I_0(t)$ and the UCS input $I(t)$ oscillate through time, these coefficients influence only the *rates* with which X and Z are influenced by θ, but not the *directions* in which X and Z can change in response to θ. It is this property that generalizes to yield the universal theorem cited above.

In particular, term $F(\theta_i - X_i)$ in A6 says that X_i approaches θ_i as learning proceeds (UCS read into STM). Term $E(Z_i - X_i)$ in A6 says X_i approaches Z_i (readout of LTM into STM). The net effect of these two terms shows how present demands of the UCS, expressed via θ, and past memories, expressed via Z, compete to change STM via X. Equation A7 shows that Z_i approaches X_i (transfer from STM to LTM). As X approaches θ, and Z approaches X, Z learns the spatial pattern θ. On later performance trials, a CS input to v_0 activates x_0, which in turn activates the signal B. Signal B reads the pattern Z into STM via the terms Bz_{0i} in A4. Since $Z \cong \theta$, A4 shows that the x_is that are activated in this fashion are proportional to the θ_is, as desired.

Many aspects of associative learning can be understood using these STM and LTM laws in more complex anatomies. In particular, the Z_is are stimulus sampling probabilities whose properties explain in a neural setting the partial successes of statistical learning models. The distributions of STM and LTM traces also mimic and predict various data about serial learning, paired associate learning, and free recall experiments. See Grossberg (1974, 1978a, 1978e) for additional discussion.

Appendix C

This section summarizes how feedforward competitive interactions solve the saturation problem using automatic gain control by inhibitory signals, and how properties such as noise suppression, pattern matching, edge enhancement, and spatial frequency sensitivity follow as special cases.

Noise-Saturation Dilemma

All cellular systems face the following dilemma. If their inputs are too small, they can get lost in noise. If the inputs are too large, they can turn on all excitable sites, thereby saturating the system and rendering it insensitive to input differences across the cells. For example, suppose that the ith cell v_i receives an input I_i that can turn on some of its B excitable sites by mass action. Let $x_i(t)$ be the number of excited sites and $B - x_i(t)$ be the number of unexcited sites at time t. The simplest mass action law for turning on unexcited sites and letting excited sites spontaneously turn off is

$$\frac{d}{dt}x_i = -Ax_i + (B - x_i)I_i, \qquad \text{(A10)}$$

$i = 1, 2, \ldots, n$. Term $(B - x_i)I_i$ says that the input I_i turns on unexcited sites $B - x_i$ by mass action. Term $-Ax_i$ says that excited sites spontaneously becomes unexcited by mass action at rate A. Hence, when $I_i \equiv 0$, x_i can decay to the equilibrium point 0.

System A10 is inadequate for the following reason: Let the inputs form a spatial pattern $I_i = \theta_i I$. Given a fixed pattern $\theta = (\theta_1, \theta_2, \ldots, \theta_n)$, choose a background intensity I and let the system reach equilibrium. This equilibrium is found by setting $(d/dt)x_i = 0$ and solving for x_i:

$$x_i = \frac{B\theta_i I}{A + \theta_i I}. \qquad \text{(A11)}$$

Now keep θ fixed and increase I. That is, process the same pattern with different background activity. Then all x_i in A11 approach B even if the relative input intensity θ_i is small. This is saturation. How can the system preserve its sensitivity to θ even as I increases? In other words, how does the ith cell v_i compute its "reflectance" θ_i in response to a spatial pattern $I_i = \theta_i I$, $i = 1, 2, \ldots, n$, of inputs? Since

$$\theta_i = I_i I^{-1} = I_i\Big(\sum_{k=1}^{n} I_k\Big)^{-1},$$

cell v_i needs to know what all the inputs $I_1, I_2, \ldots, I_n$ are in order to compute θ_i. Since

$$\theta_i = I_i\Big(I_i + \sum_{k \neq i} I_k\Big)^{-1},$$

increasing the ith input I_i "excites" v_i (increases θ_i), whereas increasing any input I_k, $k \neq i$, "inhibits" v_i (decreases θ_i). When this intuition is most simply modeled by a cellular mass action network, we find the system

$$\frac{d}{dt}x_i = -Ax_i + (B - x_i)I_i - x_i \sum_{k \neq i} I_k, \qquad \text{(A12)}$$

$i = 1, 2, \ldots, n$. In Equation A12, I_i excites v_i via term $(B - x_i)I_i$, just as in A10. The new term

$$-x_i \sum_{k \neq i} I_k$$

describes how the inputs I_k, $k \neq i$, inhibit (note the minus sign) the excited sites of v_i (which number x_i) by mass action. The *gain* of x_i is its decay rate. This is found by grouping together all the terms that multiply x_i. The sum of these terms is $A + I$, where

$$I = \sum_{k=1}^{n} I_k.$$

Thus the inputs automatically change the gain of x_i. In A10 the gain of x_i is $A + I_i$. The two gains differ by the sum

$$\sum_{k \neq i} I_k$$

of inhibitory signals. We now note how automatic gain control by the inhibitory signals overcomes the saturation problem.

Present a spatial pattern $I_i = \theta_i I$ to A12 and let each x_i reach equilibrium. Setting $(d/dt)x_i = 0$, we find

$$x_i = \theta_i \frac{BI}{A + I}. \qquad (A13)$$

In A13, x_i remains proportional to θ_i no matter how intense I is, and $BI(A + I)^{-1}$ has the form of a Weber-Fechner law. The saturation problem is hereby overcome using automatic gain control by inhibitory signals.

Noise Suppression

In A12, the passive equilibrium point, due to term $-Ax_i$, and the inhibitory saturation point, due to term

$$-x_i \sum_{k \neq i} I_k,$$

are both zero. This is not always true in vivo, where a cell potential can sometimes be actively inhibited below the passive equilibrium point. How does this fact alter pattern processing? Consider the system

$$\frac{d}{dt}x_i = -Ax_i + (B - x_i)I_i - (x_i + C) \sum_{k \neq i} I_k, \qquad (A14)$$

which differs from A12 only in that x_i can fluctuate between B and $-C$, rather than B and 0, where $-C < 0$. Often in vivo B represents the saturation point of a Na^+ channel, $-C$ represents the saturation point of a K^+ channel, and B is much larger than C.

To see how the inhibitory saturation point C influences pattern processing, let A14 equilibrate to the spatial pattern $I_i = \theta_i I$. Setting $(d/dt)x_i = 0$, we find the equilibrium activities

$$x_i = \frac{(B + C)I}{A + I}\left(\theta_i - \frac{C}{B + C}\right). \qquad (A15)$$

By A15, $x_i > 0$ only if $\theta_i > C(B + C)^{-1}$. The constant $C(B + C)^{-1}$ is an *adaptation level* that θ_i must exceed in order to excite x_i. For simplicity, suppose that the ratio CB^{-1} matches the ratio of the number of cells excited by each I_i, namely 1, to the number of cells inhibited by I_i, namely $(n - 1)$. If $CB^{-1} = (n - 1)^{-1}$, then $C(B + C)^{-1} = 1/n$. Since, in response to a uniform spatial pattern of inputs, all $\theta_i = 1/n$, no matter how intense I is, it then follows by A15 that all $x_i = 0$. This is noise suppression in its simplest form. It is due to a matched symmetry-breaking between the intracelluar excitatory versus inhibitory parameters (B, C) and the intercellular spread of off-surround versus on-center pathways.

Edge Enhancement, Spatial Frequency Detection, and Pattern Matching

The noise suppression property generalizes to systems whose excitatory and inhibitory interactions can depend on intercellular distances, as in

$$\frac{d}{dt}x_i = -Ax_i + (B - x_i) \sum_{k=1}^{n} I_k C_{ki} - (x_i + D) \sum_{k=1}^{n} I_k E_{ki}, \qquad (A16)$$

where C_{ki} (E_{ki}) is the excitatory (inhibitory) coefficient from v_k to v_i. Noise suppression at v_i (i.e., $x_i \leq 0$) occurs in response to a uniform pattern (all $\theta_i = 1/n$) in A16 if

$$B \sum_{k=1}^{n} C_{ki} \leq D \sum_{k=1}^{n} E_{ki}, \qquad (A17)$$

which generalizes $CB^{-1} = (n - 1)^{-1}$ in A15. If a rectangular pattern perturbs such a network, then a cell's activity x_i will be suppressed either if its interactions fall so far outside the rectangle or so far inside it that the pattern looks uniform to its interaction coefficients C_{ki} and E_{ki}. Consequently, only activities near the edge of the rectangle will be enhanced. More generally, the spatial gradients of activity in any input pattern are matched against the spatial gradients in each cell's interaction coefficients to enhance the activity of only those cells to whom the input pattern looks nonuniform. In recurrent networks, this property is supplemented by active contrast-enhancing, disinhibitory, and STM processes that can join together cells with similar interaction gradients into a dynamically coherent subfield that is sensitive to a band of spatial frequencies in the input patterns.

Pattern matching is illustrated as follows. Suppose in A14 that each input I_i is a sum of two inputs J_i and K_i whose patterns $J = (J_1, J_2, \ldots, J_n)$ and $K = (K_1, K_2, \ldots, K_n)$ are to be matched. If J and K mismatch each other's peaks and troughs to form an almost uniform total pattern $I = (I_1, I_2, \ldots, I_n)$, then by A15 all x_i will be inhibited if $CB^{-1} \geq (n-1)^{-1}$. By contrast, if the two patterns reinforce each other, say $J_i = \alpha K_i$, then by (A15),

$$x_i = \frac{(B+C)(1+\alpha)\bar{K}}{A+(1+\alpha)\bar{K}}\left[\theta_i - \frac{C}{B+C}\right], \quad (A18)$$

where

$$\bar{K} = \sum_{i=1}^{n} K_i$$

and $\theta_i = K_i(\bar{K})^{-1}$. In other words, matching J and K amplifies each x_i without changing the pattern θ_i.

Appendix D

This section summarizes some properties of recurrent on-center off-surround networks, including normalization, contrast enhancement, quenching threshold, and STM properties.

To see how recurrent networks normalize their STM activity, we first note by Appendix C that these networks need competitive interactions to solve the noise-saturation dilemma. The simplest recurrent on-center off-surround network is defined by

$$\frac{d}{dt}x_i = -Ax_i + (B - x_i)[f(x_i) + I_i] - x_i[\sum_{k \neq i} f(x_k) + J_i], \quad (A18)$$

$i = 1, 2, \ldots, n$. As usual, x_i is the STM activity of v_i, term $(B - x_i)f(x_i)$ describes the self-excitation of v_i via a positive feedback signal $f(x_i)$—the recurrent on-center—and term

$$-x_i \sum_{k \neq i} f(x_k)$$

describes the inhibition of v_i via negative feedback signals $f(x_k)$, $k \neq i$—the recurrent off-surround. Term I_i is the *i*th excitatory input, and term J_i is the *i*th inhibitory input, for example,

$$J_i = \sum_{k \neq i} I_k$$

in A12.

Contrast Enhancement, Normalization, and Quenching Threshold

An important problem in system A18 is to choose the feedback signal function $f(w)$ as a function of activity level w in such a way as to suppress noise but contrast enhance and store in STM behaviorally important patterns. This problem was solved in Grossberg (1973). The solution is reviewed in Grossberg (1978e, Sections 14 and 15).

To understand the simplest STM properties, A18 is transformed into pattern variables $X_i = x_i x^{-1}$ and total activity variables

$$x = \sum_{k=1}^{n} x_k$$

using the notation $g(w) = w^{-1}f(w)$ and supposing that all $I_i = J_i = 0$. Then

$$\frac{d}{dt}X_i = BX_i \sum_{k=1}^{n} X_k[g(X_i x) - g(X_k x)] \quad (A19)$$

and

$$\frac{d}{dt}x = -Ax + (B - x)\sum_{k=1}^{n} f(X_k x). \quad (A20)$$

For example, if $f(w)$ is linear, namely, $f(w) = Cw$, then $g(w) = C$ and all $(d/dt)X_i = 0$ in A19. In other words, A19 can perfectly remember *any* initial pattern of reflectances. However, by A20 if $A \geq B$, then $x(t)$ approaches zero as $t \rightarrow \infty$, whereas if $B > A$, then $x(t)$ approaches $B - A$ as $t \rightarrow \infty$, whether or not a prior input pattern occurs. Thus if STM storage is ever possible, then $B > A$, and consequently noise will be amplified as vigorously as inputs. A linear signal amplifies noise, and is therefore inadequate despite its perfect memory of reflectances.

A slower-than-linear signal $f(w)$, for example, $f(w) = Cw(D + w)^{-1}$ or more generally, any $f(w)$ such that $g(w)$ is monotone decreasing, is even worse. By A19, if $X_i > X_k$, $k \neq i$, then $(d/dt)X_i < 0$ and if $X_i < X_k$, $k \neq i$, then $(d/dt)X_i > 0$. All differences in reflectances are hereby erased by the reverberation, and noise amplification also occurs. The whole network experiences a type of seizure.

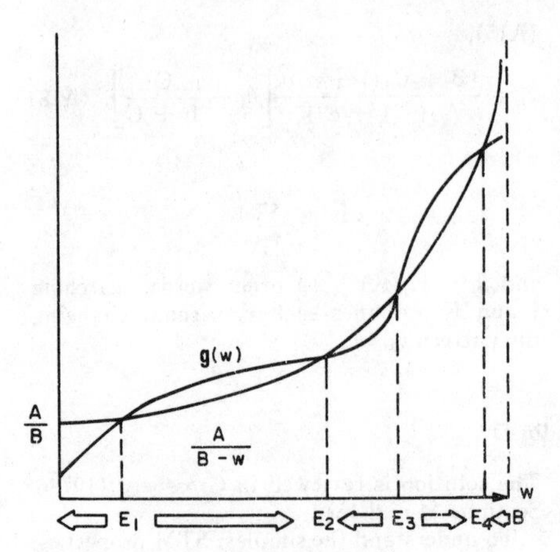

Figure A2. The even solutions E_0, E_2, ... of $g(w) = A(B - w)^{-1}$ are stable equilibrium points of $x(\infty) = \lim_{t\to\infty} x(t)$. (Since $g(w) = w^{-1}f(w)$, these points are solutions of $f(w) = Aw(B - w)^{-1}$. If $x(0) < E_1$, then $x(\infty) = 0$; thus E_1 defines the level below which $x(t)$ is treated as noise and quenched. All equilibrium points satisfy $E_i \leq B$; hence, short-term memory is normalized.)

If $f(w)$ is faster than linear, then the situation is better; for example, $f(w) = Cw^n$, $n > 1$, or more generally any $f(w)$ such that $g(w)$ is monotone increasing. In this case, if $X_i > X_k$, $k \neq i$, then $(d/dt)X_i > 0$, and if $X_i < X_k$, $k \neq i$, then $(d/dt)X_i < 0$. Consequently, this network chooses the population with the initial maximum in activity and totally suppresses activity in all other populations. This network behaves like a finite state, or binary choice machine. The same is true for total activity, since as $t \to \infty$, A20 becomes approximately

$$(d/dt)x \cong x[-A + (B - x)g(x)]. \qquad \text{(A21)}$$

Thus the equilibrium points of $x(t)$ as $t \to \infty$ are $E_0 = 0$ and all the solutions of the equation

$$g(x) = A(B - x)^{-1}. \qquad \text{(A22)}$$

If $g(0) < A/B$, then the smallest solution E_1 of A22 is unstable (Figure A2) so that small activities $x(t)$ are suppressed as $t \to \infty$. This is noise suppression due to recurrent competition. Every other solution E_2, E_4, ... of A22 is a stable equilibrium point of $x(t)$ as $t \to \infty$ (total activity quantization) and all equilibria are smaller than B (normalization).

The faster-than-linear signal contrast enhances the pattern so violently that the good property of noise suppression is joined to the extreme property of binary choice. This latter property is weakened by contructing a hybrid signal function that is chosen faster than linear at small activities to achieve noise suppression, but which levels off at high activities if only because all signal functions must be bounded. In the simplest case, $f(w)$ is a sigmoid, or S-shaped signal function. Then there exists a quenching threshold (QT). If v_is initial activity $x_i(0)$ falls below the QT, then its STM activity is quenched, or laterally masked: $x_i(\infty) = 0$. All the $x_i(0)$s that exceed the QT are contrast enhanced and stored in STM. Simultaneously, the total STM activity is normalized. Speaking intuitively, the QT exists because the faster-than-linear range starts to contrast enhance the pattern. Simultaneously, normalization shifts the activities into the intermediate linear range that stores any pattern, in particular the partially contrast-enchanced pattern. Because a QT exists, the network is a tunable filter. For example, a nonspecific arousal signal that multiplicatively inhibits all the recurrent inhibitory interneurons will lower the QT and facilitate storage of inputs in STM. Grossberg and Levine (1975) mathematically studied how such attentional shunts alter the resultant STM pattern by differentially sensitizing prescribed subfields of feature detectors that are joined together by competitive feedback interactions. The privileged subfields mask the activities in less sensitive subfields.

Such examples, either taken separately or linked together by feedback, provide insight into how interactions between continuously fluctuating quantities can sometimes generate discrete collective properties of the system as a whole. More generally, Grossberg (1978c) proves that every competitive system induces a decision scheme that can be used to globally characterize its pattern transformations as time goes on.

Appendix E

This section summarizes how the simplest transduction law realizable by a depletable chemical generates properties of antagonistic rebound due to specific cue offset and to nonspecific arousal onset when two parallel transduction pathways compete.

Transmitters as Gates

The transmitter law that we need can be derived in two ways. Originally, it was derived as the minimal law that was compatible with psychological postulates of classical conditioning (Grossberg, 1969c, Section 20; Grossberg, 1972c, Section 2). I now show that the law is the simplest transduction rule that can be computed using a depletable chemical transducer.

The simplest transduction rule converts an input I into a proportional signal S, namely,

$$S = BI, \tag{A23}$$

where $B > 0$ is some proportionality constant. Equation A23 says that I is *gated* by B to yield S. If we interpret B as the amount of transducer and BI as the rate with which transducer is released to create signal S, then A23 says that the input I activates the transducer B in a statistically independent, or mass action, way.

When the transducer is released to activate another cell, there must exist a mechanism whereby it can be replenished, so that A23 can be maintained, at least approximately, through time.

Let $z(t)$ be the amount of transducer at time t. How can we keep $z(t) \cong B$ for all $t \geq 0$ so that the transduction rule

$$S = Iz(t) \tag{A24}$$

approximately agrees with A23? This question leads to the following law for the temporal evolution of the amount $z(t)$ of available transducer

$$\frac{dz}{dt} = A(B - z) - Iz. \tag{A25}$$

The term $A(B - z)$ in A25 says that $z(t)$ accumulates until it attains level B. The term does this by accumulating transducer at rate AB, that is proportional to B, and by feedback inhibition of the production rate at a rate $-Az(t)$ that is proportional to $z(t)$. The term $-Iz(t)$ in A25 indicates that transducer is depleted at a rate proportional to its rate of elimination, which is due to gating of I by $z(t)$. When $z(t) \cong B$, term $-Iz$ is proportional to $-BI$, as required by A23. Thus A25 is the law that "corresponds" to the law $S = BI$ when depletion of transducer can occur. It describes four effects working together: production, feedback inhibition, gating, and depletion.

Rebound Due to Cue Offset

Suppose that the adaptation level is I and that the cue input is J. Consider the simplest case in which the total signal in the on-channel is $S_1 = I + J$ and in the off-channel is $S_2 = I$. Let the transmitter z_1 in the on-channel satisfy the equation

$$\frac{d}{dt}z_1 = A(B - z_1) - S_1 z_1, \tag{A26}$$

and the transmitter z_2 in the off-channel satisfy the equation

$$\frac{d}{dt}z_2 = A(B - z_2) - S_2 z_2. \tag{A27}$$

After z_1 and z_2 equilibrate to S_1 and S_2, $(d/dt)z_1 = (d/dt)z_2 = 0$. Thus by A26 and A27,

$$z_1 = \frac{AB}{A + S_1} \tag{A28}$$

and

$$z_2 = \frac{AB}{A + S_2}. \tag{A29}$$

Since $S_1 > S_2$, it follows that $z_1 < z_2$; that is, z_1 is depleted more than z_2. However, the gated signal in the on-channel is $S_1 z_1$, and the gated signal in the off-channel is $S_2 z_2$. Since

$$S_1 z_1 = \frac{ABS_1}{A + S_1} \tag{A30}$$

and

$$S_2 z_2 = \frac{ABS_2}{A + S_2}, \tag{A31}$$

it follows from $S_1 > S_2$ that $S_1 z_1 > S_2 z_2$ despite the fact that $z_1 < z_2$. Thus the on-channel gets a bigger signal than the off-channel. After the two channels compete, the cue input J produces a sustained on-response whose size is proportional to

$$S_1 z_1 - S_2 z_2 = \frac{A^2 BJ}{(A + I + J)(A + I)}. \tag{A32}$$

Now shut J off. Then the cell potentials rapidly adjust until new signal values $S_1^* = I$ and $S_2^* = I$ obtain. However, the transmitters z_1 and z_2 change much more slowly, so that A28 and A29 are approximately valid in a time interval that follows J offset. Thus the net signals are approximately

$$S_1^* z_1 = \frac{ABI}{A + S_1} \tag{A33}$$

and

$$S_2^* z_2 \cong \frac{ABI}{A + S_2}. \tag{A34}$$

Since $S_1 > S_2$, $S_1{}^*z_1 < S_2{}^*z_2$. The off-channel now gets the bigger signal, so an antagonistic rebound occurs whose size is approximately

$$S_2{}^*z_2 - S_1{}^*z_1 = \frac{ABIJ}{(A+I+J)(A+I)}. \quad (A35)$$

The rebound is transient because the equal signals $S_1{}^* = S_2{}^* = I$ gradually equalize the z_1 and z_2 levels until they both approach $AB(A + S_1{}^*)^{-1}$. Then $S_1{}^*z_1 - S_2{}^*z_2$ approaches zero, so the competition between channels shuts off both of their outputs.

Rebound due to Arousal Onset

Suppose that the on-channel and off-channel have equilibrated to the input levels I and J. Now increase I to I*, thereby changing the signals to $S_1{}^* = I^* + J$ and $S_2{}^* = I^*$. The transmitters z_1 and z_2 continue to obey A28 and A29 for awhile, with $S_1 = I + J$ and $S_2 = J$. A rebound occurs if $S_2{}^*z_2 > S_1{}^*z_1$. This inequality is true if

$$I^* > I + A, \quad (A36)$$

since

$$S_2{}^*z_2 - S_1{}^*z_1 = \frac{ABJ(I^* - I - A)}{(A+I)(A+I+J)}. \quad (A37)$$

In particular, a rebound will occur if I* exceeds I + A no matter how J is chosen. In other words, if the mismatch is great enough to increment the adaptation level by more than amount A, then all dipoles will simultaneously rebound, and by an amount that increases as a function of J, as in Equation A37. This is not true in all versions of the dipole model, since the signals S_i, $i = 1, 2$, are not always linear functions of their inputs. There exist examples in which the most active dipoles can be rebounded even though less intensely activated dipoles are amplified without being rebounded. Moreover, if the signals are sigmoid functions of input size, then inverted-U effects occur in both the on- and off-responses to cue and arousal increments (Grossberg, 1972b, 1972c, 1975).

Received February 27, 1979
Revision received August 15, 1979 ■

CHAPTER 2

SOME PHYSIOLOGICAL AND BIOCHEMICAL CONSEQUENCES OF PSYCHOLOGICAL POSTULATES

PREFACE

This note is a flashback to the time when the theory's foundation was being laid. The foundation is a synthesis of psychological, mathematical, and philosophical ideas in a series of papers that appeared between 1967 and 1969 (Papers [1] - [18] in the List of Publications at the end of this book.) The note is the tip of this theoretical iceberg.

Paper [10] carries out a thought experiment using the simplest real-time constraints on associative learning as postulates (see note items (3a)-(3h)). During my student years, I realized that associative learning was a familiar way to speak about real-time nonstationary prediction, so that the results should generalize greatly once explicit examples could be found. The simplest examples describe the nonlinear interactions of short-term memory (STM) traces with long-term memory (LTM) traces in a neural network. These examples formed the core of my undergraduate Dartmouth thesis of 1960-1961. At that time, they represented a radical way to describe associative learning. This continued to be the case until after 1970, which is why I proudly think of these equations, as well as the general approach, as my own. Contemporary information processing notions like unitized nodes and STM priming by probed LTM read-out already enjoy a quantitative formulation in these equations.

Equations (1) and (2) of the note describe the simplest version of the equations. This version disturbed me for several reasons. One reason was based on a stability question: How could I be sure that the solutions of these nonlinear systems remain bounded through time? Stability issues have continued to play a major role in guiding the theory to this day, as Chapter 1 has illustrated. In 1965, I was influenced by the psychological intuition that associative competition can prevent even large associations from manifesting themselves in observable behavior. Equations (4) - (6) embody this intuition by building competition into the quations using LTM associative strengths y_{jk}. Each y_{jk} is defined to be the LTM trace z_{jk} from the *jth* node to the *kth* node divided by the sum of all LTM traces from the *jth* node to all other nodes. This intuition paid off in several ways. Not only were the solutions easily bounded, but the use of LTM ratios y_{jk} suggested that I should change variables to study how STM ratios and LTM ratios interact ([1] - [6]). In this new system, I could discern a variety of new properties, notably how the system separates out, or factorizes, the patterned information to be learned, which can change monotonically through time, from the energy driving the learning process, which can fluctuate wildly through time. This factorization property enabled me to prove that pattern learning occurs under rather general constraints on network design. Through these theorems on pattern learning, I could see that the functional unit of LTM is a spatial pattern (a 'reflectance' pattern) distributed across the whole system, and that a stimulus sampling operation exists whereby a STM pattern could be encoded in LTM or not, depending on whether or not a sampling cell capable of activating the LTM traces fires. These insights excited me for several reasons.

For one, I could interpret the LTM ratios y_{jk} as stimulus sampling probabilities, and could thereby physiologically interpret and extend the stimulus sampling theory of Estes, which was the leading mathematical learning theory at that time. For another, I could see that a type of wave-particle duality existed in the systems: Their memory was encoded in LTM waves of probability (the y_{jk}'s), yet they could generate deterministic predictions.

Or could they? Although study of the LTM ratios y_{jk} showed how the distribution of all LTM traces should influence performance, an additional STM mechanism was needed which could sense the probed read-outs from all the slowly varying LTM traces into STM, and quickly decide which STM traces were sufficiently large to influence observable behavior. I turned to the information functional to formalize this competitive STM process. But now, although formally successful, the theory was in philosophical trouble. Neither the y_{jk} computations nor the information functional computations were local computations. Each computation required that STM or LTM traces from all across the network instantaneously communicate their values to individual network nodes or paths, respectively. This situation was philosophically intolerable to me.

I then realized how to overcome both difficulties with a single physical idea. Locality and wave-particle duality were both salvaged by invoking the existence of lateral inhibition, or competitive interactions, as in Equations (7) - (10). The LTM competition expressed by the ratios y_{jk} was hereby replaced by a real-time STM competition, which thereupon could indirectly influence LTM. Some psychologists still use the information functional to express STM competitive effects despite its nonlocality and related formal limitations.

A philosophical price was paid for this maneuver. Some system interactions were no longer directly observable by a macroscopic observer, in particular the inhibitory interactions. This step represented a decisive one away from stimulus-response psychology, and had made me sympathetic to the idea that hidden interactions are needed to stabilize self-organizing matter.

A mathematical price was also paid. If the stimulus sampling probabilities had to be abandoned, then would the property of system boundedness also collapse? This dilemma forced me to prove some theorems which guarantee boundedness; for example, boundedness is assured in many systems if STM changes sufficiently quickly and LTM changes sufficiently slowly ([16], [19]). What could be better? What about the factorization of pattern and energy on which the networks' pattern learning properties depended? Could I still prove that pattern learning occurred despite the absence of LTM ratios in the network equations?

At this point, I began to more deeply appreciate how subtle the cross pollination between vague physical intuitions and mathematical technique can be. Factorization still obtained even without the LTM ratios built in! What did this mean? What features of system design really controlled factorization? Such mathematical questions, which were motivated by philosophical issues like system observability and locality, ultimately led to the universal theorem on nonstationary prediction of Chapter 4.

The step from stimulus sampling probabilities y_{jk} to lateral inhibition showed me how to uncover finer details of system design by a 'correspondence principle', whereby spatial scales are expanded and temporal scales are slowed down by an ever-more-powerful conceptual microscope. Using this correspondence tool, I was able to uncover more microscopic interactions in the evolutionary hierarchy. I kept asking myself how the brain's quantum measurement process compares with the usual quantum theory, so I was ready to adapt Niels Bohr's idea of a correspondence principle to my own purposes.

Equations (12) - (24) describe a major step in this correspondence process. These equations describe laws for the production, feedback inhibition, mobilization, gating, and release of a chemical transmitter. These laws imply such microscopic transmitter properties as intracellular adaptation and habituation. The availability of these laws later made it possible for me to complete some work on pharmacological substrates of reinforcement, since these laws also imply antagonistic rebound properties, although I did not realize this for several years (Chapter 5). Still later, the rebound properties suggested a pharmacological substrate of the event-related potential called the P300 (Chapters 1 and 13). Just this year, the same laws enabled G. A. Carpenter and me to quantitatively fit Baylor, Hodgkin, and Lamb's exhaustive 1974 data on vertebrate photoreceptors [58].

Back in the 1960's, my interpretation of the LTM law (8) in terms of the dynamics of a chemical transmitter system led me to a prediction that still has not been tested: The chemical transactions which set LTM levels at the synaptic junctions form a chemical dipole, and interactions between the ion pairs (Na^+, K^+) and (Ca^{++}, Mg^{++}), acting as cofactors, can reset the operating level of this dipole. When I suggested an important role for Ca^{++} as a modulating signal in the 1960's, one of my Rockefeller professors remarked that this idea was so speculative that it would be best for me to leave the abstract variables uninterpreted. I thought otherwise, because the ionic labels helped me to organize and functionally interpret a variety of biochemical and physiological data. As experimental studies of cyclic AMP and cyclic GMP have intensified during the past decade, the role of ionic influences, notably the role of Ca^{++} as a gate or as a second messenger, has been increasingly understood.

The usefulness of these transmitter properties towards the explanation of such a variety of difficult interdisciplinary data argues against a philosophical position which is popular today. This is the Artificial Intelligence dogma which asserts that committing oneself to a particular instantiation of an information processing concept is not a crucial step in building a successful computational theory. This dogma ignores the fact that a particular instantiation can imply a coherent constellation of properties in whose absence the data can only be understood using a series of unrelated *ad hoc* processing tricks.

*SOME PHYSIOLOGICAL AND BIOCHEMICAL CONSEQUENCES OF PSYCHOLOGICAL POSTULATES**

Communicated by Norman Levinson, May 1, 1968

(1) *Introduction.*—This note lists some psychological, physiological, and biochemical predictions that have been derived from simple psychological postulates. These psychological postulates have been used to derive a new learning theory,[1–3] which is called the theory of embedding fields. The mathematical variables of the theory have natural psychological labels—such as "presentation of a letter or spatial pattern at time t_1," "guess of a letter at time t_2," "stimulus trace," "associational strength," etc.—due to the fact that the theory is derived on a psychological basis. *Given* the psychologically derived theory, one then observes that its mathematical variables are *already* in a form that suggests a neurophysiological, anatomical, and in some cases biochemical labeling for these variables. For example, the theory contains geometrical objects which are readily identified with cell bodies, axons, synaptic knobs, and synapses. It also contains, associated with the geometrical objects, dynamical variables that readily call to mind membrane potentials, spiking frequencies, transmitter substances, various ions, and the like. Once the mathematical variables are labeled with these suggestive physiological and anatomical labels, the *psychologically* derived laws of the theory thereupon imply functional relationships between these empirical variables, as well as a psychological rationale for the existence of these relationships in terms of how the brain might learn, remember, and recall what it has learned.

Naturally the leap from mathematical to neural variables cannot be justified in a deductive way. It is governed, as is inevitable, merely by rules of prudence and the dictates of intuition. Fortunately, the simplest neural labeling seems often to yield functional relationships which represent, at least qualitatively, known and nontrivial neural data. In other cases, the functional relationships seem never to have been measured, and therefore stand as new predictions. The strength of such predictions is, of course, no greater than the correctness of the neural labeling, and an assessment of this requires a close scrutiny of the theory's development.[1–3]

We have also begun a rigorous mathematical analysis of the learning, memory, and recall capacities of the theoretical equations in various experimental situations.[4–10]

(2) *Some Qualitative Results.*—(*a*) The equations reduce in a special case to the Hartline-Ratliff equation for lateral inhibition in the *Limulus* retina.[11] Theoretical formulas for the empirical coefficients in the H-R equation are found, and various transients can be readily studied. A new phenomenon of "enhancement of associations" or "spontaneous improvement of memory," closely related to "contour enhancement" due to lateral inhibition, is found.[2] It shares many properties with the Ward-Hovland phenomenon, or "reminiscence."[12] The

"accumulation of inhibition" postulated by Hull[12] to explain bowing in serial verbal learning is identified with lateral inhibition.[2, 13]

(*b*) A unified formal explanation is given of various serial learning phenomena,[13] such as backward learning, bowing, anchoring, chunking, response oscillation, All-or-None versus Gradualist learning, and Gestalt versus Peripheralist learning.

(*c*) A unified formal explanation of the decrease of reaction time with increased learning and of spatiotemporal masking is found.

(*d*) The level of excitatory transmitter production is controlled *jointly* by presynaptic and postsynaptic levels of membrane excitation.

(*e*) Learning needs suggest the interaction of no fewer than two pairs of antagonistic ions, say (Na^+, K^+) and (Ca^{++}, Mg^{++}).

(*f*) Na^+ and Ca^{++} are bound as synergistic cofactors on the intracellular sites, or enzymes, which activate the production of excitatory transmitter, say acetylcholine.

(*g*) There exists a spiking threshold, greater than the cell body equilibrium potential, above which average spiking frequency is proportional to cell body membrane potential (after excitatory transients subside and before saturation sets in).

(*h*) Presynaptic spiking both mobilizes and depletes transmitter. Whereas the steady-state mobilized transmitter that is *released* per unit time increases as a function of steady-state spiking frequency and saturates at a finite value, the *total* steady-state mobilized transmitter decreases as a function of spiking frequency.

(*i*) A slowly varying form of post-tetanic potentiation occurs in the synaptic knobs.

(*j*) An excitatory transient in transmitter release occurs when presynaptic spiking is resumed after a rest interval.

(*k*) The amount of intracellular acetylcholine is regulated in part by a feedback inhibition within the synaptic knob of transmitter onto a previous stage of transmitter production. This inhibition affects an intermediate or terminal stage of transmitter production, rather than an initial stage.

(*l*) K^+ is more likely to be found in unbound form within the synaptic knob than are Na^+ and Ca^{++}.

(*m*) The ionic movements suggested by learning needs are compatible with some data concerning the pattern of ion translocation in the mitochondrion, and with the assumption that these movements make adenosine 5′-triphosphate available for production of acetyl-Co A, and thereupon acetylcholine, under the guidance of choline acetylase in the synaptic vesicles (see, e.g., ref. 14).

(*n*) A mechanism is found which makes plausible the distribution of synaptic vesicles and mitochondria near the synapse of the synaptic knob, rather than (say) uniformly distributed throughout the knob.

(*o*) In response to excitatory transmitter, there exists an inward flow of Na^+ through the cell membrane which is coupled at suprathreshold values to an outward flow of K^+.

(*p*) In response to inhibitory transmitter, there exists an outward flow of K^+ through the cell membrane.

(*q*) Acetylcholine release from synaptic knobs is coupled to the intracellular K^+ concentration.

(*r*) The sensitivity of RNA activation to Mg^{++} concentration is compatible with the need to guarantee control by membrane excitation of intracellular production levels, say of proteins, and thus

(*s*) membrane excitation due to learning experiments causes systematic variations in nuclear RNA, although individual RNA strands do not encode entire behavioral memories, which are spread over many cells.

(*t*) Learning needs suggest a cell nucleus which is localized in the cell body, rather than being spread throughout the cell. More generally, various functions performed by nerves as learning mechanisms seem to determine their shape, at least qualitatively.

(*u*) A system of intracellular tubules, such as in endoplasmic reticulum, is compatible with the need to carry chemicals used in learning between cell body membrane and nucleus and from the nucleus along the axon and to the synaptic knobs.

(*v*) In an *idealized* nerve cell (say without dendrites), cell body membrane area is proportional to nuclear volume and to the membrane area of axon and end-bulbs. This is a special case of the general property of spatiotemporal self-similarity, which is apparent in many biological shapes and interactions (e.g., shape of leaves, proportionality of axon diameter, and velocity of spike along the axon).

(*w*) The size of a cell in a given idealized cell type can, in principle, be controlled by a single gene whose activity is sensitive to the average total membrane excitation.

The theory can also be used to illustrate in various cases how particular anatomical cell distributions and multiple somatotopic representations might be used to perform particular tasks of learning and performance, such as in the sensory-motor cortex, cerebellum, and retina.

(3) *Postulates and Equations.*—The psychological postulates that lead to the equations which describe our learning machines M are quite simple. The following discussion heuristically describes these postulates in the case of learning a list of "simple" letters or events, such as the alphabet $ABC\ldots Z$.

(*a*) The letter A is never decomposed into two or more parts in daily speech and listening. It is a "simple" behavioral unit. Thus we assign to every simple behavioral unit r_i a single abstract point v_i in M, $i = 1, 2, \ldots, n$. (As the theory becomes more microscopic, even simple events create a space-time trajectory of excitation and inhibition that includes many points, which are ultimately "blown up" and identified as caricatures of nerves.)

(*b*) M must react to presentation of behavioral units at specified times. Hence a real-valued function of time $x_i(t)$ is assigned to each point v_i. The value of $x_i(t)$ at any time describes how recently r_i has been presented to M.

(*c*) Consider M's response to presentation of A, then B, and then C at a speed ω. If ω is small (say $\omega \cong 2$ sec), then the influence of A and B on M's response to C is substantial. As ω increases, the influence of A and B on M's response gradually changes and ultimately becomes negligible. Since the effects of prior presentations of events wear off gradually, each $x_i(t)$ is continuous. Since our theory describes the macroscopic behavior of M, we can also readily assume that each $x_i(t)$ is differentiable.

(*d*) If r_i is never presented to M, then $x_i(t)$ remains at a fixed equilibrium value, which is (initially) set equal to zero. If r_i is presented to M at time $t = t_i$, then $x_i(t)$ must at least temporarily assume nonequilibrium values once $t > t_i$. We assume that $x_i(t)$ becomes positive after $t = t_i$, by convention. Since the effect of an event ultimately wears off, $x_i(t)$ eventually decays towards zero. (The choice of a zero equilibrium value tacitly assumes that all $x_i(t)$'s values are *observable* to a psychological experimenter. This assumption must ultimately be abandoned, for reasons that soon become clear.)

(*e*) *After M* has learned the list AB, a presentation of A to M at time t_A gives rise to the guess B by M a little while later, say at time $t_A + \tau_{AB}$, where τ_{AB} is positive. Thus a signal travels from v_A to v_B at finite velocity along a pathway e_{AB}.

(*f*) *Before M* has learned the list AB, other responses than B to A must exist, or else B would already be the only response to A. Thus a function $z_{AB}(t)$ exists which can distinguish the presentation or nonpresentation of AB and lets only B occur in response to A after AB has been learned. Since $z_{AB}(t)$ grows only if A and then B are presented to M, $z_{AB}(t)$ *correlates* (prescribed) past values of x_A with $x_B(t)$. $z_{AB}(t)$ therefore occurs at the only position at which past x_A and present x_B values exist, namely, at the end of the pathway leading from v_A to v_B.

(*g*) The list AB is not the same as the list BA. Thus $e_{AB} \neq e_{BA}$, and $z_{AB}(t) \not\equiv z_{BA}(t)$. e_{AB} is drawn as an *arrow* from v_A to v_B with arrowhead N_{AB}. By (*f*), $z_{AB}(t)$ occurs in N_{AB}.

(*h*) If C is not said, then AB can be learned in first approximation independently of CB. Thus the signals received by B combine independently.

When the postulates (*a*)–(*h*) are translated into mathematical terms, the following equations are found as, perhaps, their simplest realization.

$$\dot{x}_i(t) = -\alpha x_i(t) + \beta \textstyle\sum_{m=1}^{n} x_m(t - \tau_{mi}) p_{mi} z_{mi}(t) + I_i(t), \quad (1)$$

$$\dot{z}_{jk}(t) = -u z_{jk}(t) + \beta p_{jk} x_j(t - \tau_{jk}) x_k(t), \quad (2)$$

where $i, j, k = 1, 2, \ldots, n$; α, β, and u are positive; all τ_{jk} are positive; all p_{jk} are nonnegative; and all initial data are nonnegative and continuous. The nonnegative and continuous inputs $I_i(t)$ often have the form

$$I_i(t) = \textstyle\sum_{k=1}^{N_i} J_i(t - t_i^{(k)}), \quad (3)$$

where $t_i^{(k)}$ is the kth onset time of r_i, and $J_i(t)$ is a given nonnegative and continuous function that is positive in a finite interval of the form $(0,\lambda_i)$.

Equations (1) and (2) can be given a qualitative neural interpretation that includes cell bodies, axons, synaptic knobs, synapses, membrane potentials, spiking frequencies, and transmitter production and release.[1] These equations are not totally satisfactory because of the hypothesis (*d*) of observability. By including the following additional postulate, they can be improved without violating (*d*) in the special case that all reaction times τ_{ij} have the same value τ.

(*i*) M can learn AB perfectly by practicing AB sufficiently often. This postulate is achieved by implementing the following property. *Increasing* the strength of the choice B, given an isolated presentation of A, *decreases* the strength of the choices $C, D, E, \ldots,$ etc. In other words, a "set of response alternatives" to isolated presentations of A exists, and these alternatives *compete* with one another. This property has the effect of reducing behaviorally irrelevant background noise.

Then (1) and (2) are replaced by

$$\dot{x}_i(t) = -\alpha x_i(t) + \beta \sum_{m=1}^n x_m(t-\tau) y_{mi}(t) + I_i(t), \tag{4}$$

$$y_{jk}(t) = p_{jk} z_{jk}(t) [\sum_{m=1}^n p_{jm} z_{jm}(t)]^{-1}, \tag{5}$$

$$\dot{z}_{jk}(t) = -u z_{jk}(t) + \beta p_{jk} x_j(t-\tau) x_k(t). \tag{6}$$

Both equations (4)–(6) and (1) and (2) can be described as cross-correlated flows on networks in a manner that has been previously described in this journal.[4, 5] *Bounded* versions of both (4)–(6) and (1) and (2) can readily be given.

(4) *Lateral Inhibition and Thresholds.*—Equations (4)–(6) improve the learning of (1) and (2) formally, but introduce a conceptual difficulty; namely, by (5), the value $z_{jm}(t)$ at the arrowhead N_{jm} of e_{jm} instantaneously jumps to the arrowhead N_{jk} where $y_{jk}(t)$ is computed. This "virtual" interaction must be replaced by a finite-rate and local interaction with the same qualitative properties. Since $y_{jm}(t) \geq 0$ and $\sum_{m=1}^n y_{jm}(t) = 0$ or 1, the mapping from $p_{jk} z_{jk}(t)$ to $y_{jk}(t)$ by which (4) replaces (1) describes an *inhibition* between the associations $y_{jm}(t)$, $m = 1, 2, \ldots, n$. The finite-rate analogue of this "virtual" inhibition requires the introduction of lateral inhibitory interactions and thresholds.[2] The finite-rate analogue, in the unbounded case, is given by

$$\dot{x}_i(t) = \alpha_i^+ [P_i - x_i(t)]^+ - \alpha_i^- [x_i(t) - P_i]^+ + J_i^+(t) - J_i^-(t) + I_i^+(t) - I_i^-(t) \tag{7}$$

and

$$\dot{z}_{jk}(t) = u_{jk}^+ [Q_{jk} - z_{jk}(t)]^+ - u_{jk}^- [z_{jk}(t) - Q_{jk}]^+ + \beta_j^+ \gamma_{jk}^+ p_{jk}^+ [x_j(t - \tau_{jk}^+) - \Gamma_{jk}^+]^+ [x_k(t) - \Lambda_k^+]^+, \tag{8}$$

where

$$J_i^+(t) = \sum_{m=1}^n \beta_m^+ [x_m(t - \tau_{mi}^+) - \Gamma_{mi}^+]^+ p_{mi}^+ [z_{mi}(t) - \Omega_{mi}^+]^+, \tag{9}$$

$$J_i^-(t) = \sum_{m=1}^n \beta_m^- [x_m(t - \tau_{mi}^-) - \Gamma_{mi}^-]^+ p_{mi}^-, \tag{10}$$

$I_i^+(t)$ and $I_i^-(t)$ are known excitatory and inhibitory inputs, respectively, and the notation $[\omega]^+$ denotes

$$[\omega]^+ = \max(\omega, 0),$$

whereby various thresholds are described. P_i is the equilibrium value (or "potential") of $x_i(t)$, and Q_{jk} is the equilibrium value of $z_{jk}(t)$. The "spiking threshold" Γ_{ij} and the equilibrium value P_i satisfy $\Gamma_{ij} > P_i$. Similarly, $\Omega_{ij}^+ \geq$

Q_{ij} and $\Lambda_j^+ \geq P_j$. Equations (7)–(10) can be given a neural interpretation which is substantially more quantitative than that of (1) and (2).[2] For example, consider (7)–(10) under steady-state excitatory inputs and let all interactions be inhibitory. Then (7) reduces in the steady state to the Hartline-Ratliff equation

$$r_i = e_i - \sum_{j=1}^n K_{ij}[r_j - r_{ij}^0]^+ \tag{11}$$

for lateral inhibition in the *Limulus* retina if

$$K_{ij} = \frac{\mu_i{}^+\beta_j{}^-}{\mu_j{}^+\alpha_i{}^-} p_{ji}{}^-$$

and

$$r_{ij}{}^0 = \mu_j{}^+(\Gamma_{ji}{}^- - \Gamma_j{}^+),$$

when $\mu_i{}^+[x_i(t) - \Gamma_i{}^+]^+$ is the output from the retina to higher neural centers.

(5) *Symmetry-Breaking by Na^+ and K^+*.—The bounded analogue of (7) is

$$\dot{x}_i(t) = \alpha_i{}^+(M_i - x_i(t))(\gamma_i{}^+ + J_i{}^+(t) + I_i{}^+(t)) \\ - \alpha_i{}^-(x_i(t) - m_i)(\gamma_i{}^- + J_i{}^-(t) + I_i{}^-(t)), \tag{12}$$

where $m_i \leq x_i(t) \leq M_i$ for all $t \geq 0$. There exists an obvious symmetry between the excitatory and inhibitory terms in (12). This symmetry can be made explicit by replacing (12) with equations for a *pair* of variables $x_i{}^+(t)$ and $x_i{}^-(t)$ which are positively and negatively "polarized," respectively. This symmetrization procedure must not, however, destroy the "excitatory bias" within (7)–(10) that makes learning possible. The result is, in first approximation,

$$\dot{x}_i{}^+(t) = \alpha_i{}^{++}(M_i{}^+ - x_i{}^+(t))(\gamma_i{}^{++} + J_i{}^{++}(t) + I_i{}^{++}(t)) \\ - \alpha_i{}^{+-}\gamma_i{}^{+-}(x_i{}^+(t) - m_i{}^+), \tag{13}$$

$$\dot{x}_i{}^-(t) = \alpha_i{}^{-+}\gamma_i{}^{-+}(M_i{}^- - x_i{}^-(t)) \\ - \alpha_i{}^{--}(x_i{}^-(t) - m_i{}^-)(\gamma_i{}^{--} + J_i{}^{--}(t) + I_i{}^{--}(t)), \tag{14}$$

and

$$\chi([x_m{}^+(t) - \Gamma_{mi}{}^{+-}]^+)(\beta_m{}^{+-}[x_m{}^+(t) - \Gamma_{mi}{}^{+-}]^+ \\ - \beta_m{}^{--}[\Gamma_{mi}{}^{-+} - x_m{}^-(t)]^+) = 0, \tag{15}$$

where

$$J_i{}^{++}(t) = \sum_{m=1}^n \beta_m{}^{++}[x_m{}^+(t - \tau_{mi}{}^{++}) - \Gamma_{mi}{}^{++}]^+ p_{mi}{}^{++}[z_{mi}{}^{++}(t) \\ - \Omega_{mi}{}^{++}]^+, \tag{16}$$

$$J_i{}^{--}(t) = \sum_{m=1}^n \beta_m{}^{+-}[x_m{}^+(t - \tau_{mi}{}^{+-}) - \Gamma_{mi}{}^{+-}]^+ p_{mi}{}^{--}[z_{mi}{}^{--}(t) \\ - \Omega_{mi}{}^{--}]^+, \tag{17}$$

and

$$\chi(\omega) = \begin{cases} 1, & \omega > 0 \\ 0, & \omega \leq 0. \end{cases}$$

Equations (13)–(17) can be interpreted to yield the "symmetry-breaking" properties (*o*) and (*p*) of section (2). The condition (15) is merely of qualitative interest, and will be made more quantitive in a later paper, along a pathway that is suggested in reference 2.

(6) *Transmitter Production and Release.*—In first approximation, the bounded equation for excitatory transmitter production $z_{ij}^{++}(t)$ in the collection of synaptic knobs N_{ij}^{++} is

$$\dot{z}_{ij}^{++}(t) = (M_{ij}^{++} - z_{ij}^{++}(t))(u_{ij}^{+++} + \gamma_{ij}^{++}F_{ij}^{++}(t)R_j(t)) - u_{ij}^{++-}(z_{ij}^{++}(t) - m_{ij}^{++}), \quad (18)$$

where

$$F_{ij}^{++}(t) = \beta_i^{++}[x_i^{+}(t - \tau_{ij}^{++}) - \Gamma_{ij}^{++}]^{+}p_{ij}^{++} \quad (19)$$

and

$$R_j(t) = [x_j(t) - \Lambda_j^{+}]^{+}. \quad (20)$$

The inequalities

$$m_{ij}^{++} \leq z_{ij}^{++}(t) \leq M_{ij}^{++}$$

hold for all $t \geq 0$. All learning within (18) is due to the term

$$\gamma_{ij}^{++}(M_{ij}^{++} - z_{ij}^{++}(t))F_{ij}^{++}(t)R_j(t). \quad (21)$$

The spiking frequency term $F_{ij}^{++}(t)$ is interpreted as an antagonistic coupling between Na^{+} and K^{+} at suprathreshold values, whereas $R_j(t)$ is interpreted as an antagonistic coupling between Ca^{++} and Mg^{++}. Na^{+} and Ca^{++} act synergistically in (21) to activate $z_{ij}^{++}(t)$.

It is readily seen that the coupling between $F_{ij}^{++}(t)$ and

$$G_{ij}^{++}(t) = [z_{ij}^{++}(t) - \Omega_{ij}^{++}]^{+}$$

in (16) describes a transmitter release process in which the depleted transmitter is instantaneously replenished. The finite-rate analogue of this coupling is given by the pair of equations

$$\dot{Z}_{ij}^{++}(t) = \lambda_{ij}^{+}(\delta_{ij}z_{ij}^{++}(t) - Z_{ij}^{++}(t)) - \lambda_{ij}^{-}F_{ij}^{++}(t)[\tilde{Z}_{ij}^{++}(t) - U_{ij}^{++}]^{+} \quad (22)$$

and

$$\dot{\tilde{Z}}_{ij}^{++}(t) = \omega_{ij}^{+}(Z_{ij}^{++}(t) - \tilde{Z}_{ij}^{++}(t)) - \lambda_{ij}^{-}F_{ij}^{++}(t)[\tilde{Z}_{ij}^{++}(t) - U_{ij}^{++}]^{+} - \omega_{ij}^{-}[\tilde{Z}_{ij}^{++}(t) - V_{ij}^{++}]^{+}, \quad (23)$$

with

$$U_{ij}^{++} = \delta_{ij}\Omega_{ij}^{++} > V_{ij}^{++},$$

and

$$0 \leq \tilde{Z}_{ij}^{++}(t) \leq Z_{ij}^{++}(t).$$

$Z_{ij}^{++}(t)$ = the total amount of excitatory transmitter in the synaptic knobs N_{ij}^{++} at time t,

$\tilde{Z}_{ij}^{++}(t)$ = the total amount of *mobilized* transmitter at time t,
$z_{ij}^{++}(t)$ = the total number of *active* transmitter-producing sites at time t.

A simple physical interpretation of (22) and (23) yields properties (h), (i), (j), (k), and (q) of section (2). Equations (22) and (23) can be solved explicitly for the transient responses of $Z_{ij}^{++}(t)$ and $\tilde{Z}_{ij}^{++}(t)$ when (say) $F_{ij}^{++}(t)$ is a steady-state spiking frequency F for $t \geq 0$, $\lambda_{ij}^{+} = \omega_{ij}^{-}$, and $U_{ij}^{++} = V_{ij}^{++} = 0$. Then, ignoring slow variations of $z_{ij}^{++}(t)$,

$$\tilde{Z}_{ij}^{++}(t) = \frac{\delta_{ij} z_{ij}^{++}(0)\omega_{ij}^{+}}{\omega_{ij}^{-} + \omega_{ij}^{+}} \cdot \Big[\exp\left(-(\lambda_{ij}^{+} + \lambda_{ij}^{-}F)t\right) + \frac{\lambda_{ij}^{+}}{\lambda_{ij}^{+} + \lambda_{ij}^{-}F}\left(1 - \exp\left(-(\lambda_{ij}^{+} + \lambda_{ij}^{-}F)t\right)\right) \Big], \quad (24)$$

and the amount of mobilized transmitter which is *released* from N_{ij}^{++} at time t is $\lambda_{ij}^{-}F\tilde{Z}_{ij}^{++}(t)$, as (22) and (23) show.

* The preparation of this work was supported in part by the Office of Naval Research (N00014-67-A-0204-0016).

[1] Grossberg, S., "Embedding fields: A new theory of learning with physiological implications," *J. Math. Psych.*, in press.
[2] Grossberg, S., "On learning, information, lateral inhibition, and transmitters," *Math. Biosci.*, in press.
[3] Grossberg, S., "On the production and release of chemical transmitters and related topics in cellular control."
[4] Grossberg, S., "Nonlinear difference-differential equations in prediction and learning theory," these PROCEEDINGS, **58**(4), 1329–1334 (1967).
[5] Grossberg, S., "Some nonlinear networks capable of learning a spatial pattern of arbitrary complexity," these PROCEEDINGS, **59**(2), 368–372 (1968).
[6] Grossberg, S., "Global ratio limit theorems for some nonlinear functional-differential equations, I, II," *Bull. Am. Math. Soc.*, **74**(1), 95–105 (1968).
[7] Grossberg, S., "A prediction theory for some nonlinear functional-differential equations, I. Learning of lists," *J. Math. Anal. Applic.*, in press.
[8] Grossberg, S., "A prediction theory for some nonlinear functional-differential equations, II. Learning of spatial patterns," *J. Math. Anal. Applic.*, in press.
[9] Grossberg, S., "On the global limits and oscillations of a system of nonlinear differential equations describing a flow on a probabilistic network," *J. Diff. Eq.*, in press.
[10] Grossberg, S., "On the variational systems of some nonlinear difference-differential equations," *J. Diff. Eq.*, manuscript submitted for publication.
[11] Ratliff, F., *Mach Bands: Quantitative Studies on Neural Networks in the Retina* (New York: Holden-Day, 1965).
[12] Osgood, C. E., *Method and Theory in Experimental Psychology* (New York: Oxford, 1953), chap. 12.
[13] Grossberg, S., "On the serial learning of lists," *Math. Biosci.*, in press.
[14] Fruton, J. S., and S. Simmonds, *General Biochemistry* (New York: John Wiley, 1958).

CHAPTER 3

CLASSICAL AND INSTRUMENTAL LEARNING BY NEURAL NETWORKS

PREFACE

This article reviews some of my main theoretical advances before 1973 in a self-contained and nontechnical exposition. Among other features, the article describes some predictions which still need to be tested.

Verbal learning has always fascinated me as a goldmine of subtle philosophical issues. Consider for example the question: "What is the earliest time that you can classify the last item in a serially presented list as the end of the list?" The answer has revolutionary implications, because it implies that internal events can go backwards in time relative to the event space of a macroscopic observer. The answer also indicates that a computer which treats list items as a list on a tape does not behave like a brain. In fact, verbal learning data show that internal events *do* go backwards in time. That is what backward learning and the bowed serial position curve tell us, among many other things. I was seized by this engaging property when I took a learning course during my sophomore year at Dartmouth and it changed my life. The existence of backward effects, more than almost any other data, force one to think in terms of network geometries. Verbal learning data also vividly show that the *non*occurrence of future list items can powerfully influence the way we learn past list items. This fact made it plain that a real-time theory was indispensible to parameterize both the occurrences and non-occurrences of observable events.

Serial verbal learning is a subtle subject because forward and backward effects are asymmetric; in particular, the bowed serial position curve is skewed towards the end of the list. Here is a nonlinear many-body problem which exhibits symmetry-breaking between the future and the past! This symmetry-breaking allows us to learn a forward-pointing arrow in time among macroscopic events despite the occurrence of backward effects. Time reversal invariance holds no more in the mind than it does in particle physics.

The theory suggests an explanation of the skewed bowed curve ([13], [23]). It also makes a prediction that is a variant of the factorization of pattern and energy. This factorization is often structurally realized as an anatomical bifurcation of specific cue signals from nonspecific arousal signals, followed by convergence of these signals at a later processing stage. The theory predicts that, as the nonspecific arousal level is parametrically increased (say by amphetamine), the bow should move from its skewed position backwards towards the list middle, and the relative strengths of associations at the beginning vs. the end of the list should reverse. Network properties during this overaroused state can be compared with attentional deficits of certain schizophrenics, since they show how contextual collapse, reduction of attention span, and fuzzy response categories can be caused by overarousal of a serial processor. Thus by this stage of the theory, parametric changes within rigorous models could be used to formally explain the transition between normal behavior and certain pathological symptoms.

Some other issues of interest in nonequilibrium statistical physics are reviewed in the article. I proved that a phase transition in memory can occur in feedback anatomies but not in feedforward anatomies ([2], [3], [8]). The phase transitions describe a type of 'imprinting' behavior. A phase transition can convert a network that remembers everything into one that forgets everything [8], or into one that remembers only pictures [3]. Increasing a single nonspecific parameter can cause this global change in information processing. These results imply at least three general lessons. They emphasize the impossibility of understanding behavior by a purely grind-and-find approach. They show that some theoretical properties can be experiment-dependent, such as an interference theory of forgetting, because one experiment might excite a feedforward anatomy and another experiment might excite a feedback anatomy. They illustrate how precise control of high-dimensional information-processing changes can be controlled by a one-dimensional command, like a hormone or a burst of nonspecific arousal.

The phase transition examples also possess other properties that are interesting as nonequilibrium many-body effects [8]. These effects highlight the difficulties faced by a hypothetical experimenter who is trying to understand the system. For example, a linear system becomes nonlinear when it is perturbed by an experimental input. As the perturbation wears off, the system looks linear again. A locally and globally reversible system becomes globally irreversible when it is perturbed, but it remains locally reversible. It might or might not revert to global reversibility depending on the size of an unobservable parameter. This is the parameter that controls the phase transition in memory. All of these issues need further exploration.

Another appearance of factorization occurs in my theory of *avalanches*

([14], [19]). The avalanches are the minimal network realizations capable of learning arbitrary space-time patterns (acts). After I derived them, I realized that avalanches look like the command cells of invertebrate physiology . However, invertebrate experiments do not easily reveal the evolutionary strategy that is illustrated in a primitive way by command cells. From the outset, the theory showed that command cells act as nonspecific arousal sources which turn on or off sequentially organized activity in associated networks. The next theoretical question was then clear. How do sensory cells which are not pre-programmed at birth gain control over these arousal sources through experience? This question led to the reinforcement theory ([25], [28], [29]) that is sketched in Section VIII of the article. The anatomical relationships of command cells to reinforcement mechanisms is still unclear in invertebrate experiments. The more sophisticated forms into which the theory suggests that command cells are transmuted by evolution in higher species (Chapter 13) have not been carefully investigated by any experimental method.

The results on avalanches, verbal learning, and phase transitions filled my mind with questions about the regulation of nonspecific arousal. Was there not a more profound reason for the existence of specific vs nonspecific pathways? Such a reason was discovered by a thought experiment concerning the so-called *synchronization problem* of classical conditioning. This problem asks: How can an organism learn about unsynchronized events in real-time? How can it overcome the disorientation latent in William James' 'blooming buzzing confusion'? The solution includes explicit mechanisms for instrumental conditioning concepts, such as reward, punishment, drive, incentive motivation. Due to the derivation of these mechanisms from the synchronization problem, distinctions between classical and instrumental conditioning concepts melted away before being reworked into a new synthesis (Chapters 5 and 6).

Classical and Instrumental Learning by Neural Networks *

I. Introduction

A. Embedding Fields: A Psychophysiological Theory

This article reviews results chosen from the theory of *embedding fields*. Embedding field theory discusses mechanisms of pattern discrimination and learning in a psychophysiological setting. It is derived from psychological postulates that correspond to familiar behavioral facts. The theory tries to isolate facts which embody fundamental principles of neural design, and which therefore imply and illuminate many less evident facts and predictions. The postulates reveal their implications by being translated into rigorous mathematical expressions. On various occasions, the precision of this mathematical language has uncovered unsuspected physical properties of the postulates, or corrected erroneous conclusions of prior heuristic thinking. In particular, the mathematics can be given a natural anatomical and physiological interpretation. The neural networks hereby derived can thus be rigorously analyzed both behaviorally and neurally.

B. The Method of Minimal Anatomies

The theory introduces a particular method to approach the several levels of description that are relevant to understanding behavior. This is the method of *minimal anatomies*. At any given time, we will be confronted by laws for neural components, which have been derived from psychological postulates. The neural units will be interconnected in specific anatomies. They will be subjected to inputs that have a psychological interpretation, which create outputs that also have a psychological interpretation. At no given time could we hope that all of the more than 10^{12} nerves in a human brain would be described in this way. Even if a precise knowledge of the laws for each nerve were known, the task of writing down all the inter-

* This work was supported in part by the Alfred P. Sloan Foundation and the Office of Naval Research (N00014-67-A-0204-0051).

actions and analyzing them would be bewilderingly complex and time-consuming. Instead, a suitable method of successive approximations is needed. Given specific psychological postulates, we derive the *minimal* network of embedding field type that realizes these postulates. Then we analyze the psychological and neural capabilities of this network. An important part of the analysis is to understand what the network cannot do. This knowledge often suggests what new psychological postulate is needed to derive the next, more complex network. In this way, a hierarchy of networks is derived, corresponding to ever more sophisticated postulates. This hierarchy presumably leads us closer to realistic anatomies, and provides us with a catalog of mechanisms to use in various situations. Moreover, once the mechanisms of a given minimal anatomy are understood, variations of this anatomy having particular advantages or disadvantages can be readily imagined. The procedure is not unlike the study of one-body, then two-body, then three-body, and so on, problems in physics, leading ever closer to realistic interactions; or the study of symmetries in physics as a precursor to understanding mechanisms of symmetry-breaking; or the study of thermodynamics as a preliminary to statistical mechanical investigations.

At each stage of theory construction, natural formal analogs of nontrivial psychological and neural phenomena emerge. We shall denote these formal properties by their familiar experimental names. This procedure emphasizes at which point in theory construction, and ascribed to which mechanisms, these various phenomena first seem to appear. No deductive procedure can justify this process of name-calling, and incorrect naming of formal network properties does not compromise the formal correctness of the theory as a mathematical consequence of the psychological postulates. Nonetheless, if ever psychological and neural processes are to be unified into a coherent theoretical picture, such name-calling, with all its risks and fascinations, seems inevitable, both as a guide to more microscopic theory construction and as a tool for a deeper understanding of relevant data. The following pages will attempt to distinguish clearly between postulates, mathematical properties, factual data, and mere interpretations of network variables.

This policy of theory construction has more than practical convenience to recommend it. Even a routine behavioral act can utilize billions of nerves distributed along complexly interacting pathways that extend from sensory receptors to motor effectors. The organization of these pathways—the *global* properties of the network—powerfully influence the transformation of stimuli into responses. To the extent that these properties are ignored, one loses insight into the behavioral constraints which guide neural development and design. Even if one's neural data about individual cells are precise, the meaning of these data can remain obscure until global information about the role of these cells in behavior is obtained.

C. Overview

This article summarizes some main results that are distributed in several papers (Grossberg, 1968a,b, 1969a-f, 1970a,b, 1971a-c, 1972a-d; Grossberg and Pepe, 1971). An intuitive description of mathematical results will replace mathematical details wherever possible. Emphasis will be placed on theoretical ideas. Relevant data are discussed in the references.

The theory begins by analyzing simple facts about classical, or Pavlovian, conditioning (Grossberg, 1969a, 1971a). This form of learning is illustrated by the following experiment. A hungry dog is presented with food and thereupon salivates. A bell is rung, but the dog does not salivate. Then the bell is rung just before food presentation on several learning trials. Thereafter presentation of the bell alone yields salivation. Food is called the unconditioned stimulus (UCS), salivation is called the unconditioned response (UCR), and the bell is called the conditioned stimulus (CS). Thus Pavlovian conditioning is a problem in nonstationary prediction: The CS eventually predicts the UCR if it is paired sufficiently often with the UCS. Alternatively, this learning process can be described by considering an experimentalist, $\mathcal{E}$, who interacts with a machine, $\mathfrak{M}$, to teach $\mathfrak{M}$ to predict B given A by practicing the list AB. The sensory presentation of A is analogous to a CS, the sensory presentation of B is analogous to UCS, and the motor response, B, is analogous to a UCR.

The first derivation of the theory asks how a particular CS $\rightarrow$ UCR transition can be embedded in memory by sequential pairing of CS and UCS, and how future presentation of the CS can elicit the UCR. This derivation is reviewed in Section II.

Given the derivation, which is based on only the most rudimentary facts about classical conditioning, a number of psychophysiological and mathematical surprises ensue. For example, the mathematical systems that arise already have a natural anatomical and neurophysiological interpretation which includes cell bodies, axons, synaptic knobs, cell potentials, spiking thresholds and frequencies, and transmitter substances. In psychological terms, one finds such items as short-term memory (STM) traces, long-term memory (LTM) traces, a stimulus sampling theory, Now Print or Amplifier mechanisms, imprinting mechanisms, serial learning phenomena, a way to learn arbitrary patterns, influences of overarousal on paying attention, and a teleology for attacking problems of sensory filtering and pattern discrimination. Pattern discrimination problems will not be discussed herein.

Given this foundation, the theory proceeds in several directions. First, it studies the minimal anatomy that can learn an arbitrary space–time pattern, such as a piano sonata or a dance (see Sections IV and V). Only one command cell is needed to encode the memory of such a pattern, although there exist larger networks that can also do the job with interneurons. The main liability of such a cell is that performance is ritualistic: Once performance of the pattern begins, it cannot be terminated in midcourse even if more urgent environmental demands are made on the network. Anatomies in which sensitivity to environmental feedback exists typically encode sequences of events using many cells. This fact motivates a study of learning in arbitrary anatomies (Section VI). The goal is to find those constraints that make learning in a general context possible. Thereafter, one can specialize the anatomy to perform particular tasks. Studies of serial (or list) learning, and some related problems concerning dependence of serial learning parameters on arousal level, arise as special cases in these investigations (Section VII).

Next the theory is developed in the direction of instrumental, or operant, conditioning; namely, it approaches the question of how learning is influenced by rewards, punishments, drives, etc. (Section VIII). These questions arise naturally from a closer investigation of classical conditioning. For example, the time lags between CS and UCS presentation on successive learning trials need not be the same, since the two events are usually independent of each other. Also, after learning has occurred, the CS elicits the UCR on recall trials in the absence of the UCS. These and a few other simple facts can be used as postulates to derive networks that include mechanisms of reinforcement, drive, and incentive motivation. In short, classical and instrumental learning mechanisms are not conceptually independent. Given these networks, one imposes postulates which are aimed at preventing the network from seeking previously rewarded goals that presently lead to punishment, and which permit learned avoidance of such goals. The above postulates eventually yield networks containing rudimentary formal analogs of midbrain reinforcement centers. These analogs include the interaction of two formal transmitter systems, whose properties can be compared with data concerning cholinergic and adrenergic effects at midbrain sites. Various other facts and predictions about punishment and avoidance formally emerge in these systems.

The theoretical equations can also be refined in several directions to provide a deeper insight into possible chemical substrates of network mechanisms (Section IX). This refinement procedure uncovers various transient interactions, and suggests an important concept: that a cell capable of learning is a chemical dipole, with the two ends of the dipole existing near the cell body and synaptic knobs.

II. Classical Conditioning

The derivation below (Grossberg, 1969a, 1971a) is given in storybook form to emphasize its intuitive basis. It studies how an experimentalist, $\mathcal{E}$, can teach a machine, $\mathcal{M}$, to predict B given A by practicing the list AB.

A. Each Letter Seems Simple

In daily speech and listening, a letter is never decomposed into two parts. To maintain close contact with experience, we assume that a single state, v_A, in $\mathcal{M}$ corresponds to A. In a similar fashion, let v_B correspond to B, v_C to C, etc. We designate each v_i by a point, or vertex. (A vertex is not necessarily an individual cell, but can represent a cell population acting as a control unit.)

B. Presentation Times

The times at which letters are presented to $\mathcal{M}$ must be represented within $\mathcal{M}$. For example, presenting A and then B with a time spacing of 24 hours should yield different behavior than does presentation with a time spacing of 2 seconds. Thus various functions of time should be associated with each vertex to designate how recently a given letter has been presented. To maintain contact with the "one-ness" of each letter, and to maximize the simplicity of our derivation, we let one function $x_A(t)$ be associated with v_A, one function $x_B(t)$ be associated with v_B, etc., as in Fig. 1.

C. Continuous Vertex Functions

The functions $x_A(t), \ldots, x_Z(t)$ will be chosen continuous, and in fact differentiable. Several reasons for this exist. One reason is the following. Consider the question: What follows ABC? It is tempting to say D, but really the problem is ill-defined if the letters are presented one at a time with time spacing, w, between successive letters. If indeed w is small, say $w \cong 2$ seconds, then D might well be the correct response, but if $w \cong 24$ hours, then to the sound C (= "see") one can also reply "See what?" That is, as w varies from small to large values, the influence of A and B on

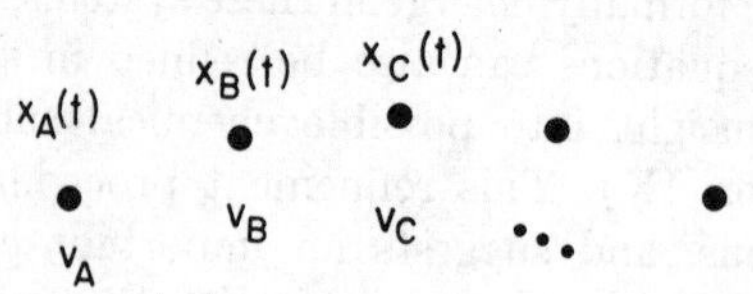

FIG. 1. Vertex functions register how recently given events occur.

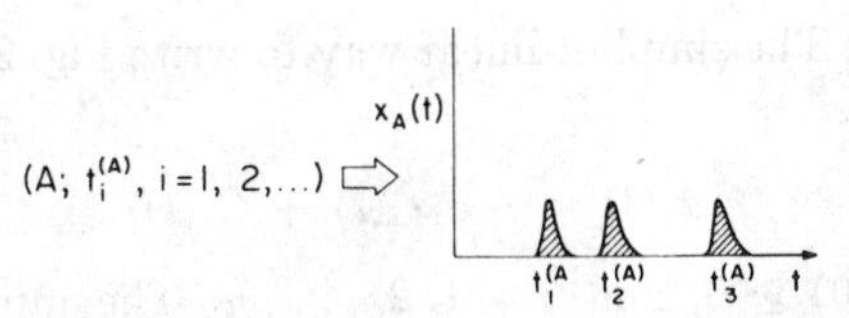

FIG. 2. Sequential presentation of an event induces sequential perturbation of its vertex function.

the prediction following C gradually wears off. Since $x_A(t)$ and $x_B(t)$ describe the relevance at time t of A and B in $\mathfrak{M}$, we conclude that these functions also vary gradually in time.

D. Perturbations Instead of Presentations

Suppose that A is never presented to $\mathfrak{M}$. Corresponding to the occurrence of "nothing" is the natural mathematical predisposition to set $x_A(t) = 0$ at all times t. (The equilibrium point, 0, can, it turns out, be rescaled ultimately relative to the signal thresholds.)

Suppose that A is presented to $\mathfrak{M}$ for the first time at time $t = t_A$. Then $x_A(t)$ must be perturbed from 0 for certain $t > t_A$, or else $\mathfrak{M}$ would have no way of knowing that A occurred. We associate the occurrence of "something" with a positive deflection in the graph of x_A. (The theory could also, in principle, be carried out with negative deflections.)

Shortly after A is presented, A no longer is heard by $\mathfrak{M}$. That is, $x_A(t)$ gradually returns to the value signifying no recent presentation of A, namely 0. In a similar fashion, if A is presented at times $t_A^{(1)} < t_A^{(2)} < \cdots < t_A^{(N_A)}$, then we find the graph of Fig. 2. The same construction holds true for all letters. In this way, we have translated the presentation of any letters A, B, C, . . . in the alphabet at prescribed times into a definite sequence of perturbations of the vertex functions $x_A(t), x_B(t), x_C(t), \ldots$.

E. Linearity

For notational convenience, we replace the alphabet A, B, C, . . . by any sequence r_i, $i = 1, 2, \ldots, n$, of n behavioral atoms; the vertices $v_A, v_B, v_C, \ldots$ by the vertices v_i, $i = 1, 2, \ldots, n$; and the vertex functions $x_A(t)$, $x_B(t)$, $x_C(t), \ldots$ by the vertex functions $x_i(t)$, $i = 1, 2, \ldots, n$. Now r_i corresponds to $[v_i, x_i(t)]$, $i = 1, 2, \ldots, n$.

What is the simplest way to translate Fig. 2 into mathematical terms? Since we are constructing a system whose goal is to adapt with as little bias as possible to its environment, we are strongly advised to make the system as linear as possible. In Section VI, we shall discuss which of these linearities is really essential. The simplest linear way to write Fig. 2 is in terms of the equations

$$\dot{x}_i(t) = -\alpha_i x_i(t) + C_i(t) \tag{1}$$

with $\alpha_i > 0$, $x_i(0) \geq 0$, and $i = 1, 2, \ldots, n$. The input $C_i(t)$ can, for example, have the form

$$C_i(t) = \sum_{k=1}^{Ni} J_i(t - t_i^{(k)}) \tag{2}$$

where $J_i(t)$ is some nonnegative and continuous function that is positive in an interval of the form $(0, \lambda_i)$. Thus $x_i(t)$ decays at the exponential rate α_i unless it is perturbed by an input pulse $J_i(t - t_i^{(k)})$.

F. After Learning

In order that $\mathfrak{M}$ be able to predict B given A after practicing AB, interactions between the vertices v_i must exist. Suppose, for example, that $\mathfrak{M}$ has already learned AB, and that A is presented to $\mathfrak{M}$ at time t_A. We expect $\mathfrak{M}$ to respond with B after a short time interval, say at time $t = t_A + \tau_{AB}$, where $\tau_{AB} > 0$. The term τ_{AB} is called the *reaction time* from A to B. Let us translate these expectations into graphs for the functions $x_A(t)$ and $x_B(t)$. We find Fig. 3. The input, $C_A(t)$, controlled by $\mathcal{E}$ gives rise to the perturbation of $x_A(t)$. The internal mechanism of $\mathfrak{M}$ must give rise to the perturbation of $x_B(t)$. In other words, after AB is learned, $x_B(t)$ gets large τ_{AB} units after $x_A(t)$ gets large.

There exists a linear and continuous way to say this; namely, v_A sends a linear signal to v_B with time lag τ_{AB}. Then Eq. (3) with i = B is replaced by

$$\dot{x}_B(t) = -\alpha_B x_B(t) + \beta_{AB} x_A(t - \tau_{AB}) + C_B(t) \tag{3}$$

with β_{AB} some positive constant. More generally, if $r_i r_j$ has been learned, we conclude that

$$\dot{x}_j(t) = -\alpha_j x_j(t) + \beta_{ij} x_i(t - \tau_{ij}) + C_j(t) \tag{4}$$

If $\beta_{ij} = 0$, then the list $r_i r_j$ cannot be learned, since a signal cannot pass from v_i to v_j.

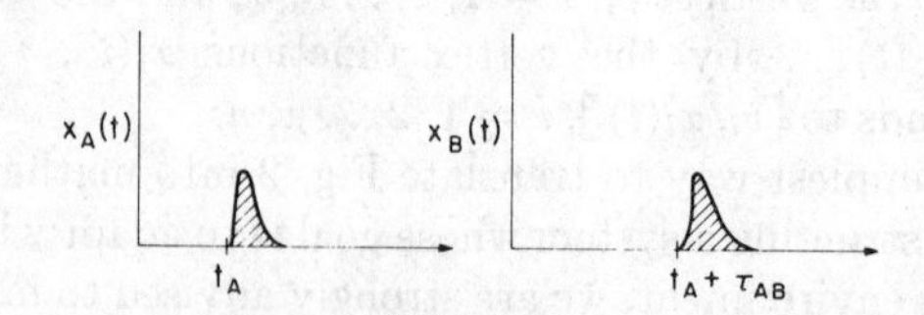

Fig. 3. Traces of sequential A-then-B presentation.

FIG. 4. Network interpretation of psychophysiological variables.

G. Directed Paths

The signal $\beta_{ij}x_i(t - \tau_{ij})$ from v_i to v_j in Eq. (4) is carried along some pathway at a finite velocity, or else the locality of the dynamics would be violated. Denote this pathway by e_{ij}. The pathways e_{ij} and e_{ji} are distinct because the lists $r_i r_j$ and $r_j r_i$ are distinct. To designate the direction of flow in e_{ij}, we draw e_{ij} as an arrow from v_i to v_j, whose arrowhead, N_{ij}, touches v_j, as in Fig. 4.

H. Before Learning

Consider the network before any learning occurs. If A leads only to B, then learning would have already occurred. The letter A must therefore also be able to lead to C, D, or some other letters. Thus the process of learning can be viewed as elimination of the incorrect pathways AC, AD, etc., while the correct pathway AB endures, or is strengthened. In other words, the connections between all vertices cannot be constant through time if learning occurs. The constant connections β_{ij} must be supplemented by time-varying connection coefficients. (This is a "connectionist" theory in the broadest sense, but, it turns out, not a traditional one.)

I. Distinguishing Order

How does $\mathfrak{M}$ know that AB and not AC is being learned? By Fig. 3, practicing AB means that x_A and then x_B become large several times. Saying A alone, or B alone, or neither A nor B should yield no learning. We seek a mathematical way to distinguish the event "A occurs and then B occurs" from all other possibilities. This can be done most simply as follows. If AB occurs with a time spacing of w, then the product $x_A(t - w)x_B(t)$ is large at suitable times $t \cong t_A^{(i)} + w$, $i = 1, 2, \ldots, N_A$. If either A or B does not occur, then the product will be small. We therefore seek a process $\mathfrak{M}$ that can compute products of past $x_A(v)$ values $(v < t)$ and present $x_B(t)$ values. Denote this process by $z_{AB}(t)$. Note that $z_{AB} \neq z_{BA}$.

Where in $\mathfrak{M}$ do past values of $x_A(v)$ and present values of $x_B(t)$ come together, so that $z_{AB}(t)$ can compute them? (Locality again!) By Fig. 4, this happens only in the arrowhead N_{AB}. Thus $z_{AB}(t)$ takes place in N_{AB}. But then the past $x_A(v)$ value received by N_{AB} at time t is the signal $\beta_{AB}x_A(t - \tau_{AB})$. The most linear and continuous way to express this rule for $z_{AB}(t)$ is the following:

$$\dot{z}_{AB}(t) = -\gamma_{AB}z_{AB}(t) + \delta_{AB}x_A(t - \tau_{AB})x_B(t)$$

with γ_{AB} a positive constant, and δ_{AB} a nonnegative constant that is positive only if β_{AB} is positive. More generally, for list r_ir_j we find in N_{ij} the process

$$\dot{z}_{ij}(t) = -\gamma_{ij}z_{ij}(t) + \delta_{ij}x_i(t - \tau_{ij})x_j(t) \tag{5}$$

Thus z_{ij} depends linearly on the crucial product $x_i(t - \tau_{ij})x_j(t)$. Note that the reaction time, τ_{ij}, replaces w in Eq. (5), owing to locality. This fact adumbrates a connection between reaction times and presentation rates optimal for learning.

J. Gating Outputs

The $z_{ij}(t)$ function can distinguish whether or not r_ir_j is practiced. But more is desired. Namely, if r_ir_j is practiced, presenting r_i should yield a delayed output from v_j. If r_ir_j is not practiced, presenting r_i should not yield an output from v_j. And even if r_ir_j is practiced, no output from v_j should occur if r_i is not presented. In other words, $x_j(t)$ should become large only if $x_i(t - \tau_{ij})$ *and* $z_{ij}(t)$ are large. Again a product is called for, and Eq. (4) is changed to

$$\dot{x}_j(t) = -\alpha_j x_j(t) + x_i(t - \tau_{ij})\beta_{ij}z_{ij}(t) + C_j(t) \tag{6}$$

Thus z_{ij}'s location in N_{ij} allows z_{ij} both to compute products of past signals from v_i and present values at v_j, and to gate the signal from v_i before it reaches v_j.

K. Independence of Lists in First Approximation

If B is not presented to $\mathfrak{M}$, then in first approximation CA should be learnable without interference from B. (Not so in second approximation, since a signal could travel from C to B to A.) Similarly, if C is not presented to $\mathfrak{M}$, then BA should be learnable without interference from C, in first approximation. In other words, it should be possible to practice particular skills without activating the entire embedded vocabulary of behavioral units. Mathematically speaking, this means that all signals to each v_j combine additively at v_j. Thus Eq. (6) becomes

$$\dot{x}_j(t) = -\alpha_j x_j(t) + \sum_{i=1}^{n} x_i(t - \tau_{ij})\beta_{ij}z_{ij}(t) + C_j(t) \tag{7}$$

The system (5) and (7) is a mathematically well-defined proposal for a learning machine that uses only such general notions as linearity, continuity, and locality, and a mathematical analysis of how a machine can learn to predict B given A on the basis of practicing AB. Note that this system is nonlinear, notwithstanding our efforts to keep it as linear as possible, owing to constraints (I) and (J).

L. Thresholds

One further modification of systems (5) and (7) is needed—namely, the introduction of signal thresholds. Here we introduce this modification directly to keep background noise down. A more fundamental analysis would introduce it by first analyzing the need in complex learning situations for inhibitory interactions—they shut off competing or irrelevant channels, among other tasks (Grossberg, 1970a, 1972a,b). Learning becomes difficult, if not impossible, without signal thresholds if inhibitory interactions exist, since the signs of all functions x_i and z_{jk} begin to oscillate in uncontrollable ways.

The modification can also be motivated by the following possible difficulty in (5) and (7). Small signals can possibly be carried around and around the network, thereby building up background noise and interfering with the processing of behaviorally important inputs. We therefore seek to eliminate the production of signals in response to small $x_i(t)$ values, in the most linear possible way. Thresholds do this for us. Letting $[\xi]^+ = \max(\xi, 0)$, we replace (5) and (7) by

$$\dot{x}_i(t) = -\alpha_i x_i(t) + \sum_{m=1}^{n} [x_m(t - \tau_{mi}) - \Gamma_{mi}]^+ \beta_{mi} z_{mi}(t) + C_i(t) \quad (8)$$

and

$$\dot{z}_{jk}(t) = -\gamma_{jk} z_{jk}(t) + \delta_{jk}[x_j(t - \tau_{jk}) - \Gamma_{jk}]^+ x_k(t) \quad (9)$$

where all Γ_{mi} are nonnegative (usually positive) thresholds, and $i, j, k = 1, 2, \ldots, n$.

III. Psychophysiological Interpretation

A. Psychological Variables

The function $x_i(t)$ is called the ith *stimulus trace*: it responds to the *stimulus* $C_i(t)$. The function $z_{jk}(t)$ is called the (j, k)th *memory trace*: it records the pairing of successive events r_j and r_k. Alternatively, $x_i(t)$ is called the ith *short-term memory trace*: it represents brief activation of the state v_i either by inputs $C_i(t)$ or by signals from other states v_j. Similarly, $z_{jk}(t)$ is called the (j, k)th *long-term memory trace*: its record of past events can endure long after the short-term memory traces have decayed. *Transfer* from short-term memory to long-term memory denotes the operation whereby the z_{jk}'s are altered by the distribution of x_i's. *Activation* of short-term memories via long-term memories denotes the operation whereby signals from a given set of v_j's, modulated in the pathways e_{jk} by the z_{jk}'s, activate a given pattern of x_k's—for example, as in the subliminal activation of learned predispositions, or "psychological sets," in response to particular sensory cues.

The term Γ_{jk} is the (j, k)th *signal threshold*: no signal is emitted by v_j into e_{jk} at time t unless $x_j(t) > \Gamma_{jk}$. The vertex v_j is said to *sample* v_k at time t if the signal received at N_{jk} from v_j at time t is positive. The *signal strength* at N_{jk} at time t is defined by $B_{jk}(t) = [x_j(t - \tau_{jk}) - \Gamma_{jk}]^+\beta_{jk}$. The constant β_{jk} is a structural parameter called the *path strength* of e_{jk}. The $n \times n$ matrix $\beta = \| \beta_{jk} \|$ determines which directed paths between vertices exist, and how strong they are. Otherwise expressed, β determines the "anatomy" of connections between all vertices.

B. Neural Variables

A natural neurological interpretation of these variables is readily noticed. This interpretation does not claim uniqueness, however, because there exist only two kinds of variables, x_i's and z_{jk}'s, at this level of theorizing, and these variables can at best represent averages of finer physiological or biochemical variables. The anatomical interpretation seems unambiguous: v_i is a cell body (population), e_{jk} is an axon (population), N_{jk} is a synaptic knob (population), and the gap between N_{jk} and v_k is a (population of) synapse(s). Part of the physiological interpretation also seems inevitable: $x_i(t)$ is an average potential taken over all units in v_i and over a brief time interval. The signal $B_{jk}(t)$ should correspondingly represent an average over individual signals in the axon(s) e_{jk}; it is therefore assumed to be proportional to the spiking frequency in e_{jk}. The interpretation of $z_{jk}(t)$ is more speculative. The process $z_{jk}(t)$ exists either in, or adjacent to, the synaptic knobs N_{jk}, and, by Eq. (8), $z_{jk}(t)$—coupled to the spiking frequency $B_{jk}(t)$—determines the signal from N_{jk} to v_k. Thus it is natural to let $z_{jk}(t)$ correspond to the rate of transmitter production in N_{jk}, or to the sensitivity of postsynaptic sites at v_k to fixed amounts of transmitter. The former interpretation is accepted herein for definiteness. Then Eq. (9) becomes a statistical law for transmitter production. Section IX shows that, even if $z_{jk}(t)$ is a presynaptic process, it is coupled to postsynaptic processes in v_k.

IV. Outstars

A. Pavlovian Choices

This section studies the smallest anatomy that can learn a choice by Pavlovian conditioning (Grossberg, 1968a, 1969b, 1970b). The anatomy is shown in Fig. 5. Figure 5*a* shows the smallest anatomy that can possibly learn AB, as opposed to the lists AC, AD, etc.; that is, it can learn the choice B given A, as opposed to C given A, D given A, etc. Figure 5*b* interprets the same anatomy using the Pavlovian concepts CS, UCS, and UCR. Figure 5*c* replaces these particularized notations by a purely abstract labeling of states using indices. The cell population with cell body v_1 emits an axon which breaks up into axon collaterals whose synaptic knobs appose the UCS-activated cells $\mathcal{B} = \{v_i;\ i = 2, 3, \ldots, n\}$. Figure 5*d* represents this system in a more symmetric fashion, which suggests the name *outstar* for it. Here v_1 is called the *source* of the outstar. Each v_i, $i \neq 1$, is called a *sink* of the outstar, and the set $\mathcal{B}$ of all sinks is called the *border* of the outstar.

The outstar equations can readily be derived from Eqs. (8) and (9). The main constraint is that only v_1 can send signals to other cells v_i. Hence $\beta_{jk} = 0$ unless $j = 1$ and $k \neq 1$. We find the equations

$$\dot{x}_1(t) = -\alpha_1 x_1(t) + C_1(t) \tag{10}$$

$$\dot{x}_i(t) = -\alpha_i x_i(t) + \beta_{1i}[x_1(t - \tau_{1i}) - \Gamma_{1i}]^+ z_{1i}(t) + C_i(t) \tag{11}$$

and

$$\dot{z}_{1i}(t) = -\gamma_{1i} z_{1i}(t) + \delta_{1i}[x_1(t - \tau_{1i}) - \Gamma_{1i}]^+ x_i(t) \tag{12}$$

where $i = 2, 3, \ldots, n$.

B. Unbiased Outstars

First we consider outstars in which no choice r_i, $i \neq 1$, is preferred above any others because of asymmetric choices of system parameters. In other words, we make the following restrictions on these parameters: (1) set all time lags τ_{1i} equal to τ; (2) set all thresholds Γ_{1i} equal to Γ; (3) set all decays rates $\alpha_i(\gamma_{1i})$ equal to $\alpha(\gamma)$; and (4) set all interaction weights $\beta_{1i}(\delta_{1i})$ equal to $\beta(\delta)$. The unbiased outstar therefore satisfies the Eqs. (10),

$$\dot{x}_i(t) = -\alpha x_i(t) + \beta[x_1(t - \tau) - \Gamma]^+ z_{1i}(t) + C_i(t) \tag{13}$$

and

$$\dot{z}_{1i}(t) = -\gamma z_{1i}(t) + \delta[x_1(t - \tau) - \Gamma]^+ x_i(t) \tag{14}$$

where $i = 2, 3, \ldots, n$.

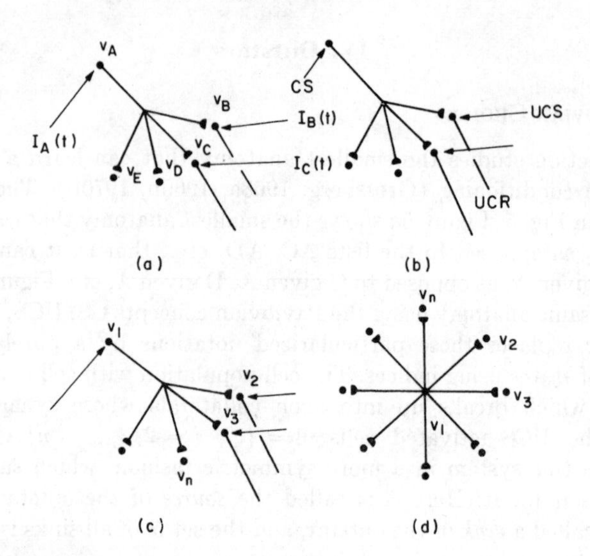

FIG. 5. Outstar: minimal network capable of classical conditioning.

Inspection of this system readily shows that it is a linear system of equations with variable coefficients. Indeed, integration of Eq. (10) yields

$$x_1(t) = x_1(0)e^{-\alpha_1 t} + \int_0^t C_1(v)e^{-\alpha_1(t-v)}\,dv$$

Hence the term $[x_1(t-\tau) - \Gamma]^+$ in Eqs. (13) and (14) is a known function of time. System (13) and (14) is therefore a special case of the following more general system of equations:

$$\dot{x}_i(t) = A(t)x_i(t) + B(t)z_{1i}(t) + C_i(t) \tag{15}$$

and

$$\dot{z}_{1i}(t) = D(t)z_{1i}(t) + E(t)x_i(t) \tag{16}$$

where $A(t)$, $B(t)$, $D(t)$, and $E(t)$ are continuous functionals of t, and moreover $B(t)$ and $E(t)$ are nonnegative. (A functional is a mapping from functions to real numbers. A functional can depend on system variables, evaluated at past times, in a complicated way.) A rigorous mathematical analysis of this class of systems has been carried out. Below we list in intuitive terminology some of the formal properties that have been found.

C. Spatial Pattern Learning

What is the most general UCS whose UCR can be reproduced by a CS after Pavlovian conditioning in an unbiased outstar? The answer is a

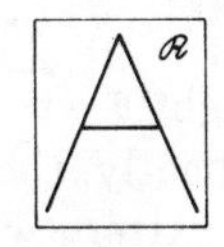

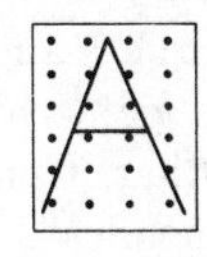
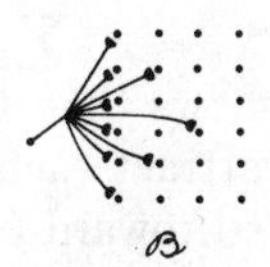

FIG. 6. Classical conditioning of a spatial pattern on a grid.

spatial pattern (or *picture*). This we define to be any UCS of the form $C_i(t) = \theta_i C(t)$, $i = 2, 3, \ldots, n$, such that $\theta_i \geq 0$ and $\sum_{k=2}^{n} \theta_k = 1$. In other words, $\theta = \{\theta_i : i = 2, 3, \ldots, n\}$ is a fixed, but otherwise arbitrary, probability distribution, and $C(t)$ is a nonnegative and continuous function of t. A spatial pattern is the unit of long-term memory in an embedding field.

The intuitive meaning of this definition will be illustrated by an example. Actually the concept of spatial pattern will arise in more varied circumstances than this example might suggest. Consider Fig. 6. Suppose that an arbitrary picture in shades of black, white, and gray is shown on the region $\mathcal{R}$. We want $\mathcal{M}$ to be able to reproduce this picture on $\mathcal{R}$, with an arbitrarily good spatial resolution, by Pavlovian conditioning.

How is the spatial resolution prescribed? Suppose that m cells of $\mathcal{M}$ are embedded in $\mathcal{R}$, and that each cell receives an input proportional to the intensity of the picture at its position. For definiteness, imagine that the m cells are arranged in a rectangular grid in $\mathcal{R}$. As m is increased to ever larger values, the density of the cells in $\mathcal{R}$ increases, as does the accuracy with which the picture is represented in $\mathcal{M}$ by these cells. We shall let these cells be the border, $\mathcal{B}$, of an outstar. The mathematical results on learning by outstars hold for any $n = m + 1$, or any spatial resolution.

Now imagine a fixed picture, such as the Mona Lisa in shades of gray, shown on $\mathcal{R}$. We can vary the *total* intensity of the light which illuminates the picture without changing the picture itself. The total intensity can be steady (and bright or dim), or can flicker between broad physiological limits, without changing our impression that the Mona Lisa is still being presented. In other words, the *relative* intensity of light, not its absolute intensity, characterizes the picture. Only the relative intensity of the picture is constant through time. The constant relative intensity at v_i is denoted by θ_i. The total intensity, which can fluctuate in time, is $C(t)$. In other words, the fact that outstars can learn the weight, θ_i, means that they can pick out the "relative figure to ground" of an input pattern. The outstar can learn such a pattern no matter how we interpret the border, $\mathcal{B}$, to which it is attached. For example, the border can consist of motor control cells, interneurons, cells in any sensory cortex, etc.

These assertions are made precise by studying the relative traces, or *pattern variables*, $X_i = x_i(\sum_{k=2}^{n} x_k)^{-1}$ and $Z_{1i} = z_{1i}(\sum_{k=2}^{n} z_{1k})^{-1}$. Mathematical analysis shows that the pattern weight θ_i attracts X_i, while X_i and Z_{1i} mutually attract each other. Consequently, the relative memory trace Z_{1i} is attracted toward ("encodes") the pattern weight θ_i. On recall trials, an input to v_1 ("presenting A") creates an equal signal to each N_{1i}. In N_{1i}, the signal is multiplied by z_{1i}, which is proportional to θ_i. Thus the input to v_i is proportional to θ_i. The learned pattern is hereby reproduced on the border $\mathcal{B}$ by an input to the source cell v_1.

Spatial pattern learning by an outstar has the following properties (Grossberg, 1968a, 1969b, 1970b).

1. *Practice Makes Perfect*

The more r_1r_i is practiced, the better can $\mathfrak{M}$ predict r_i in response to r_1. This learning can be "all-or-none"—occurring in one trial—or "gradual"—requiring several trials. In an outstar, learning rate is determined by CS and UCS input rate, intensity, relative timing, the number of response alternatives, and related factors. These factors influence both the rate with which Z_{1i} approaches θ_i and the size of z_{1i}. In more general anatomies, the learning rate of a given item in a list of events depends on list position, or more generally on the context of other events in which the item occurs (cf. Section VII). For example, in serial learning of a long list presented at a rapid rate, the items at the two ends of the list might be quickly learned, whereas the items near the middle of the list might not be learned at all on the first few trials.

2. *Overt Practice Unnecessary*

The machine $\mathfrak{M}$ can remember without overt practice. The potentials and thus the outputs from $\mathfrak{M}$ can be zero during memory intervals without destroying the memory; that is, each Z_{1i} remains constant. In fact, positive potentials (in particular, "reverberations" among the vertices) can destroy the memory in certain anatomies (Grossberg, 1968b). One must also distinguish perfect memory of pattern weights Z_{1i} from perfect performance. For example, in Eq. (14), z_{1i} can exponentially decay even if Z_{1i} remains constant. If z_{1i} decays to the level of network noise, then the memory is essentially zero.

3. *Recall Preserves Memory*

Item r_i can be recalled in response to r_1 as often as one pleases without destroying the memory of r_1r_i; that is, Z_{1i} remains constant during recall trials. In fact, recall of r_i, given r_1, can "potentiate" the memory of

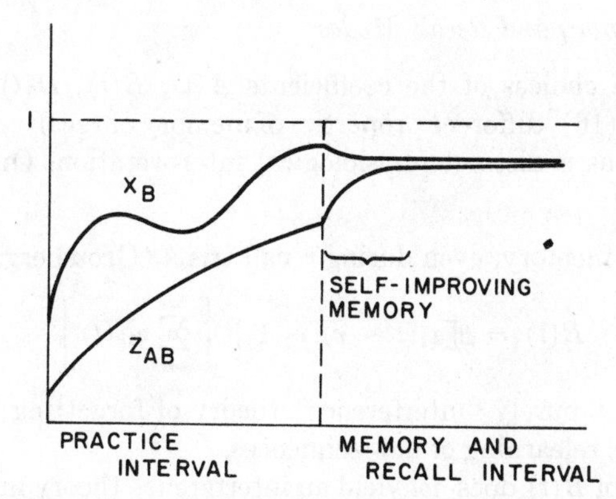

FIG. 7. Self-improving memory due to coupling of STM and LTM traces.

r_1r_i; that is, z_{1i} can grow. There exist anatomies in which this is false: the very act of recall tends to destroy the memory (Grossberg, 1968b); these anatomies usually reverberate signals in closed loops.

4. *Self-Improving Memory*

Self-improving memory, or *reminiscence*, exists. For example, let two outstars, $\mathfrak{M}_1$ and $\mathfrak{M}_2$, practice r_1r_i the same number of times. Let recall occur T_j time units after practice ceases in $\mathfrak{M}_j$, $i = 1, 2$. For certain choices of $T_1 < T_2$, recall is better in $\mathfrak{M}_2$ than in $\mathfrak{M}_1$. See Osgood (1953, pp. 509–513) for a discussion of an analogous experimental phenomenon. Figure 7 illustrates this phenomenon. It is due to a coupling between STM traces and LTM traces.

5. *Contour Enhancement*

After practice of a spatial pattern ceases, the memory of dark (bright) regions of the pattern can become darker (brighter). The mechanisms for contour enhancement and reminiscence are the same in an outstar. See Grossberg (1972b) and Ratliff (1965) for a discussion of contour enhancement due to lateral inhibition.

6. *Error Correction*

All errors can be corrected. Even after r_1r_i is learned, r_1r_j can be learned instead. The rate of learning r_1r_j can depend on such factors as the prior level of r_1r_i performance and on the total number of response alternatives.

7. *Several Memory and Recall Modes*

For suitable choices of the coefficients $A(t)$, $B(t)$, $D(t)$, and $E(t)$ in Eqs. (15) and (16), different properties of memory or recall can be achieved. Each choice has a distinct physiological interpretation. One can, for example, achieve:

(*a*) Perfect memory, even during recall trials (Grossberg, 1970b); let

$$B(t) = \beta[x_1(t - \tau) - \Gamma]^+ \left[\sum_{k=2}^{n} z_{1k}(t) \right]^{-1} \tag{17}$$

This describes a purely "interference" theory of forgetting: All forgetting is due to active relearning of new sequences.

This choice of $B(t)$ does *not* yield an interference theory in all anatomies. For example, if each v_i sends signals to all vertices v_k, or to all vertices but itself, then a "phase transition" can occur (Grossberg, 1968b, 1969c). Given suitable numerical parameters, a learned pattern will be forgotten; given other parameters, it will be remembered. Exactly what is forgotten depends on the anatomy of the network. One can pass from the forgetting phase to the remembering phase by (say) speeding up axonal signals at a critical time; such an operation can "imprint" the pattern that exists at the critical time. See Section VI,G. This example dramatizes the fact that one cannot generally infer the global properties of a network from its local properties.

(*b*) Exponential decay of memory, at any prescribed rate (Grossberg, 1970b); let $C(t) = -\gamma$. Even though z_{1i} can spontaneously decay, the relative traces Z_{1i} are changed only by "interference" due to new learning, or by reminiscence, as in property (4). The net decay rate of z_{1i} itself is not always γ. This rate can be slowed down, or even reversed, by recall trials, by "spontaneous" rhythmic inputs to v_1 during memory intervals, by reminiscence effects, etc. Again, a local property—this time a decay rate—is not necessarily the global one.

(*c*) Perfect memory until recall trials, followed by possible extinction of memory during recall if the prediction is not rewarded or retrained (Grossberg, 1970b); let $D(t) = -\gamma E(t)$. Again an interference theory of forgetting holds for the relative traces Z_{1i}, but not necessarily in all anatomies.

These examples point out that important properties of learning are invariant under changes that allow many variations in the details of learning and performance. Speaking mathematically, the pattern variables $X_i(t)$ and $Z_{1i}(t)$ have the same limiting and oscillatory possibilities given various choices of the coefficients $A(t)$, $B(t)$, $D(t)$, and $E(t)$. These coefficients determine the transient motions of the system, including learning rates.

8. *Stimulus Sampling*

Stimulus sampling theory is a purely behavioral theory that has successfully described various learning data using probability models (Atkinson and Estes, 1963). A physiological mechanism of stimulus sampling and a physiological interpretation of stimulus sampling probabilities exist in embedding fields. The relative memory traces $Z_1 = (Z_{12}, Z_{13}, \ldots, Z_{1n})$ are attracted toward the pattern weights $\theta = (\theta_2, \theta_3, \ldots, \theta_n)$ only at times when the synaptic knobs, N_{1i}, receive CS-activated spikes from v_1. This is the property of "stimulus sampling" in an outstar: v_1 samples the patterns playing on $\mathcal{B}$ by emitting signals at prescribed times. The relative memory traces, Z_1, which form a probability distribution at each time t, are the "stimulus sampling probabilities" of an outstar (Grossberg, 1970b). Whenever v_1 samples $\mathcal{B}$, the memory traces in its synaptic knobs begin to learn the spatial pattern playing on $\mathcal{B}$ at this time. If a sequence of patterns (that is, a space–time pattern) plays on $\mathcal{B}$ while v_1 is sampling, then v_1's synaptic knobs learn a weighted average of all the patterns, rather than any single spatial pattern. Thus if an outstar samples $\mathcal{B}$ while a long sequence of spatial patterns reaches $\mathcal{B}$, then after sampling terminates, the sampling probabilities, Z_1, can be different from any one of the spatial patterns. On recall trials, a CS input to v_1 creates equal signals in the axons e_{1i}. These signals flow down to the N_{1i}. In N_{1i}, the signal interacts with the memory trace z_{1i} to reproduce at the cell v_i an output proportional to Z_{1i}. In this way, recall trials reproduce at $\mathcal{B}$ the weighted average of sampled patterns that was encoded on learning trials.

9. *Oscillatory Inputs and Monotonic Response*

When r_1r_i is practiced on successive trials, the inputs $C_1(t)$ and $C_i(t)$ are highly oscillatory in time. Yet increased practice yields the impression of a steady increase in learning (see Fig. 8). The probabilities Z_{1i} bridge the gap between oscillatory inputs and monotonic learned response.

10. *Eidetic Memory*

An outstar is capable of *eidetic* memory. This remarkable phenomenon has been tested by using human subjects in the following ingenious way. Two pictures are constructed by computer from 10,000 randomly distributed black and white dots. These pictures conceal a figure in depth that can be seen only when the pictures are viewed binocularly (Julesz, 1964). An eidetic woman studies the first picture with one eye on day 1 of the experiment and returns the next day to study the second picture with the other eye. She then identifies the concealed figure (B. Julesz, personal

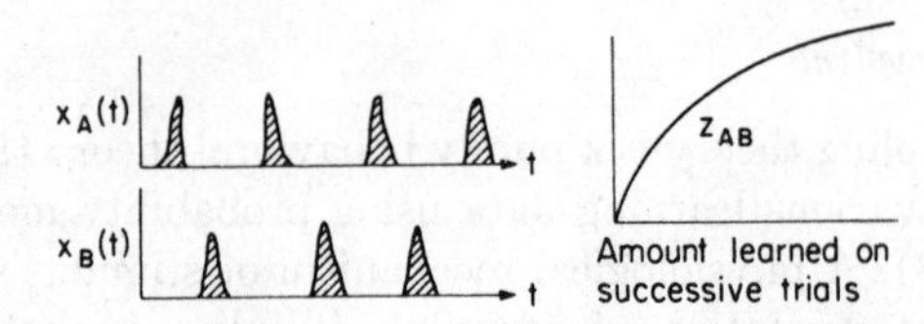

FIG. 8. Oscillatory inputs due to repetitive A-then-B presentation and monotonic response of stimulus sampling probability.

communication, 1970)! To accomplish this, she must presumably be able to conjure up in her mind's eye an almost perfect replica of the 10,000 dots shown on the previous day. In short, textural memory with an enormous storage capacity is possible.

A single cell in our networks can do this formally. Let an outstar (or a cluster of outstars that fire in unison) send axon collaterals to the correct visual representation area. If the network can activate the source cell(s) at will, then it can learn the first picture to an arbitrary degree of accuracy on day 1. On day 2, if it again activates the source cell, the internally produced representation of the first picture will interact with the externally produced representation of the second picture to produce the binocular effect of a figure in depth (see Fig. 9). Several properties of this mechanism are of interest.

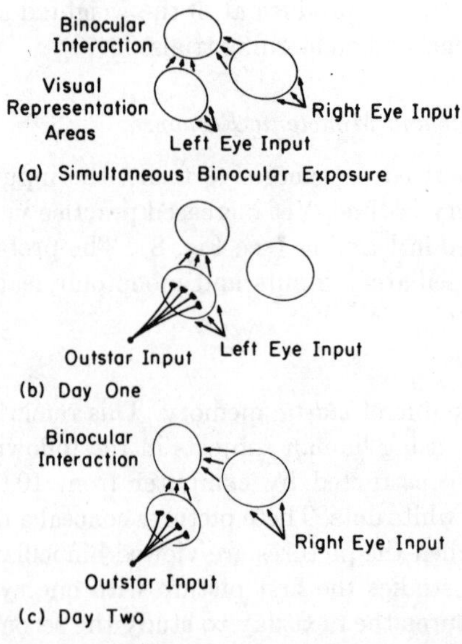

FIG. 9. Eidetic memory using outstar sampling.

(*a*) To learn 10^m pictures playing on 10^n visual cells, one needs no more than $10^m + 10^n (\leq 2 \times 10^{\max(m,n)})$ cells, not 10^{m+n} cells, as is occasionally claimed. In fact, it will later be shown that to learn 10^m moving pictures (space–time patterns, such as piano sonatas or dances) playing on 10^n cells of any kind, one needs no more than $10^m + 10^n$ cells. In principle, one could learn 10^{11} dances playing on 10^{11} motor control cells without using as many as 10^{12} cells, and our brains are thought to contain no fewer than 10^{12} cells. One could also learn a new pattern playing on 10^6 cells every second for sixty years using fewer than 2×10^{12} encoding cells. The networks contain more than enough storage capacity. None of us seems to know how to do 10^{11} complex acts, however. Hence we must ask what the extra cells are doing. Section V will begin to address this question.

(*b*) To learn eidetically as in Fig. 9, there must exist cells—other than cells leading progressively from the retina itself—that send axons to the visual representation areas. Further evidence for the existence of such cells has been acquired by studying epileptics (Penfield, 1958). An electrode in the temporal lobe of man can vividly activate a sequence of perhaps very old memories, including visual and auditory memories. Discontinuing electrode current while the sequence is being recalled can stop recall. Reapplying current at the same point can reinitiate recall of the same sequence. These data suggest that the cells being sought might project from the temporal lobes to visual and/or auditory representation areas.

(*c*) The blessing of eidetic memory also carries with it a possible liability. Suppose that the visual representation areas received a continual barrage of nonvisual inputs which were not synchronized with visual inputs to produce functionally useful results. Then hallucinations and other internal visual experiences could continually pop into our minds against our will. To prevent this, such cells should be forbidden from firing to the visual areas unless there exist functionally desirable stimulus conditions for visual learning or recall. The ability to fire the source cell at will seems to be a more remarkable phenomenon than the existence of textural memory per se, since it resembles the ability to hallucinate specific subject matter at will.

11. *Response Generalization: Variable Performance Velocities*

Suppose that the UCR sends signals to muscles which contract at a rate proportional to the signal. Let the UCR be a spatial pattern; that is, the UCR creates fixed relative contraction rates of the various muscle groups. An outstar that learns to contract these muscles at a given total velocity can also contract them—in the same pattern—at many other total velocities. This form of "response generalization" is the output version of the "stimulus generalization" property of being able to learn the "relative figure to ground" in an input pattern. Not all motor patterns are spatial patterns, however, and this property is modified when more complicated motor tasks are imposed, as in Section V.

V. Avalanches

A. Ritualistic Learning of Space–Time Patterns

This section studies the following question: What is the minimal number of cells needed to encode the memory of an arbitrarily complicated space–time pattern, such as a piano sonata or a dance? The answer is: One! What could be more "minimal"? Yet this answer creates a paradox. If one cell can encode a whole dance, and our brains contain at least 10^{12} cells, why doesn't anyone know 10^{12} (or even 10^4) dances? What do the extra cells do?

Encoding a space–time pattern with one cell has a severe limitation: Performance is ritualistic, or by rote. Once performance of the "dance" begins, the entire dance must be completed, even if the stage on which the dance is being performed is consumed by flames as the dance progresses. In other words, such a system is insensitive to environmental feedback; it cannot adapt to changing environmental demands once the performance of an act begins. Once we note how to encode a space–time pattern without feedback, we shall also readily see how to begin construction of systems that are sensitive to feedback. Such systems will require many more than one cell to encode the entire pattern.

Study of systems that perform with little feedback is not of purely academic interest, however. There exist examples of such performance throughout the phylogenetic kingdom. For example, the seagoing mollusk Tritonia has individual, large cells, with extensively branched axons, whose direct electrical stimulation causes a well-organized swimming escape response (Willows, 1968). Clearly, given such individual cells, it is crucial that they fire only at appropriate times. For example, Tritonia would starve if it "escaped" whenever it approached a source of food. Nonetheless, Tritonia can escape from predators, such as starfish, with considerable reliability. Thus certain characteristic stimuli at Tritonia's periphery can create inputs to its swimming escape cells, but inappropriate inputs cannot. Such facts motivate the construction of networks that can selectively filter environmental inputs on their way to prescribed control cells (Grossberg, 1970a, 1972a,b, Hubel and Wiesel, 1968).

Other organisms also have individual cells capable of controlling well-organized behavioral acts. These include insects (Dethier, 1968, p. 8), and crayfish (Kennedy, 1968). On a higher level, the ring dove performs a ritualistic sequence of acts during its reproductive cycle (Lehrman, 1965). Successive stages of this sequence are triggered by the previous stage and a well-defined combination of exteroceptive and interoceptive stimuli. The maternal behavior of the rat also involves a characteristic sequence of ritualistically organized acts (Thomas *et al.*, 1968, pp. 265–273). Even man is capable of performing complex sequences of acts without the benefit of continuous feedback. For example, a cadenza can be played by a skilled pianist so rapidly that motor feedback cannot possibly determine the next note to be played (Grossberg, 1969b, 1970b; Lashley, 1951). On the other hand, as one ascends the phylogenetic ladder, one finds that ever more subtle types of feedback can influence behavior. For example, the pianist can try to escape from a concert hall before it burns down, and can modify his performance of a piece in exquisitely subtle ways.

B. Sequential Sampling

Given a finite collection of cells v_i, $i \in I$ (I some set of integers), suppose that an arbitrary nonnegative and continuous input, $C_i(t)$, perturbs v_i. Consider the weights $\theta_i(t) = C_i(t)[\sum_{k \in I} C_k(t)]^{-1}$ as they fluctuate in time. Can we learn these weights to an arbitrary degree of accuracy? We can do so by using a collection of sequentially activated outstars if we invoke three mechanisms: (1) stimulus sampling; (2) brief signals from the CS-activated cell body; and (3) an anatomy in which each CS-activated outstar sends an axon collateral to each UCS-activated cell, v_i.

To see this, first note that $\theta_i(t)$, as a continuous function of t, can be arbitrarily well approximated by the discrete sequence

$$\{\theta_i(0), \theta_i(\zeta), \theta_i(2\zeta), \theta_i(3\zeta), \ldots, \theta_i(N_\zeta - 1)\}$$

of its values, if the positive number ζ is chosen sufficiently small; that is, the "moving picture" is replaced by a sequence of N_ζ "still pictures." Suppose that a sequence, $\mathfrak{M}_j$, of outstars is given, $j = 1, 2, \ldots, N_\zeta$, such that (1) each outstar sends one axon collateral to each cell v_i, $i \in I$, and (2) the synaptic knobs of $\mathfrak{M}_j$'s axon collaterals are active only during an interval of time $[(j-1)\zeta, (j-1)\zeta + \Delta\zeta]$. If $\Delta\zeta$ is sufficiently small, then the pattern weights, $\theta_i(t)$, change arbitrarily little from their values $\theta_i[(j-1)\zeta]$ during the time interval $[(j-1)\zeta, (j-1)\zeta + \Delta\zeta]$. Hence $\mathfrak{M}_j$ can learn the spatial pattern with weights $\theta_i[(j-1)\zeta]$ to an arbitrary degree of accuracy. The outstar $\mathfrak{M}_j$ samples only this pattern, by the property of stimulus sampling.

How can these sampling intervals be guaranteed? Simply let a cell body, v_1, send out a long axon, and attach the outstar $\mathfrak{M}_j$ at the axonal position which is excited by a signal emitted from v_1 at $(j-1)\zeta - \tau$ time units earlier, where τ is the time needed for signals to travel from $\mathfrak{M}_j$ to any v_i (see Fig. 10). Such a system is called an *outstar avalanche*, or an *avalanche*, by analogy with avalanche conduction in the parallel fibers of the cerebellum (Grossberg, 1969b, 1970b). Physiologically, it is a cell whose axon emits sequential clusters of axon collaterals which converge on the common cells v_i, $i \in I$. Performance of the pattern is elicited by a signal from v_1, which successively activates the outstar-encoded spatial pattern approximations to the space–time pattern on the cells v_i every ζ time units.

Note that the avalanche has the minimal number of formal degrees of freedom needed to learn the pattern perfectly, given a prescribed spatial and temporal resolution of the inputs: the number $|I|$ of cells v_i determines the spatial resolution of the inputs $C_i(t)$, and the number N_ζ of time intervals determines the temporal resolution in memory that is desired. The minimal number of formal degrees of freedom is $|I| N_\zeta$, which is also the number of axon collaterals in the avalanche.

A sample set of equations for an avalanche is stated below. Let x_1 be the potential of v_1, and let x_i be the potential of v_i, $i \in I$. Let z_{ji} be the transmitter in the axon leading from the jth outstar to the cell v_i. Then system (8) and (9) becomes

$$\dot{x}_1 = -\alpha_1 x_1 + C_1 \tag{18}$$

$$\dot{x}_i = -\alpha x_i + \beta \sum_{k=1}^{N_\zeta} [x_1(t - (k-1)\zeta) - \Gamma]^+ z_{ki} + C_i \tag{19}$$

and

$$\dot{z}_{ji} = -\gamma z_{ji} + \delta [x_1(t - (j-1)\zeta) - \Gamma]^+ x_i \tag{20}$$

where $i \in I$, $j = 1, 2, \ldots, N_\zeta$. Suppose that $[x_1(t) - \Gamma]^+$ is positive in an interval whose duration is shorter than ζ. Then at every time t, at most one term in the sum

$$\sum_{k=1}^{N_\zeta} [x_1(t - (k-1)\zeta) - \Gamma]^+ z_{ki}$$

is positive. At times when the positive term corresponds to $k = K$, then

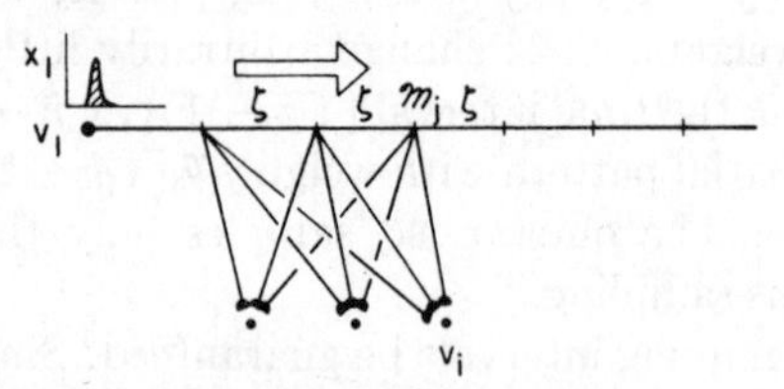

FIG. 10. Sequential sampling of a space–time pattern by an avalanche, or command cell.

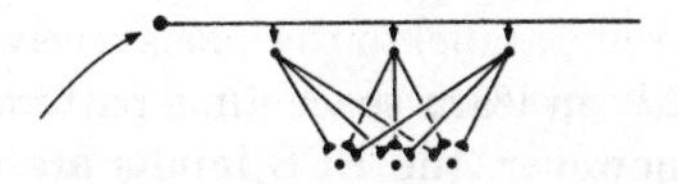

FIG. 11. A command cell that sequentially activates outstar interneurons.

Eq. (19) becomes

$$\dot{x}_i = -\alpha x_i + \beta[x_1(t - (K-1)\zeta) - \Gamma]^+ z_{ki} + C_i \tag{21}$$

and when $k = K$, the system (18), (20), and (21) is an outstar.

There exist variations on this theme. For example, a single cell, v_i, can give off sequential axon collaterals to a series of outstars. Figure 10 is then replaced by Fig. 11, in which the outstars are interneurons between v_1 and v_i, $i \in I$. In this anatomy, several different command cells can sample the same outstar. Perhaps the most abstract anatomical arrangement is that given by Fig. 12, which shows that the local anatomy alone of the system does not necessarily disclose its function. In Fig. 12*a*, a parallel series of axons gives off regular axon collaterals to a rectangular lattice of cells. What this system learns depends entirely on what inputs are sent to it. For example, in Fig. 12*b*, synchronized CS inputs reach the first three sampling cell bodies, and (perhaps differently) synchronized CS inputs reach the next three cell bodies. Figure 12*c* draws the equivalent avalanches for this case. Next one must determine the distribution of UCS inputs. If, for

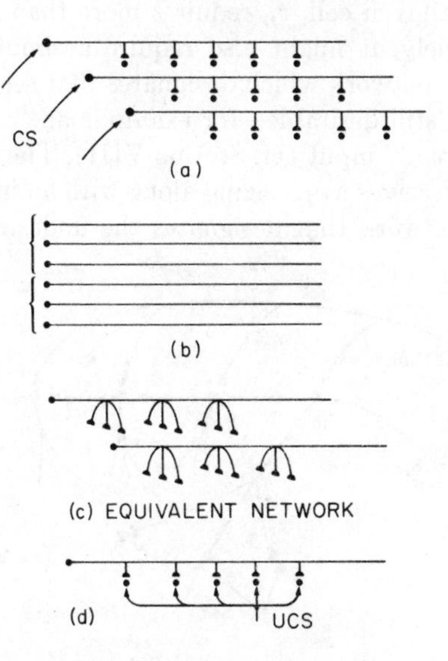

FIG. 12. An anatomy in which input symmetries determine equivalent avalanches.

example, all cells in a row parallel to the axons receive the same input, then the system of Fig. 12*d* can learn space–time patterns much as the system of Fig. 10 does. If, however, the UCS inputs are the same in each row perpendicular to the axons, then one learns only redundant copies of a sequence of perhaps uncorrelated events. In other words, the *symmetries of the input mechanisms* determine what the equivalent learning network is; the local anatomy itself need not reveal these symmetries. Various other anatomical variations are considered in Grossberg (1970b).

Avalanches of avalanches, or avalanches of avalanches of avalanches, etc., can readily be constructed. For example, a given cell population can control motions of a finger, a higher cell population in the hierarchy can control motions of all fingers in a hand, a still higher cell population can control motions of both hands, etc. Inputs can, in principle, enter this hierarchy at any level to activate a prescribed population.

C. Sensitivity to Feedback: Command Cells as Arousal Sources

How can an avalanche be modified so that sequential performance can be stopped and switched to more urgent behavioral modes? Clearly this cannot be done in Fig. 10 because the signal propagates down the entire axon once it is emitted by v_1. To prevent this, successive outstars can be separated by interpolated cells, as in Fig. 13. Immediately we have gone from one encoding cell to N_ξ such cells. These extra cells will provide no advantage unless a given cell, v_j, requires more than a signal from v_{j-1} in order to fire. Namely, it might also require a simultaneous input from another part of the network which designates that sequential performance of the given act is still desirable—for example, an "arousal" or "positive incentive motivational" input (cf. Section VIII). The cell v_j should also be unable to fire if it receives a v_{j-1} signal along with an inhibitory signal from elsewhere in the network that designates the undesirability of continued

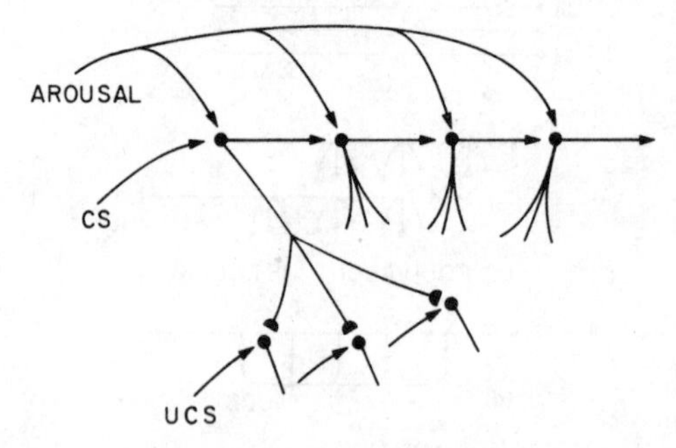

Fig. 13. A command cell as a nonspecific arousal source supporting sequential sampling.

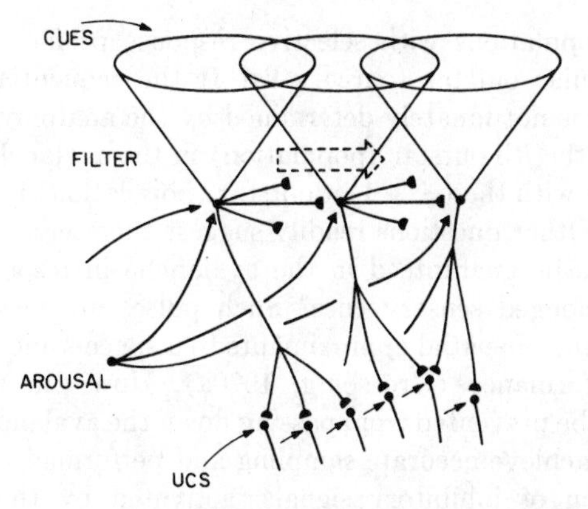

FIG. 14. An avalanche whose successive sampling sources are selected by cues and then learned.

sequential performance—for example, a "suppressor" or "negative incentive motivational" input; cf. crayfish swimmerets (Stein, 1971, p. 317).

Not every competing stimulus should be able to terminate ongoing performance. Only those inputs that have greater "significance" to the network should have this privilege. What are "significant" inputs? This question naturally leads one to discuss the question: What are rewarding or punishing inputs? In short, what is reinforcement? How does it influence the network's arousal level (Grossberg, 1971b, 1972c,d)?

Note that performance rate can be varied in Fig. 13. Each successive cell, v_j, can have its reaction time (that is, the time lag between input onset and onset of output signal) decreased, either by increasing the signal from v_{j-1} or by increasing the arousal input. Performance rate can thus be continuously modified by continuously varying the arousal level of the machine. That is, even if the avalanche-like anatomy encodes the same sequence of events (the same "information") on successive performance trials, nonetheless the arousal level of the machine (its "energy" level) can modify details of performance. The same argument holds if no learning occurs at the synaptic knobs, and the avalanche anatomy merely controls the performance of a sequence of motions. Note that modifying the arousal level does not require feedback from the avalanche outputs. Successive outstars can be sampled much faster—and at variable rates—than feedback need permit.

Until this point, we have considered avalanches whose successive outstars are predetermined by the network anatomy. In general, this need not be true. Successive links can also be determined by sensory and motor cues, including feedback cues. Then one is led to ask: How are these cues filtered through cell populations with selective response profiles to sequentially activate particular outstar source cells? If the sequential activation of outstar sources is not innately determined by the anatomy, one must also ask: How does the jth outstar (population) in the avalanche form sequential connections with the $(j + 1)$st outstar (population) in the avalanche? (See Fig. 14.) Other questions readily suggest themselves. How can brief sampling pulses be guaranteed in the avalanche in response to possibly temporally prolonged sensory cues? Such pulses are needed to achieve accurate sampling of spatial approximants to a space–time pattern, as well as precise performance (Grossberg, 1970a). How can more than one sampling pulse be prevented from passing down the avalanche at any given time, again to achieve accurate sampling and performance? This requires the introduction of inhibitory signals, activated by the outstars, and descending toward the input sources. In short, the expansion of ritualistic avalanches to achieve responsiveness to environmental feedback imposes a definite teleology on our later constructions. Some of these constructions yield mechanisms of pattern discrimination, and in particular an analysis of various uses for nonspecific inhibitory interneurons (Grossberg, 1970a, 1972a, 1973).

As learning and performance become less ritualistic in an avalanche, the complexity of the total input to each of its outstar sources increases. The total input can be a sum of a rapidly fluctuating arousal input, an input from a complex hierarchy of sensory filters, an input from a previous outstar source that was itself perturbed by a complex input, etc. Thus we seek assurances that learning can occur even if the source is perturbed by very general inputs. The next section provides such assurances in a rigorous mathematical setting. Holographic theories of memory, which depend on the existence of precisely regulated periodic sampling sources, depart heavily from the spirit of this discussion.

VI. Arbitrary Anatomies and Generalized Physiological Laws

A. One Level in a Hierarchy

When an avalanche is modified to permit feedback adaptations, the cells v_i, $i \in I$, can be sampled by many cells v_j, $j = 1, 2, \ldots, N_\xi$. Below we therefore study the following question: Under what circumstances can a collection of cells $\mathcal{A} = \{v_j, j \in J\}$ sample a collection of cells $\mathcal{B} = \{v_i, i \in I\}$ in such a fashion that simultaneous sampling of $\mathcal{B}$ by different cells in $\mathcal{A}$ does not irrevocably bias what these cells learn? We shall find that this is possible, given any finite number of cells $\mathcal{A}$ and $\mathcal{B}$, under very weak conditions. The relevant theorems (Grossberg, 1969d, 1971c, 1972b) hold even if the cells $\mathcal{A}$ fire out of phase and in response to wildly oscillatory, and mutually uncorrelated, inputs. Thus the inputs to cells $\mathcal{A}$ can be constructed from the outputs emitted by cells at a previous stage of learning or other preprocessing, and the outputs from $\mathcal{B}$ can be used to construct inputs to a later stage of cells. In this way, a hierarchy of learning cells can be constructed. The theorems study one level in such a hierarchy in detail. If such a mechanism evolved at a given time, it could be adapted to any later specialization.

B. A General Class of Systems

The equations that govern one level of this hierarchy can be substantially generalized beyond Eqs. (8) and (9) by weakening some linearities in these equations without changing their general form. These equations are defined by

$$\dot{x}_i = A_i x_i + \sum_{k \epsilon J} B_{ki} z_{ki} + C_i \tag{22}$$

and

$$\dot{z}_{ji} = D_{ji} z_{ji} + E_{ji} x_i \tag{23}$$

$i \in I, j \in J$, where A_i, B_{ji}, D_{ji}, and E_{ji} denote continuous functionals, not necessarily linear, with all B_{ji} and E_{ji} nonnegative. The input functions and initial data are chosen nonnegative and continuous. Mathematical analysis of Eqs. (22) and (23) shows that the classification of limiting and oscillatory possibilities for the pattern variables of these systems is invariant under broad changes in functionals, much as in the study of Eqs. (15) and (16). As in that situation, transient motions of the systems can be altered by changes in functionals, and a proper choice of functionals (including anatomy) must be made to guarantee efficient real-time learning of particular tasks. The invariance properties show that the systems are very stable and can be adapted to many particular situations. Below are reviewed some physically relevant choices of these functionals.

As in the case of Eqs. (15) and (16), the long-term memory decay functional, D_{ji}, can be chosen to guarantee a variety of forgetting possibilities. The choice of performance functional B_{ji}, as in Eq. (17), can also influence how decay due to D_{ji} shows up in network response to inputs. Other useful choices of these functionals are listed below.

1. *Now Print Signals of Shunting Type*

Suppose that a sequence of spatial patterns perturbs the cells $\mathcal{B}$. There exist mechanisms that can quickly accelerate learning of the patterns which arrive during prescribed time intervals. These intervals can heuristically be called Now Print intervals (Livingston, 1967, p. 132). Such mechanisms

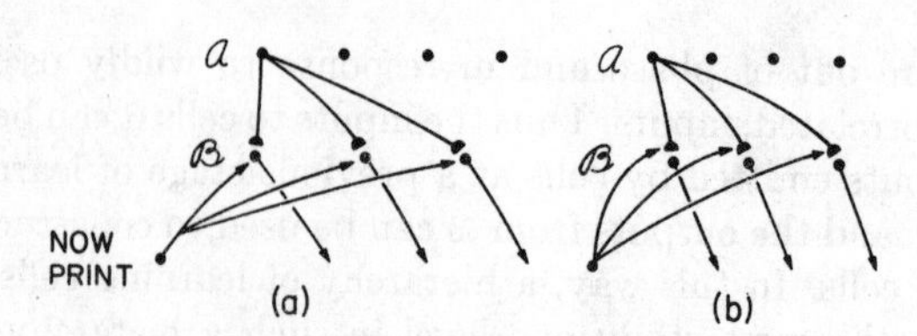

FIG. 15. Nonspecific arousal as a shunt of potentials or signals.

can be activated by arousal inputs that are turned on by the occurrence of significant events. The first mechanism works by sending synchronized signals to all cells in $\mathcal{B}$. These signals then interact multiplicatively with (or "shunt") the potentials x_i (see Fig. 15*a*).

Consider, for example, the system

$$\dot{x}_i = [-\alpha + f(t)]x_i + \theta_i C$$

where C is a constant, $0 \leq f(t) < \alpha$, and $x_i(0) = 0$. Let $f(t)$ be constant in the interval $[0, T]$. Then

$$x_i(t) = \theta_i \left[\frac{C}{\alpha - f} (1 - e^{-(\alpha - f)t}) \right]$$

for $t \in [0, T]$. The function

$$g_t(w) = \frac{1}{w} (1 - e^{-wt})$$

is, for fixed $t \geq 0$, a monotone decreasing function of $w \geq 0$. Thus, given ever-increasing values of $f \in [0, \alpha]$, $x_i(t)$ increases as well; the "shunt" f has amplified the input intensity C. This multiplicative form of Now Print mechanism is not, for some purposes, as satisfactory as the additive mechanism that will be introduced in Section VIII.

Alternatively, the nonspecific shunting signal can act directly on the synaptic knobs that deliver the inputs C_i to v_i (see Fig. 15*b*). This would have the effect of directly amplifying the inputs, as in

$$\dot{x}_i = -\alpha x_i + \theta_i f(t) C$$

The same synaptic knob shunt can influence the memory traces by amplifying the presynaptic signals that perturb the knobs. For example, let

$$\dot{z}_{ji} = -\gamma_i z_{ji} + \delta_j f(t) [x_j(t - \tau_j) - \Gamma_j]^+ x_i$$

or let

$$\dot{z}_{ji} = -\gamma_i z_{ji} + \delta_j [x_j(t - \tau_j) - \Gamma_j(f(t))]^+ x_i$$

where $f(t)$ is a nonnegative, monotone increasing function of arousal level, and $\Gamma(s)$ is a monotone decreasing function of $s = f(t)$. These laws mix exponential memory decay with a cross-correlator that can be shut on or off at will. Perfect memory until recall can also be modified in a similar fashion by letting

$$\dot{z}_{ji} = f(t)[x_j(t - \tau_j) - \Gamma_j]^+(-\gamma_j z_{ji} + \delta_j x_i)$$

or

$$\dot{z}_{ji} = [x_j(t - \tau_j) - \Gamma_j(f(t))]^+(-\gamma_j z_{ji} + \delta_j x_i)$$

Both sampling and Now Print must here be active as a precursor to learning or forgetting. Various other formal possibilities are special cases of our analysis; for example, shutting off the Now Print mechanism can prevent all memory change, whereas turning it on can permit exponential memory decay and/or new learning, as in the equation

$$\dot{z}_{ji} = f(t)\{-\gamma_j z_{ji} + \delta_j[x_j(t - \tau_j) - \Gamma_j]^+ x_i\}$$

2. *Local Flow*

The signal terms $\beta_{jk}[x_j(t - \tau_{jk}) - \Gamma_{jk}]^+$ and $\delta_{jk}[x_j(t - \tau_{jk}) - \Gamma_{jk}]^+$ in Eqs. (8) and (9), respectively, can be replaced, say, by

$$B_{jk}(t) = \beta_{jk}(t)[x_j(t - \tau_{jk}(t)) - \Gamma_{jk}(t)]^+$$

and

$$E_{jk}(t) = \delta_{jk}(t)[x_j(t - \sigma_{jk}(t)) - \Omega_{jk}(t)]^+$$

which permit different, and variable, time lags, thresholds, and path strengths in the two signal strength functionals. This includes the possibility of coupling a Now Print mechanism to these functionals, through either the variable path strengths or the thresholds. Functional $E_{jk}(t)$ describes the effect of the signal from v_j on the cross-correlational process within N_{jk} that determines z_{jk}. Functional B_{jk} describes the net signal from v_j that ultimately influences v_k after being processed in N_{jk}. It is therefore natural to physically expect that $\Gamma_{jk} \geq \Omega_{jk}$. This *local flow condition* says little more than that the signal from v_j passes through N_{jk} on its way to v_k. Such a condition is, in fact, needed to guarantee that many cells can simultaneously sample a given pattern without creating asymptotic biases in their memory (Grossberg, 1971c, 1972b). This condition has an easily realized physical interpretation, given the assumption that the process z_{jk} occurs in the synaptic knob or at postsynaptic membrane sites. Various other interpretations for z_{jk} do not yield a physical basis for the local flow condition, and could not realize the possibility of simultaneous sampling by many input channels. The local flow condition provides examples of systems that can learn patterns without performing them until later, but cannot perform old patterns without also learning new patterns that are imposed during performance.

The functionals B_{jk} and E_{jk} permit more complicated possibilities as well. For example, *in vivo*, after a signal is generated in e_{jk}, it is impossible to generate another signal for a short time afterward (absolute refractory period) and harder to generate another signal for a short time after the absolute refractory period (relative refractory period). Also, some cells emit signals in complicated bursts. Intricate preprocessing of input signals can occur in the dendrites of cells before the transformed inputs influence the cell body. All such continuous variations are, in principle, covered by our theorems, which say that, whereas such variations can influence transient motions of the system, the classification of limits and oscillatory possibilities is unchanged by them. Given that weak constraints such as local flow hold, what is learned depends on which cells sample what patterns, and how intensely, no matter how complicated the rules are for determining when a cell will sample.

It is physically interesting that those terms, such as B_{jk} and E_{jk}, which describe processes that act over a distance (such as signals flowing along e_{jk}) are the terms in Eqs. (22) and (23) that permit the most nonlinear distortion without destroying learning properties. The term x_i in Eq. (23) is not of this type. This term is computed in N_{ji} from the value x_i in the contiguous vertex v_i.

C. Local Symmetry Axes

In their final form, the theorems show that unbiased pattern learning can occur in systems with arbitrary positive path weights β_{ji} from $j \in J$ to $i \in I$. This is achieved by first restricting attention to systems of the form

$$\dot{x}_i = Ax_i + \sum_{k \epsilon J} B_k z_{ki} + C_i \tag{24}$$

and

$$\dot{z}_{ji} = D_j z_{ji} + E_j x_i \tag{25}$$

where $i \in I$ and $j \in J$. That is, all functionals A_i, B_{ji}, D_{ji}, and E_{ji} are chosen independent of $i \in I$, and the anatomy is constrained to make this possible. These constraints mean that all cells $\mathcal{B} = \{v_i : i \in I\}$ are sampled by a given cell, v_j, in $\mathcal{A} = \{v_j : j \in J\}$ without biases due to system parameters ($B_{ji} = B_j$, $D_{ji} = D_j$, $E_{ji} = E_j$), and that the inputs to all cells $\mathcal{B}$ are averaged by their cell potentials without biases due to averaging rates ($A_i = A$) (see Fig. 16*a*). Systems (24) and (25) allow each cell to have a different time lag, threshold, and axon weight, as in

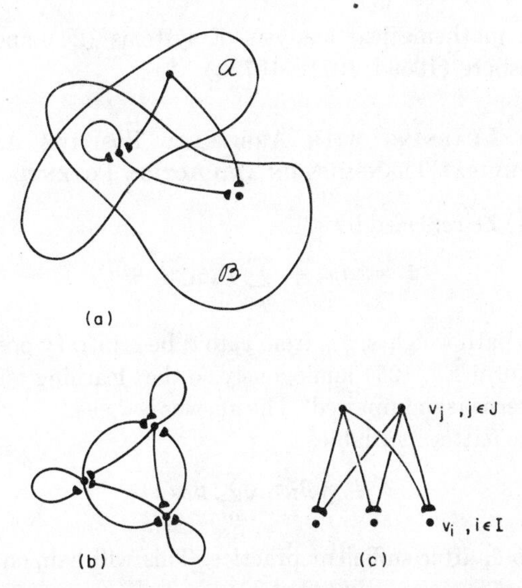

FIG. 16. Constraining an arbitrarily large set of sampling cells by imposing local symmetry axes.

$B_j(t) = \beta_j[x_j(t - \tau_j) - \Gamma_j]^+$. Even if all cells interact, as in Fig. 16*b*, no biases in asymptotic learning need occur due to these asymmetries in signal transfer among possibly billions of cells.

Figure 16, *b* and *c*, illustrates two extremal anatomies, the completely recurrent ($I = J$) and the completely nonrecurrent ($I \cap J = \phi$) cases. Generalizations of Fig. 16*a* are also possible. In these generalizations, $\mathcal{A}$ and $\mathcal{B}$ are replaced by sets $\{\mathcal{A}_k\}$ and $\{\mathcal{B}_k\}$ of subsets such that each cell in a given $\mathcal{B}_k$ is sampled by all cells in $\mathcal{A}_k$. One seeks the maximal subsets $\mathcal{B}_k$ for which this decomposition exists. For some purposes, a fixed set $\{\mathcal{B}_k\}$ is determined by structural considerations; for example, each $\mathcal{B}_k$ controls a different motor effector. It is then sometimes profitable to introduce fictitious cells into the sampling cells $\mathcal{A}$ if some cells in $\mathcal{A}$ sample two or more subsets $\mathcal{B}_k$. For example, if cell v_i in $\mathcal{A}$ samples $\mathcal{B}_1$ and $\mathcal{B}_2$, replace v_i by two cells, v_{i1} and v_{i2}, such that v_{ij} samples only $\mathcal{B}_j$, $j = 1, 2$, and each v_{ij} receives the same inputs, and has the same parameters and initial data, as the original cell, v_i, had. Otherwise expressed, suppose that a given cell (population) can sample motor controllers of both hands, but that only the left hand is used to learn a given task. We then want to study the pattern variables associated with the left hand only, not both hands. The decomposition exhibits the system in a form suitable to this analysis. The mathematical analysis of systems (24) and (25) can be found in Grossberg (1969d, 1971c, 1972b).

D. Unbiased Learning with Arbitrary Positive Axon Weights Using Chemical Transmission and Action Potentials

Let Eq. (24) be replaced by

$$\dot{x}_i = Ax_i + \sum_{k \epsilon J} B_k \beta_{ki} z_{ki} + C_i \tag{26}$$

that is, let the path weights, β_{ji}, from v_j to v_i be arbitrary positive numbers. Can we transform Eq. (25) analogously so that learning and performance of spatial patterns is unimpaired? The answer is "Yes."

We want the pattern variables

$$Z_{ji}^{(\beta)} = \beta_{ji} z_{ji} \Big(\sum_{k \epsilon I} \beta_{jk} z_{jk} \Big)^{-1}$$

to converge to θ_i after sufficient practice. This will happen if Eq. (25) is replaced by

$$\dot{z}_{ji} = D_j z_{ji} + E_j \beta_{ji}^{-1} x_i \tag{27}$$

since letting $w_{ji} = \beta_{ji} z_{ji}$, Eqs. (26) and (27) yield

$$\dot{x}_i = Ax_i + \sum_{k \epsilon J} B_k w_{ki} + C_i$$

and

$$\dot{w}_{ji} = D_j w_{ji} + E_j x_i$$

which are again of the form of Eqs. (24) and (25). A mathematical analysis shows that our goal could *not* be achieved by replacing Eq. (25) with

$$\dot{z}_{ji} = D_j z_{ji} + E_j \beta_{ji} x_i$$

which would be the natural thing to do if we supposed that $E_j \beta_{ji}$ is determined wholly by spiking frequency (Grossberg, 1972b).

How can the β_{ji}'s in Eqs. (26) and (27) be interpreted? Suppose that $\beta_{ji} = \lambda_j R_{ji}$, where $\lambda_j > 0$ and R_{ji} is the circumference of the cylindrical axon, e_{ji}. Let the signal in e_{ji} [for example, the action potential (Ruch *et al.*, 1971)] propagate along the circumference of the axon to its synaptic knob. Let the signal disperse throughout the cross-sectional area of the knob [for example, as ionic fluxes (Katz, 1966)]. Let local chemical transmitter production in the knob be proportional to the local signal density. Finally, let the effect of the signal on the postsynaptic cell be proportional to the product of local signal density and local transmitter density and the cross-sectional area of the knob.

These laws generate Eqs. (26) and (27) as follows. Signal strength is proportional to R_{ji} or to β_{ji}. The cross-sectional area of the knob is proportional to R^2_{ji}. Hence signal density in the knob is proportional to $R_{ji}R^{-2}_{ji} = R^{-1}_{ji}$, or to β^{-1}_{ji}, as in Eq. (27). Thus (signal density) $\times$ (transmitter density) $\times$ (area of knob) $\cong R^{-1}_{ji}z_{ji}R^2_{ji} = R_{ji}z_{ji} \cong \beta_{ji}z_{ji}$, as in Eq. (26).

By contrast, a mechanism whereby signals propagate throughout the cross-sectional area of the axon could not produce unbiased learning given arbitrary axon connection strengths, or at least such a mechanism is still elusive. The difficulty here is that signal strength is proportional to R^2_{ji}, signal density is proportional to one, and local transmitter production rate is then proportional to one. The postsynaptic signal is proportional to (signal density) $\times$ (transmitter density) $\times$ (area of knob) $\cong \beta^2_{ji}\, z_{ji}$. Thus we are led to the system

$$\dot{x}_i = Ax_i + \sum_{k \epsilon J} B_k\beta^2_{ki}z_{ki} + C_i$$

and

$$\dot{z}_{ji} = D_j z_{ji} + E_j x_i$$

which can be written as

$$\dot{x}_i = Ax_i + \sum_{k \epsilon J} B_k\beta_{ki}w_{ki} + C_i$$

and

$$\dot{w}_{ji} = D_j w_{ji} + E_j\beta_{ji}x_i$$

in terms of the variables $w_{ji} = \beta_{ji}z_{ji}$. This system has unpleasant mathematical properties (Grossberg, 1972b).

These observations suggest that the action potential not only guarantees faithful signal transmission over long cellular distances, as is well known, but also executes a subtle transformation of signal densities into transmitter production rates that compensates for differences in axon diameter. Note also that this transformation seems to require the chemical transmitter step. Purely electrical synapses presumably could not execute it. Thus our laws for transmitter production (and/or related processes) not only guarantee that learning occurs, but also that unbiased learning occurs, under very weak anatomical constraints. Section IX suggests another way in which the action potential contributes to unbiased learning on the level of individual cells.

The next two sections illustrate some phenomena that occur in networks with specific anatomies.

E. Threshold-Dependent Phase Transitions in Recurrent Networks

Consider Figs. 16 and 17. Figure 16*b* is a *recurrent* network: the cells send signals to each other. Figure 16*c* is a *nonrecurrent* network: the cells send signals only to different cells. Not surprisingly, under certain circumstances, the memory of recurrent and nonrecurrent networks can differ dramatically. Less intuitively, a recurrent network can sometimes behave like a nonrecurrent network. Moreover, an anatomist could not tell the difference between a recurrent network which behaves recurrently from one which behaves nonrecurrently

Figure 17 illustrates what is involved in making this distinction. Figure 17 depicts a recurrent network whose recurrent signals are carried by interneurons between the signal generating cells. Let the threshold for signals to leave the cells be Γ_1, and let the threshold of the interneurons be Γ_2. Suppose that $\Gamma_1 = \Gamma_2 = 0$. Then any input to a cell v_j will create outputs and signals to other cells v_i. These signals will, in turn, create outputs from v_j and feedback signals to v_i, and so on. As a consequence, recall trials can destroy the memory of this system. Suppose, however, that $\Gamma_2 \gg 0$. Then an output from a cell can again create signals to other cells. These signals can in turn, create outputs from these cells *without* causing feedback signals. Such a network has a nonrecurrent kind of memory: Recall need not destroy the memory of the system. During recall, each cell and its interneurons behaves like an outstar embedded in a larger, but functionally passive, anatomy in this case. The thresholds thus serve to localize the memory trace, and to provide a kind of localized "context" which a given input can activate. Whereas this argument holds during recall of a spatial pattern or during slow recall of a space–time pattern, Section VII shows that it need not hold during rapid recall of a space–time pattern.

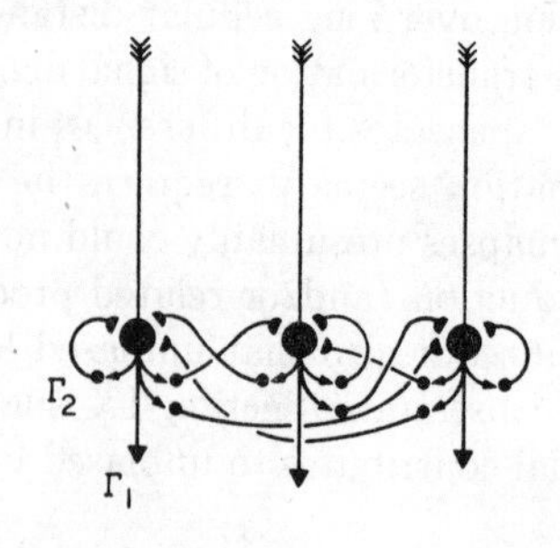

Fig. 17. Influence of interneuronal thresholds on whether a recurrent anatomy behaves recurrently or nonrecurrently.

Small inputs to the network of Fig. 17 can make it behave like a nonrecurrent network; even slightly larger inputs can make it behave recurrently, by creating signals that are sufficiently large to exceed the feedback thresholds. For example, varying the overall arousal level of the system can change its behavior in response to fixed externally controlled inputs from nonrecurrent to recurrent or conversely. The asymptotic behavior of the system is a discontinuous function of input and threshold perturbations: There is a "phase transition" at critical values of these parameters. Given this possibility, one can argue in the reverse direction. Suppose that a nonrecurrent type of memory is desired at all times. How can the total input to the cells be "normalized" so that the feedback thresholds are never exceeded? Various arrangements of nonspecific inhibitory interneurons can accomplish this task (Grossberg, 1970a, 1972d, 1973).

F. Pattern Completion and Mass Action

In Fig. 16*c*, suppose that any fraction of sampling cells is excised away. The remaining sampling cells can reproduce an entire learned pattern on the sampled cells if some of the remaining sampling cells were active when that pattern was being learned ("pattern completion"). In Fig. 16*b*, *each* vertex, v_i, can encode and perform a *different* spatial pattern at all the vertices, if the dynamics of the network are nonrecurrent in the sense of the previous section. By contrast, suppose that sampling cells can sample only a fixed fraction of sampled cells, and that the sampled cells are chosen randomly. Then, on the average, excising ever greater numbers of sampling cells will create a proportional deficit in the ability of the remaining sampling cells to reproduce a previously learned pattern spread across all sampled cells ("mass action").

G. Imprinting and Irreversibility

Mathematical analysis of systems (24) and (25) shows that, once these systems are factored into pattern variables and total energy variables, different choices of functionals influence transient motions of pattern variables, but not the possible oscillations of these variables. In particular, different functionals, or different values of fixed functionals due to particular choices of inputs, can determine different numerical limits of the pattern variables as $t \to \infty$. This section summarizes some results concerning these limits which have been proved for a particular choice of functionals, but which should hold for many other functionals chosen in the same anatomies (Grossberg, 1968b, 1969c).

This choice of functionals determines an interference theory of forgetting in the nonrecurrent outstar anatomy; for example, let $(B_j z_{ji})(t) = \beta_j[x_j(t - \tau_j) - \Gamma_j]^+ Z_{ji}(t)$ in Eq. (24). In various recurrent anatomies, however, these functionals do not determine an interference theory. Instead, there exists a phase transition in memory, such that one type of memory prevails if the network's numerical parameters have certain values, whereas a distinct type of memory prevails if the parameters take on the remaining values. Consider Fig. 18. Given the anatomy of Fig. 18*a*, there exists an example of the following type. The numerical values of the network parameters—such as α, β, γ, τ, Γ in Eqs. (8) and (9)—form two exhaustive and nonoverlapping sets, A and B. If the parameter values fall in A, then the network can remember everything; if the parameter values fall in B, then the network cannot remember anything. Thus, spontaneous forgetting occurs if parameter values fall in B, even though, speaking locally, the interaction terms describe an interference theory of forgetting. The global anatomy determines this forgetting effect. In Fig. 18*b*, if the parameter values fall in A, then the network can remember everything; if the parameter values fall in B, then the network can remember spatial patterns. For example, given A, the network can remember lists, or space–time patterns. Given B, the network forgets temporal discriminations, and its memory seeks the spatial pattern closest to what it has learned. Thus the global recurrent anatomy not only determines that two phases exist, but also what the memory characteristics of each phase will be.

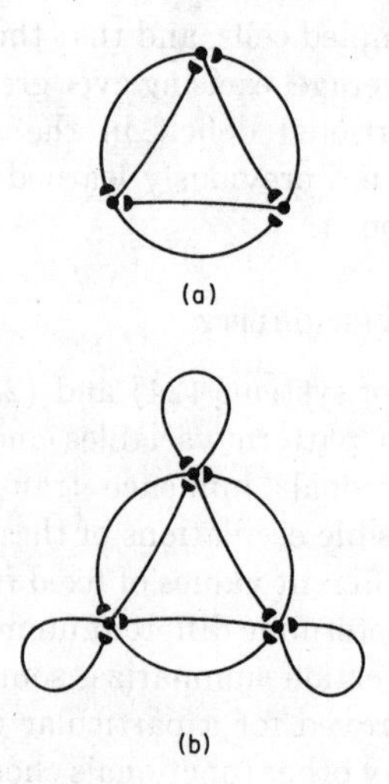

FIG. 18. Imprinting due to a phase transition in memory.

By varying network parameters, network dynamics can be transformed from phase B to phase A. Any mechanism that does this will "imprint" the memory of the input pattern that perturbs the network at the time this transition takes place. The transition from B to A can be effected, for example, by increasing the velocity of signals in the network axons. Given this formal observation, we now note various possible analogs of this phenomenon *in vivo.*

Signal velocity can be increased *in vivo* by laying down an axonal sheath around unmyelinated axons. Such a sheath can cause signals to jump along the axon in a saltatory fashion. Various strategies for imprinting a pattern of axonal connections in a particular subnetwork of a total network hereby suggest themselves. A nonspecific command signal (for example, a hormone) to this subnetwork to lay down sheaths on all subnetwork axons would suffice. Alternatively, one could imprint a pattern in the axons of particular nerves as they became active by coupling the activity of the sheath-producing cells to that of the nerves (cf. Orkand *et al.*, 1966). The *order* in which various cells imprinted patterns could be determined by such a mechanism. The interaction between external inputs and the total network anatomy could establish this order by determining which cells would reach the critical activity levels for sheath production first; cf. Grossberg (1969f, Section 19) in the light of Section IX below. Although the order in which particular nerves or subnetworks are imprinted can be developmentally predetermined by such a mechanism, the actual patterns that are imprinted depends on the choice of external inputs. If given cells do not pass from phase B to phase A, then they retain a plastic memory which can continue to spontaneously forget old patterns.

Grossberg (1969c) shows that these systems also have various properties that are of interest from the statistical mechanical point of view. For example, before such a network is probed by experimental inputs, its output might be linear, locally reversible ($z_{jk} = z_{kj}$), and globally reversible ($Z_{jk} = Z_{kj}$). An experimental input can make the output nonlinear, globally irreversible ($Z_{jk} \neq Z_{kj}$), but still locally reversible. After the effect of the input wears off, the output can become linear again. Whether the output again becomes globally reversible or not, however, depends on the sign of a function of network parameters that cannot be easily measured by an input–output analysis. Thus the (non)linearity of the system can be decoupled from its global (ir)reversibility. The decision whether the system will be become globally reversible or will remain globally irreversible after inputs cease depends on whether the network parameters fall into B or A. In all cases where this system is eventually free from inputs, its asymptotic behavior approaches that of a stationary

Markov chain. Network dynamics provide a real-time description of the transient nonstationary behavior of the system as it approaches its stationary asymptote.

VII. Serial Learning

A. Qualitative Data

This section discusses the response of a recurrent network to a particular type of space–time pattern–namely, a list, or sequence of spatial patterns, in which only one component of each spatial pattern is positive. Section VI pointed out that a recurrent network can behave nonrecurrently in response to a spatial pattern if signals from a given vertex do not create feedback signals to that vertex. Even if parameters are chosen to guarantee this, the response of the network to a space–time pattern, in particular to a list of length n, can differ significantly from that of n independent outstars to n spatial patterns.

There exists a large body of data on list learning. Some of the themes in these data are sketched below. Our analysis of these data will be heuristic and will focus only on the effects that arise in the minimal anatomies that are capable of learning a list. Proofs and extensions of these assertions are found in Grossberg (1969e) and Grossberg and Pepe (1971). A more complete phenomenological analysis of the data on a neural level would study how list items, and sequences of items, are coded by hierarchically organized fields of cells with selective response profiles, and in particular of how the field activity is sustained by short-term memory mechanisms while it is transformed and transferred to long-term memory (cf. Atkinson and Shiffrin, 1968; Grossberg, 1973). This section studies one level of recurrent interactions in such a hierarchy. The goal is to better understand the hierarchical case by first gaining insight into various one-level cases. Once this is accomplished, hierarchical anatomies can be more readily synthesized.

1. *Backward Learning*

Suppose that the list AB is sequentially presented several times to a learning subject $\mathcal{O}$. Let B alone be presented to the subject on recall trials. Other things being equal, prior practice of AB increases the probability of guessing A given B. That is, practicing AB yields at least partial learning of BA. Relative to the time scale of external events, which flows forward from A to B, learning both AB and BA, given practice of AB alone, means that the internal dynamics of $\mathcal{O}$ flow both forward (AB) and backward (BA) in time.

2. *Global Arrow in Time*

Now suppose that the list ABC is practiced with a time lag of w time units between successive presentations of each letter. After B has been presented to $\mathcal{O}$, and before C is presented, $\mathcal{O}$ has received only the list AB, and thus the association from B to A begins to form. We know, however, that ultimately ABC can be learned. Thus the forward association BC is stronger than the backward association BA, and can therefore inhibit it to yield a global arrow in time from A to B to C. In this sense, "time" is flowing both forward and backward within $\mathcal{O}$, but the forward flow is stronger and ultimately enables $\mathcal{O}$ to imitate the direction in time of external events.

3. *Bowing*

The same theme is illustrated by the phenomenon of bowing, which means that the middle of a serially learned list is harder to learn than either end, or, more familiarly, that we can often remember how a sequence of events began and ended but forget many intermediate details. If internal events in $\mathcal{O}$ flowed only forward in time, we might expect the plot of mean number of recall errors as a function of list position to be monotone nondecreasing, since at list positions ever deeper within the list, more response interference can accumulate from previously presented list items. In actuality, however, list positions near the list's middle are hardest to learn, which illustrates that the nonoccurrence of items after the last list item has somehow made items near the end of the list, which were presented earlier in time, easier to learn.

4. *Skewing*

A closely related phenomenon is skewing, which means that the list position that is hardest to learn often occurs nearer to the end than to the beginning of the list. This fact recalls the fact that learning in the forward direction (AB) is stronger than learning in the backward direction.

5. *Intratrial versus Intertrial Interval*

Many parametric studies of learning difficulty at various list positions have been reported. The intratrial interval (denoted by w) is the time between presentation of successive list items. The intertrial interval (denoted by W) is the time between two successive presentations of the

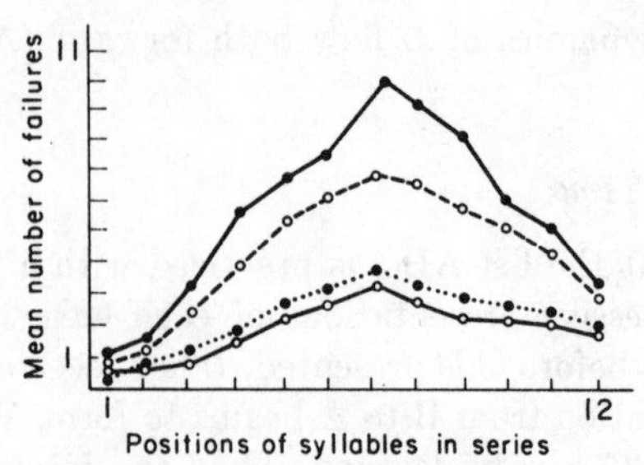

FIG. 19. Influence of intratrial interval and intertrial interval on degree of bowing. From Hovland (Osgood, 1953, p. 506). ●——● W = 6 seconds, w = 2 seconds; ○- - -○ W = 2 minutes 6 seconds, w = 2 seconds; ●· · ·● W = 6 seconds, w = 4 seconds; ○——○ W = 2 minutes 6 seconds, w = 4 seconds.

list—that is, the time between successive learning trials. Figure 19 illustrates the influence on bowing of varying w and W. Note that increasing w from 2 seconds to 4 seconds can substantially flatten the bowed curve, and that, once the curve is flattened in this fashion, increasing W has little influence on the rate of learning. Slowing the presentation rate is an example of "distributing practice." Figure 19 shows that distributing practice reduces the number of learning errors.

When the list is presented rapidly (for example, w = 2 seconds), increasing W substantially reduces the number of errors in the middle of the list. In short, increasing the rest interval after the practice trial has simplified learning of the entire list, especially at its middle. This effect also illustrates the existence of backward learning effects. Increasing W much beyond the 2-minute 6-second value does not reduce the number of errors substantially in these data.

Note that the dictum "Distributing practice improves learning" must be interpreted with caution. Letting w approach 24 hours certainly distributes practice, but makes learning of the list quite unlikely. Thus we shall seek a list presentation speed, much less than w = 24 hours but greater than w = 0, that optimizes the benefits of distributing practice.

6. *Response Oscillation and Generalization*

This phenomenon is closely related to bowing (see Fig. 20). It says that the gap between the first correct guess and the last error is largest near the middle of the list. More list intrusions interfere with the correct association near the middle of the list than at its ends. In fact, a generalization gradient exists at each list position such that the probability of guessing an item, given presentation of a fixed item, decreases as a function of the number of intervening items presented on a single trial. The shape of this generaliza-

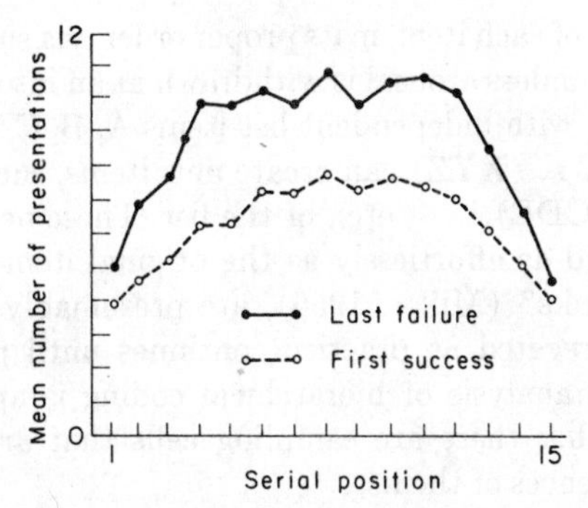

FIG. 20. Response oscillation. From Hull *et al.* (Osgood, 1953, p. 503).

tion gradient depends on list position. Given a sufficiently large intertrial interval, the gradient is skewed forward near the beginning of the list, backward near the end of the list, and in both directions near the list's middle, often with a broader span near the middle, and an advantage given to anticipactory rather than perseverative errors (Osgood, 1953), presumably as a manifestation of stronger forward than backward associations.

7. *Anchoring*

This phenomenon describes the order in which list items are learned (Atkinson and Shiffrin, 1968). Items are often learned both in the forward direction and in the backward direction around the "anchor" stimulus, A. For example, AB, then YZ, then BC and CD, then XY might be the first associations to be learned, and in the given order.

8. *Chaining*

By putting the learned fragments around the anchor together, we see that list items are often learned in growing chains around the anchor stimulus. These chains propagate from the anchor in both forward and backward directions, toward the middle of the list, and can gradually reduce the number of competing items that contribute to response oscillation at the list's middle.

9. *Chunking*

Suppose that a chain has formed. The chain can be performed—*as a unit*—given presentation of the anchor stimulus and persistent arousal, if it has an avalanche structure, in which each unit of the chain excites its motor representation as well as the next unit of the chain (Fig. 14). This yields performance of each item, in its proper order, via successive excitation of the entire chain, unless arousal is withdrawn at an intermediate point. In this sense, starting with independent list items A, B, C, . . . , Z, practicing the alphabet (ABC . . . XYZ) can create new items, such as subsequences (AB), (ABC), (BCDE), . . . , etc., of the list. These new items can eventually be performed as effortlessly as the original items were. Composite list units, or "chunks" (Miller, 1956), are presumably being continually formed and reaggregated as practice continues until perfect learning is achieved. Here an analysis of hierarchical coding is appropriate, and in particular of whether there are sampling cells that are excited only by particular subsequences of the list.

10. *Primacy versus Recency*

Typically, the beginning of a serially learned list is easier to learn than the end, as in Fig. 19; that is, the *primacy* effect is stronger than the *recency* effect, or "primary dominates recency." In the minimal network, increasing the arousal level to high values can reverse this effect. Is there a corresponding phenomenon *in vivo*?

11. *Inverted* U *in Learning*

Either too little motivation (or arousal), or too much, can hamper performance. Figure 21 illustrates this typical result in general terms. It is well described in Hebb (1955). Analogous difficulties occur in the network below. Given underarousal, there is too little energy to drive the learning process. Given overarousal, there is ample energy to drive learning, but a high level of response interference is produced by incorrect associations that are similar either in time of presentation or in meaning to the correct associations. In other words, overarousal produces "fuzzy response sets," and by impairing the network's ability to focus on the correct association interferes with "paying attention."

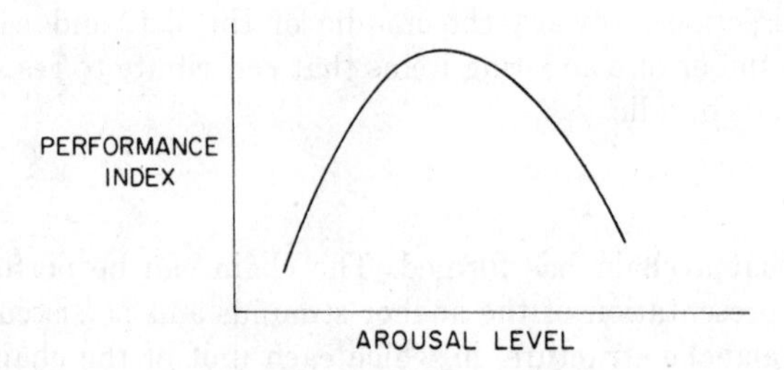

FIG. 21. Inverted U in learning.

Recent experiments (Kornetaky and Eliasson, 1969; Phillips and Bradley, 1970) have investigated the hypothesis that paying attention can be impaired by overarousal. Kornetsky and Eliasson varied the excitation level of white rats during a sustained attention task. The task chosen was for the rats to press the lever on presentation of a specific auditory stimulus. The experimenters noted any impairment in performance as a function of excitation level. High excitation was produced by electrically stimulating the rats' reticular formation. Low excitation was produced by administering a dose of chlorpromazine. Both electrical stimulation and chlorpromazine interfered with attention. The two treatments together resulted in performance indistinguishable from that seen after injections of saline alone. Presumably electrical stimulation and chlorpromazine antagonize each other and return the rat to a normal arousal level.

12. *Overarousal and Punning*

There exist networks in which overarousal weakens the strength of correct associations at the list's beginning by forcing them to compete with incorrect associations formed with later list items. Suppose that the list is a sentence. By the time the entire sentence has been presented to such an overaroused network, the earlier portions of the sentence have been washed away by a flood of competing associations. The meaning of the sentence is similarly lost. Only the last few list items survive the flood, and only these can therefore influence responses to the sentence. Structurally similar words, such as rhymes or puns, can be expected, rather than meaningful replies. Maher (1968) has discussed a phenomenon of punning in certain schizophrenics who are presumed to be in a continual state of overarousal. Various manics also pun. Chlorpromazine can improve the performance of schizophrenics at tasks that require sustained attention, presumably by lowering their arousal level. Lithium presumably has a similar effect in manics (Dally, 1967). Inspection of the networks shows that different mechanisms can produce similar symptoms of overarousal. For example, unduly large inputs from a nonspecific arousal source, such as reticular formation, can cause overarousal. Alternatively, pathological changes in the binding of ions (for example, Ca^{++}) at network cells, none of which is necessarily a nonspecific arousal source, can cause overarousal by amplifying all signals in the network. Presumably cures for similar difficulties in paying attention needed by different patients can be quite dissimilar.

Different network anatomies respond to fluctuations in arousal level in different ways. For example, in networks that describe the interaction of rewarding events with internal drives, either under- or overarousal can produce "emotional depression" by reducing the incentive motivational response of the network to emotionally charged cues (Grossberg, 1972c). The underaroused network responds "irritably" to sufficiently large increases in such cues, whereas the overaroused network is "indifferent" to these cues. In networks describing recurrent on-center off-surround interactions of shunting type, low arousal can help the network "choose" among many response alternatives, whereas high arousal tends to store many cues in short-term memory (Grossberg, 1972b). Grossberg (1975) combines such mechanisms to analyze various attentional and discrimination learning data.

The remainder of this section qualitatively describes some formal network mechanisms that behave analogously to psychological data such as that above.

B. Backward Learning

Consider the minimal anatomy that can learn AB or BA (that is, $\beta_{AB} > 0$ and $\beta_{BA} > 0$), as well as related response alternatives such as AA, AC, BB, or BC. Suppose that the network parameters are unbiased and that no association is preferred initially. The very possibility of learning BA in this context will imply that BA will be at least partially learned when AB is practiced. Thus backward learning effects can arise simply because choices exist. The greater learning of AB than of BA will be due to the existence of better cross-correlations between signals and potentials in the forward direction than in the backward direction.

Let the network be represented by the following equations for definiteness. More general functionals can also be used.

$$\dot{x}_A = -\alpha x_A + \beta[x_A(t-\tau) - \Gamma]^+ z_{AA} + \beta[x_B(t-\tau) - \Gamma]^+ z_{BA} + C_A$$

$$\dot{x}_B = -\alpha x_B + \beta[x_B(t-\tau) - \Gamma]^+ z_{BB} + \beta[x_A(t-\tau) - \Gamma]^+ z_{AB} + C_B$$

$$\dot{x}_C = -\alpha x_C + \beta[x_A(t-\tau) - \Gamma]^+ z_{AC} + \beta[x_B(t-\tau) - \Gamma]^+ z_{BC} + C_C$$

and

$$\dot{z}_{ij} = -\gamma z_{ij} + \delta[x_i(t-\tau) - \Gamma]^+ x_j$$

where $(i, j) =$ (A, A), (B, B), (A, B), (B, A), (A, C), or (B, C). Present the serial list once with an intratrial interval of w. Then $C_A(t) = C_B(t+w)$, and $C_C(t) \equiv 0$.

A particular, but noncrucial, choice of w will be made to emphasize the main effects. To maximize the possibility of learning AB, let the signal from v_A to N_{AB} arrive at N_{AB} as the input $C_B(t)$ to v_B arrives; that is, let the sampling delay from the onset time of the input C_A, namely,

$$D(\tau, \Gamma) = \tau + \min\{t: x_A(t) = \Gamma, \dot{x}_A(t) > 0\}$$

satisfy the identity

$$D(\tau, \Gamma) = w \tag{28}$$

This yields maximal overlap of the signal $\beta[x_A(t-\tau) - \Gamma]^+$ and the potential $x_B(t)$ for purposes of cross-correlation by $z_{AB}(t)$. All knobs N_{AA}, N_{AB}, and N_{AC} receive equal signals from v_A. The signal from N_{AC} to v_C is dominated at v_B by the signal from N_{AB} and the input $C_B(t)$. Thus, after learning begins, $Z_{AB} > Z_{AC}$, where $Z_{ij} = z_{ij}(\sum z_{ik})^{-1}$. The vertex v_A also receives two inputs—namely, the signal from N_{AA} and $C_A(t)$. Nonetheless, the correlation between the N_{AA} signal and $C_A(t)$ is not as good as the correlation between the N_{AB} signal and $C_B(t)$. Thus, $Z_{AB} > Z_{AA} > Z_{AC}$. A similar argument shows that $Z_{BA} > Z_{BC}$ after sampling begins at the knobs N_{BA} and N_{BC}. The correlation between the N_{BA} signal and $C_A(t)$ is not as good as that between the N_{AB} signal and $C_B(t)$. Choosing between the inequalities $Z_{BA} > Z_{BB}$ and $Z_{BA} \leq Z_{BB}$ requires a study of network parameters. This is because N_{BA} samples the decaying input $C_A(t)$ boosted by self-excitation via N_{AA}, whereas N_{BB} samples the decaying input $C_B(t)$ boosted by its own self-excitation.

C. Optimal Learning Speeds

Consider the following network anatomies for definiteness.

1. *Complete* n*-Graph without Loops.* This is the minimal anatomy that can learn any list, $r_i r_j$, of length 2 with distinct entries (see Fig. 22*a*).
2. *Complete* n*-Graph with Loops.* This is the minimal anatomy that can learn any list of length 2 (see Fig. 22*b*).
3. *Two-Layer Graph with Completely Nonrecurrent Sampling.* Each input $C_i(t)$ is delivered to two vertices, v_{1i} and v_{2i}. Each vertex v_{1i} can sample all the vertices v_{2k} (see Fig. 22*c*); e.g., each v_{1i} is a command population excited by a subsequence of the list at a uniform rate.

We shall denote a particular network corresponding to a given alphabet $\mathcal{U} = \{r_1, r_2, \ldots, r_n\}$ of behavioral units by $\mathfrak{M}(\mathcal{U})$. The graphs in Fig. 22 will be assumed to be unbiased for definiteness; that is, all vertices or edges of a given type possess the same parameters. For an example of an unbiased complete n-graph without loops consider

$$\dot{x}_i = -\alpha x_i + \beta \sum_{k \neq i} [x_k(t-\tau) - \Gamma]^+ z_{ki} + C_i \tag{29}$$

$$\dot{z}_{jk} = -\gamma z_{jk} + \delta[x_j(t-\tau) - \Gamma]^+ x_k, \qquad j \neq k \tag{30}$$

and

$$z_{jj} = 0 \tag{31}$$

where $i, j, k = 1, 2, \ldots, n$.

Let a long list $r_1 r_2 \ldots r_L$ be serially presented to an unbiased complete n-graph without loops, for definiteness. Thus $C_1(t) = C_2(t + w) = \cdots = C_L[t + (L - 1)w]$. The stimulus sampling probabilities of such a network are defined by $Z_{jk} = z_{jk}(\sum_{m \neq j} z_{jm})^{-1}$. Suppose initially that the network is at rest and that all associations are equally strong; that is, $x_i(t) = 0$ and $Z_{jk}(0) = 1/(n - 1)$, for $i = 1, 2, \ldots, n$, $j \neq k$, and $t \leq 0$.

Even if the inputs $C_i(t)$ arrive through independent input channels, no learning occurs if $w = 0$, since then all inputs are equal and, by symmetry, the memory traces remain uniformly distributed.

Suppose by contrast that $w \gg D(\tau, \Gamma)$. Then v_i begins to sample $D(\tau, \Gamma)$ time units after it is perturbed by $C_i(t)$. After $C_i(t)$ becomes zero again, these sampling signals gradually decay to zero. Only after sampling ceases does $C_{i+1}(t)$ become positive. Hence $[x_i(t - \tau) - \Gamma]^+ x_j(t) \cong 0$ for all i, j, and no learning occurs.

No learning occurs if $w = 0$ because the potentials are uniformly distributed, and therefore indistinguishable from each other. No learning

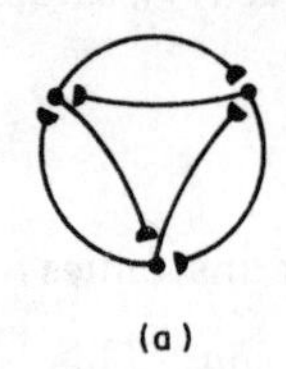
(a)

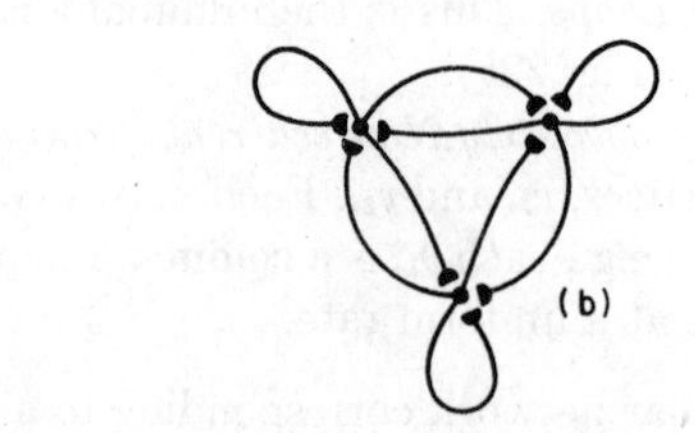
(b)

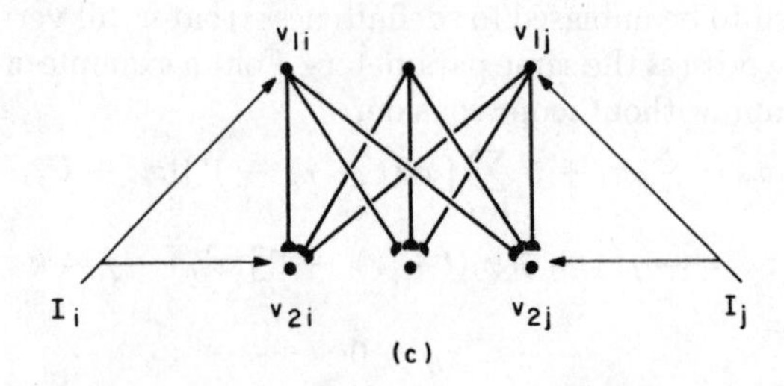

(c)

FIG. 22. Some networks in which bowing can occur.

occurs if $w \gg D(\tau, \Gamma)$ because the cross-correlations are poor. Learning is best given intratrial intervals w such that $w \simeq D(\tau, \Gamma)$, at which good distinguishability *and* good cross-correlations prevail.

D. Bare Field

The description of bowing can be approached in several stages. First, suitable anatomies must be chosen. The networks of Fig. 22, given unbiased parameters, are suitable examples. When a long serial list is presented to these graphs, bowing occurs. Thus, the analysis in Section II of the "two-body problem" of learning AB implies the existence of phenomena, such as bowing, which occur when "n-bodies" such as the alphabet ABC . . . XYZ, interact. For definiteness, we shall restrict attention to Fig. 22*a* using the simplest possible functions as in Eqs. (29)–(31). Let the inputs $C_i(t)$ be presented with intratrial interval w and intertrial interval W on N trials $\mathcal{E}_1(w, W; L), \ldots, \mathcal{E}_N(w, W; L)$ of the list $r_1 r_2 \ldots r_L$. Thus

$$C_i(t) = \sum_{m=0}^{N-1} J[t - (i-1)w - (L-1)mw - mW], \qquad i = 1, 2, \ldots, L \tag{32}$$

and

$$C_i(t) \equiv 0, \qquad i = L+1, L+2, \ldots, n \tag{33}$$

where $J(t)$ is an input pulse that is positive in the interval $(0, \lambda)$.

We seek a closed formula for $Z_{jk} = z_{jk}(\sum_{m=1}^{n} z_{jm})^{-1}$ as a functional of the serial inputs C_i. Such a formula is not available for the system (29)–(31), but one can be derived for a closely related system that embodies the main effect of the serial inputs on the sampling probabilities Z_{jk}. This system, called the *bare field* of (29)–(31), ignores the influence of the nonlinear interaction term $\beta \sum_{k \neq i} [x_k(t-\tau) - \Gamma]^+ z_{ki}$ in (29), which tends to preserve learned pattern weights except for a certain amount of smoothing when several vertices are simultaneously active, and the decay term, $-\gamma z_{jk}$, in (30), which does not change the equations for pattern variables. The bare field of a complete n-graph without loops is therefore defined by

$$\dot{x}_i = -\alpha x_i + C_i \tag{34}$$

$$\dot{z}_{jk} = \delta[x_j(t-\tau) - \Gamma]^+ x_k, \qquad j \neq k \tag{35}$$

and (31), subjected to the inputs (32) and (33). Thus, bowing can be derived from three properties taken together: (1) exponentially averaged serial inputs, from (34), (2) delayed cross-correlations of the averaged inputs, from (35), and (3) the influence of competing associations $r_j \rightarrow r_m$, $m \neq k$, on $r_j \rightarrow r_k$, from the definition of $Z_{jk} = z_{jk}(z_{jk} + \sum_{m \neq k} z_{jm})^{-1}$.

Rather than state theorems about the bare field, we first present an intuitive argument that clarifies the main effects.

E. Accumulation Sets

At what times does learning occur from r_j to r_k? That is, at what times does $Z_{jk}(t)$ grow? $Z_{jk}(t)$ grows if $z_{jk}(t)$ grows and the competing terms $z_{jm}(t)$, $m \neq k$, do not grow commensurately. By (35) this means that $Z_{jk}(t)$ grows if

$$[x_j(t-\tau)-\Gamma]^+ x_k(t) \gg [x_j(t-\tau)-\Gamma]^+ x_m(t), \qquad m \neq k \qquad (36)$$

Equation (36), in turn, can be achieved if

$$[x_j(t-\tau)-\Gamma]^+ \gg 0, \qquad x_k(t) \gg 0, \qquad \text{and} \quad x_m(t) \cong 0, \quad m \neq k \qquad (37)$$

Equation (37) shows that the growth of $Z_{jk}(t)$ will be influenced by the number of STM traces, $x_m(t)$, that are large at any given time—that is, by the distinguishability of the correct association. We therefore seek a simple way to count how many, and which, STM traces are large at any time. For simplicity, we shall constrain the input pulse, $J(t)$, from which the inputs $C_i(t)$ are constructed by the following conditions:

(1) $J(t)$ is positive only in $(0, \lambda)$, where $\lambda < D(\tau, \Gamma)$. That is, the duration of each input is less than the time needed for sampling at v_j to be induced by an input to v_i.

(2) $J(t)$ increases monotonically to a finite maximum and then decreases monotonically to zero.

(3) $w \leq W$.

Given these conditions, the following proposition holds.

Proposition 1. Suppose that $x_i(0) = 0$, $i = 1, 2, \ldots, n$. Then

$$(a) \qquad x_i(t) = x_1[t-(i-1)w], \qquad i = 1, 2, \ldots, L \qquad (38)$$

where

$$(b) \qquad x_1(t) = \begin{cases} 0 & t < 0 \\ \displaystyle\int_0^t e^{-\alpha(t-v)} J(v)\, dv, & 0 \leq t \leq \lambda \\ Ke^{-\alpha t} & \lambda < t \leq (L-1)w + W \end{cases} \qquad (39)$$

with $K = \int_0^\lambda e^{\alpha v} J(v)\, dv$. Moreover

(c) x_1 rises monotonically to its maximum in $(0, \lambda)$ and thereafter decays to zero, at the exponential rate α for $t > \lambda$.

To count how many potentials are large at time t, introduce a criterion $\epsilon > 0$ of largeness. For fixed $\epsilon > 0$, let $A_\epsilon(w, W, L; i)$ denote the collection of indices i such that $x_i(t) \geq \epsilon$. That is $A_\epsilon(w, W, L; t)$ tells us which x_i are at least as large as ϵ at time t. For simplicity, we write $A_\epsilon(w, W, L; t)$ as $A_\epsilon(t)$, and also let $| A_\epsilon(t) |$ be the number of indices in $A_\epsilon(t)$. The set $A_\epsilon(t)$ is called the *ϵ-accumulation set at time t*, since it contains the indices of all vertices, v_i, which have accumulated at least an amount ϵ of potential at time t. We always suppose in the following that ϵ is fixed in such a way that $0 < \epsilon < \max\{x_1(t) : t \geq 0\}$ to avoid trivialities. The following basic facts concerning $A_\epsilon(t)$ on the first trial $\mathcal{E}_1(w, W; L)$ are easy consequences of proposition 1.

The function $| A_\epsilon(t) |$ remains zero until the first time $t = t_\epsilon$ at which $x_1(t) = \epsilon$. Then $| A_\epsilon(t) | = 1$. The index 1 remains in $A_\epsilon(t)$ until the time $t = T_\epsilon$ at which $x_1(t) = \epsilon$ for the last time, since by proposition 1 we can also assert that $x_1(t) \geq \epsilon$ for all t in $[t_\epsilon, T_\epsilon]$. Since $x_i(t) = x_1[t - (i - 1)w]$ for all $i = 1, 2, \ldots, L$, the index 2 enters $A_\epsilon(t)$ at time $t = t_\epsilon + w$, the index 3 enters $A_\epsilon(t)$ at time $t = t_\epsilon + 2w$, and in general the index i enters $A_\epsilon(t)$ at time $t = t_\epsilon + (i - 1)w$, $i = 1, 2, \ldots, L$. Each of these indices remains in $A_\epsilon(t)$ for $T_\epsilon - t_\epsilon$ time units, and none of the indices $i = L + 1, \ldots, n$ ever enters $A_\epsilon(t)$.

The overall behavior of $A_\epsilon(t)$ as t varies within $[0, Lw]$ depends on two factors, for fixed w, W, and L. These are the amount of time $S_\epsilon = T_\epsilon - t_\epsilon$ that a single index remains in $A_\epsilon(t)$, and the number of new indices that are added to $A_\epsilon(t)$ during this time. To describe the interplay of these quantities in a precise way, we introduce the following notation.

For any $u \geq 0$, let $[\![u]\!]$ be the greatest integer that does not exceed u. Now let $G_\epsilon(w) = [\![S_\epsilon/w]\!]$. The term $G_\epsilon(w)$ measures the number of new indices that can be added to $A_\epsilon(t)$ before an old index drops out. Since S_ϵ is independent of w, $G_\epsilon(w)$ is a *monotone decreasing* function of w. We shall find that the existence or nonexistence of a bowing effect in $\mathfrak{M}(\mathfrak{U})$ during the learning of a given list $\mathcal{L} = r_1 r_2 \ldots r_L$ can be qualitatively decided by examining the absolute and relative sizes of $G_\epsilon(w)$, L, and W. To do this, we must distinguish two cases.

Case 1 $[G_\epsilon(w) < L - 1]$: In this case, $A_\epsilon(t)$ accumulates the indices $1, 2, \ldots, G_\epsilon(w) + 1$ at a linear rate at the times $t = t_\epsilon, t_\epsilon + w, \ldots, t_\epsilon + wG_\epsilon(w)$. In particular, $| A_\epsilon(t) |$ jumps by 1 every w time units from its initial value 0 until it reaches $G_\epsilon(w) + 1$. After time $t = T_\epsilon$, the "old" index 1 drops out of $A_\epsilon(t)$, but at time $t = w[G_\epsilon(w) + 1]$ the "new" index

$G_\epsilon(w) + 2$ enters $A_\epsilon(t)$. Thereafter, one old index leaves $A_\epsilon(t)$ and one new index enters $A_\epsilon(t)$ every w time units until all indices $i = 1, 2, \ldots, L$ have entered $A_\epsilon(t)$. Thus, after $|A_\epsilon(t)|$ climbs at a linear rate to its maximum value $G_\epsilon(w) + 1$, it thereafter oscillates with period w between $G_\epsilon(w) + 1$ and $G_\epsilon(w)$ until $t = Lw$.

Case 2 $[G_\epsilon(w) \geq L - 1]$: This case can be treated just as case 1 was with the following difference: The list indices $i = 1, 2, \ldots, L$ all enter $A_\epsilon(t)$ as $|A_\epsilon(t)|$ climbs to the value L. Thus, $|A_\epsilon(t)|$ climbs at a linear rate to the maximum value L, and there is no steady-state oscillatory behavior with period w.

Cases 1 and 2 exhaust all the possibilities for t in $[0, Lw]$, so that in all cases $A_\epsilon(t)$ is a *connected* set of indices of the form

$$A_\epsilon(t) = \{k_\epsilon(t), k_\epsilon(t) + 1, \ldots, k_\epsilon(t) + r_\epsilon(t)\}$$

where $k_\epsilon(t)$ and $r_\epsilon(t)$ depend on w and L. We summarize these facts in the following proposition.

Proposition 2. On trial $\mathcal{E}_1(w, W, L)$, for t in $[0, Lw]$, $A_\epsilon(t)$ is a connected set of indices such that (a) indices are added to $A_\epsilon(t)$ in chronological order at times $t = t_\epsilon, t_\epsilon + w, \ldots, t_\epsilon + (L - 1)w$, where $t_\epsilon = \min\{t\colon x_1(t) = \epsilon\}$, and (b) each index remains in $A_\epsilon(t)$ for $T_\epsilon - t_\epsilon$ time units, where $T_\epsilon = \max\{t\colon x_1(t) = \epsilon\}$. In particular, letting $G_\epsilon(w) = [\![(T_\epsilon - t_\epsilon)/w]\!]$, if $G_\epsilon(w) < L - 1$, then $|A_\epsilon(t)|$ increases in unit steps every w time units until $|A_\epsilon(t)| = G_\epsilon(w) + 1$. Thereafter $|A_\epsilon(t)|$ oscillates between $G_\epsilon(w) + 1$ and $G_\epsilon(w)$ with period w, whereas if $G_\epsilon(w) \geq L - 1$, then $|A_\epsilon(t)|$ increases in unit steps every w time units until $|A_\epsilon(t)| = L$.

We now use proposition 2 to study how changes in w and L produce changes in the associational strengths $Z_{jk}(t)$ through time.

F. Massed versus Distributed Practice

The association $Z_{jk}(t)$ from r_j to r_k grows quickly at times t for which (37) holds. This means that j is in $A_\epsilon(t - \tau)$, k is in $A_\epsilon(t)$, and all $m \neq j, k$ are not in $A_\epsilon(t)$, for some sufficiently large ϵ which we fix once and for all. In particular, $|A_\epsilon(t)|$ is a small number, since only j and k can be in $A_\epsilon(t)$.

How can we guarantee that $|A_\epsilon(t)|$ be a small number? By proposition 2, the maximum of $|A_\epsilon(t)|$ in $[0, Lw]$ is $G_\epsilon(w) + 1$. We need therefore merely require that $G_\epsilon(w)$ be small. But $G_\epsilon(w) = [\![S_\epsilon/w]\!]$, which is monotone decreasing in w. Therefore $|A_\epsilon(t)|$ will remain small for all t in $[0, Lw]$ if w is taken sufficiently large. One way of speeding up learning in $\mathfrak{M}(\mathfrak{A})$ is thus to slow down the rate with which list symbols are presented—that is, to "distribute practice."

Conversely, if the presentation rate of the list is fast, $G_\epsilon(w)$ will be large, and there will exist times t in $[0, Lw]$ at which $|A_\epsilon(t)|$ is large. Indices that are in $A_\epsilon(t)$ at these times will not be incorporated rapidly into new associations. Thus "massed practice" can slow the learning rate.

Distributing practice will not always facilitate learning. Choosing $w \gg D(\tau, \Gamma)$ yields bad learning even though such a choice of w certainly "distributes practice." The good or bad effects on learning of increasing w correspond to two different factors:

1. *Distinguishability.* Increasing w decreases $G_\epsilon(w)$ and thereby keeps $|A_\epsilon(t)|$ small in $[0, Lw]$. Thus only a few $x_i(t)$ are large at any time, and these can easily be distinguished from the many small $x_i(t)$ by the associational strengths.

2. *Correlations.* Choosing $w \gg D(\tau, \Gamma)$ means that all products $[x_j(t-\tau) - \Gamma]^+ x_k(t)$ with $k \neq j$ are always small, and thus all $Z_{jk}(t)$ remain approximately constant.

Distributing practice helps learning only if good distinguishability *and* good correlations prevail—that is, if $G_\epsilon(w)$ is small *and* $w \cong D(\tau, \Gamma)$. Since $G_\epsilon(w) = [\![S_\epsilon/w]\!]$, $G_\epsilon(w)$ is small and $w \cong D(\tau, \Gamma)$ only if $S_\epsilon \cong w \cong D(\tau, \Gamma)$.

G. Contiguity versus Connectedness

The above analysis shows that *contiguous* symbols, such as r_{j-1}, r_j, and r_{j+1}, are most likely to enter into associations with one another. This is because $A_\epsilon(t)$ is always a *connected* set. For example, in order that $Z_{jk}(t)$ grow rapidly, $x_j(t-\tau)$ and $x_k(t)$ must be large, and $|A_\epsilon(t)|$ must be small. Since this is best guaranteed when $S_\epsilon \cong w$, no index i remains in $A_\epsilon(t)$ much longer than w time units. By proposition 2, we also know that indices are added to $A_\epsilon(t)$ in chronological order. Since fast learning requires that j be in $A_\epsilon(t-\tau)$ and k be in $A_\epsilon(t)$, we conclude that $Z_{jk}(t)$ will grow fastest if $k \cong j+1$—that is, if r_j and r_k are contiguous.

When w is small [and $G_\epsilon(w)$ is large], $A_\epsilon(t)$ is still a connected set, even though there exist times, t, when it contains many indices. Once again contiguous associations are the strongest ones, but only in the weak sense that associations form best at any time t among the indices in the connected set $A_\epsilon(t)$. In particular, associations such as $Z_{j,j+2}(t)$, $Z_{j,j+3}(t)$, and $Z_{j,j-1}(t)$ can be of substantial size, thereby reducing the size of $Z_{j,j+1}(t)$.

These facts show that a decrease in w can cause a smooth change from contiguity-type conditioning to field effects among closely related items. Such field effects are closer to a Gestaltist than to a contiguity theoretical viewpoint.

H. The Beginning and the Middle of a List

We can now define the beginning and middle of a list in a way that takes into account various learning factors. For example, apply proposition 2 to the case in which w and L satisfy $L \gg 1$ and $G_\epsilon(w) < L - 1$. Then three cases can occur:

Case 1 $[G_\epsilon(w) \gg 0$ and $G_\epsilon(w) < L - 1]$: In this case, $A_\epsilon(t)$ goes through two phases:

(*a*) A *transient* phase at times t corresponding to the monotonic increase of $| A_\epsilon(t) |$ from 0 to $G_\epsilon(w) + 1 \gg 0$; and
(*b*) A *steady-state* phase at times t corresponding to the periodic oscillation of $| A_\epsilon(t) |$ between $G_\epsilon(w)$ and $G_\epsilon(w) + 1$.

Case 2 $[G_\epsilon(w) \gg 0$ and $G_\epsilon(w) \geq L - 1]$: In this case, $A_\epsilon(t)$ goes through only a transient phase, as $| A_\epsilon(t) |$ increases toward L.

Case 3 $[G_\epsilon(w) \cong 0]$: In this case, $A_\epsilon(t)$ goes through essentially only one phase, since $| A_\epsilon(t) |$ oscillates between $G_\epsilon(w) \cong 0$ and $G_\epsilon(w) + 1 \cong 1$ at *all* times t in $[0, Lw]$.

Define the (*dynamical*) *beginning* of the list $r_1 r_2 \ldots r_L$, for fixed w and L at times t in $[0, Lw]$, by the set of symbols r_i whose indices i are in the *same phase* of $A_\epsilon(t)$'s development *as the index* 1 is. The (*dynamical*) *middle* of the list is the set of symbols r_i corresponding to the *second* phase of $A_\epsilon(t)$'s development, whenever this phase exists. We denote the set of symbols in the dynamical beginning by $B_\epsilon \equiv B_\epsilon(w, L)$, and those in the dynamical middle by $M_\epsilon \equiv M_\epsilon(w, L)$. When $G_\epsilon(w)$ is large, several symbols will be in both B_ϵ and M_ϵ. This ambiguity is in the nature of the problem.

The above definitions of B_ϵ and M_ϵ have some unusual but informative consequences. For example, (*a*) the numerical length of a list's dynamical beginning is a function of w; and (*b*) there exist lists that have no dynamical middle, and all of whose symbols belong to the list's dynamical beginning. Various experimental bowing effects can be conveniently summarized in terms of these definitions. For example, (*c*) symbols in the list's dynamical middle are harder to learn than symbols in the dynamical beginning.

To see this, suppose that i enters A_ϵ when $| A_\epsilon |$ is large. Thus v_i begins to sample other v_j when $| A_\epsilon |$ is large. In particular, v_{i+1} is not readily distinguishable from the many other vertices with large potentials. Hence the association $r_i \rightarrow r_{i+1}$ will receive substantial competition from other associations $r_i \rightarrow r_k$, and $Z_{i,i+1}$ will grow slowly, if at all.

By contrast, even if $G_\epsilon(w)$ is large, $x_1(t)$ is large when $| A_\epsilon(t) |$ is small,

$x_2(t)$ is large when $|A_\epsilon(t)|$ is only slightly larger, and so on. Learning in B_ϵ is therefore faster than learning in M_ϵ if $G_\epsilon(w)$ is large.

If $G_\epsilon(w)$ is small, say $\cong 0$, learning is fast throughout B_ϵ, which now includes all symbols r_i in the list, since $|A_\epsilon(t)|$ remains small for all t in $[0, Lw]$.

The assertion that learning is faster in B_ϵ than in M_ϵ confirms the first half of the experimental bowed learning curve of Fig. 19. The assertion that case $[G_\epsilon(w) \geq L-1 \gg 0]$ is transformed into case $[L-1 > G_\epsilon(w) \gg 0]$ and finally into case $[G_\epsilon(w) \cong 0]$ as w increases agrees with the experimental fact that slowing the presentation rate flattens the first part of the bowed curve.

The definitions B_ϵ and M_ϵ illustrate the interplay of temporal and geometrical factors in determining ease of learning. The size of $|A_\epsilon(t)|$ when each v_i is sampling is the crucial fact to determine. The rigorous treatment of this system also considers transfer from STM to LTM, and how this transfer depends on the interplay of presentation rate of list items, decay rate of STM traces, arousal level, and sampling times (Grossberg, 1969e; Grossberg and Pepe, 1971).

I. Where Is the End of a List?

We have thus far considered the behavior of $A_\epsilon(t)$ only for t in $[0, Lw]$ and have based our definitions of B_ϵ and M_ϵ on this behavior. *No dynamical end exists in the list before time $t = Lw$, even though the last list item was presented at time $t = (L-1)w$.* This is not surprising, since a learning subject cannot know that r_L is the last list item until an extra intratrial interval transpires followed by no future items. The dynamical end of a list is created only *after* time $t = Lw$, and is due to the interactions of stimulus traces $x_i(t)$ and associations $Z_{jk}(t)$ before the list is presented for the second time. To see this, let us now consider $A_\epsilon(t)$ throughout trial $\mathcal{E}_1(w, W; L)$, where, as usual, $w \leq W$—that is, throughout the time interval $[0, (L-1)w + W]$. It suffices to consider the interval $[Lw, (L-1)w + W]$. This is readily done, since all $x_i(t) \equiv 0$, $i = L+1, \ldots, n$, by (34), and thus no new indices enter $A_\epsilon(t)$ for t in $[Lw, (L-1)w + W]$. Old indices continue to drop out of $A_\epsilon(t)$, however, and consequently $|A_\epsilon(t)|$ decreases in unit steps every w time units. We can distinguish two cases.

Case 1 $[G_\epsilon(w) \gg 0]$: The steady-state phase or peak of the first transient phase of $A_\epsilon(t)$ is followed by a second transient phase during the times t at which $|A_\epsilon(t)|$ decreases in unit steps at the rate w.

Case 2 $[G_\epsilon(w) \cong 0]$: Since $|A_\epsilon(t)|$ is always small in $[0, Lw]$, any decrease in $|A_\epsilon(t)|$ due to an uncompensated dropping out of indices is negligible, and so once again $A_\epsilon(t)$ has essentially only one phase.

We now define the (*dynamical*) *end* of a list as the set of symbols, if any, whose indices appear in $A_\epsilon(t)$ during its *second* transient phase. Denote the symbols in the dynamical end by $E_\epsilon \equiv E_\epsilon(w, W, L)$. We immediately conclude that:

(*a*) Learning is faster in the dynamical end of a list than in its dynamical middle. The reasoning is the same as that which showed the advantage of the beginning over the middle. Thus one can show that the middle is harder to learn than the beginning and end simply by counting the number of large stimulus traces $x_i(t)$ which the associations $Z_{jk}(t)$ must distinguish and correlate with $x_j(t - \tau)$ at any time t.

The distinction between a list's dynamical beginning, middle, and end is ambiguous when $G_\epsilon(w) \cong 0$. This is because $A_\epsilon(t)$ goes through essentially only one phase, and all one can say heuristically is that *all* list symbols are either in the beginning, or the middle, or the end, and that learning is satisfactory if also $w \cong D(\tau, \Gamma)$. [This statement can be modified to take into account the numerical length, L, of the list when the interactions, and thus accumulating noise, are introduced into the network. It is also modified when sequential STM buffers having a maximal storage capacity are included in the discussion (Atkinson and Shiffrin, 1968)]. This ambiguity therefore implies that:

(*b*) The bowed curve flattens both at its beginning and its end as the intratrial interval increases, as also occurs in Fig. 19.

J. The Dependence of a List's End on the Intertrial Interval and Associational Span

The intertrial interval, W, affects $E_\epsilon(w, W; L)$ because $|A_\epsilon(t)|$ has less opportunity to decrease when W is small. For example, suppose that $W = w$. Then trial $\mathcal{E}_2(w, w; L)$ begins right after trial $\mathcal{E}_1(w, w; L)$ ends. v_1 receives its second input pulse, $J(t - Lw)$, at time $t = Lw$, and thus the index 1 enters $A_\epsilon(t)$ on trial $\mathcal{E}_2(w, w; L)$ not longer than w time units after L enters $A_\epsilon(t)$ on trial $\mathcal{E}_1(w, w; L)$. Since each $x_i(t)$ with $i = 1, 2, \ldots, L$ satisfies $x_i(t) = x_1[t - (i - 1)w]$, the indices $1, 2, 3, \ldots, L$ enter $A_\epsilon(t)$ on trial $\mathcal{E}_2(w, w; L)$ in chronological order at rate w.

Consider the effect of increasing W step by step when the list has a middle; for example, let $0 \ll G_\epsilon(w) < L - 1$. If $W = 2w$, then $|A_\epsilon(t)|$ decreases by 1 after its steady-state phase on trial $\mathcal{E}_1(w, 2w; L)$. Trial $\mathcal{E}_2(w, 2w; L)$ then begins, and $|A_\epsilon(t)|$ quickly rises once again to its steady-state phase. The advantage to $Z_{12}(t)$ of trial $\mathcal{E}_1(w, 2w; L)$ is not entirely destroyed on trial $\mathcal{E}_2(w, 2w; L)$, but the advantage to $Z_{23}(t)$ on trial $\mathcal{E}_2(w, 2w; L)$ is slight.

Now increase W step by step. For a fixed value of W, $| A_\epsilon(t) |$ decreases step by step at a rate w for t in $[Lw, (L-1)w + W]$ to a minimum value of $\max\{0, G_\epsilon(w) + 1 - [\![W/w]\!]\}$. If W is chosen so large that $[\![W/w]\!] = G_\epsilon(w) + 1$, then $| A_\epsilon(t) |$ decreases to 0 before trial $\mathcal{E}_2(w, W; L)$ begins. Therefore $A_\epsilon(t)$ will have essentially the same phases on trials $\mathcal{E}_2(w, W; L)$ through $\mathcal{E}_N(w, W; L)$ as it had on trial $\mathcal{E}_1(w, W; L)$. In particular, symbols that are in B_ϵ, M_ϵ, or E_ϵ on one trial will be in the same dynamical part of the list on all trials [except for funneling effects (Grossberg, 1969e)], and the effects on associations which characterize a given list part will be cumulative as more and more trials occur.

These mechanisms suggest formal analogs of the major bowing effects of Fig. 19. For example:

(*a*) If $L \cong 2$ and $w \cong D(\tau, \Gamma)$, bowing does not occur, since $| A_\epsilon(t) |$ is always small.

(*b*) Bowing occurs when $0 \ll L \leq G_\epsilon(w)$ or $0 \ll G_\epsilon(w) < L$, since then $| A_\epsilon(t) |$ achieves large values.

(*c*) The bowed curve is flattened, but raised, when $0 \ll L \leq G_\epsilon(w)$ or $0 \ll G_\epsilon(w) < L$ if $W \cong w$, since then all list symbols are usually in M_ϵ.

(*d*) If for fixed $W > w$ bowing does occur, then increasing W lowers the bowed curve near its numerical middle by increasing the numerical length of B_ϵ and E_ϵ.

(*e*) Increasing W by a fixed amount has less of a lowering effect if w is large than if w is small, because $G_\epsilon(w)$ is monotone decreasing in w.

(*f*) For fixed w, increasing W beyond a W_0 such that $[\![W_0/w]\!] = G_\epsilon(w) + 1$ has little lowering effect on the bowed curve, since $| A_\epsilon(t) |$ decays to zero at the end of each trial for all such W.

(*g*) If bowing occurs but $1 \ll L - 1 < G_\epsilon(w)$, then increasing the list's numerical length L, while keeping w and W fixed, can decrease the skewness of the bowed curve by increasing the numerical length of M_ϵ.

The list's associational span and intertrial interval interact to influence the bowed curve. Consider Fig. 23. The ith *associational span* is the interval of sampling by v_i—namely, the set $\{t: x_i(t-\tau) > \Gamma\}$. By (34), this interval is $(i\tau + T_1, i\tau + T_2)$, where

$$x_1(t) = \int_0^t e^{-\alpha(t-v)} J(v)\, dv > \Gamma \qquad \text{for} \quad t \in (T_1, T_2)$$

Only those list positions whose associational span includes times when $| A_\epsilon(t) |$ is in its second transient phase are influenced by an increase in W. In Fig. 23*a*, these indices include all indices greater than j. In Fig. 23*b*,

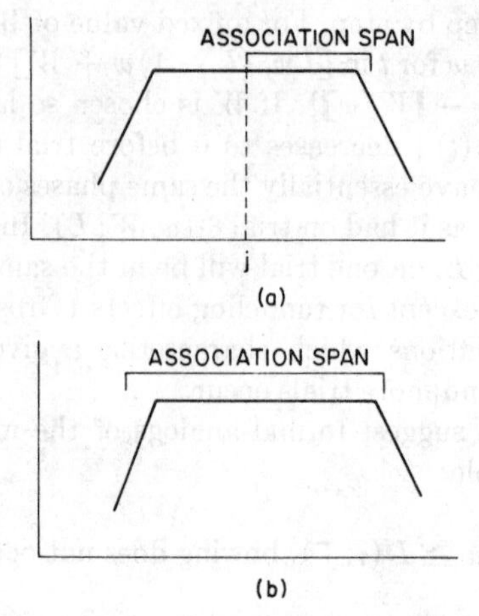

FIG. 23. Interaction of associational span and intertrial interval.

these indices include indices in B_ϵ, M_ϵ, and E_ϵ. Without all associational spans being known, the size of $|A_\epsilon|$ gives incomplete information concerning the way in which the bow changes as a function of W.

K. Response Oscillation and Remoteness

Suppose that $1 \ll G_\epsilon(w) < L - 1$ and that W is sufficiently large for some bowing to occur. Then at times t when $|A_\epsilon(t)|$ is small, the formation of new associations will be restricted to a small number of indices. Thus learning will begin to show its effects faster in B_ϵ and E_ϵ than in M_ϵ, and competing responses are restricted to a relatively small set of list symbols. By contrast, for r_i in M_ϵ, i is in $A_\epsilon(t)$ when $|A_\epsilon(t)|$ is large. Competing response tendencies to a symbol in M_ϵ are therefore broadly distributed across the list. Learning therefore takes relatively long to show its effects in M_ϵ, and a long time is needed to eliminate the large collection of competing response tendencies after learning begins. These are the main effects of Fig. 20. The analysis can be refined by studying the shape of the generalization gradients at each list position.

L. Overarousal and Inverted U in Learning

Each v_i can sample all v_k with $k \leq i - 1$, but not necessarily any v_k with $k \geq i + 1$ other than v_{i+1}. That is to say, when associations are being formed with r_i, different information is available in the network concerning the past than the future. In fact, if $J(t)$ is a rectangular input pulse of intensity J and duration λ, then the associational span has length

$$\mathcal{S} = \lambda + \frac{1}{\alpha} \log \left[\left(\frac{J}{\alpha\Gamma} - 1 \right) (1 - e^{-\alpha\lambda}) \right] \tag{40}$$

which is monotone decreasing in the signal threshold Γ. As Γ decreases, more forward associations, $r_i \to r_k$, $k > i + 1$, can form, thereby reducing the relative strength of $r_i \to r_{i+1}$. This does not mean, however, that increasing Γ always improves learning of $r_i \to r_{i+1}$. If Γ is too large, then, even though no forward associations can compete with $r_i \to r_{i+1}$, nonetheless $[x_1(t) - \Gamma]^+$ is usually zero or small in value, so that little learning of $r_i \to r_{i+1}$ occurs. Thus there exists an optimal region of threshold choice that reduces response interference without unduly diminishing the rate of learning. Alternatively expressed, this optimal region maximizes distinguishability of the correct association while providing enough energy to drive the learning process.

Notice that decreasing J in Eq. (40) has the same qualitative effect as increasing Γ. Thus all our statements concerning threshold regulation given fixed levels of physiological excitation can be transformed into corresponding statements concerning variations in the level of excitation ("arousal") as it compares with the system's fixed threshold parameters.

M. Skewing

The fact that the middle of the list is harder to learn than either end is the net result of two effects in the bare field of $\mathfrak{M}(\mathfrak{U})$. First, as list position i increases, there always exist more backward associations, $r_i \to r_k$, $k < i$, that compete with $r_i \to r_{i+1}$, thereby increasing learning difficulty. Second, there exist fewer forward associations, $r_i \to r_{i+1}$, thereby decreasing learning difficulty. However, by varying the associational span, we can guarantee that no forward association ever competes with $r_i \to r_{i+1}$ for any i. For example, choose Γ so large that $[x_i(t) - \Gamma]^+ = 0$ whenever $x_k(t) > 0$ and $k > i + 1$. Then the associations $r_i \to r_k$ never form, and consequently the major effect on the association $r_i \to r_{i+1}$ as i increases is to increase response interference due to increasing numbers of backward response alternatives. Apart from such degenerate cases, however, it can be proved that bowing always occurs in the bare field. Indeed, letting

$$\mathcal{B}(i, \Gamma) \equiv \lim_{t \to \infty} Z_{i,i+1}(t), \qquad i = 1, 2, \ldots, L - 1$$

one can prove that, for any fixed $\Gamma \geq 0$, $\mathcal{B}(i, \Gamma)$ either first decreases and then increases as i increases from 1 to L, or the degenerate case occurs in which $\mathcal{B}(i, \Gamma)$ is monotone decreasing. By definition, for fixed Γ, the bow occurs at the list position $M(\Gamma)$ for which $\mathcal{B}(i, \Gamma)$ is a minimum. If there exists more than one such position, we let $M(\Gamma)$ be the largest one, since in the presence of nonlinear interactions, background noise can only increase as more events are presented.

In the bare field, $M(\Gamma)$ is a monotone increasing function of Γ. Furthermore, $M(0) = \frac{1}{2}(L - 1)$ if L is odd and $M(0) = \frac{1}{2}L$ if L is even (Grossberg, 1969e). In the degenerate case above, $M(\Gamma) = L$ for sufficiently large Γ. Thus maximal difficulty in learning can occur at any list position greater than the list's numerical middle. Since "normal" learning requires a positive Γ, the bow will occur nearer to the end than to the beginning of the list, and the bowed curve will therefore be skewed.

At times $t < \infty$, let $\mathcal{B}(i, \Gamma, t) = Z_{i,i+1}(t)$, and suppose that $\min_i \mathcal{B}(i, \Gamma, t)$ occurs at list position $M(t, \Gamma)$ for every fixed t and Γ. Then for fixed Γ, $M(t, \Gamma)$ ultimately decreases from $M(t, \Gamma) = L$ to $M(t, \Gamma) = M(\Gamma)$ as t increases beyond the time at which r_L is presented to infinity (Grossberg and Pepe, 1971). This happens because the *non*occurrence of the events $r_{L+1}, r_{L+2}, \ldots, r_n$ gradually decreases the relative amount of response interference to $r_{L-1} \rightarrow r_L$ growth, since the future associations $r_{L-1} \rightarrow r_k$, $k > L$, never form as t increases. Thus skewing can depend both on Γ and on the intertrial interval. If Γ is very large, the intertrial interval effect will be negligible.

VIII. Instrumental Conditioning

A. Additional Postulates

The derivation of Section II can be supplemented by additional postulates that lead to mechanisms of reinforcement, drive, and incentive motivation. The first of these postulates are the following:

Postulate 1. Practice makes perfect.

Postulate 2. The time lags between CS and UCS on successive learning trials can differ.

Postulate 3. After learning has occurred, the UCR can be elicited by the CS alone on recall trials.

Postulate 4. A given CS can be conditioned to any of several drives (for example, bell → salivation if the UCS is food, or bell → fear if the UCS is a shock).

Postulate 5. Amount and/or rate of responding is influenced by the state of deprivation.

Postulate 1 is a truism that will be implemented in conjunction with postulate 2. Postulates 2 and 3 are observations about the Pavlovian conditioning paradigm. Postulates 4 and 5 are obvious facts. Such trivialities would yield little directive in a theoretical vacuum. Applied to the theory already derived, however, they are powerful guides to constructive theorizing.

B. UCS-Activated Nonspecific Arousal of CS-Activated Sampling Cells

Consider the typical situation in which a spatial pattern to be learned is embedded in a space–time pattern presented to $\mathcal{B}$, and the space–time pattern can be different on successive learning trials. Alternatively, one could let the UCS be the space–time pattern, and could consider the problem of learning a particular spatial pattern of the UCS perfectly by practicing the UCS several times. How is a particular event in a stream of events picked out as significant and learned? To simplify our notation, we suppose that the same space–time pattern is presented on each trial. Thus, on each trial a sequence $\theta^{(1)}, \theta^{(2)}, \theta^{(3)}, \ldots, \theta^{(N)}$ of spatial patterns with weights $\theta^{(k)} = \{\theta_i^{(k)}: i \in I\}$ is the UCS delivered to $\mathcal{B}$, $k = 1, 2, \ldots, N$. In this situation, an outstar anatomy does not suffice to achieve postulate 1 if postulate 2 also holds; that is, a given sampling cell, v_j, in $\mathcal{A}$ cannot learn a definite spatial pattern, $\theta^{(m)}$, chosen from the UCS sequence if the CS alone can fire v_j on successive learning trials. To see this, consider sampling by v_j of $\theta^{(1)}$ for definiteness. The sampling cell v_j can learn $\theta^{(1)}$ only if v_j fires briefly a fixed time before the onset of $\theta^{(1)}$ on *every* trial, and if the signals from v_j reach $\mathcal{B}$ only when $\theta^{(1)}$ plays on $\mathcal{B}$. This will not happen if the CS alone can fire v_j while postulate 2 holds, since signals from v_j will reach $\mathcal{B}$ on successive trials while spatial patterns $\theta^{(k)}$ other than $\theta^{(1)}$ play on $\mathcal{B}$. Thus the stimulus sampling probabilities $Z_j = (Z_{ji}: i \in I)$ will learn a weighted average of the patterns $\theta^{(k)}$ rather than $\theta^{(1)}$.

To avoid noisy sampling, the outstar must be embedded in a larger network. The sampling cell v_j must be prevented from firing unless it simultaneously receives a CS input *and* an input controlled by the UCS which signals that the UCS will arrive at $\mathcal{B}$ a fixed time interval later. This is accomplished in two steps: Let the UCS activate axons leading to v_j that deliver an input to v_j a fixed time before the UCS arrives at $\mathcal{B}$; and set the common spiking threshold, Γ_j, of all v_j's axon collaterals so high that v_j can fire only if it simultaneously receives large CS- and UCS-controlled inputs. Then, on every trial, v_j can fire and begin to sample the spatial pattern $\theta^{(1)}$ as it arrives at $\mathcal{B}$, if also the CS has been presented. Grossberg (1970a) discusses an inhibitory mechanism that guarantees brief v_j outputs in response to even prolonged CS plus UCS inputs; sampling can therefore terminate before $\theta^{(2)}$ occurs at $\mathcal{B}$.

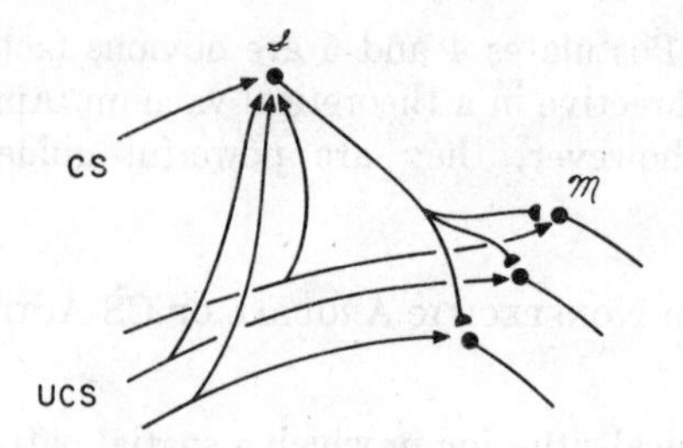

FIG. 24. UCS-activated nonspecific arousal of CS-activated sampling cells.

All cells in the network that can sample $\mathcal{B}$ receive UCS-activated axons, for the reasons given above. In other words, there exists a UCS-activated nonspecific arousal of CS-activated sampling cells. These cells are polyvalent cells, or cells that are influenced by more than one modality, such as the sound of a bell (CS) and the smell of food (UCS). The polyvalent cells fire only if the sum of CS and UCS inputs is sufficiently large. Grossberg (1971b) reviews physiological data relevant to this concept.

Some suggestive terminology is now introduced by denoting sampling cells $\mathcal{A}$ generically by $\mathcal{S}$, for "sensory cells" or "sensory representation," and sampled cells $\mathcal{B}$ by $\mathfrak{M}$ for "motor cells" or "motor representation." This distinction has no absolute significance, of course, since both $\mathcal{A}$ and $\mathcal{B}$ contribute to sensory and motor processing. It is nonetheless convenient (see Fig. 24).

C. Conditioned Reinforcers

Postulate 3 is invoked on recall trials. After learning has taken place, the CS alone can elicit performance on recall trials. Thus the CS alone can fire cells in $\mathcal{S}$ on recall trials. But $\mathcal{S}$ cells can fire only if inputs along two axon paths converge simultaneously on them. The UCS is not available on recall trials to activate one of these paths. Only the CS is available. How does CS–UCS pairing on learning trials enable the CS to gain control over the UCS $\rightarrow \mathcal{S}$ pathway on recall trials? This dilemma imposes the concept of "conditioned arousal," which will later be specialized as "conditioned incentive motivation." Namely, CS–UCS pairing during learning trials allows the CS to gain control over the nonspecific arousal channel via Pavlovian conditioning (that is, by cross-correlating presynaptic spiking frequencies and postsynaptic potentials at suitable synaptic knobs). Conditioning of nonspecific arousal at these synaptic knobs takes place while specific motor patterns are learned in the $\mathcal{S} \rightarrow \mathfrak{M}$ synaptic knobs. Consequently, on recall trials, the CS can activate two input channels:

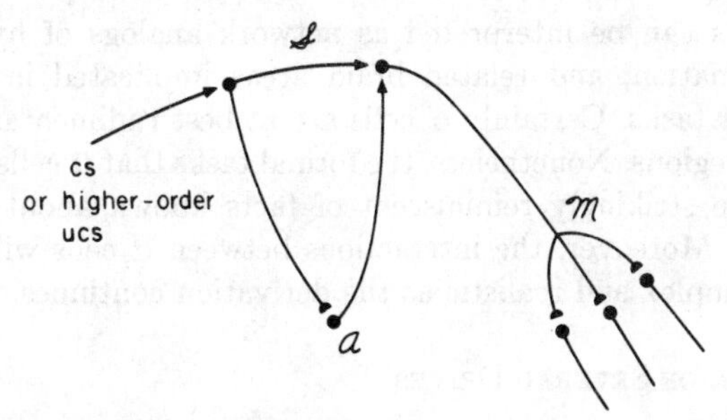

Fig. 25. Minimal nonrecurrent interaction between external cue and arousal source.

unconditioned specific inputs to $\mathcal{S}$, and conditioned nonspecific arousal inputs to $\mathcal{S}$. At cells in $\mathcal{S}$ where these two inputs converge, the cell potential can be driven above its spiking threshold. These cells can fire, yielding signals along $\mathcal{S} \rightarrow \mathcal{M}$ axons which activate the $\mathcal{S} \rightarrow \mathcal{M}$ synaptic knobs and reproduce at $\mathcal{M}$ the patterns encoded in these knobs. In this way, a CS can acquire UCS properties, and thus aspects of higher-order conditioning emerge as a consequence of postulates 2 and 3.

After a CS can activate the arousal pathway, it has UCS properties; it can serve as the UCS for a new CS in a later learning experiment. The transition from CS to UCS in these networks is effected by an alteration (not necessarily a strengthening!) of extant pathways, rather than by the creation of new pathways. Thus both CS and UCS inputs are processed in parallel pathways ("path equivalence"), except possibly the primary UCS input (for example, taste of food) on which a chain of conditioning experiments can be built. In particular, "higher-order" UCS inputs, as well as CS inputs, are delivered to $\mathcal{S}$.

D. Arousal Cells

The cells $\mathcal{A}$ at which conditioning of arousal takes place are neither $\mathcal{S}$ cells nor $\mathcal{M}$ cells. This is because the $\mathcal{S}$ cells must be aroused before they sample the activity of $\mathcal{M}$ cells, and $\mathcal{M}$ cell activation must await the onset of sampling—and thus prior firing—by $\mathcal{S}$ cells, or else $\theta^{(1)}$ cannot be learned. Similar arguments have been used to prove that at least two successive cell sites are needed in each sensory representation. The first site receives the CS input and thereupon sends signals to $\mathcal{A}$ and to the second site. The second site can fire to $\mathcal{M}$ only if it also receives a feedback signal from $\mathcal{A}$ (see Fig. 25). Sensory representations with more than two cell sites are also possible, but the theory restricts itself to the construction of minimal anatomies. As new requirements are imposed, the anatomy can be expanded to include new properties.

The $\mathcal{A}$ cells can be interpreted as network analogs of hypothalamus, reticular formation, and related brain areas implicated in arousal and reinforcement tasks. Certainly $\mathcal{A}$ cells are at best rudimentary analogs of these neural regions. Nonetheless, the formal tasks that $\mathcal{A}$ cells perform will be seen to be strikingly reminiscent of facts known about their neural counterparts. Moreover, the interactions between $\mathcal{A}$ cells will become increasingly complex and realistic as the derivation continues.

E. Existence of Several Drives

The $\mathcal{A}$ cells include drive-activated cells. For example, when a bell (CS) is conditioned to elicit salivation (UCR), it activates the $\mathcal{A}$ cells corresponding to hunger. Now invoke postulate 4. Postulate 4 directs us to further expand the minimal network to include several subsets of $\mathcal{A}$ cells, such that each subset subserves a different "drive." These $\mathcal{A}$ subsets can overlap if their corresponding drives are not mutually independent—for example, hunger and thirst. For convenience of representation, however, we draw them as individual points in Fig. 26. By postulate 4, a given sensory event can be conditioned to any of several drive contingencies. Thus, each $\mathcal{S}$ in the minimal construction will send axons to several subsets of $\mathcal{A}$ cells. Each $\mathcal{A}$ subset, in turn, sends axons nonspecifically to $\mathcal{S}$ cells; otherwise the several drives could not control nonspecific arousal signals from $\mathcal{A}$ to $\mathcal{S}$ capable of releasing signals in particular $\mathcal{S} \to \mathcal{M}$ pathways (see Fig. 26).

F. Drive Inputs

Postulate 5 imposes a new constraint on the firing of $\mathcal{A}$ cells. If an $\mathcal{A}$ cell can always fire in response to conditioned arousal inputs from $\mathcal{S}$ cells alone, then an $\mathcal{A}$ cell can always elicit (say) hunger-specific motor activity, even if $\mathcal{O}$ is not hungry, whenever food is presented. This property would kill $\mathcal{O}$. The difficulty is formally analogous to allowing an $\mathcal{S}$ cell to fire in the absence of its CS input. Maladaptive $\mathcal{A}$ cell firing of this kind can be easily

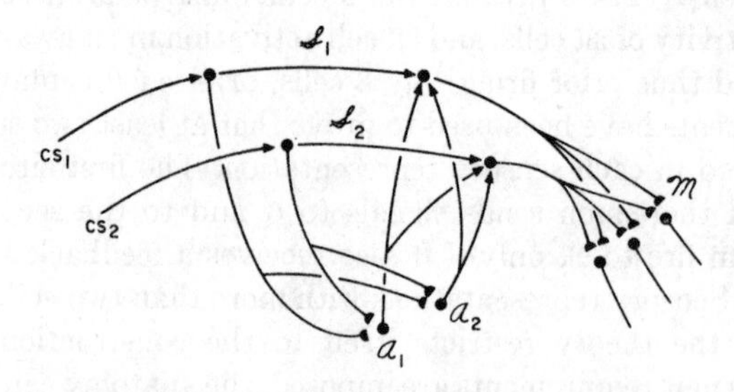

Fig. 26. Sampling of spatially distributed drive representations.

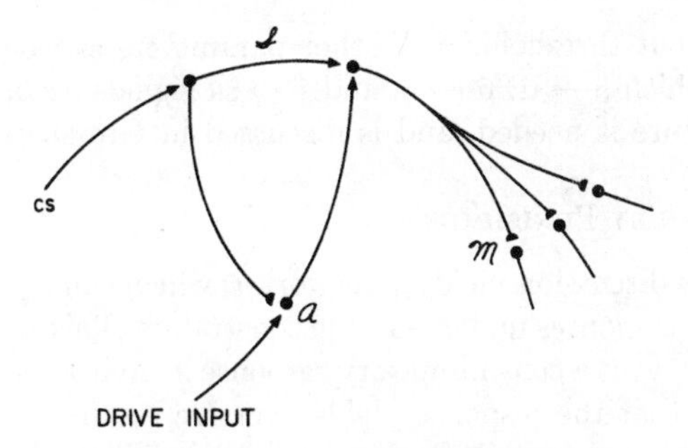

Fig. 27. Existence of drive inputs.

prevented, just as in the $\mathcal{S}$ cell case. In the $\mathcal{S}$ cell case, an $\mathcal{S}$ cell can fire to $\mathcal{M}$ only if it simultaneously receives a nonspecific input from $\mathcal{A}$ and a specific sensory input. Require analogously that an $\mathcal{A}$ cell can fire only if it simultaneously receives a nonspecific input from $\mathcal{S}$ (for example, a conditioned input from $\mathcal{S}$ or a primary UCS input) and a specific sensory input. In the $\mathcal{A}$ cell case, the sensory input is interpreted to be a drive input whose source is within $\mathcal{O}$. The size of this input indicates the level of this drive in $\mathcal{O}$ through time. This restriction on $\mathcal{A}$ cell firing is achieved by setting the spiking threshold of $\mathcal{A} \to \mathcal{S}$ axons so high that only the sum of sufficiently large inputs from $\mathcal{S}$ and from internal drive sources can fire an $\mathcal{A}$ cell (see Fig. 27). Now $\mathcal{A}$ cells are also "sensory" cells, but their sensory inputs describe the internal state of $\mathcal{O}$ rather than the external state of the world.

Grossberg (1971b) develops those ideas and cites relevant data. Noteworthy is the possibility of learning to push a lever persistently to deliver electric shocks to a (consummatory) drive representation without reducing the internal drive input (no "drive reduction"), as Olds and his collaborators have reported (Olds, 1955).

Various psychological terms can be used to describe $\mathcal{A}$ cells. They supply "incentive motivation" in support of learned sensory-motor acts encoded in $\mathcal{S} \to \mathcal{M}$ pathways. They resemble the "amplifier" elements of Estes (1969), the "Go" mechanism of Miller (1963), and the Now Print mechanism of Livingston (1967).

The foregoing construction is supported by rigorous mathematical theorems. For example, in Fig. 26, any number of cells in $\mathcal{S}$ can sample any number of cells in $\mathcal{A}$, where the $\mathcal{A}$ cells can receive primary UCS inputs, internal drive inputs, and/or conditioned inputs. This situation is covered by theorems in Grossberg (1969d, 1971c, 1972b) on nonrecurrent sampling. The same theorems cover the case of $\mathcal{S} \to \mathcal{M}$ sampling. These are the only places in Fig. 26 where learning occurs. (Actually, learning in the feedback pathway $\mathcal{A} \to \mathcal{S}$ is needed in more advanced discussions.) It remains only to guarantee that thresholds and other parameters can be set to restrict the times at which $\mathcal{S} \rightarrow \mathcal{A}$, $\mathcal{A} \rightarrow \mathcal{S}$, and $\mathcal{S} \rightarrow \mathcal{M}$ signals occur. Some further network structure is needed, and is discussed in Grossberg (1972d).

G. Suppression by Punishment

The previous discussion yields a network $\mathcal{O}$ which can learn and perform consummatory responses under suitable constraints. This construction does not suffice to prevent a consummatory response if environmental contingencies change so that the response yields aversive results. The construction will now be extended to include this crucial possibility. We shall consider the following situation for definiteness. Suppose that a CS (bell) which was once a cue for food is now a cue for shock. How does $\mathcal{O}$ prevent itself from inappropriately carrying out food-consummatory behavior in response to the CS and thereby getting shocked? To implement our construction we shall use the following postulate, which prevents $\mathcal{O}$ from indiscriminately learning unsuccessful responses.

Postulate 6. $\mathcal{O}$ does not (readily) learn escape responses that do not terminate shock.

The construction is, of course, constrained by the network that has already been derived, since the postulates from which this network emerged still hold. In Fig. 26, consummatory behavior is modifiable by two parallel conditioning processes: conditioning of nonspecific $\mathcal{A} \rightarrow \mathcal{S}$ arousal via the $\mathcal{S} \rightarrow \mathcal{A}$ synaptic knobs, and conditioning of specific motor patterns via the $\mathcal{S} \rightarrow \mathcal{M}$ synaptic knobs. Which of these conditioning processes must be supplemented to fulfill postulate 6? We proceed by asking for the minimal possible change: Can $\mathcal{O}$ recondition the $\mathcal{S} \rightarrow \mathcal{M}$ pathway without altering the $\mathcal{S} \rightarrow \mathcal{A}$ pathway? The answer will be "No" for the following reasons. The $\mathcal{S} \rightarrow \mathcal{M}$ pathway can be reconditioned in two ways:

1. *Passive Extinction*

Prevent firing of the $\mathcal{S} \rightarrow \mathcal{M}$ pathway for long time intervals. Then transmitter levels in $\mathcal{S} \rightarrow \mathcal{M}$ synapses can slowly decay to the level of network random noise. This process takes too long, however, to prevent $\mathcal{O}$ from violating postulate 6, and there exist workable transmitter laws in which no passive extinction occurs; for example, laws such as

$$\dot{z}_{jk}(t) = \{-\delta_{jk}z_{jk}(t) + \epsilon_{jk}x_k(t)\}[x_j(t - \tau_{jk}) - \Gamma_{jk}]^+$$

in which perfect memory exists until practice or recall trials, or random bursts of presynaptic spiking, occur. Also, decay can be retarded or even reversed if recall trials intermittently occur when $\mathcal{O}$ is hungry (cf. Section IV). Then the $\mathcal{S} \rightarrow \mathcal{M}$ pathway is activated and the $\mathcal{S} \rightarrow \mathcal{M}$ synaptic levels are restored to supranoise levels by transmitter potentiation, without destroying the encoded motor pattern ["post-tetanic potentiation" (Eccles, 1964)].

2. *Interference Theory of Forgetting* (Adams, 1967)

Let every occurrence of shock input generate a new UCR pattern at $\mathcal{M}$, which is incompatible with eating. If the CS also occurs at these times, and $\mathcal{O}$ is hungry, then $\mathcal{S}$ will sample the new pattern at $\mathcal{M}$, and the $\mathcal{S} \rightarrow \mathcal{M}$ synaptic knobs will encode the new UCR pattern. Thereafter, whenever the bell rings and $\mathcal{O}$ is hungry, the new motor pattern will be released, rather than eating. This mechanism has severe faults during recall trials. First, $\mathcal{O}$ cannot learn specific avoidance tasks, since the shock—and not a specific avoidance response—controls the competing UCR at $\mathcal{M}$. Second, $\mathcal{O}$ remains conditioned to the hunger $\mathcal{A}$ cells. Thus $\mathcal{O}$ will indulge in general (for example, autonomic) preparations for eating without being able to eat. Third, $\mathcal{O}$ is maladaptively fearless, since only positive consummatory drives are conditionable to the CS. Counterconditioning along a new $\mathcal{S} \rightarrow \mathcal{A}$ pathway is clearly needed. Denote the new subset of $\mathcal{A}$ cells by $\mathcal{A}_f$.

Let shock create an input at the subset $\mathcal{A}_f$. Let this input be a monotone increasing function of shock intensity. Again we are called upon to psychologically interpret a formal operation. In this case, associate activation of the cells $\mathcal{A}_f$ by shock with production within $\mathcal{O}$ of a comparable amount of fear. This interpretation introduces fear into the network using a minimum of network machinery. Given this interpretation, activating conditioned $\mathcal{S} \rightarrow \mathcal{A}_f$ synaptic knobs will yield a CER [conditioned emotional response (Estes, 1969)], both by eliciting fear in $\mathcal{O}$ and, perhaps, by activating autonomic expressions of fear through $\mathcal{A}_f$. Let $\mathcal{A}_h$ denote the subset of $\mathcal{A}$ cells that subserves hunger, and consider postulate 6 in this context.

Why is postulate 6 needed? Suppose that it does not hold. Then $\mathcal{O}$ can learn all unsuccessful escape responses. Efficient avoidance performance would therefore be unlikely, since mistakes are more likely than correct response during a period of frantic trial and error in a complex experimental chamber. At best, $\mathcal{O}$ would learn to execute the avoidance response as the terminal response in a long chain of previously learned incorrect responses. To prevent this from happening, $\mathcal{A}_h$ cells cannot be the only $\mathcal{A}$ cells that fire to $\mathcal{S}$ when the CS occurs and shock is on. For, if they were, not only could maladaptive consummatory responses be performed give the CS and sufficient hunger, but also all erroneous escape responses could be sampled and learned by $\mathcal{S} \rightarrow \mathcal{M}$ synaptic knobs with the $\mathcal{A}_h$ cells as the arousal

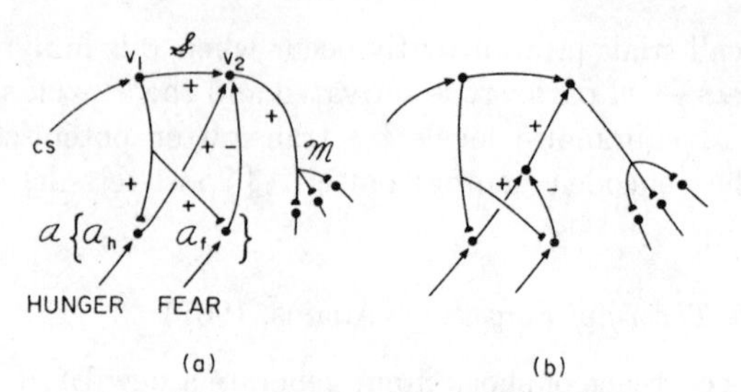

FIG. 28. Net incentive-motivational feedback.

source. The effect of $\mathcal{A}_h$ arousal on $\mathcal{S}$ must be inhibited while shock is on. The $\mathcal{A}_f$ cells are the minimal source of this inhibition. Hunger and fear arousal cells thus reciprocally inhibit each other, as Logan (1969) suggested in his discussion of net incentive motivation. Figure 28 displays two inhibitory mechanisms. Consider Fig. 28*a* when the synaptic knobs of v_1 are active. At these times, the sampling probabilities $Z_1(t)$ learn a weighted average of the spatial patterns $\theta(t) = [\theta_h(t), \theta_f(t)]$ that reach $\mathcal{A}_h$ and $\mathcal{A}_f$. Thus the probabilities learn the net balance of hunger and fear during times when v_1 samples $\mathcal{A}$. When v_1 fires and $\mathcal{O}$ is hungry, $\mathcal{A}_h$ sends excitatory feedback signals to v_2, whereas $\mathcal{A}_f$ sends inhibitory signals to v_2. Cell v_2 requires the sum of two excitatory inputs—one from v_1 and one from $\mathcal{A}_h$—in order to fire. As the relative strength of the inhibitory signal from $\mathcal{A}_f$ grows, it cancels the effect of the $\mathcal{A}_h$ input and prevents v_2 from firing. Thus v_2 cannot sample and learn the motor patterns reaching $\mathcal{A}$ at times when $\mathcal{A}_f$ feedback is active. This is true of every sensory representation.

Five conclusions follow: (1) An intense shock can suppress consummatory behavior by competing with $\mathcal{A}_h \to \mathcal{S}$ arousal via the inhibitory $\mathcal{A}_f \to \mathcal{S}$ pathway. (2) This suppression does not extinguish memory of the patterns already encoded in the $\mathcal{S} \to \mathcal{M}$ synaptic knobs. (3) Suppression can take place faster than passive extinction. (4) An intense shock can prevent new $\mathcal{S} \to \mathcal{M}$ associations from forming by inhibiting release of sampling signals from $\mathcal{S}$. (5) After $\mathcal{S} \to \mathcal{A}_f$ conditioning takes place, properties (1) through (4) can be elicited on recall trials wherever the CS input activates $\mathcal{S} \to \mathcal{A}_f$ synapses.

Similar qualitative properties hold for Fig. 28*b*. Here, however, the $\mathcal{A}_f$ and $\mathcal{A}_h$ signals compete with each other at a second stage of processing before a signal to $\mathcal{S}$ is emitted. It can be proved that only $\mathcal{A}_h$ can create an input (excitatory) to $\mathcal{S}$, and does so only if it emits a stronger signal than $\mathcal{A}_f$ does. The competitive mechanism is called a subtractive on-center off-surround field. Its mathematical properties have been discussed

(Grossberg, 1970a). Figure 28*b* requires half as many $\mathcal{A} \to \mathcal{S}$ axons as Fig. 28*a*. This represents a considerable saving of axons, since each $\mathcal{A}$ subset projects nonspecifically to numerous $\mathcal{S}$ cells. On the other hand, Fig. 28*a* requires fewer cellular processing stations.

H. Avoidance: Heuristics

The following postulate is essentially a rewording of postulate 6.

Postulate 7. $\mathcal{O}$ learns escape responses that do terminate shock faster than it learns escape responses that do not terminate shock.

This postulate also builds upon mechanisms that are already at our disposal. In particular, while shock is on, $\mathcal{S} \to \mathcal{M}$ sampling is prevented by $\mathcal{A}_f \to \mathcal{S}$ inhibition. Shock termination removes $\mathcal{A}_f \to \mathcal{S}$ inhibition, but $\mathcal{S} \to \mathcal{M}$ sampling remains impossible until *some* excitatory arousal source is activated. Postulate 7 can thus be reduced to the following question: What excitatory arousal source releases $\mathcal{S} \to \mathcal{M}$ sampling just after shock is turned off, and thereby establishes conditioned pathways from the sensory cues that are available when the avoidance response occurs to both the active arousal source and the motor controls of the avoidance response? Speaking heuristically, this arousal source provides the "motivational support" for learning the avoidance response. We suggest that an experimental analog of exciting this new arousal source is, other things being equal, an internally perceived "relief" from fear (Denny, 1971; Masterson, 1970; Reynierse and Rizley, 1970).

Denote by $\mathcal{A}_f^-$ the arousal cells which are excited by termination of shock input to the cells $\mathcal{A}_f$, which we henceforth denote by $\mathcal{A}_f^+$. Some formal requirements must be imposed on $\mathcal{A}_f^-$ and $\mathcal{A}_f^+$ to ensure that the arousals work together effectively. First, require that excitation of $\mathcal{A}_f^-$ by shock termination is transient. Transient response is needed to prevent irrelevant sensory-motor coordinations from being learned whenever shock is off. The cells $\mathcal{A}_f^+$ are *on-cells*; they are turned on by shock, and they remain on until shock is shut off. The cells $\mathcal{A}_f^-$ are *off-cells*; they are turned on temporarily by shock termination. On-cells and off-cells are familiar physiological components (Thompson, 1967, pp. 253 and 349). Second, require that the outputs from $\mathcal{A}_f^+$ to $\mathcal{A}_f^-$ reciprocally inhibit each other before they send signals to $\mathcal{S}$. Thus these outputs interact to form a consensus between "fear" and "relief." A possible behavioral analog of this rebound from $\mathcal{A}_f^+$ on-cells to $\mathcal{A}_f^-$ off-cells is the rebound in behavioral effects reported to occur after electrical hypothalamic stimulation terminates (Cox *et al.*, 1969; Grastyan, 1968; Valenstein *et al.*, 1969). This analogy receives further support from a chemical and anatomical analogy which is developed in Grossberg (1972d) between the twofold system ($\mathcal{A}_f^+$, $\mathcal{A}_f^-$) and sites in the twofold system of ventromedial and lateral hypothalamus.

The network must be expanded once again to allow $\mathcal{S}$ to become conditioned to the new arousal source. Thus let each sensory representation, $\mathcal{S}$, send axons to $\mathcal{A}_f^-$ as well as to $\mathcal{A}_f^+$, $\mathcal{A}_h$, and other $\mathcal{A}$ cell clusters. At any time, the synaptic knobs of each $\mathcal{S}$ encode a spatial pattern derived from the patterns $\theta(t) = [\theta_f^+(t), \theta_f^-(t), \theta_h(t), \ldots]$. This pattern describes the net balance of excitatory and inhibitory $\mathcal{A} \rightarrow \mathcal{S}$ feedback that this representation controls. It is determined by a weighted average of the spatial patterns $\theta(t)$ that reach $\mathcal{A}$ when the given $\mathcal{S}$ is sampling.

In summary, the classical notion that instrumental reinforcement is due to "drive reduction" when shock terminates is replaced by rebound from negative-incentive motivational on-cells to positive-incentive motivational off-cells when shock terminates. The balance of excitation of on-cells and off-cells can be classically conditioned, perhaps at different times, to any $\mathcal{S}$ representations. The net $\mathcal{A} \rightarrow \mathcal{S}$ output, and thus $\mathcal{S} \rightarrow \mathcal{M}$ firing and performance on recall trials, is determined by *all* the $\mathcal{S}$ sites that fire to $\mathcal{A}$ at such times. Even if half of $\mathcal{S}$ fires to $\mathcal{A}_f^-$, no $\mathcal{S} \rightarrow \mathcal{M}$ channel need be activated by positive $\mathcal{A} \rightarrow \mathcal{S}$ feedback if the other half fires to $\mathcal{A}_f^+$, since $\mathcal{A}_f^-$ and $\mathcal{A}_f^+$ will reciprocally inhibit each other's outputs. Similarly, shock termination yields little "relief" if it is antagonized by a switching-on of new $\mathcal{S} \rightarrow \mathcal{A}_f^+$, or "fear," channels. Shock termination per se is not necessarily "drive reducing."

Recent psychophysiological data and concepts can be qualitatively analyzed in terms of these network analogs (see Grossberg, 1972c). These concepts include aspects of the following: relaxation, or elicitation, theory, which claims that an unconditioned response of relief precedes reinforcement; the concept of "effective reinforcement," which notes that shock offset and onset of fearful situational cues can influence reward in opposite ways, as is illustrated by two-way avoidance tasks in which a rat escapes a chamber in which it is shocked by running into another chamber where it was previously shocked; classical and instrumental properties of a CS+ paired with shock, a CS− paired with no-shock, and feedback stimuli contingent on the avoidance response, including transfer of their effects from classical to instrumental conditioning experiments; autonomically nonchalant asymptotic avoidance performance originally motivated by fear; forced extinction of the CAR without fear extinction; response suppression without an avoidance response; relief without an avoidance response; opposite effects of contingent and noncontingent punishment on fear and suppression of consummatory responding; punishment hypothesis of avoidance learning, describing rewarding effects of terminating proprioceptive cues that correspond to nonavoidance responses; response (or no-response) generalization from one shock level to a different level; and rewarding effects of response-contingent reduction in frequency of shock.

The argument leading to an explicit construction of the rebound mechanism falls into eight main stages.

1. *Existence of a Tonic Input*

When shock terminates, $\mathcal{A}_f^-$ emits a transient output. Thus, by Eq. (22), the potentials x_f^- of $\mathcal{A}_f^-$ cells grow transiently to suprathreshold values. In Eq. (22), an input source is required to thusly perturb x_f^-. What input source does the job? (The concept of "input source" includes possible energy sources within the cells themselves.)

In these systems, shutting off one input (such as the shock input to $\mathcal{A}_f^+$) does not provide energy for turning on another input (such as the one driving $\mathcal{A}_f^-$ rebound). Terminating shock input can, however, unmask the effects of an *internally* driven input to $\mathcal{A}_f^-$ whose influence is inhibited by shock. The internal source of $\mathcal{A}_f^-$ input is therefore neither turned on nor off by shock offset. It is not turned off by shock onset, since then it would be off at shock offset, and could not drive $\mathcal{A}_f^-$ rebound. Finally, if it is turned on by shock onset, or is unaffected by shock onset, then it is always on. The internal input is therefore *tonic*.

2. *Existence of Accumulation–Depletion*

Output from $\mathcal{A}_f^-$ shuts off soon after it is turned on. How is this done? No externally driven input is available to do this. The $\mathcal{A}_f^-$ output is depleted by its own activity. In other words, while shock is on, an accumulation process occurs at $\mathcal{A}_f^-$. When shock is off, output from $\mathcal{A}_f^-$ is a monotone increasing function of the amount accumulated at each given time. This amount is gradually depleted when shock is off, until the $\mathcal{A}_f^-$ output vanishes. [The accumulation mechanism that is ultimately used is derived in Section IX, and is given by Eq. (44).]

3. *Consensus between Fear and Relief*

We suppose that at most one of the outputs from $\mathcal{A}_f^+$ and $\mathcal{A}_f^-$ is nonzero at any time. In other words, either fear or relief, but not both, can be "perceived" by the network at a given time. Thus the final state of processing in $\mathcal{A}_f^+$ and $\mathcal{A}_f^-$, before signals are sent to $\mathcal{S}$, is the resultant of a competition between the $\mathcal{A}_f^+$ and $\mathcal{A}_f^-$ channels due to some form of mutual inhibition.

4. *Existence of a Parallel Accumulation Process in the Fear Channel*

When shock is off for a long time, outputs from both $\mathcal{A}_f^+$ and $\mathcal{A}_f^-$ to $\mathcal{S}$ are zero. Thus the accumulation process at $\mathcal{A}_f^-$, driven by its tonic input, is balanced by a process going on at $\mathcal{A}_f^+$. The simplest idea is that a parallel process of accumulation–depletion, driven by its own tonic input which equals the $\mathcal{A}_f^-$ input, takes place in the $\mathcal{A}_f^+$ channel. When shock is on, the shock input summates with the tonic input in the $\mathcal{A}_f^+$ channel.

This idea is strengthened by the next few arguments, which elucidate the basic question: What accumulates? Is it potential or is it transmitter? Several facts favor the latter alternative. Other possibilities have been discussed by Grossberg (1972d).

5. *The Rebound Is Slow*

It lasts at least seconds rather than milliseconds. It is a slow process compared to network fluctuation rates of cell potentials in response to input changes. After shock terminates, $\mathcal{A}_f^+$ and $\mathcal{A}_f^-$ receive no externally driven inputs. Their potentials presumably equalize rapidly. Output from $\mathcal{A}_f^-$ nonetheless continues. Thus there exists a process slower than potential change that can bias output from $\mathcal{A}_f^+$ and $\mathcal{A}_f^-$ in favor of $\mathcal{A}_f^-$ after shock terminates.

6. *Both Fear and Relief Are Increasing Functions of Shock Duration and Intensity*

Data on the effect of CS and UCS intensity on the CER and CAR have been reported. Thus both channels contain slowly varying processes which parametrically depend on shock intensity and duration, and which counterbalance each other when shock is off for long intervals.

7. *The Relative Balance of Accumulation Is Changed by Shock*

What causes the $\mathcal{A}_f^-$ rebound to shut itself off? Is complete depletion of the accumulated product at $\mathcal{A}_f^-$ responsible for this? Suppose that the answer is "Yes." Then the tonic input alone can deplete $\mathcal{A}_f^-$. By symmetry, during shock, the shock input plus the tonic input to $\mathcal{A}_f^+$ could surely deplete $\mathcal{A}_f^+$. This does not occur, since then fear could not be maintained by a prolonged shock. A weaker conclusion is necessary: Shock shifts the relative balance of accumulation in the two channels by depleting the $\mathcal{A}_f^+$ channel more than the $\mathcal{A}_f^-$ channel.

8. *Signal Size Is a Joint Function of Input Size and Amount Accumulated*

This argument is crucial. During $\mathcal{A}_f^-$ rebound, both $\mathcal{A}_f^+$ and $\mathcal{A}_f^-$ receive equal tonic inputs which ultimately balance the amounts accumulated at $\mathcal{A}_f^+$ and $\mathcal{A}_f^-$, and thereby nullify $\mathcal{A}_f^+$ and $\mathcal{A}_f^-$ signals to $\mathcal{S}$. Before this happens, $\mathcal{A}_f^-$ output exceeds $\mathcal{A}_f^+$ output because $\mathcal{A}_f^-$ accumulation exceeds $\mathcal{A}_f^+$ accumulation. In other words, given a fixed input size (the equal tonic inputs to $\mathcal{A}_f^+$ and $\mathcal{A}_f^-$), output is an increasing function of accumulation level (in the two channels, $\mathcal{A}_f^+$ and $\mathcal{A}_f^-$).

When shock is on, increasing shock intensity increases $\mathcal{A}_f^+$ output, since it causes an increase in fear. Increasing shock intensity also *decreases* the amount accumulated at $\mathcal{A}_f^+$; this is the basis of the rebound at $\mathcal{A}_f^-$ when shock is turned off. Thus, output is not a function of accumulation level alone, since then increasing shock intensity would decrease $\mathcal{A}_f^+$ output by decreasing the amount accumulated at $\mathcal{A}_f^+$. Output size is a joint function of input size and accumulation level.

The terms $B_{ji}z_{ji}$ in (22) shows that output size is the product of spiking frequency and transmitter level. Spiking frequency is an increasing function of potential, which is an increasing function of input size. This leaves transmitter accumulation level as the abstract accumulation level discussed above. This argument commits us to our formalism. We could not proceed further unless: (i) the amount of accumulated transmitter is a decreasing function of input size, and (ii) output size is nonetheless an increasing function of input size. Fortunately, both (i) and (ii) are true in embedding fields, and make a construction of the rebound mechanism possible in this context.

Grossberg (1972d) carries out this construction and rigorously analyzes the resulting mechanisms. These mechanisms include an analogy with adrenergic and cholinergic interactions in series with lateral and ventromedial hypothalamic sites, dependent on phasic sensory input and tonic reticular formation input. Mechanisms emerge for such phenomena as: the lesser rewarding effect of reducing J units of shock to $J/2$ units than of reducing $J/2$ units to 0 unit; a relationship between the rewarding effect of reducing J units of shock to $J/2$ units and the possibility of releasing a conditioned avoidance response in the presence of fearful cues; two kinds of depressed emotional affect—one due to overarousal, which can also be associated with massive associational confusions and poor paying attention, and one due to underarousal, which can also be associated with overreactive fear and relief responses; persistent nonspecific fear which biases interpretation of specific cues, and can "resist" new learning or "repress" old learning; different effects of gradual and abrupt shock on response supression; response generalization from one shock level to another; reduction of pain in the presence of loud noise (analgesia); influences of drugs such as carbachol, atropine, and scopolamine on conditioned emotional and avoidance responses, and on self-stimulation via implanted hypothalamic electrodes; sensory-drive heterarchy that allows changes in situational cues to release responses compatible with any of several nonprepotent drives; feedback inhibition of adrenergic transmitter production; potentiation of adrenergic production by presynaptic spiking, and by postsynaptic spiking via a feedback loop that controls higher-order instrumental conditioning; and learning at cholinergic synapses.

IX. Possible Chemical Substrates of Network Processes

A. Refinement of Spatiotemporal Scales

Equations (22) and (23) are derived from psychological postulates and yield an abstract network anatomy whose variables are interpreted as averages over physiological variables. This section illustrates a correspondence procedure whereby spatial and temporal scales in the network are expanded to reveal possible finer processes that are compatible with Eqs. (22) and (23). Further details of this procedure can be found in Grossberg (1969f), along with additional references to relevant data. Here we develop the interpretation of z_{ji} as a transmitter variable, rather than as a measure of postsynaptic membrane sensitivity to fixed amounts of transmitter. Postsynaptic modifications nonetheless arise.

B. Coupling of K^+ to ACh Release

Consider the term $F_{ji} \equiv B_{ji}z_{ji}$ in Eq. (22). The physiological interpretation given in Section III suggests a coupling between outward flux of K^+ and of ACh (acetylcholine) from synaptic knobs. Such a coupling has been experimentally reported (Hebb and Krnjevic, 1962; Hutter and Kostial, 1955; Liley, 1956). It is approached as follows: B_{ji} increases with spiking frequency, and each spike is associated with an inward flux of Na^+ and an outward flux of K^+ (Katz, 1966). Hence an increase in B_{ji} is associated, on a microscopic level, with an increased total outward flux of K^+. The term z_{ji} describes the production of excitatory transmitter (say Ach) within N_{ji}. $F_{ji} = B_{ji}z_{ji}$ is proportional to the rate of excitatory transmitter released from N_{ji}. Hence, increasing the outward flux of K^+ increases the rate of transmitter release from N_{ji}.

The argument holds even if B_{ji} is a functional of spiking frequency or spike size. This added generality is needed to interpret B_{ji} if x_j becomes large. Since F_{ji} represents rate of transmitter release and z_{ji} is proportional to total transmitter, B_{ji} must have a finite maximum as $x_j \to \infty$; for example:

$$B_{ji} = \frac{a_{ji}[x_j(t - \tau_{ji}) - \Gamma_{ji}]^+}{b_{ji} + [x_j(t - \tau_{ji}) - \Gamma_{ji}]^+}$$

The mathematical development discussed in Section VI includes this possibility, among many others.

C. Two Pairs of Antagonistic Ions: (Na^+, K^+) and (Ca^{++}, Mg^{++})

The above interpretation of network variables can be used to suggest the existence of more speculative couplings. These couplings are also compatible with various data, but direct confirmation of their existence seems to be lacking, if only because the necessary experiments would be very hard to perform. First note that, in the presence of inhibitory interactions, Eq. (23) is changed to

$$\dot{z}_{ji} = D_{ji}z_{ji} + E_{ji}[x_i]^+ \tag{41}$$

to prevent negative values of the potential x_i from producing negative amounts of transmitter. How can the product $G_{ji} = E_{ji}[x_i]^+$ in Eq. (41) be interpreted? The term E_{ji} is, along with B_{ji}, associated with spiking frequency. The most obvious participants in the spike are the antagonistic ions Na^+ and K^+. Hence we assume that increases in E_{ji} correspond, on a microscopic level, to (a process in parallel with) an inward flux of Na^+ and an outward flux of K^+. This process will occur within N_{ji} if we associate z_{ji} with transmitter. The product G_{ji} is then also computed within z_{ji}, since it determines the rate of transmitter production, by Eq. (41). The term $[x_i]^+$ in F_{ji} corresponds, however, to a process in v_i. Thus there exists a transport of material from v_i to N_{ji}, in an amount proportional to $[x_i]^+$, that enables G_{ji} to be computed in N_{ji}. What is transported?

Product G_{ji} is a result of two processes. Process E_{ji} is in parallel with a pair of rapidly fluctuating antagonistic ion fluxes. The other process presumably occurs on a similar time scale, and involves chemical species that are known to interact with these ions. Also the two processes in G_{ji} are treated symmetrically: G_{ji} is a product of terms which, in the simplest cases, are both functionals of cell potentials cut off at a threshold (for example, $G_{ji} = \delta_{ji}[x_j(t - \tau_{ji}) - \Gamma_{ji}]^+[x_i]^+$), and it is known in the case of spike production that the threshold is produced by interaction between the pair Na^+ and K^+ of antagonistic ions. The simplest assumption is thus that $[x_i]^+$ also represents a process (in parallel with) a pair of antagonistic ion fluxes. This assumption turns out to be compatible with various data. In the following discussion of these data, the phrase *in parallel with* a pair of antagonistic ions is critical. Indeed, our macroscopic theory can do little more than suggest the symmetries of microscopic interactions, so that the pairs being sought, in principle, need not be composed of ions at all (cf. amino acids). The formal structure of the argument seems to hold, no matter how we interpret these chemicals.

The pair of ions associated with $[x_i]^+$ cannot be (Na^+, K^+). If it were, increases in $[x_i]^+$ would correspond to an influx of Na^+ and an outflux of K^+ at v_i. The process z_{ji} is, however, influenced only by those aspects of these fluxes that affect N_{ji}. These effects are a decrease in Na^+ and an increase in K^+. Process E_{ji} involves the same ions and has the opposite effect when E_{ji} increases. How then do these processes affect z_{ji} in Eq. (41) only through their product? In particular, by Eq. (41), z_{ji} cannot grow in response to even an enormous E_{ji} value if $[x_i]^+ = 0$, even though E_{ji} provides within N_{ji} all the effects that $[x_i]^+$ can trigger. Thus, if $[x_i]^+$ is in parallel with a pair of antagonistic ions, it must be a pair other than (Na^+, K^+).

In many biochemical processes, the divalent ions Ca^{++} and Mg^{++} powerfully interact with Na^{++} and K^+, and the pair (Ca^{++}, Mg^{++}) is mutually antagonistic (Dixon and Webb, 1958). We take this to be the pair being sought. In many reactions, Na^+ and Ca^{++} act synergistically (Fruton and Simmonds, 1958). We therefore consider this possibility in the present context: Let an increase in $[x_i]^+$ correspond microscopically to an increase in Ca^{++} and a decrease in Mg^{++}.

D. Binding of Na^+ and Ca^{++} as Synergistic Cofactors on Transmitter Production Sites

Now term G_{ji} says that transmitter production sites are activated at a rate proportional to the product of (processes in parallel with) Na^+ and Ca^{++} concentrations. In particular, we expect joint inward Na^+ and Ca^{++} fluxes to be created by membrane excitation and to thereby stimulate transmitter production, whereas K^+ and Mg^{++} antagonize Na^+ and Ca^{++}, respectively, in this role. Analogous fluxes have been experimentally reported (del Castillo and Engbaek, 1954; Harvey and MacIntosh, 1940; Hodgkin and Keynes, 1957). Just as inward fluxes of Na^+ and Ca^{++} presumably facilitate transmitter production, it is natural to expect that such fluxes facilitate transmitter release, so as not to cancel out one process with another. If ACh is the transmitter, then reducing Ca^{++} concentration around N_{ji} would reduce ACh release, other things being equal. If Mg^{++} is acting as a Ca^{++} antagonist, then Mg^{++} should antagonize Ca^{++} in controlling the amount of ACh release. Compatible experimental reports are found in del Castillo and Engbaek (1954), del Castillo and Katz (1954), Hubbard (1961), Jenkinson (1957), and Liley (1956).

E. A Hierarchy of Intracellular Ionic Binding Strengths

By Eq. (41), new transmitter production sites are activated only when $G_{ji} > 0$—that is, only if supraequilibrium amounts of (quantities in parallel with) Na^+ and Ca^{++} simultaneously reach these sites. When equilibrium is restored, $G_{ji} = 0$. The rate of change in z_{ji} due to G_{ji} is also zero during equilibrium; the sites remember how much transmitter to produce.

The following basic questions hereby arise. How can high concentrations of Na^+ and Ca^{++} jointly activate a process that maintains its activity even after the concentrations of these ions are reduced at equilibrium? Otherwise expressed, what keeps z_{ji} at the high values needed to produce a memory of past events even when the sources of these high values are removed as equilibrium is restored? In particular, why doesn't the high intra-end-bulb K^+ concentration at equilibrium reversibly inhibit z_{ji} growth, just as Na^+ and Ca^{++} excited z_{ji} growth at nonequilibrium?

Since z_{ji} does maintain the high values acquired during nonequilibrium, and joint coupling of Na^+ and Ca^{++} causes these values, we are led into the following conclusion: The Na^+ and Ca^{++} ions which activated the transmitter production sites are not removed from the end bulb when equilibrium is restored; a fraction of the free Na^+ and Ca^{++} ions which enter the end bulb during excitation is bound on intra-end-bulb transmitter production sites, and this binding is so strong that it cannot be displaced by the return of a high intra-end-bulb K^+ concentration as equilibrium is restored. In particular, the intracellular K^+ ions are not so strongly bound. We are hereby led to expect that most of the intracellular K^+ exists in unbound form, whereas higher proportions of intracellular Na^+ and/or Ca^{++} exist in bound form. These expectations have been experimentally reported (Brink, 1954; Ussing, 1960).

F. The Control of Cellular Production Rates by Ions: Strength of Binding versus Ion Availability

The above remarks suggest a qualitative answer to a special case of the following general question: How do cells "know" how much of a given quantity to produce in response to external environmental demands?

Our point of departure is the hypothesis that ions such as Na^+ and Ca^{++}, which presumably activate intra-end-bulb sites (or enzymes) with considerable vigor, are kept substantially out of the end bulb during equilibrium. Only in nonequilibrium periods such that $x_j(t - \tau_{ji}) > \Gamma_{ji}$

and $x_i(t) > 0$ can these ions penetrate the membrane en masse to initiate higher levels of intra-end-bulb transmitter production. Since equilibrium time intervals can, in principle, exceed nonequilibrium time intervals by a very large numerical factor, the ions Na^+ and Ca^{++}, which bind most strongly, are available least frequently within the end bulb. In other words, the process of synergistic (Na^+, Ca^{++}) binding, having a limited opportunity to occur, is made effective by guaranteeing that, whenever the opportunity does occur, the process takes place vigorously and its effects are long-lasting (cf. Brink, 1954; Quastel, 1962).

These facts suggest the following general heuristic scheme for integrating equilibrium and nonequilibrium phases in the life of a cell, which subsumes the problem of rendering the cell responsive to fluctuations in its external environment. The argument can be broken into three main steps.

1. *Coexistence of Equilibrium and Evolution*

An equilibrium phase of a cell can, in principle, be characterized by particular values of prescribed cellular parameters. For example, the equilibrium of a nerve cell can be characterized by the membrane concentrations of such parameters as Na^+ and K^+. Suppose that a cell exists whose equilibrium is characterized by particular values of all its parameters. Such a cell "forgets" all nonequilibrium values of its parameters when it returns to equilibrium. In particular, the equilibrium of such a cell cannot coexist with long-term responses of the cell to brief changes in its external environment. For convenience, we henceforth call such long-term responses evolutionary trends.

Certainly not all cells are of this type. Brains can learn! Henceforth we concern ourselves only with cells whose equilibrium phase can coexist with an evolutionary trend. We denote such a cell by C. By definition, the equilibrium phase of C does not require a specification of values for all cellular parameters. It suffices to specify the values of a fraction of these parameters. We denote these equilibrium parameters collectively by E. A particular evolutionary trend in C requires the specification of values for parameters which we denote by N. Since the parameters N control an evolutionary trend, they need not always take on the same values when the parameters E take on equilibrium values.

2. *The External Environment Perturbs the Equilibrium Parameters*

The external environment communicates its demands upon C by changing the values of parameters at C's periphery, or membrane. These parameters are, however, often the parameters E, since equilibrium is a state of C which is characterized by a particular choice of external environment. For example, a nerve cell returns to equilibrium when all excitatory and inhibitory inputs are zero. We conclude that the external environment often induces an evolutionary trend in the parameters N by perturbing the parameters E. The parameters E therefore *faithfully* communicate to the parameters N the demands of the external environment. We are hereby led to the following basic but merely ostensible paradox: If the parameters E faithfully communicate to the parameters N the external environmental demands that signal an evolutionary trend, then why don't the parameters E also faithfully communicate to the parameters N the external environmental demands that signal equilibrium, and thereby eradicate the evolutionary trend in N whenever equilibrium is restored?

3. *The Equilibrium Values Compete with the Nonequilibrium Values of the Equilibrium Parameters*

Given the natural assumption that the parameters E pass on faithfully to N *all* states of the external environment, the following resolution of this paradox seems natural: The equilibrium values of E do not eradicate the evolutionary trend in N because they cannot dislocate from N the nonequilibrium values of E that induced the trend. In the case that the parameters E are realized by ions, this means that a hierarchy of ionic binding strengths exists at the intracellular sites (or enzymes) which alter intracellular demands. The ions that are most available during equilibrium are bound least strongly to these sites. The ions introduced at these sites by the extracellular demands are strongly bound as synergistic cofactors to these sites, and thereby activate them. Proceeding in the reverse direction, suppose that the ions that bind most strongly to these sites are *not* substantially kept out of the cell during equilibrium, and are allowed freely to bind with these sites and thereby to activate them. Then essentially all sites will always be occupied, and the production rate at these sites will always be in a state of equilibrium, albeit a very active equilibrium. The evolutionary trend is hereby destroyed.

G. The Mitochondrion and Ion Translocation

Given the hypothesis that Na^+ and Ca^{++} are synergistic cofactors in the activation of sites that contribute to transmitter production, it is desirable to find candidates for these sites. A cellular system which has a strong affinity for Na^+ and Ca^{++} is the mitochondrion, whose importance as the "power plant" of aerobic cells is well known. For example, Lehninger (1965, pp. 169–171) reports a striking increase during respiration in both the relative uptake of Na^+ over K^+ and of Ca^{++} over Mg^{++}. To the extent that this fact is an example of our theoretical expectations, then ion translocation in neural mitochondria can be interpreted as a means for setting mitochondrial reaction rates at a level commensurate with the intensity and duration of a positively polarized nonequilibrium excitation phase. These rates endure long into the equilibrium phase.

H. Provision of ATP for Synaptic Vesicles by Mitochondria

Suppose that ion translocation in the mitochondrion is indeed an example of the synergism between Na^+ and Ca^{++} that contributes to transmitter production. Then mitochondria should be found clustered near regions of high transmitter density. Histological evidence suggests that transmitter is stored in synaptic vesicles, and that mitochondria can be found clustered near these vesicles (de Robertis, 1964, p. 32, and micrographs throughout the book). Perhaps the activated mitochondria supply the ATP needed to produce acetyl coenzyme A, which in turn presumably reacts with choline under the aegis of the enzyme choline acetylase to produce acetylcholine (Eccles, 1964; Fruton and Simmonds, 1958).

I. Contiguity of Synaptic Vesicles and the Synaptic Cleft

The histological investigations (Eccles, 1964; de Robertis, 1964) which have revealed the existence of synaptic vesicles also show that these vesicles are often clustered most densely along the end-bulb surface which faces the synaptic cleft. This location is well chosen for a vesicle whose supposed role is to expeditiously release transmitter into the synaptic cleft to excite the postsynaptic membrane. Yet how does the vesicle know how to choose this useful location? Such knowledge will seem mysterious in any theory that holds that transmitter production depends only on the past excitation history of the presynaptic nerve which contains the transmitter, since the excitation of just this nerve does not provide information concerning the location of the synaptic cleft relative to the end-bulb membrane. Such a theory predicts that transmitter vesicles will be found uniformly throughout the end bulb, or closer to the presynaptic source of excitation than to the synaptic cleft, or at best with uniform density along all end-bulb surfaces.

The preferential location of synaptic vesicles near the synaptic cleft is qualitatively easily understood in a theory in which transmitter production depends on both presynaptic and postsynaptic influences. Presumably the postsynaptic influence is carried over the synaptic cleft to the presynaptic end bulb, so that the region most likely to have all the ingredients needed for transmitter production lies nearest to the synaptic cleft. The postsynaptic ionic influence does not spread evenly throughout the presynaptic end bulb because the Ca^{++} influence near the synaptic cleft is presumably bound within the end bulb as soon as it reaches an appropriate site, and the amount of Ca^{++} entering the cell cannot be so large as to uniformly saturate all sites within the end bulb, or else the desired evolutionary trend will be destroyed. Indeed, one way to turn a knob capable of learning into a knob incapable of learning is to open the tight junctions for the transport from v_i to N_{ji}, and thereby bathe the presynaptic end bulb in an ionic atmosphere that is not driven by postsynaptic events.

J. Binding of Mg^{++} by RNA in the Cell Body

The Ca^{++} needed for synergistic binding of Na^{++} and Ca^{++} in N_{ji} are released into the synaptic cleft facing N_{ji} when the postsynaptic cell, v_i, is sufficiently excited. Otherwise, much of the Ca^{++} in the synaptic cleft is presumably reabsorbed into v_i.

This argument fails completely if N_{ji} can provide as much Ca^{++} as v_i, given a fixed level of excitation, since then E_{ji} would stand for essentially the same ionic fluxes as $[x_i]^+$, and the coupling F_{ji} could not be realized. Since v_i presumably can supply more Ca^{++} than N_{ji}, we must find a rationale for this fact.

Given that $[x_i]^+$ represents an antagonism between Ca^{++} and Mg^{++}, the fact that Ca^{++} is released when v_i is excited means that Mg^{++} is needed by v_i during excitation. A structure therefore exists within v_i—which is not found in N_{ji}—which selectively binds Mg^{++} ions when v_i is active and whose binding with Mg^{++} is preferred to (or antagonized by) binding with Ca^{++}. This argument does not mean that no Ca^{++} is provided by N_{ji}, but only that more Ca^{++} is provided by v_i. In a similar fashion, the fact that presynaptic excitation at N_{ji} induces coupled Na^+ and K^+ fluxes does not imply that such fluxes are absent from postsynaptic excitation at v_i.

The cell body v_i certainly has at least one prominent structure which the end bulb N_{ji} does not have—namely, the cell nucleus. If this is the structure being sought, then the cell nucleus, or processes sustained by the nucleus, ought to selectively bind Mg^{++} ions when the cell body is activated. Among the most plentiful cell body constituents of this type are the RNA's. It is also known that RNA activity depends sensitively on Mg^{++} concentration (Boedtker, 1960; Spirin, 1964; Watson, 1965).

K. Interaction of Neural Excitation and RNA

Suppose indeed, that the RNA's are among the structures that we are seeking to bind Mg^{++}. Then learning will be associated with systematic variations in the RNA's. Such variations have been reported experimentally (Hamberger and Hydén, 1963; Hydén, 1962; Koenig, 1964).

Once experiments were produced demonstrating variations in RNA activity in learning situations, it was proposed that individual RNA strands coded the content of the learning in some fashion, and that one could, in principle, recover the content of whole segments of learned experiences in such a strand if one but had the key for decoding its structure. This view seems unnecessary from the present perspective. The RNA's seem to be needed merely to keep the cell at production levels appropriate to the metabolic drains placed on the cell by the levels of excitation imposed from the external environment. Indeed, if a spatial pattern is the unit of long-term memory, then an individual cell does not have enough information to know what is being learned. Nonetheless, the cross-correlational processes presumed to occur at the cellular level do provide enough information for the cell to discriminate whether a learning type of process is occurring or not.

L. Transport down the Axon

The hypothesis that Mg^{++} is bound to nucleus-related processes is further strengthened by the following observation.

Figure 29 schematically represents a presynaptic nerve cell with nucleus N_j whose excitatory end bulb, N_{ji}, impinges upon the postsynaptic nerve cell v_i with nucleus N_i. Suppose that N_i selectively binds Mg^{++} in order to free Ca^{++} for binding within N_{ji} when both N_{ji} and v_i are vigorously excited. If v_i and v_j are of the same cell type, then Mg^{++} will also be selectively bound by N_j when v_j is vigorously excited. Since v_j is connected to N_{ji} by the axon e_{ji}, we must prevent most of the molecules that bind Mg^{++} within v_j from flowing down the axon to N_{ji}, or else N_{ji} will have too many Mg^{++}-binding molecules. Thus at least part of the Mg^{++} must be bound within v_j to structures that are so large or so well cemented within v_j that they are never carried down the axon to the end bulb. Macromolecules within N_j, such as the RNA's, are plausible candidates for such a role.

On the other hand, whenever v_j is excited to suprathreshold values, then the axon e_{ji} and the end bulb N_{ji} are also excited. The axon and the end bulb must be able to recover from this excitation. The postulated mecha-

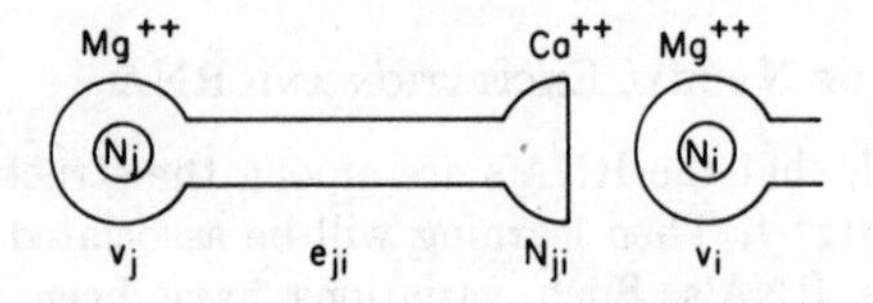

Fig. 29. Interacting chemical dipoles.

nism of recovery is activation by (processes in parallel to) Mg^{++} of the RNA's during excitation, leading to higher rates of protein synthesis, etc. However, the RNA's are substantially localized within the cell body v_j. Thus the molecules produced by RNA activation, after being produced in v_j, must be able to travel down the axon to the end bulb where they will be needed to guarantee recovery from excitation. These molecules therefore might well be lighter than the more immobile RNA's, and they might well be bound to less Mg^{++} than is bound to the activated RNA's. A transport of material from the cell body along the axon to the end bulb exists (Friede, 1959; Ochs and Burger, 1958; Waelsch and Lajtha, 1960; Weiss and Hiscoe, 1946). Various details concerning this formal transport process are considered in Grossberg (1969f).

M. Why Aren't Nerve Cells Spherical? An Intimate Bond between Neural Geometry and Neural Dynamics

It is practically a truism that the simplest geometrical objects are as homogeneous and as symmetric as possible. Thus, among the simplest three-dimensional and finite bodies are the spheres, and it is useful to think of the complexity of a three-dimensional and finite body—such as a nerve cell—in terms of its deviations from sphericity. It is also natural to suppose that a finite system in nature will assume the simplest shape that is compatible with its function. We are then readily led to ask: What features of a nerve cell's functions require that it be nonspherical?

Our speculations suggest that the role of nerve cells as mechanisms of learning requires their nonspherical shape. We link a nerve cell's ability to learn with the existence of different chemical affinities at two opposite poles of the nerve cell—namely, near the cell body and end bulbs; that is, the nerve cell is presumed to be a chemical dipole. Were the nerve cell spherical in all ways, in particular with a spherical nucleus in its center, then symmetry arguments would imply that this chemical dipole could not be realized.

Given the need for a dipole shape, the nerve cell is then confronted with the formidable problem of carrying signals from its external environment reliably from one end of the dipole to the other. This problem is formidable because the functional biases caused by the dipole might well be expected to distort the signal as it travels along the cell. The cell has solved this problem in an ingenious, but intuitively simple, way. The signals from the external environment, which first perturb the boundary, or membrane, of the cell, are transmitted reliably from one end of the dipole to the other along this boundary, whereas the chemical dipole properties of the cell are safely ensconced well within the cellular interior, where they can secondarily benefit from external environmental news without profoundly distorting the transmission of this news along the entire cell. Note by Section VI that this constraint aiming at unbiased signal transfer on the individual cell level also seems to create unbiased learning on the network level.

N. Two Main Steps in Transmitter Production

We now show that the single variable, z_{ji}, represents two processes taking place at two different rates. These two processes are the following ones.

1. *Slowly Varying Transmitter Production Rates.* Long-term memories of past network events are contained in the z_{ji} functions. These functions therefore vary more slowly than the events themselves. In particular, if $G_{ji} = \delta_{ji}[x_j(t - \tau_{ji}) - \Gamma_{ji}]^+[x_i]^+$, then z_{ji} varies more slowly than x_j and x_i.

2. *Rapidly Varying Transmitter Release.* Suppose that

$$F_{ji} = \beta_{ji}[x_j(t - \tau_{ji}) - \Gamma_{ji}]^+ z_{ji}$$

for definiteness. At suprathreshold values, F_{ji} is a linear function of $x_j(t - \tau_{ji})$, and is therefore rapidly varying compared to z_{ji}.

The physical interpretation of F_{ji} leads to the two processes represented by z_{ji}. The function F_{ji} is proportional to the rate of transmitter release from N_{ji}, and z_{ji} is the total amount of transmitter in N_{ji}. Why, then, doesn't the law (41) for z_{ji} read as follows?

$$\dot{z}_{ji} = D_{ji}z_{ji} + E_{ji}[x_i]^+ - F_{ji}$$

That is, shouldn't the total amount of transmitter in N_{ji} be reduced by the amount of transmitter that is released from N_{ji}? On formal grounds, this subtraction procedure is inadmissible; then z_{ji} would be drastically reduced in size whenever the presynaptic spiking frequency became large, and the "memory" represented by z_{ji} would quickly be destroyed. A conceptual distinction clearly must be made between z_{ji} as "memory" and z_{ji} as "releasable transmitter."

Two problems must simultaneously be resolved:

1. Distinguish z_{ji}, the rate of transmitter production, from y_{ji}, the amount of transmitter.

2. Show how z_{ji} can represent both variables in the macroscopic psychological picture; that is, show that, on the average,

$$y_{ji}(t) \cong \epsilon_{ji}z_{ji}(t) \tag{42}$$

where ϵ_{ji} is a positive constant. The relation (42) can hold at all times t

only if the transmitter lost from N_{ji} is instantly replenished until it reaches a level proportional to z_{ji}. This happens only if the rate of replenishment is infinite. This rate only seems to be infinite on the time scale of psychological events because replenishment is a rapid process on this scale. We now refine this time scale by assuming that the replenishment rate is finite, but otherwise do not change our equations. As usual, we seek the most linear way to express our intuitive ideas, while realizing that there exist variations on the linear theme.

By (42), at times when no transmitter is released from N_{ji}, y_{ji} seeks a level proportional to z_{ji}. Hence, at these times,

$$\dot{y}_{ji}(t) = \zeta_{ji}[\epsilon_{ji}z_{ji}(t) - y_{ji}(t)] \tag{43}$$

where $0 \leq y_{ji} < \epsilon_{ji}z_{ji}$ and $\zeta_{ji} > 0$.

If transmitter is released from N_{ji} at a rate H_{ji}, then $\dot{y}_{ji}$ in (43) is reduced by this amount. Thus, in general,

$$\dot{y}_{ji} = \zeta_{ji}(\epsilon_{ji}z_{ji} - y_{ji}) - H_{ji} \tag{44}$$

The term H_{ji} cannot be identified with F_{ji} because z_{ji} no longer represents the amount of transmitter. Guided by the definition of $F_{ji} = B_{ji}z_{ji}$ and (42), we let

$$H_{ji} = \eta_{ji}B_{ji}y_{ji}, \qquad \eta_{ji} = \epsilon_{ji}^{-1}$$

Thus Eq. (44) merely replaces a process with an infinite reaction rate by a qualitatively identical process with a finite reaction rate. In the special case that the transmitter is ACh, a possible interpretation of these variables is

z_{ji} = total amount of available ACh in N_{ji}

and

y_{ji} = total activity of the choline acetylase (ChAc) system which controls ACh production (Fruton and Simmonds, 1958; Krnjević, 1965; Sumner and Somers, 1953).

O. Feedback Inhibition

Equation (44) has the following chemical interpretation. Write (44) as a sum of three terms:

$$\dot{y}_{ji} = U_{ji} + V_{ji} + W_{ji} \tag{45}$$

where

$$U_{ji} = \zeta_{ji}\epsilon_{ji}z_{ji} \tag{46}$$

$$V_{ji} = -\zeta_{ji}y_{ji} \tag{47}$$

and

$$W_{ji} = -\eta_{ji} B_{ji} y_{ji} \tag{48}$$

The term (46) says that transmitter production rate is proportional to the number of active transmitter producing sites. Term (47) says that transmitter production rate is diminished by an amount proportional to the amount of transmitter—that is, by a feedback inhibition by the transmitter end product of a prior stage of transmitter production. This inhibition cannot influence those transmitter-producing sites which are activated by extracellular demands without destroying the cellular memory of these demands. Hence a later, or intermediate, stage of transmitter production is inhibited (cf. Fruton and Simmonds, 1958; Wyatt, 1964). Term (48) implies that feedback inhibition is reduced by release of transmitter from N_{ji}.

It is interesting to compute the response of Eqs. (45)–(48) to a spiking frequency that is switched to a steady-state level $B > 0$ at time $t = 0$ after a long internal of zero spiking. One finds three major effects:

1. A transient overshoot in transmitter release.
2. A progressive *decrease* in the asymptotic total available transmitter $y_{ji}(\infty)$ as a function of increasing B.
3. A progressive *increase* in the asymptotic rate of transmitter release $H_{ji}(\infty)$ as a function of increasing B.

Thus the total amount of transmitter in N_{ji} and the amount of transmitter that is released from N_{ji} do not covary as function of E (cf.). This fact makes it possible to construct the rebound mechanism using transmitter accumulation–depletion in Section VIII.

P. Transmitter Mobilization

The process of refining scales can be continued indefinitely. For example, if process z_{ji} takes place within N_{ji} but contributes transmitter for release from the N_{ji} membrane, then transmitter will be transported to the membrane. Various models for this can be contemplated. The simplest again rely on linearity wherever possible. For example, let

y_{ji} = the total amount of transmitter in N_{ji}

and

w_{ji} = the total amount of transmitter in N_{ji} at the membrane facing the synaptic cleft

The rate of transmitter release in this case involves w_{ji}, not y_{ji}, and is derived much as H_{ji} was derived from F_{ji}. Thus we find, using linearity wherever possible, that

$$\dot{y}_{ji} = \zeta_{ji}(\epsilon_{ji}z_{ji} - y_{ji}) - \eta_{ji}E_{ji}w_{ji} \tag{49}$$

and

$$\dot{w}_{ji} = \theta_{ji}(y_{ji} - w_{ji}) - \eta_{ji}E_{ji}w_{ji} - \kappa_{ji}[w_{ji} - \lambda_{ji}]^+ \tag{50}$$

Equation (50) can be understood by writing it as

$$\dot{w}_{ji} = U^*_{ji} + V^*_{ji} + W^*_{ji}$$

with

$$U^*_{ji} = \theta_{ji}(y_{ji} - w_{ji}) \tag{51}$$

$$V^*_{ji} = -\eta_{ji}E_{ji}w_{ji} \tag{52}$$

and

$$W^*_{ji} = -\kappa_{ji}[w_{ji} - \lambda_{ji}]^+ \tag{53}$$

Term (51) says that transmitter is mobilized at a rate proportional to the amount $(y_{ji} - w_{ji})$ of unmobilized transmitter. Term (52) gives the rate of releasing mobilized transmitter from N_{ji}. Term (53) says that mobilized transmitter can become spontaneously demobilized until only an amount λ_{ji} of transmitter is still mobilized.

Equation (50) has some interesting properties. If we study its transient response, then the slowly varying z_{ji} remains approximately constant. Suppose also that $\lambda_{ji} = 0$ and $\zeta_{ji} = \kappa_{ji}$. Then the equation can be explicitly integrated. Properties 1 to 3 of the previous section hold, and in addition the amount of mobilized transmitter is constant through time. The last property is not generally true if $\zeta_{ji} \neq \kappa_{ji}$.

In summary, this paper illustrates a procedure whereby the physiological equations themselves, and the network anatomy, can be successively refined to accommodate increasingly subtle psychological postulates. At each level of analysis, one finds phenomena that caution against arguing from local to global, or from linear to nonlinear, network properties.

REFERENCES

Adams, J. A. (1967). "Human Memory." McGraw-Hill, New York.

Atkinson, R. C., and Estes, W. K. (1963). *In* "Handbook of Mathematical Psychology" (R. D. Luce, R. R. Bush, and E. Galanter, eds.), Vol. II, p. 121. Wiley, New York.

Atkinson, R. C., and Shiffrin, R. M. (1968). *In* "The Psychology of Learning and Motivation" (K. W. Spence and J. T. Spence, eds.), Vol. 2, p. 98. Academic Press, New York.

Boedtker, H. (1960). *J. Mol. Biol.* **2,** 171.

Brink, F. (1954). *Pharmacol. Rev.* **6, 243,** 276 and 284.

Cox, V. C., Kakolewski, J. W., and Valenstein, E. S. (1969). *J. Comp. Physiol. Psychol.* **68,** 337.

Dally, P. (1967). "Chemotherapy of Psychiatric Disorders." Plenum, New York.

del Castillo, J., and Engbaek, L. (1954). *J. Physiol.* (*London*) **124,** 370.

del Castillo, J., and Katz, B. (1954). *J. Physiol.* (*London*) **124,** 560.
Denny, M. R. (1971). *In* "Aversive Conditioning and Learning" (F. R. Brush, ed.), p. 000. Academic Press, New York.
de Robertis, E. D. P. (1964). "Histophysiology of Synapses and Neurosecretion." Macmillan, New York.
Dethier, V. G. (1968). "Physiology of Insect Senses." Methuen, London.
Dixon, M., and Webb, E. C., eds. (1958). "Enzymes," 1st ed. Academic Press, New York.
Eccles, J. C. (1964). "The Physiology of Synapses." Springer-Verlag, Berlin and New York.
Estes, W. K. (1969). *In* "Punishment and Aversive Behavior" (B. A. Campbell and R. M. Church, eds.), p. 57. Appleton, New York.
Friede, R. L. (1959). *Exp. Neurol.* **1,** 441.
Fruton, J. S., and Simmonds, S. (1958). "General Biochemistry." Wiley, New York.
Grastyán, E. (1968). *In* "Biological Foundations of Emotion" (E. Gellhorn, ed.), p. 000. Scott, Foresman, Glenview, Illinois.
Grossberg, S. (1968a). *J. Math. Anal. Appl.* **21,** 643.
Grossberg, S. (1968b). *J. Math. Anal. Appl.* **22,** 490.
Grossberg, S. (1969a). *J. Math. Psychol.* **6,** 209.
Grossberg, S. (1969b). *J. Math. Mech.* **19,** 53.
Grossberg, S. (1969c). *J. Differential Equations* **5,** 531.
Grossberg, S. (1969d). *J. Statis. Phys.* **1,** 319.
Grossberg, S. (1969e). *Math. Biosci.* **4,** 201.
Grossberg, S. (1969f). *J. Theor. Biol.* **22,** 325.
Grossberg, S. (1970a). *J. Theor. Biol.* **27,** 291.
Grossberg, S. (1970b). *Stud. Appl. Math.* **49,** 135.
Grossberg, S. (1971a). *J. Cybernet.* **1,** 28.
Grossberg, S. (1971b). *J. Theor. Biol.* **33,** 225.
Grossberg, S. (1971c). *Proc. Nat. Acad. Sci. U.S.* **68,** 828.
Grossberg, S. (1972a). *Kybernetik* **10,** 49.
Grossberg, S. (1972b). *In* "Delay and Functional Differential Equations and Their Applications" (K. Schmitt, ed.), p. 121.
Grossberg, S. (1972c). *Math. Biosci.* **15,** 39.
Grossberg, S. (1972d). *Math. Biosci.* **15,** 253.
Grossberg, S. (1973). *Studies Appl. Math.* **52,** 213.
Grossberg, S. (1975). *Int. Rev. Neurobiol.* **18.** (To be published).
Grossberg, S., and Pepe, J. (1971). *J. Statist. Phys.* **3,** 95.
Hamberger, A., and Hydén, H. (1963). *J. Cell Biol.* **16,** 521.
Harvey, A. M., and MacIntosh, F. C. (1940). *J. Physiol.* (*London*) **97,** 408.
Hebb, C. O., and Krnjević, K. (1962). *In* "Neurochemistry" (K. A. C. Elliott, I. H. Page, and J. H. Quastel, eds.), p. 452. Thomas, Springfield, Illinois.
Hebb, D. O. (1955). *Physiol. Rev.* **62,** 243.
Hodgkin, A. L., and Keynes, R. D. (1957). *J. Physiol.* (*London*) **138,** 253.
Hubbard, J. I. (1961). *J. Physiol.* (*London*) **159,** 507.
Hubel, D., and Wiesel, T. N. (1968). *In* "Physiological and Biochemical Aspects of Nervous Integration" (F. O. Carlson, ed.), p. 153. Prentice-Hall, Englewood Cliffs, New Jersey.
Hutter, O. F., and Kostial, K. (1955). *J. Physiol. London* **129,** 159.
Hydén, H. (1962). *In* "Neurochemistry" (K. A. C. Elliott, I. H. Page, and J. H. Quastel, eds.), p. 331. Thomas, Springfield, Illinois.

Jenkinson, D. H. (1957). *J. Physiol. (London)* **138,** 434.
Jensen, A. R. (1962). *J. Psychol.* **53,** 127.
Julesz, B. (1964). *Science* **145,** 356.
Katz, B. (1966). "Nerve, Muscle and Synapse." McGraw-Hill, New York.
Kennedy, D. (1968). *In* "Physiological and Biochemical Aspects of Nervous Integration" (F. O. Carlson, ed.), p. 285. Prentice-Hall, Englewood Cliffs, New Jersey.
Koenig, H. (1964). *In* "Morphological and Biochemical Correlates of Neural Activity" (M. M. Cohen and R. S. Snider, eds.), p. 39. Harper, New York.
Kornetaky, C., and Eliasson, M. (1969). *Science* **165,** 1273.
Krnjević, J. (1965). *In* "Pharmacology of Cholinergic and Adrenergic Transmission" (G. B. Koelle, W. W. Douglass, and A. Carlson, eds.), p. 21. Macmillan, New York.
Lashley, K. S. (1951). *In* "Cerebral Mechanisms in Behavior: The Hixon Symposium" (L. P. Jeffress, ed.), p. 112. Wiley, New York.
Lehninger, A. L. (1965). "The Mitochondrion." Benjamin, New York.
Lehrman, D. S. (1965). *In* "Sex and Behavior" (F. A. Beach, ed.), p. 355. Wiley, New York.
Liley, A. W. (1956). *J. Physiol. (London)* **134,** 427.
Livingston, R. B. (1967). *In* "The Neurosciences" (F. O. Schmitt *et al.*, eds.), Vol. 2, p. 91. MIT Press, Cambridge, Massachusetts.
Logan, F. A. (1969). *In* "Punishment and Aversive Behavior" (B. A. Campbell and R. M. Church, eds.), p. 43. Appleton, New York.
Maher, B. A. (1968). *Psychol. Today* **2,** 30.
Masterson, F. A., (1970). *J. Comp. Physiol. Psychol.* **72,** 471.
Miller, G. A. (1956). *Psychol. Rev.* **63,** 81.
Miller, N. E. (1963). *In* "Nebraska Symposium on Motivation" (M. R. Jones, ed.), p. 65. Univ. of Nebraska Press, Lincoln.
Ochs, S., and Burger, E. (1958). *Amer. J. Physiol.* **194,** 499.
Olds, J. (1955). *In* "Nebraska Symposium on Motivation" (M. R. Jones, ed.) Univ. of Nebraska Press, Lincoln.
Orkand, R. K., Nicholls, J. G., and Kuffler, S. W. (1966). *J. Neurophysiol.* **29,** 788.
Osgood, C. E. (1953). "Method and Theory in Experimental Psychology." Oxford Univ. Press, London and New York.
Penfield, W. (1958). "The Excitable Cortex in Conscious Man." Liverpool Univ. Press, Liverpool.
Phillips, M. I., and Bradley, P. B. (1970). *Science* **168,** 1122.
Quastel, J. H. (1962). *In* "Neurochemistry" (K. A. C. Elliott, I. H. Page, and J. H. Quastel, eds.), p. 226. Thomas, Springfield, Illinois.
Ratliff, F. (1965). "Mach Bands: Quantitative Studies of Neural Networks in the Retina." Holden-Day, San Francisco, California.
Reynierse, J. H., and Rizley, R. C. (1970). *J. Comp. Physiol. Psychol.* **72,** 223.
Ruch, T. C., Patton, H. D., Woodbury, J. W., and Towe, A. L. (1961). "Neurophysiology." Saunders, Philadelphia, Pennsylvania.
Spirin, A. S. (1964). "Macromolecular Structure of Ribonucleic Acids," Part II, Chapter 4. Van Nostrand-Reinhold, Princeton, New Jersey.
Stein, P. S. G. (1971). *J. Neurophys.* **34,** 310.
Sumner, J. B., and Somers, G. F. (1953). "Chemistry and Methods of Enzymes," 3rd rev. ed., p. 351. Academic Press, New York.
Thomas, G. J., Hostetter, G., and Barker, D. J. (1968). *Progr. Physiol. Psychol.* **2,** 265.
Thompson, R. F. (1967). "Foundations of Physiological Psychology." Harper, New York.

Ussing, H. H. (1960). *In* "Handbuch der Experimentellen Pharmakologie," Vol. 6, Part B, p. 1. Springer-Verlag, Berlin and New York.
Valenstein, E. S., Cox, V. C., and Kakolewski, J. W. (1970). *Psychol. Rev.* 77, 16.
Waelsch, H., and Lajtha, A. (1960). *In* "Neurochemistry of Nucleotides and Amino Acids" (R. O. Brady and D. B. Tower, eds.), p. 205. Wiley, New York.
Watson, J. D. (1965). "Molecular Biology of The Gene." Benjamin, New York.
Weiss, P., and Hiscoe, H. B. (1946). *J. Exp. Zool.* **107**, 315.
Willows, A. O. D. (1968). *In* "Physiological and Biochemical Aspects of Nervous Integration" (F. O. Carlson, ed.), p. 217. Prentice-Hall, Englewood Cliffs, New Jersey.
Wyatt, H. W. (1964). *J. Theor. Biol.* **6**, 441.

CHAPTER 4

PATTERN LEARNING BY FUNCTIONAL-DIFFERENTIAL NEURAL NETWORKS WITH ARBITRARY PATH WEIGHTS

PREFACE

This paper proves the universal theorem on associative learning that culminates my 1967–1972 articles on this subject. The theorem is universal in the following sense. It says that if my associative learning laws were invented at a prescribed time during the evolutionary process, then they could be used to guarantee unbiased associative learning in essentially any later evolutionary specialization. That is, the laws are capable of learning arbitrary spatial patterns in arbitrarily many, simultaneously active sampling channels that are activated by arbitrary continuous data preprocessing in an essentially arbitrary anatomy. The learning of arbitrary space-time patterns is also guaranteed given modest requirements on the temporal regularity of stimulus sampling, as in avalanches and generalizations thereof.

The result can be described in another way. It describes the *evolutionary invariants* of pattern learning; namely, it classifies those system oscillations and limits that are not altered by evolutionary specializations. These invariants exist whenever the learning rules are computed in a canonical ordering. The learning laws describe that canonical ordering of system computations (spatial averaging, temporal averaging, preprocessing, gating, cross-correlation) which enables the system to factorize pattern from energy even if highly nonlinear feedback is operative in arbitrarily many parallel channels. This is the same factorization property which was first noticed using the LTM ratios y_{jk} in Chapter 2.

Because of the abstract form of the laws, they can be interpreted as laws for directed growth during development, as laws for receptor sensitization, or as laws for enhanced production of a chemical transmitter. The crucial thing is whether a prescribed system enjoys certain statistical properties, not whether it uses particular chemicals.

Using engineering terminology, I can describe the universal theorem as follows. It describes absolutely stable and unbiased parallel processing by a self-organizing machine. Because the result is so basic, I hoped that it would cause quite a stir. For example, computer enthusiasts should have gotten excited because of its parallel processing aspects. They did not partly because Artificial Intelligence dogma, as expressed in Minsky and Papert's book on Perceptrons, declared that associative learning was either unimportant or trivial. Also the laws are expressed by continuous systems, which were taboo in AI circles then. The self-organizing properties might have caused a stir in learning or developmental circles, but my use of global nonlinear ideas might have seemed intimidating, although these ideas really unify and simplify my analysis. In any case, an increasing number of researchers including AI enthusiasts and developmental biologists have recently begun to study associative learning or directed growth by parallel systems. Many of their models and results are special cases of my systems and theorems, or fall into errors which were side-stepped by the theory. I hope that these backward steps will be corrected soon.

These results led to some other design ideas and predictions. After understanding how the theorems could be proved, I was led to ask: "How can unbiased learning be guaranteed if each cell can communicate with other cells at variable distances, and if intercellular distances can change drastically due to development and growth? Can the system be designed so that prior learning is not distorted by developmental changes in intercellular distances?" This question reaffirmed the use of a *self-similarity principle* to design the cells. I say 'reaffirmed', because I earlier used self-similarity to suggest how the chemical dipole which I mentioned in Chapter 2 works; in particular, how a cell body knows how to produce the right amount of chemical precursors for efficient transmitter release at the cell's synaptic knobs [12]. I later realized that developmental biologists call this self-similarity property self-regulation, or invariance of form under size changes (little leaf becomes big leaf). I called the property self-similarity because I was influenced by a lecture of Benoit Mandelbrot on self-similar stochastic processes while I was working on transmitters during my student days at Rockefeller.

My theory needs self-regulation for a reason that seems not to be known, or is at least not emphasized, by developmental biologists. Self-regulation helps individual system components to arrive at globally correct conclusions from locally ambiguous data. In the present case, this general theme is specialized to suggest a rather unexpected reason why some nerve cells use chemical transmission whereas other cells use electrical transmission: Only chemical transmission seems to be capable of achieving unbiased parallel learning. My prediction on this matter has not yet been tested. Chapter 13 (Section 37) uses self-similarity to design a network capable of choosing those internal representations which are most informative, or best predictive, in a prescribed temporal context, such as a word, sentence, or piano piece. Again a design principle which arose to solve one problem found its way into the solution of a constellation of related problems.

PATTERN LEARNING BY FUNCTIONAL-DIFFERENTIAL NEURAL NETWORKS WITH ARBITRARY PATH WEIGHTS*

INTRODUCTION

The Theory of Embedding Fields studies systems of nonlinear functional-differential equations which can be derived from psychological postulates and interpreted as neural networks [1]. These systems describe cross-correlated flows on signed directed graphs. They have been applied to problems in pattern discrimination, learning, memory, and recall (e.g., [1]–[10]).

The theory is derived in several stages ([1], [4], [9], [10]). Each stage exhibits the minimal systems that are compatible with a given list of psychological postulates. Successive stages refine either the dynamical equations themselves, or the synthesis of network connections, to satisfy additional postulates. This paper reviews the derivation of stage one, for completeness, and proves two general theorems about spatial pattern learning by a suitable class of networks of Embedding Field type. Weaker versions of these theorems were announced without proof in [7]. The theorems will also be interpreted psychologically and physiologically. They describe properties of learning that are invariant under broad changes in physiological and anatomical constraints. In particular, they permit a discussion of learning in an essentially arbitrary anatomy.

DERIVATION OF SOME NETWORKS

We will globally analyse systems of the form

$$\dot{x}_i = A_i x_i + \sum_{k \in J} B_{ki} z_{ki} + C_i(t) \tag{1}$$

and

$$\dot{z}_{ji} = D_{ji} z_{ji} + E_{ji} x_i, \tag{2}$$

where $i \in I$, $j \in J$, and I and J are finite but possibly arbitrarily large sets

* Supported in part by the Alfred P. Sloan Foundation and the Office of Naval Research (N00014–67–A–0204–0051).

(First published in K. Schmitt (ed.), *Delay and Functional Differential Equations and their Applications*, Academic Press, New York and London, 1972, pp. 121–160.)

that will be subject to suitable constraints. The symbols A_i, B_{ji}, D_{ji}, and E_{ji} denote continuous functionals, not necessarily linear, with all B_{ji} and E_{ji} nonnegative. The input functions C_i and initial data are chosen nonnegative and continuous.

Systems of this type can be derived by considering an experimentalist $\mathscr{E}$ who interacts with a machine $\mathscr{M}$ to teach $\mathscr{M}$ to predict B given A by practicing AB. An alternative version of this task is described by the following experiment. A hungry dog is presented with food and thereupon salivates. A bell is rung but the dog does not salivate. Then the bell is rung just before food presentation on several learning trials. Thereafter presentation of the bell alone yields salivation. This learning process is called respondent, or Pavlovian, conditioning [11]. Food is called the unconditioned stimulus (UCS), salivation is called the unconditioned response (UCR), and the bell is called the conditioned stimulus (CS). The sensory presentation of A is analogous to a CS, the sensory presentation of B is analogous to a UCS, and the motor response B is analogous to a UCR.

Systems (1) and (2) will thus describe versions of machines $\mathscr{M}$ capable of learning complicated patterns by respondent conditioning. The inputs $C_i(t)$ will be chosen to represent a particular experiment performed on $\mathscr{M}$ by $\mathscr{E}$. The outputs of $\mathscr{M}$ will be suitable functionals of the vector function $X = (x_i : i \in I)$. The simplest version of $\mathscr{M}$ is derived below. The derivation is given in story-book form to emphasize its intuitive basis.

(A) *Each Letter Seems Simple*

In daily speech and listening, a letter is never decomposed into two parts. To maintain close contact with experience, we assume that a single state v_A in $\mathscr{M}$ corresponds to A. In a similar fashion, let v_B correspond to B, v_C to C, etc. We designate each v_i by a point, or vertex.

(B) *Presentation Times*

The times at which letters are presented to $\mathscr{M}$ must be represented within $\mathscr{M}$. For example, presenting A and then B with a time spacing of twenty-four hours should yield far different behavior than presentation with a time spacing of two seconds. Thus various functions of time should be associated with each vertex. To maintain contact with the 'one-ness' of each letter, and to maximize the simplicity of our derivation, we let one function $x_A(t)$ be associated with v_A, one function $x_B(t)$ be associated with v_B, etc., as in Figure 1.

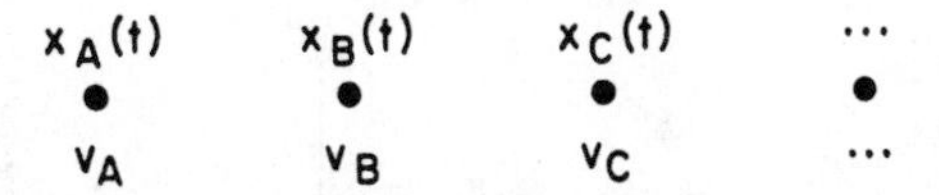

Fig. 1. Vertices and vertex functions.

(C) *Continuous Vertex Functions*

The function $x_A(t), \ldots, x_Z(t)$ will be chosen continuous, and in fact differentiable. Several reasons for this exist. The most specific reason is the following. Consider the question: What follows *ABC*? It is tempting to say *D*, but really the problem is ill-defined if the letters are presented one at a time with time spacing w between successive letters. If indeed w is small, say $w \cong 2$ seconds, then *D* might well be the correct response, but if $w \cong$ 24 hours then to the sound C (= 'see') one can also reply 'see what?' That is, as w varies from small to large values, the influence of *A* and *B* on the prediction following *C* gradually wears off. Since $x_A(t)$ and $x_B(t)$ describe the relevance at time t of *A* and *B* in $\mathscr{M}$, we conclude that these functions also vary gradually in time.

(D) *Perturbations Instead of Presentations*

Suppose *A* is never presented to $\mathscr{M}$. Corresponding to the occurrence of 'nothing' is the natural mathematical predisposition to set $x_A(t) = 0$ at all times t. (The equilibrium point 0 can, it turns out, be rescaled ultimately relative to the signal thresholds).

Suppose *A* is presented to $\mathscr{M}$ for the first time at time $t = t_A$. Then $x_A(t)$ must be perturbed from 0 for certain $t > t_A$, or else $\mathscr{M}$ would have no way of knowing that *A* occurred. We associate the occurrence of 'something' with a positive deflection in the graph of x_A. (The theory could also, in principle, be carried out with negative deflections.)

Shortly after *A* is presented, *A* no longer is heard by $\mathscr{M}$. That is, $x_A(t)$ gradually returns to the value signifying no recent presentation of *A*, namely 0. In a similar fashion, if *A* is presented at times $t_A^{(1)} < t_A^{(2)} < \cdots$

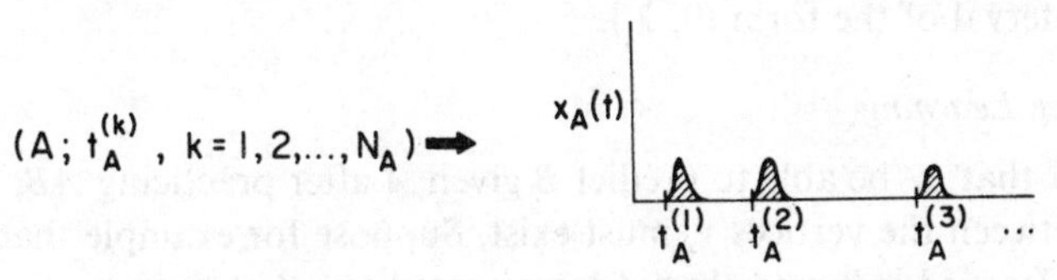

Fig. 2. Input presentations induce vertex perturbations.

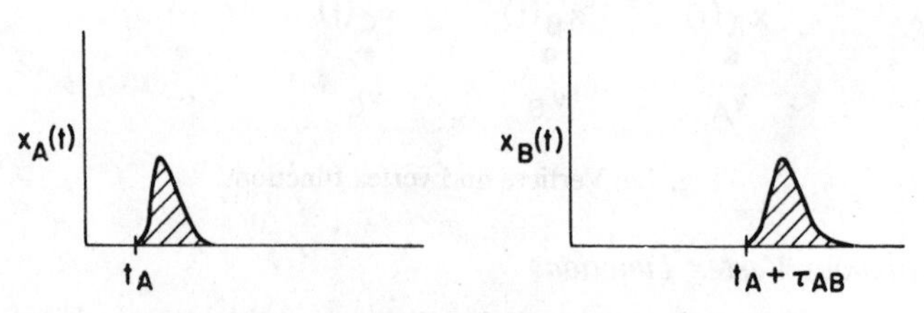

Fig. 3. Vertex translation of predicting B given A.

$< t_A^{(N_A)}$, then we find the graph of Figure 2. The same construction holds true for all letters. In this way, we have translated the presentation of any letters A, B, C, ... in the alphabet at prescribed times into a definite sequence of perturbations of the vertex functions $x_A(t)$, $x_B(t)$, $x_C(t)$, ...

(E) *Linearity*

For notational convenience, we replace the alphabet A, B, C, ... by any sequence r_i, $i = 1, 2, \ldots, n$, of n behavioral atoms; the vertices v_A, v_B, v_C, ... by the vertices v_i, $i = 1, 2, \ldots, n$; and the vertex functions $x_A(t)$, $x_B(t)$, $x_C(t)$, ... by the vertex functions $x_i(t)$, $i = 1, 2, \ldots, n$. Now r_i corresponds to $(v_i, x_i(t))$, $i = 1, 2, \ldots, n$.

What is the simplest way to translate Figure 2 into mathematical terms? Since we are constructing a system whose goal is to adapt with as little bias as possible to its environment, we are strongly advised to make the system as linear as possible. The simplest linear way to write Figure 2 is in terms of the equations

$$\dot{x}_i(t) = -\alpha_i x_i(t) + C_i(t), \tag{3}$$

with $\alpha_i > 0$, $x_i(0) \geqslant 0$, and $i = 1, 2, \ldots, n$. The input $C_i(t)$ can, for example, have the form

$$C_i(t) = \sum_{k=1}^{N_i} J_i(t - t_i^{(k)}),$$

where $J_i(t)$ is some nonnegative and continuous function that is positive in an interval of the form $(0, \lambda_i)$.

(F) *After Learning*

In order that $\mathscr{M}$ be able to predict B given A after practicing AB, interactions between the vertices v_i must exist. Suppose for example that $\mathscr{M}$ has already learned AB, and that A is presented to $\mathscr{M}$ at time t_A. We expect $\mathscr{M}$ to respond with B after a short time interval, say at time $t = t_A + \tau_{AB}$,

Fig. 4. Directed network and network processes.

where $\tau_{AB} > 0$. τ_{AB} is called the *reaction time* from A to B. Let us translate these expectations into graphs for the functions $x_A(t)$ and $x_B(t)$. We find Figure 3. The input $C_A(t)$ controlled by $\mathscr{E}$ gives rise to the perturbation of $x_A(t)$. The internal mechanism of $\mathscr{M}$ must give rise to the perturbation $x_B(t)$. In other words, after AB is learned $x_B(t)$ gets large τ_{AB} units after $x_A(t)$ gets large.

There exists a linear and continuous way to say this; namely, v_A sends a linear signal to v_B with time lag τ_{AB}. Then (3) with $i = B$ is replaced by

$$\dot{x}_B(t) = -\alpha_B x_B(t) + C_B(t) + \beta_{AB} x_A(t - \tau_{AB}),$$

with β_{AB} some positive constant. More generally if $r_i r_j$ has been learned we conclude that

$$\dot{x}_j(t) = -\alpha_j x_j(t) + C_j(t) + \beta_{ij} x_i(t - \tau_{ij}). \tag{4}$$

If $\beta_{ij} = 0$, then the list $r_i r_j$ cannot be learned since a signal cannot pass from v_i to v_j.

(G) *Directed Paths*

The signal $\beta_{ij}\, x_i(t - \tau_{ij})$ from v_i to v_j in (4) is carried along some pathway at a finite velocity, or else the locality of the dynamics would be violated. Denote this pathway by e_{ij}. The pathways e_{ij} and e_{ji} are distinct because the lists $r_i r_j$ and $r_j r_i$ are distinct. To designate the direction of flow in e_{ij}, we draw e_{ij} as an arrow from v_i to v_j whose arrowhead N_{ij} touches v_j, as in Figure 4.

(H) *Before Learning*

Before any learning occurs, if A leads only to B, then learning would have already occurred. A must therefore also be able to lead to C, D, or some other letters. Thus the process of learning can be viewed as elimination of the incorrect pathways AC, AD, etc., while the correct pathway AB endures, or is strengthened.

(I) *Distinguishing Order*

How does $\mathcal{M}$ know that AB and not AC is being learned? By Figure 3 practicing AB means that x_A and then x_B become large several times. Saying A alone, or B alone, or neither A nor B should yield no learning. This can be mathematically stated most simply as follows. If AB occurs with a time spacing of w, then the product $x_A(t - w)x_B(t)$ is large at suitable times $t \cong t_A^{(i)} + w$, $i = 1, 2, \ldots, N_A$. We therefore seek a process in $\mathcal{M}$ that can compute products of past $x_A(v)$ values $(v < t)$ and present $x_B(t)$ values. Denote this process by $z_{AB}(t)$. Note that $z_{AB} \neq z_{BA}$.

Where in $\mathcal{M}$ do past values of $x_A(v)$ and present values of $x_B(t)$ come together, so that $z_{AB}(t)$ can compute them? (Locality again!) By Figure 4, this happens only in the arrowhead N_{AB}. Thus $z_{AB}(t)$ takes place in N_{AB}. But then the past $x_A(v)$ value received by N_{AB} at time t is the signal $\beta_{AB}x_A(t - \tau_{AB})$. The most linear and continuous way to express this rule for $z_{AB}(t)$ is the following.

$$\dot{z}_{AB}(t) = -\gamma_{AB}z_{AB}(t) + \delta_{AB}x_A(t - \tau_{AB})x_B(t),$$

with γ_{AB} a positive constant, and δ_{AB} a nonnegative constant that is positive only if β_{AB} is positive. More generally, for $r_i r_j$ we find in N_{ij} the process

$$\dot{z}_{ij}(t) = -\gamma_{ij}z_{ij}(t) + \delta_{ij}x_i(t - \tau_{ij})x_j(t). \tag{5}$$

(J) *Gating Outputs*

The $z_{ij}(t)$ function can distinguish whether or not $r_i r_j$ is practiced. But more is desired. Namely, if $r_i r_j$ is practiced, presenting r_i should yield a delayed output from v_j. If $r_i r_j$ is not practiced, presenting r_i should not yield an output from v_j. And even if $r_i r_j$ is practiced, no output from v_j should occur if r_i is not presented. In other words, $x_j(t)$ should become large only if $x_i(t - \tau_{ij})$ *and* $z_{ij}(t)$ are large. Again a product is called for, and (4) is changed to

$$\dot{x}_j(t) = -\alpha_j x_j(t) + C_j(t) + x_i(t - \tau_{ij})\beta_{ij}z_{ij}(t). \tag{6}$$

(K) *Independence of Lists in First Approximation*

If B is not presented to $\mathcal{M}$, then in first approximation CA should be learnable without interference from B. (Not so in second approximation, since a signal could travel from C to B to A.) Similarly if C is not presented to $\mathcal{M}$, then BA should be learnable without interference from C, in first approximation. Mathematically speaking, this means that all signals to each v_j combine additively at v_j. Thus (6) becomes

$$\text{(7)} \qquad \dot{x}_j(t) = -\alpha_j x_j(t) + C_j(t) + \sum_{i=1}^{n} x_i(t - \tau_{ij})\beta_{ij} z_{ij}(t).$$

The system (5) and (7) is a mathematically well-defined proposal for a learning machine that uses only such general notions as linearity, continuity, and locality, and a mathematical analysis of how a machine can learn to predict B given A on the basis of practicing AB.

(L) *Thresholds*

One further modification of systems (5) and (7) is convenient; namely, the introduction of signal thresholds. Here we introduce this modification directly to keep background noise down. A more fundamental analysis would introduce it by first analysing the need in complex learning situations for inhibitory interactions, and then by pointing out that learning becomes difficult without signal thresholds if inhibitory interactions exist.

A possible difficulty in (5) and (7) is this. Small signals can possibly be carried round-and-round the network thereby building up background noise and interfering with the processing of behaviorally important inputs. We therefore seek to eliminate the production of signals in response to small $x_i(t)$ values, in the most linear possible way. Thresholds do this for us. Letting $[\xi]^+ = \max(\xi, 0)$, we replace (5) and (7) by

$$\text{(8)} \qquad \dot{x}_i(t) = -\alpha_i x_i(t) + \sum_{m=1}^{n} [x_m(t - \tau_{mi}) - \Gamma_{mi}]^+ \beta_{mi} z_{mi}(t) + C_i(t)$$

and

$$\text{(9)} \qquad \dot{z}_{jk}(t) = -\gamma_{jk} z_{jk}(t) + \delta_{jk}[x_j(t - \tau_{jk}) - \Gamma_{jk}]^+ x_k(t),$$

where all Γ_{jk} are positive thresholds, and $i, j, k = 1, 2, \ldots, n$. Systems (8) and (9) complete the derivation of this paper.

PSYCHOPHYSIOLOGICAL INTERPRETATION

The function $x_i(t)$ is called the ith *stimulus trace*: it responds to the *stimulus* $C_i(t)$. The function $z_{jk}(t)$ is called the (j, k)th *memory trace*: it records the pairing of successive events r_j and r_k. Alternatively, $x_i(t)$ is called the ith *short-term memory trace*: it represents brief activation of the state v_i either by inputs $C_i(t)$ or by signals from other states v_j. Similarly, $z_{jk}(t)$ is called the (j, k)th *long-term memory trace*: its record of past events can endure long after the short term memory traces have decayed. *Transfer* from short-term memory to long-term memory denotes the operation whereby the z_{jk}'s are altered by the distribution of x_i's. *Activation* of short-term memories via long-term memories denotes the operation whereby signals from a given set of v_j's, modulated in the pathways e_{jk} by the z_{jk}'s, activate a given pattern of x_k's.

Γ_{jk} is the (j, k)th *signal threshold*: no signal is emitted by v_j into e_{jk} at time t unless $x_j(t) > \Gamma_{jk}$. v_j is said to *sample* v_k at time t if the signal from v_j to N_{jk} is positive at time t. The *signal strength* at N_{jk} at time t is defined by $B_{jk}(t) = [x_j(t - \tau_{jk}) - \Gamma_{jk}]^+ \beta_{jk}$. The constant β_{jk} is a structural parameter called the *path strength* of e_{jk}. The $n \times n$ matrix $\beta = \|\beta_{jk}\|$ determines which directed paths between vertices exist, and how strong they are. Otherwise expressed, β determines the 'anatomy' of connections between all vertices.

A physiological interpretation of these variables in terms of cell bodies (v_i), axons (e_{jk}), synaptic knobs (N_{jk}), cell potentials ($x_i(t)$), spiking frequencies ($\propto B_{jk}(t)$), and transmitter production rates ($z_{jk}(t)$) can also be noted [1].

Mathematical analysis of system (8) and (9) shows that important properties of learning are preserved in the more general systems (1) and (2). Given the psychophysiological interpretation above, this generalization has an important physical meaning.

(A) *Short-Term Memory Decay*

Consider the replacement of the exponential decay term $- \alpha_i x_i$ in (8) by the general decay term $A_i x_i$ in (1). For example, let $A_i(t) = - \alpha_i + f_i(t)$, where $0 \leqslant f_i(t) < \alpha_i$. $f(t)$ represents a 'shunt' of cell potential at v_i that can amplify the effects of the input $C_i(t)$ at prescribed times when $f_i(t)$ is large, as the equation $\dot{x}_i = - (\alpha_i - f_i(t))x_i + C_i$ illustrates. These 'Now Print' times will presumably occur when information important to the survival of the network is delivered; e.g., the presentation of food in a prescribed situation. The inputs $f_i(t)$ are often controlled by 'nonspecific arousal' cells that are activated by biologically important events. See Figure 5a. An alternative shunt can act directly on the synaptic knobs delivering $C_i(t)$ to v_i, thereby replacing $C_i(t)$ by $f_i(t)C_i(t)$, as in the equation $\dot{x}_i = -\alpha_i x_i + f_i(t)C_i(t)$. See Figure 5b. An additive Now Print mechanism is for some purposes more useful than the shunting mechanism ([4], [9], [10]).

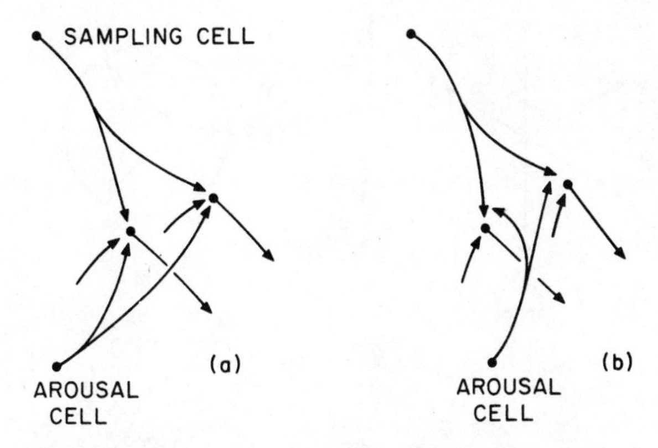

Fig. 5. Interactions between sampling and arousal cells.

The function $A_i(t)$ also permits noisy variations in decay parameters, as well as feedback from system variables, as in

$$A_i(t) = -\alpha - \sum_{k \in I} \int_0^t \{[x_k(v - \tau_k) - \Gamma_k]^+\}^2 \, g(v) \, dv.$$

(B) *Long-Term Memory Decay*

The exponential decay term $-\gamma_{jk} z_{jk}$ in (9) can be replaced by a wide variety of alternatives. For example, $D_{jk}(t) = -\gamma_{jk}[x_j(t - \tau_{jk}) - \Gamma_{jk}]^+$ implies that there is perfect memory until v_j samples v_k, say on performance trials [12], since otherwise $D_{jk}(t) = 0$. Memory is plastic on sampling trials, however. The choice $D_{jk}(t) = -\gamma_{jk} f_{jk}(t)[x_j(t - \tau_{jk}) - \Gamma_{jk}]^+$ prevents memory decay unless both general Now Print and specific sampling of v_k by v_j occurs. Again the anatomy of Figure 5b holds.

Long-term memory decay can also be altered indirectly by changing the functional B_{jk}. For example, the choice

$$(10) \qquad B_{jk}(t) = \beta_{jk}[x_j(t - \tau_j) - \Gamma_j]^+ \left(\sum_{m \in I} z_{jm}(t) \right)^{-1}$$

can have the effect of making memory perfect until new items are practiced [13]. In particular, memory can be perfect during performance trials. This is true, however, only in certain anatomies; for example, the 'outstar' anatomy of Figure 6a. There exist other anatomies, such as those depicted in Figures 6b and 6c, in which perfect memory is replaced by 'phase transi-

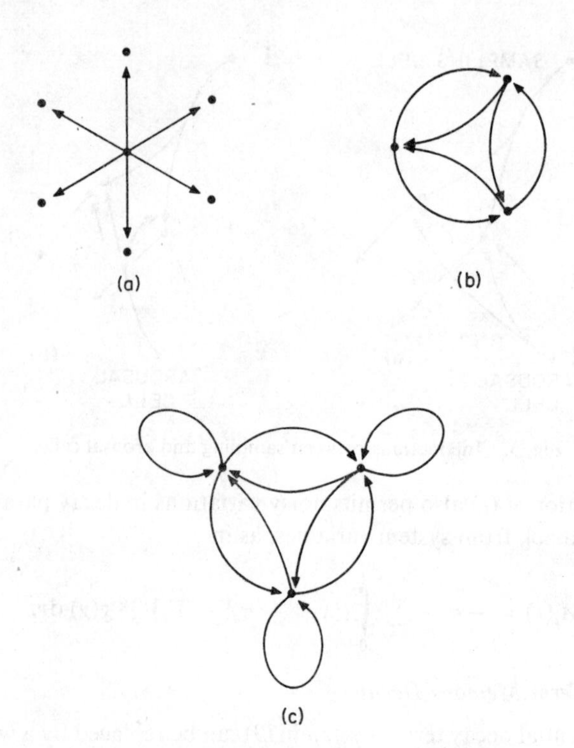

Fig. 6. Phase transitions depend on anatomy.

tions in memory' or 'imprinting' of memories ([14], [15]). That is, for certain choices of numerical parameters, memory of a given class of patterns decays; for the remaining parameter choices, memory is rigid. The particular class of patterns for which memory decays depends on the anatomy of the network ([1], p. 35). Prescribed alterations in network parameters can 'imprint' an ongoing input pattern by transforming the system from its plastic phase to its rigid phase.

(C) *Signal Strength*

The signal terms $\beta_{jk}[x_j(t - \tau_{jk}) - \Gamma_{jk}]^+$ and $\delta_{jk}[x_j(t - \tau_{jk}) - \Gamma_{jk}]^+$ in (8) and (9), respectively, can be replaced, say, by

$$B_{jk}(t) = \beta_{jk}(t)[x_j(t - \tau_{jk}(t)) - \Gamma_{jk}(t)]^+$$

and

$$E_{jk}(t) = \delta_{jk}(t)[x_j(t - \sigma_{jk}(t)) - \Omega_{jk}(t)]^+$$

which permit different, and variable, time lags, thresholds, and path strengths in the two signal strength functionals. This includes the possibility of coupling a Now Print mechanism to these functionals. Functional $E_{jk}(t)$ describes the effect of the signal from v_j on the cross-correlational process within N_{jk} that determines $z_{jk}(t)$. Functional $B_{jk}(t)$ describes the net signal from v_j that ultimately influences v_k after being processed in N_{jk}. It is therefore natural to physically expect that $\Gamma_{jk} \geqslant \Omega_{jk}$. This *local flow condition* says little more than that the signal from v_j passes through N_{jk} on its way to v_k. Such a condition is, in fact, needed to guarantee that many cells can simultaneously sample a given pattern without creating asymptotic biases in their memory. The local flow condition provides examples of systems which can learn patterns without performing them until later, but which cannot perform old patterns without also learning new patterns that are imposed during performance.

The functionals B_{jk} and E_{jk} permit more complicated possibilities as well. For example, *in vivo*, after a signal is generated in e_{jk}, it is impossible to generate another signal for a short time afterwards (absolute refractory period) and harder to generate another signal for a short time after the absolute refractory period (relative refractory period). Also, some cells emit signals in complicated bursts. All such continuous variations are, in principle, covered by our theorems, which say, that whereas such variations can influence transient motions of the system, the classification of limits and oscillatory possibilities is unchanged by them.

It is physically interesting that those terms, such as B_{jk} and E_{jk}, which describe processes that act over a distance (such as signals flowing along e_{jk}) are the terms in Equations (1) and (2) that permit the most nonlinear distortion without destroying learning properties. The term x_i in (2) is not of this type. This term is computed in N_{ji} from the value $x_i(t)$ in the contiguous vertex v_i.

LOCAL SYMMETRY AXES

In its final form, the theorem shows that unbiased pattern learning can occur in systems with arbitrary positive path weights β_{ji} from $j \in J$ to $i \in I$. This is achieved by first restricting attention to systems of the form

$$\text{(11)} \qquad \dot{x}_i = Ax_i + \sum_{k \in J} B_k z_{ki} + C_i(t)$$

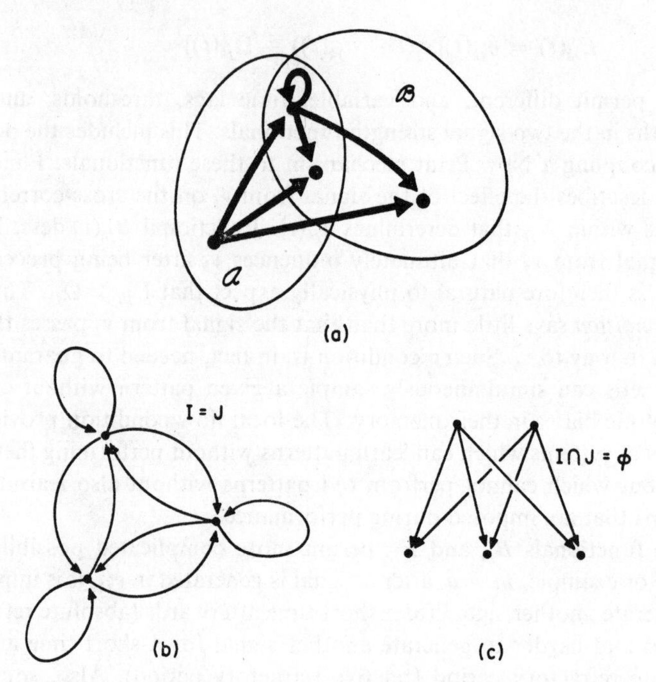

Fig. 7. Connections between sampling and sampled cells.

and

$$\dot{z}_{ji} = D_j z_{ji} + E_j x_i, \tag{12}$$

where $i \in I$ and $j \in J$. That is, all functionals A_i, B_{ji}, D_{ji}, and E_{ji} are chosen independent of $i \in I$, and the anatomy is constrained to make this possible. These constraints mean that all cells $\mathscr{B} = \{v_i : i \in I\}$ are sampled by a given cell v_j in $\mathscr{A} = \{v_j : j \in J\}$ without biases due to system parameters ($B_{ji} = B_j$, $D_{ji} = D_j$, $E_{ji} = E_j$), and that the inputs to all cells $\mathscr{B}$ are averaged by their cell potentials without biases due to averaging rates ($A_i = A$). See Figure 7a.

Systems (11) and (12) allow each cell to have a different time lag, threshold, and axon weight, as in $B_j(t) = \beta_j[x_j(t - \tau_j) - \Gamma_j]^+$. Even if all cells interact, as in Figure 7b, no biases in asymptotic learning need occur due to these asymmetries in signal transfer among possibly billions of cells.

Figures 7b and 7c illustrate two extremal anatomies, the completely recurrent ($I = J$) and completely nonrecurrent ($I \cap J = \emptyset$) cases. Generalizations of Figure 7a are also possible. In these generalizations, $\mathscr{A}$ and $\mathscr{B}$ are replaced by sets $\{\mathscr{A}_k\}$ and $\{\mathscr{B}_k\}$ of subsets such that each cell in a given $\mathscr{B}_k$ is sampled by all cells in $\mathscr{A}_k$. One seeks the maximal subsets $\mathscr{B}_k$ for which this decomposition exists. For some purposes, a fixed set $\{\mathscr{B}_k\}$ is predetermined by physical considerations; e.g., each $\mathscr{B}_k$ controls a different motor effector. It is then sometimes profitable to introduce fictitious cells into the sampling cells $\mathscr{A}$ if some cells in $\mathscr{A}$ sample two or more subsets $\mathscr{B}_k$. For example, if cell v_i in $\mathscr{A}$ samples $\mathscr{B}_1$ and $\mathscr{B}_2$, replace v_i by two cells v_{i_1} and v_{i_2} such that v_{ij} samples only $\mathscr{B}_j$, $j = 1, 2$, and each v_{ij} receives the same inputs, and has the same parameters and initial data, as the original cell v_i had.

MATHEMATICAL RESULTS

The theorems discuss learning by any number of cells in $\mathscr{A}$ of a spatial pattern presented to any number of cells in $\mathscr{B}$. Learning is by respondent conditioning. The CS's are delivered to $\mathscr{A}$. The UCS is the spatial pattern received by $\mathscr{B}$. Generalization to learning an arbitrary space-time pattern is readily accomplished [12].

A *spatial pattern* delivered to $\mathscr{B}$ is a vector function of the form $C_i(t) = \theta_i C(t)$, $i \in I$. The relative pattern weights $\theta_i (\theta_i \geqslant 0, \sum_{k \in I} \theta_k = 1)$ characterize the pattern. The total intensity $C(t)$ can fluctuate wildly in time. This definition of spatial pattern notes, for example, that recognition of a picture is invariant under wide fluctuations of background illumination. The inputs $C_i(t)$ need not, however, represent a pattern created by external events at peripheral network receptor sites. Any spatial pattern playing on any finite collection of cells (e.g., cells controlling motor outputs, or centrally located sensory cells of any type) can be learned and performed in the manner described by the following theorems. References [2] and [3] discuss other applications of this definition, especially to problems of pattern discrimination.

Since the relative intensity θ_i at each v_i characterizes the pattern, we study the limiting and oscillatory behavior of 'pattern variables'; namely, the relative stimulus traces (potentials) $X_i = x_i(\sum_{k \in I} x_k)^{-1}$ and the relative memory traces (transmitters) $Z_{ji} = z_{ji}(\sum_{k \in I} z_{jk})^{-1}$. Once behavior of these variables is established for general functionals, analysis of the 'total energy' variables $x = \sum_{k \in I} x_k$ and $z_j = \sum_{k \in I} z_{jk}$ can be carried out for particular choices of functionals. Then behavior of x_i and z_{ji} is also known. See, for example, [8].

Theorem 1 studies what happens when *all* cells in $\mathscr{A}$ receive their CS's 'sufficiently often', given also that the UCS is presented sufficiently often at times when the CS's can practice it. This theorem is not always realistic. A given CS need not activate all the cells $\mathscr{A}$ that are capable of sampling $\mathscr{B}$. Theorem 2 studies the case in which an arbitrary subset of $\mathscr{A}$ is activated 'sufficiently often' by CS's. The local flow condition is needed in this situation. Proposition 1 shows what can go wrong if the local flow condition is not imposed. Theorem 1 will be stated without proof, since the same method is used as for the more difficult Theorem 2.

Theorem 1 will be expressed in terms of the function $f(S, T) = \int_S^T C \exp(\int_t^T A \, dv) \, dt$; the functions $M(i)$: $[0, \infty) \to J$ such that $Z_{[M(i)](t),i}(t) = \max\{Z_{ji}(t)\colon j \in J\}$ and $m(i)$: $[0, \infty) \to J$ such that $Z_{[m(i)](t),i}(t) = \min\{Z_{ji}(t)\colon j \in J\}$; and the functional L defined for every piecewise constant function m: $[0, \infty) \to J$ and every continuous g on $[S, T)$ by

$$L(m, g; S, T] = \int_S^T E_m g \exp\left[-\int_S^t D_m \, dv\right] dt.$$

THEOREM 1. *Suppose that*

(i) *the system is bounded*;

(ii) *each* CS *is presented sufficiently often*; *that is*, *for every* $j \in J$,

$$L[j, x; 0, \infty] = \infty \text{ (also necessary)}; \tag{13}$$

(iii) *the* UCS *is presented sufficiently often*; *that is*,

$$\int_0^\infty C x^{-1} \, dt = \infty \text{ (also necessary)}; \tag{14}$$

and

(iv) *each* CS *and the* UCS *are practiced together sufficiently often*; *that is for some* $\varepsilon > 0$ *and each* $i \in I$, *there exist increasing divergent sequences* $\{S_{in}\}$ *and* $\{T_{in}\}$ *such that*

$$\sum_{n=1}^{\infty} \frac{L[M(i), f(S_{in}, \cdot); S_{in}, S_{i,n+1}]}{\varepsilon + L[M(i), x; S_{in}, S_{i,n+1}]} = \infty \tag{15}$$

and

$$\sum_{n=1}^{\infty} \frac{L[m(i), f(T_{in}, \cdot); T_{in}, T_{i,n+1}]}{\varepsilon + L[m(i), x; T_{in}, T_{i,n+1}]} = \infty. \tag{16}$$

Then perfect pattern learning occurs; that is, all the limits $Q_i = \lim_{t\to\infty} X_i(t)$ *and* $P_{ji} = \lim_{t\to\infty} Z_{ji}(t)$ *exist globally and*

$$(17) \qquad P_{ji} = Q_i = \theta_i.$$

Condition (i) can be removed, but leads to a physically implausible situation. Then the nth appearances of ε in (iv) are replaced by $z_{[M(i)](S_{in})}(S_{in})$ and $z_{[m(i)](T_{in})}(T_{in})$, respectively. A counterexample can be constructed if (iv) is violated.

Equation (17) means that the pattern is perfectly learned in the sense that a test input to *any* subset of cells v_j after learning occurs can reproduce the pattern at all cells v_i. The act of recall can, however, destroy the memory in special cases [14].

COROLLARY 1. *Conditions* (ii)–(iv) *are implied by the following conditions*: (i),

(v) *for every* $j \in J$,

$$(18) \qquad \int_0^\infty E_j \, dt = \infty;$$

and

(vi) *the* UCS *energy is presented, on the average, with a uniform lower bound; that is, there exist positive constants* K_1 *and* K_2 *such that for every* $T \geqslant 0$,

$$(19) \qquad f(T, T + t) \geqslant K_1 \quad \text{if} \quad t \geqslant K_2.$$

Theorem 2 uses the following functions. Let $N(i)$: $[0, \infty) \to J(1)$ be defined by $Z_{[N(i)](t),i}(t) = \max\{Z_{ji}(t) : j \in J(1)\}$, and $n(i)$: $[0, \infty) \to J(1)$ be defined by $Z_{[n(i)](t),i}(t) = \min\{Z_{ji}(t) : j \in J(1)\}$, where $J(1) = \{j \in J : \int_0^\infty B_j z_j x^{-1} \, dt = \infty\}$.

THEOREM 2. *Again suppose that the system is bounded, the* UCS *is presented sufficiently often, and*

(vii) *those* CS's *which are performed continually are also practiced with the* UCS *sufficiently often; that is, if* $J(1) \neq \emptyset$, *then condition* (iv) *holds with* $M(i)$ *and* $m(i)$ *replaced by* $N(i)$ *and* $n(i)$.

Then the potentials pick up the pattern weights and all transmitters learn the pattern at least partially; that is, all the limits Q_i *and* P_{ji} *exist with* $Q_i = \theta_i$. *If, moreover, a* CS *is practiced with the* UCS *sufficiently often, then it learns the pattern perfectly; that is, if (13) holds for some* $j \in J$, *then* $P_{ji} = \theta_i$.

The analog of Corollary 1, including a suitable version of the local flow condition, is given by Corollary 2.

COROLLARY 2. *Conditions* (iii) *and* (vii) *are implied by conditions* (i), (vi), *and*

(viii) *a local flow condition holds; that is, for every* $j \in j$, *either*

$$\int_0^\infty B_j z_j x^{-1} \, dt = \infty \quad \textit{only if} \quad \int_0^\infty E_j x \exp\left(-\int_0^t D_j dv\right) dt = \infty, \tag{20}$$

or

$$\int_0^\infty B_j \, dt = \infty \quad \textit{only if} \quad \int_0^\infty E_j \, dt = \infty. \tag{21}$$

Under these circumstances, if either $L[j, x; 0, \infty] = \infty$ *or* $\int_0^\infty E_j \, dt = \infty$, *then* $P_{ji} = \theta_i$.

Suppose for example that $B_j(t) = \beta_j[x_j(t - \tau_j) - \Gamma_j]^+$ and $E_j(t) = \delta_j[x_j(t - \sigma_j) - \Omega_j]^+$. Then condition (21) is satisfied if $\Gamma_j \geqslant \Omega_j$. In applications (21) is a constraint on the parameters of the system, rather than on its trajectories.

Condition (15) has the following intuitive meaning. The nth summand in (15) considers how much total input $C(t)$ reaches $\mathcal{M}$ during the time interval $[S_{in}, S_{i,n+1})$. The function $f(S_{in}, \cdot)$ describes the effect of averaging by functional A on C to yield C's contribution to the total potential x. For fixed $j \in J$, the functional $L[j, f(S_{in}, \cdot); S_{in}, S_{i,n+1}]$ describes the effect of averaging D_j and E_j on $f(S_{in}, \cdot)$ to yield C's contribution to the jth total transmitter z_j. $L[M(i), f(S_{in}, \cdot); S_{in}, S_{i,n+1}]$ measures the effect of C on the cell v_j whose relative memory trace Z_{ji} has been least attracted downwards towards θ_i, whenever this case occurs.

The term $L[M(i), x; S_{in}, S_{i,n+1}]$ in (15) has a similar interpretation, except that $x(t)$ replaces $f(S_{in}, t)$ to express the total effect of potential on transmitter. $x(t)$ differs from $f(S_{in}, t)$ due to the interaction term $\sum\{B_j z_{ji}: i \in I, j \in J\}$ in (11), which is also averaged by A to yield a contribution to x. These terms tend to preserve the old patterns that are already in $\mathcal{M}$'s memory (see Proposition 1). Condition (15) therefore says the following. For each $i \in I$, there exists some sequence of time intervals $[S_{in}, S_{i,n+1})$, such that enough input energy $C(t)$ is presented in each $[S_{in}, S_{i,n+1})$ to guarantee that, after averaging by potentials and transmitters, this energy suffices to overcome the stabilizing effect of interaction terms and thereby drive all the Z_{ji}'s towards the limits θ_i imposed by the new pattern.

Proposition 1 below notes that the local flow condition is not superfluous in a case of some physical interest.

PROPOSITION 1. *Suppose* (viii) *does not hold. Partition J into subsets $J(2)$ and $J(3)$ such that*

$$(22)\qquad J(2) = \{j: \int_0^\infty B_j \, dt = \infty \text{ and } \int_0^\infty E_j \, dt < \infty\} \neq \emptyset.$$

Suppose that the system is bounded, that (vi) *holds, that*

(ix) *there is perfect memory until recall in $J(2)$; that is, $D_j \geqslant -\gamma_j E_j$ for some constant $\gamma_j > 0$, $j \in J(2)$; and that*

(x) *average performance energy in $J(2)$ does not converge to zero; that is, for every $T \geqslant 0$,*

$$(23)\qquad \limsup_{t\to\infty} \sum_{k\in J(2)} \int_T^t B_k \exp\left[\int_v^t A \, d\xi\right] dv > 0.$$

Then given initial data such that $\max(Z_{ji}(0): j \in J) \geqslant X_i(0) > \theta_i$ *and* $\min(Z_{ji}(0): j \in J) > \theta_i$, *even if Q_i exists, $Q_i \neq \theta_i$, so that even if P_{ji} exists and $\int_0^\infty E_j \, dt = \infty$, $P_{ji} \neq \theta_i$.*

Theorem 2 will be proved below. The first step of the proof is to transform (11) and (12) into equations in the pattern variables $X_i = x_i x^{-1}$ and $Z_{ji} = z_{ji} z_j^{-1}$, where $x = \sum_{k\in I} x_k$ and $z_j = \sum_{k\in I} z_{jk}$.

LEMMA 1. *Suppose $x(0) > 0$ and $z_j(0) > 0$, $j \in J$. Then*

$$(24)\qquad \dot{X}_i = \sum_{k\in J} F_k(Z_{ki} - X_i) + G(\theta_i - X_i)$$

and

$$(25)\qquad \dot{Z}_{ji} = H_j(X_i - Z_{ji}),$$

where

$$(26)\qquad F_j = B_j z_j x^{-1}$$

$$(27)\qquad G = C x^{-1}$$

and

$$(28)\qquad H_j = E_j x z_j^{-1}.$$

The proof uses the standard equation

$$\left(\frac{f}{g}\right)^{\cdot} = \frac{1}{g}\left(\dot{f} - f\frac{\dot{g}}{g}\right)$$

along with the equations

$$\dot{x} = Ax + \sum_{k \in J} B_k z_k + C \tag{29}$$

and

$$\dot{z}_j = D_j z_j + E_j x. \tag{30}$$

Equations (24) and (25) are then readily transformed into equations for the difference variables $X_i^{(\theta)} = X_i - \theta_i$ and $Z_{ji}^{(\theta)} = Z_{ji} - \theta_i$.

LEMMA 2. *Given any spatial pattern, then*

$$\dot{X}_i^{(\theta)} = \sum_{k \in J} F_k(Z_{ki}^{(\theta)} - X_i^{(\theta)}) - GX_i^{(\theta)} \tag{31}$$

and

$$\dot{Z}_{ji}^{(\theta)} = H_j(X_i^{(\theta)} - Z_{ji}^{(\theta)}). \tag{32}$$

Lemmas 1 and 2 provide information concerning the oscillations of pattern variables. In order to conveniently describe these facts, we introduce the following terminology.

DEFINITION 1. If a system of inequalities, valid at some time $t = T$, is therefore valid at all times $t \geqslant T$, we say that the inequalities *propagate* in time.

The oscillations depend on whether or not the UCS is being presented. Hence let U_0 be the union of all intervals during which $C = 0$ and let U_1 be the union of all intervals during which $C > 0$. In both cases, X_i's motion is compared with that of $Y_i = \max(Z_{ji}: j \in J)$ and $y_i = \min(Z_{ji}: j \in J)$; that is, with the 'envelope' of all relative memory traces facing v_i. We will see that X_i is a kind of nonlinear 'center of mass' of these variables, and therefore attracts and is attracted towards the values in the interval $[y_i, Y_i]$. The influence of C changes the configuration to which X_i is attracted by also attracting X_i to θ_i. For $t \in U_1$, we will therefore compare $X_i^{(\theta)}$ with $Y_i^{(\theta)} = Y_i - \theta_i$ and $y_i^{(\theta)} = y_i - \theta_i$.

PROPOSITION 2. *Given any t in an interval of U_0, the following cases are exhaustive.*

(I) *The inequalities $y_i \leqslant X_i \leqslant Y_i$ propagate with y_i monotone increasing and Y_i monotone decreasing. Thus the limits $y_i(\infty)$ and $Y_i(\infty)$ exist.*

(II) *The inequality* $X_i > Y_i$ *either propagates with* X_i *monotone decreasing and all* Z_{ji} *monotone increasing, or switches into* Case (I). *Thus,* Y_i *oscillates at most once, and* y_i *is always monotone increasing. Hence either all limits exist, or* $y_i(\infty)$ *and* $Y_i(\infty)$ *exist.*

(III) *If* $X_i < y_i$ *holds, then the conclusions of* Case (II) *hold, with* y_i *replacing* Y_i *and all inequalities reversed.*

These results follow by inspection of (24) and (25) given that $C = 0$ and F_j and H_j are nonnegative. In short, either X_i is attracted to the interval $[y_i, Y_i]$ as it attracts all Z_{ji}, or X_i is trapped between y_i and Y_i as they are drawn together. If, therefore, as a result of prior practice $y_i \cong X_i \cong Y_i \cong \theta_i$, then these approximations propagate, yielding perfect memory of pattern weights.

PROPOSITION 3. *Given any t in an interval of* U_1, *the following cases are exhaustive.*

(IV) *The inequalities* $X_i^{(\theta)} \geqslant 0$ *and* $y_i^{(\theta)} \geqslant 0$ *propagate. If moreover*

(IVa) $X_i^{(\theta)} \leqslant Y_i^{(\theta)}$, *then the inequality propagates with* $Y_i^{(\theta)}$ *decreasing. Hence* $Y_i^{(\infty)}$ *exists. If however*

(IVb) $X_i^{(\theta)} > Y_i^{(\theta)}$, *then the inequality propagates as* $X_i^{(\theta)}$ *decreases and all* $Z_{ji}^{(\theta)}$ *increase until (if ever)* Case (IVa) *is entered. Hence either all limits exist or* Case (IVa) *is entered.*

(V) *The inequalities* $X_i^{(\theta)} \leqslant 0$ *and* $Y_i^{(\theta)} \leqslant 0$ *also propagate. The conclusions of* Case (IV) *hold with* y_i *and* Y_i *interchanged, and all inequalities reversed.*

(VI) *The inequalities* $Y_i^{(\theta)} \geqslant 0 \geqslant y_i^{(\theta)}$ *either propagate, or* Cases (IV) *or* (V) *are entered. If moreover,*

(VIa) $Y_i^{(\theta)} \geqslant X_i^{(\theta)} \geqslant y_i^{(\theta)}$, *then this inequality propagates with* $y_i^{(\theta)}$ *monotone increasing and* $Y_i^{(\theta)}$ *monotone decreasing. Both limits* $y_i(\infty)$ *and* $Y_i(\infty)$ *thus exist. If however*

(VIb) $X_i^{(\theta)} \notin [y_i^{(\theta)}, Y_i^{(\theta)}]$, *then* $X_i^{(\theta)}$ *and all* $Z_{ji}^{(\theta)}$ *are monotonic until* Case (VIa) *is entered. If* Case (VIa) *is never entered, then all limits exist.*

In short, X_i *is attracted towards the interval* $[y_i, Y_i]$, *but also towards* θ_i. *If both attractive forces steer* X_i *in the same direction, then the inequalities propagate.*

Proof of Theorem 2: First we consider the subcases of (IV)–(VI) in which all X_i and Z_{ji} are monotonic for large t. Suppose for example that Case (IVb) holds. Let this Case hold at all $t \geqslant 0$ for convenience. Then by (31), $\dot{X}_i^{(\theta)} \leqslant -GX_i^{(\theta)}$, and X_i decreases to the limit $Q_i \geqslant \theta_i$. Thus $\dot{X}_i^{(\theta)} \leqslant G(\theta_i - Q_i)$, or in integral form,

$$0 \leqslant X_i(t) \leqslant X_i(0) + (\theta_i - Q_i) \int_0^t G(v)\, dv$$

$$\leqslant 1 + (\theta_i - Q_i) \int_0^t G(v)\, dv,$$

for all $t \geqslant 0$. In particular, if $Q_i > \theta_i$, then

$$\int_0^\infty G(v)\, dv \leqslant (Q_i - \theta_i)^{-1} < \infty,$$

which contradicts (14). Thus $Q_i = \theta_i$. A similar procedure works in the other monotonic subcases.

To show that all P_{ji} exist in these cases, we prove the following:

LEMMA 3. *If Q_i exists, then all P_{ji} exist.*

Proof: For every $\varepsilon > 0$, there exists a T_ε such that $t \geqslant T$ implies $X_i(t) \in [Q_i - \varepsilon, Q_i + \varepsilon]$. If $Z_{ji}(t) \notin [Q_i - \varepsilon, Q_i + \varepsilon]$ for some $t \geqslant T_\varepsilon$, then $Z_{ji}(t)$ is monotonically attracted towards the interval $[Q_i - \varepsilon, Q_i + \varepsilon]$ until it enters it, by (25). If for some $\varepsilon > 0$, $Z_{ji}(t)$ never enters $[Q_i - \varepsilon, Q_i + \varepsilon]$, then $Z_{ji}(t)$ is monotonic for $t \geqslant T_\varepsilon$, and hence P_{ji} exists. If for every $\varepsilon > 0$, $Z_{ji}(t) \in [Q_i - \varepsilon, Q_i + \varepsilon]$ for all sufficiently large t, then $P_{ji} = Q_i$.

It remains in Case (IVb) to show that $P_{ji} = \theta_i$ if (13) holds. In this case, also $Y_i^{(\theta)}(t) \leqslant 0$ for $t \gg 0$. Otherwise, there will exist a T such that $Y_i(T) > \theta_i$, and since $Y_i(t)$ is monotone increasing, $X_i(t) \geqslant Y_i(t) \geqslant Y_i(T) > \theta_i$ for $t \geqslant T$. In particular, $Q_i > \theta_i$, which is impossible.

We can therefore restrict attention to the subcase in which $X_i^{(\theta)}(t) \geqslant 0 \geqslant Y_i^{(\theta)}(t)$ for $t \gg 0$. We will assume that $\theta_i > P_{ji}$ to derive a contradiction. Since X_i is monotone decreasing, and all Z_{ji} are monotone increasing, (25) implies

$$\dot{Z}_{ji} \geqslant H_j(\theta_i - P_{ji}),$$

and thus for $t \geqslant T$ and T sufficiently large,

$$1 \geqslant Z_{ji}(t) - Z_{ji}(T) \geqslant (\theta_i - P_{ji}) \int_T^t H_j\, dv.$$

Consequently $\int_0^\infty H_j\, dv < \infty$.

By (28) and (30),

$$H_j = \frac{\mathrm{d}}{\mathrm{d}t} \log \left[z_j(0) + \int_0^t E_j x \exp\left(- \int_0^v D_j \,\mathrm{d}\xi \right) \mathrm{d}v \right].$$

Hence for every $t \geqslant 0$,

$$\int_0^t H_j \,\mathrm{d}v = \log \{1 + z_j^{-1}(0) E[j, x; 0, t]\}.$$

Letting $t \to \infty$ shows that $E[j, x; 0, \infty] < \infty$, which contradicts (13).

Now we consider the nonmonotonic subcases of Proposition 3; namely

(VII) *$Y_i^{(\theta)} \geqslant X_i^{(\theta)} \geqslant 0$ and $y_i^{(\theta)} \geqslant 0$ with $Y_i^{(\theta)}$ monotone decreasing for large t*;

(VIII) *the reverse inequalities with $Y_i^{(\theta)}$ and $y_i^{(\theta)}$ interchanged*; and

(IX) *$X_i^{(\theta)} \in [y_i^{(\theta)}, Y_i^{(\theta)}]$ and $y_i^{(\theta)} \leqslant 0 \leqslant Y_i^{(\theta)}$ with $y_i^{(\theta)}$ increasing and $Y_i^{(\theta)}$ decreasing at large values of t.*

Only Case (VII) will be explicitly considered, since Cases (VIII) and (IX) can be treated by an analogous method. First, we treat the subcase in which

$$\sum_{k \in J} \int_0^\infty F_k \,\mathrm{d}t < \infty,$$

noting by (26) the relevance of F_j to (20). Then, for every $\varepsilon > 0$, there exists a T_ε such that $t \geqslant T_\varepsilon$ implies

$$\sum_{k \in J} \int_t^\infty F_k \,\mathrm{d}v \leqslant \varepsilon/2. \tag{33}$$

By (24),

$$\dot{X}_i^{(\theta)} \leqslant \sum_{k \in J} F_k - G X_i^{(\theta)}, \tag{34}$$

and thus for $t \geqslant T_\varepsilon$,

$$0 \leqslant X_i^{(\theta)}(t) \leqslant X_i^{(\theta)}(T_\varepsilon) \exp\left(- \int_{T_\varepsilon}^t G \,\mathrm{d}v \right) + \sum_{k \in J} \int_{T_\varepsilon}^t F_k \exp\left(- \int_v^t G \,\mathrm{d}\xi \right) \mathrm{d}v,$$

which by (33) implies

$$0 \leqslant X_i^{(\theta)}(t) \leqslant \exp\Big(-\int_{T_\varepsilon}^{t} G\,\mathrm{d}v\Big) + (\varepsilon/2).$$

Let $t \to \infty$ and apply (14) to conclude that $Q_i = \theta_i$. Lemma 3 now implies that all P_{ji} exist.

It remains only to consider Case (VII) – and Cases (VIII) and (IX) analogously – if

$$(35) \qquad \sum_{k \in J} \int_0^\infty F_k \,\mathrm{d}v = \infty.$$

Partition J into two sets $J(1)$ and $J(2)$ such that $j \in J(1)$ iff

$$(36) \qquad \int_0^\infty F_j \,\mathrm{d}v = \infty.$$

By (35) and the nonnegativity of F_j, $J(1) \neq \varnothing$, and we can define the function

$$\tilde{Y}_i^{(\theta)} = \max\{y_{ji}^{(\theta)} : j \in \mathrm{J}(1)\}.$$

We now show that $\tilde{Y}_i^{(\theta)}(\infty) \equiv \lim_{t\to\infty} \tilde{Y}_i^{(\theta)}(t)$ exists in Case (VII). To do this note by (32) that

$$(37) \qquad \operatorname{sign}\left[\frac{d}{\mathrm{d}t}\,\tilde{Y}_i^{(\theta)}(t)\right] = \operatorname{sign}\,[X_i^{(\theta)}(t) - \tilde{Y}_i^{(\theta)}(t)]$$

whenever $(\mathrm{d}/\mathrm{d}t)\,\tilde{Y}_i^{(\theta)}(t) \neq 0$. Thus, if there exists a T such that $X_i^{(\theta)}(t) \leqslant \tilde{Y}_i^{(\theta)}(t)$ for all $t \geqslant T$, then $\tilde{Y}_i(\infty)$ exists. There cannot exist a T such that $\tilde{Y}_i^{(\theta)}(t) \leqslant X_i^{(\theta)}(t)$ for all $t \geqslant T$, as we now show.

Define the functions $L^{(i)} = \sum_{k \in J(i)} F_k$, $i = 1, 2$. Then by (31),

$$(38) \qquad \dot{X}_i^{(\theta)} \leqslant L^{(2)} + (\tilde{Y}_i^{(\theta)} - X_i^{(\theta)})L^{(1)} - GX_i^{(\theta)}.$$

If $\tilde{Y}_i^{(\theta)} \leqslant X_i^{(\theta)}$, for $t \geqslant T$, then (38) implies

$$\dot{X}_i^{(\theta)} \leqslant L^{(2)} - GX_i^{(\theta)}$$

for $t \geqslant T$, and since by (37) $\tilde{Y}_i^{(\theta)}(t)$ is nondecreasing for $t \geqslant T$,

$$\dot{X}_i^{(\theta)} \leqslant L^{(2)} - \tilde{Y}_i^{(\theta)}(T)G \quad \text{for} \quad t \geqslant T,$$

where we can assume that $\tilde{Y}_i^{(\theta)}(T) > 0$ without loss of generality. Integrating (38) between T and ∞ readily yields the inequality $\int_0^\infty G(v)\,dv < \infty$, which contradicts (14).

Thus, either $\tilde{Y}_i^{(\theta)}(t) \geqslant X_i^{(\theta)}(t)$ for all $t \geqslant T$ and some T sufficiently large, or $\tilde{Y}_i^{(\theta)}(t) - X_i^{(\theta)}(t)$ changes sign at arbitrarily large times. We use this fact to prove that, for all $t \geqslant T$ and T sufficiently large, the inequality

$$(39) \qquad \max\,[X_i^{(\theta)}(t), \tilde{Y}_i^{(\theta)}(t)] \leqslant \tilde{Y}_i^{(\theta)}(T) + U(T, t)$$

holds, where

$$(40) \qquad U(T, t) = \int_T^t L^{(2)}(v)\,dv.$$

(39) can be used to show that $\tilde{Y}_i^{(\theta)}(\infty)$ exists in all cases. Since by definition of $J(2)$, $\int_0^\infty L^{(2)}(v)\,dv < \infty$, the function $U(T) \equiv \int_T^\infty L^{(2)}(v)\,dv$ monotonically decreases to zero as $T \to \infty$. Thus by (39) and (40), the bounded and continuous function $Y_i^{(\theta)}(t)$ is alternately monotone decreasing or increasing by an amount that approaches zero as $t \to \infty$. Hence $\tilde{Y}^{(\theta)}(\infty)$ exists.

Inequality (39) is proved as follows. It is trivial if $\tilde{Y}_i^{(\theta)}(t) \geqslant X_i^{(\theta)}(t)$ for all $t \geqslant T$ and some T sufficiently large. Suppose that $X_i^{(\theta)}(t) \geqslant \tilde{Y}_i^{(\theta)}(t)$ only in the disjoint sequence of intervals $[S_{ik}, T_{ik}]$, $k = 1, 2, \ldots$. Then $\tilde{Y}_i^{(\theta)}(S_{ik}) = X_i^{(\theta)}(S_{ik})$ for all $k = 1, 2, \ldots$, and by (38)

$$\dot{X}_i^{(\theta)}(t) \leqslant L^{(2)}(t), \quad t \in [S_{ik}, T_{ik}],$$

which implies

$$(41) \qquad \tilde{Y}_i^{(\theta)}(t) \leqslant X_i^{(\theta)}(t) \leqslant \tilde{Y}_i^{(\theta)}(S_{ik}) + U(S_{ik}, t)$$

for $t \in [S_{ik}, T_{ik}]$. Since $X_i^{(\theta)}(t) \leqslant \tilde{Y}_i^{(\theta)}(t)$ with $\tilde{Y}_i^{(\theta)}(t)$ monotone nonincreasing for $t \notin \bigcup_{k=1}^\infty [S_{ik}, T_{ik}]$, (39) readily follows by pasting these cases together.

Inequalities (38) and (39) along with the existence of $\tilde{Y}_i^{(\theta)}(\infty)$ will now be used to prove that $\tilde{Y}_i(\infty) = \theta_i$, and thus by (39) that $Q_i = \theta_i$. Then the existence of all P_{ji} follows by Lemma 3, and the proof is readily completed. Defined the functions $M = (d/dt)\log x - A - G$ and $N = (d/dt)\log x - A$. By (26) and (29), $L^{(1)} + L^{(2)} = \sum_{k \in J} F_k = M$. (38) therefore implies

$$(42) \qquad \dot{X}_i^{(\theta)} \leqslant 2L^{(2)} + M\tilde{Y}_i^{(\theta)} - NX_i^{(\theta)},$$

which can be expressed in integral form for any $t \geqslant T \geqslant 0$ as

$$(43) \qquad X_i^{(\theta)}(t) \leqslant P_i(T, t) + Q_i(T, t) + R(T, t)$$

using the notation

$$P_i(T, t) = X_i^{(\theta)}(T)x(T)x^{-1}(t)\exp\left(\int_T^t A\,\mathrm{d}\nu\right), \tag{44}$$

$$Q_i(T, t) = x^{-1}(t)\int_T^t \tilde{Y}_i^{(\theta)}(\nu)M(\nu)x(\nu)\exp\left(\int_\nu^t A\,\mathrm{d}\xi\right)\mathrm{d}\nu, \tag{45}$$

and

$$R(T, t) = 2x^{-1}(t)\int_T^t L^{(2)}(\nu)x(\nu)\exp\left(\int_\nu^t A\,\mathrm{d}\xi\right)\mathrm{d}\nu.$$

We now estimate each of these terms from above. First note that

$$R(T, t) \leqslant 2U(T, t). \tag{46}$$

This follows from (29) which shows that $\dot{x} \geqslant Ax$, and thus that

$$\exp\left[\int_\nu^t A\,\mathrm{d}\xi\right] \leqslant x^{-1}(\nu)x(t). \tag{47}$$

(39) is used to estimate (45), yielding

$$Q_i(T, t) \leqslant [\tilde{Y}_i^{(\theta)}(T) + U(T, t)]V(T, t), \tag{48}$$

where

$$V(T, t) = x^{-1}(t)\int_T^t Mx\exp\left(\int_\nu^t A\,\mathrm{d}\xi\right)\mathrm{d}\nu.$$

Note that

$$V(T, t) = x^{-1}(t)\exp\left(\int_0^t A\,\mathrm{d}\xi\right)\int_0^t(\dot{x} - Ax - C)\exp\left(-\int_0^\nu A\,\mathrm{d}\xi\right)\mathrm{d}\nu$$

$$= 1 - x(T)x^{-1}(t)\exp\left(\int_T^t A\,\mathrm{d}\nu\right)$$

$$- x^{-1}(t)\int_T^t C(\nu)\exp\left(\int_\nu^t A\,\mathrm{d}\xi\right)\mathrm{d}\nu.$$

(39) is also used to estimate (44), yielding

$$(49) \qquad P_i(T, t) \leqslant [\tilde{Y}_i^{(\theta)}(T) + U(T, t)]W(T, t),$$

where

$$W(T, t) = x(T)x^{-1}(t) \exp\left(\int_T^t A \, dv\right).$$

Combining (43), (46), (48), and (49) yields the basic inequality

$$(50) \qquad X_i^{(\theta)}(t) \leqslant 2U(T, t) + [\tilde{Y}_i^{(\theta)}(T) + U(T, t)][1 - P(T, t)],$$

where

$$(51) \qquad P(T, t) = x^{-1}(t) \int_T^t C(v) \exp\left(\int_v^t A \, d\xi\right) dv.$$

Inequality (50) is the basis for all the follows in the proof that $\tilde{Y}_i(\infty) = \theta_i$. Inequality (50) will be applied to the equation

$$(52) \qquad \frac{d}{dt} \tilde{Y}_i = N_i(X_i - \tilde{Y}_i),$$

where

$$N_i(t) = H_{[N(i)](t)}(t).$$

Interpret the derivative in (52) as a left-handed derivative at times when $N(i)$ changes value. Letting $Y_i(T, t) = \tilde{Y}_i(T) - \tilde{Y}_i(t)$, we find by (50) and (52) that

$$\begin{aligned} \frac{\partial}{\partial t} \tilde{Y}_i(T, t) &= N_i(\tilde{Y}_i - X_i) \\ &\geqslant - N_i Y_i(T, t) + \tilde{Y}_i^{(\theta)}(T) N_i P(T, t) \\ &\quad + N_i U(T, t) P(T, t). \end{aligned}$$

Integrate this inequality from $t = T$ to any $t > T$. Then

$$Y_i(T, t) \geqslant \tilde{Y}_i^{(\theta)}(T) R_i(T, t) + S_i(T, t)$$

where

$$(53) \qquad R_i(T, t) = \int_T^t P(v, T) N_i \exp\left(- \int_T^v N_i \, d\xi\right) dv$$

and

$$(54) \qquad S_i(T, t) = \int_T^t U(v, T)\,[P(v, T) - 3]N_i \exp\left(-\int_T^v N_i\, d\xi\right) dv.$$

In other words,

$$(55) \qquad \tilde{Y}_i^{(\theta)}(T)[1 - R\,(T, t)] \geqslant \tilde{Y}_i^{(\theta)}(t) + S_i(T, t).$$

Inequality (55) will now be iterated. Let $\{T_{in}\}$ be an increasing divergent sequence. Choose $T = T_{in}$ and $t = T_{i,n+1}$ and iterate (55). (We suppose for simplicity that T_{i1} is chosen sufficiently large that (55) holds for all $t \geqslant T_{i1}$.) Then iterating the inequality from $k = K$ to $k = n$, we find

$$(56) \qquad \tilde{Y}_i^{(\theta)}(T_{iK}) \prod_{k=K}^{n} [1 - R_i(T_{ik}, T_{i,k+1})] \geqslant \tilde{Y}_i^{(\theta)}(T_{i,n+1}) + G_i(K, n),$$

where

$$(57) \qquad G_i(K, n) = \sum_{k=K}^{n} S_i(T_{ik}, T_{i,k+1}) \prod_{m=k+1}^{n} [1 - R_i(T_{im}, T_{i,m+1})].$$

Since $\tilde{Y}_i^{(\theta)} \geqslant 0$ by hypothesis, (56) will suffice to prove that $\tilde{Y}_i(\infty) = \theta_i$ if it can also be shown that

$$(58) \qquad \prod_{k=1}^{\infty} [1 - R_i(T_{ik}, T_{i,k+1})] = 0$$

and that

$$(59) \qquad \lim_{K\to\infty} \limsup_{n\to\infty} G_i(K, n) = 0.$$

To prove (58), note by (28) and (30) that for every $t \geqslant T$,

$$N_i(t) = \frac{d}{dt} \log\left[z_{N(i)}(T) + \int_T^t E_{N(i)} x \exp\left(-\int_T^t D_{N(i)}\, d\xi\right) dv\right]$$

or alternatively

$$N_i(t) = \frac{d}{dt} \log\left[z_{N(i)}(t) \exp\left(-\int_T^t D_{N(i)}\, dv\right)\right].$$

Thus in (53), for any $\mu \geqslant T$ and $\nu \geqslant T$,

$$\exp\left(-\int_\mu^\nu N_i\, dt\right) = \frac{z_{N(i)}(\mu) \exp(-\int_T^\mu D_{N(i)}\, dt)}{z_{N(i)}(T) + \int_T^\nu E_{N(i)} x \exp(-\int_T^v D_{N(i)}\, d\xi)\, dv}$$

and hence

$$(60)\qquad R_i(T, t) = \frac{L[N(i), f(T, \cdot); T, t]}{z_{N(i)}(T) + L[N(i), x; T, t]} \leqslant 1.$$

Since (58) holds iff

$$\sum_{k=1}^{\infty} R_i(T_{ik}, T_{i, k+1}) = \infty,$$

(58) holds by (15) with $N(i)$ replacing $M(i)$ and $T_{in} = S_{in}$.

To prove (59), it suffices to show that

$$|G(K, n)| \leqslant \Omega U(T_{iK}, T_{i, n+1})$$

for some positive constant Ω. By (57) and (60)

$$G_i(K, n) \leqslant \sum_{k=K}^{n} |S_i(T_{ik}, T_{i, k+1})|.$$

By (54),

$$G_i(K, n) \leqslant \sum_{k=K}^{n} U(T_{ik}, T_{i, k+1}) \cdot$$

$$\int_{T_{ik}}^{T_{i, k+1}} |P(v, t) - 3| N_i \exp\left(-\int_{T_{ik}}^{v} N_i \, d\xi\right) dv.$$

By (51), $P(v, T) - 3$ is bounded. Moreover

$$\int_T^t N_i \exp\left(-\int_T^v N_i \, d\xi\right) dv = 1 - \exp\left(-\int_T^t N_i \, dv\right) \leqslant 1.$$

These estimates complete the proof.

Corollary 2 is proved as follows:

Proof of Corollary 2. (14) follows from (19) and (47) by noting that, for $t \geqslant K_2$.

$$K_1 \leqslant f(T, T + t) \leqslant x(t + T) \int_T^{t+T} G \, dv \leqslant (\sup x) \int_T^{t+T} G \, dv.$$

Thus

$$\int_0^{\infty} G \, dv = \sum_{n=0}^{\infty} \int_{nK_2}^{(n+1)K_2} G \, dv = \infty.$$

We now show that (21) implies (20). Then (20) will be used to prove the remaining assertions. It suffices to show that

$$L[j, x; 0, \infty] < \infty \quad \text{implies} \quad \int_0^\infty E_j \, dt < \infty \tag{61}$$

and that

$$\int_0^\infty B_j \, dt < \infty \quad \text{implies} \quad \int_0^\infty B_j z_j x^{-1} \, dt < \infty. \tag{62}$$

To prove (61), note by (30) that

$$\exp\left(-\int_T^t D_j \, dv\right) \geqslant z_j(T) \, (\sup z_j)^{-1},$$

and by (19) that $x(t) \geqslant K_1$ for $t \geqslant K_2$. To avoid trivialities, we assume that $z_j(T) > 0$ for some $T \geqslant K$. Then

$$L[j, x; 0, t] \geqslant R_j \int_T^t E_j \, dv,$$

where

$$R_j = z_j(T) \, (\sup z_j)^{-1} \, K_1 \exp\left(-\int_0^T D_j \, dv\right).$$

This proves (61). (62) has a similar proof.

(15) and (16) with $N(i)$ and $n(i)$ replacing $M(i)$ and $m(i)$, respectively, are proved using (18) and (19). Consider (15) for definiteness. It suffices to prove that, for any fixed $T \geqslant 0$ and t sufficiently large,

$$\frac{L[N(i), f(T, \cdot); T, t]}{\varepsilon + L[N(i), x; T, t]} \geqslant K_1 \, (2 \sup x)^{-1}. \tag{63}$$

Then define the sequence $\{S_{in}\}$ in (15) iteratively, setting $S_{in} = T$ and $S_{i,n+1} = t$ at each stage, and (15) readily follows. Consider (63). If (63) holds for $T \leqslant t < T + K_2$, we are done. If not, use condition (i) and (19) to conclude for $t \geqslant T + K_2$ that

$$\frac{L[N(i), f(T, \cdot); T, t]}{\varepsilon + L[N(i), x; T, t]} \geqslant \frac{f_i(T) + K_1 (\sup x)^{-1} \, h_i(T, t)}{g_i(T) + h_i(T, t)} \tag{64}$$

where

$$f_i(T) = L[N(i), f(T, \cdot); T, T + K_2],$$
$$g_i(T) = \varepsilon + L[N(i), x; T, T + K_2],$$

and

$$h_i(T, t) = \int_{T+K_2}^{t} E_{N(i)} x \exp\Big(-\int_{T}^{v} D_{N(i)}\, d\xi\Big)\, dv.$$

By (64), it suffices to show that

$$L[N(i), x; 0, \infty] = \infty.$$

which will be proved by contradiction. Namely, we show that

$$(65) \qquad L[N(i), x; 0, \infty] < \infty$$

implies

$$(66) \qquad L[j, x; 0, \infty] < \infty \quad \text{for some} \quad j \in J(1).$$

This contradicts the local flow condition (20).

Assume (65) and let Case (VII) hold for all $t \geqslant 0$ with $\tilde{Y}_i^{(\theta)}(0) > 0$. Then by (52), $(d/dt)\tilde{Y} \geqslant -N_i\tilde{Y}_i$. Thus there exists an $\eta > 0$ such that $\tilde{Y}_i^{(\theta)}(t) \geqslant \eta$ for all $t \geqslant 0$. Moreover, by (19) and (50) for any $T \geqslant 0$ and all $t \geqslant T + K_2$,

$$0 \leqslant X_i^{(\theta)}(t) \leqslant 2U(T) + (1 - \mu)[\tilde{Y}_i^{(\theta)}(T) + U(T)],$$

where $\mu = K_1(\sup x)^{-1}$ and $U(T)$ monotonically approaches zero as $t \to \infty$. In all, there exists a $T_1 \geqslant 0$ and a $\nu \in (0, 1)$ such that

$$(67) \qquad X_i^{(\theta)}(t) \leqslant (1 - \nu)\tilde{Y}_i^{(\theta)}(T)$$

if $T \geqslant T_1$ and $t \geqslant T + K_2$. Since trivially,

$$\tilde{Y}_i^{(\theta)}(t) - X_i^{(\theta)}(t) = \tilde{Y}_i^{(\theta)}(t) - \tilde{Y}_i^{(\theta)}(T) + \tilde{Y}_i^{(\theta)}(T) - X_i^{(\theta)}(t)$$

for any t and T, (67) shows that

$$(68) \qquad \tilde{Y}_i^{(\theta)}(t) - X_i^{(\theta)}(t) \geqslant \nu\tilde{Y}_i^{(\theta)}(T) + \tilde{Y}_i^{(\theta)}(t) - \tilde{Y}_i^{(\theta)}(T)$$

if $T \geqslant T_1$ and $t \geqslant T + K_2$. (68) and the existence of $\tilde{Y}_i(\infty)$ (which follows without involving the hypothesis to be proved) imply the existence of a time T_2 such that

$$\text{(69)} \qquad \tilde{Y}_i^{(\theta)}(t) - X_i^{(\theta)}(t) \geqslant (\nu\eta/2) > 0 \quad \text{for} \quad t \geqslant T_2.$$

(69) will now be shown to be impossible, thereby completing the proof.

By (69) there exists a T_3 such that for $t \geqslant T_3$,

$$\tilde{Y}_i^{(\theta)}(t) \geqslant \tilde{Y}_i^{(\theta)}(\infty) - (\nu\eta/8) > \tilde{Y}_i^{(\theta)}(\infty) - (3\nu\eta/8) \geqslant X_i^{(\theta)}(t).$$

Thus if for any $j \in J(1)$ and any $t \geqslant T_3$, $Z_{ji}^{(\theta)}(t) < \tilde{Y}_i^{(\theta)}(\infty) - (\nu\eta/4)$, then $Z_{ji}^{(\theta)}(t) < \tilde{Y}_i^{(\theta)}(t)$ for all $t \geqslant T_3$. In other words, every $j \in J(1)$ such that $\tilde{Y}_i^{(\theta)}(t) = Z_{ji}^{(\theta)}(t)$ at any $t \geqslant T_3$ satisfies $Z_{ji}^{(\theta)}(t) - X_i^{(\theta)}(t) \geqslant \nu\eta/8$ for all $t \geqslant T_3$. By (25) $\dot{Z}_{ji}^{(\theta)}(t) \leqslant -(\nu\eta/8)H_j(t)$ for $t \geqslant T_3$, and thus (66) holds. This completes the proof.

Now we turn to the

Proof of Proposition 1: Suppose Q_i exists. Then by Lemma 3, all P_{ji} exist. Suppose $Q_i = \theta_i$. Choose $j \in J(2)$ and let Case (VII) hold for all $t \geqslant 0$ with $y_i^{(\theta)}(0) > 0$. Then since $\dot{Z}_{ji}^{(\theta)} \geqslant -H_j Z_{ji}^{(\theta)}$, it follows from $\int_0^\infty E_j \, dt < \infty$ that $P_{ji} > \theta_i$.

Now consider (24). Let $\omega_i = \min(P_{ji} - \theta_i \colon j \in J(2)) > 0$. Then for all sufficiently large t

$$\dot{X}_i^{(\theta)} \geqslant \omega_i \sum_{k \in J(2)} F_k - (F + G) X_i^{(\theta)},$$

where $F = \sum_{k \in J} F_k$. Integrating from $t = T$ to any $t \geqslant T$ yields

$$X_i^{(\theta)}(t) \geqslant P_i(t, T),$$

where

$$P_i(t, T) = \omega_i \sum_{k \in J(2)} \int_T^t F_k \exp\left[-\int_v^t (F + G) \, d\xi\right] dv.$$

It will suffice to show that $\lim_{t \to \infty} \sup P(t, T) > 0$. By (26), (27), and (29),

$$\exp\left[-\int_v^t (F + G) \, d\xi\right] = \frac{x(v)}{x(t)} \exp\left[\int_v^t A \, d\xi\right].$$

Thus

$$P_i(t, T) = \omega_i \, x^{-1}(t) \sum_{k \in J(2)} \int_T^t B_k z_k \exp\left[\int_v^t A \, d\xi\right] dv.$$

But x is bounded. Moreover z_k has a positive lower bound, since by condition (ix) and (19),

$$\dot{z}_k \geqslant E_k(-\gamma_k z_k + x)$$
$$\geqslant E_k(-\gamma_k z_k + K_1)$$

for $t \geqslant K_2$. Thus there exists a $\lambda_i > 0$ such that

$$P(t, T) \geqslant \lambda_i \sum_{k \in J(2)} \int_T^t B_k \exp\left[\int_v^t A \, d\xi\right] dv.$$

Now apply (23) to show that $\lim_{t \to \infty} \sup X_i(t) > \theta_i$. In particular, $Q_i \neq \theta_i$. Moreover, if $\int_0^\infty E_j \, dt = \infty$, then $P_{ji} = Q_i \neq \theta_i$.

UNBIASED LEARNING WITH ARBITRARY POSITIVE AXON WEIGHTS USING CHEMICAL TRANSMISSION AND ACTION POTENTIALS

Let (11) be replaced by

$$\dot{x}_i = Ax_i + \sum_{k \in J} B_k \beta_{ki} z_{ki} + C_i; \tag{70}$$

that is, let the path weights β_{ji} from v_j to v_i be arbitrary positive numbers. Can we transform (12) analogously so that learning and performance of spatial patterns is unimpaired? The answer is yes.

We want the ratios $Z_{ji}^{(\beta)} = \beta_{ji} z_{ji} [\sum_{k \in I} \beta_{jk} z_{jk}]^{-1}$ to converge to θ_i after sufficient practice. This will happen if (12) is replaced by

$$\dot{z}_{ji} = D_j z_{ji} + E_j \beta_{ji}^{-1} x_i, \tag{71}$$

since letting $w_{ji} = \beta_{ji} z_{ji}$, (70) and (71) yield

$$\dot{x}_i = Ax_i + \sum_{k \in J} B_k w_{ki} + C_i$$

and

$$\dot{w}_{ji} = D_j w_{ji} + E_j x_i,$$

which are again of the form (11)–(12).

Our goal could *not* be achieved by replacing (12) with

$$\dot{z}_{ji} = D_j z_{ji} + E_j \beta_{ji} x_i, \tag{72}$$

which would be the natural thing to do if we supposed that $E_j \beta_{ji}$ is determined wholly by spiking frequency. That (72) is inadmissible can be seen by transforming (70) and (72) into pattern variables. Doing this yields an infinite hierarchy of equations in the variables

$$\theta_{j_1 j_2 \cdots j_m i} = \frac{\beta_{j_1 i}\beta_{j_2 i}\cdots\beta_{j_m i}\theta_i}{\sum_{k\in I}\beta_{j_1 k}\beta_{j_2 k}\cdots\beta_{j_m k}\theta_k},$$

$$X_{j_1 j_2 \cdots j_m i} = \frac{\beta_{j_1 i}\beta_{j_2 i}\cdots\beta_{j_m i}x_i}{\sum_{k\in I}\beta_{j_1 k}\beta_{j_2 k}\cdots\beta_{j_m k}x_k},$$

and

$$Z_{j_1 j_2 \cdots j_r i} = \frac{\beta_{j_1 i}\beta_{j_2 i}\cdots\beta_{j_r i}z_{j_r i}}{\sum_{k\in I}\beta_{j_1 k}\beta_{j_2 k}\cdots\beta_{j_r k}z_{j_r k}},$$

where all $j_k \in J$, $m = 0, 1, 2, \ldots$, and $r = 1, 2, 3, \ldots$. These equations have the form

$$\dot{X}_{j_1 j_2 \cdots j_m i} = \sum_{k\in J} F_{j_1 j_2 \cdots j_m k}(Z_{j_1 j_2 \cdots j_m k i} - X_{j_1 j_2 \cdots j_m i}) + G_{j_1 j_2 \cdots j_m}(\theta_{j_1 j_2 \cdots j_m i} - X_{j_1 j_2 \cdots j_m i}) \tag{73}$$

and

$$\dot{Z}_{j_1 j_2 \cdots j_m i} = H_{j_1 j_2 \cdots j_m}(X_{j_1 j_2 \cdots j_m i} - Z_{j_1 j_2 \cdots j_m i}). \tag{74}$$

Note that when X in (73) depends on m values of J, Z in (74) depends on $(m + 1)$ values of J. Thus the hierarchy of equations never ends.

Suppose that we could analyse (73) and (74) and that all X's and Z's had limits θ which were approached with sufficient regularity that all $\dot{X}$'s and $\dot{Z}$'s approached zero as $t \to \infty$. Since all the coefficients F, G, and H are nonnegative, we would expect each term on the right-hand side of (73) and (74) to also approach zero. In particular, we would find that

$$P_{j_1 j_2 \cdots j_m k i} = Q_{j_1 j_2 \cdots j_m i} \tag{75}$$

and

$$\theta_{j_1 j_2 \cdots j_m i} = Q_{j_1 j_2 \cdots j_m i}, \tag{76}$$

from (73), and

$$Q_{j_1 j_2 \cdots j_m i} = P_{j_1 j_2 \cdots j_m i} \tag{77}$$

from (74). Letting $j_1 = j_2 = \cdots = j_m = k = j$, this would mean that

$$\theta_{\underbrace{jj\cdots j}_{m\text{ times}}i} = Q_{\underbrace{jj\cdots j}_{m\text{ times}}i} = P_{\underbrace{jj\cdots j}_{m+1\text{ times}}i} = Q_{\underbrace{jj\cdots j}_{m+1\text{ times}}i} = \theta_{\underbrace{jj\cdots j}_{m+1\text{ times}}i}$$

for every $m \geqslant 0$. In particular

$$\theta_i = \frac{\beta_{ji}^m\theta_i}{\sum_{k\in I}\beta_{jk}^m\theta_k}, \qquad m \geqslant 0. \tag{78}$$

Letting $m \to \infty$ in (78) and defining $\beta_j = \max\{\beta_{ji}: i \in I\}$, we find for any $\theta_i > 0$ that

$$\theta_i = \lim_{m\to\infty} \frac{\beta_{ji}^m \theta_i}{\sum_{k\in I}\beta_{jk}^m\theta_k} = \begin{cases} 0 & \text{if } \beta_{ji} < \beta_j \\ \dfrac{\theta_i}{\sum\{\theta_k: \beta_j = \beta_{jk}\}} & \text{if } \beta_{ji} = \beta_j. \end{cases}$$

In particular, $\sum\{\theta_k: \beta_j = \beta_{jk}\} = 1$, so one can at best learn patterns which are concentrated on the cells v_i with $\beta_{ji} = \beta_j$. These cells have uniformly distributed path weights. If there exists a subset $\tilde{J} \subset J$ such that

$$\bigcap_{j\in\tilde{J}} \{i \in I: \beta_{ji} = \beta_j\} = \varnothing,$$

then no pattern can simultaneously be learned by all the cells $\tilde{J}$. This is a very inflexible system.

How can the β_{ji}'s in (70) and (71) be interpreted? Let $\beta_{ji} = \lambda_j R_{ji}$, where $\lambda_j > 0$ and R_{ji} is the circumference of the cylindrical axon e_{ji}. Let the signal in e_{ji} (e.g., the action potential [16]) propagate along the circumference of the axon to its synaptic knob. Let the signal disperse throughout the cross-sectional area of the knob (e.g., as ionic fluxes [16]). Let local chemical transmitter production in the knob be proportional to the local signal density. Finally, let the effect of the signal of the postsynaptic cell be proportional to the product of local signal density and local available transmitter density and the cross-sectional area of the knob.

These laws generate (70) and (71) as follows. Signal strength is proportional to R_{ji}, or β_{ji}. The cross-sectional area of the knob is proportional to R_{ji}^2. Hence signal density in the knob is proportional to $R_{ji}R_{ji}^{-2} = R_{ji}^{-1}$, or to β_{ji}^{-1}, as in (71). Thus (signal density) × (transmitter density) × (area of knob) $\cong R_{ji}^{-1}z_{ji}R_{ji}^2 = R_{ji}z_{ji} \cong \beta_{ji}z_{ji}$, as in (70).

By contrast, a mechanism whereby signals propagate throughout the cross-sectional area of the axon could not produce unbiased learning given arbitrary axon connection strengths, or at least such a mechanism is still elusive. The difficulty here is that signal strength is proportional to R_{ji}^2, signal density is proportional to one, and local transmitter production rate is then proportional to one. The post synaptic signal is proportional to (signal density) × (transmitter density) × (area of knob) $\cong \beta_{ji}^2 z_{ji}$. We are led to the system

$$\dot{x}_i = Ax_i + \sum_{k\in J} B_k\beta_{ki}^2 z_{ki} + C_i$$

and

$$\dot{z}_{ji} = D_j z_{ji} + E_j x_i,$$

which can be written as

$$\dot{x}_i = Ax_i + \sum_{k \in J} B_k \beta_{ki} w_{ki} + C_i$$

and

$$\dot{w}_{ji} = D_j w_{ji} + E_j \beta_{ji} x_i$$

in terms of the variables $w_{ji} = \beta_{ji} z_{ji}$. As in (70) and (72), this system yields an infinite hierarchy of equations for the pattern variables.

These observations suggest that the action potential not only guarantees faithful signal transmission over long cellular distances, as is well known, but also executes a subtle transformation of signal densities into transmitter production rates that compensates for differences in axon diameter. Note also that this transformation seems to require the chemical transmitter step. Purely electrical synapses presumably could not execute it. Thus our laws for transmitter production not only guarantee that learning occurs, but also that unbiased learning occurs, under very weak anatomical constraints.

REFERENCES

[1] Grossberg, S., 'Embedding Fields: Underlying Philosophy, Mathematics, and Applications to Psychology, Physiology, and Anatomy', *J. of Cybernetics* **1** (1971), 28.

[2] Grossberg, S., 'Neural Pattern Discrimination', *J. of Theoretical Biology* **27** (1970), 291.

[3] Grossberg, S., 'Neural Expectation: Cerebellar and Retinal Analogs of Learnable and Unlearned Pattern Classes', *Kybernetik* **10** (1972), 49.

[4] Grossberg, S., 'On the Dynamics of Operant Conditioning', *J. of Theoretical Biology* **33** (1971), 225.

[5] Grossberg, S., 'On the Serial Learning of Lists', *Math. Biosci.* **4** (1969), 201.

[6] Grossberg, S. and J. Pepe, 'Spiking Threshold and Overarousal Effects in Serial Learning', *J. of Statistical Physics* **3** (1971), 95.

[7] Grossberg, S., 'Pavlovian Pattern Learning by Nonlinear Neural Networks', *Proc. of the Natl. Acad. of Science, U.S.A.* **68** (1971), 68.

[8] Grossberg, S., 'On Learning and Energy-Entropy Dependence in Recurrent and Nonrecurrent Signed Network', *J. of Statistical Physics* **1** (1969), 319.

[9] Grossberg, S., 'A Neural Theory of Punishment and Avoidance, I. Qualitative Theory', *Math. Biosci.*, in press.

[10] Grossberg, S., 'A Neural Theory of Punishment and Avoidance, II. Quantitative Theory', *Math. Biosci.*, in press.
[11] Kimble, G. A., *Foundations of Conditions and Learning*, Appleton-Century-Crafts, New York, 1967, p. 26.
[12] Grossberg, S., 'Some Networks That Can Learn, Remember, and Reproduce any Number of Complicated Patterns, II', *Studies in Applied Math.* **XLIX** (1970), 137.
[13] Grossberg. S., 'A Prediction Theory for some Nonlinear Functional-Differential Equations, I', *J. Math. Anal. and Applics.* **21** (1968), 643.
[14] Grossberg, S., 'A Prediction Theory for Some Nonlinear Functional-Differential Equations, II', *J. Math. Anal. and Applics.* **22** (1968), 490.
[15] Grossberg, S., 'On the Global Limits and Oscillations of a System of Nonlinear Differential Equations Describing a Flow on a Probabilistic Network', *J. Diff. Eqns.* **5** (1969), 531.
[16] Ruch, T. C., H. D. Patton, J. W. Woodbury, and A. L. Towe, *Neurophysiology*, W. B. Saunders, Philadelphia, 1961.

CHAPTER 5

A NEURAL THEORY OF PUNISHMENT AND AVOIDANCE, II: QUANTITATIVE THEORY

PREFACE

This article continues where section VIII of Chapter 4 leaves off. The article is filled with psychopharmacological predictions, some of which have since been confirmed. Others still need to be tested. The article follows the conceptual path forced by the synchronization problem. It leads to a reinforcement theory whose network realizations define explicit drive, reinforcement, and incentive motivational concepts.

Few areas of psychology have generated so much divisiveness and confusion as instrumental learning. Even my simple networks suggest why this has been true: Several processes take place simultaneously on different spatial and temporal scales. My theory penetrated where Estes and Neal Miller could not, despite their awesome intuitive powers, because its real-time concepts are powerful enough to tease out some of these interactions. In particular, the theory automatically side-steps the pitfalls of a drive reduction theory, shows how simple antagonistic rebound ideas harmonize drive reduction paradoxes, and easily explains such phenomena as superconditioning, self-stimulation, vicious circle behavior, and learned helplessness. The Rescorla-Wagner model, which also appeared in 1972, explains some of these phenomena, but it is grounded in formal rather than physical variables that do not include my model's most important operations.

The core of this article is its introduction and analysis of a new *gated dipole* model. This model joins together physiological ideas about nonspecific arousal, chemical gates, and competition with psychological ideas about drives, reinforcement, and incentive motivation to suggest how reward and punishment work in real-time. For example, the dipole's antagonistic rebound from fear to relief upon shock offset provides the motivational substrate for learned avoidance behavior in the model.

The gated dipole model still seems to me like a series of minor mathematical miracles. The transmitter laws which were derived in Chapters 2 and 3 are just the right laws to cause antagonistic rebounds. The nonspecific arousal level that energizes a rebound at the offset of a specific cue can also cause a rebound, all by itself, if it is rapidly increased while a specific cue is on. This purely mathematical insight forced me to realize that unexpected events, and cognitive events generally, that trigger an arousal burst can directly influence the magnitude of reinforcement. An abnormal choice of arousal level can cause two types of emotional depression. In underaroused depression, the behavioral threshold is abnormally high, but the system is hyperactive at suprathreshold values. The threshold arousal increment that can trigger an antagonistic rebound is elevated, and the sudden halving of a phasic input might cause no rebound whatsoever. I believe that Parkinson's patients, hyperactive children and hyperphagic rats are all underaroused in this sense. In overaroused depression, the behavioral threshold is abnormally low, but the system is so hypoactive that it doesn't matter. I believe that certain schizophrenic patients are overaroused in this sense, and that certain analgesic agents work by causing an overaroused syndrome. The theory also predicts how several normal learning indices should covary as the arousal level is parametrically increased. All of these predictions are made in terms of interactions that are hypothesized to occur between two transmitter systems (cholinergic and catecholaminergic). During the last few years, cholinergic-catecholaminergic interactions have been intensively studied by neuropharmacologists and psychophysiologists. I therefore hope that some of these predictions will be tested soon.

A Neural Theory of Punishment and Avoidance, II: Quantitative Theory

Communicated by Richard Bellman

ABSTRACT

Quantitative neural networks are derived from psychological postulates about punishment and avoidance. The classical notion that drive reduction is reinforcing is replaced by a precise physiological alternative akin to Miller's "Go" mechanism and Estes's "amplifier" elements. Cell clusters $\mathscr{A}_f^+$ and $\mathscr{A}_f^-$ are introduced which supply negative and positive incentive motivation, respectively, for classical conditioning of sensory-motor acts. The $\mathscr{A}_f^+$ cells are persistently turned on by shock (on-cells). The $\mathscr{A}_f^-$ cells are transiently turned on by shock termination (off-cells). The rebound from $\mathscr{A}_f^+$ cell activation to $\mathscr{A}_f^-$ cell activation replaces drive reduction in the case of shock. Classical conditioning from sensory cells $\mathscr{S}$ to the pattern of activity playing on arousal cells $\mathscr{A}_f = (\mathscr{A}_f^+, \mathscr{A}_f^-)$ can occur. Sufficiently positive net feedback from $\mathscr{A}_f$ to $\mathscr{S}$ can release sampling, and subsequent learning, by prescribed cells in $\mathscr{S}$ of motor output controls. Once sampled, these controls can be reactivated by $\mathscr{S}$ on recall trials. This concept avoids some difficulties of two-factor theories of punishment and avoidance.

Recent psychophysiological data and concepts are analyzed in terms of network analogs, and some predictions are made. The rebound from $\mathscr{A}_f^+$ cell activation to $\mathscr{A}_f^-$ cell activation at shock termination is interpreted to be a consequence of different rates of transmitter accumulation – depletion in the parallel neural channels associated with $\mathscr{A}_f^+$ and $\mathscr{A}_f^-$. This interpretation culminates in an analogy with adrenergic and cholinergic interactions at lateral and ventromedial hypothalamic sites, dependent on phasic sensory input and tonic reticular formation input. Mechanisms are suggested for such phenomena as: the lesser rewarding effect of reducing J units of shock to $J/2$ units than of reducing $J/2$ units to 0 units; a relationship between the rewarding effect of reducing J units of shock to $J/2$ units and the possibility of releasing a conditioned avoidance response in the presence of fearful cues; two kinds of depressed emotional affect, one due to overarousal, that can also be associated with massive associational confusions and poor paying attention, and one due to underarousal, that can also be associated with overreactive fear and relief responses; persistent nonspecific fear that biases interpretation of specific cues, and can "resist" new learning or "repress" old learning; different effects of gradual and abrupt shock on response suppression; response generalization from one shock level to another; reduction of pain in the presence of loud noise (analgesia); influences of drugs, such as carbachol, atropine, and scopolamine on conditioned emotional and avoidance responses, and on self-stimulation via implanted hypothalamic electrodes; sensory-drive heterarchy that allows changes in situational cues to release responses compatible with any of several nonprepotent drives; feedback inhibition of adrenergic transmitter production; potentiation of adrenergic production by presynaptic spiking, and by postsynaptic spiking via a feedback loop that controls higher-order instrumental conditioning; learning at cholinergic synapses.

---

1. INTRODUCTION

Part I of this article [29] derived qualitative neural mechanisms of punishment and avoidance from psychological postulates, and analyzed recent psychological data using these mechanisms. This part of the article derives quantitative versions of these mechanisms. Psychophysiological data will be discussed using the quantitative results, and some predictions will be made. These include:

1. a relationship between the reinforcing effect of reducing J units of shock to $J/2$ units of shock, and the possibility of eliciting a conditioned avoidance response in the presence of fearful cues;

2. a relationship between higher-order instrumental conditioning, and postsynaptic effects on presynaptic norepinephrine production via an anatomical feedback loop, which is possibly a formal analog of medial forebrain bundle pathways;

3. a formula to decide when reducing J_1 units of shock to K_1 units of shock in the presence of I_1 units of arousal is more reinforcing than reducing J_2 units of shock to K_2 units in the presence of I_2 units of arousal, where one arousal mechanism is a possible formal analog of reticular formation inputs;

4. a mechanism for two kinds of depressed emotional affect. The first kind, due to overarousal, is stable with respect to psychological inputs; the network is "indifferent" to emotionally charged cues. This effect can be associated with poor paying attention, associational confusions, and punning behavior. The second kind of depressed affect, due to underarousal, is an unstable form of depression in the sense that, after the system's elevated thresholds are exceeded by external cues, either aversive or rewarding cues can cause overreactive fear or relief responses; network response is emotionally "irritable";

5. a relationship between administration of carbachol (an acetylcholine mimicker) at lateral and ventromedial hypothalamic sites, and the spatial distribution of fearful and rewarding cues in the experimental chamber, to tell if carbachol will enhance or depress conditioned avoidance learning.

2. BRIEF REVIEW

Part I derives neural networks whose anatomy is of the type depicted in Fig. 1, where (Fig. 1a) the ith conditioned stimulus (CS_i) among n

possible stimuli excites state U_{i1} of its sensory representation. Sensory representations will be denoted generically by $\mathscr{S}$. In response to the CS_i input, U_{i1} sends signals to stage U_{i2} as well as to arousal cell clusters $\mathscr{A} = (\mathscr{A}_h, \mathscr{A}_f^+, \mathscr{A}_f^-, \ldots)$. The arousal cells $\mathscr{A}_h$ subserve hunger, $\mathscr{A}_f^+$ subserve fear, and $\mathscr{A}_f^-$ subserve relief from fear. The cells $\mathscr{A}_h$ receive a drive input that is a monotone increasing function of hunger level, and $\mathscr{A}_f^+$ receives an input that is a monotone increasing function of shock level. Offset of shock elicits a transient excitatory response in the relief

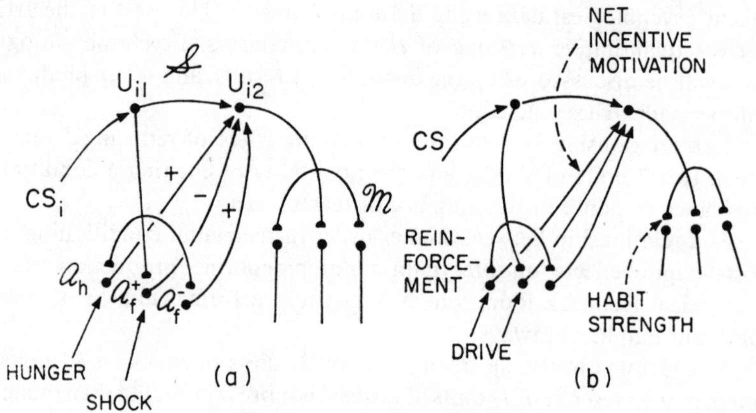

FIG. 1. Summary of network components.

center $\mathscr{A}_f^-$. Inputs from $\mathscr{A}_h$ and $\mathscr{A}_f^-$ to U_{i2} are excitatory, whereas the input from $\mathscr{A}_f^+$ to U_{i2} is inhibitory. U_{i2} can send signals to $\mathscr{M}$ only if it simultaneously receives a large signal from U_{i1} and a large excitatory signal from $\mathscr{A}$. In particular, a large excitatory $\mathscr{A}_h \rightarrow U_{i2}$ signal can be canceled by a large inhibitory $\mathscr{A}_f^+ \rightarrow U_{i2}$ signal, which thus prevents U_{i2} from firing. Moreover, if shock is terminated by an avoidance response (AR), $\mathscr{A}_f^-$ is excited and creates a large excitatory $\mathscr{A}_f^- \rightarrow \mathscr{S}$ input to all sensory representations. Feedback cues of the AR also excite particular sensory representations, which we denote by $\mathscr{S}$(AR). The U_{i2} stages of $\mathscr{S}$(AR) cells thus receive U_{i1} and $\mathscr{A}_f^-$ inputs. They can therefore fire. Cells that receive only the $\mathscr{A}_f^-$ input cannot fire.

Both $U_{i1} \rightarrow \mathscr{A}$ synaptic knobs and $U_{i2} \rightarrow \mathscr{M}$ synaptic knobs can learn ("sample") patterns of activity playing on $\mathscr{A}$ or $\mathscr{M}$, respectively, when these knobs are activated by U_{i1} or U_{i2} signals. The $U_{i1} \rightarrow \mathscr{A}$ synapses learn "motivational" patterns; the $U_{i2} \rightarrow \mathscr{M}$ synapses learn "motor" patterns playing on the motor control cells $\mathscr{M}$. Figure 1b describes the network components in more conventional psychological terms. Part I should be consulted for details.

The task of this article is to construct explicit networks for the rebound mechanism that transiently turns on the "relief" cells $\mathscr{A}_f^-$ when the shock input to the "fear" cells $\mathscr{A}_f^+$ is turned off.

Physiological laws for these networks have previously been derived [22, 26]. In their simplest form, these laws are

$$\dot{x}_i(t) = -\alpha_i x_i(t) + \sum_{k=1}^{n} [x_k(t - \tau_{ki}) - \Gamma_{ki}]^+ \beta_{ki} z_{ki}(t) - \sum_{k=1}^{n} [x_k(t - \sigma_{ki}) - \Omega_{ki}]^+ \gamma_{ki} + I_i(t) \tag{1}$$

and

$$\dot{z}_{jk}(t) = -\delta_{jk} z_{jk}(t) + \varepsilon_{jk}[x_j(t - \tau_{jk}) - \Gamma_{jk}]^+ x_k(t), \tag{2}$$

where $i, j, k = 1, 2, \ldots, n$ and $[\omega]^+ = \max(\omega, 0)$ for any real number ω. These laws describe interactions of stimulus traces (cell potentials) $x_i(t)$ at the cells (or cell clusters) v_i with memory traces (excitatory transmitter substances) $z_{jk}(t)$ at the synaptic knobs (or synaptic knob clusters) N_{jk} via the axons e_{jk}. The external event r_i excites the stimulus trace $x_i(t)$ via the input $I_i(t)$ (see Fig. 2). These equations are refined in [24]. The need for

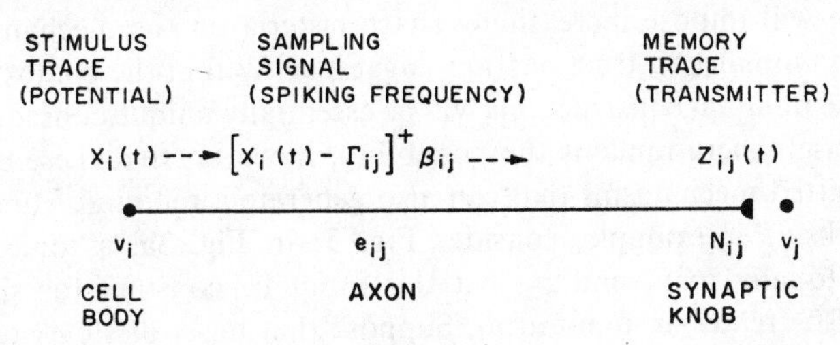

FIG. 2. Psychophysiological interpretation of network variables.

this refinement is seen by considering the terms $T_{ki}(t) = [x_k(t - \tau_{ki}) - \Gamma_{ki}]^+ \beta_{ki} z_{ki}(t)$ in (1). T_{ki} describes the rate of transmitter release from N_{ki}. The term $[x_k(t - \tau_{ki}) - \Gamma_{ki}]^+ \beta_{ki}$ is proportional to the spiking frequency reaching N_{ki} in the time interval $[t, t + dt]$, and $z_{ki}(t)$ is proportional to the amount of available transmitter in N_{ki} during $[t, t + dt]$. Given this interpretation of T_{ki}, why does not (2), which describes transmitter accumulation rate in N_{jk}, include an extra term $-T_{jk}(t)$ to describe release of transmitter? Such a term would destroy the memory of past pairings of the events r_j and r_k. This dilemma is resolved by using two variables $z_{jk}(t)$ and $Z_{jk}(t)$ instead of the single variable $z_{jk}(t)$. The new $z_{jk}(t)$ describes the rate of transmitter production. $Z_{jk}(t)$ describes the amount of available transmitter. The fact that $z_{jk}(t)$ and $Z_{jk}(t)$ can be lumped together in (1) and (2) means that $Z_{jk}(t)$ seeks a level proportional to $z_{jk}(t)$ that can transiently be reduced by transmitter release. Thus (1) is replaced by

$$\dot{x}_i(t) = -\alpha_i x_i(t) + \sum_{k=1}^{n} [x_k(t - \tau_{ki}) - \Gamma_{ki}]^+ \beta_{ki} Z_{ki}(t) + I_i(t) \tag{3}$$

and

$$\dot{Z}_{jk}(t) = \zeta_{jk}(\eta_{jk} z_{jk}(t) - Z_{jk}(t)) - \kappa_{jk}[x_j(t - \tau_{jk}) - \Gamma_{jk}]^+ Z_{jk}(t). \tag{4}$$

Note that $Z_{jk}(t)$ replaces $z_{jk}(t)$ in (3). Equation (4) describes the net effect of three processes on rate of transmitter production: (i) transmitter production rate is proportional to $z_{jk}(t)$, as in the term $\zeta_{jk}\eta_{jk}z_{jk}(t)$; (ii) production rate is decreased by feedback inhibition proportional to the amount of transmitter, via the term $-\zeta_{jk}Z_{jk}(t)$; and (iii) amount of transmitter is reduced at a rate $\kappa_{jk}[x_j(t - \tau_{jk}) - \Gamma_{jk}]^+Z_{jk}(t)$ by transmitter release. When no release occurs, $Z_{jk}(t)$ rapidly seeks the level $\eta_{jk}z_{jk}(t)$, proportional to $z_{jk}(t)$, as required by (1) and (2). In cases where no learning occurs, $\gamma_{jk} = \delta_{jk} = 0$ in (2); hence Z_{jk} seeks a constant level $\zeta_{jk}z_{jk}(0)$ in the absence of transmitter release. Some finer transients in transmitter production and release are also considered in [24].

3. REBOUND MECHANISM: IS IT UNIQUE?

How can a rebound mechanism be constructed in the context of Eqs. 2–4? We will impose increasingly sharp criteria on the mechanism until definite minimal constructions are suggested. Within the context of Eqs. 2–4, these minimal constructions will be essentially unique consequences of our criteria. There remains the possibility, however, that these equations have omitted mechanisms that can also generate a rebound from on-cells to off-cells. For example, consider Fig. 3: in Fig. 3a, a tonic input is applied equally to v_1 and v_2. Let this input be constant for simplicity. Shock also creates an input at v_1. Suppose that onset of shock creates an excitatory overshoot in x_1 and that offset of shock creates an inhibitory undershoot of x_1, as in Fig. 3b. After signals from v_1 and v_2 compete subtractively on their way to v_3 and v_4 in Fig. 1a (that is, form a subtractive on–off field), the net input to x_3 (the fear channel) is persistently turned on by shock, and the net input to x_4 (the relief channel) is transiently turned on by the inhibitory undershoot of x_1. Such overshoots and undershoots can be due (say) to ionic fluxes that can exist in a description of cell membrane dynamics which is finer than Eqs. 2–4. A rebound mechanism of this type is not chosen primarily because (i) such ionic rebounds are typically faster than the rebound process needed here. A rebound is needed that lasts on the order of seconds rather than milliseconds to enable conditioning in $U_{i1} \to \mathscr{A}$ and $U_{i2} \to \mathscr{M}$ channels to occur.

(ii) Also, the size of x_1's inhibitory undershoot should increase as a function of shock intensity, corresponding to data on the greater rewarding effects of terminating more intense shocks. In particular, the range of

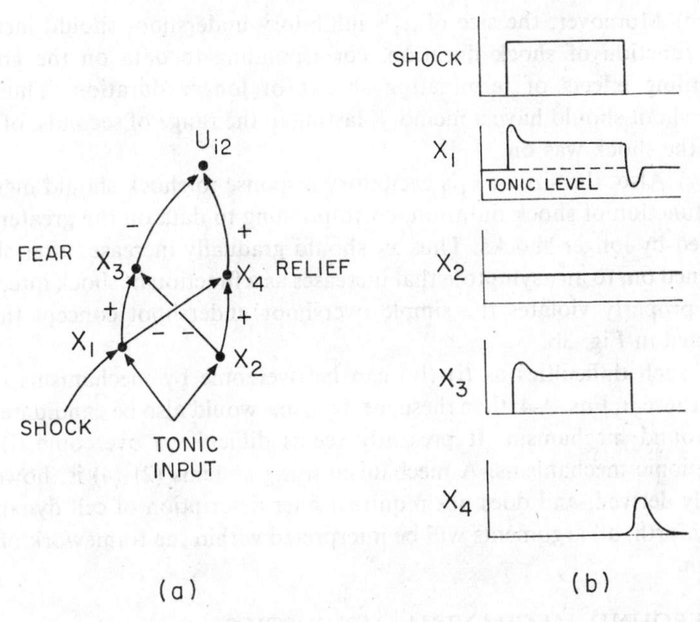

FIG. 3. Possible way to create relief after fearful shock terminates.

x_1's undershoot magnitudes as a function of shock intensity should be commensurate with the range of x_1's steady-state responses as a function of shock intensity (see Fig. 4); otherwise the parametric changes in undershoot will not generate significant changes in conditioned responding.

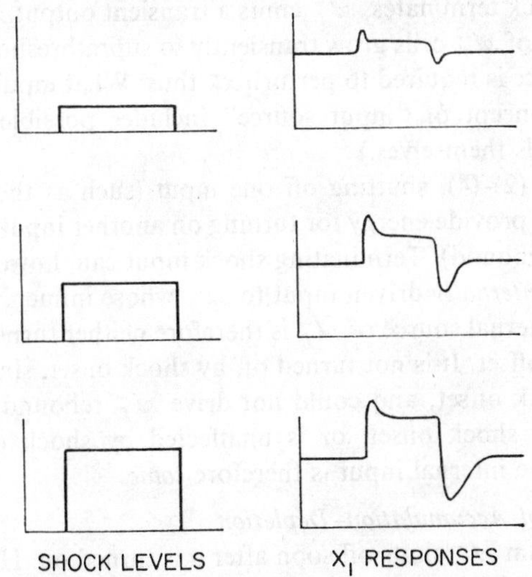

FIG. 4. Parametric dependence of potential response on shock intensity.

(iii) Moreover, the size of x_1's inhibitory undershoot should increase as a function of shock duration, corresponding to data on the greater rewarding effects of terminating shocks of longer duration. Thus the undershoot should have a memory, lasting in the range of seconds, of how long the shock was on.

(iv) Also, the size of x_1's excitatory response to shock should increase as a function of shock duration, corresponding to data on the greater fear created by longer shocks. Thus x_1 should gradually increase, after shock is turned on, to an asymptote that increases as a function of shock intensity. This property violates the simple overshoot–undershoot concept that is depicted in Fig. 3b.

If such difficulties as (i)–(iv) can be overcome by mechanisms other than those in Eqs. 2–4, then these mechanisms would also be candidates for a rebound mechanism. It presently seems difficult to overcome (i)–(iv) using ionic mechanisms. A mechanism using systems (2)–(4) is, however, readily derived, and does not require a finer description of cell dynamics. Henceforth, all arguments will be interpreted within the framework of this system.

4. REBOUND MECHANISM: HEURISTICS

The argument leading to the rebound mechanism falls into eight main stages. Some readers might prefer studying the mechanism in Section 5 before reading this section.

A. Existence of a Tonic Input

When shock terminates, $\mathscr{A}_f^-$ emits a transient output. Thus, by (3), the potentials x_f^- of $\mathscr{A}_f^-$ cells grow transiently to suprathreshold values. In (3), an input source is required to perturb x_f^- thus. What input source does the job? (The concept of "input source" includes possible energy sources within the cells themselves.)

In system (2)–(4), shutting off one input (such as the shock input to $\mathscr{A}_f^+$) does not provide energy for turning on another input (such as the one driving $\mathscr{A}_f^-$ rebound). Terminating shock input can, however, unmask the effects of an *internally* driven input to $\mathscr{A}_f^-$ whose influence is inhibited by shock. The internal source of $\mathscr{A}_f^-$ is therefore neither turned on nor turned off by shock offset. It is not turned off by shock onset, since then it would be off at shock offset, and could not drive $\mathscr{A}_f^-$ rebound. Finally, if it is turned on by shock onset, or is unaffected by shock onset, then it is always on. The internal input is therefore *tonic*.

B. Existence of Accumulation–Depletion

Output from $\mathscr{A}_f^-$ shuts off soon after it is turned on. How is this done? No externally driven input is available to do this. The $\mathscr{A}_f^-$ output is depleted by its own activity. In other words, while shock is on, an accumulation process occurs at $\mathscr{A}_f^-$. When shock is off, output from $\mathscr{A}_f^-$ is a monotone increasing function of the amount accumulated at each given time. This amount is gradually depleted when shock is off, until the $\mathscr{A}_f^-$ output vanishes.

C. Consensus between Fear and Relief

We suppose that at most one of the outputs from $\mathscr{A}_f^+$ and $\mathscr{A}_f^-$ is nonzero at any time. In other words, either fear or relief, but not both, can be "perceived" by the network at a given time. Thus the final state of processing in $\mathscr{A}_f^+$ and $\mathscr{A}_f^-$, before signals are sent to $\mathscr{S}$, is the resultant of a competition between the $\mathscr{A}_f^+$ and $\mathscr{A}_f^-$ channels due to some form of mutual inhibition, much as in Fig. 3a.

D. Existence of a Parallel Accumulation Process in the Fear Channel

When shock is off for a long time, outputs from both $\mathscr{A}_f^+$ and $\mathscr{A}_f^-$ to $\mathscr{S}$ are zero. Thus the accumulation process at $\mathscr{A}_f^-$, driven by its tonic input, is balanced by a process going on at $\mathscr{A}_f^+$. The simplest idea is that a parallel process of accumulation–depletion, driven by its own tonic input that equals the $\mathscr{A}_f^-$ input, takes place in the $\mathscr{A}_f^+$ channel. When shock is on, the shock input summates with the tonic input in the $\mathscr{A}_f^+$ channel.

This idea is strengthened by the next few arguments, which elucidate the basic question: what accumulates? Is it potential or is it transmitter? Several facts favor the latter alternative.

E. The Rebound is Slow

It lasts at least seconds rather than milliseconds. It is a slow process compared to known fluctuation rates of cell potentials in response to input changes. After shock terminates, $\mathscr{A}_f^+$ and $\mathscr{A}_f^-$ receive no externally driven inputs. Their potentials presumably rapidly equalize. Output from $\mathscr{A}_f^-$ nonetheless continues. Thus there exists a process slower than potential change that can bias output from $\mathscr{A}_f^+$ and $\mathscr{A}_f^-$ in favor of $\mathscr{A}_f^-$ after shock terminates.

F. Both Fear and Relief Are Increasing Functions of Shock Duration and Intensity

Data on the effect of CS and UCS intensity [2, 8, 17, 35–37, 45, 52] and duration [2, 6, 9, 14, 38, 58] on the CER and CAR have been reported. Thus both channels contain slowly varying processes which parametrically depend on shock intensity and duration, and which counterbalance each other when shock is off for long intervals.

G. The Relative Balance of Accumulation Is Changed by Shock

What causes the $\mathscr{A}_f^-$ rebound to shut itself off? Is complete depletion of the accumulated product at $\mathscr{A}_f^-$ responsible for this? Suppose "yes."

Then the tonic input alone can deplete $\mathscr{A}_f^-$. By symmetry, during shock, the shock input plus the tonic input to $\mathscr{A}_f^+$ could surely deplete $\mathscr{A}_f^+$. This does not occur, since then fear could not be maintained by a prolonged shock. A weaker conclusion is necessary: shock shifts the relative balance of accumulation in the two channels, by depleting the $\mathscr{A}_f^+$ channel more than the $\mathscr{A}_f^-$ channel.

H. Signal Size Is a Joint Function of Input Size and Amount Accumulated

This argument is crucial. During $\mathscr{A}_f^-$ rebound, both $\mathscr{A}_f^+$ and $\mathscr{A}_f^-$ receive equal tonic inputs which ultimately balance the amounts accumulated at $\mathscr{A}_f^+$ and $\mathscr{A}_f^-$, and thereby nullify $\mathscr{A}_f^+$ and $\mathscr{A}_f^-$ signals to $\mathscr{S}$. Before this happens. $\mathscr{A}_f^-$ output exceeds $\mathscr{A}_f^+$ output because $\mathscr{A}_f^-$ accumulation exceeds $\mathscr{A}_f^+$ accumulation. In other words, given a fixed input size (the equal tonic inputs to $\mathscr{A}_f^+$ and $\mathscr{A}_f^-$), output is an increasing function of accumulation level (in the two channels, $\mathscr{A}_f^+$ and $\mathscr{A}_f^-$).

When shock is on, increasing shock intensity increases $\mathscr{A}_f^+$ output, since it causes an increase in fear. Increasing shock intensity also *decreases* the amount accumulated at $\mathscr{A}_f^+$; this is the basis of the rebound at $\mathscr{A}_f^-$ when shock is turned off. Thus, output is not a function of accumulation level alone, since then increasing shock intensity would decrease $\mathscr{A}_f^+$ output by decreasing the amount accumulated at $\mathscr{A}_f^+$. Output size is a joint function of input size and accumulation level.

By Eq. 3, output size is the product of spiking frequency and transmitter level. Spiking frequency is an increasing function of potential, which is an increasing function of input size. This leaves transmitter accumulation level as the abstract accumulation level discussed earlier. This argument commits us to our formalism. We could not proceed further unless (i) the amount of accumulated transmitter were a decreasing function of input size; and (ii) output size were nonetheless an increasing function of input size. Fortunately, both (i) and (ii) are true in system (2)–(4), and make our construction of the rebound mechanism possible in this context.

5. REBOUND MECHANISM: NONRECURRENT CASE

The minimal nonrecurrent (that is, feed-forward) embedding field compatible with Section 4 is defined as follows. Odd (even) subscripts denote cell sites associated with $\mathscr{A}_f^+(\mathscr{A}_f^-)$. Let

(a) v_1 and v_2 be the first stage of input processing;
(b) v_3 and v_4 be the second stage of processing;
(c) v_5 and v_6 be the third stage of processing;
(d) x_i be the potential of v_i, $i = 1, 2, \ldots, 6$;
(e) z_1 and z_2 be the transmitters in N_{13} and N_{24}, respectively;
(f) $I(t)$ be the tonic internal input to v_1 and v_2;

(g) $J(t)$ be the phasic aversive input to v_1;
(h) O_i be the output of v_i, $\quad i = 5, 6.$

The following process occurs in the network of Fig. 5.

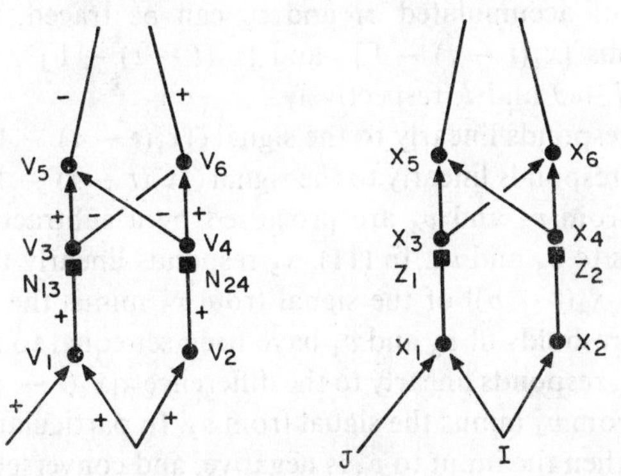

FIG. 5. Minimal nonrecurrent rebound mechanism using transmitter accumulation-depletion.

$$\dot{x}_1 = -\alpha x_1 + I + J, \tag{5}$$

$$\dot{x}_2 = -\alpha x_2 + I, \tag{6}$$

$$\dot{z}_1 = \beta(\gamma - z_1) - \delta[x_1(t-\tau) - \Gamma]^+ z_1, \tag{7}$$

$$\dot{z}_2 = \beta(\gamma - z_2) - \delta[x_2(t-\tau) - \Gamma]^+ z_2, \tag{8}$$

$$\dot{x}_3 = -\varepsilon x_3 + \zeta[x_1(t-\tau) - \Gamma]^+ z_1, \tag{9}$$

$$\dot{x}_4 = -\varepsilon x_4 + \zeta[x_1(t-\tau) - \Gamma]^+ z_2, \tag{10}$$

$$\dot{x}_5 = -\eta x_5 + \kappa[x_3(t-\sigma) - x_4(t-\sigma)], \tag{11}$$

$$\dot{x}_6 = -\eta x_6 + \kappa[x_4(t-\sigma) - x_3(t-\sigma)], \tag{12}$$

$$O_5 = \lambda[x_5 - \Omega]^+, \tag{13}$$

and

$$O_6 = \lambda[x_6 - \Omega]^+. \tag{14}$$

In (5), x_1 responds linearly to the sum $I + J$ of tonic and phasic inputs. In (6), x_2 responds linearly to the tonic input I alone. This asymmetry in inputs drives the outputs O_5 and O_6. The transmitter rules in (7) and (8) are of the form defined by (4) if no learning is possible. As in (4), the rate of transmitter change is controlled by three processes. For example, in (7), transmitter is produced at rate $\beta\gamma$. Feedback inhibition of transmitter production occurs at rate $-\beta z_1$. It is due to the action of transmitter, or a transmitter-activated substance, at an intermediate stage of transmitter production. Transmitter release from N_{13} occurs at rate $\delta[x_1(t-\tau) - \Gamma]^+ z_1$, where $[x_1(t-\tau) - \Gamma]^+$ is proportional to the spiking frequency from v_1 that reaches N_{13} at time t. When no spikes reach N_{13} in a given interval, z_1 exponentially approaches a saturation level of γ at rate β. When spikes do reach N_{13}, z_1 is depleted at a rate that increases linearly with the spiking frequency. The same process determines accumulation and depletion of z_2 in N_{24}, except v_2 sends spikes to N_{24}. Thus asymmetries in the amounts of accumulated z_1 and z_2 can be traced, via the spiking frequency terms $[x_1(t - \tau) - \Gamma]^+$ and $[x_2(t - \tau) - \Gamma]^+$, to asymmetries in the inputs $I + J$ and I, respectively.

In (9), x_3 responds linearly to the signal $\zeta[x_1(t - \tau) - \Gamma]^+ z_1$ from N_{13}, and in (10) x_4 responds linearly to the signal $\zeta[x_2(t - \tau) - \Gamma]^+ z_2$ from N_{24}. The outputs from v_3 and v_4 are processed by a subtractive on–off field yielding inputs to v_5 and v_6. In (11), x_5 responds linearly to the difference $\kappa[x_3(t - \sigma) - x_4(t - \sigma)]$ of the signal from v_3 minus the signal from v_4. The output thresholds of v_3 and v_4 have been set equal to zero for simplicity. In (12), x_6 responds linearly to the difference $\kappa[x_4(t - \sigma) - x_3(t - \sigma)]$ of the signal from v_4 minus the signal from v_3. In particular, if the input to v_5 is positive, then the input to v_6 is negative, and conversely. This guarantees Property C of Section 4. The outputs (spiking frequencies) O_5 and O_6 from v_5 and v_6 in (13) and (14) are linear functions of the potentials x_5 and x_6 above the common spiking threshold Ω.

Equations 5–14 can be explicitly integrated step by step. Three phases of network activity are considered: (i) *Before shock*: Let $J(t) = 0$ for $t \leqslant 0$. (ii) *During shock*: Fix $J(t)$ at a positive value J for $0 \leqslant t < T$. (iii) *After shock*: Let $J(t) = 0$ for $t \geqslant T$. The following constraints are imposed. (A) Inputs I and J vary slowly relative to the fluctuation rate of potentials, except when J is switched on and off. (B) I is chosen sufficiently large to fire spikes along e_{13} and e_{24} even if $J = 0$; positive J biases the pattern of firing. (C) Potentials adjust to input changes faster than transmitters accumulate. (D) T is sufficiently large to let transmitters adjust to the influence of positive J, and is large compared to the time lags τ and σ.

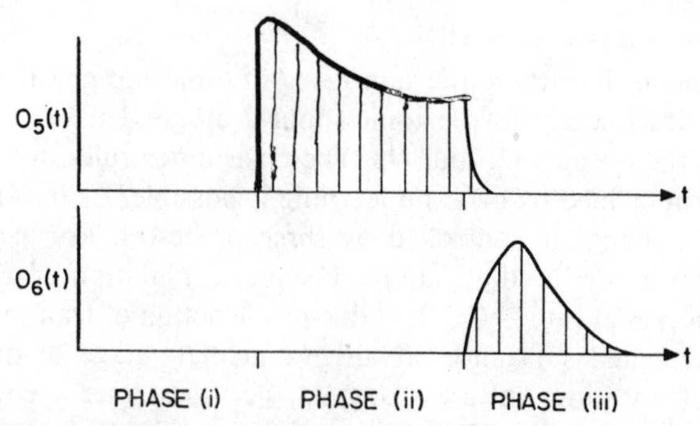

FIG. 6. Idealized persistent negative- and transient positive-incentive motivational responses to shock.

The outputs O_5 and O_6 will be studied during each phase. We will prove that in phase (i), $O_5 = 0 = O_6$. In phase (ii), O_5 becomes positive and approaches a positive asymptote, whereas $O_6 = 0$. O_5 is the basis for response suppression during phase (ii). In phase (iii), O_5 is rapidly driven to zero, but O_6 becomes positive and after achieving a positive asymptote decays to zero. O_6 during phase (iii) describes the rebound on which the CAR is based (see Fig. 6).

Consider phase (i). The inputs are symmetrically distributed; hence, the outputs are zero. To see this, note by (5) and (6) that $x_1 \cong \alpha^{-1}I \cong x_2$. Hence by (7) and (8) $z_1 \cong z_2$. Then by (9) and (10) $x_3 \cong x_4$, which in (11) and (12) yields $x_5 \cong 0 \cong x_6$. Finally, (13) and (14) imply $O_5 = 0 = O_6$.

Consider phase (ii). We compute the asymptote $x_5(T)$ of the $\mathscr{A}_f^+$ potential. $x_5(T)$ provides a measures of maximal fear and of the size of $\mathscr{A}_f^+ \rightarrow \mathscr{S}$ feedback. By Properties A and D, shortly after $t = 0$,

$$x_1 \cong x_1(T) \cong \alpha^{-1}(I + J) \tag{15}$$

and

$$x_2 \cong x_2(T) \cong \alpha^{-1}I. \tag{16}$$

Substitute (15) and (16) into (7) and (8). By Property B, $[\alpha^{-1}I - \Gamma]^+ > 0$. Thus, by (7) and Property A, z_1 is (approximately) a monotone decreasing function. By (8) and Property A, z_2 remains (approximately) constant. By Property D, z_1 and z_2 reach their new asymptotes before $t = T$. These asymptotes are

$$z_1(T) \cong \frac{\alpha\beta\gamma}{\alpha\beta + \delta(I + J - \alpha\Gamma)} \tag{17}$$

and

$$z_2(T) \cong \frac{\alpha\beta\gamma}{\alpha\beta + \delta(I - \alpha\Gamma)}. \tag{18}$$

Note that $z_1(T) < z_2(T)$: the aversive input depletes z_1 more than z_2. Nonetheless the signals $\zeta[x_1(t - \tau) - \Gamma]^+ z_1$ and $\zeta[x_2(t - \tau) - \Gamma]^+ z_2$ to v_3 and v_4 asymptotically satisfy the reverse inequality

$$\zeta[x_1(T - \tau) - \Gamma]^+ z_1(T) > \zeta[x_2(T - \tau) - \Gamma]^+ z_2(T).$$

This follows from (15)–(18) and Property D. By Property D, $x_i(T - \tau) \cong x_i(T)$. Thus by (15)–(18),

$$\zeta[x_1(T) - \Gamma]^+ z_1(T) \cong \frac{E(I + J - F)}{G + (I + J - F)} \tag{19}$$

and

$$\zeta[x_1(T) - \Gamma]^+ z_2(T) \cong \frac{E(I - F)}{G + (I - F)}, \tag{20}$$

with $E = \beta\gamma\delta^{-1}\zeta$, $F = \alpha\Gamma$, and $G = \alpha\beta\delta^{-1}$. The function

$$f(w) = \frac{E(w - F)}{G + (w - F)}$$

is monotone increasing in $w \geqslant F$ if E, F, and G are positive, so that (19) exceeds (20) if $J > 0$. Thus, the multiplicative coupling between spiking frequency and transmitter yields an output signal that is monotone increasing as a function of spiking frequency, even though the transmitter term is monotone decreasing as a function of spiking frequency. This is the crucial fact.

The asymptotic outputs O_5 and O_6 are found by substituting (19) and (20) into (9) and (10) and then $x_3(T)$ and $x_4(T)$ into (11) and (12). We find by using Property D that

$$x_5(T) \cong \frac{UJ}{(V + I)(V + I + J)} \tag{21}$$

and $x_6(T) \cong -x_5(T)$, with $U = \alpha\beta^2\gamma\delta^{-2}\zeta\varepsilon^{-1}\kappa\eta^{-1}$ and $V = \alpha(\beta\delta^{-1} - \Gamma)$. The following facts are obvious from (21).

PROPOSITION 1

For $I \geqslant \alpha\Gamma$, $x_5(T)$ is a monotone increasing function of J and a monotone decreasing function of I. Equations 13 and 21 show that $O_5(T)$ is a monotone increasing function of the aversive input J and that $O_6(T) = 0$. $O_5(T)$ is a monotone decreasing function of I. If the network $\mathcal{O}$ can control the size of I, then $\mathcal{O}$ can also reduce the CER created by an aversive input J of fixed size. The dependence of $O_5(T)$ on I will also be important when the maximum of O_5 in phase (ii) is compared with the maximum of O_6 in phase (iii). The ratio $O_5^{-1}(T) \max\{O_6(t): t \geqslant T\}$ controls the relative strength of motivational support for CAR and CER channels. The size of I influences this ratio.

To see this, consider phase (iii) using the approximation in which all potentials instantaneously adjust to the removal of J, whereas z_1 and z_2 gradually move from their asymptotes $z_1(T)$ and $z_2(T)$ to their new asymptotes. Let the notation $f(T+)$ denote the (approximate) value of the function f at a time, shortly after $t = T$, when the x_i's have readjusted but the z_i's have not. We study $x_6(T+)$, which is the maximum of the rebound at x_6 during phase (iii), given instantaneous adjustment of potentials.

First note that

$$x_1(T+) \cong \alpha^{-1}I \cong x_2(T+) \tag{22}$$

whereas

$$z_1(T+) \cong z_1(T) < z_2(T) \cong z_2(T+). \tag{23}$$

At $t \cong T+$, the signals $\zeta[x_1(t-\tau)\Gamma]^+ z_1$ and $\zeta[x_2(t-\tau)-\Gamma]^+ z_2$ from N_{13} and N_{24} satisfy the inequality

$$\zeta[x_1(T+)-\Gamma]z_1(T+) < \zeta[x_2(T+)-\Gamma]z_2(T+), \tag{24}$$

since by (17), (18), and (22),

$$\zeta[x_1(T+)-\Gamma]z_1(T+) \cong \frac{E(I-F)}{G+(I+J-\Gamma)} \tag{25}$$

and

$$\zeta[x_2(T+)-\Gamma]z_2(T+) \cong \frac{E(I-F)}{G+(I-F)}. \tag{26}$$

Inequality (24) is the basis of the rebound at v_6. Substituting (25) and (26) into (9) and (10) and then (11) and (12) yields

$$x_6(T+) \cong \frac{WJ(I-F)}{(V+I)(V+I+J)}, \tag{27}$$

and $x_5(T+) \cong -x_6(T+)$, with $W = \beta\gamma\delta^{-1}\zeta\varepsilon^{-1}\kappa\eta^{-1}$. A routine computation proves the following facts.

PROPOSITION 2

$x_6(T+)$ is a monotone increasing function of J. $x_6(T+)$ is monotone increasing in I for $F \leqslant I \leqslant F + (G^2 + JG)^{\frac{1}{2}}$ and monotone decreasing in I for $I > F + (G^2 + GJ)^{\frac{1}{2}}$. The maximum

$$\frac{WJ(G^2+JG)^{\frac{1}{2}}}{[G+(G^2+JG)^{\frac{1}{2}}][J+G+(G^2+JG)^{\frac{1}{2}}]}$$

of $x_6(T+)$ as a function of I, for fixed J, is a monotone increasing function of J. Thus there exists a value of I that maximizes the rebound driving the CAR for fixed J. Both this maximum and $x_6(T+)$ for fixed I are monotone increasing in J.

The following proposition studies the CAR/CER ratio, which estimates the relative size of O_6 and O_5 during phase (iii) and phase (ii), respectively. The importance of this ratio is shown by the following example. Let the rebound at O_6 be small relative to the size of O_5 during shock; that is, CAR/CER $\ll 1$. Let avoidance cues and cues conditioned to response suppression be simultaneously present during avoidance trials. Then the avoidance cues cannot provide enough positive feedback to $\mathscr{S} \to \mathscr{M}$ channels to overcome the CER and release the CAR. A large CAR/CER ratio means that avoidance cues can, in principle, overcome fear and induce a CAR. We study the ratio $x_5^{-1}(T)x_6(T+)$ instead of the closely related ratio $O_5^{-1}(T)\max\{O_6(t): t \geqslant T\}$.

PROPOSITION 3

$$x_5^{-1}(T)x_6(T+) = G^{-1}(I - F). \tag{28a}$$

Thus

$$x_5^{-1}(T)x_6(T+) = \left(1 + \frac{J}{G}\right)^{\frac{1}{2}} \quad \text{if} \quad I = F + (G^2 + JG)^{\frac{1}{2}}. \tag{28b}$$

Note in (28a) that the CAR/CER ratio is independent of J. In other words, the tonic arousal level determines the *relative* preference for relief over fear in the rebound mechanism. By (28a), increasing I biases the CAR/CER ratio in favor of avoidance. By (21) and (27), however, increasing I to large values drives O_5 and O_6 toward zero, and therefore eliminates both negative and positive $\mathscr{A}_f^{\pm} \rightarrow \mathscr{S}$ feedback; all "emotional" support of responding is removed. Nonetheless, letting $I = F + (G^2 + JG)^{\frac{1}{2}}$ maximizes the rebound $O_6(T+)$, by Proposition 2, and yields a CAR/CER ratio that favors avoidance, by (28b), by an amount that increases like the square root of J. Thus there exist values of I which create both a favorable CAR/CER ratio and sufficient arousal to drive the CAR.

6. ANALGESIA, DEPRESSION, NOVELTY, AND INVERTED U IN LEARNING

Equation (27) estimates the rebound created by turning off J units of shock. Denote this rebound by $\{J \rightarrow 0\}$. Equation (27) shows that the rebound caused by shutting off J units is greater than that caused by shutting off $J/2$ units; that is, $\{J \rightarrow 0\} > \{J/2 \rightarrow 0\}$. We now show that, moreover, $\{J/2 \rightarrow 0\} > \{J \rightarrow J/2\}$. This inequality can be interpreted as follows. Suppose that two aversive input sources summate at v_1. Let each source create $J/2$ units of input. Then the chain of inequalities

$$\{J \rightarrow 0\} > \left\{\frac{J}{2} \rightarrow 0\right\} > \left\{J \rightarrow \frac{J}{2}\right\} \tag{29}$$

means that it is most reinforcing to shut off both aversive sources; second most reinforcing to shut off one source in the absence of the other; and least reinforcing to shut off one source and leave the other on. This result illustrates the importance of parallel input channels on perceived fear and relief. Various pain analgesic effects [12, 20, 48] also suggest such influences. For example, loud white noise attenuates perceived pain due to drilling by a dentist. The attenuating influence of nonspecific inputs I to both $\mathscr{A}_f^+$ and $\mathscr{A}_f^-$ on perceived fear, as in (21), should be noted in this connection.

PROPOSITION 4

$\{J/2 \to 0\} > \{J \to J/2\}$. A rebound exists at v_6 in the case $\{J \to J/2\}$ if and only if $G^{-1}(I - F) > 1$; that is, if and only if the CAR/CER ratio favors avoidance. Otherwise, the transition $\{J \to J/2\}$ merely results in a reduction in fear.

PROOF

For convenience of exposition, let $x_5(T, J \to K)$ and $x_6(T+, J \to K)$ denote the fear and relief maxima created by switching J units of shock that are on throughout the interval $[0, T)$ to K units of shock at time T. A routine computation shows that for the transition $\{J \to J/2\}$,

$$x_6\left(T+, J \to \frac{J}{2}\right) = \frac{W(J/2)(I - F - G)}{(V + I)(V + I + J)}, \tag{30}$$

which is positive if and only if $G^{-1}(I - F) > 1$.

By (28a) and (30), $x_6(T+, J \to J/2) > 0$ if and only if, for any $K > 0$, $x_6(T+, K \to 0) > x_5(T, K \to 0)$. Indeed, by (28a) and (30), for all J, $K > 0$ and $I \geq \alpha\Gamma$,

$$x_6\left(T+, J \to \frac{J}{2}\right) = \frac{WG(J/2)[x_5^{-1}(T, K \to 0)x_6(T+, K \to 0) - 1]}{(V + I)(V + I + J)}. \tag{31}$$

In any case,

$$x_6\left(T+, J \to \frac{J}{2}\right) < x_6\left(T+, \frac{J}{2} \to 0\right),$$

since by (27),

$$x_6\left(T+, \frac{J}{2} \to 0\right) = \frac{W(J/2)(I - F)}{(V + I)[V + I + (J/2)]}, \tag{32}$$

which is larger than (30). ■

Equation 31 relates the amount of reward due to cutting J units of shock in half to the relative strength of fear and relief channels. Such a relationship is, in principle, testable and predicts that an individual who finds cutting shock in half unusually unrewarding should also have difficulty performing avoidance tasks in the presence of fearful cues.

Propositions 2 and 4 are special cases of the following general principle.

THEOREM 1 (*Reinforcement Ordering*)

Let two copies $\mathcal{N}_1$ and $\mathcal{N}_2$ of the network in Fig. 5 be given. Let the tonic input in $\mathcal{N}_i$ have size $I_i \geq F$. Let an aversive input of size J_i be switched on in $\mathcal{N}_i$ for a duration of T time units. Then let the aversive input level be *switched to* K_i, $i = 1, 2$. *Switching from* J_1 *to* K_1 *is more reinforcing than switching from* J_2 *to* K_2 *if and only if*

$$R(I_1, J_1, K_1) > R(I_2, J_2, K_2), \tag{33}$$

where

$$R(I, J, K) = \frac{W(J - K)(I - F) - WKG}{(V + I)(V + I + J)}. \tag{34}$$

Theorem 1 breaks up the total input to v_1 into two parts: an input I to v_1 and v_2 equally, which fluctuates slowly if at all; and an input J to v_1 alone, which can be quickly switched between various levels. In later sections, another source of inputs will be applied to v_1 and v_2; namely, the $\mathscr{S} \to \mathscr{A}_f^{\pm}$ input channels. At various times, some $\mathscr{S} \to \mathscr{A}_f^{\pm}$ channels will contribute equal inputs to $\mathscr{A}_f^+$ and $\mathscr{A}_f^-$ ("irrelevant cues"), some $\mathscr{S}$ channels will contribute a larger input to $\mathscr{A}_f^+$ than to $\mathscr{A}_f^-$ ("secondary aversive cues"), and some $\mathscr{S}$ channels will contribute larger inputs to $\mathscr{A}_f^-$ than to $\mathscr{A}_f^+$ ("secondary rewarding cues"). Because inputs summate at $\mathscr{A}_f^+$ and $\mathscr{A}_f^-$, the crucial quantity is the relative size of the *total* inputs to $\mathscr{A}_f^+$ and $\mathscr{A}_f^-$. If, for example, the total input to $\mathscr{A}_f^+$ exceeds that to $\mathscr{A}_f^-$, then the net effect is fear, even if several $\mathscr{S}$ channels project primarily to $\mathscr{A}_f^-$. If the total rapidly drops at $\mathscr{A}_f^+$, the net effect is a rebound at $\mathscr{A}_f^-$. The size of the rebound depends, as (34) shows, on both the absolute drop size and the size of the total "irrelevant" input at $\mathscr{A}_f^+$ and $\mathscr{A}_f^-$.

Both unduly small and unduly large I levels create "abnormal" learning by "emotionally depressing" network response to phasic inputs, and thereby removing the incentive motivation that controls $\mathscr{S} \to \mathscr{M}$ sampling. Thus an "inverted U" in learning exists [34]. To study the inverted U, set I at successively higher values, starting with zero. If $I < F$, no rebound is possible, since by (6), (8), and (10), no signals pass from v_2 to v_4. Also, the fear threshold is raised, in the sense that a larger value of J is needed to create signals from v_1 to v_3 than is needed when $I \geqslant F$: neither fear nor relief occur in response to J inputs of "normal" size. If $I \gg 0$, the network is again "emotionally depressed," but for a different reason. A fixed J level, or secondary aversive input, or secondary rewarding input, creates a relatively small asymmetry in the total inputs to v_1 and v_2. Thus both O_5 and O_6 are very small, as (21) and (27) show. Any unduly large source of equal inputs to v_1 and v_2 will have this effect.

The two types of depression can be distinguished by increasing the J level step by step. In the case where $I < F$, the fear threshold is high, but once J is sufficiently large to fire $v_1 \to v_3$ signals, then additional increments in J create *larger* than normal increments in fear ("hyperexcitable" with high threshold). In fact, for any $I \geqslant 0$, $\partial x_5(T)/\partial J = U(V + I + J)^{-2}$, which is quadratically decreasing in I. In a similar fashion, small rewarding inputs create larger than normal increments in relief. In the case where $I \gg 0$, the fear threshold is low, but increments in J create abnormally small increments in fear ("hypoexcitable" with low threshold). In this sense, the overaroused form of reduced affect is insensitive (is "indifferent") to emotionally charged events, whereas the underaroused form of reduced affect can overreact (is "irritable") to emotionally charged events that exceed the abnormally large threshold of this system.

In a similar fashion, either strongly conditioned $\mathscr{S} \rightarrow \mathscr{A}_f^+$ ($\mathscr{A}_{\bar{f}}$) channels that are persistently turned on, or an overaccumulation of transmitter at $\mathscr{A}_f^+$ ($\mathscr{A}_{\bar{f}}$) synaptic knobs, can produce persistent fear (relief). In the overaccumulation cases, these persistent fear or relief reactions are independent of the occurrence of particular cues, and therefore form a general emotional context which biases the interpretation of specific cues. In the case of strongly conditioned $\mathscr{S} \rightarrow \mathscr{A}_f^+$ channels, the negative feedback $\mathscr{A}_f^+ \rightarrow \mathscr{S}$ inhibits both learning of new $\mathscr{S} \rightarrow \mathscr{M}$ patterns and performance of old $\mathscr{S} \rightarrow \mathscr{M}$ patterns in response to the cues that elicit $\mathscr{S} \rightarrow \mathscr{A}_f^+$ signals. These effects can create a "resistance" to unlearning the fearful "meaning" of these cues, and a "repression" of the patterns encoded in their $\mathscr{S} \rightarrow \mathscr{M}$ synapses [27].

Another possible cause of the inverted U in learning has been studied. Grossberg and Pepe [30, 31] discuss the influence of $\mathscr{A} \rightarrow \mathscr{S}$ overarousal on serial learning by $\mathscr{S}$. Pathologically small $\mathscr{S} \rightarrow \mathscr{S}$ spiking thresholds or inadequate $\mathscr{S} \rightarrow \mathscr{S}$ lateral inhibition produce similar effects. They prove that overarousal yields associational confusions between erroneous choices that are closely related, in time or meaning, to the correct association. These "fuzzy" associational sets interfere with paying attention and subsequent learning. In this model, overarousal influences the relative strength of associations at the beginning and end of a serially learned list (primacy/recency ratio), the skewing of the bowed serial position curve, and the distribution of remote errors. Some of these effects seem not to have been studied experimentally, and if true would provide a conceptual bridge between the effects of overarousal on paying attention and other serial learning parameters.

The two inverted U's are due to different network mechanisms. The inverted U that yields depressed emotional affect will, in Section 9, be compared to the response of midbrain reward and punishment centers, such as hypothalamic sites, to variations of various arousal parameters. The inverted U that yields poor paying attention and associational confusions can be compared to responses of neocortical cells to variations in arousal level. The reticular formation supplies arousal to both regions. Our results therefore suggest that reticular formation overarousal can yield a behavioral syndrome combining poor attention, massive associational confusions, and depressed emotion affect. Other type of arousal deficits can yield other symptoms. These will be compared with clinical findings in another place.

The size of I must be carefully controlled to avoid emotional under- or overarousal. What cells control the internal tonic part of I? Possibly v_1 and v_2 contain endogenous energy sources. If not, a likely interpretation of the formal input source is the ascending reticular activating system [59, page 434]. Kornetsky and Eliasson [42] and Phillips and Bradley [51] report that under- and overarousal of the reticular formation yields an inverted U. Can reticular overarousal also yield "hypoexcitable emotional depression"? The total uniform $\mathscr{S} \to \mathscr{A}_f^{\pm}$ input also contributes to I and is therefore also carefully controlled. Section 10 discusses another reason and a mechanism for doing this.

Given such diverse sources of I input, various subtle interactions can occur. If the "novelty" or abruptness of shock creates a brief increase in total I, then by (21), more fear will follow abrupt shock than shock which is gradually switched on [13, 47]. Grossberg [25, 28] discusses some anatomies that can discriminate complex input characteristics, including velocity detectors. An excitatory transient in J due to abrupt shock can also produce more fear, by (21). Grossberg [24] describes such a transient due to prior transmitter accumulation. A sufficiently slow overshoot of the type in Fig. 3b could also have this effect. New $\mathscr{S}$ channels can also be switched on or off by rate differences in shock onset. Suppose gradual shock causes less initial response suppression than abrupt shock. Then more motor activity R is possible and more sensory representations $\mathscr{S}$, denoted by $\mathscr{S}(R)$, which are activated by feedback cues of R, can be conditioned, either to shock ("spreading the fear around $\mathscr{S}$") or to unsuppressed consummatory activity (counterconditioning). The latter conditioning can produce resistance to the suppressive effect of a later intense shock. Suppose that shock also has an $\mathscr{S}$ representation; that is, the sensory channels activated by a given shock level J_1 also project to the first stage of sensory processing in certain $\mathscr{S}$ representations denoted by $\mathscr{S}(J_1)$, and these $\mathscr{S}(J_1)$ channels can sample $\mathscr{A}_f^{\pm}$. Then $\mathscr{S}(J_1)$ sites can be conditioned to the prevailing motor activity, whether it is response suppression, given intense shock, or consummatory responding, given weak shock. If some $\mathscr{S}(J_1)$ channels are activated by a different shock level J_2 on recall trials, then the motor activity prevalent on learning trials given shock level J_1 can generalize to the J_2 level [13].

The foregoing discussion tacitly answers the question: What cell sites control "fear" and "relief" in Fig. 5? If fear and relief are mutually exclusive attributes, they are "perceived" after the operation of the subtractive on–off field. In Fig. 5, this means that "fear" is controlled by v_5 and "relief" by v_6.

7. CONDITIONING FEAR AND RELIEF

No learning occurs in Fig. 5. $\mathcal{S}$ and $\mathcal{M}$ must be included in the figure. Where do $\mathcal{S}$ axons terminate? They terminate after the rebound mechanism, to permit CER and CAR learning, and before the on–off field, since fear and relief are mutually exclusive. Thus $\mathcal{S}$ projects to v_3 and v_4, as in Fig. 7, which depicts synaptic knobs at which learning occurs as semicircles, knobs at which (slow) accumulation takes place as squares, and all other knobs as arrowheads. Two formally distinct transmitter processes now converge at v_3 and v_4.

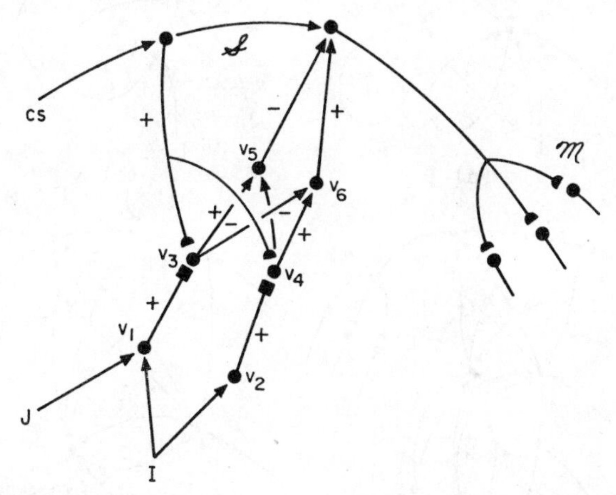

FIG. 7. Minimal synthesis of sampling and accumulation-depletion mechanisms.

Higher-order "classical" fear or relief conditioning is possible in Fig. 7. If $\mathcal{S}_1$ projects strongly to v_3 (v_4) and $\mathcal{S}_2$ projects equally to v_3 and v_4, then simultaneous activation of $\mathcal{S}_1$ and $\mathcal{S}_2$ will strengthen the $\mathcal{S}_2$ connection to v_3 (v_4), and weaken, or extinguish, the $\mathcal{S}_1$ connection to v_3 (v_4). Higher-order "instrumental," or rebound, conditioning does not occur in Fig. 7. Cessation of CS_1, when $\mathcal{S}_1$ projects strongly to v_3, does not drive a rebound at v_4, because accumulation occurs at the stage prior to the v_3 and v_4 cell sites. The next section considers the minimal extension of Fig. 7 in which higher-order instrumental conditioning is possible.

8. HIGHER-ORDER INSTRUMENTAL CONDITIONING

This section implements the following postulate.

Postulate XII. Higher-order instrumental conditioning is possible.

By Postulate XII, the network will be extended so that $\mathscr{S}$ can drive a conditioned rebound from $\mathscr{A}_f^+$ to $\mathscr{A}_f^-$. Thus $\mathscr{S}$ will send axons to a stage *prior* to the rebound, and these axons will have conditionable synaptic knobs. By Fig. 7, $\mathscr{S}$ sends axons to a stage *after* the rebound. Thus the anatomy of $\mathscr{A}_f$ is *recurrent*. The cell sites v_1 and v_3 are identified, as are v_2 and v_4. See Fig. 8 for some recurrent anatomies: In Fig. 8a, termination

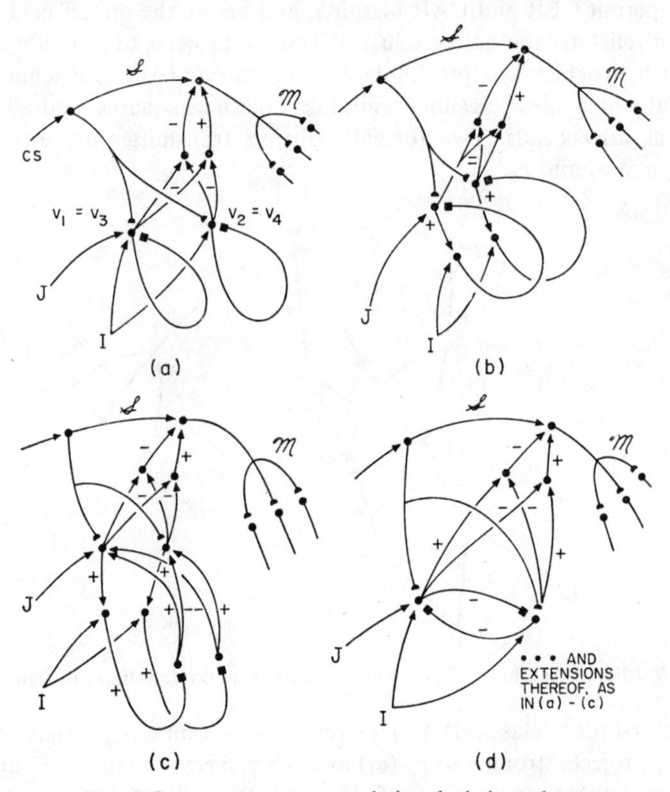

FIG. 8. Some recurrent accumulation-depletion schemes.

of J or of a conditioned aversive input from $\mathscr{S}$ to v_1 can drive a rebound at v_2. Thus higher-order instrumental conditioning is possible. Some important properties of this anatomy are listed below. Termination on a conditioned rewarding input from $\mathscr{S}$ to v_2 can also drive a rebound at v_1, yielding a brief increment in fear after abrupt termination of an interval of relief. This second-order fear conditioning effect is relatively small. Conditioned inputs from $\mathscr{S}$ to v_1 or v_2 are actively extinguished by the positive equal I inputs to v_1 and v_2 on recall trials. The recurrent "accumulation loops" $v_1 \rightarrow v_1$ and $v_2 \rightarrow v_2$ distort the relative intensity of the total $\mathscr{S}$,

J, and I inputs to v_1 and v_2 in the direction of "contour enhancement"; they produce a "contrast" effect (cf. [23]).

Figures 8b and 8c extend Fig. 7a in a way that diminishes the extinction on recall trials of conditioned $\mathscr{S} \to v_1$ and $\mathscr{S} \to v_2$ pathways due to equal I inputs at v_1 and v_2. In Fig. 8b, I only extinguishes $\mathscr{S}$ channels via the recurrent $v_1 \to v_1$ and $v_2 \to v_2$ loops, rather than directly and recurrently as in Fig. 8a. In particular, if a given $\mathscr{S}$ representation projects strongly to v_1 on recall trials, then the feedback along $v_1 \to v_1$ and $v_2 \to v_2$ will still favor v_1 in spite of I. $\mathscr{S}$ never samples the uniformly distributed I input, although the I input does indirectly reduce the relative strength of the $v_1 \to v_1$ feedback channel. In Fig. 8c, the effect of I on extinction is largely canceled out by the subtractive on–off field juxtaposed between the accumulation step and v_1 and v_2. These inhibitory interneurons cancel out the uniform part of the input to them, including the I input. $\mathscr{S}$ samples only the non-uniform part of the feedback. Extinction of $\mathscr{S} \to \mathscr{A}_f$ channels is due primarily to the influence of "irrelevant" $\mathscr{S}$ cues, namely, $\mathscr{S}$ inputs that are equal at $\mathscr{A}_f^+$ and $\mathscr{A}_f^-$, or to counter conditioning by aversive cues. In this anatomy, therefore, a CAR is very stable if the CAR removes $\mathscr{O}$ from the cues driving the CAR itself, or at least from the cues of the aversive situation. Extinction of the CAR is possible if aversive cues are not terminated; cf. one-way versus two-way avoidance tasks.

Figure 8d describes a recurrent inhibition-with-accumulation between v_1 and v_2. The accumulating transmitter is here an inhibitory transmitter rather than an excitatory transmitter, as in Figs. 8d–c. Thus, if v_1 receives a larger input than v_2, then v_1 inhibits v_2 more than conversely, and transmitter is depleted more in N_{12} than in N_{21}. When the inputs to v_1 and v_2 equalize, then $v_2 \to v_1$ inhibition exceeds $v_1 \to v_2$ inhibition. Hence the output from v_2 exceeds that from v_1, and a rebound occurs.

A useful exercise for the reader is to perform "ablation" and "nerve section" experiments on these networks at various loci and to check what happens to the fear, relief, and rebound reactions.

9. CHOLINERGIC AND ADRENERGIC TRANSMITTERS

Our networks contain two formally distinct transmitter systems which converge at $\mathscr{A}_f^{\pm}$ sites to mediate the CER and CAR: learning occurs at semicircular synapses and slow accumulation–depletion occurs at square synapses. *In vivo*, two distinct transmitter systems influence CER and CAR learning and performance at hypothalamic sites, namely, the cholinergic and adrenergic systems. This striking coincidence is interpreted below. The interpretation is necessarily tentative, since Fig. 8 describes, at best, one fragment of the several systems needed during a routine learning task. This description is also minimal and presumably does not encompass all the tasks addressed by the analogous system *in vivo*. Other systems also contain cholinergic and adrenergic components [56], and Fig. 8 provides little insight into their relative importance. Moreover, even if in the minimal description two or more synaptic knobs end at a given cellular site, which we interpret (say) as analogous to a lateral hypothalamic site, the corresponding system *in vivo* might separate these knobs by interneurons; cf. Fig. 8a–c. Hence our model might predict the existence of both cholinergic and adrenergic interactions at a single hypothalamic site, whereas *in vivo* these interactions might be occurring in parallel at different hypothalamic, or even nonhypothalamic, sites that are separated by interneurons. Nonetheless, the analogy suggests so many precise questions, possible insights concerning the relevant data, and directives for further studies that it will be given here. Some questions will be mixed with interpretive comments below.

Interpret the semicircular synapses as cholinergic synapses; they contain acetylcholine (ACh). Let the square synapses be adrenergic; they contain norepinephrine (NE). Let $\mathscr{A}_f^+$ be associated with ventromedial hypothalamus (VMH). Let $\mathscr{A}_f^-$ be associated with lateral hypothalamus (LH). Given this labeling, then both ACh and NE influence both VMH and LH sites.

Some encouraging data are readily noticed. Central monoamine neurons are tonically active [1], as are square synapses driven by the tonic internal I input. Is the *in vivo* tonic activity inhibited by blockade of reticular formation activity, or is it endogenous? Can over- and under-arousal due to this tonic activity cause emotional depression?

We now note that all the terms in Eq. (7) are qualitatively supported by data on NE regulation. The rate of NE synthesis is controlled by the amount of NE; that is, feedback inhibition of endogenous monoamine biosynthesis exists [5]. This feedback inhibits the enzyme tyrosine hydroxylase, which controls the rate-limiting step, from tyrosine to DOPA, in NE production [60, 61]. Feedback inhibition of z_1 production exists in Eq. (7), in the term $-\beta z_1$. Increased adrenergic nervous activity yields enhanced synthesis of NE from tyrosine, which is prevented by adding NE to the bath, in the isolated hypogastric nerve–vas deferens preparation of the guinea pig [60]. The terms $-\beta z_1$ and $-\delta[x_1(t-\tau)-\Gamma]^+ z_1$ in Eq. (7) have a similar effect. The term $\delta[x_1(t-\tau)-\Gamma]^+ z_1$ depletes transmitter by releasing it at a rate that increases with presynaptic spiking frequency. Feedback inhibition of transmitter production is thereby reduced. Adding transmitter increases the term $-\beta z_1$ and thereby increases feedback inhibition of transmitter production. All terms in Eq. (7) are thus qualitatively supported by data. A model of some faster transients in transmitter production is considered in [24].

NE often, but not always, has inhibitory effects on postsynaptic sites [5]. In Figs. 8a–c, the square synapses are excitatory. Nonetheless, if NE is nonspecifically applied to the $\mathscr{A}_f^{\pm}$ region, and acts equally on $v_1 \to v_1$ and $v_2 \to v_2$ loops, then the *outputs* from $\mathscr{A}_f$ to $\mathscr{S}$ can be inhibited by the subsequent increase in total I input. In the recurrent anatomy of Fig. 8d square synapses are inhibitory. Bloom and Giarman [5] discuss recurrent inhibitory NE pathways in the olfactory bulb. Is Fig. 8d an analog of some such pathways in a different anatomical locus? Can a rebound be elicited at suitable olfactory sites?

The olfactory example suggest a basic question: Why are off-cells so prevalent in sensory processing regions [59, pages 253, 349]? One formal answer seems clear in these networks: Suppose $\mathscr{O}$ can learn that offset of a given stimulus is the cue for a given response. Then cells which are reliably turned on by stimulus offset, and which can sample $\mathscr{A}$ and $\mathscr{M}$, are needed. These off-cells would appear in $\mathscr{S}$, and would be driven by the on-cells that respond to the stimulus. The stimulus replaces the aversive input J in this case. The "dipole" of on-cell and off-cell can thus discriminate both externally important cues in $\mathscr{S}$ and internally important cues in $\mathscr{A}$; the different roles for such dipoles need depend only on the interpretation of what input source drives the on-cell and on the interpretation of what cells receive the on-cell and off-cell signals. A tonic internal input would be needed to drive the dipoles in $\mathscr{S}$, just as it was needed in $\mathscr{A}$. Do such dipoles exist in cerebral cortex? The profound influence of the ascending reticular activating system on cerebral cortex is well known [59, Chapter 14]. Are cerebral dipoles, supposing that they exist, driven by tonic reticular input or by endogeneous energy sources? Are cerebral dipoles adrenergic, or are they of the type described by Fig. 3?

Reserpine depletes NE, thereby reducing both sensitivity to shock and lever-press shock avoidance responses [55, 61]. Is a formal analog of this depletion a reduction in transmitter accumulation in the $v_1 \to v_1$ and $v_2 \to v_2$ synaptic knobs, and thus of the signals driving $\mathscr{A}_f^+$ and $\mathscr{A}_f^-$ output? Self-stimulation behavior is also dependent on NE activity [62]. Is this formally analogous to varying the amount of feedback driven by the stimulating electrode in the $v_1 \to v_1$ pathway by altering the amount of accumulation in $v_1 \to v_1$ synaptic knobs? Increased shock intensities produce more rapid recovery from reserpine [61]. Is this formally analogous to a greater synthesis of transmitter driven by higher presynaptic spiking frequencies? Medial forebrain bundle (MFB) lesions in the LH cause increased sensitivity to shock [32]. Is this formally analogous to removing $\mathscr{A}_f^-$ inhibition of $\mathscr{A}_f^+$ output, and thereby generating greater fear? If this formal analogy is accurate, then the $v_1 \to v_1$ and $v_2 \to v_2$ loops, say in Fig. 8c, would be rudimentary formal analogs of MFB pathways. In particular, MFB pathways would be descending and ascending, as are the $v_1 \to v_1$ and $v_2 \to v_2$ loops [49]. This formal possibility is also compatible with the fact that NE is released by electrical stimulation of the MFB [19].

The formal "cholinergic system" of Fig. 8 encodes memory in $\mathcal{S} \to \mathcal{A}_f$ synapses. Many workers have reported that anticholinergics disrupt memory in punishment and avoidance situations [4, 7, 10, 18, 33, 39, 43, 46, 57, 61a]. Some details of these data will be cited and compared with formal network properties.

Operant behavior that is suppressed by punishment is disinhibited by cholinergic blockade of the VMH [44, 54]. Is this formally analogous to inhibition of $\mathcal{A}_f^+$ response to $\mathcal{S} \to \mathcal{A}_f^+$ channels and thus to a reduction of inhibitory $\mathcal{A}_f^+ \to \mathcal{S}$ feedback? Hypothalamic self-stimulation involves a cholinergic process, since rats learn to press a lever to self-inject carbachol (an ACh mimicker) into the LH [50]. Denote the lever-press response by R. Then the Olds *et al.* data have the following possible formal interpretation. Suppose carbachol injection is analogous to exciting v_2 in Fig. 8. Then carbachol acts as an input source to v_2 much as J does at v_1. This input brings $\mathcal{S}(R) \to \mathcal{A}_f^-$ sampling into the suprathreshold range and releases $\mathcal{A}_f^- \to \mathcal{S}$ feedback which drives $\mathcal{S}(R) \to \mathcal{M}$ sampling of the motor controls of R. Thus formal "carbachol injection" at v_2 acts much like an electrical self-stimulation pulse at v_2 [27]. This interpretation is strengthened by the fact that, *in vivo*, carbachol in LH facilitates the CAR, whereas cholinergic blockage by atropine or local depression by pentobarbital impairs the CAR [53]. Also compatible is the fact that rewarding hypothalamic stimulation reduces the aversive properties of shock [15]. This latter effect is formally analogous to inhibiting $\mathcal{A}_f^+$ response by increasing $\mathcal{A}_f^-$ input. Sepinwall [53] also reports the paradoxical result that carbachol in LH *and* in VMH facilitates the CAR. The VMH facilitation is not so large as the LH facilitation. By contrast, atropine in VMH reduces response suppression and in LH interferes with the CAR. This asymmetry in the effects of carbachol and atropine at LH and VMH would be understandable if carbachol in VMH were helping to drive a rebound in LH. This would be analogous to the following situation. Let feedback cues corresponding to nonavoidance responses during shock activate the sensory representations $\mathcal{S}$(non AR). The $\mathcal{S}$(non AR)s are conditioned to $\mathcal{A}_f^+$, since they are active when shock is on. Suppose that when the avoidance response occurs, the $\mathcal{S}$(non AR)s are shut off, and the feedback cues of the avoidance response activate the sensory representations $\mathcal{S}$(AR). In other words, we suppose that the cues of the avoidance response and of nonavoidance responses are substantially disjoint. Offset of $\mathcal{S}$(non AR)s removes input to $\mathcal{A}_f^+$, which thereupon drives a rebound at $\mathcal{A}_f^-$ that is sampled by $\mathcal{S}$(AR)s. This interpretation of CAR facilitation by carbachol in VMH can be experimentally tested. The effect should be greatest in experiments where the overlap of $\mathcal{S}$(non AR) and $\mathcal{S}$(AR) cues is minimal. Then if carbachol is present in VMH during $\mathcal{S}$(non AR) $\rightarrow \mathcal{A}_f^+$ conditioning, it will strengthen the $\mathcal{S}$(non AR)$\rightarrow \mathcal{A}_f^+$ channels. The avoidance response, by shutting off the relatively large $\mathcal{A}_f^+$ input from $\mathcal{S}$(non AR)s, then yields a relatively large rebound at $\mathcal{A}_f^-$, which is learned by $\mathcal{S}$(AR)s. Carbachol in VMH has thereby enhanced CAR conditioning by indirectly strengthening $\mathcal{S}$(AR)$\rightarrow \mathcal{A}_f^-$ channels. Carbachol in LH also enhances CAR conditioning by directly strengthening $\mathcal{S}$(AR)$\rightarrow \mathcal{A}_f^-$ channels. If, by contrast, the experiment merely calls for fear conditioning, then carbachol in VMH should enhance the rate of fear conditioning and subsequent response suppression [44].

Given this interpretation of carbachol action, the effect of atropine has the following interpretation. Atropine in VMH reduces response suppression by reducing the conditioned $\mathcal{S} \rightarrow \mathcal{A}_f^+$ input by blocking $\mathcal{S} \rightarrow \mathcal{A}_f^+$ synapses. Atropine cannot drive a rebound at $\mathcal{A}_f^-$ because it reduces the likelihood that termination of $\mathcal{S} \rightarrow \mathcal{A}_f^+$ inputs can drive a significant rebound at $\mathcal{A}_f^-$. Atropine in LH can nonetheless interfere with the CAR by reducing the positive incentive motivation controlled by $\mathcal{S} \rightarrow \mathcal{A}_f^-$ channels.

Carlton ([11; see also Khavari [40]) argues that anticholinergics act by inhibiting habituation of incorrect responses, rather than by disrupting memory of correct responses. Some of his remarks in special support of the habituation hypothesis are compatible with memory disruption effects on Fig. 8; for example, concerning the influence of carbachol and anticholinergics on VMH [11, page 324]. Carlton [11, page 305] also argues against memory disruption by showing that scopolamine (an ACH inhibitor) produces more profound deficits with negative than with positive rewards. This argument should be supplemented by dose-dependent studies, since positive and negative reward systems have different thresholds in our networks: a shock input by itself can drive a CER; food input must summate with the prevailing internal hunger input before it can release $\mathcal{S} \rightarrow \mathcal{M}$ signals. Carlton [11, page 323] suggests that the VMH contains ACh and that the LH contains NE. Our present interpretation finds ACh and NE at parallel loci in both VMH and LH, subject to the qualification that interneurons, such as those comprising the $v_1 \rightarrow v_2$ and $v_2 \rightarrow v_2$ loops in Fig. 8c, might interconnect other nuclei in parallel with LH and VMH, and thereby separate the two transmitter systems. The present interpretation also differs from the view [40, 57] that the cholinergic system mediates punishment and the adrenergic system mediates reward.

10. REGULATION OF TOTAL $\mathscr{S}\to\mathscr{A}$ AND TOTAL $\mathscr{A}\to\mathscr{S}$ INPUTS BY NONSPECIFIC INHIBITORY INTERNEURONS

Section 6 notes that the total $\mathscr{S}\to\mathscr{A}$ input must be regulated to keep the total I input away from the extremes in the inverted U. Another reason for doing this is the following. CSs can differ in sensory complexity and intensity. Thus variable numbers of $\mathscr{S}$ representations can be excited at different times, and each $\mathscr{S}$ representation can be excited to a different degree. Suppose that the representations send signals to $\mathscr{A}$ that are independent of each other. Consider two different CSs, CS_1 and CS_2. Let CS_1 excite the N $\mathscr{S}$ representations $\mathscr{S}_1$, and CS_2 excite the MN $\mathscr{S}$ representations $\mathscr{S}_2$, each to the same degree D. Suppose that CS_1 can be learned as the cue for a lever-press response for food. On recall trials, conditioned $\mathscr{S}_1\to\mathscr{A}$ signals can therefore release $\mathscr{A}\to\mathscr{S}$ feedback *if* $\mathscr{O}$ is hungry; $\mathscr{S}_1\to\mathscr{M}$ signals then drive the lever press. CS_2 can drive learning and performance of the lever press even if $\mathscr{O}$ is not hungry, if M is sufficiently large. This is because total $\mathscr{S}\to\mathscr{A}$ input increases with M until $\mathscr{S}\to\mathscr{A}$ input alone can overcome the $\mathscr{A}\to\mathscr{S}$ spiking threshold. In a similar fashion, CS_1 can drive the lever press without hunger if D is sufficiently large. To avoid these catastrophes, $\mathscr{S}\to\mathscr{A}$ signals cannot be independently delivered; the total $\mathscr{S}\to\mathscr{A}$ input must be regulated.

There exists at least two ways to do this. The general mechanism is known as *pattern normalization*, and was introduced in [25]. Two possibilities are illustrated in Fig. 9: Figure 9a illustrates a subtractive, nonrecurrent, nonspecific, inhibitory interneuron, or "horizontal cell" [25, Sections 8–10]. Such interneurons can truncate the total $\mathscr{S}\to\mathscr{A}$ input when it reaches a fixed maximum. In the simplest case, let the output of $\mathscr{S}_i$ be $I_i(t)$ and the net output after operation of an inhibitory interneuron at layer $\tilde{\mathscr{S}}$ be $f_i(t) = I_i(t) - [\Sigma_{k=1}^n I_k(t) - \Gamma]^+$, with all $I_i(t) \geqslant 0$

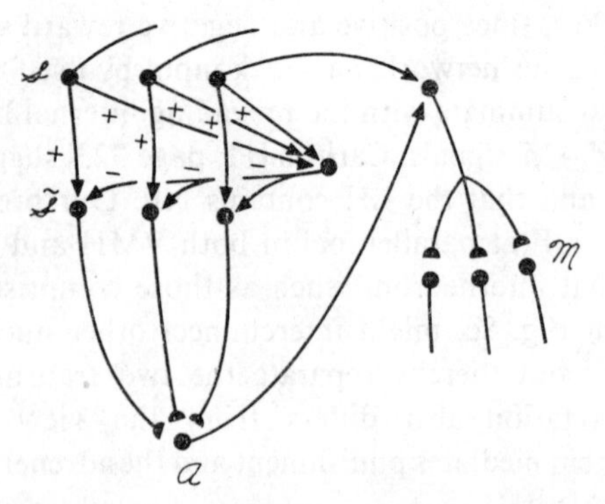

FIG. 9a. $\mathscr{S}\to\mathscr{A}$ Normalization: subtractive inhibition using nonspecific interneurons.

and $\Gamma > 0$. Clearly $f_i = I_i$ whenever $\Sigma_{k=1}^n I_k \leqslant \Gamma$. Thus $f_i = I_i \leqslant \Sigma_{k=1}^n I_k \leqslant \Gamma$ in this case. If $\Sigma_{k=1}^n I_k \geqslant \Gamma$, then $\Sigma_{k=1}^n f_k = (1-n)\Sigma_{k=1}^n I_k + n\Gamma \leqslant \Gamma$. In all cases, the total output $\Sigma_{k=1}^n f_k$ is bounded by Γ.

Figure 9b illustrates an on-center off-surround field with shunting excitation and inhibition [25, Section 14A]. In the simplest case, let the ith cell in the $\tilde{\mathscr{S}}$ layer of Fig. 9b have potential

$$\dot{x}_i = (M - x_i)I_i - \alpha x_i - x_i \sum_{k \neq i} I_k.$$

This is a passive membrane equation, with equilibrium scaled at zero for convenience, and inputs I_i representing depolarizing or hyperpolarizing conductance changes. It can be shown that the total output from $\tilde{\mathscr{S}}$ to $\mathscr{A}$ is bounded by M, and that each x_i is asymptotically proportional to the pattern weight $\theta_i = I_i(\Sigma_{k=1}^n I_k)^{-1}$. One can also study influences of different thresholds, time lags, exponential averaging rates, and axonal path weights in excitatory and inhibitory cells, variations in total output due to variations in input pattern, and so on. Each of these normalization mechanisms has particular advantages, which Grossberg [25] studies.

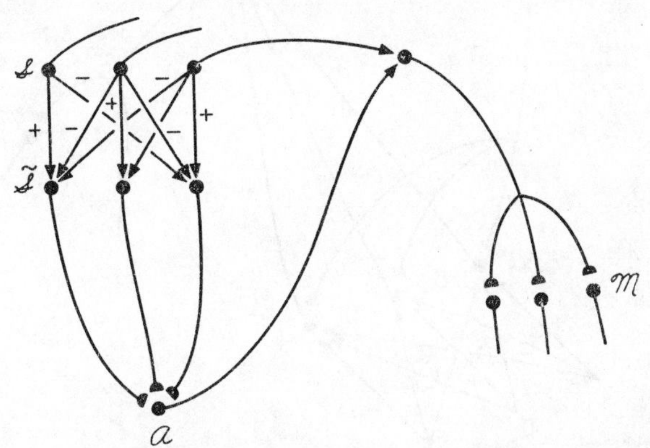

FIG. 9b. $\mathscr{S} \rightarrow \mathscr{A}$ Normalization: shunting inhibition using on-center off-surround field.

These normalization mechanisms form part of the filtering mechanism that permits only prescribed stimulus features, or classes of patterns, to excite particular $\mathscr{S}$ representations. Thus it is possible that the filtering mechanism, by creating selective $\mathscr{S}$ channels, automatically regulates the total $\mathscr{S} \rightarrow \mathscr{A}$ output.

The total $\mathscr{A} \rightarrow \mathscr{S}$ feedback input must also be regulated to prevent this input from indiscriminately firing $\mathscr{S} \rightarrow \mathscr{M}$ channels in the absence of sensory cues to these channels. Inhibitory interneurons therefore modify the outputs of the various arousal sources before they reach $\mathscr{S}$. The next section studies some anatomies that incorporate inhibitory interneurons that achieve this goal by satisfying another basic principle of network design.

11. A SENSORY-DRIVE HETERARCHY

Do the inhibitory interneurons that regulate total $\mathscr{A} \to \mathscr{S}$ feedback operate before or after the stage at which $\mathscr{S} \to \mathscr{A}$ signals combine with internal drive inputs? The answer is "after" if we accept the next postulate.

Postulate XIII. $\mathscr{O}$ can (sometimes) consummate drive D_1, even when drive D_2 is higher, if sensory cues appropriate to D_1 are available whereas cues appropriate to D_2 are not available.

For example, many $\mathscr{O}$s can eat if food is regularly available, even if their sex drives become very high in the absence of a mate. Consider Fig. 10: In Fig. 10a, the internal homeostatic inputs representing different

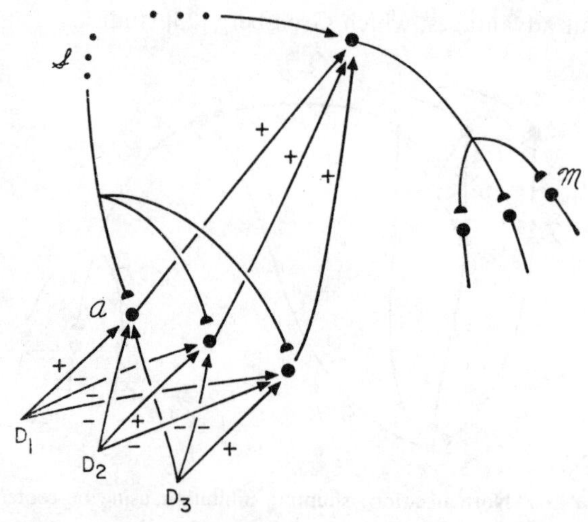

FIG. 10a. Only prepotent drive can release behavior.

drives inhibit each other before $\mathscr{S}$ can affect them. Only one drive representation receives a net positive input after operation of the nonspecific inhibitory interneurons. Only this drive can be a source for $\mathscr{A} \to \mathscr{S}$ feedback and motor output. If the $\mathscr{S}$ cues needed to release this feedback are not available, $\mathscr{O}$ will not satisfy any drive. This $\mathscr{O}$ will starve in the absence of sex. In Fig. 10b, sensorily driven $\mathscr{S} \to \mathscr{A}$ inputs summate with internal homeostatic inputs before the inhibitory interneurons operate. Thus a positive, but not prepotent, drive can release $\mathscr{A} \to \mathscr{S}$ feedback and compatible motor output if sensory cues appropriate to this drive predominate.

This $\mathcal{O}$ can eat and wait for sex. This sensory-drive heterarchy seems related to data of Cox and Valenstein [16], who show that different sensory cues can release different behavior in the presence of hypothalamic stimulation at a fixed spatial locus. Analogous data are collected by Kopa *et al.* [41], who stimulated an area dorsal to the centrum medianum nucleus of the thalamus.

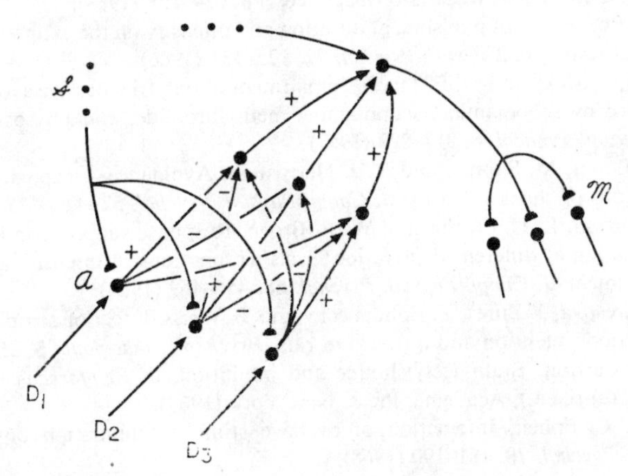

FIG. 10b. Mixing of sensory and drive cues in heterarchical anatomy.

The relative importance of $\mathscr{S}\to\mathscr{A}$ versus internal homeostatic inputs can be explicitly computed in specific cases. Note that the $\mathscr{S}\to\mathscr{A}$ input received at one drive representation is influenced by the pattern of $\mathscr{S}\to\mathscr{A}$ inputs sent to *all other* drive representations, due to the action of nonspecific inhibitory interneurons. Similarly, the distribution of $\mathscr{A}\to\mathscr{S}$ feedback is influenced by the pattern of all sensory-plus-drive combinations. This is a highly nonlocal system.

New sources of pathological $\mathscr{A}\to\mathscr{S}$ overarousal are now evident; for example, an increase in the threshold Γ of nonspecific inhibitory interneurons, as in Fig. 9a; or an increase in the saturation level M of potentials in cells of layer $\tilde{\mathscr{S}}$, as in Fig. 9b. The parametric studies in Grossberg [25] reveal still other possible sources of overarousal.

The work reported in this article was supported in part by the Alfred P. Sloan Foundation and the Office of Naval Research (N00014–67–A–0204–0051).

REFERENCES

1 N. E. Andén, A. Carlsson and J. Häggendal, Adrenergic Mechanisms, *Ann. Rev. Pharmacol.* (H. W. Elliott, W. C. Cutting and R. H. Dreisbach eds.) **9**, 119–134 (1969).
2 Z. Annau and L. J. Kamin, The conditioned emotional response as a function of intensity of the US, *J. Comp. Physiol. Psychol.* **54**, 428–432 (1961).

3 A. B. Ashton, S. C. Bitgood and J. W. Moore, Auditory differential conditioning of the rabbit nictitating membrane response: III. Effects of US shock intensity and duration, *Psychonomic Sci.* **15**, 127–128 (1969).
4 B. Berger, D. L. Margules and L. Stein, Prevention of learning of fear by oxazepam and scopolamine. *Amer. Psychol.* **22**, 492 (1967).
5 F. E. Bloom and N. J. Giarman, Physiological and Pharmacological Considerations of Biogenic Amines in the Nervous System, *Ann. Rev. Pharmacol.* (H. W. Elliott, W. C. Cutting and R. H. Dreisbach, eds.) **8**, 229–258 (1968).
6 E. E. Boe, Effect of punishment duration and intensity on the extinction of an instrumental response. *J. Exptl. Psychol.* **72**, 125–131 (1966).
7 Z. Bohdanecky and M. E. Jarvik, Impairment of one-trial passive avoidance learning in mice by scopolamine, scopolamine methylbromide, and physostigmine, *Intern. J. Neuropharmacol.* **6**, 217–222 (1967).
8 J. J. Boren, M. Sidman and R. J. Herrnstein, Avoidance, escape and extinction as functions of shock intensity. *J. Comp. Physiol. Psychol.* **52**, 420–425 (1959).
9 G. Borozci, L. H. Storms and W. E. Broen, Response suppression and recovery of responding at different deprivation levels as functions of intensity and duration of punishment. *J. Comp. Physiol. Psychol.* **58**, 456–459 (1964).
10 O. Burešova, J. Bureš, Z. Bohdanecky and T. Weiss, Effect of atropine on learning, extinction, retention and retrieval in rats. *Psychopharmacologia* **5**, 255–263 (1964).
11 P. L. Carlton, Brain-acetylcholine and inhibition, in *Reinforcement and Behavior* (J. T. Tapp, ed.), Academic Press, New York (1969).
12 B. A. Campbell, Interaction of aversive stimuli: Summation on inhibition? *J. Exptl. Psychol.* **78**, 181–190 (1968).
13 R. M. Church, Response suppression, in *Punishment and Aversive Behavior* (B. A. Campbell and R. M. Church, eds.), Appleton, New York (1969).
14 R. M. Church, G. A. Raymond and R. D. Beauchamp, Response suppression as a function of intensity and duration of punishment. *J. Comp. Physiol. Psychol.* **63**, 39–44 (1967).
15 V. C. Cox and E. S. Valenstein, Attenuation of aversive properties of peripheral shock by hypothalamic stimulation. *Science* **149**, 323–325 (1965).
16 V. C. Cox and E. S. Valenstein, Effects of stimulus intensity on behavior elicited by hypothalamic stimulation, *J. Comp. Physiol. Psychol.* **69**, 730–733 (1969).
17 M. R. D'Amato, J. Fazzaro and M. Etkin, Discriminated bar-press avoidance maintenance and extinction in rats as a function of shock intensity. *J. Comp. Physiol. Psychol.* **63**, 351–354.
18 S. S. Fox, Self-maintained sensory input and sensory deprivation in monkeys: A behavioral and neuropharmacological study, *J. Comp. Physiol. Psychol.* **55**, 438–444 (1962).
19 K. Fuxe, T. Hökfelt, and U. Ungerstedt, Morphological and functional Aspects of Central Monoamine Neurons, in *Intern. Rev. Neurobiol.* (C. C. Pfeiffer and J. R. Symythies, eds.) **13**, 93–126. Academic Press, New York (1970).
20 W. J. Gardner, J. C. R. Licklider and A. Z. Weisz, Suppression of pain by sound. *Science* **132**, 32–33 (1961).
21 S. P. Grossman, The VMH: a center for affective reactions, satiety, or both? *Physiol. Behavior* **1**, 1–10 (1966).
22 S. Grossberg, Embedding fields: A theory of learning with physiological implications. *J. Math. Psychol.* **6**, 209–239 (1969).
23 S. Grossberg, On learning, information, lateral inhibition, and transmitters, *Math. Biosci.* **4**, 255–310 (1969).

24 S. Grossberg, On the production and release of chemical transmitters and related topics in cellular control, *J. Theoret. Biol.* **22**, 325–364 (1969).
25 S. Grossberg, Neural pattern discrimination. *J. Theoret. Biol.* **27**, 291–337 (1970).
26 S. Grossberg, Embedding Fields: Underlying philosophy, mathematics, and applications to psychology, physiology, and anatomy, *J. Cybernet.* **1**, 28–50 (1971).
27 S. Grossberg, On the dynamics of operant conditioning. *J. Theoret. Biol.* **33**, 225–255 (1971).
28 S. Grossberg, Neural expectation: Cerebellar and retinal analogs of cells fired by learnable or unlearned pattern classes, *Kybernetik* (1972).
29 S. Grossberg, A neural theory of punishment and avoidance, I, Qualitative theory, *Math. Biosci* **15**, 39–68 (1972).
30 S. Grossberg and J. Pepe, Schizophrenia: Possible dependence of associational span, bowing, and primacy vs. recency on spiking threshold, *Behav. Sci.* **15**, 359–362 (1970).
31 S. Grossberg and J. Pepe, Spiking threshold and overarousal effects in serial learning, *J. Statist. Phys.* **1**, 319–350 (1971).
32 J. A. Harvey and C. E. Lints, Lesions in the medial forebrain bundle: Delayed effects on sensitivity to electric shock, *Science* **148**, 250–251 (1965).
33 R. H. Haude, Effects of scopolamine on discriminated responding in the rat. *J. Pharmacol. Exp. Therap.* **126**, 349–358 (1959).
34 D. O. Hebb, Drives and the CNS (conceptual nervous system), *Psychol. Rev.* **62**, 243–254 (1955).
35 F. W. Huff, T. P. Piantanida and G. L. Morris, Free operant avoidance responding as a function of serially presented variations in UCS intensity, *Psychonomic Sci.* **8**, 111–112 (1967).
36 J. L. Johnson and R. M. Church, Effects of shock intensity on nondiscriminative avoidance learning of rats in a shuttlebox, *Psychonomic Sci.*, **3**, 497–498 (1965).
37 L. J. Kamin, C. J. Brimer and A. H. Black, Conditioned suppression as a monitor of fear in the course of avoidance training. *J. Comp. Physiol. Psychol.* **56**, 497–501 (1963).
38 J. D. Keehn, Effect of shock duration on Sidman avoidance response rates, *Psychol. Repts*, **13**, 852 (1963).
39 K. A. Khavari, Effects of central versus intraperitoneal d-amphetamine administration on learned behavior, *J. Comp. Physiol. Psychol.* **68**, 226–234 (1969).
40 K. A. Khavari, Adrenergic-cholinergic involvement in modulation of learned behavior, *J. Comp. Physiol. Psychol.* **74**, 284–291 (1971).
41 J. Kopa, I. Szabó and E. Grastyán, A dual behavioral effect from stimulating the same thalamic point with identical stimulus parameters in different conditional reflex situations, *Acta Physiol. Acad. Sci. Hungaricae* (*Budapest*), **21**, 207–214 (1962).
42 C. Kornetsky and M. Eliasson, Reticular stimulation and chlorpromazine: an animal model for schizophrenic overarousal, *Science* **165**, 1273–1274 (1969).
43 D. Macht, A pharmacodynamic analysis of the cerebral effects of atropin, homatropin, scopolamin and related drugs, *J. Pharmacol. Exptl. Therap.* **22**, 35–48 (1923).
44 D. L. Margules and L. Stein, Neuroleptics vs. tranquillizers: Evidence from animal behavior studies of mode and site of action, in *Neuropsychopharmacology* (H. Brill, ed.), International Congress Series No. 129. Excerpta Medica Foundation, Amsterdam (1967).
45 L. K. Martin and D. Reiss, Effects of US intensity during previous discrete delay conditioning on conditioned acceleration during avoidance conditioning, *J. Comp. Physiol. Psychol.* **69**, 196–200 (1969).

46 W. R. Miles, Hyoscine vs. alcohol: Effect on the behavior of rats in the elevated maze, *Proc. 9th Intern. Congr. Psychol.* (1929), pp. 309–310.
47 N. E. Miller, Learning resistance to pain and fear: Effects of overlearning, exposure, and rewarded exposure in context, *J. Exptl. Psychol.* **60,** 137–145 (1960).
48 A. K. Myers, Effects of continuous loud noise during instrumental shock-escape conditioning, *J. Comp. Physiol. Psychol.* **68,** 617–622 (1969).
49 W. J. H. Nauta and W. Haymaker, Hypothalamic nuclei and fiber connections, in *The Hypothalamus* (W. Haymaker, E. Anderson and W. J. H. Nauta, eds.), C. C. Thomas, Springfield, Illinois (1969).
50 J. Olds, A. Yuwiler, M. E. Olds and C. Yun, Neurohumors in hypothalamic substates of reward, *Amer. J. Physiol.* **207,** 242–254 (1964).
51 M. I. Phillips and P. B. Bradley, Reticular stimulation and chlorpromazine, *Science* **168,** 1122–1123 (1970).
52 D. Reiss, Sidman avoidance in rats as a function of shock intensity and duration, *J. Comp. Physiol. Psychol.* **73,** 481–485 (1970).
53 J. Sepinwall, Enhancement and impairment of avoidance behavior by chemical stimulation of the hypothalamus, *J. Comp. Physiol. Psychol.* **68,** 393–399 (1969).
54 J. Sepinwall and F. S. Grodsky, Effects of cholinergic stimulation of the rat hypothalamus or cholinergic blockade on discrete-trial conflict. Paper presented at the meeting of the Eastern Psychological Association, Washington, April (1968).
55 M. H. Sheard, J. B. Appel, D. X. Freedman, *J. Psychiat. Res.* **5,** 237–242 (1967).
56 C. D. Shute and P. R. Lewis, Cholinergic and monoaminergic pathways in the hypothalamus, *Brit. Med. Bull.* **22,** 221–226 (1966).
57 L. Stein, Noradrenergic substrates of positive reinforcements: Site of motivational action of amphetamine and chlorpromazine, in *Neuropsychopharmacology* (H. Brill *et al.*, eds.), Excerpta Med. Foundation, Amsterdam (1967).
58 A. Strouthes, Effect of CS-onset, UCS-termination delay, UCS duration, CS-onset interval, and number of CS-UCS pairings on conditioned fear response. *J. Exptl. Psychol.* **69,** 287–291 (1965).
59 R. F. Thompson, *Foundations of Physiological Psychology*, Harper, New York (1967).
60 N. Weiner, Regulation of norepinephrine biosynthesis, in *Ann. Rev. Pharmacol.* (N. W. Elliott, W. C. Cutting, and R. H. Dreisbach, eds.) **10,** 273–290 (1970).
61 B. Weiss and V. G. Laties, Behavioral Pharmacology and Toxicology, in *Ann. Rev. Pharmacol.* (H. W. Elliott, W. C. Cutting and R. H. Dreisbach, eds.), **9,** 297–326 (1969).
62 C. Wise and L. Stein, Facilitation of brain self-stimulation by central administration of norepinephrine, *Science* **163,** 299–301 (1969).

CHAPTER 6

A NEURAL MODEL OF ATTENTION, REINFORCEMENT AND DISCRIMINATION LEARNING

PREFACE

This article discusses several important phenomena wherein present behavior depends on how temporal and geometrical relationships among past events are influenced by competitive feedback.

The overshadowing phenomenon shows that learners are minimal adaptive predictors who change their internal representations of external events when behavior based upon these representations generates unexpected environmental feedback. These data support an epistemological framework wherein each observer can possess idiosyncratic and personal definitions of objects or events, and these definitions are refined only when they are maladaptive. Many constraints for designing self-organizing measurement systems and for assigning *a priori* optimal estimators of the future are suggested by such a framework. The overshadowing phenomenon can be derived from a thought experiment concerning parallel processing of sensory cues which are conditioned to incompatible motivational meanings.

The partial reinforcement acquisition effect shows that "less is more" in the sense that intermittent reward can have more pervasive effects on behavior than continuous reward. Chronic gambling behavior is a typical example. This phenomenon can also be traced to the action of expectancies, notably to the disconfirmation of expectancies. As in serial learning, the nonoccurrences of events are no less important than their occurrences. The theory shows that partially rewarded behavior, which can seem maladaptive or even self-destructive, occurs when several adaptive mechanisms are simultaneously probed by certain environmental events. This result suggests that to understand whether behavior is 'really' adaptive, one must distinguish those pervasive environmental constraints that influence the evolution of neural designs from adventitious environmental fluctuations that merely play upon these designs. Once one has arrived at this insight, it also seems natural to ask whether the environment itself contains internal 'contradictions' which can manifest themselves in maladaptive behavior (Chapter 9).

The peak shift and behavioral contrast effect shows that 'more is less' in the sense that events which have never been rewarded can elicit more vigorous consummatory behavior than events which have been highly rewarded. If asked: "Will our learned preferences always be chosen from those past experiences which have been rewarded", the answer must in general be 'no'. Internal perceptual and cognitive maps can reorganize the net effects of rewards and punishments to favor as yet unexperienced alternatives. It seems ironic to me that Skinnerians, who in their traditional garb detest the use of hidden variables, should be the scientists who have best described these phenomena. Their data base is much richer than their philosophy.

Hidden variables also play an important role in answering questions like: "What motivates avoidance of a fearful situation when the avoidance behavior persists without the experience of renewed fear? What motivates language behavior?" This article suggests some answers.

My earlier work on reinforcement (Chapters 3 and 5) recognized that competitive interactions among the internal representations of external cues are needed to normalize the total STM activity of these representations. This finite capacity constraint on STM prevents large number of active STM representations from causing the release of incentive motivation if the corresponding drive level is too low. The decisive step in this article was the realization that incentive motivational signals can feed back into the competing internal representations of external cues, and thereby bias which cues will be attended. Attentional processing was hereby associated with a feedback network wherein adaptive resonances subserve the attentional state. This feedback network idealizes interactions between cortical and hippocampal structures, and implicates the hippocampus in evaluating stimulus-reinforcement contingencies that can activate observable behavior when a resonant match occurs between the reinforcing properties of available external cues and the motivating properties of active internal drives. A separate spatial map from the hippocampus is also postulated. When this work was finished in 1973, I had not yet explicitly derived the adaptive resonance concept. When I did so during the next two years from the standpoint of code development, I could mechanistically identify a sense in which adult attentional processing is a continuation of infant critical period phenomena along a developmental continuum.

Another basic concept, both in code development and in adult attention, is also related to competition, but competition that is acting on a more global scale of system design. This concept carries the factorization notion to a global extreme. It describes the competition between the consummatory behavior that generates expected consequences and the orienting behavior that mobilizes adaptation to unexpected consequences. Dualities between the expected and the unexpected, between structure and mobile energy, between reason and passion have been explored and debated throughout human history. A good attentional model should allow us to mechanistically contemplate how these dual principles achieve the exquisitely poised balance between stability and adaptability that, moment by moment, subserves our capacity to be human.

A NEURAL MODEL OF ATTENTION, REINFORCEMENT AND DISCRIMINATION LEARNING*

I. Introduction

This paper describes a psychophysiological model aimed at discussing how animals pay attention to and discriminate among certain cues while ignoring others, based on criteria of relevance derived from past experience or innately preprogrammed in their neural apparatus. The model builds upon previous results (Grossberg, 1969a,b, 1970, 1971a,b, 1972a–d, 1973, 1974; Grossberg and Pepe, 1971) that introduce some psychophysiological mechanisms of classical and instrumental learning, and of pattern discrimination. These results include network mechanisms of drive, reward, punishment, escape and avoidance, motivation, short-term and long-term memory, serial learning, arousal, expectation, and various perceptual constancies (e.g., hue and brightness). They will be reviewed herein as needed to motivate the present work. A previous paper (Grossberg, 1974) reviews some of them more systematically.

This collection of mechanisms comprises the theory of Embedding Fields. This theory derives neural networks from simple psychological facts that are taken as fundamental postulates. The theory tries to isolate postulates that act as guiding principles of neural design during individual development and the evolution of species. The networks that are hereby derived are capable of behavior that is far more complex and subtle than the postulates themselves, and also generate various new predictions. The theory is derived by a method of successive approximations; as more postulates are imposed, the networks become ever more sophisticated and reaslistic. At each stage of the derivation, basic mechanisms of network organization emerge, and are preserved as new postulates are imposed. Thus, each stage of the derivation ties a definite class of psychophysiological phenomena to a fixed list of elementary postulates, and successive stages of the derivation show how various phenomena of differing sophistication are interrelated.

A central theme in the present model will be that two systems are continually readjusting each other. One system (an attentional system) strives toward an ever more stable response to patterns of fluctuating cues by focusing attention on important subclasses of cues. This system is incapable of adapting to unexpected environmental changes. The second system (an arousal system) overcomes the rigidity of the attentional system when unexpected events occur, and allows the network to adapt to new reinforcement

* Supported in part by the Alfred P. Sloan Foundation, the Office of Naval Research (N00014-67-A-0204-0051), and the Advanced Research Projects Agency DAHC 15-73-0320) administered by Computer Corporation of America.

contingencies. The following psychophysiological themes, which clarify this situation, will be discussed in the model, among others.

A. Blocking and Overshadowing

This theme is elegantly discussed by Honig (1970), Kamin (1968, 1969), Trabasso and Bower (1968), and Wagner (1969a), who should be consulted for details. Below are tersely summarized some main experimental facts taken from these sources. We will consider a sequence of three classical conditioning experiments. In each experiment, two cues CS_1 and CS_2, such as a sound and a flashing light, are the conditioned stimuli that will precede a prescribed unconditioned stimulus UCS, such as food or shock. Let the UCS be a shock of prescribed duration and intensity, for definiteness.

In experiment 1, let CS_1 and CS_2 be equally salient to the learning subject $\mathcal{O}$, and suppose that both cues are always presented together before the shock. On recall trials, will $\mathcal{O}$ be afraid of CS_1 or CS_2 presented separately? The answer is "yes"; thus, cues presented together can be conditioned separately.

In experiment 2, first let CS_1 be paired alone with shock, until $\mathcal{O}$ is afraid of CS_1. Then present both CS_1 and CS_2 before shock during the second phase of the experiment. On recall trials, $\mathcal{O}$ is *not* afraid of CS_2. Somehow, prior conditioning of CS_1 to the UCS has "blocked," or "overshadowed," the possibility of conditioning CS_2 to the UCS. This happens even though $\mathcal{O}$ "notices" CS_2, and the amount of blocking depends on the amount of prior conditioning between CS_1 and the UCS. A blocking effect can also be elicited in experiment 1 if CS_1 is a more intense, or salient, cue than CS_2. In a similar direction, Bitterman (1965) discussed evidence that a CS which is paired simultaneously with a UCS does not get conditioned to the UCR.

In experiment 3, again pair CS_1 with the UCS before pairing both CS_1 and CS_2 with the UCS; however, choose the UCS intensity at two different levels in the two phases of the experiment. Then the blocking effect is at least partially eliminated: $\mathcal{O}$ is afraid of CS_2. (In general, one must also discuss whether a decrease in shock makes CS_2 a conditioned source of relief, rather than of fear.)

These experiments can be interpreted as follows. In the second phase of experiment 2, CS_1 is a perfect predictor of the event UCS that is about to follow. Since CS_2 is an irrelevant cue, $\mathcal{O}$ does not connect CS_2 with the UCR even though $\mathcal{O}$ notices CS_2. In the second phase of experiment 3, however, CS_1 is not a perfect predictor of UCS intensity. Hence some conditioning of CS_2 to the new UCR (or UCR-like response) occurs. In experiment 1, neither CS_1 nor CS_2 is initially a predictor of the UCS. Hence $\mathcal{O}$ will learn connections from each CS_1 to UCR. If CS_1 is more salient or intense than

CS_2, then faster conditioning of CS_1 to the UCS can eventually block conditioning of CS_2 to the UCR.

Such experiments suggest that various learning subjects act as minimal adaptive predictors; they enlarge the set of cues that control their behavior only when the cues that presently control their behavior do not perfectly predict subsequent events. In particular, somehow the results of 𝒪's acts can feed back in time to influence which cues will control these acts in the future. This phenomenon has broad implications, since it bears on such questions as: How do we decide which cues cause events and which are adventitious? How do we characterize the cues that define the objects with which we deal? Does the persistent unpredictability of a given source of cues increase the likelihood that this source will be treated more as a "subject" than as an "object"?

B. Frustrative Nonreward

A special case of an unpredictable event is one in which an expected reward does not occur. Suppose that 𝒪 has learned to expect food as the end result of a particular sequence of motor acts, but that food is no longer available in the expected place. Were 𝒪 to continue seeking food at this place, 𝒪 would starve to death. How does 𝒪 countercondition this erroneous expectation, and thereby release exploratory behavior aimed at finding new sources of food, before starvation occurs?

An aversive state that is activated by the nonoccurrence of expected events is "frustration" (Amsel, 1958, 1962; McAllister and McAllister, 1971; Wagner, 1969b). Frustration can motivate avoidance behavior and has properties analogous to those of fear.

Frustration can follow the nonoccurrence of expected rewards other than food. Thus if a sequence of events motivated by a given positive drive is suddenly interrupted, say by nonoccurrence of the expected reward at the end of a sequence of acts aimed at getting the reward, then a negative (frustrative) reaction can occur. We will argue that this rebound effect, from positive to negative, can be given a mechanistic interpretation that is shared by rebound effects from negative to positive, such as the relief that is felt when a prolonged shock is unexpectedly terminated (Denny, 1970), or various other punishment contrast and reinforcement contrast effects (Azrin and Holz, 1966). For example, let a pigeon be trained on a VI 1 schedule to peck for food. If a maintained level of punishment is suddenly removed, the pigeon will temporarily peck faster than it did in the absence of punishment. If the frequency of reward is suddenly increased, a temporary overshoot in pecking rate will again occur. The mechanism to be discussed herein also allows comparison with the facts that classically conditioned fear can rapidly extinguish, even though learned asymptotic avoidance behavior can be very stable (Seligman and Johnston, 1973).

C. Partial Reinforcement Acquisition Effect

Why can fearful or frustrating tasks that work out well in the end become so rewarding? What causes the extra "thrill" that some people seem to feel after successfully carrying out dangerous tasks? An analogous boost in reward value is illustrated by the following example. Consider the speed with which rats run down a straight alley to a positive goal. Compared to continuously rewarded animals, animals on a random partial reinforcement schedule run slower early in training, gradually catch up, and finally, late in training, run faster (Goodrich, 1959; Haggard, 1959). This effect has been attributed by several authors to frustration (Gray and Smith, 1969). We will suggest a property of the frustration mechanism that can formally generate this effect, and can predict a relationship between an animal's ability to carry out learned escape in the presence of fearful cues, the reinforcing effect of reducing J units of shock to $J/2$ units of shock, the size of the partial reinforcement acquisition effect, and the animal's arousal level, suitably defined.

D. Steepening of Generalization Gradients Due to Discrimination Training

Jenkins and Harrison (1960) showed that if pigeons are trained to peck a key in response to a 1000 cps tone (the S+) but not to peck in the absence of the tone (the S−), then a sharper tonal generalization gradient is found than after training to peck at the S+ without discrimination training with S−.

Newman and Baron (1965) used a vertical white line on a green key as S+ and the green key as S—. They tested generalization by tilting the line at various orientations. A generalization gradient was found, but no gradient occurred if the S— was a red key or if the S— was a vertical white line on a red key.

By contrast, Newman and Benefeld (Honig, 1970) used as S+ a vertical white line on a green key and as S— a green background, but tested and found generalization of the line orientation on a black key. They also tested generalization on a black key following training without a green S— and again found a significant generalization gradient, by contrast with the case where testing used a green key. This effect was interpreted to be one of "cue utilization during testing rather than cue selection during learning," since somehow removing green during testing unmasked prior learning on the orientation dimension.

Honig (1969) used a blue key as S+ and a green key as S—. This was followed by dimensional acquisition with three dark vertical lines on a white key. Generalization testing was on the orientation dimension. This paradigm was called a true discrimination (TD) experiment. By contrast, another group of pigeons was rewarded half the time on the blue key and half the time on the green key before dimensional acquisition with the three vertical lines and generalization testing on the orientation dimension. This paradigm was called a pseudodiscrimination (PD) experiment. The generalization gradient was marked in the TD case, but flat in the PD case.

F. Freeman (unpublished master's thesis, Kent State University, Kent, Ohio, 1967) modified this experiment by training pigeons to peck at a vertical line on a dark key (S+) but not to peck at a line tilted at 120° on the same dark background (S—). Then dimensional acquisition with the vertical line on a green background was followed by generalization testing on the dimension of color. A steeper color gradient was found than in the absence of prior discrimination training on S—. This is an example of *enhancement* due to prior discrimination training, rather than blocking. Blocking can also be achieved, as Mackintosh and Honig showed (Honig, 1970). They trained pigeons with S+ and S— as above. Then they retrained them with two spectral values (501 and 675 nm) redundantly added after the animals had reached criterion. Control groups received only the second stage of training. A generalization test on four spectral values demonstrated steeper gradients for the control group.

E. Peak Shift and Behavioral Contrast

Let a pigeon be trained to peck at a key illuminated by a 550 nm light (S+) but not to peck at a key illuminated by a light of x nm (S—), where x is chosen greater than 550 for definiteness. If the pigeon makes some errors in learning this discrimination, then it will, on test trials, peck most vigorously at a key lit by a light of $y(x)$ nm, where typically $y(x) \neq 550$, $y(x) < 550$ if $|x - 550|$ is sufficiently small, and $y(x)$ tends to increase as x increases (Hanson, 1959). This shift does not occur if the pigeon learns the discrimination without making errors (Terrace, 1966).

In the same experimental setting, the influence of error-filled training at x nm can increase the rate of pecking at 550 nm if $|x - 550|$ is sufficiently large ("behavioral contrast") (Hanson, 1959; Bloomfield, 1966). These effects do not occur if the training is errorless (Terrace, 1966), and behavioral contrast disappears after long training sessions (Terrace, 1966).

Honig (1962) has noted that the peak shift occurs only if the S+ and S− are presented successively, but not if they are presented simultaneously. Grusec (1968) has shown that after errorless discrimination training, pairing a shock with the S− will create a peak shift. Bower (1966) has suggested that such contrast effects are due to frustration. Bloomfield (1969) has attempted to unify these results by stating that an "unexpected change for the worse" yields contrast and peak shift effects. Such changes include a sudden reduction in the frequency of reinforcement, or the introduction of shock.

F. Orienting Reaction vs Discriminative Cues

The frustrative reaction is but one case of a general theme; namely, why can 𝒪's responses to a fixed unexpected, or novel, event be different in different contexts? For example, suppose that a human subject sits before a lever with no prior training and that a loud noise occurs abruptly to the left of the subject. There will ensue a strong tendency for the subject to orient toward the noise by turning his head to the left (Luria and Homskaya, 1970). By contrast, suppose that the subject is taught that the noise is a discriminative cue for rapidly pressing the lever to receive a valuable reward. Then the orienting reaction can be replaced by a rapid lever press. How does conditioning redirect the internal flow of activity that would otherwise activate the orienting reaction (Lynn, 1966)?

The orienting reaction is a form of attentional mechanism, but not the only one. For example, novel stimuli can attract more attention than non-novel stimuli even if the stimuli are presented tachistoscopically (Berlyne, 1970; P. McDonnell, unpublished doctoral thesis, University of Toronto, 1968; Trabasso and Bower, 1968). We will distinguish between the two types of reaction in the mechanisms to be described below.

G. Novel Events as Context-Dependent Reinforcers

As we noted above, frustration can follow the nonoccurrence of an expected reward; thus, if a sequence of events motivated by a given positive drive is unexpectedly interrupted, say by nonoccurrence of the reward, then a negative (frustrative) reaction can ensue. By contrast, if the expected reward is replaced by an even more valued reward, then the frustrative reaction can be mitigated; for example, a check for $1,000,000 might well eliminate the frustration one might feel after opening a refrigerator and noting the absence of an expected apple. In both cases, "surprise" might occur owing to the unexpectedness of the outcome, but this surprise is channeled differently in the two cases. Indeed, if an event is rewarding to an animal, then the effectiveness of the reward can be increased if it is also novel.

Berlyne (1969) notes that novel events per se can be positively rewarding. He shows that a response-contingent change in the intensity of light in a rat's cage can be used to reward bar pressing. We will suggest that the light change enhances the positive incentive-motivation that is motivating the rat during approach and pressing of the bar. This incentive motivation is not necessarily associated with a specific drive, such as hunger, and can merely be the motor arousal mechanism that is used for general approach behavior. Berlyne also notes that an increase in light level can be less rewarding if the animal's arousal level is too high. He suggests that the rewarding value of an indifferent stimulus is an inverted U function of its novelty. The inverted U is also a function of the animal's arousal level, so that a given novel stimulus can have different reward value if the animal's arousal level is varied. Berlyne distinguishes the existence of an optimal arousal *level* from an optimal arousal *increment* and discusses the relationship between a given arousal level and its optimal arousal increment in terms of the inverted U. Our model discusses related mechanisms of arousal with the property that various types of abnormal behavior can be elicited by overarousal; cf. a schizophrenic's difficulty in paying attention, or seizure activity.

In summary, we will suggest that the nonspecific neural activity generated by a novel event filters through all internal drive representations. The effect of this activity on behavior will depend on the pattern, or context, of activity in all these representations when the novel event occurs. Sometimes the novel event can enhance the effect of an ongoing drive, sometimes it can cause a reversal in sign (as in the frustrative reaction), and sometimes it can introduce and enhance the effect of a different drive. We will be led to assume that every novel event has the capacity to activate orienting reactions, but whether it does or not depends on competition from the drive loci which the event also activates. The nonspecific activity generated by the novel event will also be assumed to reach internal sensory representations, where it helps determine which cues will enter short-term memory to influence the pattern of internal discriminatory and learning processes.

H. Motivation and Generalization

Increasing an animal's motivation during learning and performance can flatten its gradient during performance (Bersh *et al.*, 1956; Jenkins *et al.*, 1958; Kimble, 1961). By contrast, let a pigeon be trained to peck a key for food, and then trained using a 1000 cps tone as a warning for electric shock. On testing trials, its generalization gradient for response suppression as a function of tonal frequency is steeper if the pigeon is hungrier (Hoffman, 1969). Note that in this experiment two drives (hunger and fear) compete, whereas in the experiments describing flattening of generalization gradients, only one drive is operative.

I. Predictability and Ulcers

Weiss (1971a,b,c) has carefully studied the influence of several parameters on the development of stomach ulcers in rats. In his experiments, some rats can escape tail shock by turning a wheel. Each turn of the wheel delays the next onset of shock by a fixed amount of time. In some studies, each shock is preceded by a warning signal. In other studies, each wheel turn is followed either by a tone or by a brief shock, but not both. In each study, there is a control group that is not shocked, and a yoked group that is shocked whenever the animals capable of avoiding or escaping the shock are shocked. The yoked group also hears the tone whenever the avoidance-escape group does. Weiss shows that (a) avoidance-escape subjects develop less ulceration than do the yoked animals; (b) a warning signal reduces the ulceration of both groups of rats; (c) the yoked animals develop less severe ulcers than the avoidance-escape animals if both groups receive a brief shock after each avoidance-escape response; and (d) little ulceration develops in the avoidance-escape group, even if no warning signal precedes shock, if each avoidance-escape response is followed by a feedback stimulus, such as a tone.

Weiss concludes from these results that two main factors contribute to the development of ulcers: the number of *coping responses* that an animal makes, and the amount of *relevant feedback* that these coping responses produce. As the number of coping responses increases, the tendency to ulcerate also increases; but as the relevant feedback increases, the tendency to ulcerate decreases. For example, in (d), the avoidance-escape animals can make many coping responses, but they also receive a high level of relevant feedback, since each successful response is followed by a feedback stimulus that predicts an interval free from shock. In (c), the avoidance-escape animals receive low relevant feedback, since they are shocked for coping. We will find that the magnitude of negative incentive-motivation in our model is a monotone increasing function of the amounts of ulceration that are described in (a)–(d). A rebound from a source of net positive incentive motivation to a source of net negative incentive motivation produces the frustrative reaction in our model. This positive source is capable of motivating consummatory motor activity. The negative source linked with it is not the same as the source of fear. Thus our results do not imply that amounts of fear equal to the amounts of negative incentive produced by the rebound will have the same effects on ulceration. They suggest, rather, that properties of the negative rebound source are triggered in parallel with, or themselves trigger, ulcerogenic agents.

J. Anatomy and Physiology

The networks will contain several functionally distinct regions. The interactions between these regions call to mind familiar anatomical facts. It will be apparent that the network regions are not presumed to be exact replicas of real anatomical fragments. Nonetheless, the anatomical relationships between the network regions, as well as their functional roles in total network processing, suggest natural analogs with real anatomies. These analogs will be pointed out both to suggest possible new insights about the functioning of real anatomies, and to serve as an interpretive marker for the networks that will arise in the future from additional postulates. The psychological validity of formal network interactions is, however, independent of how well we guess neuroanatomical labels for network components at this stage of theorizing, since the formal anatomy is still, at best, a lumped version of a real anatomy.

A network region of particular interest is reminiscent of the hippocampus. This region supplies motivational feedback to several other network areas (Olds, 1969). This feedback is determined by a competition between channels corresponding to different drives. Each channel is influenced by sensory and drive inputs. The sensory pathways can be strengthened or weakened by reinforcing events ("conditioned reinforcers"). If a given channel has a prepotent combination of input from conditioned reinforcers and drive, it will suppress other channels using its on-center off-surround anatomy (Anderson *et al.,* 1969; Grossberg, 1973). This feedback has at least three functions. It supplies signals to the region where the sensory pathways are being conditioned by reinforcing events. These signals help to determine the pattern of motivational activity that the sensory pathways will learn. Thus the mock-hippocampus receives input from a region that is implicated in reinforcement, and delivers feedback to this region. We therefore (undogmatically) interpret this second region as a mock-septum (Raisman *et al.,* 1966). The mock-hippocampus also supplies conditionable nonspecific feedback, in the form of a late, slow potential shift, to sensory processing areas (e.g., mock-neocortex) of the network. This feedback, which is related to the network's arousal, drive, reinforcement, and motivational mechanisms, helps to determine which cues will be attended to by the network. An analogous wave, the contingent negative variation (CNV), has been reported *in vivo* (Walter, 1964). Finally, the mock-hippocampus controls a feedback pathway that helps to regulate the degree of motor arousal or suppression.

If the mock-hippocampus is removed, then transfer of short-term memory into long-term memory is prevented, and difficulties in paying attention will ensue (Milner, 1958).

The mock-septum is influenced by a source of drive input (mock-hypothalamus) and of nonspecific arousal (mock-reticular formation). The level of nonspecific arousal is modulated by the degree of unexpectedness of external events. A mechanism whose motor command cells can be preset to fire only in response to expected events has been synthesized and has an anatomy reminiscent of cerebellar interactions (Grossberg, 1972a). This mechanism projects to the mock-reticular formation. Thus, although the arousal itself is nonspecific, its regulation can be dependent upon specific sensory cues. The nonspecific arousal filters through the drive-representing channels, and can either contrast enhance their activity, or cause a positive (negative) motivational bias to flip into a negative (positive) motivational bias. Thus nonspecific arousal can have specific effects on the pattern of motivational feedback. The nonspecific arousal also feeds into sensory processing areas (e.g., mock-neocortex), where it influences which cues will generate enough neural activity to reverberate in short-term memory, and thereupon be able to influence processes of learning and discrimination. The nonspecific arousal that is triggered by unexpected events differs from the nonspecific conditionable feedback that is related to network drive, reinforcement, and motivational levels. Indeed, these two input sources can compete with each other in overshadowing experiments.

In summary, at least two major feedback loops exist in the network. One feeds between external sensory and internal sensory (e.g., drive) processing areas (cortex → hippocampus → cortex). The other feeds within the internal sensory processing areas (septum → hippocampus → septum).

The drive representations are organized into dipoles, such that each dipole controls a positive and a negative incentive motivational channel; e.g., relief and fear, hunger and frustration. The regulation of motivational output from the dipoles, and of learning based on this output, has been interpreted as using two distinct transmitter systems, which are presumed to be analogous to adrenergic and cholinergic transmitters (Grossberg, 1972c). The need to synchronize the activity of the two parallel channels in a given dipole, and to sample the resultant activity in both dipole channels, suggests that the two transmitter systems are also organized in parallel across the two channels. The organization of drives into dipoles can induce a formal "poker-chip" organization in the input source that feeds them nonspecific arousal. A poker-chip anatomy for the reticular formation has been described (Scheibel and Scheibel, 1967).

II. Drives, Rewards, Motivation, and Habits

The model is an extension of a previous model that has been derived from psychological postulates (Grossberg, 1969a; 1971a, 1972b,c, 1974). This extension is the result of imposing more postulates. The old postulates describe basic properties of classical conditioning, yet the mechanisms that arise can also be used to discuss aspects of instrumental conditioning. The main postulates are described in Grossberg (1974). Two of these postulates are, for example, that (1) the time lags between CS and UCS on successive learning trials can differ; and (2) after learning has occurred, the CS can elicit the UCR (or UCR-like event) in the absence of the UCS. Such obvious facts seem innocent enough; yet when several of them are taken together, and are translated into a rigorous mathematical description, the ensuing neural networks are capable of surprisingly subtle behavior. A heuristic discussion of various mathematical properties of these networks can be found in Grossberg (1974). Some mathematical theorems are proved in Grossberg (1972d, 1973). A review of relevant network properties is given below in several stages. Consider Fig. 1.

In Fig. 1a, the ith conditioned stimulus (CS_i) among n possible stimuli excites the cell population U_{i1} of its sensory representation. In particular, CS_i has already been filtered on its way from the sensory periphery of the network to U_{i1}, so that it reliably excites U_{i1} but not irrelevant cells. Some mechanisms of sensory filtering (i.e., pattern discrimination) are derived in Grossberg (1970) and extended in Grossberg (1972a). Sensory representations will be denoted generically by $\mathcal{S}$. In response to the CS_i input, U_{i1} sends signals to stage U_{i2} of the ith sensory representation, as well as toward *all* the populations $\mathcal{A} = (\mathcal{A}_h, \mathcal{A}_f^+, \mathcal{A}_f^-, \ldots)$ of arousal cells. (In this presentation, we ignore effects due to spatial gradients in interaction strength.) Thus the

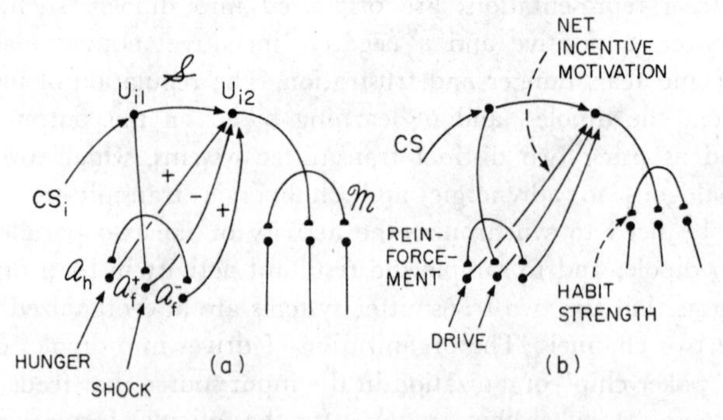

FIG. 1. Interaction of reinforcement, drive, motivation, and habit strength in minimal network.

$\mathcal{S} \rightarrow \mathcal{A}$ pathways are "nonspecific," whereas the $\mathcal{S} \rightarrow \mathcal{S}$ pathways are "specific." The arousal cells $\mathcal{A}_h$ subserve the hunger drive, cells $\mathcal{A}_f^+$ subserve fear, and cells $\mathcal{A}_f^-$ subserve relief from fear. The cells $\mathcal{A}_h$ receive an internally generated drive input that is a monotone increasing function of hunger level. The cells $\mathcal{A}_f^+$ receive an input that is a monotone increasing function of shock level. Offset of shock elicits transient excitatory activity in the relief center $\mathcal{A}_f^+$ (Denny, 1970).

Signals from the sources $\mathcal{A}$ are generated by activity at these sources and are, other things equal, monotone-increasing functions of this activity. The signals from $\mathcal{A}_h$ and $\mathcal{A}_f^-$ to all U_{i2} populations are excitatory, whereas the signals from $\mathcal{A}_f^+$ to all U_{i2} populations are inhibitory. Since a signal from a population in $\mathcal{A}$ is sent to *all* populations U_{i2}, the pathway $\mathcal{A} \rightarrow \mathcal{S}$ is nonspecific. U_{i2} can send signals to $\mathcal{M}$ only if it simultaneously receives a large signal from U_{i1} and a large net excitatory signal from $\mathcal{A}$. In particular, a large excitatory $\mathcal{A}_h \rightarrow U_{i2}$ signal can be canceled by a large inhibitory $\mathcal{A}_f^+ \rightarrow U_{i2}$ signal, which thus prevents U_{i2} from firing even if CS_i is present. In this way, consummatory activity compatible with hunger can be suppressed by shock.

Suppose that shock is terminated by an avoidance response, or AR. (Learned escape responses can become avoidance responses; hence we use only the term "avoidance" below, for simplicity.) Then $\mathcal{A}_f^-$ is excited and creates a large, but transient, excitatory $\mathcal{A}_f^- \rightarrow \mathcal{S}$ signal to all sensory representations. Sensory feedback cues of the AR also excite particular sensory representations, which we denote by $\mathcal{S}$(AR). The U_{i2} stages of $\mathcal{S}$(AR) cells thus receive U_{i1} and $\mathcal{A}_f^-$ inputs. They can therefore fire and send signals to $\mathcal{M}$. Cells in $\mathcal{S}$ that receive only the $\mathcal{A}_f^-$ input cannot fire.

Changes in long-term memory (LTM) can occur at two locations in this picture: at the $\mathcal{S} \rightarrow \mathcal{A}$ synaptic knobs, and at the $\mathcal{S} \rightarrow \mathcal{M}$ synaptic knobs. The unit of LTM is a *spatial pattern:* the relative activities of all the long-term memory traces in the synaptic knobs of a given population. The $U_{i1} \rightarrow \mathcal{A}$ synaptic knobs and the $U_{i2} \rightarrow \mathcal{M}$ synaptic knobs can learn ("sample") patterns of activity playing on the populations $\mathcal{A}$ and $\mathcal{M}$, respectively, *only* when these knobs are activated by U_{i1} or U_{i2} signals. The $U_{i1} \rightarrow \mathcal{A}$ synaptic knobs encode a weighted average of the "motivational" patterns that are sequentially presented to $\mathcal{A}$ populations when these knobs are sampling. The $U_{i2} \rightarrow \mathcal{M}$ synaptic knobs encode a weighted average of the "motor" patterns playing on the motor control cells $\mathcal{M}$ when these knobs are sampling. (The cells $\mathcal{M}$ need not be motor controllers; they can also be an arbitrary collection of sensory cells, for example.) Thus when the cells in the $\mathcal{S}$(AR) representations are active after an avoidance response, their U_{i1} stages can sample "relief" at $\mathcal{A}_f^-$, and their U_{i2} stages can sample the motor pattern corresponding to the AR. On a recall trial, the presence of avoidance or escape cues, such as seeing the wheel that delays or terminates shock when turned, activates $\mathcal{S}$(AR). Signals $\mathcal{S}(\text{AR}) \rightarrow \mathcal{A}_f^- \rightarrow \mathcal{S}$ generate positive feedback which combines with $\mathcal{S}(\text{AR}) \rightarrow \mathcal{S}(\text{AR})$ signals to activate $\mathcal{S}(\text{AR}) \rightarrow \mathcal{M}$ synaptic knobs and reproduce (a frame of) the AR. In this sense, the rebound from fear to relief when shock terminates provides the positive motivation for learning escape or avoidance motor acts.

Figure 1b describes network variables in more conventional terms. Rewarding or punishing events change the pattern of activity across populations in $\mathcal{A}$. These reinforcing events are encoded in LTM by the $\mathcal{S} \rightarrow \mathcal{A}$ synapses that are sampling at the crucial times. Drive inputs also perturb $\mathcal{A}$. $\mathcal{A}$ cells such as $\mathcal{A}_h$ can fire to $\mathcal{S}$ only if they receive a sufficiently large drive input *and* an input from a UCS or a conditioned reinforcer. Otherwise, the network would emit persistent eating behavior in response to food in the absence of hunger. In this context, a conditioned reinforcer corresponding to a given drive is a cue whose $\mathcal{S} \rightarrow \mathcal{A}$ pathway to that drive locus has become relatively large owing to prior conditioning. The $\mathcal{A} \rightarrow \mathcal{S}$ signals provide incentive motivation for modulating the activity in U_{i2}. If the net incentive motivation to U_{i2} is sufficiently positive *and* CS_i is present, then signals from U_{i2} to $\mathcal{M}$ can occur, sampling at the $U_{i2} \rightarrow \mathcal{M}$ synaptic knobs is initiated, and the habit strengths of these knobs are influenced by the patterns playing on $\mathcal{M}$ during the sampling interval.

Given this basic network structure, several refinements must be made to guarantee that the network works well. These refinements eventually lead to mechanisms that are relevant to the phenomena described in the Introduction. The next three sections review refinements that have, at least partially, been discussed in previous papers.

III. The Rebound from Fear to Relief

How does the offset of shock at $\mathcal{A}_f^+$ generate a transient rebound of activity at $\mathcal{A}_f^-$? A previous paper (Grossberg, 1972c) analyzes this question and constructs explicit rebound mechanisms. The simplest nonrecurrent (feed-forward) version of this mechanism is shown in Fig. 2, and is described mathematically in the Appendix.

An internally generated, tonic (i.e., persistent) input I derives both the $\mathcal{A}_f^+$ and the $\mathcal{A}_f^-$ channels. This input provides the activity that drives the rebound at $\mathcal{A}_f^-$ when the phasic shock-derived input J is shut off. The inputs I and J add up at v_1 and create signals along e_{13}. A smaller signal is created by I in e_{24}. At the synaptic knobs N_{13} and N_{24}, transmitter is produced at a fixed rate. This rate is inhibited by the transmitter end product at a rate proportional to transmitter concentration. The two processes taken together

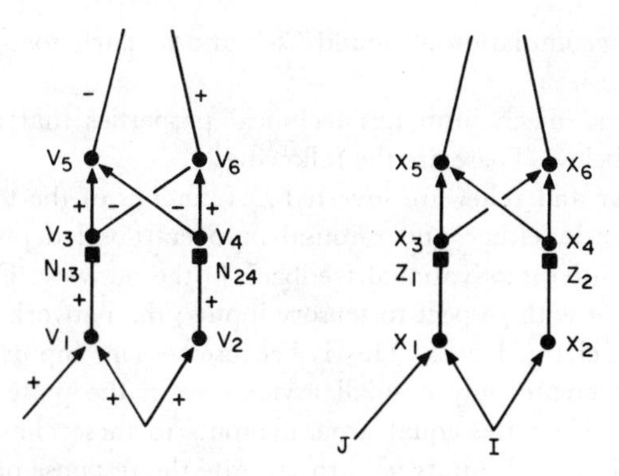

FIG. 2. Nonrecurrent rebound mechanism from fear to relief.

create the tendency for transmitter level in both N_{13} and N_{24} to approach a fixed finite upper bound. Transmitter is also removed from N_{13} at a rate jointly proportional to the signal strength in e_{13} and the amount of transmitter in N_{13}. A similar release of transmitter occurs at N_{24}. Since the signal in e_{13} is larger than the signal in e_{24} whenever $J > 0$, less transmitter exists in N_{13} than in N_{24} when $J > 0$. On the other hand, it has been proved that the multiplicative coupling of signal strength to transmitter produces a larger output from N_{13} than from N_{24} when $J > 0$. Thus when shock is on, the signals emitted by v_3 exceed those emitted by v_4. Since these signals compete subtractively at v_5 and v_6, only the output from v_5 is positive when shock is on. That is, only the fear channel supplies incentive motivation to $\mathcal{S}$. When shock is turned off, both v_1 and v_2 are driven by the equal tonic input I. The potentials at v_1 and v_2 rapidly equalize, as do the signals in e_{13} and e_{24}. By contrast, the transmitters in N_{13} and N_{24} only slowly begin to readjust to the new input levels. The input to v_3 is, however, determined by a multiplicative coupling of the signal in e_{13} with the transmitter level in N_{13}. A similar coupling of e_{24} signal with N_{24} transmitter determines the input to v_4. Since the transmitter level in N_{24} exceeds that in N_{13}, the input to v_4 exceeds the input to v_3, and hence only v_6 generates an output. Thus, after shock terminates, only the relief channel is active. Gradually the equal tonic input to v_1 and v_2 equalizes the amount of transmitter in N_{13} and N_{24}. The two channels then annihilate each other's equal signals, and no outputs arise from either channel. The relief response is transient because the imbalance in transmitter accumulation caused by shock is gradually eliminated by the uniformly distributed tonic input.

The identification of the accumulation-depletion-release substances in e_{13} and e_{24} as transmitters is speculative at present. Grossberg (1972c) cites compatible data. Any process with the same formal properties could do the job, however; cf., accumulation of bound Ca^{2+} and its participation in transmitter release.[2]

This rebound mechanism has technical properties that are relevant to the discussion below. These are the following:

1. Both fear and relief are inverted U functions of the tonic input level I. In other words, either underarousal or overarousal depresses emotional affect and incentive motivational feedback in the network. Overaroused depression is stable with respect to sensory inputs; the network is "indifferent" to emotionally charged cues. This is because sensory inputs to the fear or relief channels create only a small asymmetry in the pattern of inputs to these channels when the equal arousal inputs to these channels are large; the large equal arousal inputs tend to saturate the response of the two parallel channels. Thus, after subtractive competition between these channels, their net output is small. Underaroused depression is unstable in the sense that, after the system's elevated thresholds are exceeded by external cues (i.e., there is not enough I input to exceed threshold in response to small J inputs), either aversive or rewarding cues can cause overreactive fear or relief responses; network response is emotionally "irritable." This is because sensory inputs to the fear or relief channels create an unusually large asymmetry in the total input to these channels, since the background arousal level is smaller than usual. This phenomenon formally illustrates the paradoxical fact that underarousal can be unusually aversive in some situations and unusually rewarding in others (Berlyne, 1969).

2. There exist levels I such that maximal relief is greater than maximal fear in response to a prolonged, but then abruptly terminated, fearful cue. In fact, the ratio of maximal relief to maximal fear grows as I increases. By property (1), however, unduly small or large I levels create small absolute values of relief or fear. There exist intermediate I values, however, such that the (relief:fear) ratio is large, and the absolute size of these reactions is also large. This property is needed to make learned avoidance or escape behavior possible in the presence of fearful cues. One needs the guarantee that, although the fear channel is on, the relief channel can be so strongly activated by avoidance or escape dues that it can generate a positive *net* incentive motivational input to $\mathcal{S}$, and thereby release motor activity that leads to avoidance or escape.

3. Once the transmitter levels have adjusted to a fixed input level, either a sudden decrease in arousal input or a sudden increase in fearful cue input will cause an increase in fear. Similarly, a sudden decrease in input from irrelevant cues (i.e., cues that send equal signals to the two channels) will

[2] *Note added in proof:* C. D. Wise, B. D. Berger, and L. Stein, *Biol. Psychiatry*, 6, 1 (1973) present data suggesting that a norepinephrine reward system and a serotonin punishment system compete in parallel for relative dominance.

cause an increase in fear. By contrast, a sudden increase in arousal and/or irrelevant cue input, or a sudden decrease in fearful cue input, will tend to create a relief rebound.

4. More relief is generated by shutting off J units of shock than $J/2$ units of shock. More relief is generated by shutting off $J/2$ units of shock than by cutting the shock level from J units to $J/2$ units. A relationship exists between the rewarding effect of cutting the shock level in half, the size of the (relief:fear) ratio, and the arousal level (Grossberg, 1972c). This will be extended in Section XI to include the size of the partial reinforcement acquisition effect.

This rebound mechanism is coupled to the learning mechanism in Fig. 3. Sampling by $\mathcal{S}$ channels occurs at v_3 and v_4 for two reasons: (a) it must occur after the accumulation-depletion step to be able to sample the rebound; (b) it must occur before the subtractive stage in order to ensure that not both fear and relief control behavior at any instant of time. Another reason is given in Section V.

A noncurrent rebound mechanism is not capable of higher-order instrumental conditioning, i.e., of instrumentally motivated "chaining" (Kelleher, 1966). For example, the offset of a cue that was previously paired with shock could not be used to reward escape behavior (Maier *et al.*, 1969). To make higher-order instrumental conditioning possible, the network must be modified so that offset of activity in a conditioned $\mathcal{S} \to \mathcal{A}_f^+$ channel can drive a rebound from $\mathcal{A}_f^+$ to $\mathcal{A}_f^-$. In the above example, a cue that was paired with shock has a strong $\mathcal{S} \to \mathcal{A}_f^+$ pathway. Offset of this cue will reward escape behavior if it elicits a rebound at $\mathcal{A}_f^-$. Thus $\mathcal{S}$ must send axons to a stage *prior* to the rebound, and these axons will have conditionable synaptic knobs.

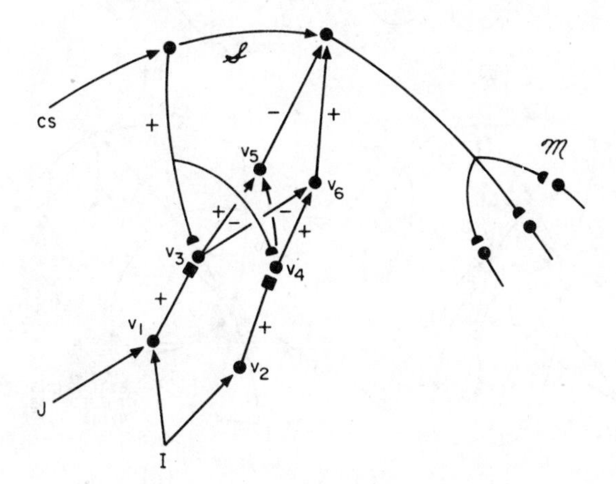

Fig. 3. Interaction of learning and rebound mechanism.

$\mathcal{S}$ must also send its axons to a stage *after* the rebound, so that cues can sample the fear and relief reactions. Thus the anatomy of $(\mathcal{A}_f^+, \mathcal{A}_f^-)$ is *recurrent* to guarantee that the rebound occurs both before and after the stage where $\mathcal{S}$ samples the rebound. See Fig. 4 for some recurrent anatomies.

In Fig. 4a, the cell sites v_1 and v_3 of Fig. 2 are identified, as are the sites v_2 and v_4. Figure 4c is particularly interesting, since it permits the learning of a stable conditioned-avoidance response that is motivated on performance trials by relief rather than fear (Maier *et al.*, 1969). The tonic source I is moved downstream from the $\mathcal{S} \to \mathcal{A}$ sampling axons. There it can drive the rebound, which occurs still further downstream, but it does not countercondition patterns in the $\mathcal{S} \to \mathcal{A}$ axons whenever sampling by $\mathcal{S}$ occurs. The outputs from the rebound stage compete before they are fed back to be sampled by $\mathcal{S}$. This feedback is positive *only* if one of the channels is stronger than the other. Thus the tonic input I alone cannot generate any feedback, and therefore does not countercondition patterns encoded in $\mathcal{S} \to \mathcal{A}$ synaptic knobs. Irrelevant cues in $\mathcal{S}$ can, however, countercondition

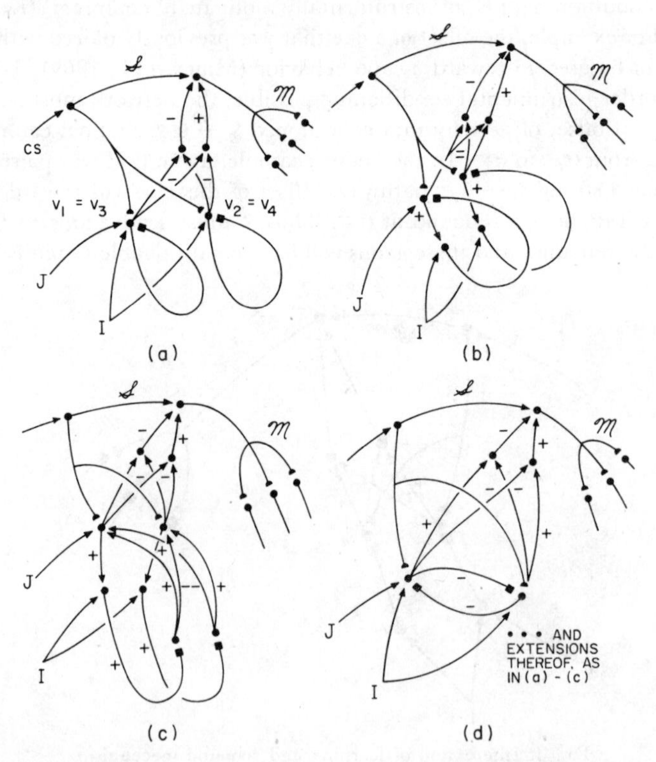

Fig. 4. Some recurrent rebound mechanisms coupled to the learning mechanism.

conditioned $\mathcal{S} \to \mathcal{A}$ pathways. Such cues send equal signals to $\mathcal{A}_f^+$ and $\mathcal{A}_f^-$, and these signals can extinguish the pattern in other active $\mathcal{S} \to \mathcal{A}$ pathways by contiguity. Even this source of extinction can be removed by a slight modification of network design: require that *only* the feedback signal along the recurrent loop can cause changes in the long-term memory of the $\mathcal{S} \to \mathcal{A}$ synaptic knobs. Then neither irrelevant cues nor the tonic input can countercondition $\mathcal{S} \to \mathcal{A}$ knobs. Section XX discusses this modification in greater detail. Section IV discusses ways to minimize the possibility of saturating the feedback loop with irrelevant cue and tonic inputs. Of course, if the avoidance or escape response does not remove $\mathcal{O}$ from sources of fearful cues, then the strong $\mathcal{S} \to \mathcal{A}_f^+$ connections which these cues control can countercondition the $\mathcal{S} \to \mathcal{A}_f^-$ channels that motivate avoidance.

In Grossberg (1972c), the inputs I and J add up at v_1, and their influence decays exponentially through time. Denote the response of v_1 (its activity, stimulus trace, or short-term memory trace) by x_1. There exists a signal threshold Γ in e_{13}, such that the signal strength in e_{13} is zero if $x_1(t) \leq \Gamma$ and is a linear function of $x_1(t) - \Gamma$ if $x_1(t) > \Gamma$. The inverted U in the relief channel, as a function of arousal level I, does not depend on the threshold Γ, but the inverted U in the fear channel does; in fact, the amount of fear is a decreasing function of I once $x_1 \geq \Gamma$.

In vivo, signal functions are not always linear functions of activity above a threshold cutoff. Often they are sigmoid functions of activity (Kernell, 1965a,b; Rall, 1955). Section IV discusses the importance of this property for the processing of neural signals in noise when the network is recurrent. Figure 4 shows that rebound mechanisms often have a recurrent anatomy. We therefore consider how the rebound mechanism is altered by making the output signals in e_{13} and e_{24} sigmoid functions of the activity levels in v_1 and v_2, respectively. The main new effect for present purposes is that, given a fixed level J of shock, an increase in the arousal level I can potentiate the fear reaction, even if the activity of e_{13} is suprathreshold (see the Appendix). This does not happen if the e_{13} signal is linear, since equal linear increments in the activity of the parallel fear and relief channels do not cause a greater than linear response to J, but they do tend to saturate the transmitter channels. Figure 5 depicts the main difference for the case of sigmoid signal functions.

Comparing Fig. 5a with Fig. 5b, we see that even if the equal arousal inputs in the parallel channels cancel at v_5 and v_6, nonetheless the response to J can be larger when it accompanies a larger arousal level. Figure 5c shows that the J response can also be smaller if the network is "overaroused." This overarousal effect can become confounded with the saturation of transmitter levels in N_{13} and N_{24}. Saturation yields an overarousal effect even if the signals in e_{13} and e_{24} are linear. Both effects cause depression of incentive motivation due to subtractive competition of signals at v_5 and v_6.

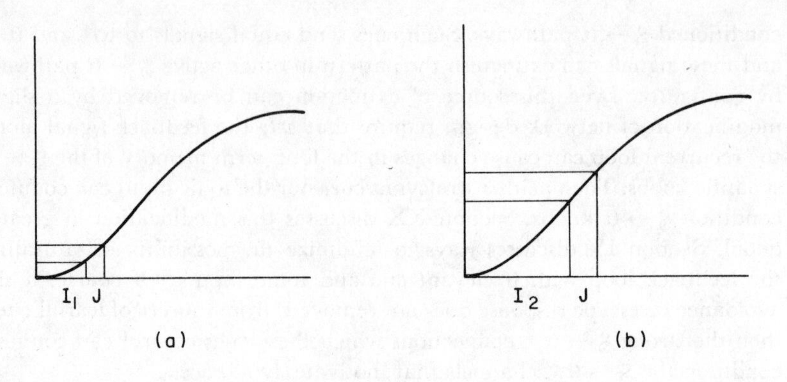

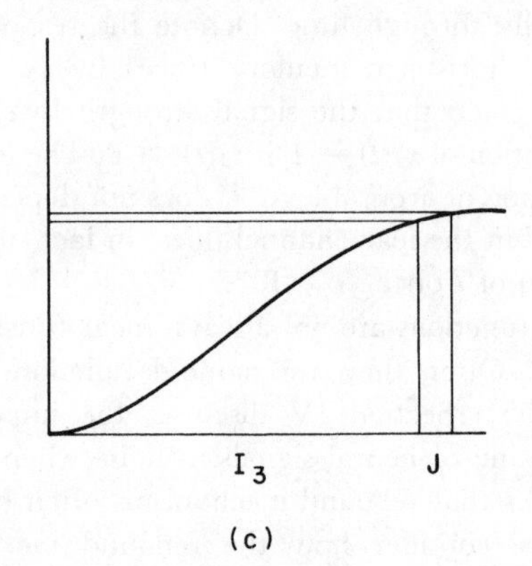

FIG. 5. Potentiation or compression of response to phasic input as arousal level varies.

The interaction of arousal level I and phasic increment J in Fig. 5 recalls Berlyne's (1969) discussion of the interaction between arousal level and optimal arousal increment. In Section XV, the effect will be used to discuss how a novel light change can reinforce lever pressing.

IV. Short-Term Memory and Total Activity Normalization

Several constraints must be placed on $\mathcal{S} \rightarrow \mathcal{A}$ signals. All these constraints can be guaranteed by the same mechanism, which is analyzed in Grossberg (1973). These constraints are listed below. They amount to predictions about the global behavior of recurrent on-center off-surround anatomies undergoing shunting interactions.

1. The *total* $\mathcal{S} \to \mathcal{A}$ output must have an upper bound that is independent of the number and intensity of $\mathcal{S} \to \mathcal{A}$ channels that are active at any time. Otherwise the $\mathcal{S} \to \mathcal{A}$ channels, in the absence of drive inputs, could activate $\mathcal{A} \to \mathcal{S}$ feedback. In such a network, persistent feeding could occur in the absence of hunger. Similarly, a *single* active $\mathcal{S} \to \mathcal{A}$ channel, combined with a prepotent drive input, should be able to activate $\mathcal{A} \to \mathcal{S}$ feedback, even though the drive input alone should not be able to do so. Thus the total activity of $\mathcal{S}$, rather than just its upper bound, must be carefully regulated. Such regulation can also prevent the feedback loop in the rebound mechanism of Section III from being saturated by irrelevant cues. We call such regulation of total activity *normalization*.

2. Consider an experiment in which lever pressing preceded by a tone is rewarded, whereas lever pressing in the absence of the tone is punished. What mechanism keeps the internal representations of the tone and the lever press active in $\mathcal{S}$ until the reward or punishment can alter the probability of lever pressing in response to the tone-plus-cues of the lever, as opposed to cues of the lever alone? A mechanism for storing these representations in short-term memory (STM) is needed.

3. Consider the processing of a pattern of inputs delivered to an ensemble of noninteracting cell populations. Suppose that neural noise exists in these populations. In many neural systems, noise cannot be avoided, if only because they operate near the quantum range, as in the case of sensory systems. In this context, if the input signals are too small, they can be lost in the noise. If they are too large, they can saturate their respective populations, thereby creating a uniform pattern of excitation across populations and destroying all information about the input pattern. To avoid these extremes in the noninteracting case, input intensities would have to be restricted to a narrow range, and the ability to process arbitrary patterns with fluctuating input intensities would be lost.

The following mechanism regulates total network activity, is capable of STM, quenches network noise, and permits the effective processing of arbitrary input patterns without saturation. The simplest example of this mechanism will be reviewed below. It is a network with a recurrent on-center off-surround anatomy whose interactions are of shunting type. The network is defined by the system

$$\dot{x}_i = -Ax_i + (B - x_i)f(x_i) - x_i\Sigma_{k \neq i}f(x_k) + I_i \tag{1}$$

where $i = i, 2, \ldots, n$, and $x_i(\leq B)$ is the mean activity of the ith cell, or cell population, v_i of the network. Four effects determine this system: (a) exponential decay, via the term $-Ax_i$; (b) shunting self-excitation, via the term $(B - x_i)f(x_i)$; (c) shunting inhibition of other populations, via the term $-x_i\Sigma_{k \neq i}f(x_k)$; and (d) externally applied inputs, via the term I_i. The function $f(w)$ describes the mean output signal of a given population as a function of its mean activity w. *In vivo*, $f(w)$ is often sigmoid function of w (Kernell, 1965a,b; Rall, 1955). This is an important property of the above model for the effective processing of signals in noise.

The system in (a) can be formally motivated as follows. Consider n states v_i whose responses $x_i(t)$ to inputs $I_i(t)$ are linear, return to equilibrium (say 0) in the absence of inputs, and have a finite maximum (say B). Then $\dot{x}_i = -Ax_i + (B - x_i)I_i$. Term $(B - x_i)I_i$ says that inactive sites become activated at a rate jointly proportional to the number of inactive sites and the excitatory input size. At equilibrium, $x_i = BI_i(A + I_i)^{-1}$, which approaches B as I_i becomes large. This system saturates and is not normalized. Both these difficulties vanish if an off-surround with shunting interaction exists. Then

$$\dot{x}_i = -Ax_i + (B - x_i)I_i - x_i\Sigma_{k\neq i}I_k$$

The new term says that active sites become deactivated at a rate jointly proportional to the number of active sites and the inhibitory input size. At equilibrium, $x_i = BI_i(A + I)^{-1}$, where $I = \Sigma_k I_k$. Letting $\theta_i = I_i I^{-1}$, we find $x_i = \theta_i BI(A + I)^{-1}$. The activity x_i is proportional to θ_i no matter how large I becomes, and the total activity $x = \Sigma_k x_k$ never exceeds B. Such a system, however, is not capable of STM. The anatomy is made recurrent by replacing each $I_i(t)$ by $f(x_i)$. External inputs are then added on to get *(a)*.

The STM capabilities of recurrent networks carry with them possible difficulties. If these networks can reverberate patterns imposed by external inputs, then why do they not also reverberate their own noise indefinitely, thereby flooding the network with its own noise? The answer is that they do, if the signal function $f(w)$ is improperly chosen. For example, if $f(w)$ is a linear function of w, or a function that grows slower than linearly, such as $f(w) = w(1 + w)^{-1}$, then noise, even in the absence of signals, will be amplified and reverberated. If $f(w)$ grows faster than linearly, such as $f(w) = w^2$, then this problem is avoided. Sufficiently small noise values will dissipate through time. If a brief, but sufficiently intense, input pattern is imposed on the noise, however, then two things happen.

First, all populations which receive the largest input in the pattern will suppress the activity in all other populations, including the noise. Second, if the function $g(w) = w^{-1}f(w)$ is convex, then *normalization* occurs: the total activity $x(t) = \Sigma_{k=1}^{n} x_k(t)$ of all the populations approaches a fixed positive limit through time. The first property shows that an extreme form of contrast enhancement occurs: only the peaks of the input pattern survive. If one population of the network receives more input than any other, then the network "chooses" this population and quenches all others. The second property shows that the system precisely regulates its total activity and can store the activity of certain populations indefinitely in STM by reverberating their activity through excitatory recurrent interneuronal loops.

The first property is too strong: too much of the pattern is suppressed in the attempt to suppress the noise. How can this be avoided? The way is to choose $f(w)$ so that it grows faster than linearly for small values of w, and (approximately) linearly at larger values of w. Then noise dissipates, and there exists a *quenching threshold*. This means that, given a sufficiently energetic pattern of inputs, the activities of populations which fall below the threshold are quenched (including noise) and those which fall above the threshold are contrast enhanced and stored in STM. Speaking intuitively, this is due to three interacting effects. The pattern is contrast enhanced for activities at which $f(w)$ grows faster than linear, but is preserved for activities at which $f(w)$ is linear. Normalization of activity can drive the system from the first to the second region of $f(w)$ growth. Thus, a pattern that is partially contrast-enhanced at small activity levels is preserved after it is normalized in the linear $f(w)$ range. Various applications of this system, and its mathematical properties, are found in Grossberg (1973). An important property of the system is shown in Fig. 6.

The function $g(w)$ alluded in Fig. 6 is defined by $g(w) = w^{-1}f(w)$; $g(w)$ tests how the signal $f(w)$ deviates from a linear function as the activity level w increases. For example, if $g(w)$ increases for small values of w and decreases for large values of w, then $f(w)$ will be a sigmoid function of w. The arrows in Fig. 6 depict the direction in which the total activity $x(t) = \Sigma_{k=1}^{n} x_k(t)$ will change if it falls in a given region between two adjacent vertical dotted lines. In Fig. 6a, $g(w)$ is convex, and only one stable equilibrium point exists. Either $x(t) \to 0$ as $t \to \infty$ if $x(0)$ is small, or $x(t)$ approaches the unique limit E_1. In Fig. 6b, $g(w)$ is not convex, and two positive limits E_1 and E_2 for $x(t)$ exist. This property motivates the following possibility, which is illustrated in Fig. 7.

In Fig. 7, a nonspecific arousal input J_A combines with a specific input J_i at each population v_i. Two important cases arise. In case 1, J_A and J_i combine multiplicatively to influence the activity level x_i. Input J_A is said to shunt the activity level (Grossberg, 1973b). In case 2, J_A and J_i combine additively to influence the activity level x_i. Consider case 1 for definiteness. Then the input J_A does not change the *relative* input levels to the various populations. (In case 2, a large J_A tends to uniformize any pattern of J_i's.) Let J_A be parametrically increased to ever higher levels. One hereby increases the number of populations that receive enough input to exceed the quenching threshold and are stored in STM. Conversely, reducing J_A decreases the number of populations that will be stored. Thus, given an input pattern in which many inputs are close to each other in relative size, one way to "make a choice" between populations is to lower the arousal level

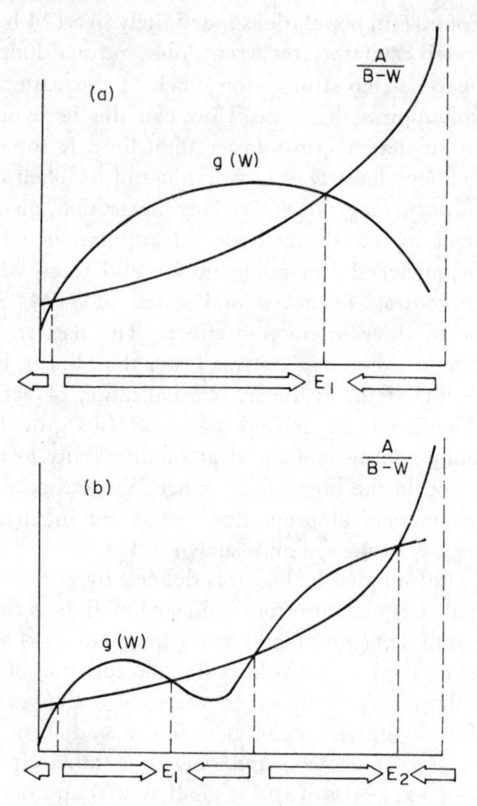

Fig. 6. Stable limit points of total activity.

of the input until only one population exceeds the quenching threshold; in common parlance, put the network in a quiet place. By contrast, one way to make as many cues as possible relevant to further network processing is to substantially increase the arousal level. Thus, suppose that a "novel" stimulus excites the network's nonspecific arousal source. Then all recently presented cues can have their network representations brought into STM to play a part in further network processing, including the sampling and subsequent learning of motor responses (Grossberg, 1972b). In this way, novel or unpredictable events can bring all possible information about presently available cues into an active state, to enhance the network's ability to deal with the unexpected situation.

A similar effect could be achieved if an increase in arousal lowered the quenching threshold by, say, decreasing the signal strength in the inhibitory

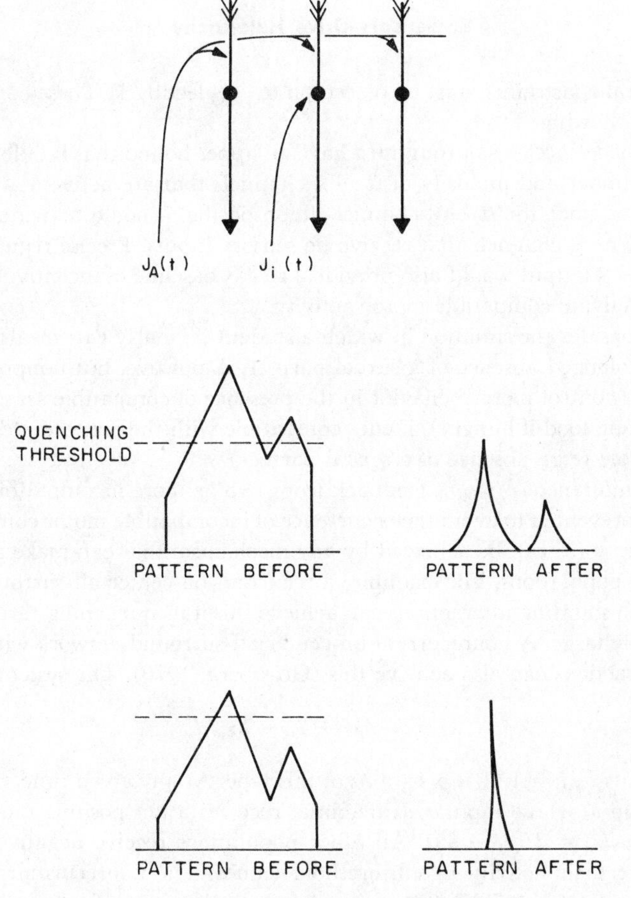

Fig. 7. Influence of arousal on final pattern stored in short-term memory.

off-surround using arousal-initiated inhibition of the inhibitory interneurons. Grossberg (1973) discusses how such a network can be thrown into "seizure" under abnormal circumstances.

Thus, we suppose that the cell populations in $\mathcal{S}$ are interconnected by a recurrent on-center off-surround anatomy whose interactions are of shunting type. This anatomy does not, of course, accomplish all the tasks that living sensory processors undertake. For example, it does not generate a sequential STM buffer (Atkinson and Shiffrin, 1968); it has no hierarchical structure. A later paper will investigate this extension in a setting that preserves properties (1)–(3) above.

V. Sensory-Drive Heterarchy

Several constraints must be placed on $\mathcal{A} \to \mathcal{S}$ feedback. These constraints are the following:

1. The *total* $\mathcal{A} \to \mathcal{S}$ output must have an upper bound that is independent of the number and intensity of $\mathcal{A} \to \mathcal{S}$ channels that are active at any time. Otherwise, since the $\mathcal{A} \to \mathcal{S}$ channel is nonspecific, it could activate $\mathcal{S} \to \mathcal{M}$ sampling by $\mathcal{S}$ channels that receive no sensory inputs. Precise regulation of total $\mathcal{A} \to \mathcal{S}$ output would also provide a steady baseline of incentive motivation to activate compatible motor activity.
2. Consider the situation in which a student regularly eats meals in spite of the prolonged absence of a sexual partner. A positive, but nonprepotent, drive can control motor behavior in the presence of compatible sensory cues (e.g., eating food if hungry), if cues compatible with the prepotent drive are unavailable (e.g., absence of a sexual partner).
3. Simultaneous $\mathcal{A} \to \mathcal{S}$ feedback from two or more incompatible drives must be prevented to avoid the occurrence of incompatible motor commands.

Property (3) can be achieved by any mechanism that can make a choice among n populations. For example, a recurrent on-center off-surround network with shunting interactions can achieve this if its quenching threshold is sufficiently large. A nonrecurrent on-center off-surround network with additive interactions can also achieve this (Grossberg, 1970). The system

$$\dot{x}_i = -Ax_i + I_i - \Sigma_{k \neq i} I_k$$

with inputs I_i, $i = 1, 2, \ldots, n$, is of this type. At any given time, only the population v_i whose input I_i is maximal receives a net positive total input $(I_i - \Sigma_{k \neq i} I_k) = 2I(\theta_i - \frac{1}{2})$. All other populations receive negative inputs that drive their activity to subthreshold values. The nonrecurrent mechanism is incapable of STM. It is driven, however, by signals from $\mathcal{S}$ which are capable of STM. Thus, the outputs from a nonrecurrent mechanism that chooses among $\mathcal{A} \to \mathcal{S}$ signals can be sustained by the STM reverberation in $\mathcal{S}$ that drives $\mathcal{S} \to \mathcal{A}$ signals.

Property (2) requires that signals from sensory cues and from drive inputs combine before the choice mechanism of property (3) determines $\mathcal{A} \to \mathcal{S}$ feedback. $\mathcal{A} \to \mathcal{S}$ feedback is then determined by the dominant combination of sensory-plus-drive cues (Fig. 8a) rather than by the dominant drive level (Fig. 8b). Figure 8a shows that a sensory-drive heterarchy can be established if the normalizer occurs after the stage at which sensory and drive inputs mix, after the stage at which $\mathcal{S} \to \mathcal{A}$ sampling occurs, after the stage at which a relief rebound occurs, and before $\mathcal{A} \to \mathcal{S}$ feedback signals can influence $\mathcal{S}$. The normalizer determines a consensus between *all* the possible $\mathcal{A} \to \mathcal{S}$ feed-

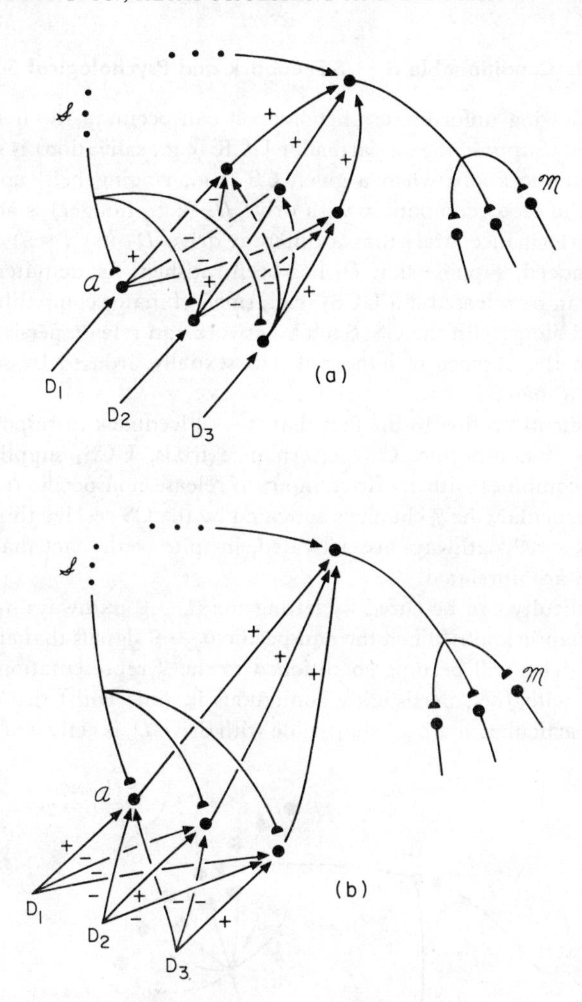

FIG. 8. Sensory-drive heterarchy vs drive hierarchy.

back channels, and thus its inhibitory interactions cut across the channels controlled by different drives.

Property (1) can be achieved by a recurrent on-center off-surround network with shunting interactions. This mechanism also regulates total activity of $\mathcal{A} \to \mathcal{S}$ feedback, and therefore provides a steady baseline of incentive motivation to activate compatible motor activity. Given a nonrecurrent choice mechanism, total activity is bounded from above, but its precise regulation can be achieved only indirectly by the regulation of total $\mathcal{S} \to \mathcal{A}$ output.

VI. Conditionable $\mathcal{A} \to \mathcal{S}$ Feedback and Psychological Set

The following unfortunate phenomenon can occur in the network thus far discussed. Suppose that a particular UCR (e.g., salivation) is encoded in $\mathcal{S} \to \mathcal{M}$ synaptic knobs when a given CS (e.g., ringing bell) and a UCS_1 (smell of food) compatible with drive D_1 (e.g., hunger) is active. Suppose on performance trials that a different drive D_2 (e.g., sex) is stronger than D_1; indeed, suppose that D_1 has been satisfied, for definiteness. Then the UCR can be released if a UCS_2 (e.g., smell of mate) compatible with D_2 is presented along with the CS. Such a network can release persistent eating behavior in the absence of hunger if it is sexually aroused by sexual cues other than a mate.

This difficulty is due to the fact that $\mathcal{A} \to \mathcal{S}$ feedback in response to any given drive is nonspecific. On performance trials, UCS_2 supplies sensory input that combines with D_2 drive input to release nonspecific $\mathcal{A} \to \mathcal{S}$ feedback. In particular, the $\mathcal{S}$ channels activated by the CS receive this feedback, and their $\mathcal{S} \to \mathcal{M}$ pathways are activated, in spite of the fact that drive D_2 and the CS are unrelated.

This difficulty can be cured by letting the $\mathcal{A} \to \mathcal{S}$ pathways have conditionable synaptic knobs. Then the nonspecific $\mathcal{A} \to \mathcal{S}$ signals that are released by a given drive will become conditioned to the $\mathcal{S}$ representations that are compatible with (i.e., persistently contiguous in time with) this drive. See Fig. 9. In particular, if a cue compatible with drive D_j is activated and drive

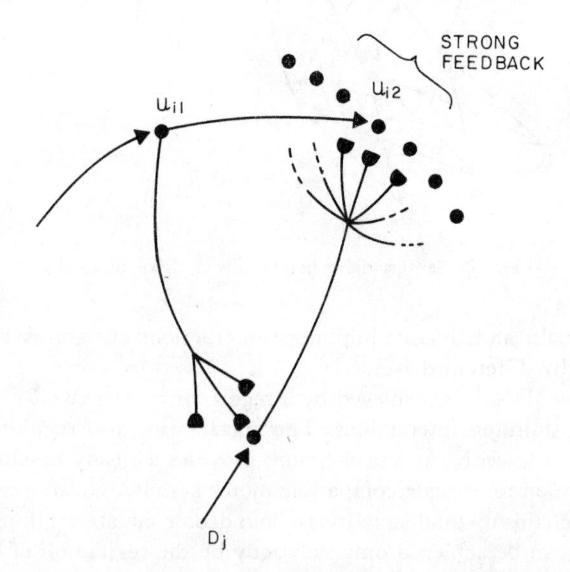

FIG. 9. Conditionable $\mathcal{A} \to \mathcal{S}$ feedback establishes psychological set.

D_j is unsatisfied, then $\mathcal{A} \to \mathcal{S}$ feedback controlled by this cue will activate the stages U_{i2} of all $\mathcal{S}$ representations that are compatible with D_j. In this way, motor activity compatible with the cue can be released, and a "psychological set," or predisposition to fire, is established in all the U_{i2} stages that are compatible with D_j.

Making $\mathcal{A} \to \mathcal{S}$ synapses conditionable increases the symmetry of the network. Now both $\mathcal{A} \to \mathcal{S}$ and $\mathcal{S} \to \mathcal{A}$ synapses are conditionable, and these pathways fire only in response to a specific cue plus a nonspecific arousal input. Do both types of cells have a common phylogenetic ancestor?

We want conditioning of synaptic knobs to occur in both excitatory and inhibitory $\mathcal{A} \to \mathcal{S}$ pathways. Then inhibitory and excitatory psychological sets can be switched on separately by different cues, but an $\mathcal{S}$ representation that often samples when both sets are on will control a mixture of excitatory and inhibitory feedback, and hence its net feedback can be small. The next section shows how to accomplish this without preventing attentiveness to fearful cues.

VII. The Persistence of Learned Meanings

Now we are ready to begin a study of attentional factors in these networks. The networks have the following unfortunate property.

Let CS_i be a conditioned reinforcer for drive D_i, $i = 1,2$. For example, let CS_1 be a roast turkey, CS_2 be one's lover, D_1 be hunger, and D_2 be sex. Consider the situation of having roast turkey for dinner with one's lover. At dinner, both CS_1 and CS_2 are scanned intermittently in rapid succession, or even simultaneously. In daily life, we do not come away from the dinner table labeling our lover as a source of food and the turkey as a source of sex, as would happen if all contiguous cues were always mutually conditioned to their respective responses. Fortunately, the learned meanings of cues can endure in spite of parallel presentation of cues with different drive representations. Of course, if the turkey is persistently and consistently paired with all of our sexual encounters, then turkey can become a discriminative cue for sex, just as pairing turkey which shock can make us afraid of turkey. The above example distinguishes the forced pairing of events from the free reorganization of attention through time.

How can persistence of learned meanings be achieved in these networks? Let the $\mathcal{S}$ representations that are activated by CS_i be denoted by $\mathcal{S}_i$, $i =$ 1, 2. We want $\mathcal{O}$ to be able to "notice" each CS_i as it is scanned. Thus each CS_i should be able to activate its $\mathcal{S}_i$. We also want to prevent *sustained simultaneous* sampling of $\mathcal{A}$ by $\mathcal{S}_1$ and $\mathcal{S}_2$. Otherwise, $\mathcal{S}_1$ would activate the $\mathcal{A}_1$ channel, and $\mathcal{S}_2$ would sample this channel and strengthen its $\mathcal{A}_1$ connection. Simultaneously $\mathcal{S}_2$ would activate $\mathcal{A}_2$, and $\mathcal{S}_1$ would strengthen its $\mathcal{A}_1$ connection. In effect, the turkey would become a cue for sex and the lover would become a cue for eating.

Our task is to prevent sustained simultaneous sampling of $\mathcal{A}$ by $\mathcal{S}_1$ and $\mathcal{S}_2$ if $\mathcal{S}_1$ and $\mathcal{S}_2$ project to incompatible drives. To achieve this in the present context, at least three stages of processing are needed:

1. $\mathcal{S}_1$ and $\mathcal{S}_2$ send signals to $\mathcal{A}$ in order to test which $\mathcal{A}$ channels they control (e.g., *do* $\mathcal{S}_1$ and $\mathcal{S}_2$ control incompatible drives?).
2. The $\mathcal{A} \rightarrow \mathcal{S}$ feedback measures which $\mathcal{A}_i$ channel is stronger at any time, via the sensory-drive heterarchy.
3. $\mathcal{S}_i \rightarrow \mathcal{A}$ sampling is shut off in the weaker channel.

Stage 1 is accomplished when the CS_i inputs activate $\mathcal{S}_i \rightarrow \mathcal{A}_i$ signals. Stage 2 is accomplished by the sensory-drive heterarchy. How is stage 3 accomplished? Suppose for definiteness that the $\mathcal{S}_1$ channel is stronger; that is, the strength of sensory-plus-drive inputs to $\mathcal{A}_1$ exceeds that to $\mathcal{A}_2$, so that only $\mathcal{A}_1 \rightarrow \mathcal{S}$ feedback is positive. Somehow this feedback must suppress $\mathcal{S}_2 \rightarrow \mathcal{A}$ sampling. By Section VI, $\mathcal{A}_1 \rightarrow \mathcal{S}$ feedback will be received only by the $\mathcal{S}_1$ representation. How does this feedback suppress $\mathcal{S}_2$ reverberation and sampling? An answer is suggested by Section IV. The total activity of the $\mathcal{S}$ representations is normalized, and a quenching threshold exists. We want strong $\mathcal{A}_1 \rightarrow \mathcal{S}_1$ feedback to enhance the activity of the $\mathcal{S}_1$ representation and, as a consequence of normalization, to thereby, at least partially, suppress the activity, and hence the sampling, of the $\mathcal{S}_2$ representation. The minimal way to accomplish this is to require that *specific* $U_{i2} \rightarrow U_{i1}$ signals exist in each $\mathcal{S}$ representation (see Fig. 10a). In Fig. 10a, strong $\mathcal{A}_1 \rightarrow U_{12} \rightarrow U_{11}$ feedback increases the strength of activity in the U_{11} population relative to the activity in the U_{21} population. The U_{21} activity is thereupon suppressed by inhibitory signals from U_{11} to U_{21}.

The above argument holds if the drives in question control positive $\mathcal{A} \rightarrow \mathcal{S}$ feedback. The case of drives, such as fear and frustration, which control negative feedback requires further argument. The problem is this. If the conditioned feedback is negative, then it will tend to differentially suppress activity in the controlling $\mathcal{S}$ representation, rather than to enhance it. This would have the following maladaptive effect on behavior. Increasing the learned fearfulness of a given cue, in a fixed context of other cues, would decrease the attention paid to it. Jumping ahead in our discussion for a moment, we also would note that fearful cues could not overshadow or block learning in response to other cues, which is false (Kamin, 1968, 1969).

Hence a distinction must be made between mechanisms for learned persistence of negative meanings and for negative incentive motivation. See Figs. 10b and 10c. The former feedback channel helps to focus attention on particular cues. The latter feedback channel suppresses motor activity. The

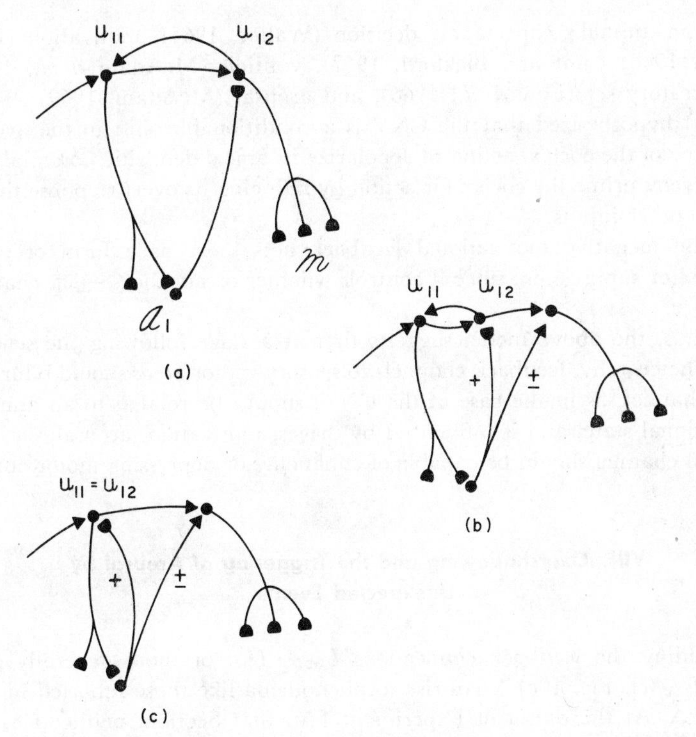

Fig. 10. Attentional feedback and motor arousal occupy different channels.

attentional feedback is always positive, even if the drive in question controls negative incentive motivation. The synapses of this feedback channel are, moreover, conditionable, so that the feedback can enhance the activity of particular representations on which attention will focus.

Given that attentional feedback is conditionable, is it also necessary to make the $\mathcal{A} \rightarrow \mathcal{S}$ incentive motivational synapses conditionable? At any given time, the conditionable attentional feedback will guarantee that only certain U_{i1} stages will send signals to their respective U_{i2} stages. Can an irrelevant drive release $\mathcal{S} \rightarrow \mathcal{M}$ sampling in the manner described by Section VI? If the irrelevant drive creates conditioned attentional feedback to its "psychological set" in $\mathcal{S}$, then this set will tend to quench other sources in $\mathcal{S}$, and therefore to prevent their firing. Thus the $\mathcal{A} \rightarrow \mathcal{S}$ incentive motivational feedback is not necessarily conditionable, although making it conditionable could only improve network efficiency.

The two kinds of feedback can be interpreted as slow potential shifts. The conditionable attentional feedback is reminiscent of the contingent negative variation, or CNV (Cohen, 1969). Such a wave has been associated with an animal's expectancy, decision (Walter, 1964), motivation (Irwin *et al.*, 1966; Cant and Bickford, 1967), volition (McAdam *et al.*, 1966), preparatory set (Low *et al.*, 1966), and arousal (McAdam, 1969). Walter (1964) hypothesized that the CNV is a conditionable shift in the average baseline of the cortex, acting to depolarize its apical dendritic potentials and to thereby prime the cortex for action by reducing its overt response threshold to other inputs.

The incentive motivational feedback acts more as a form of motor arousal or suppression, since it controls whether or not the $\mathcal{S} \rightarrow \mathcal{M}$ channels will fire.

Thus, the above model suggests that at a stage following the sensory-drive heterarchy, feedback channels to sensory-motor areas should bifurcate; one channel, as in the case of the CNV, should be related to an animal's attentional state, and is influenced by drives, motivation, arousal, etc. The second channel should be capable of enhancing or depressing motor output.

VIII. Overshadowing and the Triggering of Arousal by Unexpected Events

Adding the feedback connections $U_{i2} \rightarrow U_{i1}$, or more generally from $\mathcal{A}$ to U_{i1} (cf. Fig. 10c), gives rise to phenomena like those reported in Section I,A. At the outset of Experiment 1 in that Section, neither CS_1 nor CS_2 projects to any particular drive representation in $\mathcal{A}$. Thus both CS_1 and CS_2 can sample the fear representation when shock is on. Since the total $\mathcal{S} \rightarrow \mathcal{A}$ output is normalized, the strength of $\mathcal{S}_i \rightarrow \mathcal{A}$ signals depends on how many $\mathcal{S}_i$ channels are active at any time. Thus learning by $\mathcal{S}_1$ and $\mathcal{S}_2$ activated together will be slower than learning by $\mathcal{S}_1$ activated alone, unless there exists more than one limit point for $x(t)$, as in Fig. 6b.

In experiment 2, CS_1 first becomes conditioned to fear, which we will call the $\mathcal{A}_1$ channel. The channels $\mathcal{S}_1 \rightarrow \mathcal{A}_1$ and $\mathcal{A}_1 \rightarrow \mathcal{S}_1$ both become conditioned during the first phase of this experiment. When CS_1 and CS_2 are presented in phase 2 of the experiment, $U_{11} \rightarrow \mathcal{A}_1 \rightarrow U_{12} \rightarrow U_{11}$ feedback suppresses sampling in the $U_{21} \rightarrow \mathcal{A}_1$ channel before CS_2 can become conditioned to fear. CS_2 is hereby overshadowed by prior fear conditioning to CS_1.

Suppose in experiment 1 that CS_1 is more salient than CS_2. Then the sampling signals from $\mathcal{S}_1$ to $\mathcal{A}$ will initially be larger than those from $\mathcal{S}_2$ to $\mathcal{A}$. Consequently learning in $\mathcal{S}_1 \rightarrow \mathcal{A}_1$ synaptic knobs will occur faster than learning in $\mathcal{S}_2 \rightarrow \mathcal{A}_2$ synaptic knobs. Similarly, learning in the feedback channel $\mathcal{A}_1 \rightarrow \mathcal{S}_1$ will occur faster than learning in the $\mathcal{A}_2 \rightarrow \mathcal{S}_2$ channel. The $U_{11} \rightarrow \mathcal{A}_1 \rightarrow U_{12} \rightarrow U_{11}$ feedback therefore grows faster than the $U_{21} \rightarrow$

$\mathcal{A}_1 \to U_{22} \to U_{21}$ feedback. Sampling by U_{21} is hereby gradually suppressed as learning trials proceed, and CS_2 is gradually overshadowed by CS_1. Similarly, if a CS and UCS are simultaneously presented, then the UCS can overshadow the CS via $\mathcal{S} \to \mathcal{A} \to \mathcal{S}$ feedback. If the CS occurs shortly before the UCS, then its sampling channels are active in the time interval after the UCS occurs and before $\mathcal{S} \to \mathcal{A} \to \mathcal{S}$ feedback can quench their activity. Hence CS $\to$ UCR conditioning is possible in this latter case.

Experiment 3 is not so easily approached. Somehow, the occurrence of an unexpected UCS must prevent CS_1 from overshadowing CS_2. Either the $U_{11} \to \mathcal{A}_1 \to U_{21} \to U_{11}$ feedback must be weakened, or an independent nonspecific (e.g., "arousal") input to $\mathcal{S}$ must keep activity at $\mathcal{S}_2$ in the supra-threshold range.

Weakening $U_{11} \to \mathcal{A} \to U_{21} \to U_{11}$ feedback does not seem to be a physically plausible way to overcome overshadowing. To see this, change experiment 3 as follows: in phase 1 of the experiment ($CS_1 \to$ shock), use 40 units of shock, and in phase 2 of the experiment ($CS_1 + CS_2 \to$ shock), use 80 units of shock. The increase in shock level is unexpected, but it should surely be accompanied by an *increase* in $\mathcal{A}_1 \to U_{12}$ feedback. Indeed, the very survival of an animal can depend on its ability to process the reinforcing characteristics of unexpected events. The increase in $\mathcal{A}_1 \to U_{12}$ feedback would increase the overshadowing of CS_2 by CS_1, other things equal, but just the reverse occurs *in vivo*.

Overshadowing can be eliminated, or at least reduced, if unexpected events transiently increase the nonspecific arousal of $\mathcal{S}$, and thus the number of $\mathcal{S}$ representations whose activity exceeds quenching threshold. This increase in overall arousal of $\mathcal{S}$ competes with overshadowing tendencies controlled by motivational $\mathcal{A} \to \mathcal{S}$ channels. Alternatively, it is possible that unexpected events transiently decrease the quenching threshold of $\mathcal{S}$. The latter effect could be achieved, say, by letting an unexpected event trigger shunting inhibition of the inhibitory interneurons in the off-surround of each population in $\mathcal{S}$. The triggering of arousal by unexpected events will be seen to be a basic feature of the model for dealing with a variety of phenomena (see Fig. 11). For example, the Appendix derives a formula showing that CS_2 can become a learned source of relief, rather than of fear, if the shock level that follows $CS_1 + CS_2$ is lower than the shock level that follows CS_1. This can be achieved using the increase in tonic arousal input to $\mathcal{A}$ that accompanies the unexpected change in shock level (cf Section IX). The increase in arousal at $\mathcal{A}$ enhances the tendency for a relief rebound to occur, whereas the increase of arousal at $\mathcal{S}$ overcomes overshadowing and enables $\mathcal{S} \to \mathcal{A}$ sampling of this rebound to occur. By contributing an increase in irrelevant cue input to $\mathcal{A}$, the increase in arousal at $\mathcal{S}$ can also enhance the relief rebound at $\mathcal{A}$.

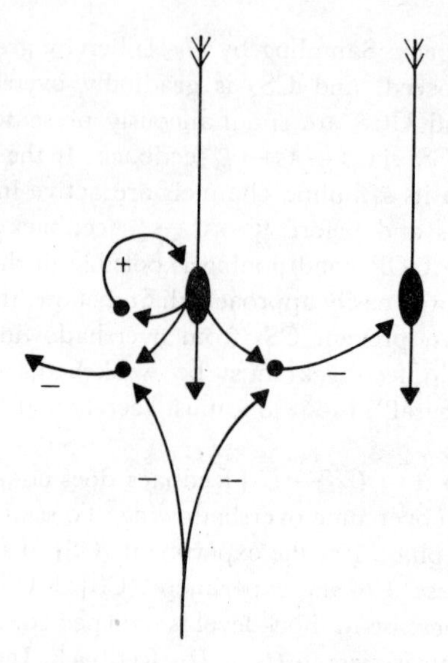

FIG. 11. Arousal-initiated inhibition of inhibitory interneurons.

Section XXI notes some possible clinical differences that would arise due to overactivity of nonspecific arousal to $\mathcal{S}$ vs overactivity of conditionable $\mathcal{A} \rightarrow \mathcal{S}$ feedback, especially with regard to the hypothesis that certain schizophrenic symptoms are due to imbalances in catecholamine production.

The above conclusions can be phrased in a way that emphasizes the adaptability of a network to changing environmental demands as a fundamental principle of its design. The mechanism for preserving learned meanings of cues is an adaptive attentional mechanism that permits parallel processing of cues without spurious cross-conditioning of the learned meanings of all cues. Overshadowing is a consequence of this mechanism. Overshadowing can, however, yield maladaptive network performance if the environment changes, or is only partially understood, since then the cues that presently control network output will be imperfect predictors of environmental response to this output. The property of persistence, by itself, creates too rigid a network. Taken together with the liberating effect of unexpected events on nonspecific arousal (or the quenching threshold), it can achieve a stable, but adaptively changeable, attentional mechanism.

The above discussion reduces the overshadowing problem to the problem of how arousal is triggered by an unexpected, but *not* by an expected, event. This latter problem can be restated in an informative way: how does a network *habituate* (Grossman, 1967) to a repetitively presented, and therefore increasingly expected, event? A mechanism whereby network output is regulated by the expectedness of an event is described in Grossberg (1972a), and will be applied to the present case in Section XIX.

IX. Pavlovian Fear Extinction vs Persistent Learned Avoidance

The above results suggest a mechanism for the fact that classically conditioned fear can rapidly extinguish, whereas learned asymptotic avoidance behavior can be very stable. An explanation that uses the concept of expectation in a descriptive psychological theory has been given by Seligman and Johnston (1973). Our neural explanation will use the arousing effect of unexpected events on the fear–relief dipole of Section III. Figure 4c illustrates a mechanism in which avoidance is stable, if it does not confront the network with a new source of fearful cues. To approach the fear extinction problem we suppose that an unexpected event transiently increases not only the arousal input to $\mathcal{S}$, but also the arousal input I to the fear–relief dipole; e.g., imagine that both regions receive arousal from a common source, such as reticular formation. Using this hypothesis, a mechanism of fear extinction is the following.

Suppose that a CS_1 (e.g., bell) has persistently been paired with a shock UCS. Eventually $\mathcal{S}_1$ will project strongly to the fear channel $\mathcal{A}_f^+$, and will be capable of generating a conditioned emotional response (cf. Grossberg, 1972b). If on a performance trial, the CS_1 is not followed by the expected shock, then a transient increase in I occurs and causes a rebound at $\mathcal{A}_f^-$. This rebound is sampled by $\mathcal{S}_1$. The $\mathcal{S}_1 \rightarrow \mathcal{A}_f^+$ channel is hereby counterconditioned by the increase in relative strength of the $\mathcal{S}_1 \rightarrow \mathcal{A}_f^-$ channel, since the *net* positive feedback controlled by $\mathcal{S}_1$ decreases. If the fear has been suppressing consummatory activity based on a positive drive, then spontaneous recovery of this activity can occur; the $\mathcal{S} \rightarrow \mathcal{M}$ synapses which encoded the activity were not counterconditioned by fear suppression, and the positive incentive motivation that originally activated these synapses is no longer inhibited by fear (Grossberg, 1972b).

A similar rebound effect, triggered by arousal subsequent to an unexpected event, can be used to approach a neural mechanism of frustration.

X. Frustration

Let a CS_1 (e.g., bell) supported by drive $\mathcal{D}_1$ (e.g., hunger) be conditioned to a response (e.g., lever press) to satisfy $\mathcal{D}_1$ (e.g., with food that appears after the lever is pressed). Suppose that the expected food does not appear. How does the network prevent itself from persistently responding to this

CS_1 with lever pressing for food? This problem can be phrased in a more general way as follows: How does an organism stop persistently performing learned motor acts which no longer satisfy its needs, and free itself to seek new sources of gratification before it suffers irreversible damage due to prolonged deprivation?

To accomplish this in the networks which have already been derived, we want the nonoccurrence of the expected event to create a negative incentive-motivational output that can be sampled by $\mathcal{S}_1$. Thereafter, the occurrence of CS_1 will create signals from $\mathcal{S}_1$ both to the positive incentive-motivational source that used to support the motor act, as well as to the negative incentive-motivational source. The *net* incentive-motivation will decrease until CS_1 no longer elicits the erroneous response.

Clearly this situation is analogous to that involving fear and relief. This analogy is depicted graphically in Fig. 12. In Figure 12a, a sudden reduction in shock or $\mathcal{S} \to \mathcal{A}_f^+$ input, or a sudden increase in irrelevant $\mathcal{S} \to \mathcal{A}$ or I input tend to cause a rebound at $\mathcal{A}_f^-$. In Fig. 12b, suppose that the network is engaged in a sequence of behaviors compatible with hunger. Then a persistent, large $\mathcal{S} \to \mathcal{A}_h^+$ input drives positive $\mathcal{A}_h^+ \to \mathcal{S}$ feedback that supports this behavior. Regulating this input through time is one of the tasks of the recurrent normalizer in $\mathcal{S}$. Suppose that the nonoccurrence of the expected event follows this sequence of acts. Because the expected event does not occur, $\mathcal{S} \to \mathcal{A}_h^+$ input can suddenly decrease. (Temporarily ignore the case in which an unexpected event projects to $\mathcal{A}_h^+$.) Simultaneously the nonoccurrence creates an increase in *I*. Both these factors conspire to create a negative incentive-motivational rebound at $\mathcal{A}_h^-$ that $\mathcal{S}_1$ can sample. Also note that an increase of arousal to $\mathcal{S}$ can also create an increase in the total $\mathcal{S} \to \mathcal{A}$ input due to cues that are "irrelevant" with respect to the $(\mathcal{A}_h^+, \mathcal{A}_h^-)$ dipole. This input can also contribute to the rebound at $\mathcal{A}_h^-$.

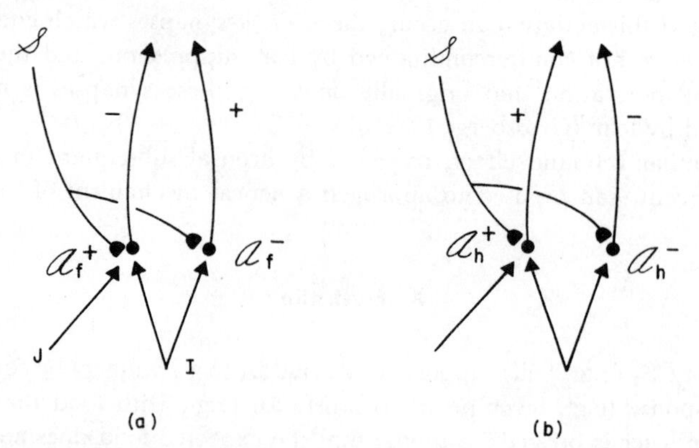

FIG. 12. Fear–relief (a) and hunger–frustration (b) dipoles.

We hereby assume, as in the case of fear and relief, that *every* drive capable of generating incentive motivational feedback has a complementary drive capable of generating incentive motivational feedback of the opposite sign. In short, drives are organized in dipoles. The properties of the fear–relief dipole are assumed to occur also in all other dipoles; e.g., a sudden reduction in expected reward can cause frustration, just as a sudden reduction in expected shock can cause relief. We also imagine that all negative incentive motivational sources are grouped together anatomically, and that all positive incentive motivational sources are grouped together. The dipole organization then becomes a *universal* feature of network design. The different behavioral meanings of particular dipoles are determined by the particular input sources that perturb them (e.g., shock at a given negative source, or metabolic levels parameterizing deprivation states at various positive sources). This anatomical grouping into positive and negative sources is reminiscent of the organization of lateral hypothalamus as a reward center and of ventromedial hypothalamus as a punishment center (Grossman, 1967) (see Fig. 13).

In Fig. 13, each dipole receives a tonic arousal input whose size through time is influenced by the unexpectedness of events. If the dipoles are arranged in a regular fashion (e.g., a row), then these arousal sources can also be regularly laid out in the network. A plausible candidate for these arousal sources is the reticular formation (Thompson, 1967). Perhaps the "poker chip" organization of the dipoles is one reason for the "poker chip" organization of reticular formation anatomy that has been so elegantly investigated by the Scheibels (1967).

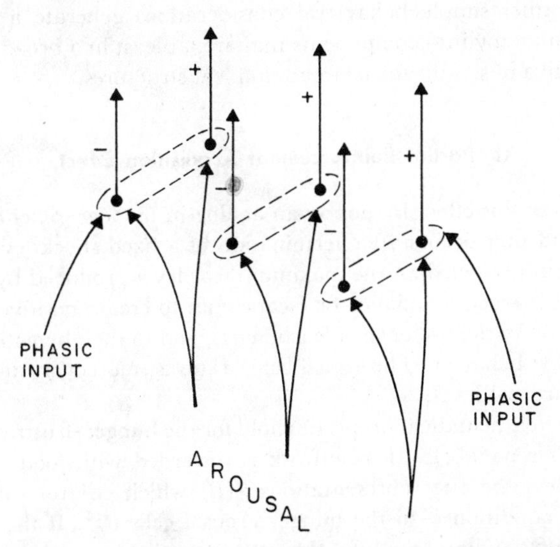

Fig. 13. Regular organization of dipoles and supportive arousal sources.

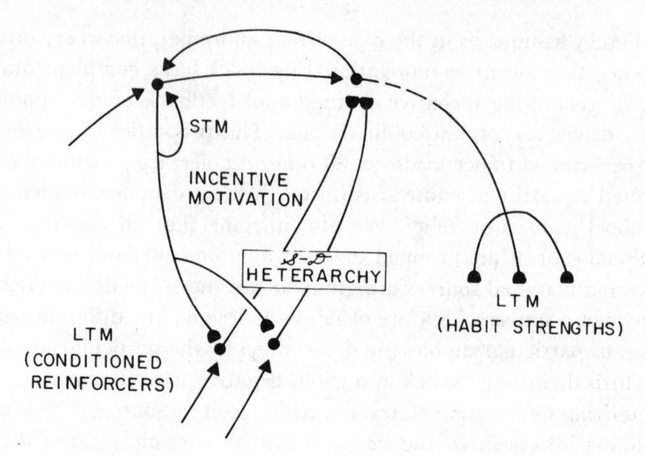

FIG. 14. Influence of motivational feedback on transfer of short-term memory (STM) to long-term memory (LTM).

The normalizer that determines the final sensory-drive heterarchical feedback to $\mathcal{S}$ must come after the dipole stage. Various data suggest the plausibility of interpreting this normalizer as an idealized analog of hippocampus. It can, for example, maintain a baseline of learned incentive motivational feedback (Olds, 1969), its elimination can prevent transfer of short-term memory to long-term memory (Milner, 1958), and it is organized as a recurrent on-center off-surround anatomy (Anderson *et al.*, 1969) (see Fig. 14). Thus rather simple behavioral considerations generate a subdivision of network anatomy into components that are, at least in a broad qualitative way, suggestive of significant neuroanatomical structures.

XI. Partial Reinforcement Acquisition Effect

We analyze this effect by noting an analog in the fear–relief dipole. The relief rebound that occurs after termination of a fixed shock level typically has a maximum larger than the maximum fear level produced by the shock. This permits learned avoidance or escape cues to create positive net incentive motivation in the presence of fearful cues, and to thereby activate avoidance or escape behavior. The (relief: fear) ratio is, moreover, an increasing function of arousal level.

The same mathematical properties hold for the hunger–frustration dipole, but with different labels. If a network is rewarded with food for running down an alley, the cue representations $\mathcal{S}(R)$ which control this behavior will become conditioned to the hunger arousal cells $\mathcal{A}_h^+$. If the network is rewarded on a random schedule, then the nonoccurrence of expected food can create a frustrative rebound at $\mathcal{A}_h^-$, both by decreasing $\mathcal{S} \rightarrow \mathcal{A}_h^+$ input and by increasing nonspecific arousal due to the unexpected event. The $\mathcal{S}(R)$ and other $\mathcal{S}$ representations that are active during the $\mathcal{A}_h^-$ rebound will therefore be conditioned to $\mathcal{A}_h^-$. The net feedback from $(\mathcal{A}_h^+, \mathcal{A}_h^-)$ to $\mathcal{S}$ activated by these representations will therefore be smaller than in the case of a continuously rewarded network, for which only $\mathcal{S}(R) \rightarrow \mathcal{A}_h^+$ conditioning occurs. In particular, early in training, while the $\mathcal{S}(R) \rightarrow \mathcal{A}_h^+$ connection and the $\mathcal{A}_h^-$ rebound develop, the partially rewarded network will find the goal box less attractive than the continuously rewarded network. As training proceeds, an ever stronger $\mathcal{S}(R) \rightarrow \mathcal{A}_h^-$ projection develops. The frequency of reward must be adjusted to prevent the $\mathcal{S}(R) \rightarrow \mathcal{A}_h^+$ channel from dominating the $\mathcal{S}(R) \rightarrow \mathcal{A}_h^+$ channel; otherwise the network will eventually stop seeking the goal. Suppose that the frequency of reward has been suitably adjusted. Then what happens on a reward trial late in the training of a randomly rewarded network? First, the usual boost in $\mathcal{S} \rightarrow \mathcal{A}_h^+$ connection strength will occur. Second, there will be a sudden reduction in activity of the cues that are conditioned to frustration. Third, owing to the partial unexpectedness of reward, there can be a transient increase in arousal. The second and third effects both tend to create a rebound from $\mathcal{A}_h^-$ to $\mathcal{A}_h^+$ due to "frustration reduction," just as reduction of shock intensity tends to produce a relief rebound. This rebound combines with the usual rewarding effect of food to produce an enhancement of the desirability of the goal late in training. In short, the enhancement can be analyzed as a double rebound, with conditioning in between, from $\mathcal{A}_h^+$ to $\mathcal{A}_h^-$, and then back to $\mathcal{A}_h^+$.

It is ironic that the persistence of this kind of frustrating behavior can be analyzed as a composite of two effects that have a manifestly adaptive biological value. The rebound from positive to negative allows an organism to countercondition erroneous expectations before they do irreversible damage. The rebound from negative to positive allows an organism to learn escape from or avoidance of noxious events. The rebound from positive to negative to positive can, however, generate a maladaptive persistence in seeking an unlikely goal; e.g., gambling. Moreover, while cues that are conditioned to frustration are active, they can negatively bias the interpretation of other cues as possible alternatives, and can suppress exploratory behavior by inhibiting positive incentives. The network can "fall into a rut," and might tenaciously await the elusive frustration reduction that can give it some relief.

XII. Generalization Gradients in Discrimination Learning

Our discussion of discrimination learning will try to show how various mechanisms fit together to qualitatively generate empirical effects. A more quantitative analysis will appear in another place. The main new assumption is that a sensory input will excite not only its own internal representation in $\mathcal{S}$, but also the representations of closely related inputs, e.g., the auditory system in which peripheral auditory cells have "tuning curves" that include a connected band of frequencies (see Fig. 15). Given this elementary fact, the model already at our disposal generates various nontrival effects.

The populations in $\mathcal{S}$ will also be interconnected by a recurrent on-center off-surround field of shunting type for the reasons cited in Section IV. There exist variations on this theme which will not be considered here, but which are being studied. For example, let each input perturb only its own $\mathcal{S}$ representation, but let the on-center and off-surround of each representation have a generalization gradient that includes related populations; or let the inputs, the on-center, and the off-surround have such a generalization gradient. Before considering particular experiments, we now note various general properties of this system, Let CS_i be an X_i cps tone for definiteness. Suppose that CS_i activates $\mathcal{S}_i$ while $\mathcal{A}_i^+$ is the dominant active arousal source. Then all representations in $\mathcal{S}$ which are sufficiently excited by CS_i to exceed the quenching threshold of $\mathcal{S}$ will sample $\mathcal{A}_i^+$ with an intensity that increases as a function of input strength. If $i = h$, then $\mathcal{S}_h$ eventually controls positive $\mathcal{A} \rightarrow \mathcal{S}$ feedback. If $i = f$, then $\mathcal{S}_f$ eventually controls negative $\mathcal{A} \rightarrow \mathcal{S}$ feedback (see Fig. 16).

Let a test CS (CS_T) have an $\mathcal{S}$ representation ($\mathcal{S}_T$) that lies within the generalization gradients of both $\mathcal{S}_h$ and $\mathcal{S}_f$. $\mathcal{S}_T$ will become conditioned to both $\mathcal{A}_h^+$ during presentations of CS_h or CS_f, with a relative strength that depends in part on how close $\mathcal{S}_T$ lies to these foci of input activity. Thus the net $\mathcal{A} \rightarrow \mathcal{S}$ feedback controlled by $\mathcal{S}_T$ will be a mixture of positive and negative signals. How strong will the feedback be when CS_T is presented on test trials?

CS_T also has a generalization gradient that excites a band of $\mathcal{S}$ representations, including $\mathcal{S}_h$ and $\mathcal{S}_f$. Some of these $\mathcal{S}$ representations lie in the generalization gradients of $\mathcal{S}_f$ and/or $\mathcal{S}_h$, and will therefore be conditioned to

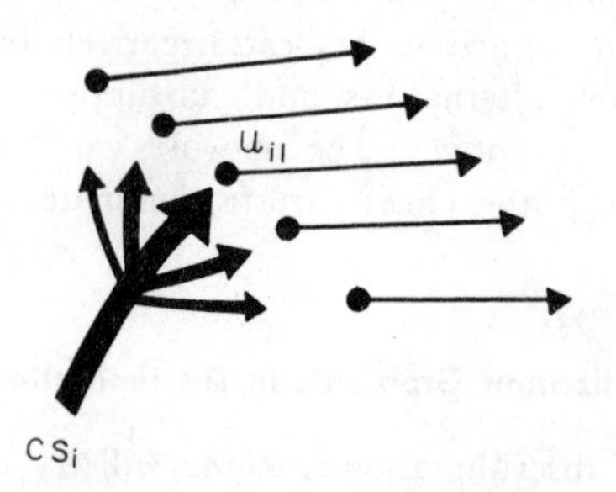

Fig. 15. Tuning curves underlying generalization gradients.

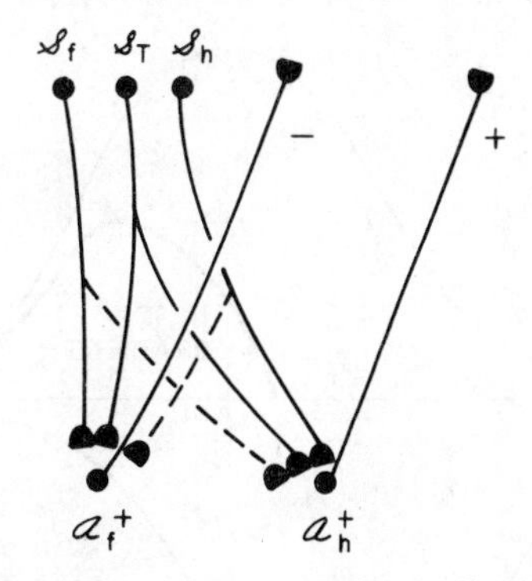

FIG. 16. Net feedback varies along generalization gradient.

$\mathcal{A}_f^+$ and/or $\mathcal{A}_h^+$ with a relative strength depending on how close they lie to these foci. Thus when CS_T is presented, it will excite a band of $\mathcal{S}$ representations that project with differing patterns to $\mathcal{A}_f^+$ and $\mathcal{A}_h^+$. The net $\mathcal{A} \rightarrow \mathcal{S}$ feedback is a composite of all these patterns after they filtered through the sensory-drive heterarchy. Recall also, however, that there is conditionable $\mathcal{A} \rightarrow \mathcal{S}$ feedback, specific $U_{i2} \rightarrow U_{i1}$ feedback, and normalization within $\mathcal{S}$ to contend with.

To understand what happens in a qualitative way, we first make an unsatisfactory approximation, and then improve it step-by-step. First, ignore the generalization gradient in $\mathcal{S}$ of CS_T and compute the net feedback that would occur in response to activating any $\mathcal{S}_T$ when this feedback is just the resultant of the relative $\mathcal{S}_T \rightarrow \mathcal{A}_h^+$ and $\mathcal{S}_T \rightarrow \mathcal{A}_f^+$ path strengths, and the total path strength is normalized (see Fig. 17a).The boldface curve in Fig. 17a shows the resultant at any $\mathcal{S}$ representation $\mathcal{S}_T$ of the gradients centered at $\mathcal{S}_f$ and $\mathcal{S}_h$. Note that the resultant gradient always is less than the $\mathcal{S}_h$ gradient, but that its slope is steeper than the $\mathcal{S}_h$ gradient. What is the effect of normalization by the on-center off-surround field? The normalized $\mathcal{S}_T$ gradient is shown in Fig. 17b. Its maximum is higher than the $\mathcal{S}_h$ gradient because the positive part of the $\mathcal{S}_T$ gradient in Fig. 17a is narrower and steeper than the $\mathcal{S}_h$ gradient. Thus normalization of the resultant gradient produces behavioral contrast. Also there is a peak shift away from $\mathcal{S}_f$, and a steepening of the generalization gradient due to discrimination training. The need for normalization, in turn, can be traced back to the need to prevent $\mathcal{S} \rightarrow \mathcal{A}$ signals from creating $\mathcal{A} \rightarrow \mathcal{S}$ feedback in the absence of supporting drives. The various other mechanisms at work in the network can now be switched in without changing these qualitative conclusions.

Why does a pronounced peak shift not occur if the training is errorless? In our networks, errorless training means that there is no fear or frustration (Bower, 1966), hence no negative gradient to interact with the positive

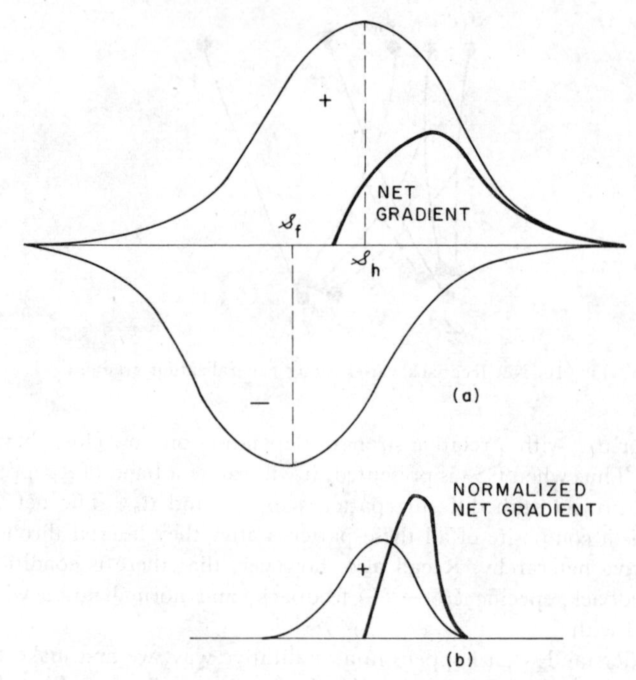

FIG. 17. Normalization yields behavioral contrast based on net generalization gradient.

gradient to cause a shift. If, however, a shock is paired with the S- after errorless discrimination training, then S$-\to \mathcal{A}_f^+$ conditioning will occur (along with $\mathcal{S} \to \mathcal{A}_f^+$ conditioning of all the $\mathcal{S}$ representations in the S$-$ gradient) and a peak shift will develop (Grusec, 1968). Bloomfield's (1969) remarks about "an unexpected change for the worse" can readily be interpreted in this context. A sudden reduction in reinforcement frequency generates a frustrative rebound, and hence sampling by $\mathcal{S}$ of a negative feedback source, as does introduction of shock, etc. Honig's (1962) suggestion that the peak shift occurs only if the S+ and S$-$ are presented successively, but not if they are presented simultaneously, can also be discussed. One possible reason for the latter fact is that rewarding *S*+ gradually gives it control over a powerful positive $\mathcal{S} \to \mathcal{A} \to \mathcal{S}$ feedback channel. This feedback enbles S+ to overshadow S$-$, so that as the animal's expectation of reward develops, the tendency to approach S+ simultaneously dominates the tendency to approach S$-$. Moreover, before this expectation develops, responding to S$-$ will not generate a large negative rebound. In the successive paradigm, approach to S$-$ is not inhibited by the presence of S+, so that more occasions occur after the expectation of reward develops that generate a frustrative negative rebound. Consequently, in the simultaneous paradigm S− does not gain (as much) control over a negative feedback channel, and the peak shift does not develop.

Normalization can also be used to interpret the Newman and Baron (1965) and Newman and Benefeld (Honig, 1970) studies. These studies suggest that the color dimension can mask the orientation dimension in pigeons, but that some conditioning is occurring in the orientation dimension nonetheless. In the Newman and Benefeld study, a vertical line on a green background during training is replaced by a vertical line on a black background during testing, and a generalization gradient is demonstrated even in the absence of discrimination training with an S−. We suppose that removal of the green background has three effects. It (1) eliminates the strong color → orientation inhibition due to the off-surround. Thus (2) the vertical orientation representation, and its generalization gradient, become more active owing to a shift of activity in the normalized field. This gradient can thereupon sample $\mathcal{A}$ and release its learned (but previously weak) $\mathcal{A} \to \mathcal{S}$ feedback. Removal of the green background also (3) causes surprise by changing the expected line-color combination used during training, and thereby enhances activity both of the orientation representation in $\mathcal{S}$ and of the sampled drive representations in $\mathcal{A}$ by triggering a transient increase in nonspecific arousal.

XIII. Habituation and the Hippocampus

The Newman and Baron study shows that a white line on a green key as S+ and a green key as S− produces a generalization gradient on the orientation dimension. Why does color not overshadow orientation in this situation? A phenomenon of this type exists in our networks. It is due to the interaction of several mechanisms, namely, (1) normalization and the quenching threshold, (2) conditioned arousal, (3) conditionable net incentive motivation, and (4) feedback of net incentive motivation to the normalization stage. On the initial S+ trials, color will partially overshadow orientation via the off-surround in $\mathcal{S}$. Denote the relevant color representations in $\mathcal{S}$ by $\mathcal{S}(C)$ and the orientation representations by $\mathcal{S}(O)$. Both $\mathcal{S}(C)$ and $\mathcal{S}(O)$ will sample $\mathcal{A}_h^+$ on these trials, but $\mathcal{S}(C)$ will build up a stronger connection since its activity is greater. As this occurs, $\mathcal{A}_h^+ \to \mathcal{S}$ feedback paths will become conditioned to $\mathcal{S}(C)$ and $\mathcal{S}(O)$ with a similar difference in relative strength. In the usual overshadowing experiment, this initial advantage of $\mathcal{S}(C)$ over $\mathcal{S}(O)$ will be progressively enhanced as training continues until $\mathcal{S}(C)$ completely overshadows $\mathcal{S}(O)$. In the present context, however, S− trials occur. On these trials, $\mathcal{S}(O)$ is inactive. $\mathcal{S}(C)$ is active, but the expected reward does not occur. A frustrative rebound is therefore generated. $\mathcal{S}(C)$ thereupon samples $\mathcal{A}_h^+$. Simultaneously, $\mathcal{A}_h^+$ samples the second stages U_{i2} of $\mathcal{S}(C)$. Thus the *net* incentive motivation controlled by $\mathcal{S}(C)$ is progressively diminished by frustrative nonreward. On S+ trials, the net feedback from $(\mathcal{A}^+_h, \mathcal{A}_h^-)$ to $\mathcal{S}(C)$ is cut down, owing to competition between these two channels before they release incentive motivation, but the feedback to $\mathcal{S}(O)$ comes only from $\mathcal{A}_h^+$, and increases through time. Asymptotically, the $\mathcal{S}(O)$ activity, bolstered by $\mathcal{S}(O) \rightarrow \mathcal{A}_h^+ \rightarrow \mathcal{S}(O)$ feedback, can dominate the $\mathcal{S}(C)$ activity. In this limited sense, the network has habituated to $\mathcal{S}(C)$, even as it grows ever more attentive to $\mathcal{S}(O)$. This habituation mechanism has several interesting properties. First, the sensory channel itself does not habituate; habituation is an active process based on interpretive feedback of sensory information via the drive representations (Grossman, 1967; Sharpless and Jasper, 1956). Second, suppose that the normalizer which creates a sensory-drive heterarchy is indeed interpreted as a simplified hippocampus. Then the hippocampus becomes involved in attentional control and the habituation of attention, but only indirectly via its determination of which motivational channel will be active in response to particular cues.

Section XX describes another habituation mechanism with the property that increasingly expected, and in particular repetitively presented, events elicit progressively smaller orienting reactions.

Why does a generalization gradient not occur if the S— is a red key or if the S— is a vertical white line on a red key? Then the color dimension is not habituated by frustrative rebound. Indeed, in the latter case, the orientation dimension might habituate, although perhaps at a slow rate because it is overshadowed by the dominant color dimension.

Note that expectation mechanisms can interact with habituation mechanisms in two opposing ways in the above experiment. First, they contribute to the frustrative rebound during S− trials, by altering the uniform (e.g., arousal) input to $\mathcal{A}_h^+$ and $\mathcal{A}_h^-$ on these trials. Second, they work against habituation by creating nonspecific arousal in $\mathcal{S}$ that tends to overcome the reduction of $\mathcal{S} \rightarrow \mathcal{A} \rightarrow \mathcal{S}$ feedback in particular channels, and allows them to once again reverberate in STM.

XIV. Overshadowing vs Enhancement

We now interpret and contrast the Honig (1969) experiments with the F. Freeman (unpublished master's thesis, 1967) experiment. Honig used TD and PD training sessions, followed by dimensional acquisition and finally testing on an orientation dimension. In the TD experiment, the pigeons were trained to make a discrimination on a dominant (namely, color) dimension. In particular, lesser dimensions were overshadowed, and the pigeon acquired a strong expectation and a positive conditioned arousal path in response to the S+ color cues. On dimensional acquisition trials, the color cues were not present, so that the orientation dimension, no longer overshadowed, could be trained, given that the pigeon still maintains general approach tendencies to the lever.

The PD training session, by contrast, frustrates the pigeon on the dominant color dimension. Yet the reinforcement schedule has been chosen so as to overcome frustration and yield a net approach tendency. The cues that elicit learned approach are not the frustrated cues of the color dimension. These cues develop their own powerful positive feedback paths. It is reasonable to assume that these cues are also present on the orientation dimensional acquisition trials. If they are, they will (at least partially) overshadow the orientation dimension both on orientation training and testing trials.

By contrast, Freeman trained pigeons to peck at a vertical line on a dark key (S+) but not to peck at a line tilted at 120° on the same dark background (S−). A generalization gradient is hereby established on the orientation dimension. Then dimensional acquisition occurs with the vertical line on a green background, and one finds a nontrivial chromatic generalization gradient on testing trials. Why does the orientation dimension not overshadow the green background during dimensional acquisition? One wants to say that surprise, and hence arousal, is triggered by changing the black background to green. Then the green $\mathcal{S}(C)$ representations will be able to sample the positive net incentive motivation controlled by the vertical $\mathcal{S}(O)$ representations. This explanation works, however, only if one first can answer the question: why does the orientation dimension not overshadow the dark background during discrimination training? And if the dark background is overshadowed, and therefore irrelevant, why is the pigeon surprised if it is removed on dimensional acquisition trials? The importance of these questions is perhaps better seen when they are phrased as follows: If the pigeon *does* get surprised when the dark background is replaced by green, then why does this not happen in *all* overshadowing experiments when the CS_1 is replaced by the $CS_1 + CS_2$, thereby preventing overshadowing from occurring?

To answer these questions, we seek differences in how the expectation mechanism (and thus arousal) responds in the Freeman experiment as opposed to the usual overshadowing experiment. We want to say that introducing green in the Freeman experiment is more surprising, say, than introducing a tone as CS_2 after prior CER training with a flashing light as CS_1 in a Kamin-type overshadowing experiment. A difference of degree is sought in the two experiments, rather than the operation of different mechanisms. We suggest that this difference exists, in part at least, because the pigeon can develop an expectation of a vertical line on a particular visual background more easily than it can develop an expectation of a flashing light in a prescribed combination of events involving nonvisual modalities. In other words, a learned expectation can be at least partially localized to a given cluster of features or events, and features which stream into the same modality in close physical contiguity can be more easily grouped together as a coherent expectation than features which enter through different modalities, other things equal. If this is true, then it might be easier to eliminate overshadowing of CS_2 by CS_1 in a Kamin-type experiment if the CS_1 is a vertical line and the CS_2 is a green background, than if the two events involve different modalities. This kind of prediction is hard to analyze completely because inputs to two different modalities are hard to equate psychophysically, and can activate orienting reactions that need not be activated by two inputs to the same modality.

The Freeman experiment demonstrates enhancement due to prior discrimination training. The closely related Mackintosh and Honig experiment (Honig, 1970) demonstrates blocking. We suggest that blocking occurs because the surprise that is triggered during redundant spectral discrimination training, after orientation discrimination training has been completed, only partly overcomes overshadowing. When no prior orientation discrimination exists, and only spectral discrimination training is given, there is no overshadowing to overcome.

If the above analysis is accurate, then one might be able to create a transition from overshadowing to enhancement in a given experimental setup by varying the relative strength of the attentional and surprise channels, say by drugs.

XV. Novelty and Reinforcement

Berlyne (1969) showed that a novel light change, contingent on lever pressing, can reinforce lever pressing. We suggest that the novelty of the light change, as usual, triggers nonspecific arousal which, as usual, filters through all drive representations. If a positive incentive motivational source is active when arousal occurs, and this source dominates other drive representations at that time, then the arousal will enhance the amount of positive motivation. The lever press cues $\mathcal{S}(L)$ can become differentially conditioned to the positive source, which also supplies enough incentive to trigger $\mathcal{S}(L) \rightarrow \mathcal{M}$ sampling of the motor commands that control the lever press. We suggest that the source of positive incentive in this case is the motor arousal source for exploratory approach and pressing of the lever, rather a specific drive representation.

We can now provide an answer to a related question: Why is the approach incentive motivation not *usually* the motivational source for learned goal objects? One reason is that, unless the approach source is *differentially* strengthened by arousal enhancement or other means, then *all* meaningless objects in the environment can be approached, and none will be approached more frequently than any other, other things equal. A second reason is that, when a specific drive is rewarded, then the source of positive incentive tends to shift from general exploration and approach to the specific drive representation that was rewarded.

The enhancing effect of arousal on the pattern of activity at drive representations can also generate incentive motivational feedback to sensory representations in the absence of external sensory cues. For example, if the hunger drive is prepotent, and all drive representations are aroused, then $\mathcal{A}_h^+$ can generate feedback to its psychological set $\mathcal{S}_h$ in $\mathcal{S}$, leading, say, after further enhancement through the feedback loop $\mathcal{S}_h \rightarrow \mathcal{A}_h^+ \rightarrow \mathcal{S}_h$, to the motor output "I want food." More generally, the network can ask itself "how it feels" by arousing its drive representations. The resulting motivational feedback from $\mathcal{A}$ to $\mathcal{S}$ can establish a psychological set that is capable of generating compatible motor activity. This possibility is a special case of the "two-thirds rule" discussed in Section XXII.

XVI. Motivation and Generalization

How does increased drive flatten generalization gradients? A formal answer exists in the networks. Increasing the drive increases positive incentive motivational signals in $\mathcal{A} \rightarrow \{U_{i2}\}$ synapses. Increasing these signals has two effects. It speeds up conditioning in the $\mathcal{A} \rightarrow \{U_{i2}\}$ synapses, and it increases the signals from $\{U_{i2}\}$ to $\{U_{i1}\}$. At $\{U_{i1}\}$, the increased input allows more $\mathcal{S}$ representations to exceed quenching threshold, and faster conditioning occurs in the $\mathcal{S} \rightarrow \mathcal{A}$ synapses of these representations.

How does this mechanism affect generalization gradients? If a particular $\mathcal{S}$ representation is activated by external cues, its generalization gradient of $\mathcal{S}$ representations will also be activated, albeit to a lesser extent. Increasing $\mathcal{A} \rightarrow \{U_{i2}\} \rightarrow \{U_{i1}\}$ signals permits more of these representations to sample drive representations in $\mathcal{A}$. Thus, on testing trials, more cues in the generalization gradient can generate the type of feedback that was elicited on training trials. The generalization gradient is hereby flattened. Section IV also shows that if the nonspecific feedback is additive at $\mathcal{S}$ sites, then it will tend to flatten the gradient by uniformizing the pattern of activity in $\mathcal{S}$.

If two drives compete, then increasing one drive can steepen the generalization gradient on the other drive (Hoffman, 1969). Let the two drives be hunger and fear for definiteness, and consider the generalization of fear conditioned to a 1000 cps tone. Let the $\mathcal{S}$ representation $\mathcal{S}(X)$ of an X cps tone be activated by the tone on testing trials. Choose X in the generalization gradient of the 1000 cps tone. On training trials, $\mathcal{S}(X)$ sampled $\mathcal{A}_f^+$, and possibly also $\mathcal{A}_h^+$. Thus the strength of $\mathcal{A}_h^+ \rightarrow \mathcal{S}$ feedback is increased relative to the strength of $\mathcal{A}_f^+ \rightarrow \mathcal{S}$ feedback, which is driven by $\mathcal{S}(X) \rightarrow \mathcal{A}_f^+$ signals. The suppressive effect of fear is hereby reduced by increasing the hunger level.

Why does this mechanism steepen the fear generalization gradient? A formal reason is that a fixed increment in positive feedback can totally overcome the suppressive effect of a sufficiently small amount of negative feedback, but has only a small relative effect on large amounts of negative feedback. An X cps generalization gradient controls large amounts of negative feedback, but tones near the edge of the 1000 cps generalization gradient control only small amounts of negative feedback. Hence the increase in hunger narrows and steepens the fear gradient.

XVII. Predictability and Ulcers

If Weiss's experiments (1971a,b,c) on the development of stomach ulcers in rats are performed on our networks, then the net incentive motivation in the networks is a monotone increasing function of the degree of ulceration in his experiments. This analysis does not give a physiological explanation of the ulcerogenic process, but it does suggest that the frustrative sources of negative incentive are also triggered at the same time as sources of ulcer-inducing agents.

Why do avoidance–escape networks develop less ulceration than yoked networks? Avoidance–escape networks have been trained to respond to cues which activate positive incentive motivation that supports avoidance and/or escape activity. The positive incentive competes with the negative incentive generated by shock, and thereby reduces the net negative incentive motivation.

Why does a warning signal reduce the ulceration of both groups of networks? It can do so by reducing the novelty of the shock. By Section VIII, this will reduce the arousal level that accompanies the shock, and thus the net negative incentive that the shock produces. In the avoidance–escape networks, the warning signal can also be used as a discriminative cue for activating avoidance–escape cues that switch on positive incentive motivation.

Why do the yoked networks develop less severe ulcers than the avoidance–escape networks if both groups receive a brief shock after each avoidance–escape response? Three effects in the network conspire to produce this result. First, the network is motivated by positive incentive in making the avoidance–escape response; this motivational source is abruptly terminated. Second, the network expects relief after performing the response, but does not get it; this unexpected event triggers nonspecific arousal. Third, a negative, or punishing, event occurs instead of the expected relief. The first effect tends to produce a positive-to-negative rebound. The third effect creates a second source of negativity. And the second effect enhances the total negative tendency. The first and second effects are absent, or at least much weaker, in yoked networks.

Why does little ulceration develop in avoidance–escape networks if each avoidance–escape response is followed by a feedback stimulus, such as a tone? Three effects are operating in our networks. First, the avoidance–escape response produces relief, as in Section III. Second, the novel tone, of itself, produces nonspecific arousal. As in the analyses of the fear–relief dipole in Section III, and of the Berlyne (1969) experiments in Section XV, this arousal enhances the relief rebound that is produced. Third, these effects speed up the conditioning of avoidance–escape cues to the positive incentive motivational source, and therefore reduce the net negative incentive that is produced even before the coping response is made.

Is this analysis compatible with Weiss's idea that no ulcers can develop in the absence of a coping response? It is compatible with a weaker statement: that coping responses can enhance or suppress ulceration, but that any mechanism that produces negative incentive in the rebound mechanism creates a predisposition to ulcerate. A deeper analysis of the way in which positive and negative incentive actually regulate muscular contraction might refine this view at a later time.

XVIII. Orienting Reaction

We will show below that some properties of this reaction can formally be represented within the networks that are already at our disposal. We will invoke psychophysical examples to illustrate the formal meaning of the mechanisms, but do not presume that they are given a complete physiological explanation. Consider Fig. 18.

In Fig. 18, different paths P_i are differentially excited by different peripheral events, e.g., retinal loci, positions on the skin, auditory inputs. Suppose that U_{i2} can fire only if orienting arousal combines with a signal from U_{i1}. Let the axon collaterals from U_{i1} to $\mathfrak{M}$ have relative strengths that determine a final orienting position for the muscles that they control. Different P_i paths will determine different orienting positions by having dif-

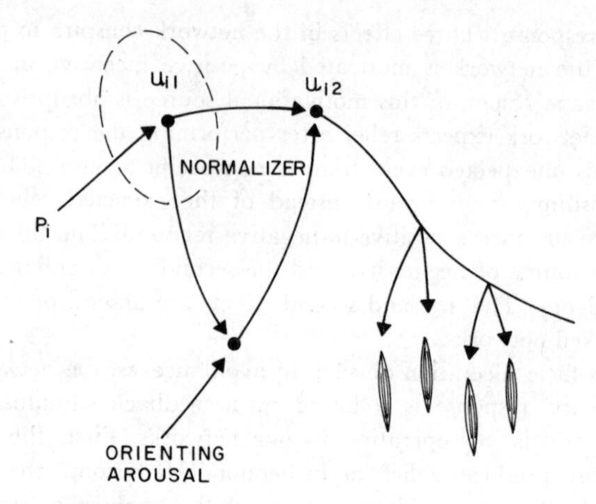

Fig. 18. Orienting arousal activates position codes for motor control.

ferent patterns of axon strength. Let a recurrent normalizer interconnect the $\{U_{i1}\}$ populations. This establishes a "position code" in the field $\{U_{i1}\}$ of populations by normalizing the total activity of the field (e.g., normalizing the effect of variations in the total light energy hitting the retina), and letting the $U_{i2} \rightarrow \mathfrak{M}$ axon strengths determine the terminal muscle positions. For example, if two U_{i1} are simultaneously and equally active, then a position will be determined that lies between the positions determined by each U_{i11} separately. As the relative activity of one U_{i1} increases, the terminal position will approach the position controlled by this U_{i1} alone. In the case of vision, for example, if the arousal level of the field is tuned so that only one population can reverberate in STM, then only one retinal light source can attract the eyes. If arousal permits several populations simultaneously to reverberate, then weighted averages of the retinal positions can attract the eyes.

Withdrawing orienting arousal prevents the release of signals from any U_{i2}. We assume that this arousal can be inhibited by activity from competing arousal sources, such as drive representations. Thus, before training, a loud noise in the direction of a subject's left side can elicit an orienting reaction toward the left. Suppose, however, that the noise is used in learning trials as a discriminative cue for rapid lever pressing for food. Then on testing trials, the noise can differentially excite the $\mathfrak{A}_h{}^+$ representation, which can inhibit the source of orienting arousal via, say, the sensory-drive heterarchy.

The source of orienting arousal is triggered by unexpected events. Minimality bids that we identify this arousal source with the arousal source, also triggered by unexpected events, that overcomes blocking and triggers enhancement or rebound in the various drive representations. A plausible candidate for this arousal source is the reticular formation.

There exist variations on this anatomical theme, such as an orienting arousal source supplying *shunting* excitation that permits the cells which carry the position code to fire. Such an arousal source can also act at the synaptic knobs, or to inhibit tonic presynaptic inhibition of these knobs (disinhibition). In all the above anatomies, excitation and disinhibition can have similar functional effects. Disinhibition has the disadvantage of requiring an extra processing step, but it has the advantage that it permits sustained activity of cells, which prevents them from undergoing a chemical degradation due to disuse.

XIX. A Learned Expectation Mechanism

An expectation mechanism is described below to help fix ideas in the above discussion. We wish to prevent orienting arousal if an expected event occurs, and to permit it if an unexpected event occurs. The first part of the construction synthesizes a network which can learn to expect a given event subsequent to the occurrence of another event. Several variations of this construction appear in Grossberg (1972a). This construction will be supplemented herein to guarantee additional properties of the expectation mechanism. The output cells U of the network will fire only if the learned expected event occurs. The construction in Grossberg (1972a) is reviewed below for completeness.

The learned input pattern (or class of patterns) which can fire the cell (or cells) U is controlled by presetting cells P. The cells P send axons to the filtering mechanism (e.g., inhibitory interneurons and dendrites) that processes inputs to U. Each P cell can learn a particular pattern that will bias U's filter when P is active. For example, consider an animal $\mathcal{O}$ that learns to lever press for food. On a testing trial, $\mathcal{O}$ "expects" food when it lever presses in response to hunger. We suppose that lever press cues also preset consummatory controls which can be released by expected sensory cues of the food reward. Similarly, suppose that one goes to the refrigerator expecting to find orange juice, which one loves, in a transparent container, but instead one finds tomato juice, to which one is indifferent. The same motor sequence of reaching, pouring, and drinking suffices for imbibing either the orange juice or the tomato juice. The orange fluid releases this sequence, but the red fluid does not; indeed, the red fluid can release a frustrative rebound. The consummatory controls have been preset by the expectation of an orange fluid.

How do the P cells learn the patterns on training trials that will bias the U-cell filter on testing trials? Consider the anatomy of Fig. 19, in which interacting signals combine additively. In Fig. 19, the cells $V_1 = \{v^{(j)}$:

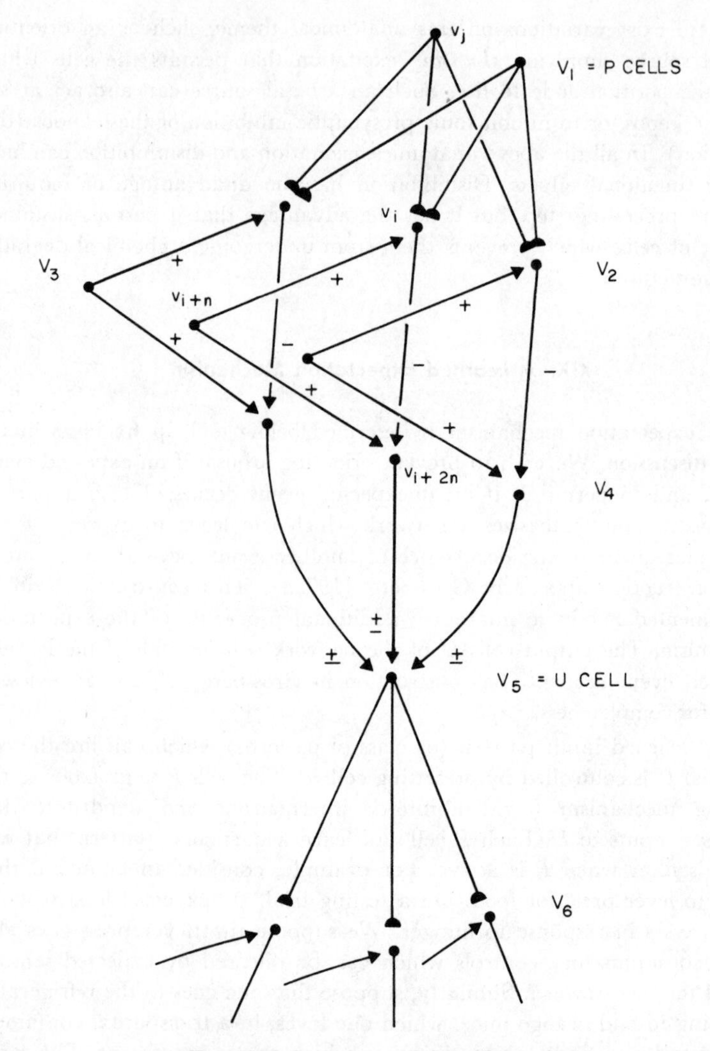

FIG. 19. Subtractive preset mechanism.

$j = 1, 2, \ldots, n\}$ are P cells. These cells sample patterns playing on the cells $V_2 = \{v_i: i = 1, 2, \ldots, m\}$ when they are active. The cells $V_3 = \{v_{i+m}: i = 1, 2, 3, \ldots, m\}$ receive the test patterns that will fire the cell U if they are expected by P. On training trials, $V_3 \rightarrow V_2$ signals reproduce these test patterns at V_2, where they can be sampled by V_1. Signals in $V_3 \rightarrow V_4$ axons, where $V_4 = \{v_{i+2m}: i = 1, 2, \ldots, m\}$, also reproduce the test patterns at V_4. On testing trials, activity in a P cell generates a pattern θ of activity in V_2, which is transferred to V_4 as inhibitory signals by $V_2 \rightarrow V_4$

axons. The test pattern $\tilde{\theta}$ at V_3 is also sent along to V_4, where it can be compared with θ. Simultaneously, $\tilde{\theta}$ is sent to V_2 to be sampled by V_1. At V_2, $\tilde{\theta}$ is transferred to V_4 as inhibitory signals, just as the $V_1 \rightarrow V_2$ presetting signals are. The inhibitory $V_3 \rightarrow V_4$ signals are chosen weaker than the excitatory $V_3 \rightarrow V_4$ signals, so that the test pattern appears at V_4 with a net excitatory strength. The inhibitory preset pattern θ and the excitatory test pattern $\tilde{\theta}$ are thereupon compared at V_4.

The above transformations can be defined in greater detail as follows. Let the strength of the excitatory $v_{i+m} \rightarrow v_{i+2m}$ signal be $\tilde{\theta}_i I$, and of the inhibitory $v_{i+m} \rightarrow v_i \rightarrow v_{i+2m}$ signal be $\tilde{\theta}_i \eta I$, $0 < \eta < 1$. Then the net signal to v_{i+2m} from v_{i+m} is $\tilde{\theta}_i(1 - \eta)I$, which is nonnegative. Let the $v^{(j)} \rightarrow v_{i+n} \rightarrow v_{i+2m}$ inhibitory signal from the jth preset cell be $-\theta_i K$. If only $v^{(j)}$ in V_1 is active, the total signal to v_{i+2m} is $\tilde{\theta}_i(1 - \eta)I - \theta_i K$. Under these circumstances, v_{i+2m} will fire only if

$$\tilde{\theta}_i(1 - \eta)I > \theta_i K \tag{2}$$

This constraint shows that all cells in V_4 can fire only if all the relative pattern activities in $\tilde{\theta}$ are not too much smaller than the relative pattern activities in θ. Since $\Sigma_k \theta_k = \Sigma_k \tilde{\theta}_k = 1$, simultaneous firing in all channels is possible only if $(1 - \eta)I > K$. Thus the *total* activities of V_1 and of V_3 must be carefully regulated.

Inequalities (2) do not suffice to prevent firing of a discriminative cell further downstream to patterns some of whose $\tilde{\theta}_i$ are much larger than θ_i (Grossberg, 1970, 1972a). To prevent this, the output signal from each v_{i+2m} in V_4 excites both an excitatory and an inhibitory pathway. The inhibitory pathway (which can, in principle, be just a high threshold inhibitory ionic channel in the same axon pathway as the excitatory channel) overcomes the excitatory pathway if the signal from v_{i+2m} is too large. When this happens, the net output from v_{i+2m} to V_5 is negative, so that not all channels are simultaneously excitatory. Thus the net signal from v_{i+2m} to V_5 is derived from two successive inhibitory mechanisms. It is positive at v_{i+2m} only if the relative pattern activity $\tilde{\theta}_i$ is not too much smaller than the relative pattern activity θ_i. This positive activity is inhibited, however, if $\tilde{\theta}_i$ is too much larger than θ_i. *All* channels in V_4 contribute a positive signal to V_5 only if the pattern $\tilde{\theta}$ is close, in every component, to the pattern θ. The signal threshold of V_5 is adjusted once and for all so that V_5 will fire only if it receives (nearly) simultaneous positive signals from all V_4 channels. Hence the cell $U = V_5$ fires only in response to the expected pattern. Grossberg (1972a) shows that this anatomy has formal properties that are reminiscent of cerebellar anatomy, and thereby illustrates the anatomical pausibility of this expectation mechanism.

The same principles have been used to synthesize a class of networks with a suggestive retinal analog (Grossberg, 1972a). These networks are capable of discriminating the relative figure-to-ground of spatial patterns (i.e., their θ_i's) but do not have a learnable expectation mechanism. Here also two successive inhibitory mechanisms are needed. If the receptor cells of this network are interpreted as light receptors, then the first inhibitory layer is reminiscent of retinal horizontal cells. Speaking functionally, this layer produces a form of light adaptation; cf. unicellular recordings in the mudpuppy retina (Werblin, 1971). The second layer is reminiscent of amacrine cells. The output cells (cf. ganglion cells) are then capable of hue constancy (including a lightness scale), brightness constancy, velocity detection, etc., depending on which receptors are hooked into the network, and on how the anatomical connection coefficients are chosen.

The expectation mechanisms defined above has two deficiencies: (1) It does not automatically regulate the total activities of V_1 and V_3; and (2) if no presetting cell in P is active, then *every* pattern presented to V_3 can fire V_5, since no net inhibitory signal is produced at V_4. The first deficiency can be overcome by introducing recurrent on-center off-surround anatomies with shunting interactions into V_1 and V_3 Section IV indicates the need for such mechanisms within sensory processors, so that their use here does not impose a new constraint. The second deficiency can be overcome by assuming that uniformly distributed *tonic* inhibitory signals are somehow generated from V_2 to V_4 in the absence of presetting signals, and that the onset of presetting signals supplants the tonic inhibition with learned patterns of $V_2 \rightarrow V_4$ inhibition. A simple way to do this is to assume that tonically active cells exist in V_1 and send uniformly distributed inputs to V_2; V_2, in turn, generates inhibitory signals to V_4 that prevent inputs to V_3 from firing V_4. When a presetting cell in V_1 becomes sufficiently active, it suppresses the activity in the tonic cells via the recurrent off-surround in V_1, and substitutes its own patterned signals to V_2. Tonically active cells that are suppressed by the onset of phasic afferents are known to exist in various neural structures; in the frog retina, for example, there are dimming cells whose tonic activity in the dark is suppressed by light (Chung *et al.*, 1970). Note also that the distribution of tonic inputs to V_2 can be uniform even if no tonic cell is connected to all cells in V_2; only the *distribution* of activity across all tonically active cells needs to be uniform, and this distribution can be suppressed uniformly by widely dispersed off-surround signals within V_1.

XX. Regulation of Orienting Arousal

The output cells U fire only if their expected event occurs. If any unexpected event occurs, we want it to generate orienting arousal. It seems very unlikely that a brain contains internal models of the infinitely many events that are unexpected at any time, and that it generates orienting arousal whenever there is a match between one of these events and its internal model. By contrast, given the above construction, it is easy to devise a network that inhibits orienting arousal only if the expected event occurs. Thus, we assume that *every* event which is processed by the network's sensory mechanisms can, in principle, activate orienting arousal using as a source the neural activity which it generates as it is processed. The output from the expectation mechanism can, however, inhibit orienting arousal (cf. Sokolov, 1960) (see Fig. 20). In Fig. 20, the output from the cells U bifurcates. One channel inhibits orienting arousal and the other channel samples the drive representations in $\mathcal{A}$. For example, suppose that the expected event is a loud noise to the left of the network, and that the noise has been trained as a discriminative cue for lever pressing. When the noise occurs, it generates activity that can drive the orienting reaction. This activity is, however, inhibited by the output from U. The U output also generates positive $\mathcal{S} \rightarrow \mathcal{A} \rightarrow \mathcal{S}$ feedback that elicits the lever press. The orienting reaction can be inhibited by this mechanism even if U controls no other motor reaction. The construction can be modified to change this conclusion. If the orienting arousal channel is included in the on-center off-surround anatomy of the

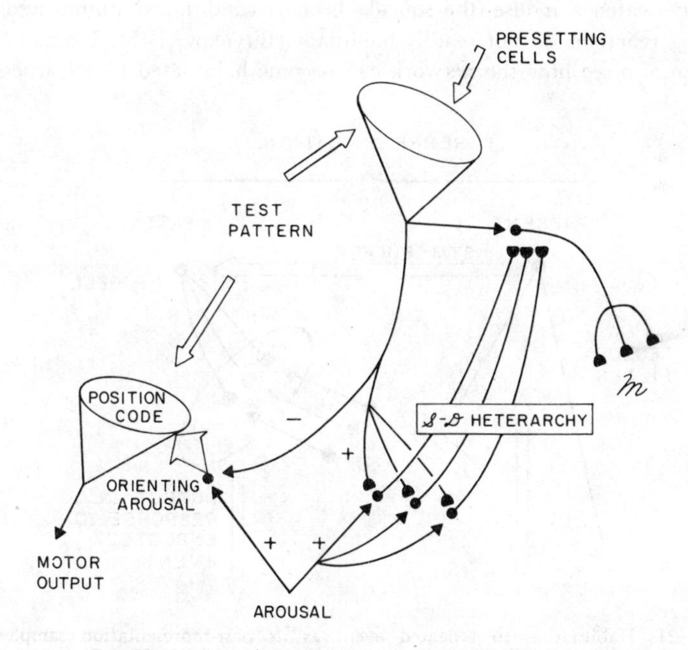

FIG. 20. Inhibition of orienting arousal by expected events.

sensory-drive heterarchy, then the orienting arousal can occur unless it is supplanted by strong competing $\mathcal{S} \rightarrow \mathcal{A} \rightarrow \mathcal{S}$ feedback in a specific drive channel.

The relationship between presetting inputs and test inputs will be more completely studied in another place. In particular, one must note that the events which excite P cells and those which excite V_3 cells need differ only in their onset times; the P events occur earlier than the V_3 events. Thus, V_3 events gain control over P cells as new events intervene. This shift in the spatial locus of an event's internal representation can be subsumed under the study of sequential short-term memory buffers (Atkinson and Schiffrin, 1968). It is schematically represented in Fig. 21. Given such a shift in representational locus, one can see how this network becomes habituated to a repetitively presented event. As the event is repeated, it serves as a source of P-cell activity in its "past" mode, and as a source of test inputs in its "present" mode. The event samples itself, in short. As the event is repeated, it samples itself repeatedly via $P \rightarrow V_2$ axons which build up the strength of the expectation. As the event becomes more expected, the output from U increases and progressively inhibits orienting arousal, but does not prevent conditioned responses from occurring. For example, young foxes quickly habituate orienting reactions to the sound of mouse squeaking, but once they have eaten a mouse, the squeaks become conditioned stimuli and the orienting reactions do not readily habituate (Biryukov, 1958; Lynn, 1966). One can also see how the network can become habituated to a learned set

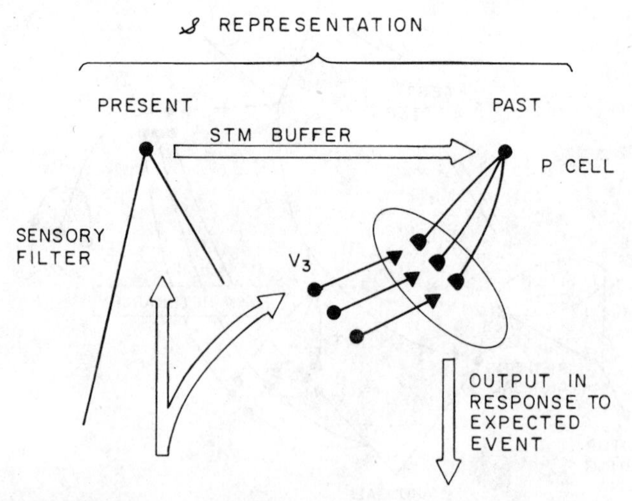

FIG. 21. Habituation to repeated event as its past-representation samples its present-representation in the expectation mechanism.

of events via conditioning of the $P \rightarrow V_2$ synapses of the P representations corresponding to this set. Indeed, if the sensory filter is capable of grouping peripheral events in classes that fall along unconditioned or conditioned generalization gradients, then these gradients will be transferred to the P cells via the sequential STM buffer.

XXI. Hippocampal Feedback, Conditioning, and Dendritic Spines

This section provides a way to implement three formal requirements in the network using a common mechanism. The mechanism has a suggestive anatomical analog in terms of hippocampus, septum, hypothalamus, reticular formation, and neocortex. In this analog, the hippocampus receives input from neocortex (*in vivo*, via the entorhinal cortex) and septum (Raisman *et al.*, 1966). The mock-hippocampal output trifurcates and eventually feeds back to septum as signals conditionable at $\mathcal{A} \rightarrow \mathcal{A}$ synapses, and to neocortex as nonspecific attentional or motor feedback, possibly via the anterior thalamic nuclei (Raisman *et al.*, 1966). The mock-hypothalamus prepares drive inputs to this system, and the reticular formation provides nonspecific arousal, which can be triggered by specific events, and which is filtered through the sensory and drive representations to enhance or rebound their activity.

The three formal requirements are these:

1. Consider Fig. 22. In the figure, $\mathcal{S}_1$ is conditioned to $\mathcal{A}_1$, but all $\mathcal{S}_i$, $i \neq 1$, are irrelevant cues; i.e., they project equally to $\mathcal{A}_1$ and $\mathcal{A}_2$. Suppose that many of these irrelevant cues are active when $\mathcal{S}_1$ is active. Then the $\mathcal{S}_1 \rightarrow \mathcal{A}_2$ synapses will become progressively stronger and eventually $\mathcal{S}_1$ will approach irrelevancy also. That is, the act of performing in response to a relevant event can countercondition the event simply because irrelevant events exist. Part of this difficulty can be overcome by $\mathcal{A} \rightarrow \mathcal{S}$ attentional feedback, which tends to quench irrelevant cues. This does not, however, prevent counterconditioning of the $\mathcal{A}_1 \rightarrow \mathcal{S}_1$ channel by irrelevant cues that are active before the feedback occurs, however, just as a CS that is presented

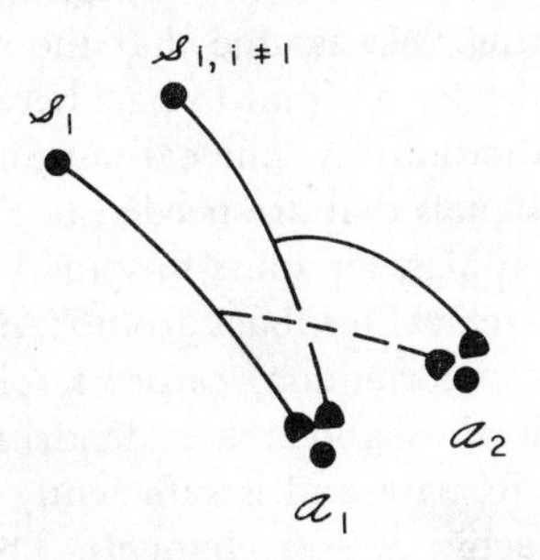

FIG. 22. Counterconditioning by irrelevant cues.

before a UCS can sample the UCS-controlled representations without being totally overshadowed by the UCS.

To prevent counterconditioning by irrelevant cues, the uniform part of the total input to a dipole's channels must be inhibited before substantial $\mathcal{S} \to \mathcal{A}$ sampling of these channels occurs. This can occur only at a stage after the $\mathcal{S} \to \mathcal{A}$ synaptic knobs, because one only knows that the $\mathcal{S} \to \mathcal{A}$ signals are uniform after they are emitted at their respective synapses. Also the resultant of this inhibition must feed back in a form that *can* be sampled by $\mathcal{S} \to \mathcal{A}$ synaptic knobs. How can the bulk of the conditionable signal be due to $\mathcal{A} \to \mathcal{A}$ feedback, and not to the $\mathcal{S} \to \mathcal{A}$ signals that are in spatial contiguity with other $\mathcal{S} \to \mathcal{A}$ synapses?

2. The existence of higher-order instrumental conditioning implies that $\mathcal{S} \to \mathcal{A}$ sampling can occur both before and after the stage of drive rebounds; hence there exists a recurrent loop from sampling mechanism, to rebound mechanism, to choice-among-drives mechanism, and back to sampling mechanism.

3. What kind of feedback should be conditionable? Should the feedback be from the resultant of each dipole separately, or from the resultant of all competing drives? In the latter case, conditioning is possible only with respect to the drive that supplies incentive motivation for regulating attention, motor performance, and the transfer of STM into LTM. We exhibit a system of the latter type for definiteness. Variations on the theme are then readily constructable.

To achieve (2) and (3), we use a mechanism as in Fig. 23. Note that the output of the sensory-drive heterarchy trifurcates: it is fed back to "neocortex" as attentional feedback and as motor arousal, and it is fed back to "septum" as conditionable signals.

To achieve (1), we must somehow allow $\mathcal{S} \to \mathcal{A}$ signals to influence events further downstream without allowing these signals to be substantially conditioned to anything but sensory-drive heterarchical output. One way to do this is suggested in Fig. 24. The $\mathcal{S} \to \mathcal{A}$ signals reach "dendritic spines." Here they produce local potentials that propagate to the cell body where they influence axonal firing. We assume that the resistances in spines are such that it is much harder for a signal to pass between spines than from a spine to the cell body. Alternatively, one can assume that the threshold for the post- to presynaptic signals that are needed to change transmitter levels in $\mathcal{S} \to \mathcal{A}$ synapses are too high for spine-to-spine interactions to overcome them. By contrast, heterarchical feedback from $\mathcal{A}$, energized by nonspecific arousal (e.g., from reticular formation) causes a spike potential, or similar global potential change, throughout the dendritic column. This spike invades all the spines in its path and is sufficiently strong to induce transmitter level changes in active $\mathcal{S} \to \mathcal{A}$ channels. Thus a mechanism using

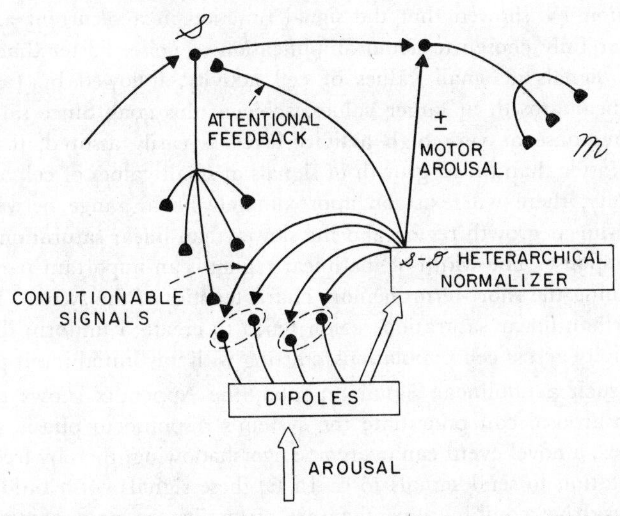

FIG. 23. Conditionable heterarchical feedback signals sampled by $\mathcal{S}$.

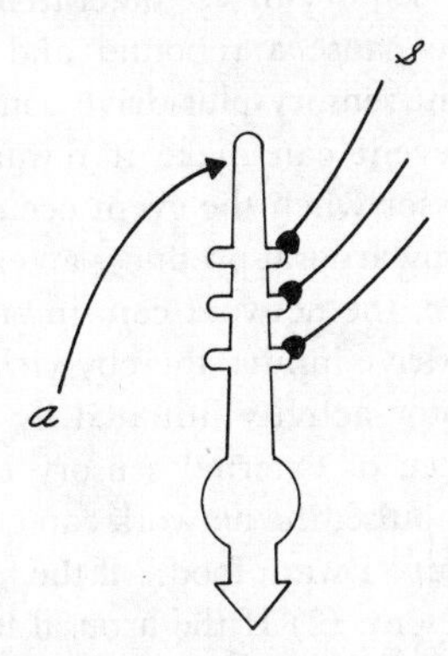

FIG. 24. Heterarchical feedback causes global potential change that invades individual conditionable $\mathcal{S}$ channels.

dendritic spines and dendritic spike generators (or some formally analogous mechanism) can allow $\mathcal{S} \to \mathcal{A}$ signals to occur without major changes in $\mathcal{S} \to \mathcal{A}$ synaptic transmitter levels unless feedback invades the entire dendritic apparatus.

XXII. Nervous Eating and Attentional Deficits Modulated by Arousal

Section III pointed out that the existence of higher-order instrumental conditioning implies the existence of feedback loops in the rebound mechanism. Section IV showed that the signal function in a recurrent anatomy must be carefully chosen to avoid amplification of noise. Faster than linear growth of signals at small values of cell activity, followed by (approximately) linear growth at larger values, achieves this goal. Since saturation of signal response at very high activity levels is easily assured, it suffices to impose faster than linear growth of signals at small values of cell activity; by continuity, there will exist an approximately linear range between the faster-than-linear growth region and the slower-than-linear saturation region of signal response. The width of the linear region is an important parameter in determining the short-term memory characteristics of the system. Indeed, the slower-than-linear saturation region tends to create a uniform distribution of activity across cell populations, starting with any initial input pattern.

Given such a nonlinear signal function, the Appendix shows that an increase in arousal can potentiate the system's response to phasic sensory inputs. Thus, a novel event can overcome overshadowing, thereby freeing its $\mathcal{S}$ representation to send signals to $\mathcal{A}$. In $\mathcal{A}$, these signals contribute to the sensory-plus-drive combinations that are struggling to gain control over $\mathcal{A} \to \mathcal{S}$ feedback via the sensory-drive heterarchy. Simultaneously, the arousal triggered at $\mathcal{A}$ can cause a rebound and/or enhance the $\mathcal{A} \to \mathcal{S}$ feedback from the dominant sensory-plus-drive combination. Similarly, the novelty of an indifferent event can make it rewarding, if the network is engaged in approach behavior when the event occurs (Berlyne, 1969).

In effect, enhancement by arousal produces an extra source of input to $\mathcal{A}$. Using this new input source, the network can, in principle, generate $\mathcal{A} \to \mathcal{S}$ feedback in the absence of drive inputs, thereby yielding the following interesting possibilities: (1) Motor activity initiated by an internally generated arousal pulse in the absence of external sensory cues; e.g., by testing its drive states with an arousal pulse, the network can generate a hunger-related output, such as the statement "I want food," if the hunger drive is dominant but no cues of food are present. (2) If the arousal level is high, it can elicit consummatory activity compatible with sensory cues or drives that are too low to otherwise initiate consummation; e.g., "nervous eating." If, for example, there is a damming up of motor activity in the absence of an appropriate goal, and this activity feeds into $\mathcal{A}$ as arousal, then potentiation by arousal can discharge the motor activity through the heterarchical channel that is dominant at that time. (3) An unexpected event, even a frustrating one, can elicit transient motor activity via heterarchical feedback, even if the heterarchical feedback is not related in a simple way to the unexpected event.

These remarks illustrate a so-called "two-thirds rule." Namely, at least two channels from among $\mathcal{S}$, $\mathcal{D}$, and $\mathcal{A}$ are needed to elicit $\mathcal{A} \to \mathcal{S}$ feedback: $\mathcal{S} + \mathcal{D}$ is the usual heterarchical constraint on $\mathcal{A} \to \mathcal{S}$ firing; $\mathcal{D} + \mathcal{A}$ is illustrated by the "I want food" example; and $\mathcal{S} + \mathcal{A}$ is illustrated by "nervous eating." One can also imagine, in principle, the perhaps pathological case in which intense $\mathcal{S} + \mathcal{D} + \mathcal{A}$ activity allows more than one heterarchical channel to be active at a time, by driving the activity of more than one channel above quenching threshold; cf. the remarks about quenching threshold relevant to determining the asymptotic "eye position" in Section XVIII.

The mixing of channel activities at the $\mathcal{A}$ cells has an analog at $\mathcal{S}$ cells. Here converge arousal inputs that are triggered by unexpected events, and which consequently tend to overcome overshadowing, as well as arousal inputs from $\mathcal{A} \to \mathcal{S}$ channels that contribute motivational feedback, and can sustain overshadowing. Under pathological circumstances, either channel can become persistently overaroused. One possible consequence of overarousal is "seizure activity" (cf. Grossberg, 1973). Another is the inability to pay attention (cf. Grossberg and Pepe, 1971). Either of the two arousal sources can cause such difficulties, in principle, but the overall "clinical syndrome" that the network would undergo could be quite different in the two situations. $\mathcal{A} \to \mathcal{S}$ overarousal can cause, in addition, emotional depression (Grossberg, 1972c), as well as pathological changes in the network's "psychological sets" (Section VI). If, for example, one heterarchical $\mathcal{A} \to \mathcal{S}$ feedback channel became dominant, then it could bias all the network's sensory processing in a direction that is compatible with the dominant drive, or more precisely, the dominant "psychological set." Such an effect need not occur if the source of novelty-bound arousal is overaroused, since this arousal source is truly nonspecific. Previous work (Grossberg, 1972c) suggests than an analogy can be drawn between $\mathcal{A} \to \mathcal{S}$ channels and midbrain channels influenced by catecholamine production. Thus, imbalances in catecholamine production might produce an overaroused syndrome ("simple schizophrenia"?) which is different from an overaroused syndrome produced by malfunction of the reticular formation, both in symptomatology and in proper treatment.

SUBMITTED FOR PUBLICATION: March, 1973.

Appendix

CS_2 can become a learned source of relief, rather than of fear, if the shock level that follows $CS_1 + CS_2$ is sufficiently small compared to the shock level that follows CS_1. This happens if the increase in tonic input I that follows the unexpected change in shock level J is sufficiently great compared to the change in J. Moreover, for sufficiently small values of I and J, an increase in I can potentiate the fear produced by a fixed level of J, if the signal function is sigmoid.

Consider the following system for definiteness (see Grossberg, 1972c, for details).

$$\begin{aligned}
\dot{x}_1 &= -\alpha x_1 + I + J \\
\dot{x}_2 &= -\alpha x_2 + I \\
\dot{z}_1 &= \beta(\gamma - z_1) - \delta f(x_1(t-\tau))z_1 \\
\dot{z}_2 &= \beta(\gamma - z_2) - \delta f(x_2(t-\tau))z_2 \\
\dot{x}_3 &= -\epsilon x_3 + \zeta f(x_1(t-\tau))z_1 \\
\dot{x}_4 &= -\epsilon x_4 + \zeta f(x_2(t-\tau))z_2 \\
\dot{x}_5 &= -\eta x_5 + \kappa[x_3(t-\sigma) - x_4(t-\sigma)] \\
\dot{x}_6 &= -\eta x_6 + \kappa[x_4(t-\sigma) - x_3(t-\sigma)]
\end{aligned}$$

First, constant levels I_1 and J_1 of tonic input and shock are switched on until after the potentials x_i and transmitters z_j adjust to their new levels. Then new levels I_2 and J_2 are imposed. The potentials adjust much quicker than the transmitters. Hence a measure of the maximum response to the change in levels in computed by maintaining the transmitters at the steady-state levels determined by I_1 and J_1, and the potentials at the new steady-state levels imposed by I_2 and J_2. Fear is produced if $x_5 > 0 > x_6$; relief is produced if $x_5 < 0 < x_6$. The function $f(w)$ computes the signal in response to the potential w. In Grossberg (1974), the function $f(w) = \max(0, w - \Gamma)$ is used, where Γ is a signal threshold. Fear is produced only if

$$f\left(\frac{I_2 + J_2}{\alpha}\right)\left[1 + \mu f\left(\frac{I_1}{\alpha}\right)\right] > f\left(\frac{I_2}{\alpha}\right)\left[1 + \mu f\left(\frac{I_1 + J_1}{\alpha}\right)\right]$$

where $\mu = \delta\beta$, and relief is produced only if the reverse inequality holds. For example, if $f(w) = w$, then fear is produced only if

$$J_2/J_1 > I_2/(v + I_1)$$

where $v = \alpha\beta/\delta$.

The steady-state fear response to constant I and J is given by

$$x_5(\infty) = \frac{\omega\left[f\left(\frac{I+J}{\alpha}\right) - f\left(\frac{I}{\alpha}\right)\right]}{\left[1 + \mu f\left(\frac{I}{\alpha}\right)\right]\left[1 + \mu f\left(\frac{I+J}{\alpha}\right)\right]}$$

where $\omega = \kappa\zeta\gamma\eta^{-1}\epsilon^{-1}$. A sigmoid f yields potentiation of $x_5(\infty)$ in response to an increase of I, if I and J are sufficiently small. To see this is the case that f'' exists, we compute $[\partial x_5(\infty)]/\partial I$, and note that this function is positive if and only if the function

$$h(w) = f'(w)/[\mu f(w) + 1]^2$$

is strictly monotone increasing in the desired region of I and J values. If f'' exists, this condition becomes

$$[\mu f(w) + 1]f''(w) > 2\mu[f'(w)]^2$$

which is true for small values of w, since $f(0) = f'(0) = 0 < f''(0)$.

References

Amsel, A. (1958). *Psychol. Bull.* **55,** 102.

Amsel, A. (1962). *Psychol. Rev.* **69,** 306.

Anderson, P., Gross, G. N., Lomo, T., and Sveen, O. (1969). *In* "The Interneuron" (M. Brazier, ed.), p. 415. Univ. of California Press, Los Angeles.

Atkinson, R. C., and Shiffrin, R. M. (1968). *In* "The Psychology of Learning and Motivation," (K. W. Spence and J. T. Spence, eds.), Vol. 2, p. 89. Academic Press, New York.

Azrin, N. H., and Holz, W. C. (1966). *In* "Operant Behavior" (W. K. Honig, ed.), p. 380. Appleton, New York.

Berlyne, D. E. (1969). *In* "Reinforcement and Behavior" (J. T. Tapp, ed.), p. 179. Academic Press, New York.

Berlyne, D. E. (1970). *In* "Attention: Contemporary Theory and Analysis" (D. E. Mostofsky, ed.), p. 25. Appleton, New York.

Bersh, P. J., Notterman, J. M., and Schoenfeld, W. N. (1956). Air University, School of Aviation Medicine, U.S.A.F., Randolph AFB, Texas.

Biryukov, D. A. (1958). The nature of orienting reactions. *In* "The Orienting Reflex and Orienting-Investigating Activity" (L. G. Voronin *et al.*, eds.), Acad. Pedag. Sci., Moscow.

Bitterman, M. E. (1965). *In* "Classical Conditioning" (W. F. Prokasy, ed.), p. 1. Appleton, New York.

Bloomfield, T. M. (1966). *J. Exp. Anal. Behav.* **9,** 155.

Bloomfield, T. M. (1969). *In* "Animal Discrimination Learning" (R. M. Gilbert and N. S. Sutherland, eds.), p. 215. Academic Press, New York.

Bower, G. H. (1966). *In* "Theories of Learning" (E. R. Hilgard and G. H. Bower, eds.), Appleton, New York.

Cant, B. R., and Bickford, R. G. (1967). *Electroencephalogr. Clin. Neurophysiol.* **23,** 594.

Chung, S.-H., Raymond, S. A., and Lettvin, J. V. (1970). *Brain Behav. Evol.* **3,** 72.

Cohen, J. (1969). *In* "Average Evoked Potentials" (E. Donchin and D. B. Lindsley, eds.), p. 143. Nat. Aeron. Space Admin., Washington, D.C.

Goodrich, K. P. (1959). *J. Exp. Psychol.* **57,** 57.

Grastyan, E. (1959). *In* "The Central Nervous System and Behavior" (M. A. Brazier, ed.) Josiah Macy, Jr. Found., New York.

Gray, J. A., and Smith, P. T. (1969). *In* "Animal Discrimination Learning" (R. M. Gilbert and N. S. Sutherland, eds.), p. 243. Academic Press, New York.

Grossberg, S. (1969a). *J. Math. Psychol.* **6,** 209.

Grossberg, S. (1969b). *Math. Biosci.* **4,** 201.

Grossberg, S. (1970). *J. Theor. Biol.* **27,** 291.

Grossberg, S. (1971a). *J. Theor. Biol.* **33,** 225.

Grossberg, S. (1971b). *Proc. Nat. Acad. Sci. U.S.* **68,** 828.

Grossberg, S. (1972a). *Kybernetik* **10**, 49.

Grossberg, S. (1972b). *Math. Biosci.* **15**, 39.

Grossberg, S. (1972c). *Math. Biosci.* **15**, 253.

Grossberg, S. (1972d). *In* "Delay and Functional Differential Equations and their Applications" (K. Schmitt, ed.), p. 121. Academic Press, New York.

Grossberg, S. (1973). *Stud. Appl. Math.* **52**, 213.

Grossberg, S. (1974). *In* "Progress in Theoretical Biology" (F. M. Snell, ed.), p. 51. Academic Press, New York.

Grossberg, S., and Pepe, J. (1971). *J. Statist. Phys.* **1**, 319.

Grossman, S. P. (1967). "A Textbook of Physiological Psychology." Wiley, New York.

Grusec, T. (1968). *J. Exp. Anal. Behav.* **11**, 239.

Haggard, D. F. (1959). *Psychol. Rec.* **9**, 11.

Hanson, H. M. (1959). *J. Exp. Psychol.* **58**, 321.

Hoffman, H. S. (1969). *In* "Animal Discrimination Learning" (R. M. Gilbert and N. S. Sutherland, eds.), p. 63. Academic Press, New York.

Honig, W. K. (1962). *J. Exp. Psychol.* **64**, 239.

Honig, W. K. (1969). *In* "Animal Discrimination Learning" (R. M. Gilbert and N. S. Sutherland, eds.), p. 35. Academic Press, New York.

Honig, W. K. (1970). *In* "Attention: Contemporary Theory and Analysis" (D. I. Mostofsky, ed.), p. 193. Appleton, New York.

Irwin, D. A., Rebert, C. S., McAdam, D. W., and Knott, J. R. (1966). *Electroencephalogr. Clin. Neurophysiol.* **21**, 412.

Jenkins, H. M., and Harrison, R. H. (1960). *J. Exp. Psychol.* **59**, 246.

Jenkins, W. O., Pascal, G. R., and Walker, R. W., Jr. (1958). *J. Exp. Psychol.* **56**, 274.

Kamin, L. J. (1968). *In* "Miami Symposium on the Prediction of Behavior 1967: Aversive Stimulation" (M. R. Jones, ed.), Univ. of Florida Press, p. 9. Coral Gables.

Kamin, L. J. (1969). *In* "Punishment and Aversive Behavior" (B. A. Campbell and R. M. Church, eds.), p. 279. Appleton, New York.

Kelleher, R. T. (1966). *In* "Operant Behavior" (W. K. Honig, ed.), p. 160. Appleton, New York.

Kernell, D. (1965a). *Acta Physiol. Scand.* **65**, 65.

Kernell, D. (1965b). *Acta Physiol. Scand.* **65**, 74.

Kimble, G. A. (1961). "Conditioning and Learning." Appleton, New York.

Low, M. D., Borda, R. P., Frost, J. D., and Kellaway, P. (1966). *Neurology* **16**, 771.

Luria, A. R., and Homskaya, E. D. (1970). *In* "Attention: Contemporary Theory and Analysis" (D. I. Mostofsky, ed.), p. 303. Appleton, New York.

Lynn, R. (1966). "Attention, Arousal, and the Orientation Reaction." Pergamon, Oxford.

McAdam, D. W. (1969). *Electroencephalogr. Clin. Neurophysiol.* **26**, 216.

McAdam, D. W., Irwin, D. A., Rebert, C. S., and Knott, J. R. (1966). *Electroencephalogr. Clin. Neurophysiol.* **21**, 194.

McAllister, W. R., and McAllister, D. E. (1971). *In* "Aversive Conditioning and Learning" (F. R. Brush, ed.), p. 105. Academic Press, New York.

Maier, S. F., Seligman, M. E. P., and Solomon, R. L. (1969). *In* "Punishment and Aversive Behavior" (B. A. Campbell and R. M. Church, eds.), p. 299. Appleton, New York.

Milner, B. (1958). *In* "The Brain and Human Behavior" (H. C. Solomon, S. Cobb, and W. Penfield, eds.), p. 244. Williams & Wilkins, Baltimore, Maryland.

Newman, F. L., and Baron, M. R. (1965). *J. Comp. Physiol. Psychol.* **60,** 59.

Newman, F. L., and Benefield, R. L. (1968). *J. Comp. Physiol. Psychol.* **66,** 101.

Olds, J. (1969). *Amer. Psychol.* **24,** 114.

Raisman, G., Cowman, W. M., and Powell, T. P. S. (1966). *Brain* **89,** 83.

Rall, W. (1955). *J. Cell. Comp. Physiol.* **46,** 413.

Scheibel, M. E., and Scheibel, A. B. (1967). *In* "The Neurosciences: A Study Program" (G. C. Quarton, T. Melnechuk, and F. O. Schmitt, eds.), p. 577. Rockefeller Univ. Press, New York.

Seligman, M. E. P., and Johnston, J. C. (1973). *In* "Contemporary Prospectives in Learning and Conditioning," Scripta, Washington.

Sharpless, S., and Jasper, H. (1956). *Brain* **79,** 655.

Sokolov, E. N. (1960). *In* "The Central Nervous System and Behavior" (M. A. Brazier, ed.), Josiah Macy, Jr. Found., New York.

Terrace, H. S. (1966). *In* "Operant Behavior" (W. K. Honig, ed.), p. 271. Appleton, New York.

Thompson, R. F. (1967). "Foundations of Physiological Psychology." Harper, New York.

Trabasso, T., and Bower, G. H. (1968). "Attention in Learning: Theory and Research." Wiley, New York.

Wagner, A. R. (1969a). *In* "Punishment and Aversive Behavior" (B. A. Campbell and R. M. Church, eds.), p. 157. Appleton, New York.

Wagner, A. R. (1969b). *In* "Animal Discrimination Learning" (R. M. Gilbert and N. S. Sutherland, eds.), p. 83. Academic Press, New York.

Walter, W. G. *Arch. Psychiat. Nervenkr.* **206,** 309.

Weiss, J. M. (1971a). *J. Comp. Physiol. Psychol.* **77,** 1.

Weiss, J. M. (1971b). *J. Comp. Physiol. Psychol.* **77,** 14.

Weiss, J. M. (1971c). *J. Comp. Physiol. Psychol.* **77,** 22.

Werblin, F. S. (1971). *J. Neurophysiol.* **34,** 228.

CHAPTER 7

NEURAL EXPECTATION: CEREBELLAR AND RETINAL ANALOGS OF CELLS FIRED BY LEARNABLE OR UNLEARNED PATTERN CLASSES

PREFACE

This article contains a confluence of several streams of theoretical ideas. Earlier mathematical work on associative learning showed that the unit of long term memory is a spatial pattern, or a pattern of 'reflectances'. After noticing this, I applied minimality, or Occam's razor, when I asked: "If the system can learn only reflectance patterns, then does it not discriminate only reflectance patterns? Would not discriminatory capabilities atrophy due to disuse if they could not abet later recognition acts by being learned?" This 'equal tuning principle' for learning and perception led me to construct the minimal networks capable of discriminating reflectance patterns.

I began to classify these networks in my 1970 article on neural pattern discrimination [20]. To my surprise, the minimal networks look like retinas. The networks require two successive inhibitory layers that are analogous to horizontal cell and amacrine cell layers. They also possess analogs of receptors, bipolar cells, and ganglion cells. Several months elapsed after deriving these networks before I realized their intuitive meaning. The conceptual road from associative learning to retinas, although smoothly paved by mathematics, took my slower, more reluctant intuition a while to traverse.

These minimal networks were decisive for my later work in several respects. The first stage of competition in the network had to satisfy a normalization rule that regulates the total activity or operating level of the network. In vision, this operating level helps to control the network's light adaptational state. The same normalization rule also appeared shortly thereafter while I was studying how variable numbers of conditioned reinforcing cues can be processed in parallel without destroying the decision rules for eliciting incentive motivational signals (Chapter 5). When I confronted this decision problem as I studied motivation, I knew the answer due to my prior work on retinas. I then realized that a frontal attack on the mathematics of competitive systems was needed to better understand reinforcement no less than perception. In particular, no one seemed to know much about how competitive feedback systems work, despite the ubiquitous appearance of these systems in Nature. Fortunately my first mathematical results on this problem (Chapter 8) provided me with the properties that I needed to extend my reinforcement theory to phenomena about attention and discrimination learning (Chapter 6).

At around the same time that these 'retinal' networks were derived, I published some results from my 1964 monograph in a 1969 article about the cerebellum [15]. This article describes a minimal network design for learning sequences of motor commands among linearly ordered motor organs. The article's main point was to show how different input-output constraints could yield different network realizations using the same associative principles. The marriage of minimality and linear ordering to the associative laws led to networks which strikingly resembled cerebellar anatomy, along with a prediction about possible associative learning at the parallel fiber-Purkinje cell synapse.

David Marr also published a paper in 1969 about the cerebellum and made a similar associative prediction. Marr's approach was, in essence, to directly interpret cerebellar anatomy. I call this theoretical approach The Fallacy of Misplaced Concreteness. Changing an interneuron here or there does not matter if the network's behavioral meaning remains correct, especially since species-specific anatomical variations will surely exist. By contrast, a direct labelling of all variables, disconnected from behavioral meanings, creates an illusion of concrete fabrication without advancing a corresponding scientific understanding. This Fallacy is very popular today, possible because too many people thinker with computers and watch TV rather than think.

My own philosophy claims instead: "The minimal solution is formally correct even if it is not the cerebellum". Incorrect labelling has no secure old age because it generates no conceptual momentum. These observations do not free us from the responsibility of making contact with real anatomies. Rather, they stimulate such contacts by requiring us to derive a principle of neural design in multiple ways to reflect the multiple uses of this design in the control of behavior. The important role of expectancies in my work on reinforcement around 1970–71 therefore led me to ask: "How is the minimal network which is capable of reflectance processing altered when it can be preset to learn an expected event?" Instead of locking like a retina, the minimal solution again looks like a cerebellum. A cerebellar-like anatomy thus can learn sequences of commands among linearly-ordered motor components as well as motor expectancies. As an exercise in minimality, the article illustrates how variations on an associative theme can generate strikingly different network realizations.

Christopher von der Malsburg immediately used the equations in this article to do a clever thing. My previous articles had shown how to construct feature filters, how to learn patterns, and how to bias the filters to respond to prescribed pattern features by using learned expectancies. In a 1973 *Kybernetik* article, Marlsburg showed that the same equations could be used to train a network to discriminate features which regularly occur in the environment. I then realized that the problem of training filters is a dual problem to the problem of learning patterns, but this is another story (Chapter 13). Of more immediate interest is the fact that Marlsburg needed a normalization rule to get his computer program to run, so he simply built one into his program. By this time, I had already proved that you get normalization for free if you use shunting competition instead of additive competition, which was the type of competition that I used in my 1972 article, and the type of competition that Marlsburg adapted to his purposes. How this observation influenced the progress of my work on code development is discussed in Chapter 12. For now, we can wonder why normalization occurs in learning, perception, motivation, as well as in code development.

Neural Expectation: Cerebellar and Retinal Analogs of Cells Fired by Learnable or Unlearned Pattern Classes*

Abstract. Neural networks are introduced which can be taught by classical or instrumental conditioning to fire in response to arbitrary learned classes of patterns. The filters of output cells are biased by presetting cells whose activation prepares the output cell to "expect" prescribed patterns. For example, an animal that learns to expect food in response to a lever press becomes frustrated if food does not follow the lever press. It's expectations are thereby modified, since frustration is negatively reinforcing. A neural analog with aspects of cerebellar circuitry is noted, including diffuse mossy fiber inputs feeding parallel fibers that end in Purkinje cell dendrites, climbing fiber inputs ending in Purkinje cell dendrites and giving off collaterals to nuclear cells, and inhibitory Purkinje cell outputs to nuclear cells. The networks are motivated by studying mechanisms of pattern discrimination that require no learning. The latter often use two successive layers of inhibition, analogous to horizontal and amacrine cell layers in vertebrate retinas. Cells exhibiting hue (in)constancy, brightness (in)constancy, or movement detection properties are included. These results are relevant to Land's retinex theory and to the existence of opponent- and nonopponent-type cell responses in retinal cells. Some adaptation mechanisms, and arousal mechanisms for crispening the pattern weights that can fire a given cell, are noted.

1. Introduction

This paper describes neural networks containing cells U that fire in response to arbitrary learnable

* Supported in part by the Alfred P. Sloan Foundation and the Office of Naval Research (N00014-67-A-0204-0051).
Address: Massachusetts Institute of Technology, 2-382, Cambridge, Massachusetts 02139.

classes of input patterns. These input patterns need not be precoded in the anatomy or physiology of the networks at their birth, and can be changed at will throughout their life. The cells U can also learn to control arbitrary output patterns. U cells are thus "universal" cells whose input and output properties are entirely flexible and changeable by classical and instrumental conditioning.

The learned class of input patterns which can fire a cell U is controlled by presetting cells P. The cells P send axons to the filtering mechanism (e.g., inhibitory interneurons and dendrites) that processes inputs to U. Each P cell can learn a particular pattern that will bias U's filter when P is active. A given P cell can bias several U cells, each with a different learned—or unlearned—pattern. If more than one P cell is active, then U's filter is biased by a pattern which is a weighted average of the patterns controlled separately by each active P cell. The weighting pattern can be changed by altering the relative firing intensities of the active P cells.

The need for such presetting mechanisms is suggested by learning experiments in which an animal O's expectations of future events are learned. For example, let O learn to lever press for food. On a recall trial, O "expects" food when it lever presses in response to hunger. If food is not delivered, O can become "frustrated". Frustration is negatively reinforcing (Kimble, 1961; Wagner, 1969) and can modify O's future expectations. Frustration is biologically useful, since it permits O to eliminate learned, but later unsuccessful, instrumental behavior before ireversible damage is done by the absence of primary reinforcement. On recall trials in the lever press example, lever press cures are presumed to also preset consummatory controls which can be released by expected sensory cues of the food reward. The presetting cells are analogous to the cells P and the consummatory control cells are analogous to the cells U. This paper approaches the study of neural expectation and its frustration by sketching some mathematical features of presetting mechanisms. Presetting mechanisms will later be merged with mechanisms of "negative incentive motivation", as developed in Grossberg (1972), to create a more global picture of the expectation-frustration mechanism.

Given a catalog of preset mechanisms, one naturally seeks analogies with known anatomies *in vivo*. This search is complicated by two factors. First, seemingly different anatomies can share common preset mechanisms; their differences might only be due to differences in arrangement ("symmetries") of P and U cells, and in the distribution of $P \rightarrow U$ axons. Second, the same preset mechanism can form part of different total processing schemes in different cell groups. A striking analogy with aspects of cerebellar anatomy is nonetheless discernible. This analogy includes the following cerebellar facts (Bell and Dow, 1967; Eccles, Ito, and Szentogothai, 1967): (1) existence of a diffuse excitatory mossy fiber input; (2) existence of a localized excitatory climbing fiber input that branches off from excitatory axons to cerebellar nuclear cells; (3) existence of an inhibitory Purkinje cell output to nuclear cells; (4) possibility of firing the Purkinje cell in response to either, or both, input sources; (5) possibility of plastic changes at parallel fiber—Purkinje spine contacts.

The network anatomies use inhibitory interneurons in several configurations. A particular configuration is partly determined by the physiology of its individual cells; for example, by whether the inhibition is subtractive or shunting (multiplicative) (Creutzfeldt, Sakmann, Scheich, and Korn, 1970; Sperling, 1970; Sperling and Sondhi, 1968). Each configuration determines a characteristic time response of the network to test inputs; for example, transient "on-off" responses, or saturated "on" responses, or graded responses to rectangular inputs can occur. Alternatives to inhibitory interneurons exist in some cases; for example, blockade of postsynaptic potential response at higher than prescribed spiking frequencies, or switch-over from net excitation to net inhibition of postsynaptic sites as presynaptic spiking frequency exceeds a critical value (Bennett, 1971; Blackenship, Wachtel, and Kandel, 1971; Wachtel and Kandel, 1971). The present mechanisms include as special cases some mechanisms of sensory adaptation, and the possibility of "crispening" the pattern weights that can elicit a response from U by altering the ambient arousal level.

U cell anatomies are related to the anatomies of cells R which can discriminate patterns without prior learning or presetting. R cell anatomies were introduced in Grossberg (1970a). They also have neural analogs, for example in vertebrate retinas. Such cells can, for example, exhibit essentially perfect hue constancy or brightness constancy (Cornsweet, 1970). Two successive stages of lateral inhibitory interactions, reminiscent of horizontal cell and amacrine cell interactions (Dowling and Werblin, 1969; Werblin and

Dowling, 1969), prepare the inputs to R cells to achieve this constancy. R cell response can also exhibit shifts in perceived hues and brightnesses as background illumination increases. Perfect constancy cells need differ from imperfect constancy cells only in the spatial distribution of the inhibitory interneurons that prepare their outputs. Logarithmic transformation of peripheral inputs is unnecessary to achieve hue or brightness constancy of the R cell variety.

2. Theoretical Review

A previous paper (Grossberg, 1970a) introduced neural networks which can discriminate arbitrary input patterns and learn to release arbitrary output patterns in response to prescribed input patterns. These discrimination mechanisms are ritualistic in the sense that a given output cell responds to the same class of patterns at all times. The ritualistic discrimination and learning mechanisms can be modified to construct the cells U. They are therefore reviewed below.

Grossberg discusses the following situation. Let n cells v_i be given, $i=1, 2, \ldots, n$. Let the input to v_i be $C_i(t)$. A *spatial pattern* is an excitatory input to $V=\{v_i: i=1, 2, \ldots, n\}$ of the form $C_i(t)=\tilde{\theta}_i\, C(t)$, where $\tilde{\theta}_i$ is the fixed relative intensity of the input at v_i (hence $\tilde{\theta}_i \geqq 0$ and $\sum_{k=1}^{n} \tilde{\theta}_k = 1$), and $C(t)$ is the total input intensity. This concept of spatial pattern notes that recognition of a picture is invariant under considerable fluctuations in background illumination.

Grossberg considers the following problem: what is the *minimal* anatomy and physiology of an input filtering device, fed by V, that fires an output cell R if and only if

$$\theta_i - \varepsilon < \tilde{\theta}_i < \theta_i + \varepsilon \tag{1}$$

for all $i = 1, 2, \ldots, n$; that is, if and only if the pattern $\tilde{\theta} = (\tilde{\theta}_1, \ldots, \tilde{\theta}_n)$ differs by less than ε from a prescribed pattern $\theta = (\theta_1, \ldots, \theta_n)$? It will be shown that such a cell R can exhibit perfect hue constancy or brightness constancy, within an error of ε.

The above definition of spatial pattern is compatible with mathematical properties of neural networks *(embedding fields)* derived from psychological postulates about classical conditioning (Grossberg, 1969a; Grossberg, 1971). In their simplest form, these networks are defined as follows.

$$\begin{aligned} \dot{x}_i(t) = & -\alpha_i\, x_i(t) + \sum_{k=1}^{N} [x_k(t - \tau_{ki}) - \Gamma_{ki}]^+ \beta_{ki}\, z_{ki}(t) \\ & - \sum_{k=1}^{N} [x_k(t - \sigma_{ki}) - \Omega_{ki}]^+ \gamma_{ki} + C_i(t) \end{aligned} \tag{2}$$

and

$$\dot{z}_{jk}(t) = -\delta_{jk}\, z_{jk}(t) + \varepsilon_{jk}\, [x_j(t - \tau_{jk}) - \Gamma_{jk}]^+ x_k(t), \tag{3}$$

where $i, j, k = 1, 2, \ldots, N$ and $[\xi]^+ = \max(\xi, 0)$ for any real number ξ. $x_i(t)$ denotes the stimulus trace (or average membrane potential) at time t of the cell body (or cell body cluster) v_i, and $z_{jk}(t)$ denotes the memory trace (or associational strength, or excitatory transmitter production activity) at time t of the synaptic knob (or knobs) N_{jk} found at the end of the axon(s)

e_{jk} from v_j to v_k. The term $[x_k(t-\tau_{ki})-\Gamma_{ki}]^+\beta_{ki}$ in (2) is proportional to the spiking frequency released into e_{ki} in the time interval $[t-\tau_{ki}, t-\tau_{ki}+dt]$. Γ_{ki} is the spiking threshold, β_{ki} is proportional to the anatomical connection strength from v_k to N_{ki}, and τ_{ki} is the time required for spikes to travel from v_k to N_{ki}. The term $\sum_{k=1}^{N}[x_k(t-\tau_{ki})-\Gamma_{ki}]^+\beta_{ki}z_{ki}(t)$ in (2) is the total excitatory input from other cells to v_i at time t. At an excitatory synapse ($\beta_{ki}>0$), spiking frequency couples multiplicatively to transmitter to release transmitter that perturbs $x_i(t)$, and all such signals combine additively at v_i. The term

$$\sum_{k=1}^{N}[x_k(t-\sigma_{ki})-\Omega_{ki}]^+\gamma_{ki}$$

is the total inhibitory input from other cells to v_i at time t. The term $C_i(t)$ is the experimental input (or stimulus) to v_i at time t. See Fig. 1.

In (3), the memory trace cross-correlates the presynaptic spiking frequency which reaches N_{jk} from v_j at time t with the value of average potential at v_k at this time.

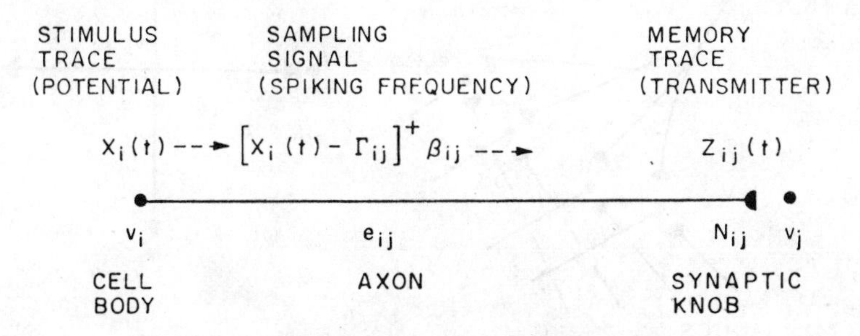

Fig. 1. Network interpretation

To illustrate how the spatial pattern concept relates to learning in (2) and (3), consider the simplest embedding field that can learn by Pavlovian conditioning; namely, an *outstar* (Grossberg, 1969b, 1970b). Let one CS-activated cell v_{n+1} send equal signals to its synaptic knobs $N_{n+1,i}$, which about the UCS-activated cells $V = \{v_i: i = 1, 2, \ldots, n\}$, see Fig. 2. v_1 can learn and perform at V a spatial pattern; that is, a UCS input to V of the form $C_i(t) = \tilde{\theta}_i C(t)$. The synaptic knobs $N_{n+1,i}$ encode the pattern, or "relative figure-to-ground", $\tilde{\theta} = (\tilde{\theta}_1, \tilde{\theta}_2, \ldots, \theta_n)$ in their transmitters $z_{n+1,i}$ at a rate which depends on $C(t)$, among other factors. The total amount $\sum_{k=1}^{n} z_{n+1,k}$ of transmitter accumulation is not constant, however. It can be potentiated either by presynaptic spikes from v_{n+1} to the knobs $N_{n+1,i}$, or by a combination of pre- and postsynaptic activity. It can also spontaneously decay. These fluctuations can occur in the absence of learning.

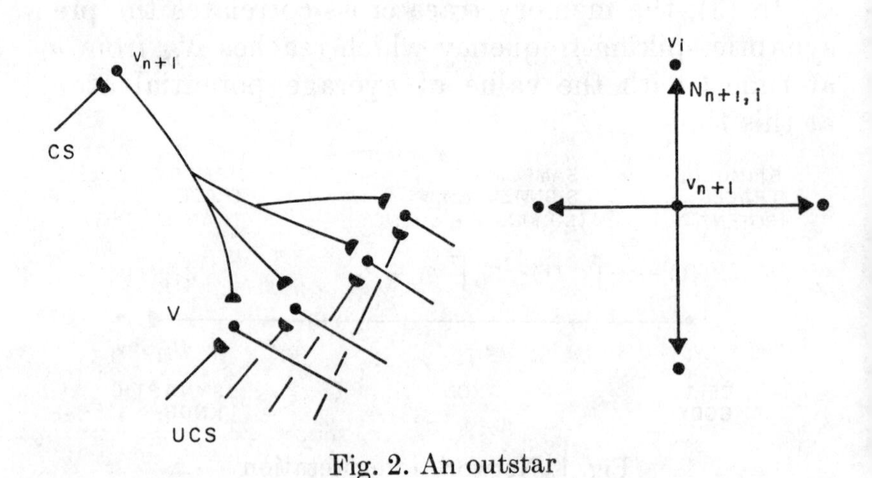

Fig. 2. An outstar

Learning alters the relative sizes of the $z_{n+1,i}$, which are attracted to the pattern weights $\tilde{\theta}_i$.

The pattern $\tilde{\theta}$ at V is learned only at times when the synaptic knobs $N_{n+1,i}$ receive CS-activated spikes from v_{n+1}. This is the property of "stimulus sampling". The relative transmitters $Z_{n+1,i} = z_{n+1,i}\left(\sum_{k=1}^{n} z_{n+1,k}\right)^{-1}$ are the "stimulus sampling probabilities" of an outstar (Grossberg, 1970b). Whenever v_{n+1} samples V, its synaptic knobs begin to learn the spatial pattern playing on V at this time. If a sequence of spatial patterns plays on V while v_{n+1} is sampling, then v_{n+1} learns a pattern which is a weighted average of all the patterns, rather than any single pattern in the sequence. Any nonnegative continuous input to V can be written as such a sequence of patterns.

During recall trials, an input to v_{n+1} reproduces at V the pattern which the outstar has learned. The total output to V can, however, vary wildly through time. For example, fluctuations in spiking frequency from v_{n+1} to the knobs $N_{n+1,i}$ can change the total output either directly, by altering the term

$$[x_{n+1}(t-\tau_{n+1}) - \Gamma_{n+1}]^+$$

in the v_{n+1}-to-v_i signals

$$\beta[x_{n+1}(t-\tau_{n+1}) - \Gamma_{n+1}]^+ z_{n+1,i}(t),$$

or indirectly, by changing the total transmitter level $\sum_{k=1}^{n} z_{n+1,k}(t)$. Such changes in total output do not destroy the encoded memory of pattern weights or the relative output to v_i (Grossberg, 1969b, 1970b). Nonetheless, v_i cannot discern what its pattern weight is if the total output fluctuates. It is therefore natural to seek anatomies, fed by outstar signals, which can discriminate the pattern that the outstar has learned in spite of fluctuations in total outstar output. This is the problem summarized by Eq. (1).

In the study of ritualistic pattern discrimination, no learning occurs. Hence $\delta_{jk} = \varepsilon_{jk} = 0$ in (3) and $z_{ki}(t)$ is set equal to 1 in (2) for notational convenience. Two stages of inhibitory processing intermediate between V and a discriminative output cell R are needed to achieve the criterion in (1) (Grossberg, 1970a, Section 8):

I. Pattern Normalization and Low-Band Filtering

Pattern normalization guarantees that as the pattern intensity $C(t)$ becomes arbitrarily large, the potential $x_i(t)$ in the i-th filtering channel remains bounded without losing the pattern weight due to saturation at the maximal potential. In other words, the first inhibitory layer limits the range of dynamical response to inputs without saturating pattern weights. For example, normalization occurs if $x_i(t) = \tilde{\theta}_i \Gamma(t)$ where $0 < \Gamma(t) \leqq \Gamma < \infty$.

Given a mechanism of pattern normalization, a suitably chosen positive spiking threshold can prevent an output from the i-th channel unless $\tilde{\theta}_i > \theta_i - \varepsilon$. For example, if $x_i(t) = \tilde{\theta}_i \Gamma(t)$, choose the spiking threshold of the i-th channel to equal

$$\Gamma_i = \Gamma(\theta_i - \varepsilon). \tag{4}$$

Then $x_i(t) \leq \Gamma_i$ unless $\tilde{\theta}_i > \theta_i - \varepsilon$. One half of the inequalities (1) is hereby achieved. By (4), given any pattern normalization device, it is necessary that

$$\max \sum_{k=1}^{n} x_k > \sum_{k=1}^{n} \Gamma_k,$$

in order that all low-band filtered channels be able to fire simultaneously.

II. High-Band Filtering

A second processing step is needed to guarantee the inequalities $\tilde{\theta}_i < \theta_i + \varepsilon$ in (1). This step inhibits the i-th low-band filtered signal when it exceeds a size corresponding to an input whose pattern weight $\tilde{\theta}_i$ exceeds $\theta_i + \varepsilon$. The doubly inhibited output from the i-th channel is therefore large only if (1) holds.

All n of these doubly inhibited signals summate at R. R's threshold is chosen so high that R can emit a signal only if it receives large signals almost simultaneously from all n channels; that is, only if (1) holds for all $i = 1, 2, \ldots, n$.

The pattern normalization and high-band filtering steps do not determine which pattern will pass through the filter. The choice of spiking threshold as in (4) does this. A U cell filter differs from an R cell filter only because active P cells can change the "threshold pattern" which must be overcome to fire U, and can learn arbitrary threshold patterns. The network that accomplishes this does not, however, use variable spiking thresholds. Rather it uses a physiologically more plausible mixture of excitatory and inhibitory signals. This mixing process will be motivated by a review of some concrete ways to realize steps (I) and (II) of the ritualistic filtering process.

Pattern normalization and low-band filtering can be achieved in at least two ways.

(i) Subtractive Nonspecific Nonrecurrent Interneuron (see Fig. 3). Let the net signal in the i-th channel after operation of the inhibitory interneuron be

$$J_i(t) = [C_i(t) - \Gamma_i]^+ - \zeta \left[\sum_{k=1}^{n} C_k(t) - \Gamma \right]^+, \qquad (5)$$

with $\Gamma > \sum_{k=1}^{n} \Gamma_k$ and $\zeta \geqq 1$. Since $\Gamma > \sum_{k=1}^{n} \Gamma_k$, Γ_i can be written in the form $\Gamma_i = \Gamma(\theta_i - \varepsilon)$, as in (4). Also write $C_i = \tilde{\theta}_i C$ where $C = \sum_{k=1}^{n} C_k$. By (5), $J_i(t) \leqq 0$ for all $t \geqq 0$ unless $\tilde{\theta}_i > \theta_i - \varepsilon$. Thus no output occurs unless $\tilde{\theta}_i > \theta_i - \varepsilon$. This is because $\tilde{\theta}_i \leqq \theta_i - \varepsilon$ implies that C

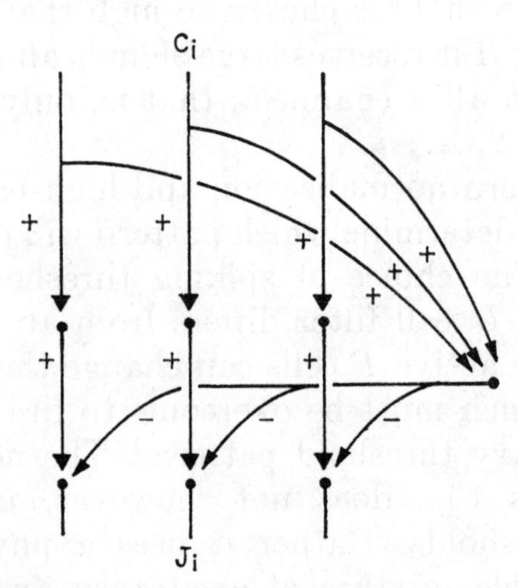

Fig. 3. Subtractive nonspecific nonrecurrent interneuron

exceeds Γ before C_i exceeds Γ_i, and that $dJ_i(t)/dt \leqq 0$ whenever $C > \Gamma$. In the case $\tilde{\theta}_i > \theta_i - \varepsilon$, Grossberg (1970a, Section 10) studies the function J_i as C oscillates, and extends the results to cases in which time lags and exponential averaging rates in the excitatory and inhibitory terms of (5) differ. For example, unless $\tilde{\theta}_i = \zeta = 1$ in (5), $J_i(t)$ responds to growth of C from zero to large values with a transient "on" response, and to decay of C from large values to zero with a transient "off" response. If $\tilde{\theta}_i = \zeta = 1$, then $J_i(t)$ saturates at a constant value when C takes on large values.

(ii) Multiplicative On-Off Field. The subtractive form of the interaction in (5) depends on inputs and thresholds which are small compared to the saturation level of cell body potentials. Given large inputs that drive the potentials towards saturation, a different inhibitory anatomy can preserve input pattern weights in the response of cell body potentials. Consider Fig. 4.

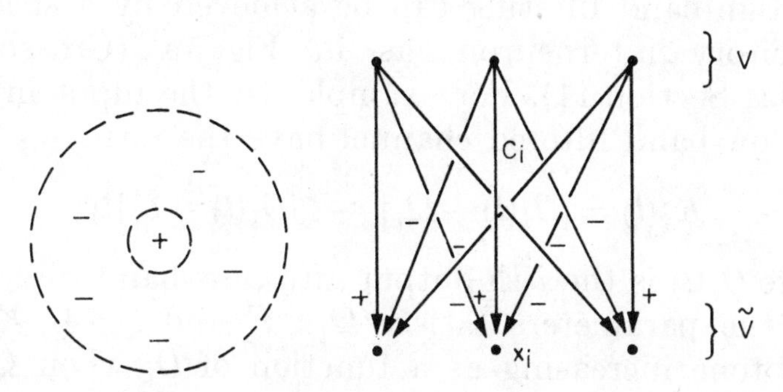

Fig. 4. Shunting on-center off-surround field

The i-th cell in V excites the i-th cell in $\tilde{V}$ and equally inhibits all other cells in $\tilde{V}$ by interacting multiplicatively with ("shunting") cell body potential. The potential x_i of the i-th $\tilde{V}$ cell thus obeys the equation

$$\dot{x}_i = (M - x_i)\, C_i - \alpha x_i - x_i \sum_{k \neq i} C_k. \tag{6}$$

This is a passive membrane equation with equilibrium scaled to zero for convenience, and inputs C_i representing depolarizing or hyperpolarizing conductance changes (Sperling, 1970; Sperling and Sondhi, 1968). Grossberg (1970a, Section 14A) proves that each x_i is bounded and is attracted to $\tilde{\theta}_i$. If, in fact, C varies slowly relative to x_i, then

$$x_i \cong \tilde{\theta}_i \frac{MC}{\alpha + C}, \tag{7}$$

so that x_i is monotone increasing as a function of C. Contrast mechanism (i). Eq. (7) accomplishes pattern normalization without low-band filtering. The next positive spiking threshold in the i-th channel accomplishes low-band filtering.

High-band filtering can be achieved by a specific inhibitory interneuron, as in Fig. 5a (Grossberg, 1970a, Section 11). For example, let the input in the i-th low-band filtered channel have the form

$$K_i(t) = [O_i(t) - \Omega_i]^+ - \zeta\, [O_i(t) - \Gamma_i]^+, \tag{8}$$

where $O_i(t)$ is the i-th output after low-band filtering, and the parameters satisfy $\Omega_i < \Gamma_i$ and $\zeta > 1$. K_i is monotone increasing as a function of O_i when $\Omega_i < O_i < \Gamma_i$, but decreases in O_i as soon as O_i exceeds Γ_i.

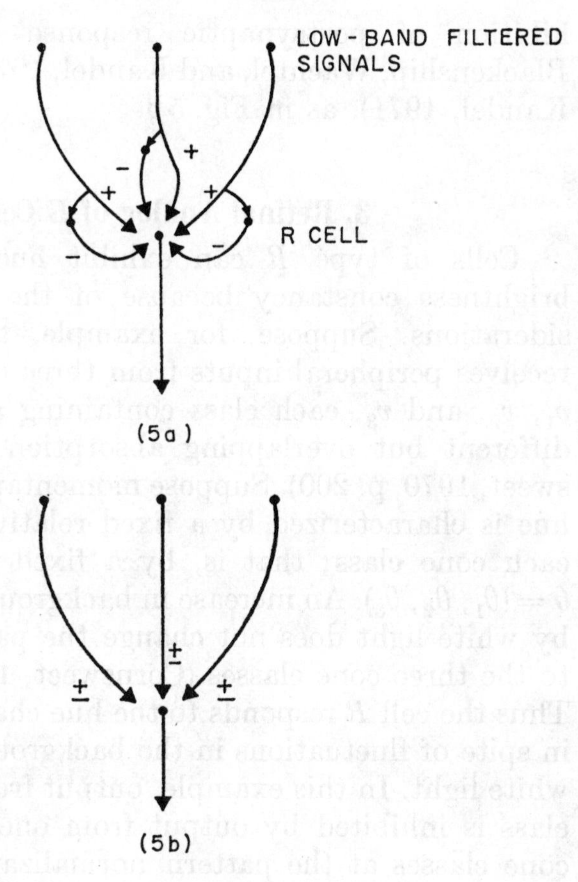

Fig. 5. Some high-band filters

In the present case, letting $\Gamma_i = \Omega_i + \eta$ suffices, where $O_i = \Gamma_i$ before $\tilde{\theta}_i = \theta_i + \varepsilon$. Other mechanisms accomplish the same functional goal; for example, blockade of postsynaptic potential response if presynaptic spiking frequency exceeds the level corresponding to $\theta_i = \theta_i + \varepsilon$, or switch-over as spiking frequency reaches the $\tilde{\theta}_i = \theta_i + \varepsilon$ level from net excitation to net inhibition of postsynaptic response (Bennett, 1971; Blackenship, Wachtel, and Kandel, 1971; Wachtel and Kandel, 1971), as in Fig. 5b.

3. Retinal Analog of *R* Cells

Cells of type R can exhibit hue constancy or brightness constancy because of the following considerations. Suppose, for example, that an R cell receives peripheral inputs from three classes of cones, v_1, v_2, and v_3, each class containing a pigment with different but overlapping absorption spectra (Cornsweet, 1970, p. 200). Suppose momentarily that a given hue is characterized by a fixed relative excitation of each cone class; that is, by a fixed spatial pattern $\theta = (\theta_1, \theta_2, \theta_3)$. An increase in background illumination by white light does not change the pattern of inputs to the three cone classes (Cornsweet, 1970, pp. 243f.). Thus the cell R responds to the hue characterized by θ in spite of fluctuations in the background intensity of white light. In this example, output from a given cone class is inhibited by output from one or more other cone classes at the pattern normalization step. This fact is compatible with the existence of spectrally opponent cells at the ganglion cell layer of the macaque retina (Abramov, 1968).

A similar minimal construction exists for cells R exhibiting brightness constancy. It suffices to change the source cells in the above example. Thus let the pattern $\theta = (\theta_1, \theta_2, \ldots, \theta_n)$ describe the relative intensities of illumination at center cells v_1 and surround cells $v_2, \ldots, v_n$, such that the source cells are either all rods, or are all members of a single cone class. The

R cells responding to these source cells will yield a brightness constancy, respectively, to either white light or to a fixed wavelength to which the cone class responds. The brightness constancy dissolves as soon as the surround is no longer excited. The existence of inhibitory interactions between cells fed by cones and between cells fed by rods has been reported in the goldfish retina. Stell (1967) showed that the external horizontal layer is fed by cones and the intermediate horizontal layer is fed by rods. Kaneko (1970) recorded luminosity- and chromaticity-type S-potentials from the external and internal horizontal layers of the goldfish retina. Naka and Rushton (1966) reported that S-potentials from color units in the fish retina are influenced by at least one stage of shunting inhibition; cf. Eq. (7). A shunt acts also at the level of the cones in the turtle retina (Baylor and Fuortes, 1970). In all these cases, the absolute range of dynamical response in retinal cells can be small, because relative responses at different spatial loci are the crucial quantities.

More sophisticated forms of color constancy are known, and some can be approached by modifying cell connections. The above simple form of color constancy is achieved by pattern normalization among cone classes, followed by low- and high-band filtering. Land (1964), in his retinex theory, notes data suggesting that each cone class establishes its own "lightness" scale before the cones interact. See also Land and McCann (1971). This idea is related to the following modified anatomy, which is included more as a directive for further studies than as a purported explanation of all lightness phenomena. First let each cone class interact within itself to yield pattern normalization. Choose a multiplicative on-off field as in Section 2(ii) for definiteness. Let the off-surround be of much broader spatial extent than the on-center. Each on-off field can, for example, feed the bipolar cell layer of the network. The normalization step reduces the variability of response due to variations in background illumination that excite each cone class, as Eq. (7) illustrates. It can be followed by low- and high-band filtering to yield R cells exhibiting brightness constancy within each cone class. Even if it isn't, however, some degree of constancy is assured. The intracone class normalization step (perhaps followed by low- and high-band filtering) provides a "retinex" for our network. Each cone class is given its own retinex, as Land suggested. Then let the different retinexes interact "classically"—namely, as in the original color constancy construction—to yield spectrally-opponent cell types in the ganglion cell layer. This construction, were it to exist in vivo, could employ bipolar cells which respond to inputs derived from only one cone class in an on-center off-surround configuration. A more local color theory, without a "lightness" scale, could require less bipolar cell selectivity; namely, bipolars could respond to more than one cone class in an on-center off-surround configuration.

The mudpuppy retina exhibits analogs with the minimal mechanisms of Section 2. Since our constructions say little about the *global* properties of neuronal fields of discriminative cells, no more than qualitative similarities will be mentioned. An analog exists between mechanisms of pattern normalization followed by low-band filtering and processing in the horizontal cell layer of the mudpuppy retina (Dowling and Werblin, 1969). Consider the multiplicative on-off field in Section 2(ii) for definiteness. An on-center and off-surround response can be measured intracellularly from bipolar cells, at which mudpuppy receptor and horizontal cell inputs converge (Werblin and Dowling, 1969). The horizontal cells hyperpolarize bipolar cells and integrate receptor signals from a broad retinal area, as is also true of a pattern normalizer. Werblin and Dowling (p. 347–348) stress the importance of the ratio of center-to-surround illumination in determining the bipolar cell response; cf., Eq. (7). They note that horizontal cell inhibition counteracts the depolarizing effects of receptor input on bipolar cell response without hyperpolarizing the bipolar cell (p. 347), that the horizontal cell apparently operates in a nonrecurrent manner relative to its receptor input sources (p. 347), and that the bipolar potential can be maintained at graded sizes in response to graded inputs (p. 347–348). These facts also hold in Eq. (6). In principle, however, Eq. (6) can also yield hyperpolarization if a tonic input term is added. Bipolar responses resemble color-coded C-type S-potentials (p. 347). This is compatible with the use of pattern normalization and low-band filtering to prepare inputs to cells R whose responses exhibit color constancy.

High-band filtering mechanisms can be compared with amacrine cell responses in the mudpuppy retina. Werblin (1970, p. 348) notes that an amacrine cell response can be elicited from a limited number of bipolar cells, much as a specific mixed excitatory-inhibitory connection forms a high-band filter. Werblin and Dowling (1969, p. 351) strengthen this interpretation by noting that the amacrine cell is activated by the initial part of the transient in the bipolar cell response. They hypothesize that a return synapse from amacrine cell to bipolar cell is inhibitory, and that the amacrine cell inhibits the excitation which it receives from the bipolar cell. Such a mechanism could suffice to high-band filter the amacrine cell input. Given this horizontal cell and amacrine cell analogy of R-cell filters, ganglion cells, receiving amacrine-bipolar output, would be a natural class of cells in which R cells, if they exist, might be sought in the mudpuppy retina. Grossberg (1970a) shows that other types of discriminative cells than hue or brightness constancy detectors can be constructed using similar mechanisms; e.g., movement detectors (Werblin, 1970).

The above constancies can be weakened by changing the relative strengths of inhibitory signals. For example, replacing (6) by

$$\dot{x}_i = (M - x_i)\, C_i\, \gamma_{ii} - \alpha\, x_i - x_i \sum_{k \neq i} C_k\, \gamma_{ki}, \tag{9}$$

where the γ_{ki} determine the strengths of signals (excitatory if $k = i$, inhibitory if $k \neq i$) from the k-th to the i-th cone class, yields an asymptotic response to the pattern $C_i(t) = \tilde{\theta}_i\, C(t)$ of the form

$$x_i \cong \frac{\tilde{\theta}_i\, \gamma_{ii}\, MC}{\alpha + \left(\sum_k \tilde{\theta}_k\, \gamma_{ki}\right) C}. \tag{10}$$

Unless the sums $\sum_k \tilde{\theta}_k\, \gamma_{ki}$ are independent of i, the pattern recorded in the relative potentials $x_i \left(\sum_{k=1}^{n} x_k\right)^{-1}$ shifts as a function of total intensity C. This shift in pattern excites those R cells which are sensitive to the new pattern, and yields an analogous shift in perceived hue. Transformations of the form $A(C) = K_1 C(K_2 + C)^{-1}$ have been used to explain properties of hue and brightness inconstancies as a function of background illumination (Cornsweet, 1970, p. 252). The choices of coefficients in these examples differ from that in (10), which illustrates only one possible cause of hue shifts. The problem of choosing coefficients in $A(C)$ is complicated by the appearance of shunting terms at retinal layers beyond the cone and horizontal layers. Creutzfeldt, Sakmann, Scheich, and Korn (1970) use such a transformation to describe ganglion cell responses in the cat retina. They also reference earlier efforts.

4. A Learnable Preset Mechanism: Subtractive Case

To motivate the construction of U cells, consider, as in Section 1, the problem of learning to lever press for food. Activity of P cells is associated with the lever press response. Activity of U cells releases consummatory motor activity. During training trials, P cells learn the patterns which unconditionally activate the U cells. During recall trials, P cells prevent the U cell filter from firing U unless the learned input patterns reoccur. Thus we must answer three questions: (A) How does P learn input patterns at U? (B) Before learning, how do the input patterns at U unconditionally drive U cell firing? (C) After learning, how does P activity bias the U cell filter to prevent all but the learned patterns from activating U? We seek the *minimal* anatomies that answer these questions. Only the U cell analog of Eq. (4) need be studied;

all other properties of R-filters carry over to the U-filter case without change.

Consider question (A). Denote the collection of P cells generically by $V_1=\{v_j: j=1, 2, \ldots, m\}$. Let each P cell be the source cell of an outstar whose axon collaterals terminate at the cells $V_2=\{v_i: i=m+1, \ldots, m+n\}$. Let the cells V_2 receive the input patterns which unconditionally activate U from cells

$$V_3=\{v_{i+n}: i=m+1, \ldots, m+n\}.$$

The $V_1 \rightarrow V_2$ synaptic knobs can learn these input patterns by Pavlovian conditioning.

Consider question (C). Denote by

$$V_4=\{v_{i+2n}: i=m+1, \ldots, m+n\}$$

the cells at which the input pattern is low-band filtered by the output from P cells. As in (5), the difference between the input pattern and the threshold pattern must be computed. Hence excitatory $v_{i+n} \rightarrow v_{i+2n}$ axons deliver the input pattern from V_3 to V_4. Since the threshold pattern is reproduced at V_2 by active V_1 cells, inhibitory $v_i \rightarrow v_{i+2n}$ axons carry the threshold pattern from V_2 to V_4. Note that the input pattern is multiply represented at V_2 and V_4, and that two excitatory input sources converge at V_2 cells.

Consider question (B). In the absence of P cell activity, input patterns can unconditionally activate U. The i-th input channel activates a net excitatory signal $v_{i+n} \rightarrow v_{i+2n}$ and a net inhibitory signal $v_{i+n} \rightarrow v_i \rightarrow v_{i+2n}$ from V_3 to V_4. Thus the absolute value of the excitatory signal exceeds the absolute value of the inhibitory signal. Moreover the spiking thresholds of

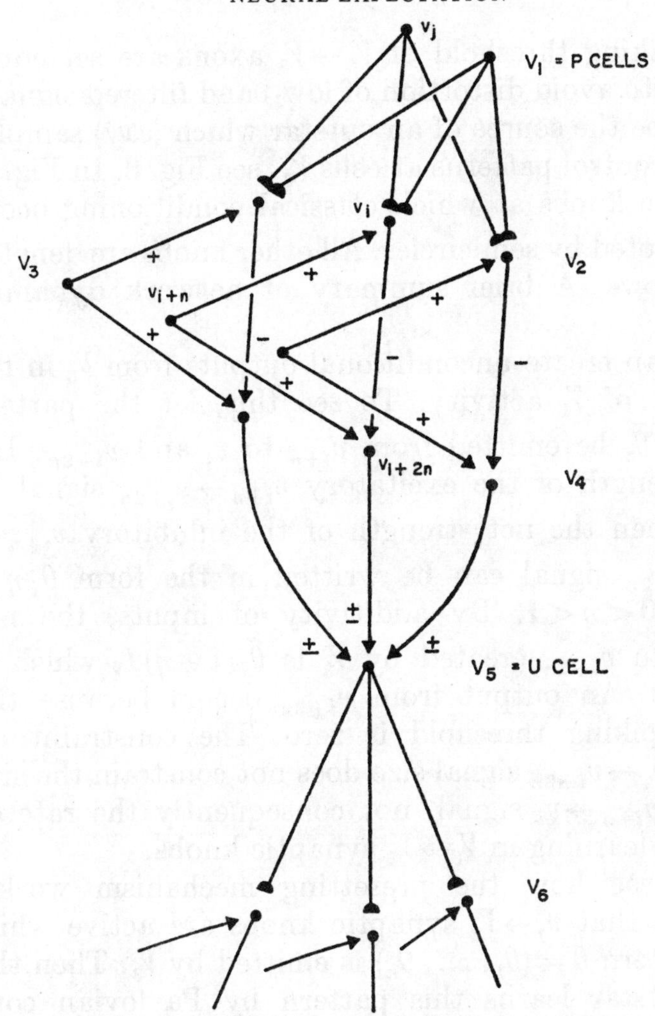

Fig. 6. A learnable preset mechanism: subtractive case

$v_i \rightarrow v_{i+2n}$ axons are set equal to zero to avoid distortion of threshold pattern weights.

Output from V_4 cells is high-band filtered on the way to the output cell $V_5 = v_{m+3n+1}$, which is of type U.

The spiking threshold of $V_4 \to V_5$ axons are set equal to zero to avoid distortion of low-band filtered signals. V_5 can be the source of an outstar which (say) samples motor control patterns at cells V_6, see Fig. 6. In Fig. 6, synaptic knobs at which classical conditioning occur are denoted by semicircles. All other knobs are denoted by arrows. A brief summary of network dynamics follows.

V_3 can create unconditional outputs from V_5 in the absence of V_1 activity. To see this, let the pattern weight $\tilde{\theta}_i$ be emitted from v_{i+n} to v_i and v_{i+2n}. Let the strength of the excitatory $v_{i+n} \to v_{i+2n}$ signal be $\tilde{\theta}_i I$. Then the net strength of the inhibitory $v_{i+n} \to v_i \to v_{i+2n}$ signal can be written in the form $\tilde{\theta}_i \eta I$, where $0 < \eta < 1$. By additivity of inputs, the net signal to v_{i+2n} created by V_3 is $\tilde{\theta}_i(1-\eta)I$, which is positive. An output from v_{i+2n} occurs because the v_{i+2n} spiking threshold is zero. The constraint on $v_{i+n} \to v_i \to v_{i+2n}$ signal size does not constrain the size of the $v_{i+n} \to v_i$ signal, nor consequently the rate of pattern learning in $V_1 \to V_2$ synaptic knobs.

To see how the presetting mechanism works, suppose that $v_j \to V_2$ synaptic knobs are active while the pattern $\theta = (\theta_1, \ldots, \theta_n)$ is emitted by V_3. Then the j-th outstar learns this pattern by Pavlovian conditioning. On recall trials, let v_j be the only active cell in V_1. It reproduces the pattern θ at V_2. V_2 communicates this pattern as inhibitory signals to V_4. Let the net $v_j \to v_i \to v_{i+2n}$ inhibitory signal be $-\theta_i K$. Now let V_3 emit the test pattern $\tilde{\theta} = (\tilde{\theta}_1, \ldots, \tilde{\theta}_n)$. The $V_1 \to V_2 \to V_4$, $V_3 \to V_2 \to V_4$, and $V_2 \to V_4$ signals combine additively at V_4. The net signal to v_{i+2n} is

$$J_i = \tilde{\theta}_i(1-\eta)I - \theta_i K. \tag{11}$$

Since the spiking threshold of v_{i+2n} is zero, (11) implies that v_{i+2n} will fire only if

$$\tilde{\theta}_i > \theta_i K(1-n)^{-1} I^{-1}. \tag{12}$$

Eq. (12) can hold for all i only if $(1-\eta)I > K$; that is, only if the V_3 channel is stronger than the V_1 channel. This achieves low-band filtering by the conditioned pattern θ. High-band filtering of *any* low-band filtered input is then automatically achieved by $V_4 \rightarrow V_5$ signals. Hence, when V_1 presets the U-filter with pattern θ, the U cell only fires if the test pattern emitted by V_3 is θ, within an error of ε. Note that increasing the total $v_j \rightarrow V_2$ input increases K, and thus the minimal pattern weights that can fire V_4. By contrast, increasing the total V_3 output increases I, and thereby decreases the minimal pattern weights that can fire V_4. This "crispening" effect can thus be controlled by varying the arousal, or adaptation, levels of V_1 and V_3, respectively.

Suppose that more than one cell in V_1 fires to V_2 at a given time. Let the $v_j \rightarrow v_i$ synaptic knob encode the pattern weight θ_{ji}. Then the net signal from v_j to v_{i+2n} has the form $-\theta_{ji} K_j$, where K_j depends on the spiking frequency in $v_j \rightarrow V_2$ axons. The total input to v_{i+2n} in response to all active V_1 cells and to V_3 output is

$$J_i = \tilde{\theta}_i(1-\eta) I - \sum_{j=1}^{m} \theta_{ji} K_j.$$

Thus v_{i+2n} fires only if

$$\tilde{\theta}_i > \sum_{j=1}^{m} \theta_{ji} K_j (1-\eta)^{-1} I^{-1}. \qquad (13)$$

The right-hand side of (13) is a weighted average of all pattern weigths in knobs abutting v_i. The weights are determined by the intensity of signals from each cell v_j. The above conclusions can be mathematically extended to consider the influence of time lags and exponential averaging rates using the analysis of the analogous ritualistic case in Grossberg (1970a).

5. Cerebellar Analog of *U* Cells

By redrawing the network in Fig. 6, a striking analogy with aspects of cerebellar anatomy emerges (Bell and Dow, 1967; Eccles, Ito, and Szentagothai, 1967). This analogy includes the following identifications: V_1 = mossy fiber glomeruli; V_2 = Purkinje cells; V_3 = inferior olive cells; V_4 = cerebellar nuclear cells; $V_1 \to V_2$ axons = excitatory parallel fibers; $V_2 \to V_4$ axons = inhibitory Purkinje cell axons; $V_3 \to V_2$ axons = excitatory climbing fibers; $V_3 \to V_4$ axons = excitatory collaterals of climbing fibers.

This analogy becomes more evident when Fig. 6 is redrawn as follows. The $V_1 \to V_2$ outstar axons can be drawn as in Fig. 7a, rather than as in Fig. 7b. The mossy fiber ends in a glomerulus (rosette) that feeds the dendrites of a band of contiguous granule cells. The granule cell axons are parallel fibers, which activate Purkinje cell dendrites. The abstract outstar anatomy of Fig. 7b is functionally identical with

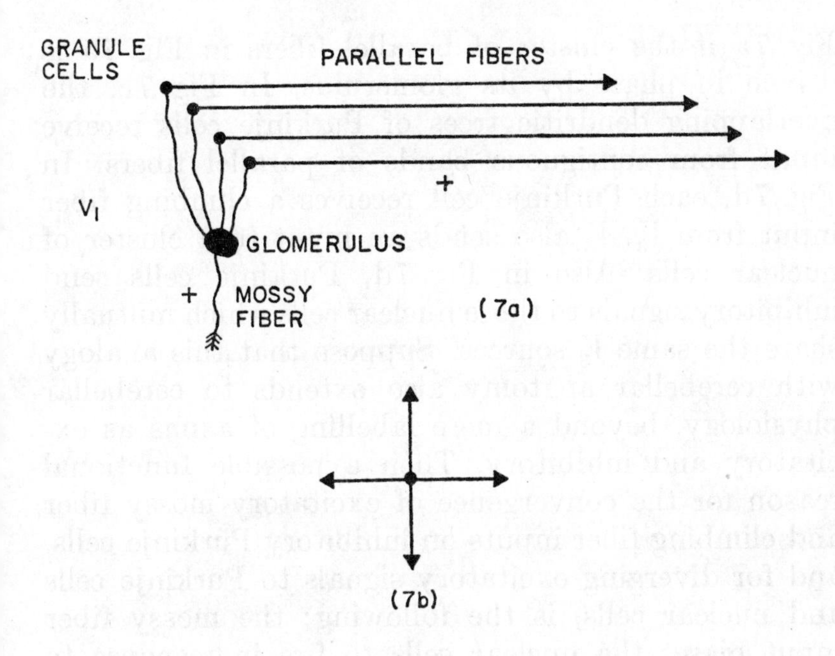

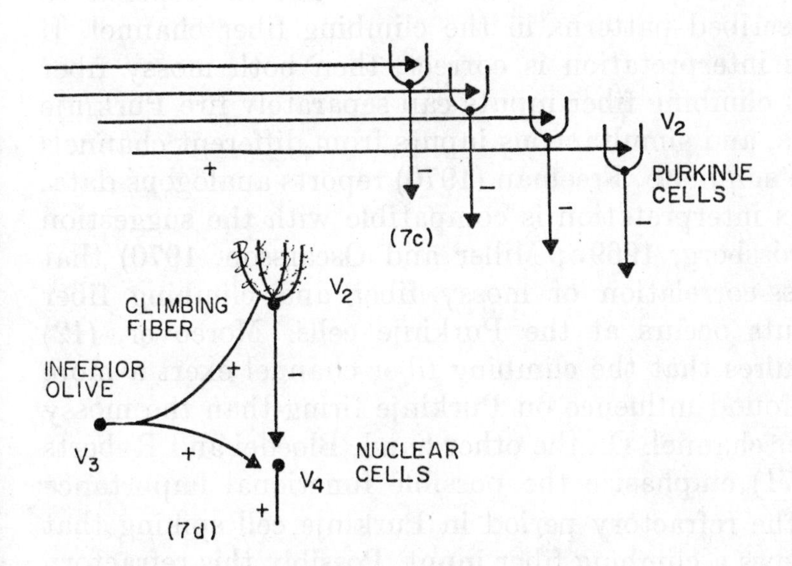

Fig. 7. Cerebellar analog of preset mechanism

Fig. 7a if the cluster of parallel fibers in Fig. 7a is driven in phase by its glomerulus. In Fig. 7c, the overlapping dendritic trees of Purkinje cells receive input from contiguous bands of parallel fibers. In Fig. 7d, each Purkinje cell receives a climbing fiber input from V_3. V_3 also sends an input to a cluster of nuclear cells. Also in Fig. 7d, Purkinje cells send inhibitory signals to those nuclear cells which mutually share the same V_3 sources. Suppose that this analogy with cerebellar anatomy also extends to cerebellar physiology, beyond a mere labelling of axons as excitatory and inhibitory. Then a possible functional reason for the convergence of excitatory mossy fiber and climbing fiber inputs on inhibitory Purkinje cells, and for diverging excitatory signals to Purkinje cells and nuclear cells, is the following: the mossy fiber input biases the nuclear cells to fire in response to prescribed patterns in the climbing fiber channel. If this interpretation is correct, then both mossy fiber and climbing fiber inputs can separately fire Purkinje cells, and simultaneous inputs from different channels can summate. Freeman (1970) reports analogous data. This interpretation is compatible with the suggestion (Grossberg, 1969c; Miller and Oscarsson, 1970) that cross-correlation of mossy fiber and climbing fiber inputs occurs at the Purkinje cells. Moreover, (12) requires that the climbing fiber channel exert a more profound influence on Purkinje firing than the mossy fiber channel. On the other hand, Bloedel and Roberts (1971) emphasize the possible functional importance of the refractory period in Purkinje cell spiking that follows a climbing fiber input. Possibly this refractory period helps to break up the temporal processing of cerebellar inputs into sequences of spatial patterns (Grossberg, 1969b, Section 12). Quantization of temporal processing seems to occur in some sensory systems. For example, exploratory sniffing and tactile input from facial vibrissae seem to be synchronized with the theta rhythm and heart beat in rats (Komisaruk, 1970). After the rat's head is fixed in position, the vibrissae twitch forward and a brief inhalation sniff occurs. The vibrissae are then retracted and the head moves to a nearby fixation point. Then the cycle of coordinated vibrissae motion and inhalation repeats itself. This mechanism seems to break up the sensory input into sequences of spatial patterns. Different sensory channels admit their next spatial pattern in phase with each other. Thus patterns in different modalities can be filtered simultaneously and correlated with each other. This interpretation of Purkinje refractoriness is at best speculative at the present time. Nonetheless, the anatomical analogy in Fig. 7 clearly shows that the abstract minimal anatomy in Fig. 6 is constructed using plausible anatomical principles.

6. A Learnable Preset Mechanism: Multiplicative Case

A mechanism for low-band filtering and pattern normalization using shunting inhibition, such as that in Section 2 (ii), will now be sketched. Consider Fig. 8. V_1 consists of outstar sources. V_2 receives outstar signals. V_3 sends test inputs to V_2 and V_4. V_5 (V_6) receives signals from V_2 (V_4) that have been preprocessed by a multiplicative on-off field. V_6 sends excitatory signals to V_7, whereas V_5 sends inhibitory signals to V_7. The low-band comparison between the input patterns controlled by

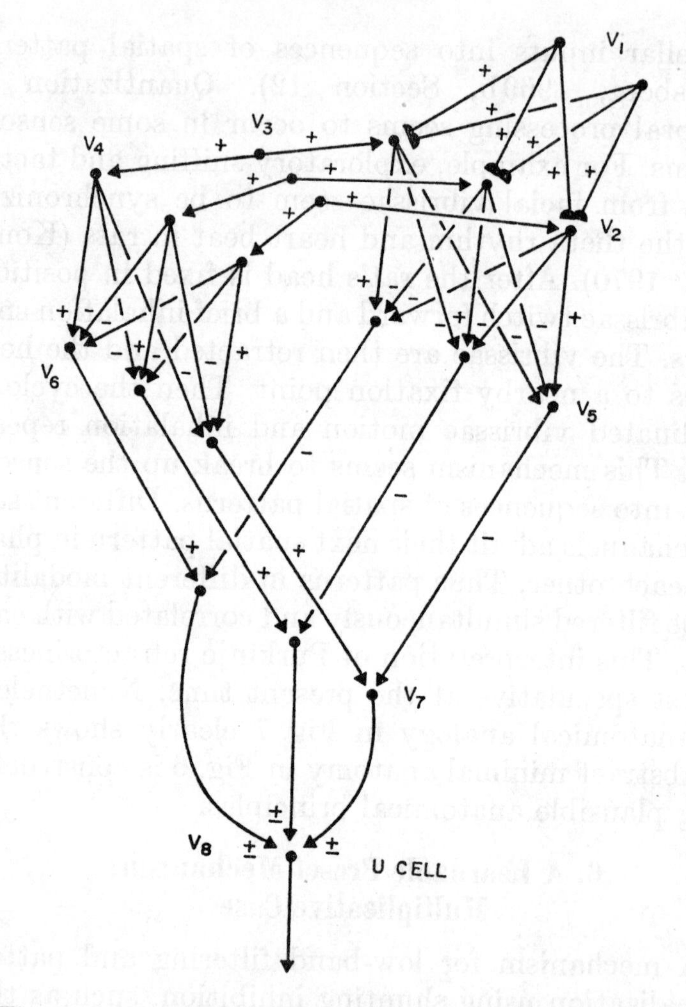

Fig. 8. A learnable preset mechanism: shunting case

V_3 and the threshold patterns controlled by V_1 thus occurs at V_7. Signals from V_7 to V_8 are high-band filtered. V_8 is a cell of type U. The inhibitory signal $V_3 \to V_2 \to V_5 \to V_7$ must be weaker than the excitatory signal $V_3 \to V_4 \to V_6 \to V_7$. This can be accomplished in several ways; for example, let the saturation level of V_5 potentials be smaller than the saturation level of V_6 potentials. The spiking thresholds of $V_2 \rightarrow V_5$, $V_4 \rightarrow V_6$, $V_5 \rightarrow V_7$, $V_6 \rightarrow V_7$, and $V_7 \rightarrow V_8$ axons are set equal to zero to avoid biasing filtered pattern weights. This completes the construction. Note that in all low-band filters of this paper, the statistical dispersion of signals in the excitatory channels is the same as the statistical dispersion of signals in the parallel inhibitory channels. This yields decision rules for cellular firing that retune themselves as the statistics of the input change; cf., Sperling (1970).

More elaborate variations on these themes readily suggest themselves. Given the existence of such striking retinal and cerebellar analogs to the minimal anatomies, it is to be hoped that some of these variations will have more quantitative neural analogs, whose functional meaning will be evident from an inspection of their psychologically derived counterparts. At the very least, the minimal anatomies show how different anatomies can be, even if they carry out similar discrimination tasks, one ritualistically and one with a learnable, or unlearned, preset mechanism.

References

Abramov, I.: Further analysis of the responses of LGN cells. J. Opt. Soc. Amer. 58, 574 (1968).

Baylor, D.A., Fuortes, M.G.F.: Electrical responses of single cones in the retina of the turtle. J. Physiol. (Lond.) 207, 77 (1970).

Bell, C.C., Dow, R.S.: Cerebellar circuitry. In: Neurosciences Research Symposium Summaries, vol. 2 (Schmitt, F.O., Melnechuk, T., Quarton, G. C., and Adelman, G., eds.). Cambridge, Mass.: M.I.T. Press 1967.

Bennett, M.V.L.: Analysis of parallel excitatory and inhibitory synaptic channels. J. Neurophysiol. **34**, 69 (1971).

Blackenship, J.E., Wachtel, H., Kandel, E.R.: Ionic mechamisms of excitatory, inhibitory, and dual synaptic actions mediated by an identified interneuron in abdominal ganglion of Aplysia. J. Neurophysiol. **34**, 76 (1971).

Bloedel, J.R., Roberts, W.J.: Action of climbing fibers in cerebellar cortex of the cat. J. Neurophysiol. **34**, 32 (1971).

Cornsweet, T.N.: Visual perception. New York: Academic Press 1970.

Creutzfeldt, O.D., Sakmann, B., Scheich, H., Korn, A.: Sensitivity distribution and spatial summation within receptive-field center of retinal on-center ganglion cells and transfer function of the retina. J. Neurophysiol. **33**, 654 (1970).

Dowling, J.E., Werblin, F.S.: Organization of retina of the mudpuppy Necturus maculosus. I. Synaptic structure. J. Neurophysiol. **32**, 315 (1969).

Eccles, J.C., Ito, M., Szentagothai, J.: The cerebellum as a neuronal machine. Berlin-Heidelberg-New York: Springer 1967.

Freeman, J.A.: Responses of cat cerebellar Purkinje cells to convergent inputs from cerebral cortex and peripheral sensory systems. J. Neurophysiol. **33**, 697 (1970).

Grossberg, S.: Embedding fields: A theory of learning with physiological implications. J. Mathematical Psychology **6**, 209 (1969a).

— Some networks that can learn, remember, and reproduce any number of complicated space-time patterns, I. J. Math. Mech. **19**, 53 (1969b).

— On learning of spatiotemporal patterns by networks with ordered sensory and motor components, I. Excitatory components of the cerebellum. Studies in Applied Mathematics **48**, 105 (1969c)

— Neural pattern discrimination. J. theor. Biol. **27**, 291 (1970a).

— Some networks that can learn, remember, and reproduce any number of complicated space-time patterns, II. Studies in Applied Mathematics **49**, 135 (1970b).

— Pavlovian pattern learning in nonlinear neural networks. Proc. nat. Acad. Sci. (Wash.) **68**, 828 (1971).

— A neural theory of punishment and avoidance. Mathematical Biosciences 1972, submitted for publication.

Kaneko, A.: Physiological and morphological identification of horizontal, bipolar and amacrine cells in goldfish retina. J. Physiol. (Lond.) **207**, 623 (1970).
Kimble, G.A.: Conditioning and learning. New York: Appleton-Century-Crofts 1961.
Komisaruk, B.R.: Synchrony between limbic system theta activity and rhythmical behavior in rats. J. comp. physiol. Psychol. **70**, 482 (1970).
Land, E.H.: The retinex. Amer. Scientist **52**, 247 (1969).
— McCann, J.J.: Lightness theory. J. Opt. Soc. Amer. **61**, 1 (1971).
Miller, S., Oscarsson, O.: Termination and functional organization of spino-olivocerebellar paths. In: The cerebellum in health and disease (Fields, W.S., and Willis, W.D., eds.). St. Louis: W.H. Green 1970.
Naka, K.I., Rushton, W.A.H.: *S*-potentials from color units in the retina of fish (*Cyprinidae*). J. Physiol. (Lond.) **185**, 536 (1966).
Sperling, G.: Model of visual adaptation and contrast detection. Perception and Psychophysics **8**, 143 (1970).
— Sondhi, M.M.: Model for visual luminance discrimination and flicker detection. J. Opt. Soc. Amer. **58**, 1133 (1968).
Stell, W.K.: The structure and relationship of horizontal cells and photo-receptor-bipolar synaptic complexes in goldfish retina. Amer. J. Anat. **121**, 401 (1967).
Wachtel, H., Kandel, E.R.: Conversion of synaptic excitation to inhibition at a dual chemical synapse. J. Neurophysiol. **34**, 56 (1971).
Wagner, A.R.: Frustrative nonreward: A variety of punishment. In: Punishment and aversive behavior (Campbell, B.A., and Church, R.M., eds.). New York: Appleton-Century-Crofts 1969.
Werblin, F.S.: Response of retinal cells to moving spots: Intracellular recording in Necturus maculosus. J. Neurophysiol. **33**, 342 (1970).
— Dowling, J.E.: Organization of the retina of the mudpuppy, Necturus maculosus. II. Intracellular recording. J. Neurophysiol. **32**, 339 (1969).

Received July 24, 1971

CHAPTER 8

CONTOUR ENHANCEMENT, SHORT TERM MEMORY, AND CONSTANCIES IN REVERBERATING NEURAL NETWORKS

PREFACE

This article is the first of a series to globally analyse competitive dynamical systems. The article suggests that competition solves a sensitivity problem that confronts all cellular systems: the *noise-saturation dilemma.* Low energy input patterns can be registered poorly by cells due to their internal noise. High energy input patterns can be registered poorly by cells because their sensitivity approaches zero when all their sites are turned on. How do cells balance between the two equally deadly, but complementary, extremes of noise and saturation? How do cells achieve a Golden Mean?

The article shows how automatic gain control by competitive signals can retune the cells to accurately transform fluctuating input patterns without noise or saturation. The article once again illustrates the importance of framing theoretical questions on a conceptual level which has evolutionary and behavioral significance. I consider the processing of patterns by cellular systems in a continuously fluctuating environment. If I had considered single cells rather than patterns (as do many experimental neurophysiologists), or discrete rather than continuous inputs (as do many computer scientists), or systems with an infinite number of sites rather than cells (as do linear system theorists), my attack would have failed. Quite unexpectedly, I was able to achieve a global mathematical analysis of these networks *because* they are so nonlinear! To me this means that the dynamics of mass action, the geometry of competition, and the statistics of competitive feedback signals work together to define a unified network module whose several parts are designed in a coordinated fashion through development.

Various other brain theorists, such as Wilson and Cowan or James Anderson, have shied away from nonlinearities in the name of 'simplicity'. I contend that Nature has carefully balanced useful nonlinearities against each other to achieve important physical properties. Throwing away some or all of these nonlinearities upsets this balance, thereby sacrificing important physical properties and making a global mathematical theory harder, rather than easier, to achieve. For example, the model of Wilson and Cowan avoids a sum of nonlinear sigmoid (S-shaped) signals $\Sigma_k f(x_k)$ by using instead a sigmoid of a sum of linear signals $f(\Sigma_k x_k)$. Their theory hereby became 'more linear'. This maneuver led them to also discard automatic gain control by competitive signals, so their system cannot retune itself (see Section 5). 'Simplicity' hereby caused the unnecessary abandonment of a fundamental physical property. James Anderson totally embraces linearity except to stop his variables from exploding to infinity by an *ad hoc* threshold rule. His networks therefore possess quite different properties than the networks derived herein. Each of these models admits certain short-run conveniences, but soon leads to paradoxes or unwieldy extra hypotheses insofar as it does not accurately express a principle of neural design.

The mathematical results include some interesting surprises, which can be viewed as global bifurcation properties of dissipative systems. For example, power law feedback signals cause these continuous systems to behave like finite state choice machines. Sigmoid feedback signals cause the systems to behave like tunable filters. I needed this latter property to understand overshadowing in Chapter 6. This is just the type of property that would have been lost in a linear model. Sigmoid signals, by the way, are often found in neural data taken from sensory processing cells. In all the networks, a normalization property constrains the storage of patterns in short term memory. This nonlinear property was also needed to study attentional processing.

The theorems illustrate several ways whereby positive feedback within the network can be destabilized to generate seizure-like or hallucinatory activity. A 1975 paper with Sam Ellias [35] extended this analysis to investigate how standing waves or travelling waves of neural activity can be triggered across networks whose interaction strengths depend on the distances between cells. Papers in 1975 and 1976 ([34], [37]) with Dan Levine showed how developmental or attentional biasing of the parameters in competitive networks can generate masking phenomena and visual illusions like line neutralization, tilt aftereffect, and angle expansion. These results began to classify some of the transformations that can be executed by competitive systems, and to hereby show how phenomenally distinct perceptual phenomena can be generated by similar underlying dynamic laws. This profusion of special properties eventually led me to ask: "Is there a general mathematical method whereby these results could all be unified?" The next chapter suggests that this question has an affirmative answer.

Contour Enhancement, Short Term Memory, and Constancies in Reverberating Neural Networks

A model of the nonlinear dynamics of reverberating on-center off-surround networks of nerve cells, or of cell populations, is analysed. The on-center off-surround anatomy allows patterns to be processed across populations without saturating the populations' response to large inputs. The signals between populations are made sigmoid functions of population activity in order to quench network noise, and yet store sufficiently intense patterns in short term memory (STM). There exists a quenching threshold: a population's activity will be quenched along with network noise if it falls below the threshold; the pattern of suprathreshold population activities is contour enhanced and stored in STM. Varying arousal level can therefore influence which pattern features will be stored. The total suprathreshold activity of the network is carefully regulated. Applications to seizure and hallucinatory phenomena, to position codes for motor control, to pattern discrimination, to influences of novel events on storage of redundant relevant cues, and to the construction of a sensory-drive heterarchy are mentioned, along with possible anatomical substrates in neocortex, hypothalamus, and hippocampus.

1. Introduction

Recent experimental studies of the hippocampus (Anderson *et al*, 1969) have suggested that its cells are arranged in a recurrent on-center off-surround anatomy. The main cell type, the pyramidal cell, emits axon collaterals to interneurons. Some of these internueurons feed back excitatory signals to nearby pyramidal cells. Other interneurons scatter inhibitory feedback signals over a broad area. Recurrent on-center off-surround networks are found in a variety of neural structures other than hippocampus; for example, neocortex (Stefanis, 1969) and cerebellum (Eccles *et al*, 1967). What does this fundamental principle of neural design accomplish? What can a recurrent, or reverberating, network do that a non-recurrent, or feed-forward, network cannot? In the special case of the hippocampus, one can in particular ask: How does this anatomy contribute to seizure

* Supported in part by the Alfred P. Sloan Foundation and the Office of Naval Research (N00014-67-A-204-0051).

activity in response to topical application of either strychnine or penicillin crystals (Anderson *et al*, 1969)? Can one functionally interpret the suggestion that afferent fibers to the hippocampus excite the inhibitory interneurons directly (Anderson *et al*, 1969), thereby creating a feed-forward inhibitory action, in addition to the recurrent inhibition activated by pyramidal cell output?

This paper describes mathematical results that seem to be relevant to these issues. We study a model that emphasizes the properties of interacting populations of cell sites. These populations can be interpreted either as populations of small membrane patches on individual cells, or as populations of whole cells. The model is perhaps more general since it is defined by mass action laws involving excitatory and inhibitory processes. As in the paper of Wilson and Cowan (1972), we assume that the cell sites in a given population are distributed in such a fashion that their interactions are spatially random and densely distributed within each population and between population pairs. Our equations differ from those of Wilson and Cowan, however. Their excitatory and inhibitory interactions combine additively before they are further processed; our interactions are of shunting type (Hodgkin, 1964; Sperling, 1970; Sperling and Sondhi, 1968). Differences in the applicability of these equations are discussed in Section 5.

Denote the average excitation at time t of the ith population v_i by $x_i(t)$, $i = 1, 2, \ldots, n$. We will study how these averages are transformed through time by recurrent on-center off-surround interactions (Figure 1); that is, each population excites itself and inhibits other populations via the system of equations

$$\dot{x}_i = -Ax_i + (B - x_i)f(x_i) - x_i \sum_{k \neq i} f(x_k) + I_i,$$

where $i = 1, 2, \ldots, n$, and $x_i(\leq B)$ is the mean activity of the ith cell, or cell population, v_i of the network. Four effects determine this system: (1) exponential decay, via the term $-Ax_i$; (2) shunting self-excitation, via the term $(B - x_i)f(x_i)$; (3) shunting inhibition of other populations, via the term $-x_i \sum_{k \neq i} f(x_k)$; and (4) externally applied inputs, via the term I_i. The function $f(w)$ describes the mean output signal of a given population as a function of its activity w. In vivo, $f(w)$ is

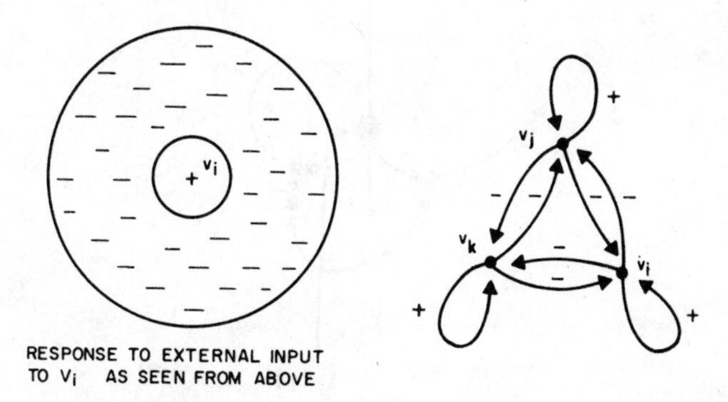

Figure 1. Recurrent on-center off-surround network.

often a sigmoid function of w (Kernell, 1965a, b; Rall, 1955a–c). The mathematical results below will show that this is an important property of the above model for the effective processing of signals in noise.

First, why is an on-center off-surround anatomy needed at all? It has been noted that such an arrangement permits contour enhancement of sensory information (Ratcliff, 1965). We will show that a more basic property can be achieved as well. In many neural systems, noise cannot be avoided, if only because they operate near the quantum range, as in the case of sensory systems. Also cells, and therefore cell populations, have finite saturation levels in response to external inputs. Given these facts, consider the processing of a pattern of input signals delivered to an ensemble of noninteracting cell populations. If the signals are too small, they can be lost in the noise. If they are too large, they can saturate their respective populations, thereby creating a uniform pattern of excitation across populations and destroying all information about the input pattern. In short, noninteracting cell populations are caught between two unsatisfactory extremes. To avoid these extremes in the noninteracting case, input intensities must be restricted to a very narrow range, and one loses the ability to process arbitrary patterns with fluctuating input intensities. On-center off-surround interactions solve this problem: they permit effective processing of arbitrary input patterns across populations, without saturation, even if the inputs are large.

Recurrent on-center off-surround anatomies are capable of short term memory (STM); that is, they can reverberate a pattern of activity distributed over cell populations for an indefinite interval of time. This reverberation can also be switched off rapidly by inhibitory inputs if a new pattern is delivered by external sources; the decay rates of individual cells can be large after the excitatory reverberating loop is broken by inhibition, even if the reverberation through an active excitatory loop is long lived. See Figure 2. A single layer of nonrecurrent on-center off-surround network has limited STM capabilities. Such a network can store a pattern only if it has small decay rates. It will therefore also recover slowly

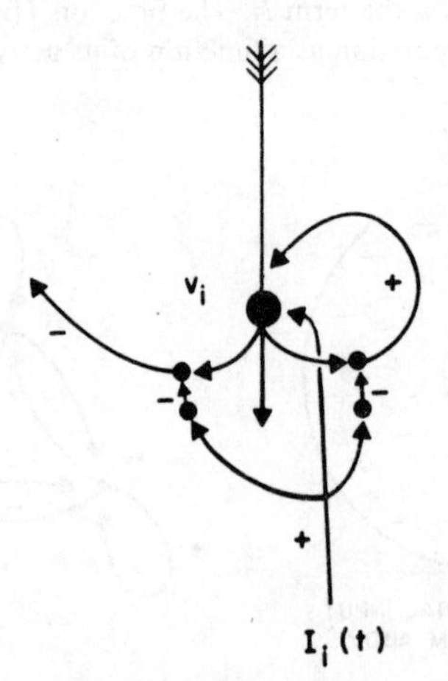

Figure 2. Input inhibits old reverberation as it imposes a new pattern to be reverberated.

from inhibition aimed at shutting the pattern off. Consequently, its response to new inputs will be biased by the lingering traces of old inputs. In a psychological context, the use of reverberation as a mechanism of STM has been suggested (Estes, 1972; Grossberg 1971a). For example, from operant conditioning experiments, one is led to seek reverberatory processes that can maintain in short term storage internal representations of sequences of external events until later rewards or punishments occur and transfer the memory of these sequences to long term storage (Grossberg, 1971a). The interplay of reverberatory and arousal processes on this transfer process has been discussed on various levels, for example neurophysiologically (John 1966) and psychologically (Grossberg, 1971a, 1972a, 1972b).

The STM capabilities of recurrent networks carry with them possible difficulties. If these networks can reverberate patterns imposed by external inputs, then why don't they also reverberate their own noise indefinitely, thereby flooding the network with its own noise? The answer is that they do, if the signal function $f(w)$ is improperly chosen. For example, if $f(w)$ is a linear function of w, or a function that grows slower than linearly, such as $f(w) = w(1 + w)^{-1}$, then noise will be amplified and reverberated. Note that if $f(w)$ is linear, then no contour enhancement will occur; the $f(w)$ that does provide contour enhancement is chosen, first and foremost, to prevent amplification and reverberation of noise. If $f(w)$ grows faster than linearly, such as $f(w) = w^2$, then this problem is avoided. Sufficiently small noise values will dissipate through time. If a brief, but sufficiently intense, input pattern is imposed on the noise, however, then two things happen. First, all populations which receive the largest input in the pattern will suppress the activity in all other populations, including the noise. Second, *normalization* occurs: the total activity $x(t) = \sum_{k=1}^{n} x_k(t)$ of all the populations approaches a fixed positive limit through time. The first property shows that an extreme form of contour enhancement occurs: only the peaks of the input pattern survive. If one population of the network receives more input than any other, then the network "chooses" this population and quenches all others. The second property shows that the system precisely regulates its total activity, and can store the activity of certain populations indefinitely in STM by reverberating their activity through excitatory recurrent interneuronal loops.

The first property is too strong: too much of the pattern is suppressed in the attempt to suppress the noise. How can this be avoided? The way is to choose $f(w)$ so that it grows faster than linearly for small values of w, and (approximately) linearly at larger values of w. Then noise dissipates, and there exists a *quenching threshold*. This means that, given a sufficiently energetic pattern of inputs, the activities of populations which fall below the threshold are quenched (including noise) and those which fall above the threshold are contour enhanced and stored in STM.

In the subsequent discussion, let the existence of a constant quenching threshold be assumed. Then the determination of which populations will be quenched, in the presence of sustained inputs, depends on the total strength $I = \sum_{k=1}^{n} I_k$ of the input to all populations. Consider Figure 3. In Figure 3, a nonspecific arousal input J_A combines with a specific input J_i at each population v_i. Two important cases arise. In Case 1, J_A and J_i combine multiplicatively to influence the activity level x_i. Input J_A is said to *shunt* the activity level (Grossberg, 1973). In Case 2,

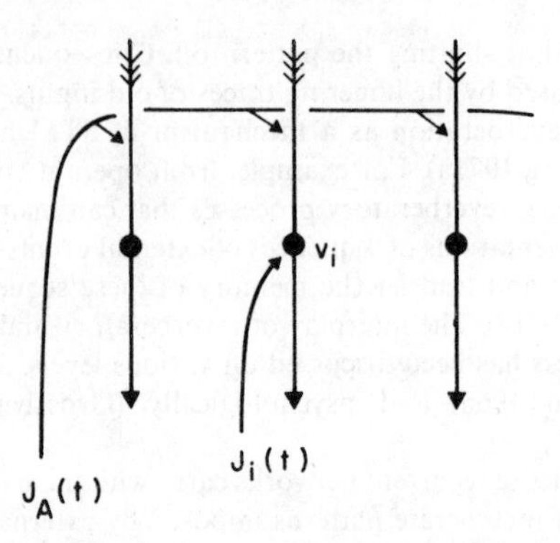

Figure 3. Interaction of arousal and specific inputs.

J_A and J_i combine additively to influence the activity level x_i. Consider Case 1 for definiteness. Then the input J_A does not change the *relative* input levels to the various populations. (In Case 2, a large J_A tends to uniformize any pattern of J_i's.) Let J_A be parametrically increased to ever higher levels. One hereby increases the number of populations that receive enough input to exceed the quenching threshold and are stored in STM. Conversely, reducing J_A decreases the number of populations that will be stored. Thus, given an input pattern in which many inputs are close to each other in relative size, one way to "make a choice" between populations is to lower the arousal level of the input until only one population exceeds the quenching threshold; in common parlance, put the network in a quiet place. By contrast, one way to make as many cues as possible relevant to further network processing is to substantially increase the arousal level. Thus, suppose that a "novel" stimulus excites the network's nonspecific arousal source. Then all recently presented cues can have their network representations brought into STM to play a part in further network processing, including the sampling and subsequent learning of motor responses (Grossberg, 1973). In this way, novel or unpredictable events can bring all possible information about presently available cues into an active state, to enhance the network's ability to deal with the unexpected situation. Using this mechanism, one can approach the problem of how redundant relevant cues are learned (Trabasso and Bower, 1968).

A particularly interesting case arises when the input J_i, unbolstered by a sufficiently large value of J_A, is too small to drive x_i above the quenching threshold. Then any mechanism that inhibits the action of J_A at a given population can prevent this population from reverberating in STM. Figure 4 provides two examples that illustrate this concept. In Figure 4a, the inhibitory input prevents arousal from activating x_i, but x_i's excitatory recurrent collateral bypasses the inhibition. Thus, if population v_i is already reverberating, it continues to do so when the inhibitory input is activated. By contrast, suppose that a new input pulse to v_i occurs simultaneously with inhibition of arousal. Then the afferent inhibition controlled by the new input briefly inhibits the reverberation to allow the new input to begin reverberating without bias due to the previously reverberated input. The new input cannot reverberate, however, because inhibition of arousal prevents

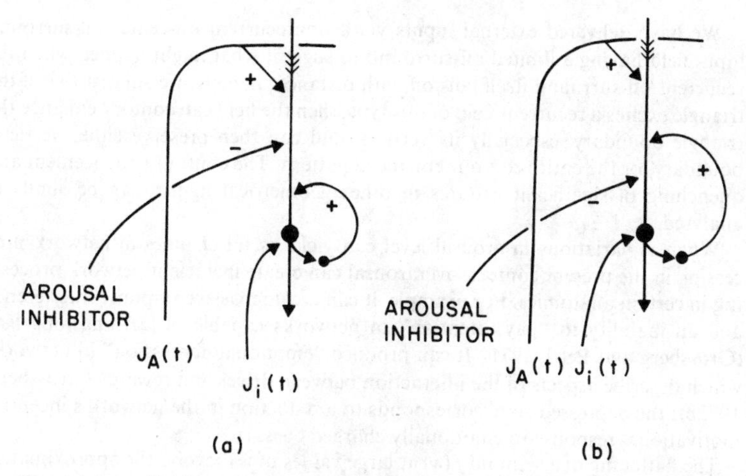

Figure 4. Arousal inhibitors can preserve old reverberations but prevent new reverberations.

it from exceeding the quenching threshold. In short, this type of arousal-inhibition can prevent transfer of new inputs into STM, but permits storage of old inputs in STM. By contrast, in Figure 4b, inhibition of arousal also inhibits STM of old and new inputs.

The above properties have many possible interpretations. For example, suppose that each population in the network responds to lines (Hubel and Wiesel, 1968), orientations (Blakemore and Campbell, 1969), or other geometrical features of external objects. Then varying the arousal level and/or arousal-inhibitors can determine whether a unique geometrical feature of the visual scene, or some particular combination of features, will control motor behavior.

In a similar fashion, particular features of a spatial pattern, such as its boundary, can be stored by the network, while other parts of the pattern are quenched. Suppose for example that the n interacting populations v_i form a rectangular grid in a plane. Choose n very large, and pack the populations closely together to achieve a good spatial resolution of external inputs. Let external inputs be delivered to the populations as follows. If an excitatory input is delivered to v_i, then inhibitory inputs are delivered to all v_k in a small circular region around v_i (nonrecurrent on-center off-surround input field). Suppose that the strength of inhibition depends on the distance of v_k from v_i, and let the same be true for all $i = 1, 2, \ldots, n$. Let a filled triangle be presented to the field. One readily computes that the populations that are excited by the triangle's vertices receive the largest net excitatory input, the populations that are excited by the remainder of the triangle's boundary receive lesser excitatory inputs, and the populations excited by the deepest parts of the triangle's interior receive the smallest excitatory inputs. If the arousal level is sufficiently high, this pattern can be preserved as delivered to the network, apart from the occurrence of normalization. Smaller arousal levels can, however, either quench the interior of the triangle and contour enhance its boundary, or can quench all but the triangle's vertices.

We have delivered external inputs via a nonrecurrent on-center off-surround input field having a limited off-surround to suggest what might happen when the recurrent off-surround itself falls off with distance; namely, we suggest that if the triangle excites a recurrent field of this type, then the field can contour enhance the triangle boundary, especially its vertices, and can then preserve either vertices, boundary, or the entire contour enhanced pattern. The contour enhancement and quenching of significant features in other geometrical figures can be similarly analysed.

Whereas variations in arousal level can yield useful changes in network processing in the present context, overarousal can create inefficient network processing in certain anatomies. For example, it can create massive response interference and an inability to "pay attention" in networks capable of learning long lists (Grossberg and Pepe, 1971). It can produce "emotional depression" in networks which describe aspects of the interaction between drives and rewards (Grossberg, 1972b); the depressed state corresponds to a reduction in the network's incentive motivational response to emotionally charged cues.

The flattening of a sigmoid $f(w)$ at large values of w (beyond the approximately linear range) can, in principle, cause amplification of noise, if the network is overaroused. Such a flattening cannot be avoided *in vivo* because cells have finite maximal firing rates and other bounded constraints on their operating characteristics. It is proved below, however, that robust choices of parameters exist for which the flattening of the sigmoid does not deleteriously affect network processing. The function $f(w)$ is determined by such parameters as the distribution of signal (or spiking) thresholds and of afferent synapses per cell within each population (Wilson and Cowan, 1972). The above results show that varying the function $f(w)$ can dramatically change the pattern features that are stored by the network in STM. Thus, by changing the relative number of cells having a given threshold *within* each population, one can change the pattern features that will be stored by interactions *between* populations.

A variant of the overarousal theme is embodied by the question: how can such a network go into seizure? *Any* operation that creates enough activity in a population to exceed its quenching threshold will cause the population activity to be amplified and maintained in STM. This can be done by creating a sufficiently large excitatory signal (or other perturbation of the population), or by reducing spiking thresholds thereby indirectly increasing noise levels), or by removing inhibitory feedback. If, for example, such cell populations subserve particular sensory impressions, such as in the visual cortices, then these impressions can be created in the absence of external sensory cues if the quenching threshold is exceeded by any other mechanism. If such cell populations control the elicitation of sensory memories, such as in the temporal cortices (Penfield, 1958), then such memories, or memory fragments, can be elicited in the absence of external sensory cues whenever the quenching threshold is exceeded. These "hallucinatory" effects (West, 1962) can be created (say) if sensory deprivation or drugs create a reduction in inhibitory controls, an increase in arousal level, or a decrease in cell spiking thresholds.

The property of normalization creates stable overall activity levels at which the network normally operates in its suprathreshold range. This property can be used to accomplish a variety of tasks by hooking up the network as a component in different overall input processing schemes. For example, it can establish position codes for motor control. This use addresses the question: how does one prevent overshoots and undershoots of orienting responses to localized lights and sounds using our eyes, head, and neck, when these cues have fixed positions but variable intensities? An idealized example is sketched below to convey the basic idea without a pretense of physiological completeness. Consider a network of n populations whose inputs differentially excite a given subset of populations in response to a particular pattern of sensory excitation. For example, suppose that a spot of light in a given retinal position excites a particular population preferentially. Let each population send axonal connections to the various eye muscles, and let the strength of each connection depend on the retinal position represented by the population. The problem is to construct connections which will guarantee that the eye moves towards an arbitrary, but fixed, peripheral spot and fixates on the spot. In this context, normalization prevents undershoots or overshoots in response to a spot of fixed position but variable suprathreshold luminance by factoring out fluctuations in total input intensity. The position code for eye movements is then established by differential *relative* excitation of populations and by the strength of their axonal connections to the eye muscles.

In a similar fashion, such a mechanism can, in principle, maintain a fixed posture in agonist and antagonist muscle pairs. See Figure 5 for an idealized example. In Figure 5, v_i sends a fixed input to the (abstract) muscle M_i, $i = 1, 2$. The relative sizes of the inputs can be changed by descending inputs I_i that move the muscles. In the absence of such descending inputs, the pattern of $v_i \rightarrow M_i$ signals is fixed. In the absence of descending inputs, the fixed total output from v_1 and v_2 can maintain a fixed total muscle length in agonist plus antagonist during maintained postures. The muscle spindles can prevent external forces from altering the muscular position imposed by the signals from v_i (Matthews, 1971).

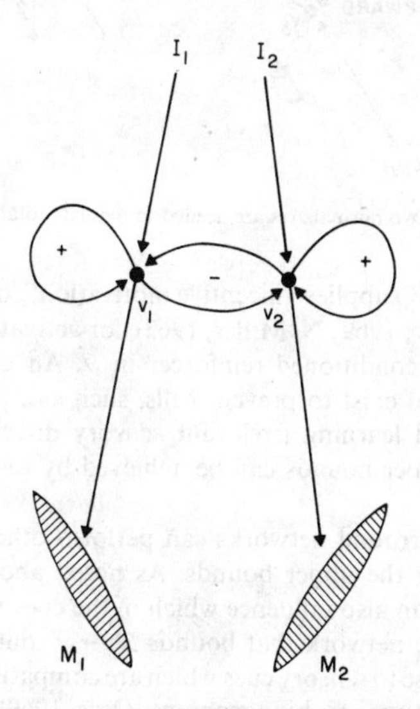

Figure 5. An idealized mechanism for maintaining a posture in the absence of continual inputs I_i.

Normalization has other uses as well. An analysis of instrumental conditioning (Grossberg, 1971a; 1972a, 1972b) shows that the total input from sensory processing areas $\mathscr{S}$ (such as neocortex) to internal drive processing areas $\mathscr{D}$ (such as hypothalamus) should have an upper bound independent of the number of sensory channels which are active at any time. This upper bound is needed to prevent the firing of cells from $\mathscr{D}$ to $\mathscr{S}$ except when they receive a suitable combination of inputs. See Figure 6. In Figure 6, a sufficiently large input from internal homeostats designating that a particular drive needs satisfaction *and* an input from a conditioned reinforcer in $\mathscr{S}$ that is compatible with this drive must combine at cells such as v_3 in $\mathscr{D}$ before these cells can fire. If inputs from $\mathscr{S}$ alone could fire v_3, then the network would seek to persistently satisfy an already satiated drive; hence the bound on total $\mathscr{S} \to \mathscr{D}$ input.

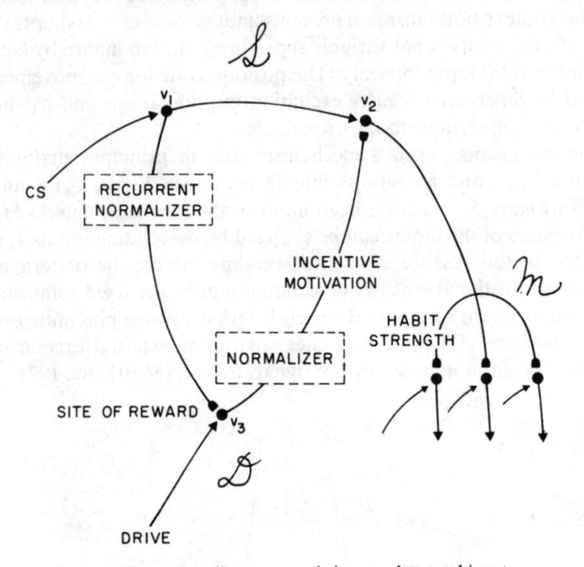

Figure 6. Two normalizers are needed to regulate total input.

The output from $\mathscr{D}$ to $\mathscr{S}$ supplies "incentive motivation", or a "Go" mechanism (Grossberg, 1971a; Logan, 1969; N. Miller, 1963), for activating the motor output at $\mathscr{M}$ controlled by the conditioned reinforcer in $\mathscr{S}$. An upper bound on total $\mathscr{D} \to \mathscr{S}$ output must also exist to prevent cells, such as v_2, in $\mathscr{S}$ from firing at unappropriate times and learning irrelevant sensory discriminations or motor acts at $\mathscr{M}$. These two upper bounds can be achieved by recurrent on-center off-surround networks.

These on-center off-surround networks can perform other important tasks in addition to guaranteeing the upper bounds. As noted above, the network that bounds $\mathscr{S} \to \mathscr{D}$ output can also influence which of the cues represented by $\mathscr{S}$ will reverberate in STM. The network that bounds $\mathscr{D} \to \mathscr{S}$ output can also prevent learning except in response to sensory cues which are compatible with the network's drive needs at any given time; cf., hippocampus (Olds, 1969). Such a network can create a sensory-drive heterarchy (Grossberg, 1972b). Consider the situation in which a student regularly eats meals in spite of the prolonged absence of a sexual partner. A positive, but nonprepotent drive can control motor behavior in the presence of compatible sensory cues (e.g., eating food if hungry), if cues compatible with the prepotent drive are unavailable (e.g., absence of sexual partner). The combination of sensory cues and drive level which controls behavior at a given time can be normalized and stored in STM by the recurrent network. A steady baseline of incentive motivation to activate compatible motor output can hereby be achieved. Interruption of $\mathscr{D} \rightarrow \mathscr{S}$ feedback by ablation or other means can prevent transfer from STM to LTM by preventing the sampling by cells in $\mathscr{S}$ of the patterns to be learned at $\mathscr{M}$ (Milner, 1958).

Normalization can also be used as one stage in the construction of anatomies whose terminal cells respond only to prescribed features of a sensory pattern (Hubel and Wiesel, 1968; Grossberg, 1970, 1972c). It does so by averaging away fluctuations in total network activity and allowing the network to process a pattern's relative weights. In special cases, this construction yields cells whose responses exhibit color or brightness constancies (Grossberg, 1972c), sensitivity to particular velocities (Grossberg, 1970), etc. These examples illustrate that an on-center off-surround anatomy has properties which take on significant, and sometimes surprising, meanings when the network is hooked up at different locations in the overall processing of neural information.

We note in passing that the systems herein are examples of "dissipative structures" (Nichols, 1971), and contribute to the discussion of how patterns of activity can develop and be self-sustained within an interactive system.

In Section 2, the equations that define our networks are presented. Section 3 qualitatively outlines the main phenomena to be reported. Section 4 states the theorems that justify the comments in Section 3. These theorems are proved in the Appendix. Section 5 compares the equations of Section 2 with those of Wilson and Cowan.

2. Network equations

In general, each population v_i contains both excitatory (v_i^+) and inhibitory (v_i^-) subpopulations of cells. See Figure 7. Consider the excitatory cells v_i^+ for definiteness. Suppose on the average that the cell sites in v_i^+ receive randomly distributed afferent pathways from within each subpopulation of the network. Let there be b_i excitable sites in v_i^+, and let $x_i(t)$ be the number of active sites at time t. Three effects determine our equations:

(*1*) *Spontaneous decay of activity:* Active sites become inactive at a fixed rate. Hence $x_i(t)$ decreases at a rate proportional to $x_i(t)$, say $a_i x_i(t)$.

(*2*) *Shunting inhibition:* Active sites are inhibited at a rate jointly proportional to the number of active sites and to the total (randomly distributed!) inhibitory input $I_i^-(t)$. This rate is proportional to $x_i(t)I_i^-(t)$.

(*3*) *Shunting excitation:* Inactive sites are excited at a rate jointly proportional to the number of inactive sites and to the total (randomly distributed!) excitatory input $I_i^+(t)$. This rate is proportional to $(b_i - x_i(t))I_i^+(t)$. In all,

$$\dot{x}_i = -(a_i + I_i^-)x_i + (b_i - x_i)I_i^+, \tag{1}$$

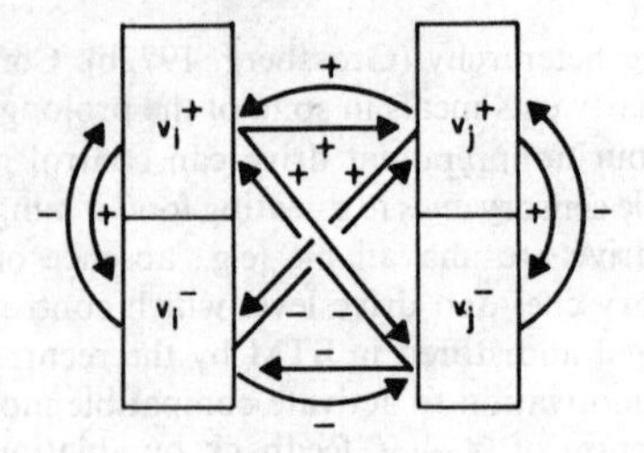

Figure 7. Interactions between excitatory and inhibitory subpopulations.

$i = 1, 2, \ldots, n$. The initial data satisfy the inequalities

$$0 \leq x_i \leq b_i, \qquad i = 1, 2, \ldots, n. \tag{2}$$

Inspection of (1) shows that the inequalities (2) then hold for all $t \geq 0$.

A similar analysis applies to inhibitory cells. Let $y_i(t)$ be the number of active sites in the inhibitory subpopulation v_i^- at time t. Let the total excitatory (inhibitory) input to v_i^- at time t be $J_i^+(t)(J_i^-(t))$. Then y_i is governed by an equation of the form

$$\dot{y}_i = -(A_i + J_i^-)y_i + (B_i - y_i)J_i^+, \tag{3}$$

$i = 1, 2, \ldots, n$, subject to the constraints

$$0 \leq y_i \leq B_i, \qquad i = 1, 2, \ldots, n. \tag{4}$$

The above equations have the same form as passive membrane equations (Hodgkin, 1964; Sperling, 1970; Sperling and Sondhi, 1968); in this context, the inputs I_i^+, I_i^-, J_i^+ and J_i^- represent (average) conductance changes. Thus our analysis formally applies to suitable interactions either between individual cells or between cell populations.

Total inputs are often sums of inputs from other cells (or cell populations) and external influences. For example, let

$$I_i^+ = \sum_{k=1}^{n} F_{ki}^+(x_k) + K_i^+(t), \tag{5}$$

$$I_i^- = \sum_{k=1}^{n} F_{ki}^-(y_k) + K_i^-(t), \tag{6}$$

$$J_i^+ = \sum_{k=1}^{n} G_{ki}^+(x_k) + L_i^+(t) \tag{7}$$

and

$$J_i^- = \sum_{k=1}^{n} G_{ki}^-(y_k) + L_i^-(t). \tag{8}$$

The functions $K_i^+(t)$, $K_i^-(t)$, $L_i^+(t)$ and $L_i^-(t)$ are external inputs. The signal strength functionals F^+_{ki}, F^-_{ki}, G^+_{ki}, and G^-_{ki} determine how mean activities within the excitatory and inhibitory subpopulations of v_k are converted into mean excitatory and inhibitory signals to the excitatory and inhibitory subpopulations of v_i. For example, one can choose

$$[F^+_{ki}(x_k)](t) = \int_0^t x_k(v) \exp\left[- \int_v^t u_k(\xi)\, d\xi \right] dv,$$

or

$$[F^-_{ki}(y_k)](t) = \frac{v_{ki} y_k(t - \tau_{ki})}{w_{ki} + y_k(t - \tau_{ki})},$$

etc.

This paper studies influences of varying signal strength functionals in a setting that minimizes other effects. Hence we consider the special case in which the excitatory and inhibitory subpopulations of each population have the same parameters and receive the same inputs. That is, the excitatory and inhibitory subpopulations of a given population are indistinguishable with respect to every input source, and contain the same number of membrane sites constructed from similar materials. Then $a_i = A_i$, $b_i = B_i$,

$$F^+_{ki}(w) = G^+_{ki}(w),\ F^-_{ki}(w) = G^-_{ki}(w),\ K^+_i = L^+_i\,,\ \text{and}\ K^-_i = L^-_i\,.$$

In this situation one readily proves that the differences $(x_i - y_i)(t)$ converge exponentially to zero as $t \to \infty$, given otherwise arbitrary inputs. Hence the excitatory and inhibitory subpopulations can be lumped together.

We furthermore impose a recurrent on-center off-surround anatomy on the lumped model. See Figure 1. This anatomy is made as homogeneous and simple as possible by imposing the following assumptions:

(1) all numerical parameters are independent of population;

(2) all signals are transmitted instantly; the signal strength functionals are functions.

These constraints lead to the system

$$\dot{x}_i = -(A + I^-_i)x_i + (B - x_i)I^+_i, \tag{9}$$

$i = 1, 2, \ldots, n$, where

$$I^+_i = f(x_i) + K^+_i(t) \tag{10}$$

and

$$I^-_i = \sum_{k \neq i} f(x_k) + K^-_i(t). \tag{11}$$

To study reverberations of system (9)–(11), we always set the external inputs K^+_i and K^-_i equal to zero, yielding the nonlinear system

$$\dot{x}_i = -\left[A + \sum_{k \neq i} f(x_k) \right] x_i + (B - x_i) f(x_i), \tag{12}$$

$i = 1, 2, \ldots, n$. Once reverberations are understood, the inputs K^+_i and K^-_i can be switched on during a finite time interval $[-T, 0]$. Given prescribed initial data at $t = -T$, these inputs will determine a particular distribution of terminal values $x_i(0), i = 1, 2, \ldots, n$. The results on reverberations can then predict how the values $x_i(0)$ will be transformed as $t \to \infty$.

System (12) says that each state v_i excites itself and inhibits all other states with equal weight. This situation can arise even if the inhibitory fields of all populations do not coincide. For example, consider Figure 8. In Figure 8, only the populations

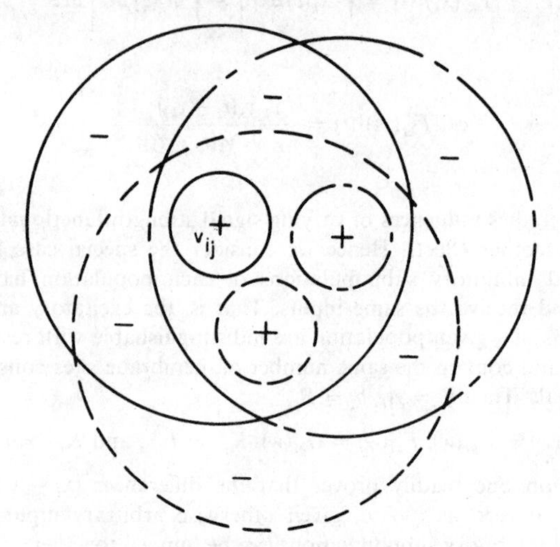

Figure 8. Overlapping inhibitory surrounds.

v_{i_j}, $j = 1, 2, \ldots, m$, receive excitatory inputs $I_{i_j}^+$ at times $[-T, 0]$. Before time $t = -T$, all populations in the network have returned to their zero equilibrium values. The inhibitory fields of each excited population v_{i_j} inhibit all other excited populations v_{i_k}. Inhibited populations which do not receive excitatory inputs can be deleted from the network, since they start out with essentially zero activity and are inhibited thereafter. Thus system (12) includes anatomies in which inhibitory fields of different populations are not the same, but those populations which are excited by external inputs in a given time interval all inhibit each other. System (12) also includes cases in which the strength of inhibitory interactions decreases as a function of distance, if we assume that the excited populations are sufficiently close to each other that their mutual inhibitory interactions are approximately equally strong. Effects of inhomogeneous anatomies on widely separated populations will be considered in another place.

The results derived for system (12) carry over, with small modifications, to the more general system

$$\dot{z}_i = -A(z_i - U) + (V - z_i)I_i^+ - (z_i - W)I_i^- \tag{13}$$

where $W \le U < V$, $I_i^+ = F(z_i)$, and $I_i^- = \sum_{k \neq i} F(z_k)$. Passive membrane equations generally contain the extra parameters U and W. Defining $x_i = z_i - W$, $B = V - W$, $C = A(U - W)$, and $f(x_i) = F(W + x_i)$, (13) becomes

$$\dot{x}_i = -\left[A + \sum_{k \neq i} f(x_k)\right] x_i + (B - x_i) f(x_i) + C. \tag{14}$$

System (14) differs from (12) only in the terms $C \geq 0$, which act like a uniformly distributed tonic input. These tonic terms tend to uniformize the distribution of random noise across populations (compare Theorem 8). The uniformizing effect can be overcome by sufficiently large external inputs (compare Theorem 9). The size of external inputs needed to drive the total activity $x(t) = \sum_{i=1}^{n} x_i(t)$ above the uniformizing range depends on the size of C, and in turn on the size of $U - W$. The term $U - W$ is generally much smaller than $B = V - W$, which is the maximum possible value of $x(t)$.

3. Summary of results

We will study how the choice of $f(w)$ influences the answers to two main questions:

(i) Under what circumstances is the reverberation persistent? transient?

(ii) How is the initial pattern of activity, that was laid down by previous external inputs, transformed as time goes on? Is there a limiting pattern of activity, or does the pattern oscillate indefinitely?

These concepts are made precise by the following definitions:

DEFINITION 1. The *total activity* is the function $x = \sum_{i=1}^{n} x_i$.

DEFINITION 2. The *i*th *pattern variable* is the function

$$X_i = x_i x^{-1}$$

DEFINITION 3. The reverberation is *persistent* if there exists an $\varepsilon > 0$ such that $x(t) \geq \varepsilon$ for all $t \geq 0$.

DEFINITION 4. The reverberation is *transient* if $\lim_{t\to\infty} x(t) = 0$.

If the limit $\lim_{t\to\infty} X_i(t)$ exists, it will be denoted by Q_i. If the limit $\lim_{t\to\infty} x(t)$ exists, it will be denoted by E. Below we define the major limiting distributions and tendencies that will arise in our discussion, and thereby set the stage for this discussion.

DEFINITION 5. The limiting distribution is *fair* if

$$Q_i = X_i(0), \qquad i = 1, 2, \ldots, n.$$

DEFINITION 6. The limiting distribution is *uniform* if

$$Q_i = \frac{1}{n}, \qquad i = 1, 2, \ldots, n.$$

DEFINITION 7. The limiting distribution is *locally uniform* if $P_{i_j} = 1/m$, $j = 1, 2, \ldots, m$, where $1 < m < n$.

DEFINITION 8. The limiting distribution is *0 – 1* if $Q_i = 1$ for some i.

DEFINITION 9. The limiting distribution is *trivalent* if each Q_i assumes one of three values.

DEFINITION 10. The limiting distribution exhibits *quenching* if $Q_{i_j} = 0$ for $j = 1, 2, \ldots, m$.

DEFINITION 11. Let $M(t) = \max\{X_i(t) : i = 1, 2, \ldots, n\}$ and $m(t) = \min\{X_i(t) : i = 1, 2, \ldots, n\}$. The limiting distribution exhibits *contour enhancement* if $\dot{M}(t) \geq 0$ and $\dot{m}(t) \leq 0$, and neither of these derivatives is identically zero.

DEFINITION 12. The limiting distribution is *uniformized* if $\dot{M}(t) \leq 0$ and $\dot{m}(t) \geq 0$, and neither of these derivatives is identically zero.

Definition 13. The reverberation is normalized if there exists a unique positive E_1 such that $E = E_1$.

The following paragraphs illustrate these definitions.

(*A*) *Fair Distribution.* Suppose that $f(w)$ is a linear function of w, as in Figure 9. Then $X_i(t)$ is constant, $i = 1, 2, \ldots, n$. Moreover, the reverberation is either transient or normalized. The conditions under which the reverberation is persistent are independent of the initial data $x_i(0)$. In other words, given persistence, the network can preserve an *arbitrary* pattern indefinitely. Moreover, if $x(0)$ is too small, the network will amplify the total activity until E_1 is reached, whereas if $x(0)$ is too large, activity will dissipate until E_1 is reached.

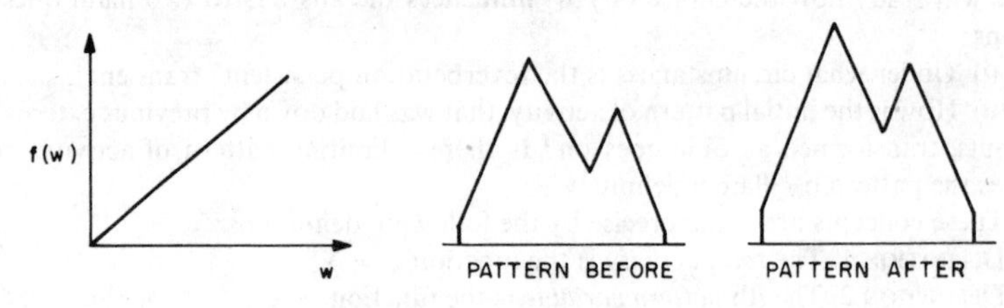

Figure 9. Fair distribution.

The fair signal function unfortunately amplifies noise in the absence of signals as vigorously as it amplifies signals.

The existence of normalization in recurrent networks constrains the possible sensory codes that these networks can sustain. Measurement of the absolute sizes of spiking frequencies given off by a cell, or cell population, in a recurrent network can be misleading. A code based on relative sizes of spiking frequencies across populations focuses on pattern transformations. To determine such a code, an experimentalist must simultaneously measure from a sample of populations. Fluctuations in signals from a single population need not be due to changes in $X_i(t)$, as this example shows; only $x(t)$ need be changing. In some of the examples below, both $x(t)$ and $X_i(t)$ can change through time, although the limits E and Q_i are ultimately approached, with E determined independently of the pattern Q_i. Thus, later readings of the relative spiking frequencies are often functionally more revealing than readings which are taken immediately after the offset of external input pulses. Macrides and Chorover (1972) describe results in the olfactory bulb which are in the spirit of this approach. The olfactory system is known to contain recurrent interactions (Freeman, 1969).

Previous papers (Grossberg, 1971b, 1972a) show that the learning capabilities of various networks are compatible with such a relative code. These networks can learn the pattern of relative excitation across an ensemble of cells, or cell populations, by classical or instrumental conditioning. They can reliably reproduce the learned pattern with an absolute intensity that depends on a complex interplay of various factors.

Deviations from a fair limiting distribution are due to whether $f(w)$ grows more slowly or more rapidly than linearly for various values of w; that is, whether the function $g(w) = w^{-1}f(w)$ is monotone decreasing or increasing.

(*B*) *0–1 or locally uniform distribution.* Suppose that $f(w)$ grows faster than a linear function, as in Figure 10. Again the reverberation is either transient or persistent. It is not necessarily normalized, however, unless $g(w)$ is convex. If the reverberation is persistent, then the limiting distribution exhibits an extreme form of contour enhancement and quenching whenever the initial pattern $X_i(0)$ is nonuniform: All pattern variables such that $X_i(0) < M(0)$ satisfy $Q_i = 0$, while the maximal pattern weights ($X_i(0) = M(0)$) receive all the weight asymptotically.

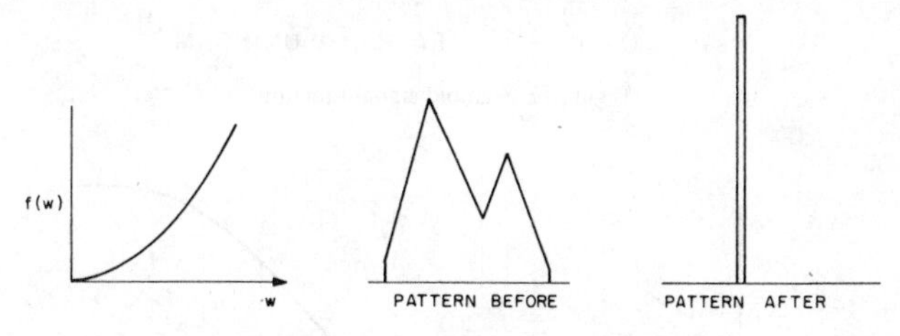

Figure 10. 0–1 Distribution.

In this example, when noise alone is present in the network, it continually dissipates (the reverberation is transient). If a sufficiently energetic pattern is imposed upon the noise, then the highest peaks of the pattern actively suppress both the noise and lesser pattern weights. Simultaneously, these peaks are accentuated, and the total energy of the pattern approaches a positive limit, which is unique if $g(w)$ is convex.

(*C*) *Uniform distribution.* Let $f(w)$ grow slower than linearly, as in Figure 11. Then pattern uniformization occurs. The reverberation is either transient or normalized. In the latter case, the limiting distribution is uniform.

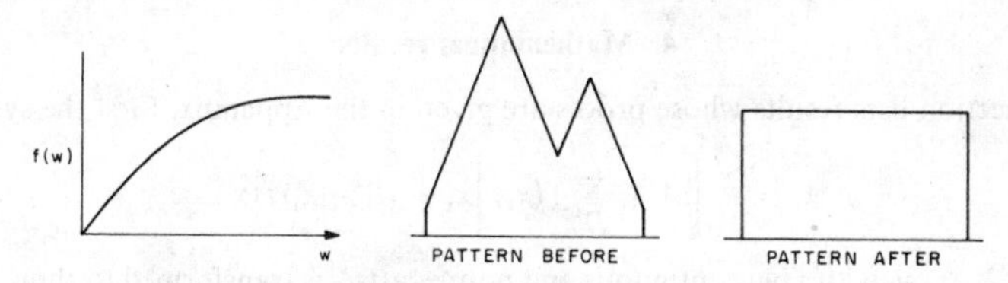

Figure 11. Uniform distribution.

Pattern uniformization can have unfortunate consequences in the presence of noise. Then all states which receive either external inputs or random noise will asymptotically have equal importance.

Functions $f(w)$ exist which combine all the three tendencies listed above; for example, the sigmoid function in Figure 12. Such a function $f(w)$ gives rise to a quenching threshold. Uniformly distributed tonic signals produce yet another uniformizing region. This region tends to uniformize the distribution of noise across populations, and thus to reduce the probability that noise can accumulate in a given population, and thereby create a persistent reverberation in the absence of signals. See Figure 13. Section 4 provides a rigorous discussion of how these regions interact to determine limiting distributions that are combinations of 0–1, fair, and uniform tendencies.

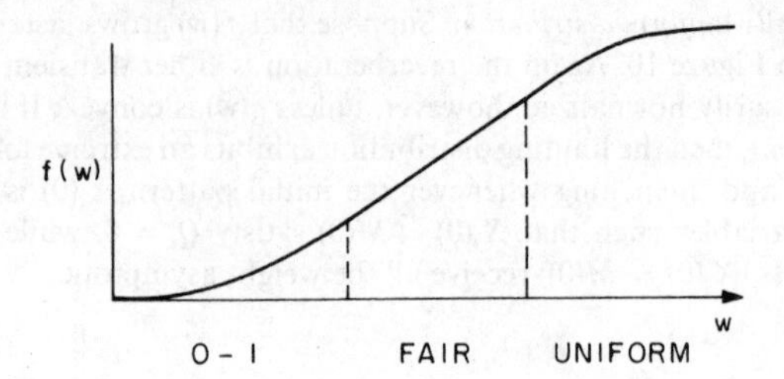

Figure 12. Sigmoid signal function.

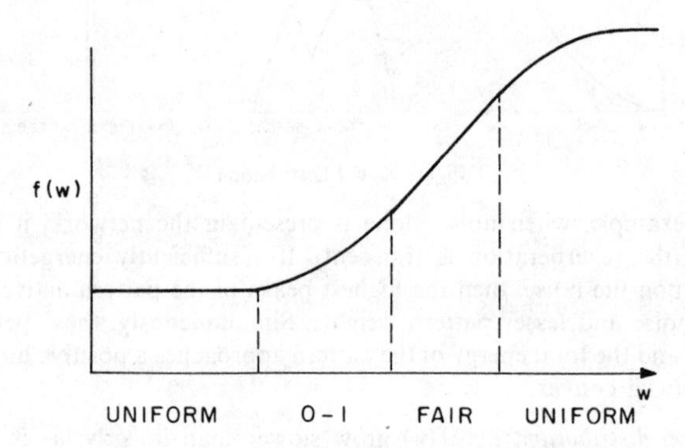

Figure 13. Tonically based Sigmoid signal function.

4. Mathematical results

This section lists results whose proofs are given in the Appendix. First the system

$$\dot{x}_i = -\left[A + \sum_{k \neq i} f(x_k)\right] x_i + (B - x_i) f(x_i), \tag{12}$$

$i = 1, 2, \ldots, n$, with $f(w)$ continuous and nonnegative, is transformed to show how the total activity $x = \sum_{i=1}^{n} x_i$ and the pattern variables $X_i = x_i x^{-1}$ interact. Below we assume that $x(0)$ and all $X_i(0)$ are positive to avoid trivialities. These assumptions imply that $x(t)$ and all $X_i(t)$ are positive for $t \geq 0$. If some $X_i(0) = 0$, then $X_i(t) = 0$ for $t \geq 0$, since v_i receives only inhibition in this case. Such a v_i can be deleted from the network without loss of generality. The notations $g(w) = w^{-1} f(w)$, $g_k = g(x_k)$, and $G = \sum_{k=1}^{n} X_k g_k$ will be used below.

PROPOSITION 1. (*Pattern Variables*). The following equations hold.

$$\dot{X}_i = BX_i \sum_{k=1}^{n} X_k (g_i - g_k), \tag{15}$$

$i = 1, 2, \ldots, n$, and

$$\dot{x} = x(B - x)\left(G - \frac{A}{B - x}\right), \tag{16a}$$

or alternatively

$$\dot{x} = xG(B - AG^{-1} - x). \tag{16b}$$

Alternatively, (15) can be written as

$$\dot{X}_i = BX_i(g_i - G) \tag{17}$$

or as

$$\dot{X}_i = BX_i \sum_{k=1}^{n} X_k[g(X_i x) - g(X_k x)]. \tag{18}$$

Remarks: (1) By (15), the influence of v_k on v_i, namely $X_i X_k(g_i - g_k)$ is the negative of the influence of v_i on v_k, namely $X_k X_i(g_k - g_i)$. Thus the interactions between pattern variables are antisymmetric.

(2) By (17) and (16a), the direction of change of each X_i and x depends on the size of g_i and $A/(B - x)$, respectively, compared to $G = \sum_{k=1}^{n} X_k g(X_k x)$, which is a weighted average of all pattern variables and x. For example, suppose that $g(w)$ has the graph in Figure 14. If $g_i \geq G$, then by (17), $\dot{X}_i \geq 0$. This is depicted in Figure 14 by the arrows facing right. If $g_i < G$, then $\dot{X}_i < 0$, which yields arrows

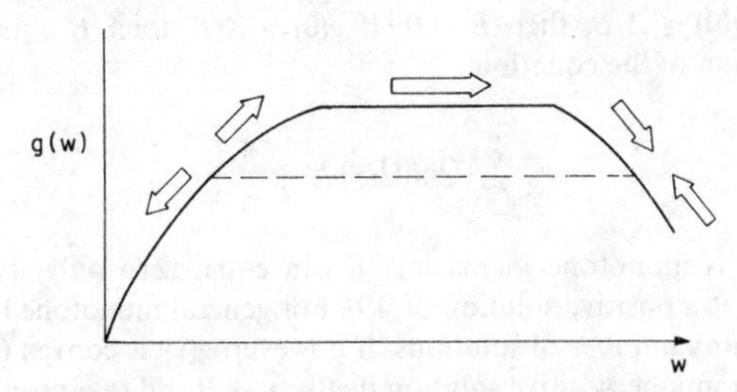

Figure 14. Convex $g(w)$ with fair intermediate range.

facing left. The collision of arrows to the right tends to produce a uniform distribution at large values of X_i. The parting of arrows to the left tends to produce contour enhancement of intermediate X_i values and quenching of small (for example, noisy) X_i values. The mathematics is complicated by the fact that an increase in X_i does not necessarily imply an increase in x_i, since x can be decreasing rapidly. In particular, even if X_i is increasing rapidly, x can be decreasing so rapidly that x_i is dragged down into a region where X_i begins to decrease. Thus the interaction between total activity and pattern variables can produce oscillations, as Proposition 4 illustrates. The results below study how these oscillations can be controlled.

PROPOSITION 2. (*Preservation of Order*). Suppose the states v_i are labelled in such a way that $X_1(0) \leq X_2(0) \leq \cdots \leq X_{n-1}(0) \leq X_n(0)$. Then $X_1(t) \leq X_2(t) \leq \cdots \leq X_{n-1}(t) \leq X_n(t)$ for all $t \geq 0$.

Remark: Consider the pattern depicted in Figure 15a. Proposition 2 says that no matter how the relative sizes of pattern weights are transformed, say as in Figure 15b, their ordering is preserved. This property does not hold in arbitrary anatomies. Henceforth, states will be labelled so that the inequalities $X_1 \leq X_2 \leq X_{n-1} \leq X_n$ hold.

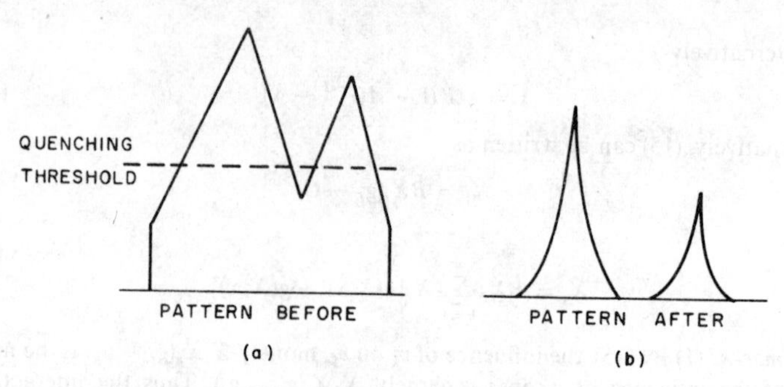

Figure 15. Preservation of order.

The next proposition describes an important condition under which limits of pattern and total activity variables always exist.

PROPOSITION 3. (*Pattern Limits and Energy Normalization*). Let all x_i, $i = 1, 2, \ldots, n$, vary in a region where $g(w)$ is monotonic. Then all the limits $Q_i = \lim_{t\to\infty} X_i(t)$ and $E = \lim_{t\to\infty} x(t)$ exist. Suppose $g(w)$ is monotone decreasing or constant. If $g(0) \leq A/B$, then $E = 0$. If $g(0) > A/B$, then E equals the unique positive solution of the equation

$$\sum_{k=1}^{n} Q_k g(Q_k x) = \frac{A}{B - x}. \tag{19}$$

Suppose $g(w)$ is monotone increasing. E can equal zero only if $g(0) < A/B$. If $E \neq 0$, then E is a positive solution of (19). For general monotone increasing $g(w)$, (19) can have any number of solutions. If however, $g(w)$ is convex (as in Figure 9) then (19) has a unique positive solution if $g(0) \geq A/B$ and two positive solutions if $g(0) < A/B$. The smaller solution is unstable; the larger solution is stable.

Remarks: (1) If $g(w)$ is increasing, then (19) can have any number of solutions unless $g(w)$ is convex. In physical situations, convexity (or near convexity) is a likely property, since $f(w)$ is often sigmoid (Kernell, 1965a, b; Rall, 1955a, b, c; Wilson and Cowan, (1972) and the simplest $g(w)$ that can achieve this shape is convex, as in Figure 14. (2) The case $g(0) \geq A/B$ is undesirable, since even small noise values can be amplified and preserved indefinitely by the network. The inequality $g(0) < A/B$ allows noise to dissipate, but sufficiently large signals in the noise can persist. Theorems 2 and 4 describe systems in which signals can quench noise, and can use the noise to accentuate the contours of the pattern that is imposed on the noise. This contour enhanced pattern can then be preserved indefinitely by the network.

The following results show how particular choices of $f(w)$ determine the limiting distribution Q_i. The crucial fact is whether $f(w)$ grows faster or slower than linearly, or linearly, for particular values of w; that is whether $g(w)$ is increasing, decreasing, or constant. There exist $f(w)$'s with the following property: given a fixed initial pattern $X_i(0)$ and fixed $f(w)$, more than one limiting pattern Q_i can occur. The particular pattern Q_i that occurs depends on $x(0)$, or the initial "arousal" level; varying the arousal level can change the type of information processing that occurs. For example, one can either preserve a given pattern or induce contour enhancement and quenching of this pattern, simply by varying $x(0)$:

THEOREM 1. (*Fair Distribution*). Let $f(w) = Cw$ for some $C > 0$. Then $Q_i = X_i(t) = X_i(0)$ for all $t \geq 0$. Let $D = BC - A$. If $D > 0$, the reverberation is persistent. If $D \leq 0$, the reverberation is transient. In fact, if $D \neq 0$,

$$x_i(t) = \frac{x_i(0)\exp(Dt)}{1 + x(0)CD^{-1}[\exp(Dt) - 1]}, \tag{20}$$

whereas if $D = 0$,

$$x_i(t) = \frac{x_i(0)}{1 + x(0)Ct}. \tag{21}$$

In particular, if $D > 0$, then $\lim_{t\to\infty} x_i(t) = X_i(0)(B - AC^{-1})$, and thus energy normalization occurs as $t \to \infty$.

Remark: Given a linear $f(w)$, if any pattern can reverberate persistently, then even small values of noise will reverberate albeit with small relative weight in the presence of large signals. This can be a liability in such systems, since in the absence of signals, noise will be amplified, and will receive a large relative weight.

The next theorem shows that if $f(w)$ grows faster than linearly, then noise can dissipate, and large values can quench small values before they are amplified and maintained. To discuss this situation, we again use the notation $M(t) = \max\{X_i(t): i = 1, 2, \ldots, n\}$ and $m(t) = \min\{X_i(t): i = 1, 2, \ldots, n\}$. By Proposition 2, if $X_i(t_0) = M(t_0)$, then $X_i(t) = M(t)$ for all $t \geq t_0$. Similarly for $m(t)$.

THEOREM 2. (*0–1 Distribution*). Let $f(w) = wg(w)$, where $g(w)$ is continuous, nonnegative, and strictly monotone increasing. If $M(0) = m(0) = 1/n$, then $M(t) = m(t) = 1/n$ for all $t \geq 0$. Otherwise, $M(t)$ is monotone increasing faster than any function $X_i < M$, and $m(t)$ is monotone decreasing. Suppose moreover that the reverberation is persistent. (It is if $g(0) \geq A/B$, or if $g(0) < A/B$ and $x(0) \geq \hat{x}$, where $\hat{x}$ is the smaller root of

$$(n - K)M(0)g(M(0)x) = \frac{A}{B - x} \tag{22}$$

(g convex) and $X_K(0) < M(0) = X_{K+1}(0)$.) Then the limiting distribution is 0–1 or locally uniform, and satisfies $Q_1 = Q_2 = \cdots = Q_K = 0$ and $Q_{K+1} = \cdots = Q_n = (n - K)^{-1}$.

A wide variety of functions are special cases of Theorem 2; for example,

$$f(w) = \sum_{k=1}^{\infty} a_k w^k$$

with all $a_k \geq 0$ and $0 < \sum_{k=2}^{\infty} a_k B^k < \infty$;

$$f(w) = \frac{Cw}{D + \exp(-Ew)}, \tag{23}$$

with $C, D, E > 0$; and

$$f(w) = \frac{w(A + Bw + Cw^2)}{D + Bw + Cw^2} \tag{24}$$

with $A, B, C, D > 0$ and $D > A$.

If $f(w)$ increases slower than linearly, then the opposite tendency occurs; the initial distribution is uniformized.

THEOREM 3. (*Uniform Distribution*). Let $f(w) = wg(w)$, where $g(w)$ is continuous, nonnegative, and strictly monotone decreasing. Then the function $M(t)$ is monotone decreasing and $m(t)$ is monotone increasing. Suppose moreover that the reverberation is persistent (that is, $g(0) > A/B$). Then all $Q_i = 1/n$, and E equals the unique positive solution of

$$g\left(\frac{x}{n}\right) = \frac{A}{B - x}. \tag{25}$$

Some special $f(w)$'s are listed below, for definiteness. An important class of functions is defined by

$$f(w) = \frac{w}{\sum_{k=0}^{\infty} b_k w^k}$$

where $b_0 > 0$, $b_k \geq 0$ and $0 < \sum_{k=1}^{\infty} b_k B^k < \infty$.
For example,

$$f(w) = \frac{Cw}{D + w},$$

with $C, D > 0$; or

$$f(w) = \frac{Cw}{D + \exp(Ew)}, \tag{26}$$

with $C, D, E > 0$ (contrast (23)); or

$$f(w) = \frac{Cw\exp(-Dw^m)}{E + Fw^n},$$

with $C, D, E, F > 0$ and $m, n \geq 1$.

Remark: Not all of the above $f(w)$'s are monotonic; nonetheless Theorem 3 holds. For example, $f(w)$ in (26) increases at small values of w and decreases to zero at large values of w.

Theorems 1–3 suggest how to construct functions $f(w)$ that will combine 0–1, fair, and uniform tendencies. For example, define a continuous, positive $g(w)$ that is strictly increasing at small values of w and is strictly decreasing at large values of w. Theorems 2 and 3 suggest that 0–1 and uniform tendencies will be included in this way. A "fair" intermediate region can be constructed by choosing $g(w)$ constant (or, for all practical purposes, approximately so) between its increasing and decreasing values, since then $f(w)$ is linear in this range. See Figure 14. More complex combinations of these three tendencies can be included by defining a $g(w)$ that oscillates finitely many times. This procedure can also be reversed. *Given* a function $f(w)$, define $g(w) = w^{-1}f(w)$ and test where $g(w)$ increases, decreases, and is constant to get an idea of $f(w)$'s 0–1, uniform, and fair tendencies. The next theorems discuss various combinations of these possibilities. First we consider an $f(w)$ that combines 0–1 and fair tendencies. In this situation, three possibilities occur. The reverberation can be transient or persistent. If the reverberation is persistent, the limiting distribution can be fair; in both cases the limiting distribution can combine contour enhancement and quenching tendencies. The choice between fair or contour enhancing and quenching tendencies can be controlled by $x(0)$.

THEOREM 4. (*Fair, or Contour Enhancing and Quenching*). Let $f(w) = wg(w)$, where $g(w)$ is continuous, non-negative, and strictly monotone increasing for $0 \leq w \leq x^{(1)}$, and $g(w) = C$ for $w \geq x^{(1)}$. (See Figure 16). Then all limits Q_i and E exist. The function $M(t)$ increases monotonically and no slower than any $X_i < M$, and the function $m(t)$ is monotone decreasing. If $x_1(t) \geq x^{(1)}$, then all $\dot{X}_i(t) = 0$. If $g_i(t) = g_j(t) = C$, then $(d/dt)(X_i X_j^{-1})(t) = 0$.

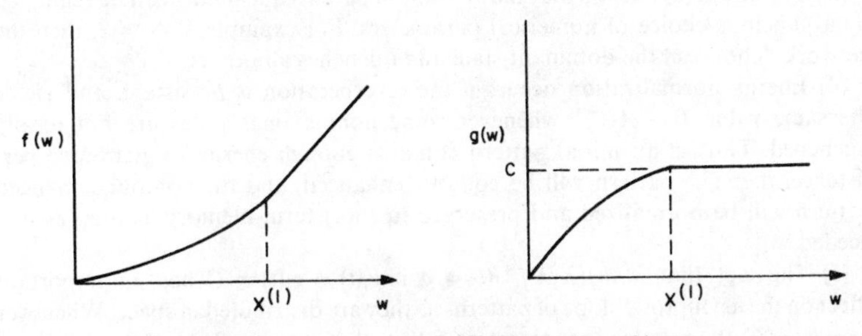

Figure 16. Fair, or contour enhancing and quenching.

Suppose moreover that the reverberation is persistent. (It is under the conditions given in Theorem 2.) Define K by $X_K(0) < M(0) = X_{K+1}(0)$. Then either $Q_i = 0$ or $g(Q_i E) = C, i = 1, 2, \ldots, K$, and $Q_i = 1/(n - K)$ or $g(Q_i E) = C, i = K + 1, \ldots, n$. In particular, if

$$X_L(0) \min\left(B - \frac{A}{\sum_{i=L}^{n} X_i(0)C}, x(0) \right) \geq x^{(1)}, \tag{27}$$

then $\dot{X}_i \geq 0$ and $(X_i X_j^{-1})^{\bullet} = 0$ for $t \geq 0$ and $i, j \geq L$. If

$$X_1(0) \min(B - AC^{-1}, x(0)) \geq x^{(1)}, \tag{28}$$

then $Q_i = X_i(t) = X_i(0)$ for $t \geq 0$ and $i = 1, 2, \ldots, n$. If however

$$X_1(0)(B - AC^{-1}) < x^{(1)}, \tag{29}$$

then $Q_1 = 0$. If

$$X_i(t_i)(B - AC^{-1}) < x^{(1)} \tag{30}$$

for some sufficiently large time $t = t_i$, then $Q_i = 0$. Moreover if

$$B - AC^{-1} < Nx^{(1)} \tag{31}$$

with $1 < N \leq n$, then $Q_1 = Q_2 = \cdots = Q_{n-N+1} = 0$.

If the limiting distribution is 0–1 or locally uniform, then E satisfies the equation

$$g(Q_n x) = \frac{A}{B - x}. \tag{32}$$

If not, then $E = B - AC^{-1}$.

Remarks:

(1) Condition (27) provides a condition under which contour enhancement occurs without quenching all but the highest pattern weight.

(2) Condition (28) shows that all patterns whose weights satisfy $X_i(0) \geq \theta_1$ can be preserved by choosing the initial arousal level $x(0)$ sufficiently high if $\theta_1 = x^{(1)}(B - AC^{-1})^{-1}$. Condition (29) shows that θ_1 is a threshold value for preserving patterns, since if (29) holds, then some pattern quenching and contour enhancement occurs. If the inequality $X_i < \theta_1$ persists, then by (30), X_i is treated as noise and is quenched.

(3) Condition (31) shows that the amount of pattern quenching can be regulated by a judicious choice of numerical parameters. For example, if $N = 2$, then the network "chooses" the dominant state and quenches all others.

(4) Energy normalization occurs if the reverberation is persistent, and yields the same value $B - AC^{-1}$ whenever some nonmaximal states are not totally quenched. Thus, if an initial pattern contains enough energy to guarantee persistence, then the pattern will be contour enhanced, and the contour enhanced pattern will be normalized and preserved in short term memory as long as it is needed.

(5) The fact that $(d/dt)(X_iX_j^{-1})(t) = 0$ if $g_i(t) = g_j(t) = C$ has an important effect on the asymptotic slope of patterns as they are distributed in space. Whenever $g_i = g_j = C$, the *relative* growth rates of X_i and X_j remain fixed. If this happens, then the slope of a pattern in space is steepened as more pattern quenching occurs, but pattern shape of unquenched states is otherwise unchanged. Not all indices i are equally likely to satisfy the equation $g_i = C$, however. Since $X_1 \leq X_2 \leq \cdots \leq X_{n-1} \leq X_n$, the identity $g_n = C$ holds most often, $g_{n-1} = C$ holds next most often (and only if $g_n = C$), and so on. If $g_i < C$ while $g_{i+1} = C$, then $(d/dt) \cdot (X_iX_{i+1}^{-1}) < 0$. The *relative* growth rate of X_{i+1} as compared to that of X_i is increasing. This creates effects such as those in Figure 15. Note that straight lines in Figure 15a become curved inwards in Figure 15b due to the greater relative growth rates of larger pattern values.

The next theorem describes the possibility of mixing fair and uniform tendencies.

THEOREM 5. (*Fair or Uniformizing*). Let $f(w) = wg(w)$ where $g(w)$ is continuous, nonnegative, and $g(w) = C$ for $0 \leq w \leq x^{(2)}$ whereas $g(w)$ is strictly decreasing for $w > x^{(2)}$. (See Figure 17). Then all limits Q_i and E exist.

The function $M(t)$ is monotone decreasing, and $m(t)$ is monotone increasing. If $x_n(t) \leq x^{(2)}$, then all $\dot{X}_i(t) = 0$. If $g_i(t) = g_j(t) = C$, then $(d/dt)(X_iX_j^{-1})(t) = 0$.

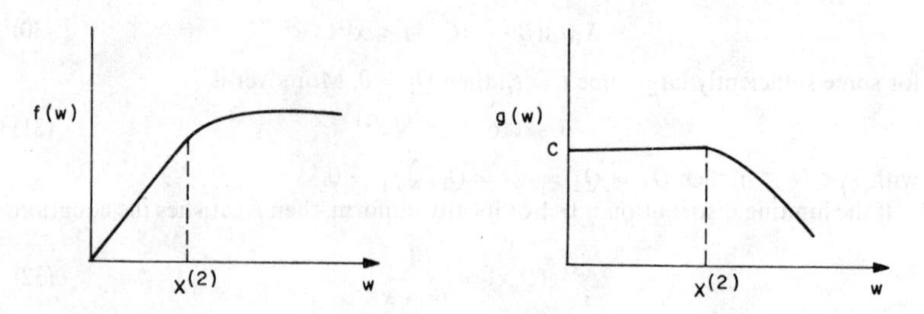

Figure 17. Fair or uniformizing.

Suppose moreover that the reverberation is persistent (that is, $g(0) > A/B$). Then either all $Q_i = 1/n$ or all $g(Q_i E) = C$. In the former case, E equals the unique root of

$$g\left(\frac{x}{n}\right) = \frac{A}{B - x}. \tag{25}$$

In the latter case, $E = B - AC^{-1}$. In particular, if

$$X_n(0) \max(B - AC^{-1}, x(0)) \leq x^{(2)}, \tag{33}$$

then $Q_i = X_i(t) = X_i(0)$ for $t \geq 0$ and $i = 1, 2, \ldots, n$. If however

$$X_1(0)\hat{E} > x^{(2)}, \tag{34}$$

where $\hat{E}$ is the unique root of (25) and g is convex, then all $Q_i = 1/n$. Indeed if

$$\hat{E} > nx^{(2)} \tag{35}$$

and g is convex, then all $Q_i = 1/n$.

Remarks: As in Theorem 4, there is a condition, namely (33) guaranteeing that all patterns that satisfy a given constraint will be preserved. In this case, if all $X_i(0) \leq \theta_2 \equiv x^{(2)}(B - AC^{-1})^{-1}$, and the initial arousal level is sufficiently small, then the patterns will be preserved. Energy normalization also occurs. By contrast, a proper choice of numerical parameters can guarantee a uniform limiting distribution.

Now we consider functions $f(w)$ that combine 0–1, fair, and uniform tendencies; for example, sigmoid functions. See Figure 18. The influence of these $f(w)$'s on the limiting distribution depends on particular choices of the parameters $x^{(0)}$ defined by $g(x_0) = g(B)$, $x^{(1)} = \min\{w : g(w) = C\}$, and $x^{(2)} = \max\{w : g(w) = C\}$. Before making such choices, we note the following proposition.

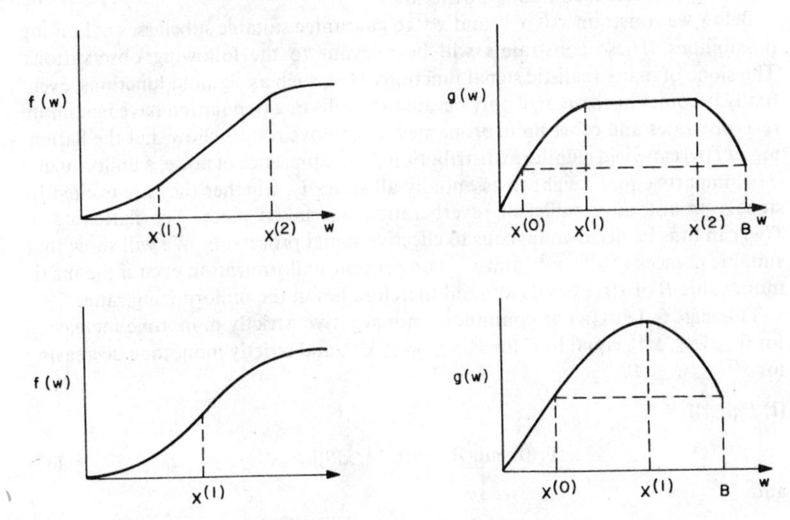

Figure 18. Important numerical parameters in $g(w)$.

PROPOSITION 4. Let $f(w) = wg(w)$, where $g(w)$ is continuous, nonnegative, strictly monotone increasing for $0 \leq w \leq x^{(1)}$, equal to C for $x^{(1)} \leq w \leq x^{(2)}$, and strictly monotone decreasing for $x^{(2)} \leq w \leq B$. If all limits Q_i exist, then the limit E also exists. Moreover there exists a $K \geq 0$ such that $Q_i = 0$, $i = 1, 2, \ldots, K$, and $g(Q_i E) = g(Q_j E)$ if $i, j > K$. If $g(0) \geq A/B$ and g is convex, then E is the unique solution of

$$g(Q_n x) = \frac{A}{B - x}. \tag{32}$$

If $g(0) < A/B$ and g is convex, then E equals 0 or one of the two solutions of (32). The smaller solution is unstable; the larger solution is stable.

In particular, if $x^{(1)} = x^{(2)}$ in Figure 13, then there exist integers L and M, $L \geq 0$, $M \geq 0$, $L + M \leq n$, such that

$$Q_i = \begin{cases} 0 & \text{if } i \leq L \\ \dfrac{\xi}{M\xi + (n - L - M)\eta} & \text{if } L + 1 \leq i \leq L + M \\ \dfrac{\eta}{M\xi + (n - L - M)\eta} & \text{if } L + M < i \leq n \end{cases} \tag{36}$$

where ξ and η satisfy $g(\xi) = g(\eta)$. Moreover $E = 0$ or $[M\xi + (n - L - M)\eta]$.

Remark: The proposition shows that, if a limiting distribution exists, then, supposing that $x^{(1)} < x^{(2)}$, it is either fair ($g_i = C$ for $i = 1, 2, \ldots, n$ and $t \geq 0$), or uniform ($Q_i = Q_j$, $i, j = 1, 2, \ldots, n$), or $0 - 1$($K = n - 1$), or contour enhancing and quenching ($0 < K < n - 1$ and $g(Q_i E) = C$, $i > K$), or trivalent (as in (36)). If $x^{(1)} = x^{(2)}$, then only trivalence is possible, including the uniform, $0 - 1$, or locally uniform cases. The existence of values w for which $f(w)$ is linear substantially enriches the limiting possibilities.

Below we constrain $x^{(0)}$, $x^{(1)}$, and $x^{(2)}$ to guarantee suitable subclasses of limiting possibilities. These constraints will be relevant to the following observations. The slope of many realistic signal functions $f(w)$, such as sigmoid functions, eventually becomes horizontal, if only because the cells in a population have maximum response rates and other finite properties. The above results show that the flattening of $f(w)$ can yield a uniform distribution. In the presence of noise, a uniformizing $f(w)$ imparts equal weight to essentially all states v_i, whether they are excited by signals or not, after sufficient reverberation has taken place. The flattening of $f(w)$ can thus be disadvantageous to effective signal processing. We will show that suitable choices of $x^{(0)}$, $x^{(1)}$, and $x^{(2)}$ can prevent uniformization even if the maximum value B of $x(t)$ exceeds $x^{(2)}$ and therefore lies in the uniformizing range.

THEOREM 6. Let $g(w)$ be continuous, nonnegative, strictly monotone increasing for $0 \leq w \leq x^{(1)}$, equal to C for $x^{(1)} \leq w \leq x^{(2)}$, and strictly monotone decreasing for $x^{(2)} \leq w \leq B$.

(I) *Fair:* If

$$X_1(0) \min(B - AC^{-1}, x(0)) \geq x^{(1)} \tag{37}$$

and

$$X_n(0) \max(B - AC^{-1}, x(0)) \leq x^{(2)}, \tag{38}$$

then $Q_i = X_i(t) = X_i(0)$, $i = 1, 2, \ldots, n$, and $x(t)$ approaches $E = B - AC^{-1}$ monotonically.

(II) *Fair, or Contour Enhancing and Quenching*: Let

$$x^{(0)} + x^{(2)} \geq \max(B - AC^{-1}, x(0)), \tag{39}$$

hold throughout this section. Then all Q_i exist, $M(t)$ is monotone increasing faster than all $X_i < M$, and $m(t)$ is monotone decreasing. The limiting distribution is either fair, 0–1, locally uniform (only if several $X_i(0) = M(0)$), or contour enhancing and quenching; no uniformization occurs. If moreover, for some $L < n$,

$$X_L(0) \min\left(B - \frac{A}{\sum_{i=L}^{n} X_i(0)C}, x(0) \right) \geq x^{(1)}, \tag{27}$$

then $\dot{X}_i \geq 0$ and $(X_i X_j^{-1})^{\boldsymbol{\cdot}} = 0$ for $t \geq 0$ and $i, j \geq L$, so that contour enhancement occurs. If

$$X_1(0) \min(B - AC^{-1}, x(0)) \geq x^{(1)}, \tag{28}$$

then $Q_i = X_i(t) = X_i(0)$, for $t \geq 0$ and $i = 1, 2, \ldots, n$. If, however, the reverberation is persistent, and

$$B - AC^{-1} < Nx^{(1)} \tag{31}$$

with $1 < N \leq n$ and $X_{n-N+1}(0) < X_n(0)$, then $Q_1 = Q_2 = \cdots = Q_{n-N+1} = 0$, so that quenching occurs. If

$$X_1(0)(B - AC^{-1}) \leq x^{(0)}, \tag{40}$$

the reverberation is persistent, and $X_1(0) < X_n(0)$, then $Q_1 = 0$. If

$$X_i(t_i)(B - AC^{-1}) \leq x^{(0)} \tag{41}$$

for a sufficiently large $t = t_i$, then $Q_i = 0$ if $X_i(t_i) < X_n(t_i)$ and the reverberation is persistent. If $x^{(1)} = x^{(2)}$ and the reverberation is persistent, then the limiting distribution is 0–1 if $X_n(0) > X_{n-1}(0)$ and locally uniform otherwise.

(III) *Quenching*: If

$$X_1(0) \max(B - AC^{-1}, x(0)) \leq x^{(0)}, \tag{42}$$

then $\dot{X}_1 \leq 0$ for $t \geq 0$. If moreover $X_1(0) < X_n(0)$ and the reverberation is persistent, then $Q_1 = 0$. If (41) holds for t_i sufficiently large and $i < n$, then $\dot{X}_i \leq 0$ for $t \geq t_i$, and $Q_i = 0$ if the reverberation is persistent and $X_i(t_i) < X_n(t_i)$.

(IV) *Quenching*: If

$$(n - 1)x^{(0)} + x^{(2)} > \max(B - AC^{-1}, x(0)), \tag{43}$$

then $\dot{X}_i \leq 0$ for $t \geq 0$. If moreover $X_1(0) < X_n(0)$ and the reverberation is persistent, then $Q_1 = 0$.

(V) *Uniformizing*: If for some K, $0 < K < n$,

$$(n - K + 1)x^{(2)} \geq \max(B - AC^{-1}, x(0)) \tag{44}$$

and

$$X_1(0)\min\left(B - \frac{A}{\sum_{i=1}^{K} X_i(0)C}, x(0)\right) \geq x^{(1)}, \tag{45}$$

then all Q_i exist with $Q_1 < 1/n < Q_n$, even though $\dot{X}_i \geq 0$ and $\dot{X}_n \leq 0$ for $t \geq 0$.

Remarks: (1) Theorem 6 provides readily computable conditions under which a given $f(w)$ will not uniformize or contour enhance. For example, consider the sigmoid function

$$f(w) = \frac{Dw^2}{E + w^2}.$$

By (39), if

$$E + B\sqrt{E} > B\max(x(0), B - 2AE^{1/2}D^{-1}),$$

then uniformization is prevented.

(2) The results in (II) hold for large t, rather than all $t \geq 0$, if $x^{(0)}$ is replaced by $x^{(3)}$, which is defined by $g(x^{(3)}) = g(B - AC^{-1})$. In making this definition, we assume that $x^{(2)} < B - AC^{-1}$. Otherwise all x_i will fall into the monotone nondecreasing range of $g(w)$ for large t, and Theorem 4 holds for large t.

Although Theorem 6 provides practical constraints on $x^{(0)}, x^{(1)}, x^{(2)}$ that guarantee functionally useful behavior, it has not yet been proved in general that all Q_i exist in the absence of constraints. Such a theorem would be of particular interest in pathological conditions where $x^{(0)}$, $x^{(1)}$, and $x^{(2)}$ might deviate from normal values. Can sustained oscillations occur in pathological cases? Interaction between x and $(X_1, X_2, \ldots, X_n)$ can produce oscillations, but whether these oscillations always dissipate remains to be proved. Such an oscillation is described in the following proposition.

PROPOSITION 5. (*Oscillation*). Let

$$x_1(0) > x^{(2)} > B - AC^{-1} > 0 \tag{46}$$

and suppose that $X_1(0) < X_n(0)$. Then $\dot{X}_1(0) > 0$ and $\dot{X}_n(0) < 0$, but asymptotically $\dot{X}_1 \leq 0$ and $\dot{X}_n \geq 0$ with $\dot{X}_1 < 0$ and $\dot{X}_n > 0$ unless all $g_i = C$.

An important class of functions $f(w)$ such that $f(0) = 0$ are those that can be written as ratios of absolutely convergent power series. Given such an $f(w)$, it is instructive to expand (15) in the form

$$\dot{X}_i = \sum_{k=1}^{n} L_{ik}(X_i - X_k), \tag{47}$$

and to note the influence of the coefficients L_{ik} on the limiting distribution in special cases. Thus we introduce the class $\mathcal{R} = \{f\}$ of functions defined by $f(w) = N(w)D^{-1}(w)$, such that

$$N(w) = \sum_{m=1}^{\infty} a_m w^m \geq 0, \qquad D(w) = \sum_{m=0}^{\infty} b_m w^m > 0,$$

$$\sum_{m=1}^{\infty} |a_m| B^m < \infty, \quad \text{and} \quad \sum_{m=0}^{\infty} |b_m| B^m < \infty.$$

Introducing the notation $N_i = N(x_i)$ and $D_i = D(x_i)$, we find the following theorem.

THEOREM 7. If $f(w)$ is in $\mathscr{R}$, then

$$\dot{X}_i = \sum_{k=1}^{n} L_{ik}(X_i - X_k) \tag{47}$$

with $L_{ik} = U_{ik}V_{ik}$, where

$$U_{ik} = BxX_iX_k(D_iD_k)^{-1}, \tag{48}$$

$$V_{ik} = \sum_{m=1}^{\infty} a_m W_m(x_i, x_k), \tag{49}$$

$$W_m(y, z) = \sum_{r=0}^{m-2} b_r(yz)^r S_{m-r-1}(y, z) - \sum_{r=m}^{\infty} b_r(yz)^{m-1} S_{r-m+1}(y, z), \tag{50}$$

and

$$S_p(y, z) = \begin{cases} 0 & \text{if } p = 0 \\ 1 & \text{if } p = 1 \\ y^{p-1} + y^{p-2}z + \cdots + yz^{p-2} + z^{p-1} & \text{if } p > 1. \end{cases}$$

Thus if $L_{ik} \leq -\varepsilon$, $k \neq i$, for some $\varepsilon > 0$, then the limiting distribution is uniform. Suppose that

$$L_{ij}X_j^{-r} \geq \varepsilon \tag{51}$$

for some $\varepsilon > 0$, $r \geq 1$, and all i and j such that $X_i(0) = M(0) > X_j(0)$. If $M(0) > \frac{1}{2}$, then the limiting distribution is $0 - 1$. If $L_{ik} \geq L_{jk}$ whenever $X_i = M > X_j$ and $k \neq i, j$, then the limiting distribution is $0 - 1$ or locally uniform even if $M(0) < \frac{1}{2}$.

Remark: Theorem 7 shows that the limiting distribution is determined essentially by the sign of each L_{ik}, and thus by the signs of the summands $W_m(x_i, x_k)$. In (50), the summands $W_m(x_i, x_k)$ can be composed of positive and negative terms. Herein lies the main sources of mathematical difficulty in studying arbitrary functions in $\mathscr{R}$.

Uniformly distributed tonic signals (as well as uniformly distributed excitatory tonic inputs) tend to create a uniform limiting distribution. For example, we have

THEOREM 8. (*Tonic Signals Uniformize*). Let

$$f(w) = K + wg(w), \qquad K > 0 \tag{52}$$

where $g(w)$ is a continuous, nonnegative, monotone nonincreasing (not necessarily decreasing) function. Then the limiting distribution is uniform and energy normalization occurs, such that E is the unique positive solution of

$$A + nK = \frac{nBK}{x} + (B - x)g\left(\frac{x}{n}\right). \tag{53}$$

Remarks: (1) Uniformly distributed tonic signals can uniformize the distribution of random noise, and thereby prevent fluctuations in noise from unduly favoring any given population of cells. A price is paid for this additional stability, however: the tonic level of activity never dissipates. This activity can be prevented from sending signals to cells further downstream by interpolating a nonrecurrent on-center off-surround field between the recurrently interacting populations and the cells downstream (Grossberg, 1970).

(2) Functions $f(w)$ exist that are not manifestly of type (52); for example, the linear fractional transformations

$$f(w) = \frac{A + Bw}{C + Dw} \tag{54}$$

with $A, B, C, D > 0$ and $BC \geq AD$ are of type (52).

(3) If in (52), $g(w)$ is strictly monotone increasing, then the tonic signal K and the phasic signal $wg(w)$ create opposing limiting tendencies. Given small values of w (or of $x(t)$), uniformization is favored, whereas for large values of w, contour enhancement is favored.

The following theorem illustrates this competition between uniformizing and contour enhancing tendencies in a special case.

THEOREM 9. (*Tonic vs. Phasic Signals*). Let $n = 3$ and

$$f(w) = a_0 + a_1 w + a_2 w^2 \tag{55}$$

with $a_i > 0$, $i = 0, 1, 2$. Choose $x_i(0) = x_j(0)$ and let Y be the common value of X_i and X_j at every time t. Then

$$\operatorname{sign} \dot{Y} = \operatorname{sign}(\tfrac{1}{3} - Y)(Y - U)(Y - V), \tag{56}$$

where

$$U = \tfrac{1}{4}[1 + \sqrt{1 - 8a_0 a_2^{-1} x^{-2}}] \tag{57}$$

and

$$V = \tfrac{1}{4}[1 - \sqrt{1 - 8a_0 a_2^{-1} x^{-2}}]. \tag{58}$$

Thus if $x \leq \sqrt{8a_0 a_2^{-1}}$, the system tends towards a uniform distribution. In the limiting case $x = \infty$, the system tends towards a $0 - 1$ distribution with $Q_3 = 1$ if $Y < \frac{1}{3}$, and towards a locally uniform distribution with $Q_1 = Q_2 = \frac{1}{2}$ if $Y > \frac{1}{3}$. If $\sqrt{8a_0 a_2^{-1}} < x \leq B$, the system exhibits mixed $0 - 1$ and uniform tendencies.

Figure 19 illustrates the flow patterns that can be achieved given various values of $x(t)$. A point on the triangle codes particular values of the three functions X_1, X_2, X_3. The system is at the ith vertex V_i of the triangle at time t if $X_i(t) = 1$. The system is at the midpoint of the edge L_i opposite V_i at time t if $\frac{1}{2} = X_j(t) = X_k(t)$, where $\{i, j, k\} = \{1, 2, 3\}$. Note that $X_j(t_0) = X_k(t_0)$ implies $X_j(t) = X_k(t)$ for $t \geq t_0$. Thus if the system starts out on the line through V_i and the midpoint of L_i, then it remains on this line. The distance from V_i on this line increases as X_i decreases. All three bisecting lines interact at the point where $X_1 = X_2 = X_3 = \frac{1}{3}$. Arrows along these lines denote the direction in which the system flows given various values of x. Closed circles denote stationary points of the system (that is, points where all $\dot{X}_i = 0$).

Consider Figure 19a for definiteness. Note that distributions close to the uniform distribution are attracted towards the uniform distribution; distributions a little further away are attracted towards the 0–1 distribution but never reach it; and distributions close to the 0–1 distribution tend to be uniformized somewhat.

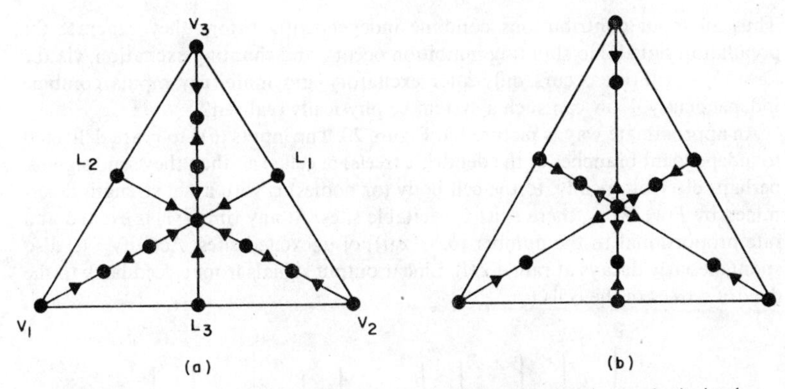

Figure 19. Interaction of uniformizing tonic signal and contour enhancing quadratic signal.

Thus there are two uniformizing regions separated by a contour enhancing region. In Figure 19b, this situation is reversed: patterns close to the uniform distribution are contour enhanced, whereas those close to the 0–1 distribution are uniformized.

The structure of these regions becomes correspondingly more complex as the degree of the polynomial

$$f(w) = \sum_{k=0}^{m} a_k w^k$$

is increased beyond the value $m = 2$ of (55). For example, if $m = 3$, (56) is replaced by a polynomial of degree 4, with a corresponding more complicated diagram replacing Figure 19.

5. Comparison with the Wilson–Cowan equations

The Wilson–Cowan equations have the form

$$\dot{x}_i = -a_i x_i + (b_i - x_i) F\Big(\sum_k x_k c_{ki} - \sum_k y_k d_{ki} + e_i\Big) \tag{59}$$

and

$$\dot{y}_i = -A_i y_i + (B_i - y_i) G\Big(\sum_k x_k C_{ki} - \sum_k y_k D_{ki} + E_i\Big). \tag{60}$$

The function $x_i(y_i)$ describes the activity of the ith excitatory (inhibitory) subpopulation. Consider the right-hand side of (59) for definiteness. The activity x_i decays at a spontaneous rate $a_i x_i$. The term

$$(b_i - x_i) F\Big(\sum_k x_k c_{ki} - \sum_k y_k d_{ki} + e_i\Big)$$

has the following interpretation. $F(w)$ is a sigmoid signal function. It sums up excitatory inputs ($\sum_k x_k c_{ki}$), inhibitory inputs ($-\sum_k y_k d_{ki}$), and the external input (e_i) before computing the signal $F(w)$ as a function of the resultant

$$w = \sum_k x_k c_{ki} - \sum_k y_k d_{ki} + e_i. \tag{61}$$

Thus all input contributions combine independently before they generate the population signal. No shunting inhibition occurs, and shunting excitation, via the term $(b_i - x_i)F(w)$, occurs only after excitatory and inhibitory inputs combine independently. How can such a system be physically realized?

An approximate way is pictured in Figure 20. The inputs (61) to v_i are delivered to independent branches of the dendritic tree(s) of cell(s) v_i; then they send signals, perhaps electrotonically, to the cell body (or bodies) v_i with a net strength determined by $F(w)$. At v_i, there exist b_i excitable sites. At any time t, v_i is excited at a rate proportional to the number $[b_i - x_i(t)]$ of unexcited sites. Activity $x_i(t)$ also spontaneously decays at rate $a_i x_i(t)$. Linear output signals from v_i feed back to the dendritic trees of the cells v_k.

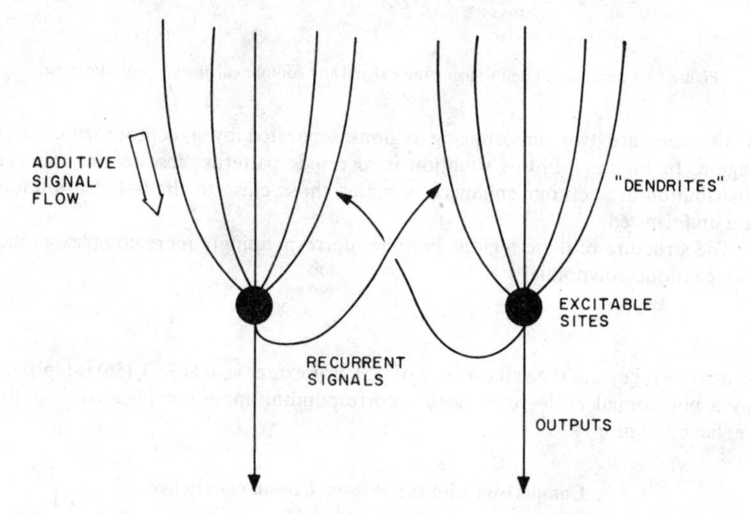

Figure 20. Graphical interpretation of Wilson–Cowan equations.

The mixture of independently combining inputs and shunting excitation seems to require a formal diagram such as that in Figure 20, whether or not we interpret the input receiving stations as dendrites. This diagram means that the interactions between excitatory and inhibitory inputs are so "weak" that they can be approximated by independent increments, without mutual shunting terms. Correspondingly, individual inputs will have a small effect on the cell body. Also, since the summands in (61) are linear functions of population activity, the outputs from each population are linear functions of population activity. Only after these outputs combine independently is a sigmoid function of their resultant computed in (59) and (60). This fact does not seem to be compatible with the interpretation that the output from each population is a sigmoid function of that population's activity.

The systems studied herein contain both shunting excitation and inhibition, such that the output of each population can be a sigmoid function of its activity. These systems thus permit "strong interactions" between excitatory and inhibitory inputs, whether at the cell body, or between cell body and dendrites. A marriage between two experimentally verified phenomena—passive membrane equations and recurrent on-center off-surround anatomies—is hereby demonstrated. Whereas the present systems are, in a naive sense, "more nonlinear" than those of Wilson and Cowan, their particular nonlinearities blend harmoniously with an on-center off-surround anatomy, thereby making possible the rigorous mathematical theory presented herein.

Appendix: Proofs of results

The notation $f_i = f(x_i)$, $F = \sum_{k=1}^{n} f_k$, and $F_i = f_i F^{-1}$ will be used below.

Proof of Proposition 1: First we show that

$$\dot{X}_i = P(F_i - X_i), \tag{62}$$

where $P = BFx^{-1}$. By (12),

$$\dot{x}_i = -(A + F)x_i + Bf_i, \tag{63}$$

which when summed over i shows that

$$\dot{x} = -(A + F)x + BF. \tag{64}$$

Apply the identity

$$(UV^{-1})^{\bullet} = V^{-1}(\dot{U} - U\dot{V}V^{-1})$$

to $U = x_i$ and $V = x$ and find

$$\dot{X}_i = x^{-1}(\dot{x}_i - X_i\dot{x}). \tag{65}$$

Substituting (63) and (64) into (65) yields (62) after cancellation and rearrangement of terms. Now $F_i - X_i$ is computed as follows:

$$\begin{aligned} F_i - X_i &= x_i g_i\Big(\sum_k x_k g_k\Big)^{-1} - x_i\Big(\sum_k x_k\Big)^{-1} \\ &= x_i(Fx)^{-1}\sum_k x_k(g_i - g_k). \end{aligned}$$

Substituting this result into (62) yields (15). Equation (17) follows from (15) and the fact that $\sum_k X_k = 1$. Equation (18) is obvious.

To derive (16), write (64) as

$$\dot{x} = -Ax + (B - x)F$$

and note that $F = xG$. QED.

Proof of Proposition 2: Suppose for definiteness that $X_i(0) < X_{i+1}(0)$. By the continuity of the functions X_i and X_{i+1}, the inequality $X_i(t) > X_{i+1}(t)$ cannot hold at any time t unless $X_i(t_0) = X_{i+1}(t_0)$ at some time $t_0 < t$. By (15), the identity $X_i(t_0) = X_{i+1}(t_0)$ implies $X_i(t) = X_{i+1}(t)$ for $t \geq t_0$. Hence ordering is preserved. QED.

Proof of Proposition 3: First we prove the existence of all limits Q_i. Recall the definitions $M(t) = \max\{X_i(t): i = 1, 2, \ldots, n\}$ and $m(t) = \min\{X_i(t): i = 1, 2, \ldots,$

$n\}$. Suppose that $g(w)$ is monotone increasing. Then by (15), $\dot{M}(t) \geq 0$ and $\dot{m}(t) \leq 0$ for $t \geq 0$. If $g(w)$ is monotone decreasing, then (15) implies $\dot{M}(t) \leq 0$ and $\dot{m}(t) \geq 0$ for $t \geq 0$. In both cases, the limits Q_1 and Q_n exist, and $Q_1 + Q_n > 0$. If $g(w)$ is monotone increasing, then $Q_n > 0$. If $g(w)$ is monotone decreasing, then $Q_1 > 0$. Consider the former case for definiteness.

Using the fact that Q_n exists, we will prove that Q_{n-1} exists. Using the existence of Q_n and Q_{n-1}, we will prove that Q_{n-2} exists, and so on.

Integrate (15) from $t = S$ to $t = T$. Then

$$X_n(T) - X_n(S) = B \int_S^T X_n \sum_{k=1}^{n} X_k |g_n - g_k|\, dt \tag{66}$$

Let $T \to \infty$ and note that for all $t \geq 0$, $X_n(t) \geq X_n(0)$. Then (66) implies

$$Q_n - X_n(S) \geq X_n(0) \sum_{k=1}^{n} h_{n,k}(S), \tag{67}$$

where

$$h_{m,k}(S) = B \int_S^\infty X_k |g_m - g_k|\, dt. \tag{68}$$

Letting $S \to \infty$ in (67) shows that

$$\lim_{S\to\infty} h_{n,k}(S) = 0, \qquad k = 1, 2, \ldots, n. \tag{69}$$

Now consider X_{n-1}. By (15), for any $T \geq S \geq 0$,

$$|X_{n-1}(T) - X_{n-1}(S)| \leq B \int_S^T \left(\sum_{k=1}^{n-1} X_k |g_{n-1} - g_k| + X_{n-1} |g_n - g_{n-1}| \right) dv. \tag{70}$$

By (68) and Proposition 2,

$$h_{n,k}(S) \geq h_{n-1,k}(S) \geq 0, \qquad k = 1, 2, \ldots, n-1.$$

Thus (69) implies that

$$\lim_{S\to\infty} h_{n-1,k}(S) = 0, \qquad k = 1, 2, \ldots, n-1,$$

which by (69) and (70) implies the existence of Q_{n-1}.

Use the existence of Q_n and Q_{n-1} to prove the existence of Q_{n-2} by showing that

$$\lim_{S\to\infty} h_{m,k}(S) = 0, \qquad m = n, n-1 \quad \text{and} \quad k = 1, 2, \ldots, n$$

implies that

$$\lim_{S\to\infty} h_{n-2,k}(S) = 0, \qquad k \neq n-2.$$

Iterate the argument until the existence of all Q_i is proved.

Now the existence of E is proved. Consider the function $H(x) = \sum_{k=1}^{n} Q_k g(Q_k x)$. $H(x)$ is monotonic since $g(w)$ is monotonic. Since $G = \sum_{k=1}^{n} X_k g(X_k x)$, $\lim_{t\to\infty} (G - H) = 0$. Thus (16) can be written in the form

$$\dot{x} = x(B - x)\left(H(x) - \frac{A}{B - x} \right) + \varepsilon(t)$$

where $\lim_{t\to\infty} \varepsilon(t) = 0$. As $t \to \infty$, the sign of $\dot{x}$ becomes essentially equal to that of $H(x) - A/(B - x)$. This situation is graphed in Figure 21 for various choices of $g(w)$. The arrows indicate the direction in which x moves at various of its values. Clearly E equals zero or is a solution of the equation $H(x) = A(B - x)^{-1}$, which is (19). The distribution of E's values, given specific choices of $g(w)$, can be read off from graphs such as those in Figure 21. Figure 16c uses the fact that if $g(w)$ is increasing and convex, then $H(x)$ is also increasing and convex. QED.

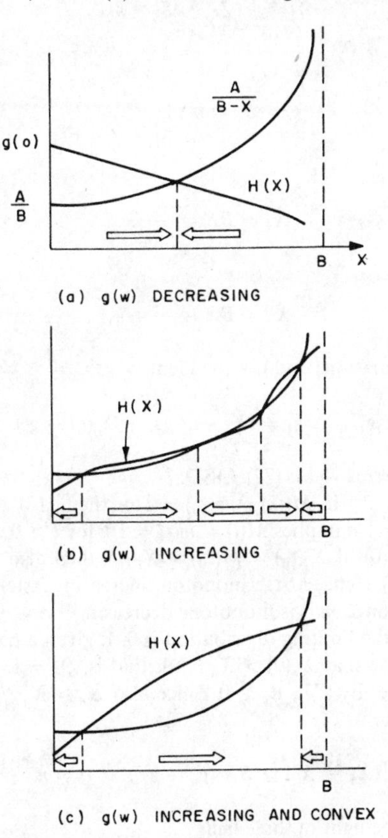

Figure 21. Equilibrium points of $x(t)$ as $t \to \infty$.

Proof of Theorem 1: Since $f(w) = Cw$, $g(w) = C$, and (15) implies that $\dot{X}_i \equiv 0$, and that every $X_i(t)$ is constant for $t \geq 0$. Thus $x_i(t) = X_i(0)x(t)$, and it suffices to study $x(t)$. By (16),

$$\dot{x} = x(D - Cx) \qquad (71)$$

where $D = BC - A$. Equation (71) is of Riccati type. It can be routinely solved using the change of variables $x = \dot{y}(Cy)^{-1}$, yielding (20) and (21) (Bellman, 1967).

The proofs of Theorem 2 and 3 make use of the following Lemma.

LEMMA 1. The following equations hold.

$$(X_i - X_j)^{\bullet} = R_i(X_i - X_j) + S_j(g_i - g_j) \tag{72}$$

with

$$R_i = B \sum_{k=1}^{n} X_k(g_i - g_k) \tag{73}$$

and

$$S_j = BX_j. \tag{74}$$

Proof: By (17),

$$\dot{X}_i = BX_i(g_i - G)$$

and

$$\dot{X}_j = BX_j(g_j - G).$$

Subtract these two equations and use the identity

$$X_i g_i - X_j g_j = (X_i - X_j)g_i + X_j(g_i - g_j).$$

A rearrangement of terms yields (72). QED.

Proof of Theorem 2: If $M(t_0) = m(t_0) = 1/n$, then (15) implies $\dot{X}_i(t_0) = 0$. Hence $M(0) = m(0) = 1/n$ implies $M(t) = m(t) = 1/n$ for $t \geq 0$.

Suppose $M(0) > m(0)$. If $X_j(t_0) = M(t_0) > X_i(t_0)$, then also $g_j(t_0) > g_i(t_0)$, and by (15), $\dot{X}_j(t_0) > \dot{X}_i(t_0)$. Hence $M(t)$ is monotone increasing faster than any $X_i < M$. By (15) and Proposition 2, $m(t)$ is monotone decreasing.

We will show that the limiting distribution is 0–1, given a persistent reverberation, in the special case that $X_n(0) > X_{n-1}(0)$; that is, $Q_n = 1$. The general proof is essentially the same. By (73), $R_n \geq 0$. Since also $X_n > X_j$, $j \neq n$, (72) and (74) imply that

$$(X_n - X_j)^{\bullet} \geq BX_j(g_n - g_j), \qquad j \neq n. \tag{75}$$

Consider $g_n - g_j$ in the light of three facts:

(i) $g(w)$ is *strictly* monotone increasing;

(ii) $X_n - X_j \geq X_n(0) - X_j(0) > 0, j \neq n$;

and

(iii) $x(t)$ varies in a positive closed interval.

Thus there exists a $\delta > 0$ such that for any $j \neq n$,

$$g_n - g_j = g(X_n x) - g(X_j x) \geq \delta.$$

By (75),

$$(X_n - X_j)^{\bullet} \geq \delta BX_j, \qquad j \neq n.$$

Integrating this inequality from $t = 0$ to ∞, and using the fact that all X_i satisfy $0 \leq X_i \leq 1$, yields the inequalities

$$\infty > (\delta B)^{-1} \geq \int_0^\infty X_j dt, \qquad j \neq n.$$

The function X_j is also a nonnegative function which, by (15), has a bounded first derivative. Hence $Q_j = 0$ for every $j \neq n$, and thus $Q_n = 1$.

We now prove that the reverberation is persistent if $g(0) < A/B$ and $x(0) \geq \hat{x}$ where $\hat{x}$ is the smaller root of (22). If $X_K(0) < M(0) = X_{K+1}(0)$, then $X_n(t) = X_{n-1}(t) = \cdots = X_{K+1}(t) = M(t) > X_K(t)$, and $\dot{M}(t) > 0$. In particular,

$$G \geq \sum_{i=K+1}^{n} X_i g(X_i x) = (n - K)Mg(Mx) \geq (n - K)M(0)g(M(0)x).$$

Thus if $\hat{x}$ is a root of (22), then by (16), $\dot{x} > 0$ if $x = \hat{x}$. Hence if $x(0) \geq \hat{x}$, then $x(t) \geq \hat{x} > 0$ for $t \geq 0$, which proves persistence. QED.

The function in (24) can be written as $f(w) = wg(w)$ with $g(w)$ strictly increasing by defining

$$g(w) = \frac{1}{1 + (D - A)(A + Bw + Cw^2)^{-1}}$$

Proof of Theorem 3: By (15) and Proposition 2, $M(t)$ is monotone decreasing and $m(t)$ is monotone increasing. Thus the limits $M(\infty)$ and $m(\infty)$ exist. We now show that $M(\infty) = m(\infty)$, and thus that all $Q_i = 1/n$, if the reverberation is persistent. By Proposition 2, $M \equiv X_n$ and $m \equiv X_1$. If $X_1(0) = X_n(0)$, we are done. Suppose that $X_n(0) > X_1(0)$. By (73), $R_n \leq 0$. Thus by (72),

$$(X_n - X_1)^{\cdot} \leq S_1(g_n - g_1).$$

By (74),

$$S_1 = BX_1 \geq BX_1(0) = \varepsilon > 0$$

Thus

$$(X_n - X_1)^{\cdot} \leq -\varepsilon(g_1 - g_n).$$

By the monotone decrease (increase) of $X_n(X_1)$,

$$(X_n - X_1)^{\cdot} \leq -\varepsilon[g(m(\infty)x) - g(M(\infty)x)].$$

Suppose $M(\infty) > m(\infty)$. Then since $g(w)$ is strictly monotone decreasing, and x varies in a positive closed interval, there exists a $\delta > 0$ such that

$$(X_n - X_1)^{\cdot} \leq -\delta < 0.$$

This implies the contradiction $1 \geq Q_1 = \infty$. Hence $M(\infty) = m(\infty)$.

To prove that E satisfies (25) if $g(0) > A/B$, it suffices to note by Proposition 3 that the reverberation is persistent if $g(0) > A/B$, and thus that all $Q_k = 1/n$. Hence

$$\sum_{k=1}^{n} Q_k g(Q_k x) = g\left(\frac{x}{n}\right).$$

Substitution of this expression into (19) yields the desired result. QED.

Proof of Theorem 4: The statements about monotone increase of $M(t)$ and decrease of $m(t)$ follow as in the proof of Theorem 2.

If $x_1(t) \geq x^{(1)}$, then all $g_i = C$, and by (15), all $\dot{X}_i(t) = 0$. If $g_i(t) = g_j(t) = C$, then by (17),

$$(\dot{X}_i X_i^{-1})(t) = C - G(t) = (\dot{X}_j X_j^{-1})(t).$$

Since

$$(X_i X_j^{-1})^{\cdot} = X_i X_j^{-1}(\dot{X}_i X_i^{-1} - \dot{X}_j X_j^{-1}),$$

it follows that $(d/dt)(X_i X_j^{-1})(t) = 0$.

Suppose that the reverberation is persistent. By (18), if $X_n(0) > 1/n$, then

$$\dot{X}_n \geq BX_n(0) \sum_{i=1}^{n} X_i[g(X_n x) - g(X_i x)] > 0.$$

Integrate this inequality and use the inequalities $0 \leq X_n \leq 1$ to conclude that

$$\int_0^\infty H_i \, dt < \infty \tag{76}$$

$i = 1, 2, \ldots, n$, where

$$H_i = X_i[g(X_n x) - g(X_i x)] \geq 0.$$

Inequality (76) implies that H_i approaches zero arbitrarily closely at arbitrarily large times. Since all Q_i and E exist,

$$Q_i[g(Q_n E) - g(Q_i E)] = 0.$$

Either $Q_i = 0$ or $g(Q_i E) = g(Q_n E)$. Suppose that $i \leq K$. Then $Q_i < Q_n$. Since $g(w)$ is strictly monotone increasing until $w = x^{(1)}$, the identity $g(Q_i E) = g(Q_n E)$ implies that $g(Q_i E) = C$. Suppose that $i > K + 1$. Then $g(Q_i E) = g(Q_n E)$ because $Q_i = Q_n$. If moreover $Q_i \neq 1/(n - K)$ then $Q_K > 0$. Thus $g(Q_K E) = g(Q_i E)$, which implies that $g(Q_K E) = C$. Since $Q_i \geq Q_K$ for $i \geq K$, also $g(Q_i E) = C$ for $i \geq K$.

Suppose that (27) holds. By (17), to show that $\dot{X}_i \geq 0$ for $t \geq 0$ it suffices to show that $x_i \geq x^{(1)}$ and thus that $g_i = C \geq G$ for $t \geq 0$. Suppose that $i \geq L$, where L is defined by (27). At any fixed time $t = T$, the inequalities

$$G \geq \sum_{i=L}^{n} X_i g(X_i x)$$

$$\geq \sum_{i=L}^{n} X_i(0) C$$

hold if $x_L(t) \geq x^{(1)}$ for $t \leq T$. Let $t = T$ be the first time that

$$x(t) = B - \frac{A}{\sum_{i=L}^{n} X_i(0) C}. \tag{77}$$

By (27), then

$$x_L(t) = X_L(t) x(t) \geq X_L(0) x(t) \geq x^{(1)} \tag{78}$$

for $t \leq T$. Consequently

$$G \geq \sum_{i=L}^{n} X_i(0)C \geq \frac{A}{B - x}$$

at this time, and by (16), $\dot{x}(T) \geq 0$. The same argument is valid at every time t such that (77) holds, and thus $x(t)$ is increasing whenever (77) holds. This shows that

$$x(t) \geq \min\left(B - \frac{A}{\sum_{i=L}^{n} X_i(0)C}, x(0)\right) \tag{79}$$

for $t \geq 0$. Inequality (77) implies that (78) is true for $t \geq 0$. Since $x_i \geq x_L \geq x^{(1)}$ for all $i \geq L$, also $\dot{X}_i \geq 0$ for $t \geq 0$ and $i \geq L$.

A similar argument shows that if (28) holds, then $x_1 \geq x^{(1)}$ for $t \geq 0$; thus all $g_i = C$ for $t \geq 0$, and by (15), all X_i are constant for $t \geq 0$.

Suppose that (29) holds. Since $G \geq C$, (16) shows that for every $\varepsilon > 0$, there exists a T_ε such that

$$x(t) \leq B - AC^{-1} + \varepsilon, \qquad t \geq T_\varepsilon. \tag{80}$$

By (15), $\dot{X}_1 \leq 0$ for $t \geq 0$. Thus for sufficiently large t, (80) implies that

$$x_1(t) = X_1(t)x(t) \leq X_1(0)(B - AC^{-1} + \varepsilon) < x^{(1)};$$

thus g is bounded away from C for $t \gg 0$; $g(Q_1E) < C$; and finally $Q_1 = 0$.

This argument can be successively applied to $X_1, X_2, \ldots, X_i$ to show that $Q_1 = Q_2 = \cdots = Q_i = 0$ if (30) holds. Suppose that we have already shown that $Q_1 = Q_2 = \cdots = Q_{i-1} = 0$. Then the terms $X_k(g_i - g_k)$, $k \leq i - 1$, in (15) approach zero as $t \to \infty$. The terms $X_k(g_i - g_k)$ with $k > i$ are nonpositive. The term $X_n(g_i - g_n)$ is moreover bounded away from zero at $t = t_i \gg 0$, since by (30) and (80),

$$x_i(t_1) \leq X_i(t_i)(B - AC^{-1} + \varepsilon) < x^{(1)}$$

for some $\varepsilon > 0$, and hence $C - g_i \geq \delta$ for some $\delta > 0$, while the gap between X_n and X_i increases as $t \to \infty$. Thus $\dot{X}_i(t_i) < 0$. This argument can be repeated at all times $t \geq t_i$ to show that $x_i(t) \leq X_i(t_i)(B - AC^{-1} + \varepsilon) < x^{(1)}$; thus $g(Q_iE) < C$; and finally $Q_i = 0$.

Suppose that (31) holds. If for arbitrarily large t, $x_{n-N+1}(t) \geq x^{(1)}$, then

$$x(t) > \sum_{i=n-N+1}^{n} x_i(t) \geq Nx^{(1)} > B - AC^{-1},$$

which contradicts (80). Hence for sufficiently large t, $x_i(t) \leq x^{(1)} + \delta$ for some $\delta > 0$ and all $i \leq n - N + 1$; thus $g(Q_iE) < C$; and finally $Q_i = 0, i \leq n - N + 1$.

If the limiting distribution is locally uniform, then E satisfies (32) because $\lim_{t\to\infty}[G - g(Q_n x)] = 0$. If the limiting distribution is not locally uniform, then some Q_K such that $X_K(0) < M(0)$ exceeds zero. Thus $g(Q_K E) = g(Q_{K+1}E) = \cdots = g(Q_n E) = C$. This is true for every such K. Hence $\lim_{t\to\infty} G = C$, and $E = B - AC^{-1}$. QED.

Proof of Theorem 5: The first few statements of the Theorem follow from Proposition 3 and arguments in the proof of Theorem 3. If $x_n(t) \leq x^{(2)}$, then all

$\dot{X}_i(t) = 0$ because all $g_i(t) = C$. If $g_i(t) = g_j(t) = C$, then $(d/dt)(X_i X_j^{-1})(t) = 0$ by the same reasoning that was used in Theorem 4.

We now show that, given a persistent reverberation, either all $Q_i = 1/n$ or all $g(Q_i E) = C$. We use the facts $\dot{X}_1 \geq 0$, $\dot{X}_n \leq 0$, and the existence of all limits Q_i and E. By (15),

$$\begin{aligned} B^{-1}\dot{X}_1 &= X_1 \sum_{k=1}^{n} X_k(g_1 - g_k) \\ &\geq X_1 X_n(g_1 - g_n) \\ &\geq X_1^2[g(X_1 x) - g(X_n x)] \\ &\geq X_1^2(0)[g(Q_1 x) - g(Q_n x)] \geq 0. \end{aligned} \tag{81}$$

If $Q_1 = Q_n$, then all $Q_i = 1/n$ and we are done. Suppose $Q_1 < Q_n$. Integrate inequality (81) from $t = 0$ to $t = \infty$. Since $0 \leq X_1 \leq 1$ and $X_1(0) > 0$,

$$\int_0^\infty [g(Q_1 x(t)) - g(Q_n x(t))]\, dt < \infty.$$

Thus the nonnegative function $[g(Q_1 x) - g(Q_n x)]$ approaches zero arbitrarily closely at arbitrarily large times. Since E exists, $g(Q_1 E) = g(Q_n E)$. Since $Q_1 \leq Q_i \leq Q_n$, $g(Q_i E) = g(Q_n E)$ for $i = 1, 2, \ldots, n$.

If all $Q_i = 1/n$, then $\lim_{t\to\infty}[G - g(x/n)] = 0$. Hence E satisfies (25). If all $g(Q_i E) = C$, then $\lim_{t\to\infty} G = C$, and $E = B - AC^{-1}$.

Suppose that (33) holds. Since $G \leq C$, (16) shows that

$$x(t) \leq \max(B - AC^{-1}, x(0)) \tag{82}$$

for $t \geq 0$. Since $\dot{X}_n(t) \leq 0$ for $t \geq 0$,

$$x_n(t) = X_n(t)x(t) \leq X_n(0)\max(B - AC^{-1}, x(0)) \leq x^{(2)};$$

thus all $x_i(t) \leq x^{(2)}$ for $t \geq 0$; all $g_i \equiv C$; and finally all X_i are constant.

Suppose that (34) holds. If g is convex, then

$$\frac{1}{n}\sum_{k=1}^{n} f(x_k) \geq f\left(\frac{\sum_{k=1}^{n} x_k}{n}\right),$$

$$\frac{1}{n}\sum_{k=1}^{n} x_k g(x_k) \geq \frac{x}{n} g\left(\frac{x}{n}\right),$$

and finally

$$G = \sum_k X_k g_k \geq g\left(\frac{x}{n}\right). \tag{83}$$

By (16) and (83), if the reverberation is persistent, then for every $\varepsilon > 0$, there exists a T_ε such that

$$x(t) \geq \hat{E} - \varepsilon \quad \text{for} \quad t \geq T_\varepsilon.$$

Since $\dot{X}_1(t) \geq 0$ for $t \geq 0$,

$$x_1(t) \geq X_1(t)x(t) \geq X_1(0)(\hat{E} - \varepsilon) \tag{84}$$

for $t \geq T_\varepsilon$. By (34) and (84), there exists a $\delta > 0$ such that $x_1(t) \geq x^{(2)} + \delta > 0$ for all sufficiently large t. Thus $g(Q_1E) < C$, and all $Q_i = 1/n$.

Suppose that (35) holds. To prove that all $Q_i = 1/n$, we argue by contradiction. Let $Q_n > 1/n$. Then

$$x_n(\infty) = Q_nE > (1/n)\hat{E} > x^{(2)};$$

hence $g(Q_nE) < C$, and all $Q_i = 1/n$. QED.

Proof of Proposition 4: The proof imitates that of Proposition 3 as much as possible. The limit E exists because $\lim_{t\to\infty}(G - H) = 0$, where $H(x) = \sum_{k=1}^{n} Q_i \times g(Q_kx)$. Thus $E = 0$ or is a solution of (19).

By (15), for every $i = 1, 2, \ldots, n$,

$$\lim_{t\to\infty} \dot{X}_i(t) = BQ_i[g(Q_iE) - \sum_{k=1}^{n} Q_kg(Q_kE)].$$

These limits must all equal zero, since otherwise some $X_i(t)$ will be unbounded as $t \to \infty$. Either $Q_i = 0$ or

$$g(Q_iE) = \sum_{k=1}^{n} Q_kg(Q_kE).$$

In particular, if $Q_iQ_j > 0$, then $g(Q_iE) = g(Q_jE)$. Since $Q_i \leq Q_{i+1}$, then exists a K, possibly zero, such that $Q_1 = Q_2 = \cdots = Q_K = 0$ and $g(Q_iE) = g(Q_jE)$ if $i, j > K$. In particular $\lim_{t\to\infty}[G - g(Q_nx)] = 0$.

Suppose that $g(0) \geq A/B$ and let g be convex. Then $G(0) \geq A/B$, and since $\lim_{t\to\infty}[G - g(Q_nx)]$, E is a solution of (32). Since $g(Q_nx)$ is convex and $A/(B - x)$ is concave, (32) has a unique solution. If $g(0) < A/B$ and g is convex, then (32) has two solutions.

If $x^{(1)} = x^{(2)}$, then $g(w)$ has no constant interval. Hence the equality $g(Q_iE) = g(Q_jE)$ can occur only if $Q_i = Q_j$, or if $Q_iE = \xi$ and $Q_jE = \eta$, for some ξ and η such that $g(\xi) = g(\eta)$. This readily yields the trivalent distribution of (36). QED.

Proof of Theorem 6: This proof uses ideas similar to those used before; hence we merely sketch the main arguments below.

Impose (37) and (38). The main effect is that

$$\min(B - AC^{-1}, x(0)) \leq x(t) \leq \max(B - AC^{-1}, x(0)), \tag{85}$$

for $t \geq 0$. At time $t = 0$, $x^{(1)} \leq x_1(0) \leq x_n(0) \leq x^{(2)}$; thus all $g_i(0) = C$, and all $\dot{X}_i(0) = 0$. The bounds (85) on $x(t)$ cause these inequalities to propagate to all $t \geq 0$. Thus $G \equiv C$, and $E = B - AC^{-1} > 0$.

Consider (39). This condition implies that $g_i \leq g_n$ for $t \geq 0$ and $i = 1, 2, \ldots, n$. To see this, note by (39) and (82) that

$$x^{(0)} + x^{(2)} \geq x(t),\ t \geq 0. \tag{86}$$

If for any $T \geq 0$, $x_n(T) \geq x^{(2)}$, then all $x_i(T) \leq x^{(0)}$, $i = 1, 2, \ldots, n - 1$. Suppose not. Then some $x_i(T) > x^{(0)}$, and

$$x(T) \geq x_i(T) + x_n(T) > x^{(0)} + x^{(2)},$$

which contradicts (86). Consequently, by definition of $x^{(0)}$ and $x^{(2)}$, $g_i(T) \leq g_n(T)$ for $i = 1, 2, \ldots, n$. If for any $T \geq 0$, $x_n(T) \leq x^{(2)}$, then since $g(w)$ is monotone increasing for $0 \leq w \leq x^{(2)}$, and since all $x_i(T) \leq x_n(T)$, again $g_i(T) \leq g_n(T)$ for $i = 1, 2, \ldots, n$. In all cases $g_i(T) \leq g_n(T)$, whence by (15), $\dot{X}_n \geq 0$ and $\dot{X}_1 \leq 0$ for $t \geq 0$. The arguments of Proposition 3 can therefore be used to show that all limits Q_i and E exist with $M(t)$ monotone increasing faster than all $X_i < M$, and $m(t)$ monotone decreasing.

Condition (27) is treated here much as it was in the proof of Theorem 4, but its use here is more subtle. By (27), $x_L(0) \geq x^{(1)} > x^{(0)}$. Thus by (39), $x_n(0) < x^{(2)}$. By interpolation, for every $i \geq L$, $x^{(2)} > x_i(0) \geq x^{(1)}$, and consequently $g_i(0) = C$. By the continuity of the functions x_i, there exists a time interval $[0, T]$ such that $g_i(t) = C$ if $0 \leq t \leq T$. For $0 \leq t \leq T$, therefore, $\dot{X}_i(t) \geq 0$ and $(d/dt)(X_i X_j^{-1})(t) = 0$. We now show that $T = \infty$.

For $0 \leq t \leq T$,

$$G \geq \sum_{i=L}^{n} X_i g_i \geq \sum_{i=L}^{n} X_i(0) C.$$

In particular, if for any t, $x(t)$ satisfies (77), then $\dot{x}(t) \geq 0$, so that, by (27), (79) holds.

Moreover, by (27) and (79), for every $i \geq L$,

$$x_i(t) = X_i(t)x(t) \geq X_i(0)x(t) \geq x^{(1)} > x^{(0)}. \tag{87}$$

Consequently, the inequality $x_n(t) > x^{(2)}$ is impossible, since when the function $R(t) = x_n(t) - x^{(2)}$ changes sign from negative to positive, all $x_i(t)$, $L \leq i \leq n - 1$, would have to instantaneously jump from values $\geq x^{(1)}$ to values $\leq x^{(0)}$ in order to satisfy (39). This they cannot do, since they are continuous. Thus the inequalities $\dot{X}_i(t) \geq 0$, $i \geq L$, (79), and (87) maintain each other for $t \geq 0$. A similar argument shows that all X_i are constant if (28) holds.

Suppose that (31) holds. To show that $Q_{n-N+1} = 0$, we argue by contradiction. If $Q_{n-N+1} > 0$, then by Proposition 4, $g(Q_{n-N+1}E) = g(Q_n E)$. By (39), $g(Q_{n-N+1}) = C$. Thus for $i \geq n - N + 1$,

$$x_i(\infty) \geq x_{n-N+1}(\infty) = Q_{n-N+1}E \geq x^{(1)},$$

and

$$E \geq \sum_{i=n-N+1}^{n} x_i(\infty) \geq N x^{(1)} > (B - AC^{-1}).$$

This contradicts (80).

The statements involving (40) and (41) are proved as in Theorem 4.

Suppose that $x^{(1)} = x^{(2)}$ and that (39) holds. Then by Proposition 4, either $Q_i = 0$ or $g(Q_i E) = g(Q_n E)$. The latter can hold only if $Q_i = Q_n$. Hence the limiting distribution is 0–1 or locally uniform.

The assertions in (III) based on (41) and (42) are proved as in Theorem 4.

Condition (43) implies that $\dot{X}_1 \leq 0$ for $t \geq 0$, since either all $x_i(t) \leq x^{(2)}$, or $x_n(t) > x^{(2)}$, which implies that $x_1(t) \leq x^{(0)}$ by (43). The other assertions of (IV) follow readily from this.

Condition (44) implies that $(n - K + 1)x^{(2)} > x(t)$, $t \geq 0$, and thus that at most $(n - K)x_i$'s can exceed $x^{(2)}$ at any time. In particular, if then $x_i(t) \le x^{(2)}$, $i = 1, 2, \ldots, K$. Condition (45) guarantees that X_1 will increase and X_n will decrease just so long as $x^{(1)} \le x_i(t) \le x^{(2)}$, and thus $g_i(t) = C$, for $i = 1, 2, \ldots, K$. By (88), these inequalities hold for $t \ge 0$. Using the monotone increase of X_1, the existence of all limits can be proved as in Proposition 3. The inequalities $Q_1 < 1/n < Q_n$ hold because $x_1(t)$ is bounded away from $x^{(2)}$ by (44). QED.

Proof of Proposition 5: By (46), $x_i(0) \ge x_1(0) > x^{(2)}$, $i = 1, 2, \ldots, n$; all x_i begin in the uniformizing region, so that $\dot{X}_1(0) > 0$ and $\dot{X}_n(0) < 0$. By (46) and (80), for t sufficiently large,

$$x_i(t) \le x(t) < B - AC^{-1} < x^{(2)};$$

all x_i end up in the contour enhancing region, so that $\dot{X}_1 < 0$ and $\dot{X}_n > 0$ unless all $g_i = C$. QED.

Proof of Theorem 7: First we prove equations (47)–(50). By (62), we must compute $F_i - X_i$.

$$\begin{aligned} F_i - X_i &= \frac{N_i D_i^{-1}}{\sum_{k=1}^n N_k D_k^{-1}} - \frac{x_i}{\sum_{k=1}^n x_k} \\ &= \frac{1}{Fx} \sum_{m=1}^{\infty} a_m \sum_{k=1}^{n} (x_k x_i^m - x_i x_k^m D_i D_k^{-1}) \\ &= \frac{1}{F x D_i} \sum_{m=1}^{\infty} a_m \sum_{k=1}^{n} x_i x_k D_k^{-1} \left[x_i^{m-1} \sum_{r=0}^{\infty} b_r x_k^r - x_k^{m-1} \sum_{r=0}^{\infty} b_r x_i^r \right] \\ &= \frac{1}{F x D_i} \sum_{m=1}^{\infty} a_m \sum_{k=1}^{n} x_i x_k D_k^{-1} \left[\sum_{r=0}^{m-2} b_r (x_i x_k)^r (x_i^{m-r-1} - x_k^{m-r-1}) \right. \\ &\quad \left. - \sum_{r=m}^{\infty} b_r (x_i x_k)^{m-1} (x_i^{r-m+1} - x_k^{r-m+1}) \right]. \end{aligned}$$

The identity

$$y^p - z^p = (y - z) \sum_{q=0}^{p-1} y^q x^{p-1-q}$$

is now applied with $y = x_i$, $z = x_k$, and $p = \pm(m - r - 1)$. Then this expression is multiplied by $P = BFx^{-1}$, as required by (62), to yield (47).

The proofs yielding uniform, 0–1, and locally uniform distributions are much like those in Theorems 1–6. Consider the 0–1 case for definiteness; thus let $X_n(0) > X_{n-1}(0)$. By (47) and (51),

$$\dot{X}_n \ge \varepsilon \sum_{k=1}^{n-1} X_k^r (X_n - X_k). \tag{89}$$

Suppose that $X_n(0) > \frac{1}{2}$. Since $\dot{X}_n \ge 0$ and $\sum_{k=1}^n X_k = 1$, there exists a $\delta > 0$ such that

$$\dot{X}_n \ge \delta \sum_{k=1}^{n-1} X_k^r \ge \delta \left(\frac{1 - X^n}{n - 1} \right)^r, \tag{90}$$

which implies $Q_n = 1$.

Suppose that $L_{nj} \geq L_{kj}$ for all $j \neq k, n$. By (48)–(50), $L_{nk} = L_{kn}$. Thus by (47) applied successively with $i = n$ and $i = k$, we conclude that $\dot{X}_n \geq \dot{X}_k$, $k \neq n$. The positive functions $X_n - X_k$, $k \neq n$, are therefore monotone increasing. Define

$$\delta = \varepsilon^{-1} \min\{(X_n - X_k)(0) : k \neq n\} > 0.$$

Then (89) implies (90), which implies $Q_n = 1$. QED.

Proof of Theorem 8: The proof is essentially the same as that of Theorem 3. Equation (15) is modified by adding an extra term $D(1/n - X_i)$ to its right hand side, where $D = nBKx^{-1} > 0$. This term pulls X_i towards $1/n$ even if $g(w)$ is constant.

To prove (53), one computes

$$\dot{x} = x[-(A + nk) + (B - x)G + nBKx^{-1}],$$

notes that $\lim_{t \to \infty} [G - g(x/n)] = 0$, and checks that $\dot{x} = 0$ only if (53) holds. (53) has one solution because its right hand side is a monotone decreasing function of x. QED.

Equation (54) defines a function of type (52) if $K = AC^{-1}$ and

$$g(w) = (BC - AD)C^{-1}(C + Dw)^{-1}.$$

Proof of Theorem 9: By (62), we must compute $F_i - X_i$, premultiply by $P = BFx^{-1}$ and find an expression equal to $\dot{X}_i$. The result is, for any $n > 1$,

$$\dot{X}_i = U(1 - nX_i) + VX_i \sum_{k=1}^{n} X_k(X_i - X_k), \tag{91}$$

where $U = a_0Bx^{-1}$ and $V = a_2Bx$. Let $n = 3$. Suppose $X_i(0) = X_j(0)$ for some i and j, $i \neq j$. Then $X_i(t) = X_j(t)$, $t \geq 0$. Denote the common value of $X_i(t)$ and $X_j(t)$ by $Y(t)$. Then $X_k(t) = 1 - 2Y(t)$, where $k \neq i, j$. Letting $W = 6a_2Bx$, equation (91) becomes

$$\begin{aligned}\dot{Y} &= U(1 - 3Y) + VY(1 - 2Y)(3Y - 1) \\ &= W(\tfrac{1}{3} - Y)[Y^2 - \tfrac{1}{2}Y + a_0(2a_2x^2)^{-1}]\end{aligned}$$

or

$$\dot{Y} = W(\tfrac{1}{3} - Y)(Y - U)(Y - V), \tag{92}$$

Since $W > 0$, (92) implies (56). Equations (56), (57), and (58) show that the value of x determines the limits to which Y converges. In particular, letting $L = (8a_0a_2^{-1})^{1/2}$, (56) implies that

$$\begin{aligned}&\text{sign } \dot{Y} = \text{sign}(\tfrac{1}{3} - Y) \quad \text{if } 0 \leq x < L, \\ &\text{sign } \dot{Y} = \text{sign}(\tfrac{1}{3} - Y)(Y - \tfrac{1}{4})^2 \quad \text{if } x = L, \\ &\text{sign } \dot{Y} = \text{sign}(\tfrac{1}{3} - Y)(Y - U)(Y - V)\end{aligned}$$

with

$$0 < V < \tfrac{1}{4} < U < \tfrac{1}{2} \quad \text{if } L < x \leq B,$$

and

$$\text{sign } \dot{Y} = \text{sign}(Y - \tfrac{1}{3}) \quad \text{if } x = \infty.$$

These changes due to progressive increase in x are pictured in Figure 22. If $x < L$, the limiting distribution is uniform. If $x = L$, $Y = \frac{1}{4}$ is an unstable critical point; hence $Y(\infty) = \frac{1}{3}$ or $\frac{1}{4}$. If $B \geq x > L$, this unstable critical point branches and creates two stable and one unstable critical points. Either U or $\frac{1}{3}$ is the unstable critical point, depending on which is smaller. As x increases, the limiting case of $x = \infty$ is approached. Here, if $Y > \frac{1}{3}$, then $Y(\infty) = \frac{1}{2}$, which defines the locally uniform distribution $P_i = P_j = \frac{1}{2}$, whereas if $Y < \frac{1}{3}$, then $Y(\infty) = 0$, which defines the 0–1 distribution $P_k = 1$. QED.

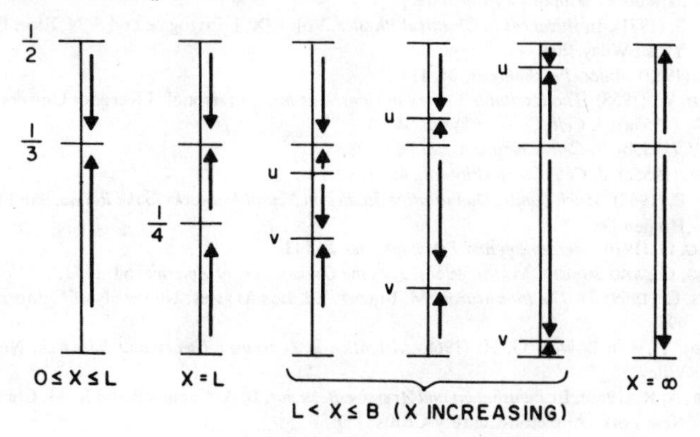

Figure 22. Influence of x on limiting distribution.

References

ANDERSON, P., GROSS, G. N., LOMO, T., AND SVEEN, O. (1969). In *The Interneuron*, M. Brazier, Ed. Los Angeles: Univ. of California Press, 415.

BELLMAN, R. (1967). *Introduction to the Mathematical Theory of Control Processes*, Vol. I, New York: Academic Press.

BLAKEMORE, C., AND CAMPBELL, F. W. (1969). *J. Physiol.*, **203**, 237.

ECCLES, J. C., ITO, M., AND SZENTAGOTHAI, J. (1967). *The Cerebellum as a Neuronal Machine*, New York: Springer.

ESTES, W. K. (1972). In *Coding Processes in Human Memory*, A. W. Melton and E. Martin, Eds. Washington, D.C.: V. H. Winston and Sons.

FREEMAN, W. J. (1969). *J. of Biomedical Systems*, **1**, 3.

GROSSBERG, S. (1970). *J. Theoret. Biol.*, **27**, 291.

GROSSBERG, S. (1971a). *J. Theoret. Biol.*, **33**, 225.

GROSSBERG, S. (1971b). *Proc. Nat'l. Acad. Sci. USA*, **68**, 828.

GROSSBERG, S. (1972a). *Math. Biosci.*, **15**, 39.

GROSSBERG, S. (1972b). *Math. Biosci.*, **15**, 253.

GROSSBERG, S. (1972c). *Kybernetik*, **10**, 49.

GROSSBERG, S. (1973). Classical and instrumental learning by neural networks. To appear in *Progress in Theoretical Biology*.

GROSSBERG, S., AND PEPE, J. *J. of Statistical Physics*, **3**, 95.

HODGKIN, A. L. (1964). *The Conduction of the Nervous Impulse*, Springfield: C. C. Thomas.

HUBEL, D. H., AND WIESEL, T. N. (1968). In *Physiological and Biochemical Aspects of Nervous Integration*, F. D., Carlson, Ed. Englewood Cliffs: Prentice-Hall, 153.

JOHN, E. ROY (1966). In *Frontiers in Physiological Psychology*, R. W. Russell, Ed., New York: Academic Press, 149.

KERNELL, D. (1965a). *Acta. Physiol. Scand.*, **65**, 65.
KERNELL, D. (1965b). *Acta. Physiol. Scand.*, **65**, 74.
LOGAN, F. A. (1969). In "*Punishment and Aversive Behavior*," New York: Appleton-Century-Crofts.
MACRIDES, F., AND CHOROVER, S. L. (1972). *Science*, **175**, 85.
MATTHEWS, P. B. C. (1972). *Mammalian Muscle Receptors and their Central Actions*, London: E. Arnold, Ltd.
MILLER, N. E., (1963). In *Nebraska Symposium on Motivation*, M. R. Jones, Ed. Lincoln: University of Nebraska Press.
MILNER, B. (1958). In *The Brain and Human Behavior*, H. C. Solomon, S. Cobb, and W. Penfield, Eds. Baltimore: Williams and Wilkins.
NICOLIS, G. (1971). In *Advances in Chemical Physics*, Vol. XIX. I. Prigogine and S. N. Rice, Eds. New York: Wiley, 209.
OLDS, J. (1969). *Amer. Psychologist*, **24**, 114.
PENFIELD, W. (1958). *The Excitable Cortex in Conscious Man*, Liverpool: Liverpool University Press.
RALL, W. (1955a). *J. Cell. Comp. Physiol.*, **46**, 3.
RALL, W. (1955b). *J. Cell. Comp. Physiol.*, **46**, 373.
RALL, W. (1955c). *J. Cell. Comp. Physiol.*, **46**, 413.
RATLIFF, F. (1965). *Mach Bands: Quantitative Studies on Neural Networks in the Retina*, San Francisco: Holden Day.
SPERLING, G. (1970). *Perception and Psychophysics*, **8**, 143.
SPERLING, G., AND SONDHI, M. M. (1968). *J. of the Optical Soc. of America*, **58**, 1133.
STEFANIS, C. (1969). In *The Interneuron*, M. Brazier, Ed. Los Angeles: University of California Press, 497.
TRABASSO, T., AND BOWER, G. H. (1968). *Attention in Learning: Theory and Research*, New York: Wiley.
WAGNER, A. R. (1969). In *Punishment and Aversive Behavior*, B. A. Campbell and R. M. Church, Eds. New York: Appleton-Century-Crofts, 157.
WEST, L. J. (1962). *Hallucinations*, New York: Grune and Stratton.
WILSON, H. R., AND COWAN, J. D. (1972). *Biophysical Journal*, **12**, 1.

MASSACHUSETTS INSTITUTE OF TECHNOLOGY

(Received May 9, 1973)

CHAPTER 9

BIOLOGICAL COMPETITION: DECISION RULES, PATTERN FORMATION, AND OSCILLATIONS

PREFACE

This article summarizes some of the new mathematical and physical ideas about competition that have emerged during the past eight years. Each of these ideas can be expressed in several ways. For example, every competitive system induces a decision scheme that can be used to analyze its global dynamics. Otherwise expressed, you learn a lot about a competition by keeping track of who is winning it! Otherwise expressed again, you can understand more about certain nonequilibrium systems by measuring where they change fastest rather than where they achieve equilibrium. Still otherwise expressed, you can sometimes learn a lot about a continuous parallel process by embedding a discrete serial process into it, even though you couldn't guess which serial process to embed without referring to the parallel process.

This article suggests that mass action competition is a universal design principle in its own right, which solves its own universal environmental problem and can be approached by its own unified mathematical method. In fact, the method has recently taken on a life of its own. In mathematical parlance, it describes a theory of Liapunov functionals which are integrals of maximum functions. All population models whose total populations change monotonically through time admit such functionals whether or not they are competitive. Thus the mathematics makes a statement about systems which induce decision schemes, whether or not they are competitive. This notion of decision is a new idea which is well-suited to the study of fast parallel processors. In particular, the theorems show that some systems contain internal contradictions in their decision schemes which can force them to oscillate forever, as in the voting paradox, whereas other systems possess such consistent decision schemes that they can always arrive at global decisions in response to arbitrary input patterns and store these decision until they are reset.

Biological competition: Decision rules, pattern formation, and oscillations

ABSTRACT Competition solves a universal problem about pattern processing by cellular systems. Competition allows cells to automatically retune their sensitivity to avoid noise and saturation effects. All competitive systems induce decision schemes that permit them to be classified. Systems are identified that achieve global pattern formation, or decision-making, no matter how their parameters are chosen. Oscillations can occur due to contradictions in a system's decision scheme. The pattern formation and oscillation results are extreme examples of a complementarity principle that seems to hold for competitive systems. Nonlinear competitive systems can sometimes appear, to a macroscopic observer, to have linear and cooperative properties, although the two types of systems are not equivalent. This observation is relevant to theories about the evolutionary transition from competitive to cooperative behavior.

1. Biological signal processing and competitive decisions

Darwin's classic work (1) on the survival of the fittest emphasized the importance of competition as a universal principle of biological organization. Darwin's theory was formulated in terms of macroscopic variables such as competing species. More recently, competitive interactions have been shown (2–5) to solve a universal dilemma concerning the processing of patterned information by any noisy system with finitely many excitable sites. All cellular systems are of this type. This di-

The publication costs of this article were defrayed in part by page charge payment. This article must therefore be hereby marked "*advertisement*" in accordance with 18 U. S. C. §1734 solely to indicate this fact.

lemma, called the noise-saturation dilemma, notes that small signals to the system can get lost in noise whereas large signals can saturate system response by exciting all of its sites and thereby reducing to zero its sensitivity to signal fluctuations. The dilemma describes a fundamental problem concerning the transmission of information by biological systems because, by trying to avoid noise, the system might amplify the signals so much that saturation occurs, and conversely. The noise-saturation dilemma is solved by competitive systems, and the solution shows how such systems can automatically retune themselves to avoid both noise and saturation. This fact supplies a basic reason for the universal existence of competition on both the microscopic and the macroscopic level and throws a new light on Darwin's concept of biological competition.

Given that competitive systems are ubiquitous, we need a general method for classifying some of the rich variety of their dynamical possibilities. Smale (6) has shown that essentially any dynamical behavior can be embedded in a suitably defined competitive system. The present method reverses his approach by providing a tool for designing, analyzing, and classifying competitive systems that have desirable biological behavior.

This note announces that every competitive system induces a decision scheme that can be used for global analysis of the competition as it evolves through time. The method has been used to explicate radically different types of dynamical behavior within competitive systems. Section 4 below summarizes a result concerning global pattern formation by systems possessing any number of competing populations. This result describes a principle of system design that guarantees the *absolute stability* of pattern formation; that is, pattern formation occurs no matter how system parameters are chosen within this class of systems. This principle of design means intuitively that the system possesses an *adaptation level*. Any such competitive system is capable of resolving essentially arbitrary irregularities in local system design into a global consensus or decision among the system's components by balancing these irregularities against the adaptation level. Some systems that arise within Eigen's theory of macromolecular evolution are of this type (7). Neural networks and other cellular and chemical mass action systems are often of this form (3, 4, 8). The systems also suggest new models of stable economic markets (M. W. Hirsch, personal communication).

Section 3 below illustrates how the method can be used to prove global oscillation theorems for systems of arbitrarily many competing populations (9). The oscillations reflect a system's inability to arrive at a global decision. The two types of theorems—pattern formation and oscillations—are extreme examples of a complementarity principle that seems to hold in competitive systems.

In this idea of decision, the decisions are defined by structures that exist far from equilibrium. The method hereby shows that measures of the nonequilibrium behavior of competitive systems often provide a deeper insight into their design than does the traditional local analysis of their equilibrium points.

2. Decisions in competitive systems

Suppose that a system is defined by n quantities $x = (x_1, x_2, \ldots, x_n)$ evolving through time. For example, $x_i(t)$ might be the population size, or activity, or concentration, etc. of the ith species v_i in the system, $i = 1, 2, \ldots, n$. A system

$$\frac{dx}{dt} = f(x), \quad x \in R, \qquad [1]$$

is said to be *competitive* if its ith component

$$\frac{dx_i}{dt} = f_i(x) \qquad [2]$$

satisfies

$$\frac{\partial f_i(x)}{\partial x_j} \leq 0 \text{ if } i \neq j \text{ and } x \in R, \qquad [3]$$

and the system remains in a bounded region R of Euclidean n space. In other words, increasing x_j can only decrease x_i's rate of change, but might not change it at all, for all $i \neq j$.

The new concept of decision can be motivated in the following fashion. Suppose that an experimentalist is looking at a petri dish filled with an unknown material. What does the experimentalist notice? Usually, one's attention goes to those regions of the petri dish where something is changing. If suddenly a change occurs in a new region of the dish, attention is focused on the new region. Our attention hereby jumps from region to region as new changes appear. Just as our attention jumps to follow the most noticeable system changes, we can formalize the maximal changes in system activity as decisions within the system that regulate which of its regions are active.

Actually, the intuitive notion of decision can be explicated in several directions by using competitive systems as a guide. The decisions to be described below are "local" decisions that might, or might not, terminate as time goes on. When they terminate in pattern formation, the entire system has made a more global decision based on the series of local decisions. If this competitive system is embedded as a component in a hierarchy of competitive subsystems, linked together by adaptive feedback pathways, then the patterns at each level in the hierarchy sometimes mutually reinforce and amplify each other, thereby locking each other into a global activity pattern that represents a functional unit of the entire system and which can thereupon drive adaptive changes in system structure. Such *adaptive resonances* define a yet higher sense in which competitive systems participate in biological decision-making (5, 10). Each of these levels of decision-making acts on different spatial and temporal scales, and each must be analyzed before decision-making by an organism as a whole can be understood.

To see how jumps, or local decisions, are formalized, a

competitive system often can be written in the form

$$\dot{x}_i = a_i(x_i)M_i(x), \quad i = 1, 2, \ldots, n. \qquad [4]$$

For example, in the classical Volterra–Lotka systems

$$\dot{x}_i = A_i x_i \left(1 - \sum_{k=1}^{n} B_{ik} x_k\right), \qquad [5]$$

$$a_i(x_i) = A_i x_i, \text{ and}$$

$$M_i(x) = 1 - \sum_{k=1}^{n} B_{ik} x_k$$

Intuitively, $M_i(x)$ defines the competitive balance at the ith state v_i, and $a_i(x_i)$ is an amplification coefficient that converts the competitive balance into the growth rate dx_i/dt of x_i. In particular, $a_i(x_i) > 0$ unless $x_i = 0$, $a_i(0) = 0$, and

$$\frac{\partial M_i(x)}{\partial x_j} \leq 0 \text{ if } i \neq j \text{ and } x \epsilon R. \qquad [6]$$

To track which states are changing fastest and slowest, we define

$$M^+(x) = \max_k M_k(x) \text{ and } M^-(x) = \min_k M_k(x). \qquad [7]$$

One then proves that there exists a *competition threshold;* namely, if $M^+[x(T)] \geq 0$ at some time $t = T$, then $M^+[x(t)] \geq 0$ at *all* times $t \geq T$. Thus, if some state v_i is being enhanced at time $t = T$ [that is, $d/dt\, x_i(\mathrm{T}) \geq 0$], then at *every* future time $t \geq T$, some state v_j will be enhanced, but possibly different states at different times. In other words, if the competition "ignites" at some time $t = T$, then it thereafter never turns off. Set

$$S^+ = [x \in R: M^+(x) = 0] \qquad [8]$$

is thus a competition threshold, and once the *positive ignition region*

$$R^{+} = [x \in R: M^{+}(x) \geq 0] \qquad [9]$$

is entered, it never can be left. In mathematical parlance, R^{+} is a positively invariant region.

Similarly, if $M^{-}[x(T)] \leq 0$ at some time $t = T$, then $M^{-}[x(t)] \leq 0$ at *all* times $t \geq T$. Thus, if some state v_i is being suppressed at time $t = T$ [that is, $d/dt\ x_i(t) \leq 0$], then at *every* future time $t \geq T$, some state v_j will be suppressed, but possibly different states at different times. Set

$$S^{-} = [x \in R: M^{-}(x) = 0] \qquad [10]$$

defines the threshold at which suppression sets in, and the *negative ignition region*

$$R^{-} = [x \in R: M^{-}(x) \leq 0] \qquad [11]$$

is a positively invariant region.

How are the regions R^{+} and R^{-} used? Suppose $x(t)$ never enters R^{+}. Then, by Eq. 4, each $d/dt\ x_i(t) \leq 0$ at all times $t \geq 0$. Consequently, $x_i(t)$ monotonically decreases to a limit. In this case, the competition never gets started. The interesting behavior in any competitive system occurs within the invariant region R^{+}, and really within the smaller invariant region $R^{*} = R^{+} \cap R^{-}$ because, if x is in R^{+} but not R^{-}, all x_is are increasing and the limit $x(\infty) = \lim_{t\to\infty} x(t)$ trivially exists.

After ignition takes place [i.e., $x(t)$ is in R^{*}], we keep track of which state v_i is being *maximally* enhanced at any time. That is, we pay attention to the biggest rates of change. If, for example, $M^{+}[x(t)] = M_i[x(t)]$ for $S \leq t < T$ but $M^{+}[x(t)] = M_j[x(t)]$ for $T \leq t < U$, then we say that the system *jumps* from i to j at time $t = T$. These jumps are the local decisions in a competitive system. A jump from i to j can only occur on the *jump set*

$$J_{ij} = [x \in R^{*}: M_i(x) = M_j(x) = M^{+}(x)]. \qquad [12]$$

Because this set is defined where the x_is are changing at a maximal rate, it defines a hypersurface that is far away from the equilibrium points x such that $dx/dt = 0$ of the system. By studying the geometrical relationships of the jump sets within R^*, global results have been proved about pattern formation and oscillations in nonlinear systems with any number $n \geq 2$ of competing states. Below, I briefly summarize two applications of the theory to systems that are, on the surface, very different, although both are amenable to the present method.

3. Oscillations and the voting paradox

In ref. 9, n-dimensional generalizations

$$\dot{x}_i = a_i(x)[1 - \sum_{k=1}^{n} B_{ik}(x)f_k(x_k)], \qquad [13]$$

$i = 1, 2, \ldots, n$, of the three-dimensional Volterra–Lotka system

$$\left.\begin{aligned} \dot{x}_1 &= x_1(1 - x_1 - \alpha x_2 - \beta x_3) \\ \dot{x}_2 &= x_2(1 - \beta x_1 - x_2 - \alpha x_3) \\ \dot{x}_3 &= x_3(1 - \alpha x_1 - \beta x_2 - x_3) \end{aligned}\right\} \qquad [14]$$

are globally analyzed. May and Leonard (11) studied system **14** to illustrate the *voting paradox*. Namely, if $\alpha + \beta \geq 2$ and $\beta > 1 > \alpha$, then in *pairwise* competition in system **14**, v_1 beats v_2, v_2 beats v_3, and v_3 beats v_1. When all three populations interact, a global "contradiction" is produced because, if the winning relationship were transitive, v_1 could beat itself, which is absurd. May and Leonard showed that this contradiction produces sustained oscillations in system dynamics. They used computer simulations and local analytic estimates to do this. The present method provides a global analysis of system decisions and hereby shows why the system is forced to jump infinitely often in the cyclic order $v_1 \rightarrow v_2 \rightarrow v_3 \rightarrow v_1$, thereby producing sustained oscillations.

Below are intuitively summarized some of the ideas that yield pattern formation and oscillation theorems. First, one observes that given initial data $x(0)$ such that

$$\int_0^\infty M^+[x(t)]dt < \infty, \qquad [15]$$

then the limit $x(\infty)$ exists, and we say that pattern formation occurs in response to $x(0)$. It is also shown that if, starting at $x(0)$, there ensue only finitely many jumps, then relationship **15** holds, and consequently $x(\infty)$ exists. Intuitively this means that, after all local decisions have been made, the system can form a well-defined pattern $x(\infty)$. For example, suppose, starting at $x(0)$, all jumps are partially ordered so that no jump cycles (e.g., $v_1 \to v_2 \to v_3 \to v_1$) exist. Then, only finitely many jumps can occur, so pattern formation occurs. Moreover if relationship **15** holds, because $M^+[x(t)] \geq 0$ at all large times, it follows that $\lim_{t\to\infty} M^+[x(t)] = 0$. Consequently, $x(t)$ approaches an equilibrium point that lies on S^+. For example, in Volterra–Lotka systems **5** with $n = 3$, to find the jump sets J_{ij} defined by Eq. **12**, one first notes if the planes $M_i(x) = 0$ and $M_j(x) = 0$ intersect on S^+. The intersection is a line segment L_{ij} except in trivial cases. Then, one defines the planar region interpolated between L_{ij} and the point $x = 0$ and intersects this planar region with R^* to find J_{ij}. Because $n = 3$, no jump cycle exists unless there are three line segments $L_{i_1i_2}$, $L_{i_2i_3}$, and $L_{i_3i_1}$ on S^+ with i_1, i_2, and i_3 distinct. If not, given *any* $x(0)$, the limit $x(\infty)$ exists; that is, *global* pattern formation occurs. Moreover the limit $x(\infty)$ lies on S^+. Fig. 1 illustrates some Volterra–Lotka systems that undergo global pattern formation.

The starting point for studying oscillations is the converse statement:

$$\int_0^\infty M^+[x(t)]dt = \infty \qquad [16]$$

implies that infinitely many jumps occur. For example, the

jump sets of the Volterra–Lotka system **14** are depicted in Fig. 2. One finds that, if $\beta > 1 > \alpha$, jumps must cycle in the order $v_1 \to v_2 \to v_3 \to v_1$ if they occur at all. To show that this jump cycle recurs infinitely often, we must prove relationship **16**, given prescribed initial data $x(0)$. To do this, one studies the ignition surface S^+ defined by [8] to test which $x(0)$s generate trajectories that penetrate S^+ and which of these trajectories are then repelled away from S^+ into R^*. Such trajectories will satisfy $M^+[x(t)] \geq \epsilon$ for some $\epsilon > 0$ and all times t that are sufficiently large. Then [**16**] readily follows. In system **14**, all trajectories penetrate S^+ except those that have uniform initial

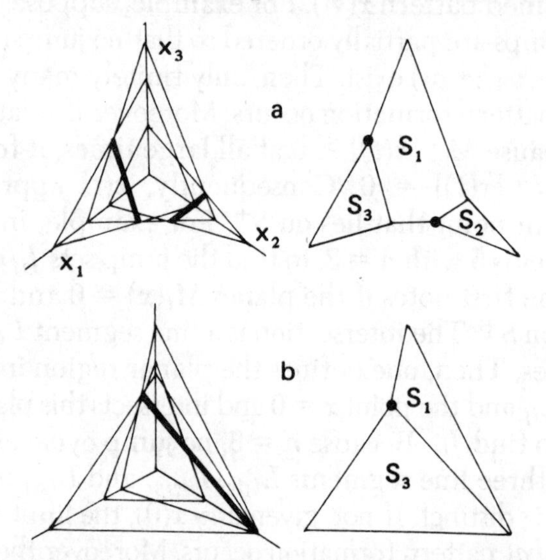

FIG. 1. The sets $S_i = [x \in \mathbf{R}_+^3: M_i(x) = 0]$ are planar segments in the case of three-dimensional Volterra–Lotka systems. (*a*) All three S_i, $i = 1,2,3$, form part of S^+. Because only two intersections $S_1 \cap S_3$ and $S_1 \cap S_2$ are nonempty in S^+, there are no jump cycles. Global pattern formation therefore occurs. Moreover, there are two equilibrium points (filled circles) on S^+, but only the one on $S_1 \cap S_3$ is stable. (*b*) Only one intersection $S_1 \cap S_3$ is nonempty in S^+. Again, global pattern formation occurs. The equilibrium point on S^+ is stable.

data $x_1(0) = x_2(0) = x_3(0)$. The latter trajectories remain uniformly distributed and approach the equilibrium point $P = (1 + \alpha + \beta)^{-1}(1, 1, 1)$ that lies at the intersection of the dark lines in Fig. 2*b*. What prevents other trajectories from approaching P after they penetrate S^+? The condition $\alpha + \beta \geq 2$ guarantees that P is an unstable equilibrium point with respect to the directions lying within R^*.

After one is sure that $x(0)$ generates infinitely many jumps, how does one know which x_i oscillate persistently as $t \to \infty$; that is, which x_i oscillate at arbitrarily large times and in such a way that the limit $x_i(\infty)$ does not exist? To study this, one defines an *asymptotic graph* that decomposes the jumps that reoccur infinitely often into a collection of jump cycles among certain of the states v_i. In system **14,** all of the v_i, $i = 1, 2, 3$, are in the

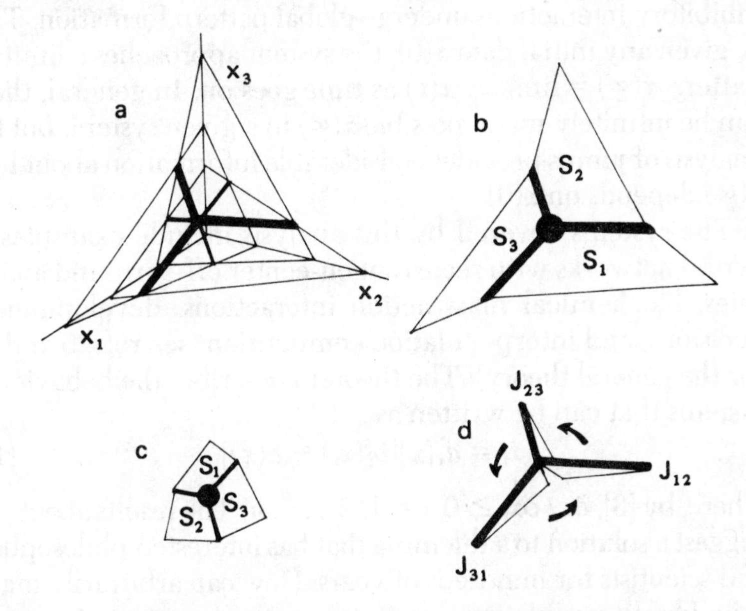

FIG. 2. (*a*) Sets $S_1 \cap S_2$, $S_2 \cap S_3$, and $S_3 \cap S_1$ are nonempty in S^+. (*b*) Positive ignition surface with equilibrium point $P = (1 + \alpha + \beta)^{-1}(1,1,1)$. (*c*) Negative ignition surface with equilibrium point P. (*d*) Jump sets form a jump cycle $v_1 \to v_2 \to v_3 \to v_1$ if $\beta > 1 > \alpha$.

asymptotic graph if $\alpha + \beta \geq 2$ and $\beta > 1 > \alpha$. One then shows how those x_i whose v_i are in the asymptotic graph cannot stop oscillating as $t \to \infty$ without contradicting [**16**].

It does not follow that persistently oscillating x_is approach a periodic solution as $t \to \infty$. For example, May and Leonard (11) numerically demonstrated oscillations of ever-increasing period in system **14** when $\alpha + \beta = 2$, and Grossberg (9) showed that such oscillations can occur when the trajectory approaches a union of heteroclinic orbits (namely, orbits between two equilibrium points) as $t \to \infty$.

4. Absolute stability of global pattern formation

In ref. 8 it is shown how a large class of systems defined by mass action, or kinetic laws, and subjected to feedback excitatory and inhibitory interactions undergo global pattern formation. That is, given any initial data $x(0)$, the system approaches a limiting pattern $x(\infty) = \lim_{t\to\infty} x(t)$ as time goes on. In general, there can be infinitely many possible $x(\infty)$ in a given system, but the analysis of jumps provides considerable information about how $x(\infty)$ depends on $x(0)$.

The systems covered by this analysis include examples of neural networks with recurrent on-center off-surround anatomies, biochemical mass action interactions, developmental decisions, and interpopulation competition (see refs. 5 and 12 for the general theory). The theorem describes the behavior of systems that can be written as

$$\dot{x}_i = a_i(x)[b_i(x_i) - c(x)] \qquad [\mathbf{17}]$$

where, by [3], $\partial c/\partial x_i \geq 0$, $i = 1, 2, \ldots, n$. The results about [**17**] suggest a solution to a dilemma that has interested philosophers and scientists for hundreds of years: How can arbitrarily many individuals, populations, or states, each obeying unique and personal laws, ever interpret each other's signals or communications well enough to ever agree about anything? Leibniz has met this dilemma by developing his theory of monads (13). The theorem suggests a different solution. In [17], each v_i can have an essentially arbitrary signal function $b_i(x_i)$ as well as an arbitrary amplification function $a_i(x)$. Global consensus, or pattern formation, can be achieved despite these local irregularities because there exists a commonly shared *adaptation level* $c(x)$ against which to evaluate local irregularities. The adaptation level $c(x)$ defines a type of symmetric long-range order that is shared by the populations.

There seems to exist a complementarity, or trade-off, between how freely one can choose local parameters ("individual differences") and how global the adaptation level ("communal understanding") must be chosen to achieve global consensus (8, 9). For example, in the Volterra–Lotka system **14** there is no adaptation level, and even linear feedback signals can produce sustained oscillations. By contrast, a large class of generalized Volterra–Lotka systems [**13**] do undergo global pattern formation. These are the systems whose interaction coefficients $B_{ik}(x)$ are determined by statistically independent factors at v_i and v_k—namely, $B_{ik}(x) = g_i(x_i)h_k(x_k)$. Then [**13**] can be written in the form of [**17**] and hence undergoes global pattern formation. Thus, within Volterra–Lotka systems, only deviations from statistically independent interactions can produce sustained oscillations.

An important class of mass action systems undergoing competitive feedback can be written in the form of [**17**]. To illustrate this, consider the usual voltage law that underlies the circuit diagrams of nerve cell membranes (14, 15):

$$C\frac{\partial V}{\partial t} = (V^+ - V)g^+ + (V^- - V)g^- + (V^P - V)g^P \qquad [18]$$

in which C is capacitance, the constants V^+, V^-, and V^P are excitatory (usually Na^+), inhibitory (usually K^+), and passive saturation points, respectively; and g^+, g^-, and g^P are excitatory, inhibitory, and passive conductances, respectively. The voltage $V(t)$ is variable, and stays between V^+ and V^- because $V^- \le V^P < V^+$. Let $y_i(t)$ be the voltage of the ith cell (or cell population) v_i. Let $C = 1$ (that is, rescale the time variable), and introduce the notation $V^+ = B$, $V^P = 0$, and $V^- = -D$, in which $B > 0$ and $D \geq 0$ because $V^- \leq V^P < V^+$. Suppose that the ith excitatory conductance g_i^+ is influenced by a constant, or tonic, external input I_i and by a positive feedback signal $f_i(y_i)$ from v_i to itself. Thus, $g_i^+ = f_i(y_i) + I_i$. Let the ith inhibitory conductance g_i^- be influenced by a constant, or tonic, input J_i and by competitive or inhibitory signals $f_k(y_k)$ from all cells v_k, $k \neq$ i. Thus, $g_i^- = \sum_{k \neq i} f_k(y_k) + J_i$. Actually, one can think of I_i and J_i as varying slowly compared to the reaction rate of x_i. Let the passive conductance g_i^P equal the constant A. In all, the feedback interactions define a recurrent on-center (v_i excites itself) off-surround (v_k inhibits v_i, $i \neq k$) anatomy. Eq. **18** then becomes

$$\dot{y}_i = -Ay_i + (B - y_i)[f_i(y_i) + I_i] - (y_i + D)\left[\sum_{k \neq i} f_k(y_k) + J_i\right]. \quad [19]$$

Now generalize [**19**]. Let each v_i have an arbitrary decay rate A_i, an arbitrary excitatory saturation point B_i, and an arbitrary inhibitory saturation point D_i. Then [**19**] becomes

$$\dot{y}_i = -A_i y_i + (B_i - y_i)[f_i(y_i) + I_i] - (y_i + D_i)\left[\sum_{k \neq i} f_k(y_k) + J_i\right] \quad [20]$$

which is an n-dimensional mass action, or kinetic, network with arbitrary parameters, tonic inputs, and feedback signals $f_k(y_k)$. To write [**20**] in the form of [**17**], first let $x_i = y_i + D_i$ and $h_i(x_i) = f_i(x_i - D_i)$. Then,

$$\dot{x}_i = -A_i x_i + (B_i + D_i - x_i)[h_i(x_i) + I_i] - x_i\left[\sum_{k \neq i} h_k(x_k) + J_i\right] + A_i D_i. \quad [21]$$

System **21** can be written in the form of [**17**] by using the definitions $a_i(x) = x_i$,

$$b_i(x_i) = -A_i - I_i - J_i + x_i^{-1}[A_i D_i + I_i + (B_i + D_i)h_i(x_i)], \text{ and } c(x) = \sum_{k=1}^{n} h_k(x_k).$$

Clearly [**17**] is vastly more general than [**20**]; for example, it permits nonlinear combinations of the signals, rather than merely additive ones, as well as state-dependent changes in the parameters. Because global pattern formation obtains given *any* choice of parameters in [**17**], I call the system *absolutely stable*. Any mechanism that changes system parameters can cause dramatic changes in the underlying decision scheme without destroying the system's ability to reach a new global consensus. The theorem thus constrains possible bifurcations in the space of decision schemes. This subject should be studied further.

For example, the system

$$\dot{x}_i = -Ax_i + (B - x_i)f(x_i) - x_i \sum_{k \neq i} f(x_k) \qquad [22]$$

describes the simplest competitive mass action feedback network. If the signal function $f(w)$ is chosen so that $b(w) = w^{-1}f(w)$ is strictly increasing, then the system chooses the population v_i possessing the maximal initial data and concentrates all system activity at v_i. By contrast, if $b(w)$ is a concave function with a flat plateau between its increasing and decreasing values, as when $f(w)$ is a sigmoid or S-shaped signal function, then a quenching threshold exists: initial activities that are smaller than the quenching threshold are suppressed, whereas the spatial pattern of initial activities that exceed the quenching threshold is contrast-enhanced and stored (3, 16). These results illustrate how a competitive system can sometimes, but not always, behave, like a finite state machine. In particular, a "hill" or "hump" in the graph of $b(w)$ can significantly alter system dynamics. Mimura and Murray (17) have noted the importance of hills in determining the qualitative behavior of prey–predator reaction–diffusion systems. Their goal was to understand spatial heterogeneity, or patchiness, in these systems. In a neural context, analogous effects occur and are called disinhibition or lateral masking (18, 19).

Global pattern formation in [**17**] is proved by first analyzing how the hills in the functions $b_i(w)$ influence system dynamics. It is shown how the decision rules sense these hills by causing a nested series of *decision boundaries* to be laid down as time goes on. These decision boundaries suddenly appear at prescribed times and, after they appear, each $x_i(t)$ can fluctuate only within the intervals that are defined by the boundaries. Once all the boundaries are laid down, the decision process is essentially complete, except for a possible series of minor system adjustments. The concept of decision boundary is reminiscent of the compartmental boundaries that Kauffman *et al.* (20) have modeled for the development of the *Drosophila* embryo. However, the Kauffman *et al.* model describes a linear threshold phenomenon that is due to the existence of a physical boundary—in their case, an elliptic boundary. A decision boundary is caused by nonlinear suprathreshold interactions even if no physical boundary effects occur.

To illustrate how hills are related to decision boundaries, choose all $b_i(w) \equiv b(w)$ and let $b(w)$ possess finitely many local maxima and minima. Consider the abscissa values of the hill peaks of highest height. There exists a time T_1 after which each $x_i(t)$ is trapped within an interval between a pair of such abscissa values. These abscissa values are the first decision boundaries to appear. To prove this fact, the ignition property is used. If at any time t, $x_i(t)$ equals one of these abscissa values, then $M_i[x(t)] = M^+[x(t)] \geq 0$. Consequently $\dot{x}_i \geq 0$, so that once x_i crosses an abscissa value, it can never cross back. Further analysis shows that there exists a time $T_2 > T_1$ after which no $x_i(t)$ can cross the abscissa values between either the highest or the next-highest hill peaks. This process of laying down decision boundaries continues until each $x_i(t)$ is trapped in the "bowl" between a pair of successive hill peaks. The first stage of pattern formation is then complete.

The second stage is analyzed by keeping track of that $x_i(t)$ whose hill height $b[x_i(t)]$ is maximal. Denote the maximal hill height by $B[x(t)]$; that is, $B[x(t)] = \max_i b[x_i(t)]$. By [**17**], $B(x) = M^+(x) + c(x)$. After all dynamic boundaries are laid down, jumps can occur among descending slopes of the hills (slopes to the right of hill peaks) as $B[x(t)]$ decreases monotonically through time; or jumps can occur among ascending slopes of the hills (slopes to the left of hill peaks) as $B[x(t)]$ increases monotonically through time; or a jump can occur from a descending slope to an ascending slope but not conversely; or the variable $x_i(t)$ such that $B[x(t)] = b[x_i(t)]$ can increase continuously as $B[x(t)]$ moves from a descending slope to an ascending slope, but not conversely, because the $x_i(t)$ variables are trapped within their bowls. In all, $B[x(t)]$ can oscillate at most once after the dynamic boundaries are laid down. Consequently, the limit $B[x(\infty)] = \lim_{t\to\infty} B[x(t)]$ exists. Using this fact, it is then shown that the limit $c[x(\infty)] = \lim_{t\to\infty} c[x(t)]$ of the adaptation level also exits and equals $B[x(\infty)]$. In other words, the local decisions among system components ultimately lead to the choice of a set-point or asymptotic adaptation level $c[x(\infty)]$. Then each x_i adjusts itself via [**17**] to this set-point as the limit $x(\infty)$ is approached.

The above analysis reveals that $B[x(t)]$ is monotonic at large times; that is, it is an *asymptotically Liapunov function*. Function $B[x(t)]$ only becomes Liapunov, however, after the decision boundaries have been laid down. Thus, the system approaches a "classical limit" only after its initially nonstationary dynamics of decision-making is over. A similar trend often occurs in learning networks: *after* the nonstationary phase of learning is over, the system settles down to a memory phase, which is described by a stationary Markov chain (21).

5. Evolutionary switch from competition to cooperation?

By analogy with [3], a *cooperative system* is one in which

$$\frac{\partial f_i}{\partial x_j}(x) \geq 0 \text{ if } i \neq j \text{ and } x \in R. \quad [23]$$

A competitive system can sometimes appear to be cooperative. For example, Grossberg (3) proved that competitive schemes, such as [22], can amplify the activities of all the competing populations, thereby making it appear that an increase in one population's activity has increased other populations' activities. This property can drive all system activities into the range where they are most sensitive to each other's signals. Grossberg (5, 22) argued that this self-tuning, or normalization, property lies behind a wide variety of biological phenomena such as sensory adaptation and self-regulation. Nonlinear interactions are required to achieve self-tuning, but the system's properties can look linear to a macroscopic observer (see ref. 5, section 8). A similar dilemma can occur in learning systems (21). Thus, a system that looks linear and cooperative to an untutored observer can, in reality, be nonlinear and competitive. Such an observer will not be able to understand how the system automatically tunes its sensitivity to match fluctuating external demands, among other properties, and efforts to model the system out of linear components can lead to unphysical instabilities (23).

This situation can create major conceptual difficulties when one considers the evolution of biological order. How do components that compete at early stages of evolution ultimately cooperate to establish a more complex structure, such as an organ? Does this switchover imply that the laws of interaction change from a condition like [3] to a condition like [**23**]? Or do the components compete throughout all the evolutionary stages, but in such a fashion that earlier stages of competition alter system parameters so that later stages can yield ostensibly cooperative macroscopic properties? For example, if [**22**] has a sigmoid signal $f(w)$ and system sensitivity is modulated by a variable arousal or enzymatic level, the system can choose a winning population at low arousal levels and amplify all activity levels at higher arousal levels (5). Interacting competitive subsystems can also begin to resonate when their feedback signals match and amplify each other (5, 10, 12).

Such considerations make it plain that the collective properties that define the evolutionary success of biological systems often cannot be reliably guessed from a study of their isolated components and indicate an important role for mathematical analysis in understanding the principles of design on which evolutionary success is founded.

This work was supported in part by National Science Foundation Grant MCS 77-02958.

1. Darwin, C. (1859) *On the Origin of Species* (London).
2. Grossberg, S. (1970) *J. Theor. Biol.* **27**, 291–337.
3. Grossberg, S. (1973) *Stud. Appl. Math.* **52**, 217–257.
4. Grossberg, S. (1977) *J. Math. Biol.* **4**, 237–256.
5. Grossberg, S. (1978) in *Progress in Theoretical Biology*, eds. Rosen, R. & Snell, F. (Academic, New York), pp. 183–232.
6. Smale, S. (1976) *J. Math. Biol.* **3**, 5–7.
7. Eigen, M. & Schuster, P. (1978) *Naturwissenschaften* **65**, 7–41.
8. Grossberg, S. (1978) *J. Math. Anal. Appl.* **66**, 470–493.
9. Grossberg, S. (1978) *J. Theor. Biol.* **73**, 101–130.
10. Grossberg, S. (1980) *Psychol. Rev.* **87**, 1–51.
11. May, R. M. & Leonard, W. J. (1975) *SIAM (Soc. Ind. Appl. Math.) J. Appl. Math.* **29**, 243–253.
12. Grossberg, S. (1978) in *Progress in Theoretical Biology*, eds. Rosen, R. & Snell, F. (Academic, New York), pp. 233–374.

13. Leibniz, G. W. (1925) *The Monadology and Other Philosophical Writings*, translated by Latta, R. (Oxford Univ. Press, London).
14. Hodgkin, A. L. (1964) *The Conduction of the Nervous Impulse* (Thomas, Springfield, IL).
15. Katz, B. (1966) *Nerve, Muscle, and Synapse* (McGraw-Hill, New York).
16. Grossberg, S. & Levine, D. S. (1975) *J. Theor. Biol.* **53,** 341–380.
17. Mimura, M. & Murray, J. D. (1978) *J. Theor. Biol.* **75,** 249–262.
18. Ellias, S. A. & Grossberg, S. (1975) *Biol. Cybernetics* **20,** 69–98.
19. Levine, D. S. & Grossberg, S. (1976) *J. Theor. Biol.* **61,** 477–504.
20. Kauffman, S. A., Shymko, R. M. & Trabert, K. (1978) *Science* **199,** 259–270.
21. Grossberg, S. (1969) *J. Differential Equations* **5,** 531–563.
22. Grossberg, S. (1980) *Bull. Math. Biol.*, in press.
23. Grossberg, S. (1978) *Psychol. Rev.* **85,** 592–596.

Communicated by Louis N. Howard, December 26, 1979

CHAPTER 10

COMPETITION, DECISION, AND CONSENSUS

PREFACE

This chapter proves that all competitive systems which admit an adaptation level are absolutely stable. This theorem suggests an approach to an old philosophical problem: How can you design systems of communicators wherein each communicator is characterized by arbitrary individual differences, or personal parameters, each communicator knows about other communicators only through locally perceived signals, yet the communication system as a whole can generate a global consensus? How can the system as a whole achieve coherence even if its parts are carelessly thrown together? One answer is: "Balance the individual differences against an adaptation level". In other words, if you design part of the system very carefully, you can let the rest go wild without sacrificing system stability. It seems to me that this type of insight should be generally better understood, notably in discussions of free market forces.

Once the concept of an adaptation level system was clearly defined, I could identify examples of this concept in various fields other than psychology and physiology. Examples are found, for example, in Eigen's theory of macromolecular evolution, in the Volterra–Lotka equations of population biology, in Willshaw and Malsburg's theory of retinotectal development, and in Lacker's theory of control of ovulation number in mammals. These examples illustrate *descriptive* appearances of adaptation level systems. *Prescriptive* appearances are also useful, since they suggest behavioral rules whereby absolutely stable interactions can be guaranteed, by the consent of competing individuals, even if these individuals know very little about each other's behavior. For example, due to Moe Hirsch's interest in economic applications of these models, I have defined a class of absolutely stable production strategies for an economic market. If all competitors produce the same product, and each competitor chooses a strategy from this class, then even without knowledge of the other competitors' choices, each competitor will realize his expected profit without disturbing the absolute stability of the market. These applications are just starting to be studied now, but it is already quite fascinating to realize how the existence of a rapidly communicated, competititve price index can act as an adaptation level that tends to stabilize a market.

Why is *absolute* stability needed? This means that the system can reach a global decision no matter how its parameters are chosen. In self-organizing machines, one often cannot predict in advance how system parameters will change through time. Absolute stability guarantees that the global decision-making capability is left invariant by self-organization.

Competition, Decision, and Consensus*

Submitted by G. C. Rota

1. Introduction

The following problem, in one form or another, has intrigued philosophers and scientists for hundreds of years: How do arbitrarily many individuals, populations, or states, each obeying unique and personal laws, ever succeed in harmoniously interacting with each other to form some sort of stable society, or collective mode of behavior? Otherwise expressed, if each individual obeys complex laws, and is ignorant of other individuals except via locally received signals, how is social chaos averted? How can local ignorance and global order, or consensus, be reconciled? This paper considers a class of systems in which this dilemma is overcome.

We begin by asking what design constraints must be imposed on a system of competing populations in order that it be able to generate a global limiting pattern, or decision, in response to arbitrary initial data? This paper proves that global pattern formation occurs in systems of the form

$$\dot{x}_i = a_i(x)\,[b_i(x_i) - c(x)] \tag{1}$$

where $x = (x_1, x_2, ..., x_n)$ and $i = 1, 2, ..., n$. Such systems can have any number of competing populations ($n \geqslant 2$), any interpopulation signal functions $b_i(x_i)$, any mean competition function, or adaptation level, $c(x)$, and any state-dependent amplifications $a_i(x)$ of the competitive balance. Systems of type (1), which can be highly nonlinear, arise in many biological applications, such as pattern formation in development [1, 2], the transformation and short-term storage of sensory data in psychophysiology [3–6], competitive interactions among groups or communities is ecology and sociology [1, 7], decision-making in a parallel processor [1, 3, 4], and related areas. Recently considerable interest has been focused on the question: How simple can a system be and still generate "chaotic" behavior? This question is motivated both by a desire to understand turbulence in fluids and by a desire to understand how organized biological interactions can break down under parametric changes [8, 9]. This paper considers the converse

* Supported in part by the National Science Foundation (NSF MCS 77-02958).

question: How complicated can a system be and still generate order? The results herein hold because, despite essentially arbitrary irregularities and nonlinearities in local system design, there exists a powerful symmetry in the global rules that bind together the interacting populations. This symmetry is expressed by the existence of a state-dependent mean competition function, or adaptation level, $c(x)$. It can be caused by the existence of long-range interpopulation interactions that have comparable effects on all populations, but otherwise represent an essentially arbitrary competition. The results herein therefore suggest that a breakdown of symmetry in competitive systems, say due to the existence of asymmetric biases in short-range interpopulation interactions, is a basic cause of oscillations and chaos in these systems; cf. [10, 11], where this fact is illustrated by the voting paradox in Volterra–Lotka systems. There appears to exist a complementary, or trade-off, between how global the adaptation level ("communal understanding") is and how freely local signals ("individual differences") can be chosen without destroying global consensus.

The main result is proved by explicating as a mathematical method a main theme about competitive systems; namely, who is winning the competition? The method keeps track of which population is being maximally enhanced as time goes on. When a different population starts to be maximally enhanced, the system "decides" to enhance the new population, or "jumps" between populations. These jumps are a source of system oscillations. Were the jumps never to cease, approximately periodic or even chaotic behavior could ensue. The theorem guarantees, however, that after a time interval of perhaps very complicated, and even seemingly random oscillations, the decision process is essentially completed, and the system approaches the final pattern in an orderly fashion, even if the jumps do not cease. Reference [12] applies this method to a less general problem and reviews earlier work in this direction.

By studying system "jumps" or "decisions", three themes of general interest emerge. First, one analyses the continuous nonlinear system by studying the discrete series of jumps that it induces. Second, although the continuous system describes parallel interactions, it can be analysed in terms of its serial jumps. Third, the analysis of jumps shows that there exists a sequence of nested "dynamic boundaries" that appear as the system evolves. By this is meant the following. Suppose that $x_i(t) \in [0, B]$ for all $t \geqslant 0$. There exists a sequence of nested partitions $E_{j1}^{(i)} \oplus E_{j2}^{(i)} \oplus \cdots \oplus E_{jm_j}^{(i)}$ of $[0, B]$ into half-open intervals $E_{jk}^{(i)}$, $j = 1, 2, \ldots$, such that after time T_1, $x_i(t)$ remains in some interval $E_{1k_1}^{(i)}$, after time T_2, $x_i(t)$ remains in some interval $E_{2k_2}^{(i)} \subset E_{1k_1}^{(i)}$, and so on. The endpoints of each interval define a "dynamic boundary" beyond which $x_i(t)$ cannot migrate. As the jumps continue, the system "decides" to restrict $x_i(t)$ to ever finite intervals, until as $t \to \infty$ a definite limiting value for $x_i(\infty)$ is established. The existence of these dynamic boundaries is a purely nonlinear effect that arises from the interaction of a nonlinear signal function and a nonlinear mass action law within a competitive geometry.

2. Competitive Systems

The simplest competitive feedback interaction among n populations v_i with activities $x_i(t)$ that obey mass action dynamics is

$$\dot{x}_i = -Ax_i + (B - x_i)[f(x_i) + I_i] - x_i \left[\sum_{k \neq i} f(x_k) + J_i \right], \tag{2}$$

$i = 1, 2,..., n$. System (1) has the following interpretation. Let each population v_i have B excitable sites, of which $x_i(t)$ are excited and $B - x_i(t)$ are unexcited at time t. Let a signal $f(x_i(t))$ be generated by the excited sites of v_i. Then term $-Ax_i$ describes the spontaneous switching-off of excitation at rate A; term $(B - x_i) f(x_i)$ describes the switching on of unexcited sites by a positive feedback signal from v_i to itself; term $-x_i f(x_k)$ describes the switching off of excited sites at v_i by a competitive (or negative) feedback signal from v_k to v_i, $k \neq i$; and terms $(B - x_i) I_i$ and $-x_i J_i$ describe the effects of excitatory input I_i and inhibitory input J_i to v_i. This system was first analysed in [3] in a psychophysiological content. In neural terminology, (2) describes the simplest recurrent (feedback) on-center (excite v_i) off-surround (inhibit all v_k, $k \neq i$) interaction of shunting, or passive membrane (or mass action) dynamics and was used to understand aspects of how input patterns to fields of neocortical feature detectors are transformed before they are stored in short-term memory. The results classify ways the choice of the signal function $f(w)$ influences this transformation. The problem studied was as follows: Suppose that the inputs $(I_1, I_2, ..., I_n)$ and $(J_1, J_2, ..., J_n)$ act before time $t = 0$ to establish an initial pattern of activity $x = (x_1, x_2, ..., x_n)$ at $t = 0$. If these inputs are switched off at time $t = 0$, how does the network

$$\dot{x}_i = -Ax_i + (B - x_i) f(x_i) - x_i \sum_{k \neq i} f(x_k) \tag{3}$$

determine the behavior of $x(t)$ as $t \to \infty$? In particular, do there exist choices of $f(w)$ such that system (3) stores biologically important patterns, yet prevents noise amplification via its positive feedback loops?

This latter problem arose because systems such as (2) solve an ubiquitous biological problem: the noise-saturation dilemma. This dilemma asks how a system of noisy populations with finitely many excitable sites can process continuously fluctuating input patterns? When the input patterns are small, they can get lost in the noise. When the inputs are large, they can saturate the system by exciting all of its excitable sites. Competitive systems such as (2) elegantly solve this problem, by balancing between the two extremes of noise and saturation. The choice of $f(w)$ helps to establish this balance; in particular, sigmoid or S-shaped signal functions $f(w)$ balance between too little vs too much noise suppression. When the competitive balance breaks down, either too much or too little noise suppression can occur, thereby leading to various pathologies, such as "seizures" [3, 4, 11].

Not all competitive systems are as simple as (2). A problem of classification is hereby suggested: How do competitive systems that differ in terms of their mass action dynamics, competitive geometry, and statistics of interpopulation signaling generate different transformations of their initial data while trying to overcome the noise-saturation dilemma in their own way? Papers [1] and [12] discuss this classification problem and review some of the transformations that have already been studied.

Systems (1) are a significant generalization of (3) and of the systems studied in [12]. For example, (1) includes systems of the form

$$\dot{x}_i = -A_i x_i + (B_i - x_i)\,[f_i(x_i) + I_i] - x_i \left[\sum_{k \neq i} f_k(x_k) + J_i\right], \tag{4}$$

in which each population v_i can have different decay rates A_i, different numbers of excitable sites B_i, different signal functions $f_i(x_i)$, and different constant (or tonic) inputs I_i and J_i. System (4) becomes (1) given

$$a_i(x) = x_i, \tag{5}$$

$$b_i(x_i) = x_i^{-1}[B_i f_i(x_i) + I_i] - A_i - I_i - J_i, \tag{6}$$

and

$$c(x) = \sum_{k=1}^{n} f_k(x_k). \tag{7}$$

System (1) also includes generalized Volterra–Lotka systems

$$\dot{x}_i = D_i(x) \left[1 - \sum_{k=1}^{n} f_k(x_k)\, E_{ki}(x)\right], \tag{8}$$

given state-dependent competition coefficients of the form $E_{ki}(x) = F_k(x_k)\, G_i(x_i)$ [7, 10]. Such competition coefficients describe statistically independent couplings between populations v_k and v_i via the statistically independent factors $F_k(x_k)$ and $G_i(x_i)$. An alternative description of this system is that the vector function

$$G(x) = (G_1(x_1), G_2(x_2), \ldots, G_n(x_n))$$

describes a state-dependent preference order among the populations. System (8) reduces to (1) given the identifications

$$a_i(x) = D_i(x)\, G_i(x_i), \tag{9}$$

$$b_i(x_i) = G_i^{-1}(x_i), \tag{10}$$

and

$$c(x) = \sum_{k=1}^{n} f_k(x_k) F_k(x_k). \tag{11}$$

The theorem also holds for such complex nonlinear examples as

$$a_i(x) = x_i^{A_i} \exp\left(\sum_{k=1}^{n} x_k^{B_{ki}}\right), \tag{12}$$

$$b_i(x_i) = \sin(C_i x_i^{D_i} - E_i), \tag{13}$$

and

$$c(x) = \sum_{k=1}^{n} \exp(G_k x_k^{H_k}), \tag{14}$$

where all the coefficients $A_i, \ldots, H_i$ are positive. Indeed, the theorem holds for essentially any physically meaningful choice of the functions $a_i(x)$, $b_i(x_i)$, and $c(x)$, and thereby describes a robust design that guarantees global pattern formation by competitive systems.

3. Global Consensus Theorem

Below are considered systems of the form

$$\dot{x}_i = a_i(x) [b_i(x_i) - c(x)], \tag{1}$$

where $x = (x_1, x_2, \ldots, x_n)$, $i = 1, 2, \ldots, n$, and n is any integer greater than 1. To state the main theorem, the following hypotheses will be needed:

(I) *Smoothness:*

(a) $a_i(x)$ is continuous for $x \geqslant 0$;

(b) $b_i(x_i)$ is either continuous with piecewise derivatives for $x_i \geqslant 0$, or is continuous with piecewise derivatives for $x_i > 0$ and $b_i(0) = \infty$;

(c) $c(x)$ is continuous with piecewise derivatives for $x \geqslant 0$.

(II) *Nonnegativity:*

$$a_i(x) > 0 \quad \text{if} \quad x_i > 0 \text{ and } x_j \geqslant 0, \ j \neq i, \tag{15a}$$

$$a_i(x) = 0 \quad \text{if} \quad x_i = 0 \text{ and } x_j \geqslant 0, \ j \neq i. \tag{15b}$$

Moreover, there exists a function $\bar{a}_i(x_i)$ such that, for sufficiently small $\lambda > 0$, $\bar{a}_i(x_i) \geqslant a_i(x)$ if $x \in [0, \lambda]^n$ and

$$\int_0^{\lambda} \frac{dw}{\bar{a}_i(w)} = \infty. \tag{16}$$

(III) *Boundedness:*

$$\limsup_{w\to\infty} b_i(w) < c(0, 0, \ldots, \infty, \ldots, 0, 0) \tag{17}$$

where "∞" occurs in the ith entry, $i = 1, 2, \ldots, n$.

(IV) *Competition:*

$$\frac{\partial c}{\partial x_k} \geqslant 0, \qquad k = 1, 2, \ldots, n. \tag{18}$$

Given essentially any functions that satisfy (15)–(18), we prove that any initial data $x(0) \geqslant 0$ generates an asymptotic pattern, or decision, $x(\infty)$ such that $0 \leqslant x(\infty) < \infty$. In general, there can exist nondenumerably many limit values that $x(\infty)$ might assume, but the analysis of jumps provides considerable information about the dependence of $x(\infty)$ on $x(0)$. There exists a highly degenerate and unlikely situation, however, in which the possibility of oscillations as $t \to \infty$ has bot been ruled out. Even in this rare case, however, all the signals $b_i(x_i(t))$ have limits as $t \to \infty$. These signals are the only observable data that the states about one another, so that global consensus of observables is always reached. Moreover, even if oscillations in certain $x_i(t)$ persist, they become arbitrarily slow as $t \to \infty$, so that for all practical purposes (e.g., measurements taken over one "generation" at large values of t), limits are always achieved. Whether these slow oscillations ever do occur remains an open problem. To state the theorem in its present form, three further concepts will be introduced.

DEFINITION 1. System (1) is said to obey the *oscillation condition* if there exists a constant b^* and three signal functions, labelled $b_1(w)$, $b_2(w)$, and $b_3(w)$ for definiteness, such that

(V) $b_1(w) = b^*$ for all $w \in W_1$, where W_1 is an interval of positive length within the range of x_1;

(VI) there exist increasing infinite sequences $\{p_{2i}\}$ and $\{v_{2i}\}$ converging at w_2^*, and all in the range of x_2, such that each p_{2i} is a local maximum of b_2, each v_{2i} is a local minimum of b_2, each $b_2(p_{2i}) > b^*$; and $\lim_{k\to\infty} b_2(p_{2k}) = \lim_{k\to\infty} b_2(v_{2k}) = b^*$; and

(VII) there exists a decreasing infinite sequence $\{q_{3i}\}$ converging at w_3^*, and all in the range of x_3, such that $b_3(q_{3i}) < b^*$ for every $i = 1, 2, \ldots$ and $\lim_{k\to\alpha} b_3(q_{3i}) = b^*$.

DEFINITION 2. System (1) achieves *weak global consensus* (or *weak global pattern formation*) if, given any $x(0) \geqslant 0$, all the limits $b_i(x_i(\infty)) = \lim_{t\to\infty} b_i(x_i(t))$ exist, $i = 1, 2, \ldots, n$.

DEFINITION 3. System (1) achieves *strong global consensus* (or *strong global pattern formation*) if, given any $x(0) \geqslant 0$, all the limits $x_i(\infty) = \lim_{j\to\infty} x_i(t)$ exist, $i = 1, 2, \ldots, n$.

THEOREM 1 (Global Consensus). *Any system of form* (1) *whose functions satisfy properties* (I)–(IV) *achieves weak global consensus. Moreover, since each* $b_i(x_i(\infty) = c(x(\infty))$, *any oscillations that might occur become arbitrarily slow as* $t \to \infty$. *Any system of form* (1) *whose functions satisfy properties* (I)–(IV), *and do not satisfy the oscillation conditions* (V)–(VII), *achieves strong global consensus.*

Remarks. Since the oscillation condition requires at least three signals, any 2-dimensional system of type (1) achieves strong global consensus. Moreover, since the oscillation condition requires b_2 to oscillate infinitely often in a compact interval, and b_2^* to identically equal $b_1(w)$ for all $w \in W_1$, essentially any biologically interesting system of type (1) achieves strong global consensus. For example, any system whose signals are built up from arbitrary finite numbers of random factors within each population achieves strong global consensus; cf. [12, Section 2]. Strong global consensus is a generic property. The main facts are summarized by the following corollary.

COROLLARY 1. *Any system of type* (1) *which satisfies properties* (I)–(IV), *and whose signal functions* b_i *possess finitely many local maxima, or intervals of local maxima, within the range of* x_i, *achieves strong global consensus. In particular, if the signals are real-analytic functions, then strong global consensus is achieved.*

The following corollaries are found when Theorem 1 is applied to competitive mass-action networks and to Volterra–Lotka systems.

COROLLARY 2. *Let system* (4) *possess signal functions* $f_i(x_i)$ *that are continuous, monotone nondecreasing, and have piecewise derivatives for* $x_i \in [0, B_i]$, $i = 1, 2, \ldots, n$. *Then weak global consensus is achieved. If moreover,* $x_i^{-1} f_i(x_i)$ *has finitely many local maxima, or intervals of local maxima for* $x_i \in [0, B_i]$, $i = 1, 2, \ldots, n$, *then strong global consensus is achieved.*

Remark. Corollary 2 generalizes the limit theorems in [4].

COROLLARY 3. *Let system* (8) *with* $E_{ki}(x) = F_k(x_k)\, G_i(x_i)$ *have a continuous* $D_i(x)$ *which is positive unless* $x_i = 0$; *continuous functions* $G_i(x_i)$ *that are positive except possibly at* $x_i = 0$, *and which possess piecewise derivatives; continuous functions* $f_k(x_k)$ *and* $F_k(x_k)$ *such that* $f_k(x_k)\, F_k(x_k)$ *is monotone nonincreasing with piecewise derivatives; and let* (15)–(17) *hold with the identifications* (9)–(11). *Then weak global consensus is achieved. If, moreover* $G_i(x_i)$ *has finitely many local minima, or intervals of local minima, within the range of* $x_i(t)$, *then strong global consensus is achieved.*

Proof of Theorem. The theorem will first be proved for the case that all $b_i \equiv b$. This proof can then be adapted to the case of arbitrary b_i. First one notes by (15) and (16) that if $x_i(0) > 0$ then $x_i(t) > 0$ for $t \geqslant 0$ [7]. If $x_i(0) = 0$, population v_i can be deleted from the network without loss of generality. Hence we restrict attention below to the case of positive initial data. The proof consists of three stages: I. Ignition, II. Jump Sequence (or Iterated Local Decisions), and III. Coda (or Global Consensus).

I. Define the functions

$$M_i(t) = b(x_i(t)) - c(x(t)) \tag{19}$$

and

$$M(t) = \max\{M_k(t) : k = 1, 2, \ldots, n\}. \tag{20}$$

"Ignition" means that either $M(t) \leqslant 0$ for all $t \geqslant 0$, or that there exists a $t = T$ such that

$$M(T) \geqslant 0 \qquad \text{implies} \qquad M(t) \geqslant 0 \qquad \text{for } t \geqslant T. \tag{21}$$

To prove (21) it suffices to show that if at any time $t = S$, $M(S) = 0$, then $\dot{M}(S) \geqslant 0$. By (19), if $M(S) = M_i(S)$, then

$$\dot{M}(S) = b'(x_i(S))\, \dot{x}_i(S) - \sum_{k=1}^{i} \frac{\partial c}{\partial x_k}(x(S))\, \dot{x}_k(S).$$

Since $\dot{x}_i(S) = 0 \geqslant \dot{x}_k(S)$, $k = 1, 2, \ldots, n$, (1) and (18) imply that $\dot{M}(S) \geqslant 0$.

By the ignition property, either all $\dot{x}_i \leqslant 0$ for $t \geqslant 0$, or there exists a time $t = T$ after which some x_i, perhaps a different one at different times, is always increasing. In the former case, all $x_i(\infty)$ exist, since all x_i are monotone decreasing and, by (16), all x_i are bounded below by 0. It remains only to consider the latter case. Below we therefore assume that $M(0) \geqslant 0$ without loss of generality.

II. By (16) and (17), there exists a $B > 0$ such that $x_i(t) \in [0, B]$ for all $i = 1, 2, \ldots, n$ and $t \geqslant 0$. Consider the graph of $h(w)$ in the interval $[0, B]$. Decompose the graph into *ascending slopes* A_i and *descending slopes* D_i as follows. Consider successively larger w values, $w \geqslant 0$, until for some $w = W$, $b'(W) \neq 0$. Suppose for definiteness that $b'(W) > 0$. Then the ascending slope A_1 is the maximal connected set of w values, including $w = 0$, wherein $b'(w) \geqslant 0$. The descending slope D_1 is the maximal connected set in $[0, B] \backslash - A_1$ that is contiguous to A_1 wherein $b'(w) \leqslant 0$. The ascending slope A_2 is the maximal connected set in $[0, B] - (A_1 \cup D_1)$ that is contiguous to D_1 wherein $b'(w) \geqslant 0$. And so on. Also define $H_j = A_j \cup D_j$, to be the jth *hill* in the graph of $b(w)$. Let $p_j = \max\{w : w \in A_j\}$ be the *peak* of H_j, and $v_j = \max\{w : w \in D_j\}$ be the *valley* of H_j. Also let $P_j = b(p_j)$ be the *height* and $V_j = b(v_j)$ be the *depth* of H_j. Speaking intuitively, $b(x_i(t))$ is the height of x_i at time t, and P_j is the height of the jth hill peak.

A *jump* is said to occur from i to j at time $t = T$ if there exist times S and U such that $M(t) = M_i(t)$ for $S \leqslant t < T$ and $M(t) = M_j(t)$ for $T \leqslant t < U$. The set of *jump variables* $J = \{i: M(t) = M_i(t)$ for some $t \geqslant 0\}$. The set of *persistent jump variables* $J^\infty = \{i: M(t) = M_i(t)$ for some $t = t_{ik}$, $k = 1, 2, \ldots$, where $\lim_{k \to \infty} t_{ik} = \infty\}$. The set of *j-persistent jump variables* $J_j^\infty = \{i: M(t) = M_i(t)$ and $x_i(t) \in H_j$ for some $t = t_{ik}$, $k = 1, 2, \ldots$, where $\lim_{k \to \infty} t_{ik} = \infty\}$. When we say that a jump occurs from an $i \in J_j^\infty$ at time $t = T$, we imply that $x_i(T) \in H_j$ at time T. Otherwise expressed, let $I(t)$ be the index i such that $M_i(t) = M(t)$; that is, $M(t) = M_{I(t)}(t)$ for $t \geqslant 0$. Also define $y(t) = x_{I(t)}(t)$ for $t \geqslant 0$. To say that a jump occurs from an $i \in J_j^\infty$ at time $t = T$ means that $y(t) \in H_j$ just before $I(t)$ changes value at time $t = T$.

To test when a jump will occur from i to j at time $t = T$, suppose that $M_i(T) = M_j(T) = M(T)$. A jump will occur from i to j if $\dot{M}_j(T) > \dot{M}_i(T)$. Since

$$\dot{M}_j(T) = b'(x_j(T))\, a_j(x(T))\, M(T) - \dot{c}(T) \tag{22}$$

and

$$\dot{M}_i(T) = b'(x_i(T))\, a_i(x(T))\, M(T) - \dot{c}(T), \tag{23}$$

where $M(T) \geqslant 0$, a jump occurs from i to j if

$$b'(x_j(T))\, a_j(x(T)) > b'(x_i(T))\, a_i(x(T)). \tag{24}$$

Since $a_i(x(T))$ and $a_j(x(T))$ are nonnegative, a jump can never occur from an ascending slope to a descending slope.

Case 1. *Finitely Many Jumps*. If only finitely many jumps occur, then after a finite amount of time goes by, there exists some i, say $i = 1$, such that thereafter $M_1 = M \geqslant 0$. By (1), (15), and (16), $\dot{x}_1 \geqslant 0$, so that x_1 is monotone increasing. By (17), x_1 is also bounded above. Hence the limit $x_1(\infty)$ exists. This limit lies on some ascending or descending slope. Suppose that it lies on an ascending slope, say A_K for definiteness. We will now prove that $\lim_{t \to \infty} c(x(t))$ exists and equals $b^* = b(x_1(\infty))$.

Suppose not. Whenever $\dot{x}_1 \geqslant 0$, $b(x_1) \geqslant c(x)$. Thus if $c(x(t))$ does not converge to b^*, there exists a sequence of increasing times t_k with $\lim_{k \to \infty} t_k = \infty$ such that

$$b^* - c(x(t_k)) \geqslant 2\epsilon, \qquad k = 1, 2, \ldots. \tag{25}$$

Now note that $c(x(t))$ is uniformly continuous for $t \geqslant 0$. This is true because $x(t)$ remains in a compact set R; $c(x)$ is continuous, and hence uniformly continuous for $x \in R$; and, by (1), there exists a constant M, $0 < M < \infty$, such that $|\dot{x}_i(t)| \leqslant M$ for all $t \geqslant 0$ and $i = 1, 2, \ldots, n$. By the uniform continuity of $c(x)$, for every $\epsilon > 0$ there exists a $\delta > 0$ such that $x, y \in R$ and $|x - y| < \delta$ implies $|c(x) - c(y)| < \epsilon$. By the mean-value theorem, $|x(t) - x(s)| \leqslant M\,|t - s|$. Consequently for $|t - s| \leqslant \delta M^{-1}$, $|c(x(t)) - c(x(s))| < \epsilon$, which proves the assertion. Thus by (25) there exists a $\beta > 0$ and a sequence of nonoverlapping intervals $[Y_k, Z_k]$ such that $Z_k - Y_k \geqslant \beta$ for $k = 1, 2, \ldots$, and

$$b^* - c(x(t)) \geqslant \epsilon, \qquad t \in \bigcup_{k=1}^{\infty} [Y_k, Z_k]. \tag{26}$$

Furthermore, since x_1 is monotone increasing, (15) along with the continuity of $a_1(x)$ for $x \in [0, B]^n$, shows that there exists a $\gamma > 0$ such that

$$a_1(x) \geqslant \gamma \tag{27}$$

when x_1 is close to $x_1(\infty)$. By (1), (26), and (27)

$$\dot{x}_1(t) \geqslant \epsilon\gamma \qquad \text{for } t \in \bigcup_{k=1}^{\infty} [Y_k, Z_k]. \tag{28}$$

Since x_1 is monotone increasing, (28) implies that $x_1(\infty) = \infty$, which is impossible. This contradiction proves that

$$\lim_{t\to\infty} c(x(t)) = b^*. \tag{29}$$

Lemma 1 below completes the proof.

Case 2. *Infinitely Many Jumps.* If jumps do not cease after a finite amount of time, delete from consideration all hills H_j whose J_j^∞ sets are empty; that is, let the process continue until no jumps ever again occur from hills with empty J_j^∞ sets. Relabel the time scale so that $t = 0$ after all such jumps have occurred. Relabel the hills H_j with nonempty J_j^∞ sets so that $j \leqslant k$ iff $P_j \geqslant P_k$.

The idea of the argument below is to show that given any hill, after a sufficient amount of time goes by, all variables get trapped either to the right or to the left of its peak. Since this is true for any hill, eventually each variable gets trapped in the "bowl" between a contiguous descending slope and ascending slope; in particular, eventually no jump variable can cross over a peak from an ascending slope to a descending slope. In general, the x_i do not get trapped in their bowls all at once. First they get trapped in some interval $E_{1k_1}^{(i)} = [u_{1k_1}^{(i)}, v_{1k_1}^{(i)})$ whose boundary values have the largest peak heights $P_1 = b(u_{1k_1}^{(i)}) = b(v_{1k_1}^{(i)})$; later they get trapped in some interval $E_{2k_2}^{(i)} = [u_{2k_2}^{(i)}, v_{2k_2}^{(i)})$ whose boundary values have the largest or the next-to-largest peak heights $b(u_{2k_2}^{(i)})$ and $b(v_{2k_2}^{(i)})$; and so on. These intervals are the nested dynamic boundaries that were discussed in Section 1. The "bowls" are the final set of dynamic boundaries that are established. After the variables get trapped in their bowls, the decision process is essentially complete. Thereafter, $(d/dt)b(y(t))$ changes sign at most once, so that at all large times $b(y(t))$ is monotonic. This Lyapunov-like behavior is then used to complete the proof.

The above heuristic description tacitly assumes that there exists a finite series of peak heights in the graph of $h(w)$. In any "physical" example, this will be true. In general, however, certain peak heights can be limit points of other peak heights. The proof is extended to such cases by using the fact that these "infinitely wiggly hills are arbitrarily small."

First we will consider the physically important case. Suppose that the successive peak heights $P_1 > P_2 > P_3 > \cdots > P_L$ form a discrete series such that only finitely many hills attain any given height. Below we will assume for definiteness that one hill H_i has height P_i, but the argument immediately generalizes to the case in which finitely many hills share the same height.

Consider hill H_1. To start, suppose that there exists a $t = S_1$ when

$$x_i(S_1) \in [p_1, B] \qquad \text{for all } i \in J_1^{\infty}. \tag{30}$$

Then

$$x_i(t) \in [p_1, B] \qquad \text{for all } i = 1, 2, \ldots, n \quad \text{and} \quad t \geqslant S_1. \tag{31}$$

This is true because, if some $x_i(t) = P_1$ at a time $t > S_1$, then $b(x_i(t)) = P_1 \geqslant b(x_j(t))$ for all $j \neq i$. Thus $M_i(t) = M(t) \geqslant 0$ and $\dot{x}_i(t) \geqslant 0$, which keeps $x_i(t) \in [p_1, B]$. Given that (31) holds, we now show that either jumps occur only among $i \in J_1^{\infty}$, while $x_i \in D_1$, or there exists a $t = T_1$ such that

$$b(x_i(t)) \leqslant P_2 \qquad \text{for all } i = 1, 2, \ldots \quad \text{and} \quad t \geqslant T_1. \tag{32}$$

This alternative is true because no jump can occur from an $i \in J_1^{\infty}$ to any hill H_j, $j \neq 1$, while $b(x_i) \geqslant P_2$. Since the $i \in J_1^{\infty}$ are *persistent* jump variables, either jumps continue to occur among the $i \in J_1^{\infty}$ on D_1, or eventually $y(t)$ enters the set $\bigcup_{j \geqslant 2} H_j$. In the former case, the heights $b(x_i)$ at successive jump times are monotone decreasing, since $b'(w) \leqslant 0$ for $w \in D_1$, and whenever $M_i(t) = M(t) \geqslant 0$, $\dot{x}_i(t) \geqslant 0$. Thus there exists a limiting height b_1^* to which the jump heights converge. In the latter case, there exists some time $t = T_1$ at which $y(T_1) \in \bigcup_{j \geqslant 2} H_j$. Consequently

$$b(x_i(T_1)) \leqslant P_2 \qquad \text{for all } i = 1, 2, \ldots, n. \tag{33}$$

It follows from (33) that (32) holds. To see this, suppose that time $t = U_1$ is the first time that some $b(x_i(t)) = P_2$. Then $M_i(U_1) = M(U_1) \geqslant 0$, so that $\dot{x}_i(U_1) \geqslant 0$. Moreover, by (30), either $x_i(U_1) = p_2$ or $x_i(U_1) \in D_1$, so that $b'(x_i(U_1)) \leqslant 0$. In both cases, $(d/dt)\, b(x_i(U_1)) = b'(x_i(U_1))\, \dot{x}_i(U_1) \leqslant 0$, which proves (33).

To summarize the above argument: If (30) holds at some $t = S_1$, then either b_1^* exists, or there exists a $t = T_1 \geqslant S_1$ such that (32) holds. It will be seen below how to complete the proof if b_1^* exists. Hence suppose that (31) and (32) hold.

Now consider H_2 at times $t \geqslant T_1$. Suppose that there exists a $t = S_2 \geqslant T_1$ such that

$$x_i(S_2) \in [p_2, B] \qquad \text{for all } i \in J_2^{\infty}. \tag{34}$$

Then

$$x_i(t) \in [p_2, B] \quad \text{for all } i = 1, 2, \ldots, n \quad \text{and} \quad t \geqslant S_2. \tag{35}$$

Property (35) follows from (34), since if some $x_i(t) = p_2$ at a time $t > S_2$, then by (32) $b(x_i(t)) = P_2 \geqslant b(x_j(t))$ for all $j \neq i$. Thus $M_i(t) = M(t) \geqslant 0$ and $\dot{x}_i(t) \geqslant 0$, which keeps $x_i(t) \in [p_2, B]$. From properties (31) and (35), we will conclude that either a limiting height b_2^* analogous to b_1^* exists, or there exists a time $t = T_2$ such that

$$b(x_i(t)) \leqslant P_3 \quad \text{for all } i = 1, 2, \ldots, n \quad \text{and} \quad t \geqslant T_2. \tag{36}$$

This alternative is true, because by (31) and (35), so long as jumps occur at heights that exceed P_3, they can only occur in $D_1 \cup D_2$. The heights at successive jump times are then monotone decreasing, whence b_2^* exists. If a jump occurs at a height $\leqslant P_3$ at some time $t = T_2$, then (36) holds. This is seen by considering the first time $t = U_2$ at which some $b(x_i(U_2)) = P_3$. Then $M_i(U_2) = M(U_2) \geqslant 0$, and by (32) and (35), either $x_i(U_2) = p_3$ or $x_i(U_2) \in D_1 \cup D_2$. In both cases, $(d/dt)\, b(x_i(U_2)) \leqslant 0$, thereby proving (36).

This argument can now be continued on hills of successively shorter heights to derive the following alternative after considering hill H_m: Either a limiting height b_m^* exists, or there exists a $t = T_m$ such that

$$b(x_i(t)) \leqslant P_{m+1} \quad \text{for all } i = 1, 2, \ldots, n \quad \text{and} \quad t \geqslant T_m. \tag{37}$$

The argument is continued until we reach the first hill H_r (possibly $r = 1$) on which there is no time at which all $x_i(t) \in [p_r, B]$ for all $i \in J_r^\infty$. In this case, we will conclude that there exists a time $t = S_r$ such that, for each $i = 1, 2, \ldots, n$,

$$x_i(t) \in [0, p_r] \quad \text{for } t \geqslant S_r \tag{38a}$$

or

$$x_i(t) \in [p_r, B] \quad \text{for } t \geqslant S_r. \tag{38b}$$

This follows from the fact that if $x_i(S_r) \in [p_r, B]$, then $x_i(t) \in [p_r, B]$ for $t \geqslant S_r$. This conclusion is due to (37), since if $x_i(S_r) = p_r$ with $S_r \geqslant T_{r-1}$, then $M_i(S_r) = M(S_r) \geqslant 0$ and thus $\dot{x}_i(S_r) \geqslant 0$.

Given that (38) is true, it follows that either a limiting height b_r^* exists, or there exists a $t = T_r$ such that

$$b(x_i(t)) \leqslant P_{r+1} \quad \text{for } i = 1, 2, \ldots, n \quad \text{and} \quad t \geqslant T_r. \tag{39}$$

This can be shown as follows. Consider any jump sequence that starts with $y(t) \in A_r$ and $b(y(t)) > P_{r+1}$ at some time $t \geqslant S_r$. By (37) and (38), if such a jump sequence does not terminate, then $y(t) \in A_r$ at all future times. This is because, by (38), the jump variables on A_r cannot cross to D_r, and the only jump variables with $b(x_i(t)) > P_{r+1}$ that are not on A_r are on $\bigcup_{j=1}^{r} D_j$, and no jump can go from an ascending slope to a descending slope, by (24). If the jump sequence does not terminate, then a limiting height exists since the successive jump heights are monotone increasing on A_r. If the jump sequence does terminate, then there is no persistent jump variable $x_i(t) \in A_r$ such that $b(x_i)) > P_{r+1}$ after some time elapses. In this latter case, continue the argument on the set $\bigcup_{j=1}^{r} D_j$ to show that either a limiting height exists, or (39) holds.

Now consider hill H_{r+1}. It is clear that the above argument can be repeated to conclude that either a limiting height exists, or there exists a $t = T_{r+1}$ such that

$$b(x_i(t)) \leqslant P_{r+2} \qquad \text{for all } i = 1, 2, \ldots, n \quad \text{and} \quad t \geqslant T_{r+1}. \tag{40}$$

If there are only finitely many hills on the interval $[0, B]$, then the above argument can be applied to each successively shorter hill until all hills are exhausted. Thus, after a finite amount of time goes by, either a limiting height exists, or no jump variable can cross a peak. In the latter case, each variable eventually gets trapped in a "bowl" between a contiguous descending slope and ascending slope.

Consider the case in which all variables eventually get trapped in their bowls. After this happens, what kinds of jumps can occur? Jumps can occur among descending slopes. The successive jumps then occur at successively lower heights. If this goes on indefinitely, then a limiting height exists. Jumps could instead occur among ascending slopes. The jumps then occur at successively higher heights. If this goes on indefinitely, then a limiting height again exists. Finally jumps can occur first among descending slopes until a jump occurs to some ascending slope. Thereafter only jumps among ascending slopes can occur, because no jump variable can continuously cross over a peak from an ascending slope to a descending slope, and no jump from an ascending slope can ever occur to a descending slope. In all cases, after a finite amount of time goes by, the successive jump heights converge monotonically to a limiting height. Indeed $b(y(t))$ is monotonic at all large times.

III. To complete the proof, given that only finite many hills exist, we must consider the case in which a limiting height is approached either on ascending slopes only, or on descending slopes only. Suppose for definiteness that there is a jump series among ascending slopes as $t \to \infty$; then $b(y(t))$ is monotone nondecreasing, and the successive heights at which jumps occur monotonically increase to b^*. We will now use this fact to prove that $\lim_{t\to\infty} c(x(t))$ exists and equals b^*. Then the proof can be completed by using Lemma 1 below.

The argument proceeds by contradiction. Suppose that $c(x(t))$ does not converge to b^*. Since $M(t) \geqslant 0$, it follows that

$$b^* \geqslant b(y(t)) \geqslant c(x(t)). \tag{41}$$

Thus there exists an $\epsilon > 0$ and a sequence of increasing times $\{t_k\}$ with $\lim_{k\to\infty} t_k = \infty$ such that

$$b^* - 3\epsilon \geqslant c(x(t_k)), \qquad k = 1, 2, \ldots \tag{42}$$

Since $c(x(t))$ is uniformly continuous for $t \geqslant 0$, there exists a $\beta > 0$ and a sequence of nonoverlapping intervals $[Y_k, Z_k]$ such that $Z_k - Y_k \geqslant \beta$ for $k = 1, 2, \ldots$ and

$$b^* - 2\epsilon \geqslant c(x(t)), \qquad t \in \bigcup_{k=1}^{\infty} [Y_k, Z_k]. \tag{43}$$

Given any k, choose a j such that $y(Y_k) = x_j(Y_k)$. Denote this j by $j = J(k)$; that is, $y(Y_k) = x_{J(k)}(Y_k)$. For k sufficiently large and $j = J(k)$,

$$b^* - b(x_j(Y_k)) \leqslant \epsilon/2. \tag{44}$$

Each function $b(x_j(t))$ is uniformly continuous in $t \geqslant 0$ for the same reason that $c(x(t))$ is. Consequently there exists a γ, $0 < \gamma \leqslant \beta$, such that for *all* large k and $j = J(k)$,

$$0 \leqslant b^* - b(x_j(t)) \leqslant \epsilon \quad \text{if} \quad t \in [Y_k, Y_k + \gamma]. \tag{45}$$

Thus by (43) and (45),

$$b(x_j(t)) - c(x(t)) \geqslant \epsilon, \qquad t \in [Y_k, Y_k + \gamma]. \tag{46}$$

Since also each $x_j(t)$ is bounded away from zero for $t \in [Y_k, Y_k + \gamma]$, (15) implies that there exists a $\delta > 0$ such that

$$a_j(x(t)) \geqslant \delta, \qquad t \in [Y_k, Y_k + \gamma]. \tag{47}$$

Putting (46) and (47) together shows, by (1), that

$$\dot{x}_j(t) \geqslant \delta\epsilon, \qquad t \in [Y_k, Y_k + \gamma], \tag{48}$$

or that

$$x_j(t) \geqslant x_j(Y_k) + \delta\epsilon(t - Y_k), \qquad t \in [Y_k, Y_k + \gamma]. \tag{49}$$

Thus $x_j(t)$ increases at an at least linear rate, that is independent of k, across an interval of length at least $\delta\epsilon\gamma$, when $t \in [Y_k, Y_k + \gamma]$. Denote this interval by Q_k.

Now choose an x_j such that $j = J(k)$ at infinitely many values of k. This x_j traverses infinitely many of the intervals Q_k. However, x_j is bounded for all $t \geqslant 0$. Thus the sets Q_k such that $j = J(k)$ must overlap as $k \to \infty$. More precisely, there exists an infinite subsequence $k_1 < k_2 < k_3 < \cdots$ of k's such that $j = J(k)$ and the closed interval $R_j = \bigcap_{m=1}^{\infty} Q_{k_m}$ has positive length. Since $\lim_{t\to\infty} b(y(t)) = b^*$, (45) holds for *any* $\epsilon > 0$ if k is chosen sufficiently large Consequently given any $\epsilon > 0$,

$$0 \leqslant b^* - b(w) \leqslant \epsilon \qquad \text{if} \qquad w \in R_j. \tag{50}$$

That is, $b(w) = b^*$ if $w \in R_j$. Thus at any sufficiently large time Y_k such that $j = J(k)$, it follows that

$$b(y(Y_k)) = b(x_j(Y_k)) = b^*. \tag{51}$$

In other words, the maximal variable attains the maximal height b^* at a *finite time* Y_k. Thereafter, no further jumps can occur, since $b(y(t))$ is monotone increasing. Thus all jumps cease after a finite time, and the proof can be completed as in Case 1, which shows that, indeed, $\lim_{j\to\infty} c(x(t)) = b^*$. This completes the proof when all $b_i \equiv b$, except for the application of Lemma 1.

Given arbitrary b_i, the same proof goes through because the original proof is "local" and "exhaustive". By this we mean the following. The proof when all $b_i \equiv b$ orders the hills by height and considers each hill in its turn. The position in $[0, B]$ of any given hill relative to other hills in the graph is immaterial. One simply has to define sets of persistent jump variables *on the hill*. In this sense, the proof is "local", because one can worry about one hill at a time. The proof is "exhaustive", because by considering all the hills, one can prove either that a limiting height exists, or that the variables eventually get trapped in their bowls. The latter case then also implies that a limiting height exists. When the behavior near the limiting height is considered, all that matters is that each limiting variable is on an ascending slope, or that each limiting variable is on a descending slope. *Where* these slopes are to be found is irrelevant—again, a "local" condition is studied.

To prove the theorem given arbitrary b_i, it suffices to consider all the hills on the graphs of all the b_i. Order these hills by height, and then proceed exactly as in the $b_i \equiv b$ case. After one exhausts all these hills, one automatically exhausts all the hills in the graph of each b_i, so the proof can be completed in the same fashion.

Now we indicate how the above arguments can be adapted to cases in which there are infinitely many hills on $[0, B]$. Order the hills so that $P_1 \geqslant P_2 \geqslant P_3 \cdots$. By hypothesis $b(w)$ is either continuous on $[0, B]$, hence uniformly continuous; or $b(0) = \infty$ and $b(w)$ is uniformly continuous on $[\delta, B]$ no matter how small $\delta > 0$ is chosen. Thus given any interval $[\delta, B]$ and any $\epsilon > 0$, there exists in $[\delta, B]$ only finitely many hills H_i such that $P_i - V_i \geqslant \epsilon$.

First consider the case in which no height P_k or depth V_k is a limit point of other heights or depths (including the case where $P_1 = +\infty$). Then $\lim_{k\to\infty}(P_k - V_k) = 0$. Since also the sequence $\{P_k\}$ is monotone decreasing and bounded below, $\lim_{k\to\infty} P_k = \lim_{k\to\infty} V_k$. Consequently, as $t \to \infty$ either the jumps are restricted to finitely many hills, which can be treated as above, or (40) holds for all $r \geqslant 0$, and

$$\lim_{t\to\infty} b(y(t)) = \lim_{k\to\infty} P_k. \tag{52}$$

By(52), a limiting height exists, and certain x_i must keep moving onto shorter hills as $t \to \infty$. Since these hills approach a limiting width of zero, and since the x_i are continuous, certain limits $x_i(\infty)$ exist. The proof can now be continued by adapting the argument in (25)–(29). The only change is that (27) holds because $x_i(t)$ is forced away from zero by being driven onto ever shorter hills. Given the existence of $c(x(\infty))$, it is proved that the limits of the remaining x_i exist by using the fact that no P_k or V_k is a limit point to drive all these x_i onto a definite ascending or descending slope, but not a peak or valley, as $t \to \infty$.

Now consider a case in which the maximal height P_1 is a limit point of other heights. Given an infinite sequence $\{w_k\}$ such that P_1 is a limit point of $\{b(w_k)\}$, then a cluster point $\bar{w}$ of $\{w_k\}$ exists and satisfies $b(\bar{w}) = P_1$ due to the continuity of $b(w)$. Consider the case in which finitely many such cluster points $\bar{w}_{11}$, $\bar{w}_{12}$,..., $\bar{w}_{1r_1}$ exist. By the uniform continuity of $b(w)$, each $\bar{w}_{1k}$ is a limit of hills that become arbitrarily small as $w \to \bar{w}_{1k}$. Intuitively speaking, $\bar{w}_{1k}$ is the peak of "a hill with infinitely small wiggles" near $\bar{w}_{1k}$. In particular, the depths of the hills close to any $\bar{w}_{1k}$ approach P_1 at a uniform rate as $w \to \bar{w}_{1k}$. Thus if $b(y(t))$ gets close to P_1 in value, it remains close to P_1 in value unless $y(t)$ eventually crosses some $\bar{w}_{1k}$ value and descends a sequence of hills. More precisely, either $\lim_{t\to\infty} b(y(t)) = P_1$ or there exists an $\epsilon > 0$ and a T_ϵ such that $b(x_i(t)) \leqslant P_1 - \epsilon$ for $i = 1, 2,..., n$ and all $t \geqslant T_\epsilon$. In the latter case, let P_2 be the maximal hill height $\leqslant P_1 - \epsilon$ among all the hills on which $y(t)$ sits at some $t \geqslant T_\epsilon$. Repeat the above argument, assuming that there are finitely many cluster points $\bar{w}_{21}$, $\bar{w}_{22}$,..., $\bar{w}_{2r_2}$ such that $b(\bar{w}_{2k}) = P_2$. Again either $\lim_{t\to\infty} b(y(t)) = P_2$ or there exists an $\epsilon > 0$ and a T_ϵ such that $b(x_i(t)) \leqslant P_2 - \epsilon$ for $i = 1, 2,..., n$ and $t \geqslant T_\epsilon$. This argument is now repeated iteratively. The main point is that finitely many cluster point peaks can be treated like finitely many peaks without cluster points. In the case where infinitely many cluster points to a given height exist, one notes that only in regions where these cluster points are isolated can a jump variable possibly escape to a distinct height; in effect, since only finitely many $P_k - V_k$ values are not smaller than any prescribed $\epsilon > 0$, the argument can again be reduced essentially to the finite case. More precisely, given any limiting height b^*, use the boundedness and uniform continuity of b_j to cover all cluster points of b_j with finitely many intervals in which $b^* \geqslant b_j \geqslant b^* - \epsilon$. These intervals surround the isolated cluster points of b_j as well as finitely many cluster points around which other cluster points of b_j cluster. Using such a finite covering at every stage of the argument, argue as in the case of finitely many cluster points to conclude that the limit $b^* = \lim_{t\to\infty} b_{I(t)}(y(t))$ exists in the general case.

It remains only to prove that $\lim_{t\to\infty} c(x(t))$ exists and equals b^*, since then the proof can be completed using Lemma 1. For convenience, the notation $B(t) = b_{I(t)}(y(t))$ will be used to denote the maximum height at any time. The proof proceeds by supposing that $c(x(t))$ does not approach b^*. Then (42)–(51) follow as before. In particular, there exists a j and a sequence of times $t_m = Y_{k_m}$ with $\lim_{m\to\infty} t_m = \infty$ such that

$$b_j(x_j(t_m)) = B(t_m) = b^*, \qquad m = 1, 2,... \tag{53}$$

The main new difficulty is that the x_i's need not be trapped on only descending slopes or only ascending slopes, so $B(t)$ is not necessarily monotonic at all large times; that is, $b^* - B(t)$ can change sign at arbitrarily large times. Suppose that this happens. Then there exists a sequence $\{s_m\}$ with $t_m < s_m < t_{m+1}$ such that

$$B(s_m) > b^*, \qquad m = 1, 2, \ldots \tag{54}$$

despite the fact that (53) holds. How can this happen? By (53) and (54), $B(t_m) < B(s_m)$ and $B(t_{m+1}) < B(s_m)$ despite the fact that $t_m < s_m < t_{m+1}$. In order for the maximal height to increase and then decrease, *some* variable x_i must go over a hill: The increase requires the maximal variable to be on an ascending slope, and the decrease requires the maximal variable to be on a descending slope; since no jump can occur from an ascending slope to a descending slope, the maximal variable must go over a hill. Moreover, since $\lim_{t\to\infty} B(t) = b^*$, given any $\epsilon > 0$ there exists a T_ϵ such that

$$b^* + \epsilon \geqslant B(t) \qquad \text{for all } t \geqslant T_\epsilon. \tag{55}$$

Consequently, the maximal variable cannot go over a hill whose height exceeds $b^* + \epsilon$ after time $t = T_\epsilon$. This forces the maximal variable to go over infinitely many hills as $t \to \infty$, since no matter how much higher the hill is than b^*, after some finite time goes by, ϵ can be chosen so small that the hill is too high to be the one that the maximal variable goes over after that time. Since there are only finitely many hills whose width and depth exceed any fixed size $\delta > 0$, by waiting sufficiently long, the maximal variable is driven to hills of arbitrarily small width and depth. Now choose an i such that $b_i(x_i(s_m)) = B(s_m)$ at infinitely many values of m. The above argument shows that $x_i(t)$ eventually gets driven onto, and trapped within, arbitrarily small hills as $t \to \infty$. Consequently, the limit $x_i(\infty)$ exists and $b_i(x_i(\infty)) = b^*$. The argument in Case 1 can now easily be adapted to prove that $\lim_{t\to\infty} c(x(t)) = b^*$.

In all cases, it has now been proved that

$$\lim_{t\to\infty} c(x(t)) = b^*. \tag{29}$$

This fact is now used to complete the proof using the following Lemma.

LEMMA 1. *If system* (1) *obeys properties* (I)–(IV), *and limit* (29) *exists, then the limits* $b_i(x_i(\infty))$ *exist and equal* $c(x(\infty))$, $i = 1, 2, \ldots, n$. *If, moreover, the oscillation condition does not hold, then the limits* $x_i(\infty)$ *exist*, $i = 1, 2, \ldots, n$.

Proof. By (29), given any $\epsilon > 0$, there exists a T_ϵ such that

$$|\, c(x(\infty)) - c(x(t))| < \epsilon \quad \text{if} \quad t \geqslant T_\epsilon. \tag{56}$$

Suppose that

$$b_i(x_i(t)) > c(x(\infty)) + \epsilon \tag{57}$$

at some time $t \geqslant T_\epsilon$. Then, by (56), $\dot{x}_i(t) > 0$. Consequently, $x_i(t)$ monotonically increases towards the limit $x_i(\infty)$ unless there is a time $t \geqslant T_\epsilon$ at which

$$b_i(x_i(t)) = c(x(\infty)) + \epsilon. \tag{58}$$

There must be such a time, since otherwise $b_i(x_i(\infty)) - c(x(\infty)) \geqslant \epsilon$, and thus $x_i(\infty) = \infty$, which is impossible. Consider the first time $t = T \geqslant T_\epsilon$ at which (58) holds after an interval of times during which (57) holds. At this time,

$$0 \geqslant \frac{d}{dt} b_i(x_i(T)) = b_i'(x_i(T))\, \dot{x}_i(T). \tag{59}$$

By (56), $\dot{x}_i(T) \geqslant 0$, and thus $b'(x_i(T)) \leqslant 0$, so that $x_i(T)$ is on a descending slope, or plateau, of b_i. Because (56) holds for all $t \geqslant T$, it follows that $x_i(t) \geqslant x_i(T)$ for all $t \geqslant T$, since whenever $x_i(t) = x_i(T)$, (56) and (58) imply that $\dot{x}_i(t) \geqslant 0$. The above argument can now be iterated. After (57) holds during some time interval, there must be a time when (58) holds while x_i is on a descending slope or plateau. Letting T_{ij} be the jth time at which (58) holds after (57) holds, it follows that $x_i(t) \geqslant x_i(T_{ij})$ for $t \geqslant T_{ij}$, and that $x_i(T_{i1}) < x_i(T_{i2}) < \cdots$. If there are only finitely many T_{ij}'s, then there exists a time U_ϵ such that

$$b_i(x_i(t)) \leqslant c(x(\infty)) + \epsilon \qquad \text{for all } t \geqslant U_\epsilon. \tag{60}$$

Otherwise, there must exist infinitely many hills in the graph of b_i. Since x_i is bounded, and bounded away from 0, and b_i is continuous on the compact set within which x_i fluctuates, b_i is also uniformly continuous. Consequently, given any δ, there exist only finitely many hills in the graph of b_i whose width or depth is greater than δ. If x_i traverses infinitely many hills on which (58) holds after (57) holds, it is eventually forced onto arbitrarily small hills whose heights P_k and depths V_k both approach $c(x(\infty)) + \epsilon$, by (58). Thus at *all* large times, $b_i(x_i(t)) - c(x(t)) \geqslant \epsilon/2$, which again implies the impossible conclusion that $x_i(\infty) = \infty$. Consequently, given any $\epsilon > 0$, there exists a time U_ϵ such that (60) holds. A similar argument with reversed inequalities allows us to conclude that there exists a time V_ϵ such that

$$b_i(x_i(t)) \geqslant c(x(\infty)) - \epsilon \qquad \text{for all } t \geqslant V_\epsilon. \tag{61}$$

Since both (60) and (61) hold for any $\epsilon > 0$,

$$\lim_{t\to\infty} b_i(x_i(t)) = c(x(\infty)). \tag{62}$$

The same argument holds for all $i = 1, 2, \ldots, n$. Consequently, system (1) achieves weak global consensus. Moreover, since all the limits $b_i(x_i(\infty))$ exist and equal $c(x(\infty))$, it follows that all the limits $\dot{x}_i(\infty)$ exist and equal 0. Thus whatever oscillations occur in the $x_i(t)$ become arbitrarily slow as $t \to \infty$.

From (62), it follows that the limit $x_i(\infty)$ exists unless there exists a nontrivial interval W_i of values throughout which $b_i(w) = b^*$. If $x_i(t) \in W_i$ at all large times, then $x_i(t)$ might oscillate back and forth across W_i as $t \to \infty$ without contradicting the fact that $b_i(x_i(\infty)) = b^* = c(x(\infty))$. By (1), $x_i(t)$ can oscillate back and forth across W_i at arbitrarily large times only if $c(x(t))$ oscillates above and below b^* at arbitrarily large times. Since $M(t) = B(t) - c(x(t)) \geqslant 0$, $c(x(t))$ can only oscillate above b^* at arbitrarily large times if $B(t)$ also exceeds b^* at arbitrarily large times. In particular, if $B(t)$ gets trapped on ascending slopes at all large times, then this cannot happen, since then $B(t)$ monotonically increases to b^* while $B(t) \geqslant c(x(t))$. Consequently, all limits $x_i(\infty)$ exist in this case.

Moreover, if every b_i has only finitely many hills above the height b^*, then again all limits $x_i(\infty)$ exist. To see this, wait until all the x_i have crossed over all the peaks of those hills that they shall ever cross. Suppose that this occurs before time $t = T$. Because $b_i(x_i(\infty)) = b^*$, the following alternative holds: Either $b_i(x_i(t)) \geqslant b^*$ for all $t \geqslant T$, or $b_i(x_i(t)) < b^*$ for all $t \geqslant T$. Only those $x_i(t)$ for which $b_i(x_i(t)) \geqslant b^*$ can ever equal $y(t)$ at arbitrarily large times. Henceforth we restrict attention to these persistent jump variables. Since all persistent $x_i(t)$ have crossed their last hill before time $t = T$, and $b_i(x_i(t)) \geqslant b^* = b_i(x_i(\infty))$, it follows that all persistent $x_i(t)$ are on descending slopes for $t \geqslant T$; that is, $b_i'(x_i(t)) \leqslant 0$ for $t \geqslant T$. Using this fact, we will prove that $\int_0^\infty M(t)\,dt < \infty$. This latter inequality implies that all limits $x_i(\infty)$ exist; see [11, Theorem 1].

Since $y(t)$ is restricted to descending slopes for all $t \geqslant T$, $B(t)$ is monotone decreasing for $t \geqslant T$. Consider the trajectory of a given $x_i(t)$ at all times when $y(t) = x_i(t)$, $t \geqslant T$. Suppose for definiteness that there is a sequence U_{i1}, U_{i2},... of nonoverlapping intervals of time, whose union is U_i, such that $y(t) = x_i(t)$ only if $t \in U_i$. Suppose moreover that $U_{ik} = [S_{ik}, T_{ik})$. Whenever $y(t) = x_i(t)$, it follows that $\dot{x}_i(t) \geqslant 0$. Consequently $x_i(T_{ik}) \geqslant x_i(S_{ik})$. It is also true that $x_i(S_{i,k+1}) \geqslant x_i(T_{ik})$. This follows from the fact that x_i is trapped on a descending slope, and that $\dot{B}(t) \leqslant 0$ for all $t \geqslant T$. Thus the nonoverlapping intervals of time U_{ik} generate nonoverlapping intervals $[x_i(S_{ik}), x_i(T_{ik}))$ in the range of x_i. Since x_i is bounded, the total length of these intervals, namely $\Sigma_k[x_i(T_{ik}) - x_i(S_{ik})]$, is finite. This total length can also be written as $\int_{U_i} \dot{x}_i\,dt$, which can be written as $\int_{U_i} a_i M\,dt$. Since each x_i is bounded away from zero, it follows that $\int_{U_i} M\,dt$ is finite for every persistent x_i. However, $\int_T^\infty M\,dt$ is the sum over a finite number of these integrals, and thus $\int_0^\infty M\,dt < \infty$.

Each limit $x_j(\infty)$ therefore exists unless b_j has infinitely many hills H_1, H_2,... whose peak heights P_1, P_2,... exceed b^* and $\lim_{m\to\infty} P_m = b^*$. Moreover, $x_j(t)$ must reach each of these hills as $t \to \infty$, and $x_j(t) = y(t)$ for some time at which $x_j(t)$ is on each hill. Otherwise $y(t)$ would get trapped on descending slopes at all large times. Also, by the uniform continuity of b_j, the depths V_1, V_2,... of these hills also satisfy $\lim_{m\to\infty} V_m = b^*$, and there exists a w^* such that the peaks and valleys of the hills converge to w^* as $m \to \infty$.

First consider the case in which $b_j(x_j(t)) \geqslant b^*$ at all large times, despite the fact that b_j has infinitely many hills. Consider times $t = T$ at which $c(x(T)) = b^*$. There must exist infinitely many such times, approaching infinity, at which $\dot{c}(x(T)) < 0$, so that $c(x(t))$ can oscillate around b^* infinitely often. At every such $t = T$, some variable, say x_j, satisfies $y(T) = x_j(T)$. Thus $\dot{x}_j(T) \geqslant 0$. Moreover the x_i for which $b_i(x_i(T)) = b^*$ satisfies $\dot{x}_i(T) = 0$. Since

$$\dot{c}(x(T)) = \sum_{m=1}^{n} \frac{\partial c}{\partial x_m}(x(T))\, \dot{x}_m(T)$$

where all

$$\frac{\partial c}{\partial x_m}(x(T)) \geqslant 0,$$

there must exist an x_k, with $k \neq i, j$, such that $\dot{x}_k(T) < 0$ at infinitely many of the times T. This justifies the oscillation condition.

In the remaining case, there can exist b_j which oscillate above and below b^* on infinitely many hills. Then a similar argument holds: In order for x_j to get across infinitely many hills, there must exist infinitely many $t = T$, approaching infinity, at which $b_j(x_i(T)) > b^*$, $c(x(T)) = b^*$, and $\dot{c}(x(T)) < 0$. Since $\dot{x}_j(T) \geqslant 0$, there must exist an x_k such that $\dot{x}_k(T) < 0$ at infinitely many values of T. Thus all $x_i(\infty)$ exist, except possibly in those cases wherein the oscillation condition holds.

4. Finite Jump Condition

The proof of Theorem 1 does not rule out the possibility that infinitely many jumps occur, say if a limiting height exists. Theorem 1 of [12] describes systems of the form

$$\dot{x}_i = a(x)\, g(x_i)\, [b(x_i) - c(x)], \tag{63}$$

in which only finitely many jumps occur, and in which the jump trends through time can be analyzed. This theorem depends on two properties that do not generally hold in (1): First, because of the form of equation (63), the variables are *ordered* in time; that is, they can be labeled so that $x_1(t) \leqslant x_2(t) \leqslant \cdots \leqslant x_n(t)$ for $t \geqslant 0$. Second, a *self-similarity* condition is assumed to hold between the hills of the graph of $b(w)$. This condition requires that the highest hills of the graph are also the speepest hills. Self-similarity explicates the intuitive idea that each hill is due to averaging over some random factor that is distributed across a subpopulation of each population, and that the averaging process will automatically produce a correlation between the steepness and height of the hills in many cases. Theorem 1 above indicates that neither the ordering nor the self-similarity is necessary to produce global limits.

The question of when a system has only finitely many jumps is of considerable physical interest, since after all jumps cease the system has "decided" on its asymptotic pattern. There exist systems more general than (63) in which only finitely many jumps can occur even if self-similarity does not hold. In these systems, the infinite sequences of jumps towards a limiting height are ruled out by imposing a dominance condition on the possible jumps between slopes. This dominance condition is weaker than self-similarity because there need not be any relationship between the relative height and steepness of a hill.

THEOREM 2 (Finite Jump Sequence). *Given any $n \geqslant 2$, consider the systems*

$$\dot{x}_i = a(x)\, g_i(x_i)\, [b_i(x_i) - c(x)], \tag{64}$$

where $x = (x_1, x_2, \ldots, x_n)$ and $i = 1, 2, \ldots, n$. Let the following hypotheses hold:

1. *Smoothness:*

(a) *$a(x)$ is continuous for $x \geqslant 0$;*

(b) *$g_i(x_i)$ is continuous for $x_i \geqslant 0$;*

(c) *$b_i(x_i)$ is either continuous with piecewise derivatives for $x_i \geqslant 0$, or is continuous with piecewise derivatives for $x_i > 0$ and $b_i(0) = \infty$;*

(d) *$c(x)$ is continuous with piecewise derivatives for $x \geqslant 0$.*

2. *Nonnegativity:*

$$a(x) > 0 \quad \text{if} \quad x \geqslant 0, \tag{65}$$

$$g_i(x_i) > g_i(0) = 0, \quad x_i > 0, \tag{66}$$

and

$$\int_0^\lambda \frac{dw}{g_i(w)} = \infty. \tag{67}$$

3. *Boundedness:*

$$\limsup_{w \to \infty} b_i(w) < c(0, 0, \ldots, \infty, \ldots, 0, 0), \tag{68}$$

where "∞" is in the ith entry, $i = 1, 2, \ldots, n$.

4. *Competition:*

$$\frac{\partial c}{\partial x_k} \geqslant 0, \qquad k = 1, 2, \ldots, n; \tag{69}$$

5. *Slope Dominance:*

Let there exist finitely many ascending slopes A_{ik} and descending slopes D_{ik} on the graph of each function $b_i(w)$, $w \in [0, B]$. Given any pair A_{jk} and A_{lm} of ascending slopes, let the slope functions $s_i(w) = g_i(w)\, b_i'(w)$ satisfy either

$$s_j(w_j) \geqslant s_l(w_l) \quad \text{if} \quad b_j(w_j) = b_l(w_l) \quad \text{and} \quad w_j \in A_{jk}, \quad w_l \in A_{lm} \tag{70a}$$

or

$$s_j(w_j) \leqslant s_l(w_l) \quad \text{if} \quad b_j(w_j) = b_l(w_l) \quad \text{and} \quad w_j \in A_{jk}\,, \quad w_l \in A_{lm}\,. \tag{70b}$$

Given any pair D_{jk} and D_{lm} of descending slopes, let the slope functions satisfy either

$$s_j(w_j) \geqslant s_l(w_l) \quad \text{if} \quad b_j(w_j) = b_l(w_l) \quad \text{and} \quad w_j \in D_{jk}\,, \quad w_l \in D_{lm} \tag{71a}$$

or

$$s_j(w_j) \leqslant s_l(w_l) \quad \text{if} \quad b_j(w_j) = b_l(w_l) \quad \text{and} \quad w_j \in D_{jk}\,, \quad w_l \in D_{lm}\,. \tag{71b}$$

Then given any nonnegative initial data $x(0)$, finite nonnegative limits $x(\infty)$ are approached after finitely many jumps occur.

Proof. The proof proceeds as in Theorem 1 until the case of a limiting height is considered. Jumps then occur between finitely many variables on (say) ascending slopes. By (24) and (64), a jump can occur from i to j at time T only if

$$s_j(x_j(T)) > s_i(x_i(T)) \qquad \text{and} \qquad b_i(x_i(T)) = b_j(x_j(T)). \tag{72}$$

Thus by (70), once a jump occurs from a variable on a given ascending slope to a different ascending slope, a jump can never return to the original variable. Since these are only finitely many variables, only finitely many jumps are possible, and the proof can be completed as in Case 1 of Theorem 1.

When Theorem 2 is applied to a generalized Volterra–Lotka system of the form

$$\dot{x}_i = D_i(x_i)\left[1 - \sum_{k=1}^{n} f_k(x_k)\, E_{ki}(x)\right], \tag{73}$$

with $E_{ki}(x) = F_k(x_k)\, G_i(x_i)$, the slope function

$$s_i(w) = \frac{-D_i(w)\, G_i'(w)}{G_i(w)}\,. \tag{74}$$

By (70) and (71), $s_j(w_j)$ and $s_l(w_l)$ are compared at values w_j and w_l such that $b_j(w_j) = b_l(w_l)$. Since $b_i(w) = G_i^{-1}(w)$ in this case, the relative sizes of the functions $S_j(w_j) = D_j(w_j)\, G_j'(w_j)$ and $S_l(w_l) = D_l(w_l)\, G'(w_l)$ must be compared at values of w_j and w_l such that $G_j(w_j) = G_l(w_i)$. This observation leads to the following corollary.

COROLLARY 3. *Let system (73) with $E_{ki}(x) = F_k(x)\, G_i(x_i)$ satisfy the conditions of Corollary 2. In addition, suppose that there exist finitely many ascending slopes A_{ik} and descending slopes D_{ik} of the functions $G_i(w)$, $w \in [0, B]$. Given any pair A_{jk} and A_{lm} of ascending slopes, let the slope functions $S_i(w) = D_i(w)\, G_i'(w)$ satisfy either*

$$S_j(w_j) \geqslant S_l(w_l) \quad \text{if} \quad G_j(w_j) = G_l(w_l) \quad \text{and} \quad w_j \in A_{jk}\,, \quad w_l \in A_{lm} \tag{75a}$$

or

$$S_j(w_j) \leqslant S_l(w_l) \quad \text{if} \quad G_j(w_j) = G_l(w_l) \quad \text{and} \quad w_j \in A_{jk}, \quad w_l \in A_{lm}. \tag{75b}$$

Given any pair D_{jk} and D_{lm} of descending slopes, let the slope functions satisfy either

$$S_j(w_j) \geqslant S_l(w_l) \quad \text{if} \quad G_j(w_j) = G_l(w_l) \quad \text{and} \quad w_j \in D_{jk}, \quad w_l \in D_{lm} \tag{76a}$$

or

$$S_j(w_j) \leqslant S_l(w_l) \quad \text{if} \quad G_j(w_j) = G_l(w_l) \quad \text{and} \quad w_j \in D_{jk}, \quad w_l \in D_{lm}. \tag{76b}$$

Then given any nonnegative initial data $x(0)$, finite nonnegative limits $x(\infty)$ are approached after finitely many jumps occur.

5. Maximizing Preference and Contrast

Since $b_i(x_i) = G_i^{-1}(x_i)$ in the Volterra–Lotka systems (73), local minima of G_i are local maxima of b_i. Thus the fact that dynamical boundaries are switched in earliest at the abscissas of the highest peaks of b_i translates into the fact that dynamical boundaries are switched in earliest at the lowest valleys of G_i. Each G_i can be interpreted as a preference function, since the vector function $G(x) = (G_1(x_1), G_2(x_2),..., G_n(x_n))$ rank-orders the strength of signals from any population v_k to all the populations $v_1, v_2, ..., v_n$ when the system is in state x. Thus the above results proves that the dynamical boundaries are switched in at successively highly values of preference as $t \to \infty$. Once x_i crosses the lowest valleys of the preference function G_i, it can never cross them again. This defines a statistical tendency for the system to try to achieve the largest preference values that are compatible with its initial data $x(0)$ and the structure of the state-dependent preference order $G(x)$. Thus these Volterra–Lotka systems tend to maximize preference, just as the analogous neural networks (4) tend to maximize contrast, other things equal. It would appear to be wrong, however, to assume that a maximization principle could be used to express this trend in these nonstationary systems, although the search for such a principle is always a tempting adventure. Such a principle is often associated with a Liapunov function in classical examples. In the present examples, the maximum function $b(y(t))$ is not a Liapunov function at all values of $t \geqslant 0$. However, where only finitely many hills exist, $b(y(t))$ becomes a Liapunov function *after* all the dynamical boundaries have been laid down; that is, after all the decisions have already been made. This is true because $b(y(t))$ is then either restricted to descending slopes at all large times, or after one jump to an ascending slope, is restricted thereafter to ascending slopes. In the former case, $b(y(t))$ is a Liapunov function at large times; in the latter case, $-b(y(t))$ is a Liapunov function at large times. Thus, after the initially nonstationary dynamics of decision-making is over, the system then settles down towards a "classical limit". A similar trend occurs in learning networks; *after* the nonstationary phase of learning is over, the network settles down to a stationary memory phase, which is described by a stationary Markov chain [13]. Such examples suggest that global insights into the nonstationary processes suggested by biology require concepts and methods that genuinely transcend those that have proved so useful toward understanding essentially stationary phenomena.

References

1. S. Grossberg, Communication, memory, and development, *in* "Progress in Theoretical Biology" (R. Rosen and F. Snell, Eds.), Vol. 5. Academic Press, New York, 1978.
2. S. Grossberg, On the development of feature detectors in the visual cortex with applications to learning and reaction–diffusion systems, *Biol. Cybernetics* **21** (1976), 145–159.
3. S. Grossberg, Contour enhancement, short term memory, and constancies in reverberating neural networks, *Studies in Appl. Math.* **52** (1973), 213–257.
4. S. Grossberg and D. S. Levine, Some developmental and attentional biases in the contrast enhancement and short term memory of recurrent neural network, *J. Theoret. Biol.* **53** (1975), 341–380.
5. S. A. Ellias and S. Grossberg, Pattern formation, contrast control, and oscillations in the short term memory of shunting on-center off-surround networks, *Biol. Cybernetics* **20** (1975), 69–98.
6. D. S. Levine and S. Grossberg, Visual illusions in neural networks: Line neutralization, tilt aftereffect, and angle expansion, *J. Theoret. Biol.* **61** (1976), 477–504.
7. S. Grossberg, Preference order competition implies global limits in *n*-dimensional competition systems (1977), submitted for publication.
8. E. N. Lorenz, The problem of deducing the climate from the governing equations, *Tellus* **16** (1964), 1–11.
9. T.-Y. Li and J. A. Yorke, Period three implies choas, *Amer. Math. Monthly* **82** (1975), 985–992.
10. R. M. May and W. J. Leonard, Nonlinear aspects of competition between three species, *SIAM J. Appl. Math.* **29** (1975), 243–253.
11. S. Grossberg, Decisions, patterns, and oscillations in nonlinear competitive systems, with applications to Volterra–Lotka systems, *J. Theoret. Biol.* **73** (1978), 101–130.
12. S. Grossberg, Pattern formation by the global limits of a nonlinear competitive interaction in *n* dimensions, *J. Math. Biol.* **4** (1977), 237–256.
13. S. Grossberg, On the global limits and oscillations of a system of nonlinear differential equations describing a flow on a probabilistic network, *J. Differential Equations* **5** (1969), 531–563.

CHAPTER 11

BEHAVIORAL CONTRAST IN SHORT TERM MEMORY: SERIAL BINARY MEMORY MODELS OR PARALLEL CONTINUOUS MEMORY MODELS?

PREFACE

This article uses the free recall paradigm to discuss several philosophical and scientific issues, and to make some predictions. First the article shows that popular computer models of free recall data imply erroneous predictions and at best paradoxical neural implementations. I have since been occasionally told that these models were never intended to be taken literally, but that provides scant comfort to their adherents. I undertook this exercise to counter the prevalent belief that a computational theory of mind can be advanced without regard to its implementation. Different realizations rule out different phenomena and imply different logical implications among the possible phenomena.

One of the article's main insights is that temporal order information in STM can be encoded by parallel processing mechanisms. There need not exist a serial buffer. The article achieves this insight by deriving all the codes for temporal order information in STM that satisfy two simple, but basic, postulates. The first postulate is called an LTM Invariance Principle. It says that future events should not contradict the occurrence of past event sequences, even if these future events alter the relative importance of past events. This is a real-time postulate about code stability. It was introduced for the first time in my human memory article of Chapter 13, which derived a somewhat less general code from it. The second postulate says that the network's carrying capacity is finite and is independent of the number of active cells. This is just a normalization rule. Both of these postulates are satisfied by mass action (or shunting) competitive networks, but not by additive competitive networks. Once again, a nonlinear property was needed to drive the theory forward.

The fact that LTM invariance and STM competition can coexists in this fashion seemed quite natural to me at the time that this article was written, since I had already known about adaptive resonances for four years. However, the mathematical results were quite surprising, because they can imply the existence of a primacy effect in STM. Such primacy effects had been experimentally reported without mechanistic interpretation, but the present dogma was that they did not exist, since they could not be seen in interference experiments. I argue herein that STM primacy effects can exist even when interference experiments do not easily measure them. The main insight is to realize that the STM recency can mask the STM primacy effect due to STM normalization before it is measured in observable behavior.

This article continues the classification of competitive properties by showing that the primacy effect in STM is a behavioral contrast effect that develops in time. I explain behavioral contrast in time using the same mechanisms whereby I explained behavioral contrast during discrimination learning experiments in Chapter 6. Both of these behavioral contrast effects are manifestations of a normalization rule.

The article also points out how the normalization and multiplicative properties of mass action competitive networks can behave like a probabilistic calculus. Chapter 13 carries this observation a step further by suggesting that various physical systems *seem* to be probabilistic because they execute competitive computations. These results begin to address the fundamental and perplexing question of why probability theory describes various natural phenomena so well. Chapters 1 and 13 suggest an alternative to probability theory which seems to be more powerful than probabilistic rules for the explanation of hypothesis testing by self-organizing systems.

The article also reviews how mass action competitive rules can generate visual properties like reflectance processing, a Weber law, and a shift in the intensity range wherein cells are maximally sensitive. These results are deceptively simple. They work so well only because the theoretical questions are posed on a conceptual level that has behavioral meaning. For example, Cornsweet's excellent 1970 book on visual perception also discusses Weber-like rules $J(A+I)^{-1}$, but Cornsweet's approach to these rules omits questions of pattern processing. Instead, Cornsweet introduces a separate theoretical discussion in which he uses logarithms to discuss reflectance processing, although logarithmic singularities at low and high input intensities have no physical meaning. Logarithms are very tempting to use in vision, because the shift property (among others) comes out so well in logarithmic coordinates. The article points out, however, that the shift property does not depend on logarithmic mechanisms.

Behavioral Contrast in Short Term Memory: Serial Binary Memory Models or Parallel Continuous Memory Models?*

This paper develops a model wherein STM primacy as well as recency effects can occur. The STM primacy effects can be used to generate correct immediate recall of short lists that have not been coded in LTM. The properties of the model are interpreted in terms of explicit neural mechanisms. The STM primacy effect is a behavioral contrast effect that is analogous to the behavioral contrast that can occur during discrimination learning. The adaptational mechanism that accounts for these effects is also implicated in data on reaction time, retinal adaptation, ratio scales in choice behavior, and von Restorff-type effects. Its ubiquitous appearance is due to the fact that it solves a universal problem concerning the parallel processing of patterned data by noisy cells with finitely many excitable sites. It is argued that the STM primacy effect is not measured in interference experiments because it is masked by competitive STM interactions. These competitive interactions do not prevent the LTM primacy effect from influencing performance. The paper criticizes recent models of STM that use computer analogies to justify binary codes, serial STM buffers, and serial scanning procedures. Several deficiencies of serial models in dealing with psychological and neural processing are overcome by a model in which continuous STM activities and parallel real-time operations play an important role.

1. Introduction: Serial and Binary Memory Processes or Parallel and Continuous Memory Processes?

A great deal of experimental and theoretical work (e.g., Melton & Martin, 1972 Restle, *et al.*, 1975; Tulving & Donaldson, 1972) has been done on problems; relating to how learning subjects store data in short-term memory (STM) before it is transcribed into long-term memory (LTM) or otherwise transformed. Many experimental findings have been interpreted, either explicitly or implicitly, in terms of computer-like constructs, such as binary codes (Anderson & Bower, 1974; Atkinson & Shiffrin, 1968), serial buffers (Atkinson & Shiffrin, 1968), and serial scanning procedures (Sternberg, 1966). This paper suggests that the computer analogy has led to several basic difficulties. It also suggests an alternative theory to explain how order information in STM evolves in real time. This theory predicts a new experimental phenomenon, behavioral contrast in time, analogous to the phenomenon of behavioral contrast in space that occurs during discrimination learning (Bloomfield, 1969), and explains both phenomena using collective properties of well-known neural mechanisms. The theory is

* Supported in part by the National Science Foundation (NSF MCS 77-02958).

illustrated by examples concerning free recall, discrimination learning, reaction time, perceptual adaptation, and von Restorff-type effects.

Discrete and serial memory models have an immediate appeal in situations where behavioral responses are counted as they occur one at a time. However, discrete and serial behavioral properties do not imply that the processes which control them are also discrete and serial. Townsend (1974) has, for example, noted that the reaction times found in the Sternberg paradigm do not imply a serial process by describing statistical parallel processing completion times that are indistinguishable from their serial processing counterparts. It can furthermore be argued that accepting a discrete serial model precludes the study of some basic processes of learning and perception. Even in simple behavioral tasks, both continuous and discrete elements are evident. Many perceptions seem to be continuous; for example, colors or sounds seem to vary continuously in intensity and quality. Yet the language with which we describe them seems to be much more discrete; for example, letters such as A or B seem, in daily speech and listening, to be indecomposable units of behavior, and all of our language utterances seem to be built up from finitely many such units. To understand the process of seeing a color and describing it by language, we must face the problem of how seemingly continuous representations can interact, or be transformed, into seemingly discrete representations. We must be able to discuss the "degree of continuity" at all levels of this transformation.

The relationship between seemingly continuous and seemingly discrete events is a deep one especially because the same behavior can seem to have either type of representation depending on how familiar it is to us. The process of learning to walk or to talk is illustrative. Before we can walk, attention is paid almost continuously to the complex coordinations that are required. Yet after we know how to walk, much of this coordination is automatic, so that we can simply start to walk, pay attention to other things, and then decide to stop walking. The control of walking eventually approximates a binary on–off switch, except for some steering and object avoidance. Thus the process of learning can alter the control of walking from a relatively continuous representation to a relatively discrete one. A similar process occurs in many learning tasks wherein some form of "abstraction" occurs. Yet it would be wrong to believe that, after such a task is learned, its representation is "really" discrete rather than "seemingly" or "relatively" discrete,since the brain waves that occur during familiar speech or walking fluctuate continuously through time across billions of cells (Donchin & Lindsley, 1969). Moreover the sound spectrogram of familiar speech is an almost continuous flow of sound despite our impression that it is a series of discrete words (Lenneberg, 1967). In fact, an unfamiliar foreign language does sound like an almost continuous flow of sound. The process of learning makes the sounds seem to be discrete by perceptually grouping them into learned units. Thus if one accepts a binary representation of familiar events by fiat, then one must in principle miss vital ingredients of the learning process. In effects, the consensual impression of the event then blinds us to its functional representation.

Similar considerations make it clear that even in tasks that appear serial, such as serial learning, important underlying control processes are parallel processes. For example, the code that controls performance after a serial list is repetitively practiced are not just the individual list items. As Young (1959, 1961, 1963, 1968) noted, if they were, prior serial practice of a list A–B–C–D–... should yield marked positive transfer for later learning of the paired associates A–B, C–D,..., but it does not. Horowitz and Izawa (1963) suggested that more than one item can be the functional stimulus for a given response in a serial list, in particular that several items preceding the response serve as its functional stimulus. This viewpoint illustrates the familiar idea that a series of items can be chunked together (Miller, 1956) to form a new code whereby a series of behaviors can be more efficiently performed. Such a chunking process is based on the simultaneous availability of all the individual units, and is thus a parallel process.

2. Bowed Serial Position Curve in Free Recall

A basic datum about STM is the bowed serial position curve that is found in free recall experiments (Fig. 1). When a subject repeats a sufficiently long standard list of items in any order after hearing it once, the items near the beginning and end of the list are performed earliest and with the highest probability (Atkinson & Shiffrin, 1971). The advantage of items near the list beginning is called a *primacy* effect, that of items near the list end is called a *recency* effect. A computer analogy to explain these effects can be developed as follows; cf. Atkinson and Shiffrin (1968). Let a list $r_1, r_2, \ldots, r_n$ of behaviorally matched items be presented to a subject. It is supposed that each item is either in an STM buffer, or is not in the buffer, at a given time. That is, assume that a binary code exists such that 1 is assigned to r_i if r_i is in the buffer, and 0 is assigned to r_i if r_i is not in the buffer. If $k > 1$ items are in the buffer at time t, one cannot determine the order in which they entered the buffer by looking at their 0's and 1's, since all k items that are in the buffer have the value 1. Thus a binary code carries no order information. If there did not exist any internal trace of the order in which items occurred, there would be no way to encode this information in LTM. Given a binary code, some mechanism other than an item's activity (0 or 1) is needed to code order information. A serial STM buffer is therefore assumed to exist. Suppose that this buffer contains m serially organized slots $s_1, s_2, \ldots, s_m$. The first item r_1 enters s_1. When r_2 occurs, it enters s_1 and displaces r_1 from s_1 to s_2. Then r_3 displaces r_2 from s_1 to s_2, and r_1 from s_2 to s_3.

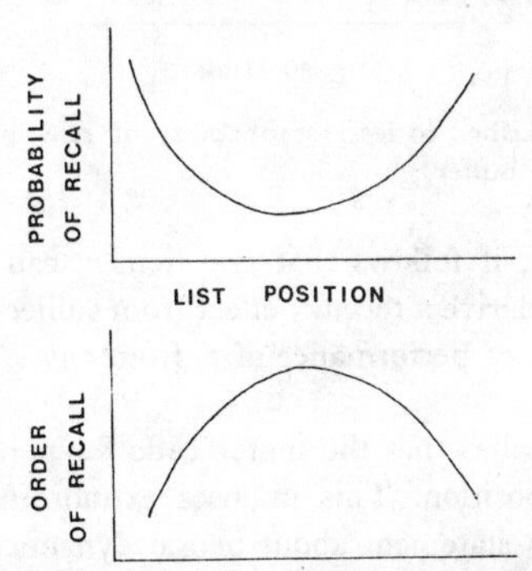

Fig. 1. Probability and order of recall in free recall experiment.

One can then tell which items occurred first by testing their relative positions in the buffer. The above process continues until m items are presented. Item r_{m+1} knocks r_1 out of the buffer from slot s_m, and each successive item eliminates the earliest remaining item from the buffer at slot s_m. In all, at any time there will be a block of successive items in the buffer, each with activity 1. Thus, given a binary code, a serially organized STM buffer is needed to store order information.

The binary buffer concept does not, however, explain the data in Fig. 1. If the buffer worked in a deterministic fashion, then each item could be perfectly performed, and each item would be performed in its correct order. The two bowed curves in Fig. 1 would be replaced by a horizontal line and an increasing straight line, respectively. Consequently, the buffer cannot work in a deterministic fashion. It must work probabilistically, if it exists at all. Introducing probabilities brings continuous variables back into the model, and creates a hybrid mixture of computer and probabilistic ideas. To explain the recency effect, this hybrid model makes two more related assumptions, both of which say that the buffer works badly in a prescribed way. First, one says that the buffer is *leaky* in the sense that an item can fall out of the buffer even before it reaches s_m. Since the probability of falling out increases as a function of how long an item is in the buffer, this makes it most probable that the most recent items are still in the buffer (Fig. 2). A recency effect for the probability of being in the buffer is achieved by averaging across subjects. In each subject, however, items that remain in the buffer all still have activity 1. Probabilistic models of STM usually stop at this point. They fail to ask a crucial question whose answer casts doubt on the binary code assumption. How is the probability distribution of being in the STM buffer translated into the real-time performance of individual items?

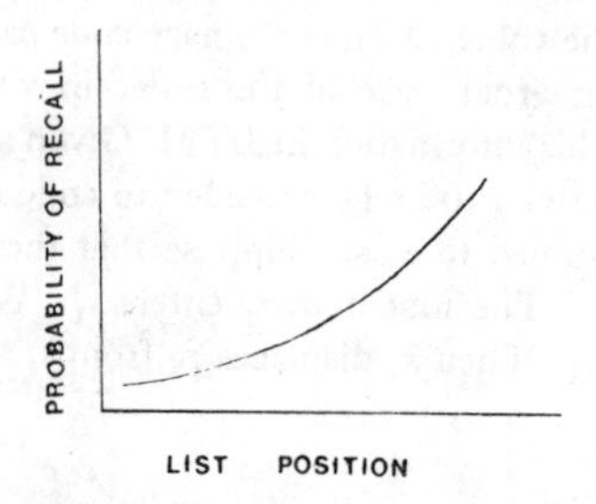

FIG. 2. Recency effect is ascribed to lesser probability of remaining in a serial STM buffer as a function of duration in the buffer.

Given the above framework, it follows that *any* item r_i can be performed from *any* buffer position s_j in order to derive a recency effect from subject performance. Otherwise there would be 0 probability of performance of r_i from any s_j from which r_i could not be elicited.

This latter assumption implies that the motor code for performing any item can be read out from any buffer position. This imposes extraordinary demands on system design, and makes a definite statement about neural dynamics. It says that the motor codes for eliciting an item, although not initially learned at all positions s_j, can be performed at all positions s_j. The code is ***shift-invariant***. An even more demanding implication is that all codes that are hierarchically built up from the STM buffer must also be shift-invariant. To predict the recency effect from a binary code model, one is hereby led to conclude that the buffer is so poorly designed that it leaks, but is so exquisitely designed that its entire hierarchy of codes is shift-invariant. Usually this implication is ignored, possibly because it is so disturbing, and possibly because probabilistic modelers often overlook real-time constraints on performing individual items when they construct their models.

Given the considerable machinery that is needed to produce the recency effect using a binary code, we ask whether it is necessary? In particular, the enormous amount of neural data on continuously fluctuating potentials, spiking frequencies, and the like, leads one to question the binary assumption itself. If the binary code is abandoned, then all of the above difficulties evaporate. Items r_i can then have fixed internal representations v_i that are innate or built up by learning; their codes need not move around in a buffer. Thus there need be no shift-invariant code. Moreover, the v_i need not be leaky, and each subject can possess a recency gradient, rather than assuming the recency gradient is a statistical property of a pool of subjects, as in a binary theory.

In a continuous theory, a recency gradient exists if the most recent items have the greatest STM activity in their representations v_i, say because their STM activity has had less opportunity to decay, either spontaneously or due to interference. If greater STM activities translate into faster reaction times of item performance, then a recency effect in performance can be achieved without the need to move items around in a serially organized buffer. In other words, if continuous STM activities exist, then they already carry order information. Below we suggest some STM interactions that can occur at a single level of input processing. Section 7 discusses how LTM feedback from a higher network level can modify these properties. The paper Grossberg (1978a) develops a more complete theory in which several levels of STM and LTM processing are needed to self-organize complex behavioral codes, maps, and plans.

How does a continuous mechanism work? Suppose for definiteness that each item r_i has an internal representation v_i with STM activity x_i. If the most recently presented items have the largest STM traces, and if r_i is the last item to have occurred, then $x_1 < x_2 < x_3 < \cdots < x_i$. The storage of these STM activities must be distinguished from their overt rehearsal. How does rehearsal translate differential STM activity into a prescribed order of performance, in particular the order $r_i, r_{i-1}, r_{i-2}, \ldots, r_1$, in the case when $x_1 < x_2 < x_3 < \cdots < x_i$? In many neural examples, a nonspecific rehearsal, or arousal, wave can accomplish this. Such a mechanism simultaneously amplifies all STM activities so that they can exceed an output threshold, or alternatively lowers the output threshold until it is exceeded by the STM activities (Fig. 3). The largest STM activity x_i exceeds the output threshold first, and thereby elicits the fastest output signal. This output signal controls performance of item r_i. If this signal was not self-terminating, then perseverative performance of item r_i would occur. Under normal circumstances, the output signal generates feedback inhibition that self-inhibits, or resets, its STM activity. Then the state v_{i-1} is most active, so that its output signal can elicit performance of item r_{i-1}. This process of STM arousal and reset continues until all of the items are

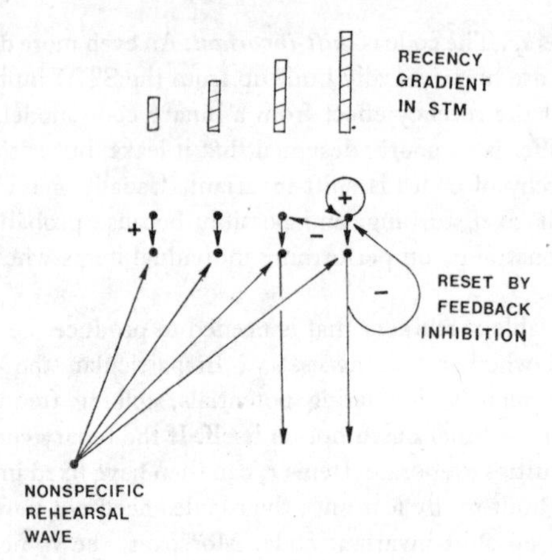

FIG. 3. Readout of STM by a nonspecific rehearsal wave, and reset of STM by feedback inhibition.

performed in the order $r_i, r_{i-1}, \ldots, r_1$. The role of nonspecific arousal and reset as rehearsal mechanisms are further discussed in Grossberg (1977a). This paper also describes how a list can be grouped into parts which can all be performed in their correct order.

In the above example, the nonspecific rehearsal mechanism is a parallel operation that simultaneously influences all representations v_i, despite the fact that item performance is serial. Serial properties do not imply serial mechanisms.

3. Primacy Effect

The binary model also implies questionable conclusions to explain the primacy effect of Fig. 1. The binary model assumes that a further operation is activated by items while they are in the STM buffer. This operation is described as coding the item in LTM. Although the binary model does not describe the LTM coding mechanism, clearly more coding can occur the longer an item is in the buffer, other things equal. Consequently the earliest items can be coded in LTM better than more recent items. This LTM process is claimed to produce the primacy effect. Thus early items produce a primacy effect via LTM, whereas late items produce a recency effect via STM.

Several types of evidence are compatible with this view. For example, if STM storage is inhibited by an interfering task, then the primacy effect remains but the recency effect is obliterated (Atkinson & Shiffrin, 1971). The similarity of primacy effects with or without interference is a main source of the belief that STM does not contribute to primacy. This is only indirect evidence, however, and it is argued in Section 7 that competitive interactions acting in parallel across internal representations can effectively mask any STM primacy effects that might exist, leaving the impression that only LTM influences performance. Other experiments are based on the premise that rehearsal should strengthen

LTM, so that a good correlation between the number of rehearsals at different list positions and the size of the LTM contribution to recall at these positions should argue for LTM as a basis for the primacy effect (Rundus, 1971). This argument does not help if STM primacy effects are masked. Moreover, it has been shown that the size of the LTM effect can depend on the type of rehearsal (maintenance vs elaborative), and on whether performance is measured by recognition or recall (Craik & Watkins, 1973; Woodward, Bjork, & Jongeward, 1973). In Grossberg (1978a, Sects. 31 and 47) a coding mechanism is described wherein mere repetition of items, improved recognition, and improved recall are distinguished. Improved recognition can result from new code formation and sustained STM activity of these new codes even when individual item codes are rapidly reset in STM. Improved recall can result from the formation of new motor associations using the new codes as sampling sources. Because the sampling sources must be synthesized before the motor associations can be learned, recognition often improves before recall does.

Using the LTM primacy hypothesis, the binary model can fit some of the interference and rehearsal data, but is also forced into the counterintuitive idea that items near the beginning of the list can only be performed in their correct order by being read out of LTM. This idea overlooks the fact that a telephone number can be perfectly repeated immediately after hearing it, yet it could have been obliterated from memory by a distracting event before performance occurred, so presumably was not stored in LTM. Indeed, amnesic patients with Korsakoff syndrome have no LTM capability, but exhibit essentially normal digit span performance (Baddeley & Warrington, 1970; Milner, 1956). These examples question whether LTM is *necessary* to produce a primacy effect. The data used to support the LTM contribution to the primacy effect do not disprove that STM also contributes to primacy, and sometimes without a large LTM contribution.

4. STM Primacy Effect

How can performance of a telephone number due to STM but not LTM be achieved? This is easy in the continuous model if the earliest items have the largest STM activities, since these items will be performed first when the STM field is amplified by a nonspecific rehearsal wave. However, if the earliest items have the largest STM strengths in a short list, then how can the most recent items also have large STM strengths in longer lists to produce a bowed STM pattern across list representations, and thus the bowed order of recall in Fig. 1? We will illustrate below how an STM primacy effect can be generated in short lists, but becomes an STM bow in longer lists, such that the STM recency effect becomes progressively stronger as list length increases. The list length at which the bow appears is called the transient memory span (TMS). The TMS can depend on such factors as a subject's attentional and motivational state, but it can be proved that the TMS is no longer than the more familiar immediate memory span (Miller, 1956) under rather general circumstances (Grossberg, 1978a, Sect. 32).

Perhaps the belief that the STM activities of earlier items should always be weaker supported the idea that only LTM can ever generate a primacy effect, despite its unfortunate implications for the immediate recall of short lists. To counter this belief, we note that an STM primacy effect has already been found in some free recall data. Baddeley and Warrington (1970) study amnesic Korsakoff patients whose STM is intact, but whose LTM is nonfunctional. In free recall taks, these patients produce a bowed probability of recall curve that is due to STM alone. Hogan and Hogan (1975) theoretically disentangle STM and LTM contributions in their free recall data for normal subjects, and find an STM primacy effect which they mention without mechanistic interpretation. Furthermore, we will suggest that STM primacy is a temporal analog of a phenomenon which is more familiar experimentally, but which until recently was theoretically paradoxical; namely, behavioral contrast in discrimination learning experiments (Bloomfield, 1969).

5. Behavioral Contrast in Space

A typical example of behavioral contrast is this. If a pigeon is rewarded on errorless discrimination trials for pecking on a key illuminated by a light of wavelength λ, then during extinction trials, when the pigeon is allowed to peck in responses to keys illuminated by various wavelengths, a generalization gradient of pecks centered at λ is generated (Fig. 4a). By contrast, if the pigeon is rewarded for pecking a key illuminated at wavelength λ_1, and punished for pecking at a nearby wavelength $\lambda_2 < \lambda_1$, then during extinction, the pigeon pecks most vigorously at wavelength $\lambda_3 > \lambda_1$ (i.e., a peak shift occurs). Remarkably, the pigeon pecks λ_3 more vigorously than it would have pecked λ_1 if λ_2 had not occurred (Fig. 4b); that is, behavioral contrast occurs. Behavioral contrast is

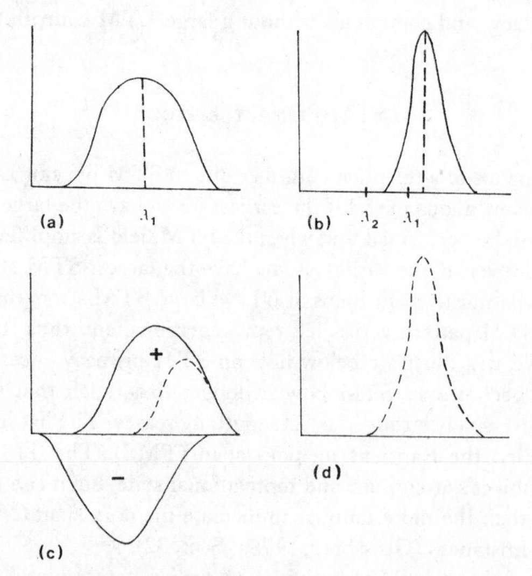

Fig. 4. (a) Generalization gradient on extinction trials if wavelength λ is rewarded; (b) peak shift and behavioral contrast; (c) net gradient produces shift but no contrast; (d) normalized net gradient produces contrast.

paradoxical because the punishing λ_2 causes the pigeon to peck at the unrewarded λ_3 more than it would have pecked at the rewarded λ_1 in the absence of λ_2. The difficulty in explaining behavioral contrast is this: Suppose that reward at λ_1 generates a positive generalization gradient centered at λ_1, and punishment at λ_2 generates a negative generalization gradient centered at λ_2. If performance at λ_3 is due to the net gradient, then a peak shift will occur, but pecking at λ_3 should be less vigorous than pecking at λ_2 (Fig. 4c). What then causes behavioral contrast?

Grossberg (1975) suggests that behavioral contrast follows from a property of cell populations that undergo mass action interactions in recurrent on-center off-surround anatomies (Fig. 5). Grossberg (1973) derives networks of this type as a solution to a

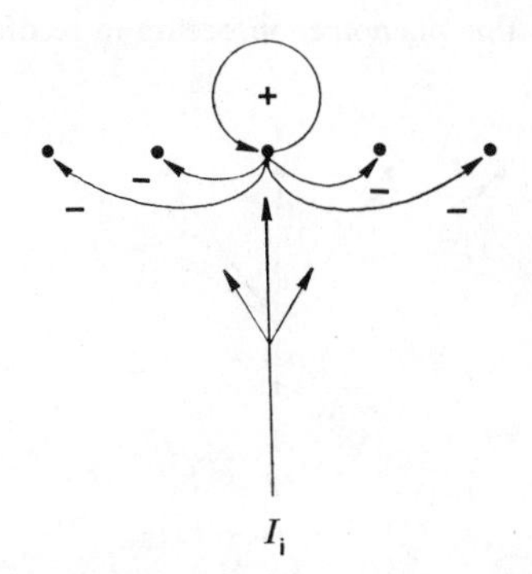

FIG. 5. Recurrent on-center off-surround anatomy can suppress noise, contrast enhance suprathreshold patterns, and store them in STM without saturation.

universal problem concerning how patterned information can be processed by noisy cells with finitely many excitable sites; cf., Levine and Grossberg (1976) or Grossberg (1977b) for a review. In the present context, the populations are sensitive to different hues. The on-center (excitatory feedback) defines an excitatory generalization gradient to nearby hues, and makes possible STM storage after external inputs cease. The off-surround (inhibitory feedback) maintains network sensitivity to relative input sizes even in response to large inputs; otherwise expressed, automatic gain control, driven by the off-surround, prevents cell saturation by adapting network responses to different background activity levels. The mass action laws reduce to the familiar equations (Hodgkin, 1964; Katz, 1966)

$$C\frac{\partial V}{\partial t} = (V_+ - V)g_+ + (V_- - V)g_- + (V_p - V)g_p \tag{1}$$

for a cellular potential $V(t)$ influenced by a capacitance C and by three conductances g_+, g_-, and g_p which change $V(t)$ insofar as it deviates from the three saturation potentials V_+, V_-, and V_p. The notation "+" designates the excitatory channel (usually the Na^+ channel); "–" designates the inhibitory channel (usually K^+), and "p"

designates a passive channel. The convention is also accepted that depolarization makes $V(t)$ more positive.

It has been proved that such networks tend to conserve the total potential of all cells in the network. This property is called *normalization*, and is a form of network adaptation due to automatic gain control by its off-surround. In Fig. 4c, the net gradient is narrower than the gradient in Fig. 4a. The total potential, or area under the curve in Fig. 4a, is approximately conserved when the excitatory gradient is replaced by the narrower net gradient. Normalization therefore amplifies the net gradient to produce the higher and steeper normalized net gradient of Fig. 4d. Thus behavioral contrast can be explained as the result of a net gradient normalized by a network that is capable of storing cues in STM without saturation.

The simplest example of this phenomenon occurs in feedforward networks (Fig. 6).

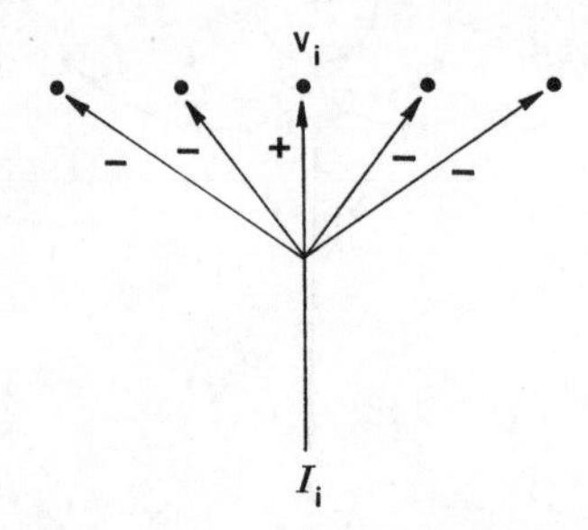

FIG. 6. Nonrecurrent on-center off-surround anatomy.

Let n cells (or cell populations) v_i, $i = 1, 2, \ldots, n$, be given, and let $x_i(t)$ be the potential of v_i. In (1), replace $V(t)$ by $x_i(t)$ and choose the constant parameters $C = 1$, $V_+ = B > 0$, $V_- = V_p = 0$, and $g_p = A > 0$ for simplicity. The conductance g_+ is influenced by an on-center input $g_+ = I_i$ and the conductance g_- is influenced by an off-surround input $g_- = \sum_{k \neq i} I_k$. Then at the cell v_i, (1) becomes

$$\frac{d}{dt} x_i = -Ax_i + (B - x_i) I_i - x_i \sum_{k \neq i} I_k . \tag{2}$$

If the inputs remain constant for awhile, then the system approaches equilibrium. At equilibrium $(d/dt)\, x_i = 0$ so that the equilibrium potential of v_i is

$$x_i = BI_i/(A + I), \tag{3}$$

where $I = \sum_{k=1}^{n} I_k$. Letting $\theta_i = I_i I^{-1}$ be the relative input to v_i, (3) can be written as

$$x_i = \theta_i\, BI/(A + I), \tag{4}$$

which shows that each v_i retains its sensitivity to θ_i even as I is parametrically increased. The dependence of x_i on the ratio θ_i is a form of adaptation to changing background activity levels. Without such adaptation, each x_i would rapidly saturate at B as I_i increased. Ratio theories have often been suggested by perceptual or learning data. For example, Zeiler (1963) developed an adaptation-level theory in which the subject's perception of a stimulus depends on its ratio with respect to an internal norm, or adaptation level. Luce (1959) developed a theory in which choice behavior depends on the ratio of two reaction tendencies. Color theories are often based on ratios that represent the reflectances of external objects (Cornsweet, 1970). I suggest that ratios appear in such a great variety of situations to deal with the ubiquitous saturation problem. Even in the simplest case of Eq. (4), however, the ratio influence is modified by a term $BI(A + I)^{-1}$ which is of Weber–Fechner form (Cornsweet, 1970, p. 249). In other examples of on-center off-surround networks, only ratios above an adaptation level can cause positive x_i values (Grossberg, 1978b), or there can be complicated hysteresis, normative, decision, and related types of behavior (Grossberg, 1977, 1978a).

Wherever ratios appear in individual activities x_i, the total activity $x = \sum_{k=1}^{n} x_k$ obeys a normalization rule. In (4), $x = \sum_{k=1}^{n} x_k = BI(A + I)^{-1}$ is always less than B. It thus has an upper bound that is independent of the number n of cells and the total input I. This is normalization in a feedforward network. In a recurrent, or feedback, network the normalization property is strengthened. The normalized inequality $x \leqslant B$ is replaced by normalized equalities.

The law (4) has another important property; namely, x_i, plotted as a function of the logarithm of its on-center $K = \ln I_i$ and its off-surround $L = \sum_{k \neq i} I_k$ obeys

$$x_i(K, L) = Be^K/(A + e^K + L). \tag{5}$$

Thus, if the off-surround input is shifted from $L = L_1$ to $L = L_2$, the whole curve (5) is shifted by an amount $S = \ln[(A + L_1)(A + L_2)^{-1}]$, since

$$x_i(K + S, L_1) = x_i(K, L_2) \qquad \text{for all} \quad K \geqslant 0. \tag{6}$$

A similar shift occurs, for example, in bipolar cell responses in the mudpuppy retina (Werblin, 1971; Grossberg, 1977b, c). The shift relocates where x_i is most sensitive.

The above properties are summarized herein to emphasize three points. First, in the laws (1) and (2), inputs exercise their effects by *multiplying* potentials. Hence these laws are called *shunting* laws. Additive models cannot generate these effects. Second, in (2), all cells v_k, $k \neq i$, inhibit v_i with equal strength. If this is not true, say because inhibitory interactions become weaker as the distance between populations increases, then the normalization effect becomes partial, and the total potential can grow to a finite asymptote as more populations are excited; that is, the adaptation effect is only partial, and saturation starts to set in as the background input becomes large. Third, the normalization rule helps to clarify from a neurophysiological perspective why probability axioms often model behavioral data so well. The normalization rule plays the role of summing all the probabilities to 1, and the shunting laws play the role of multiplying the probabilities of independent events. However, even in Eq. (4), ratios do not appear alone, and the normalization rule can often hold only partially, as we will see below.

6. Behavioral Contrast in Time

Bowing in STM will now be explained as a behavioral contrast effect that evolves as items are presented in time, rather than across space. Before developing the ideas in general, consider the simplest example as an illustration. Suppose that total activity is normalized. Set it equal to 1, for definiteness, when some item is active in STM. Also suppose that when a new item occurs, the old item's STM activity is reduced by a multiplicative factor w due to shunting inhibition. When item r_1 occurs, its activity x_1 equals 1. When item r_2 occurs, x_1 is changed to w. By normalization, $x_1 = w$ and $x_2 = 1 - w$. If $w > \frac{1}{2}$, then $x_1 > x_2$; that is, an STM primacy effect occurs. A large value of w means that the reverberating STM activity x_1 of v_1 can substantially inhibit v_2 when v_2 is receiving an input due to presentation of r_2. When r_3 occurs, the old STM activities are again multiplied by w, so that $x_1 = w^2$ and $x_2 = w(1 - w)$. By normalization, $x_3 = 1 - w$. Note that $x_1 > x_2$ and $x_3 > x_2$. A bow has occurred at v_2. As new items r_i are presented, $i > 3$, the bow remains at position v_2, but a pronounced recency effect develops due to normalization. In particular, given any list of length $k > 1$, the last item to enter STM always has STM activity $1 - w$. Below we will show that a bow can arise at any list position if network parameters are properly chosen.

The behavioral contrast mechanism can be derived from three concepts. The first concept is operationally described by saying that new items change the STM activities of old items by a multiplicative factor. This mechanism is the simplest rule for making rigorous the idea that shunting interactions join the network populations together. There exists a deeper justification for using the multiplicative rule. Grossberg (1977a) develops a theory of neural coding, in which it is shown (Sect. 25) that the multiplicative rule leaves invariant the codes of old items as new items occur and activate new codes. This concept is needed to prevent each new item from destabilizing the internal representations of all the old items. It says that new items do not deny the fact that old items occurred, even if they alter their importance, or even totally inhibit them. The rule is therefore called a *Principle of Code Invariance.* The theory hereby establishes a conceptual bridge between statements about STM interactions—via laws describing cellular potentials and signals—and statements about LTM interactions—via laws describing cellular learning. This bridge shows in a precise formal way how each type of law is adapted to the needs of the other.

The multiplicative operations of the Invariance Principle have the following effect. Let item r_i enter the network with STM activity u_i. Let the ith item multiplicatively modify the STM activity of all previous items $r_1, r_2, \ldots, r_{i-1}$ by a factor w_i. Suppose r_1 enters with weight $x_1 = u_1$. After r_2 occurs, $x_1 = w_2 u_1$ and $x_2 = u_2$. After r_3 occurs, $x_1 = w_3 w_2 u_1$, $x_2 = w_3 u_2$, and $x_3 = u_3$. And so on. The total STM strength S_i after r_i occurs is thus

$$S_i = \sum_{m=1}^{i} \prod_{r=m+1}^{i} w_r u_m . \tag{7}$$

The second hypothesis is the *Normalization Rule.* This says that total STM strength grows in a negatively accelerated way from a minimum of u_1, when only r_1 occurs, to some finite maximum M. The case $u_1 = M$ characterizes complete normalization. An analogous experimental phenomenon is that pupil diameter increases in a negatively accelerated way as a function of the number of items presented to a subject (Kahneman & Beatty, 1966), and one might try to use this paradigm to estimate u_1 and M in particular cases. Mathematically, the Normalization Rule says that

$$S_i = u_1\theta^{i-1} + M(1 - \theta^{i-1}), \tag{8}$$

where $M \geqslant u_1 \geqslant 0$ and $0 \leqslant \theta \leqslant 1$. The parameters u_1, M, and θ can depend on the geometry of the network as well as on attentional and motivational factors that can retune network interactions (Grossberg, 1976a, b). Our goal is to solve for the weights w_k, $k \geqslant 2$, in terms of the parameters θ, M, and u_i, $i = 1, 2, \ldots$. That this can be done is summarized in the next statement.

(I) Suppose that the Invariance Principle and the Normalization Rule both hold; that is, let both (7) and (8) hold. Then the shunting parameters can be explicitly determined. They are

$$w_k = \frac{u_1\theta^{k-1} + M(1 - \theta^{k-1}) - u_k}{u_1\theta^{k-2} + M(1 - \theta^{k-2})}, \qquad k > 1. \tag{9}$$

Since the STM activity of v_i after item r_j occurs is, by the Invariance Principle,

$$x_i = u_i \prod_{k=i+1}^{j} w_k, \qquad i \leqslant j, \tag{10}$$

the STM code can be completely solved by specifying the STM weight u_i of the most recent item r_i, $i = 1, 2, \ldots$. We assume that u_i estimates how much attention is paid to item r_i as it is presented. Then (9) and (10) show that, once u_1, M, and θ estimate the code geometry, and (presumably constant) performance variables, it suffices to specify how much attention is paid to each item as it is presented.

The following result also holds if each u_i depends only on r_i and possibly events $r_1, r_2, \ldots, r_{i-1}$ that have preceded it.

(II) Suppose that the Invariance Principle holds. If an STM bow occurs at position J in a list of length K, then it also occurs at position J given a list of any length $k \geqslant K$. This strong property notes that the factors w_k in (10) change the relative sizes of past STM strengths, but not where local maxima or minima occur in the STM pattern across old items. It is this property that allows us to define a TMS for lists composed of matched items presented under fixed performance conditions.

To derive further information about the code, we now impose some natural constraints on the u_i. These constraints do not hold if attentional conditions vary in an arbirary fashion as new items are presented. They summarize in mathematical terms various stable attentional conditions. Intuition suggests that if we pay equal attention to each item as it is presented, then $u_1 \geqslant u_2 \geqslant u_3 \geqslant \cdots$ (equality might be destroyed by negative feedback acting on later items) and that the u_i equilibrate at some positive value u_∞ as longer lists are used. This idea simply says that the STM strength of the last item should always be positive, if one always pays attention to the most recent item, and can only get weaker as i increases due to greater total inhibition from the larger numbers of items that are already in STM. For example, if all $u_i = u > 0$, then a fixed amount of STM strength is always given to the last item. In this case, if $M > u$, it can be proved that the bow in STM occurs at item r_J, where J is the maximal index j such that

$$(u^{-1}M - 1)(1 - \theta)\theta^{j-2} > 1, \tag{11}$$

Thus a bow can occur at any list position if u, M, and θ are suitably chosen.

Another plausible rule for the u_i sequence is: set $u_1 = u$ and $u_k = u_\infty$, $k \geqslant 2$, where $u_\infty < u$. In other words, if r_1 occurs alone, then no inhibition occurs since no others items are reverberating in STM to supply it. Once more than one item occurs, the most recent item always has a fixed amount of STM strength due to a balancing out of excitation and inhibition across all items. This occurs in the special case where $u_1 = M$ and all $w_i = w < 1$. Both of these rules are a special case of the rule that

$$u_i = u_1\phi^{i-1} + u_\infty(1 - \phi^{i-1}), \tag{12}$$

$i \geqslant 1$, with $u_1 \geqslant u_\infty > 0$ and $0 \leqslant \phi \leqslant 1$. In other words, u_i is negatively accelerated function of i. Choosing $\phi = 1$ yields the rule $u_i = u_1$, $i \geqslant 1$; choosing $\phi = 0$ yields the rule $u_1 > u_k = u_\infty$, $k \geqslant 2$. The analysis can be generalized further by letting a twice-differentiable function $u(t)$, $t \geqslant 1$, interpolate the sequence $u_1, u_2, \ldots$; that is, let $u(k) = u_k$, $k \geqslant 1$, and by placing hypotheses on $u(t)$. A natural generalization of (12) is

$$\frac{d}{dt}u(t) \leqslant 0, \qquad \frac{d^2}{dt^2}\ln u(t) \geqslant 0, \qquad t \geqslant 1, \tag{13}$$

or that $u(t)$ is a nonincreasing logarithmically convex function. Using these constraints on the u_i, we can prove the following general statements. The existence of a limiting u_∞ is sufficient to prove that a recency effect always occurs in sufficiently long lists:

(III) If the Invariance Principle and Normalization Rule both hold, and u_∞ exists, then in all sufficiently long lists, a recency effect develops. This follows from (9) and (10). By (10), a recency effect develops if the function

$$G(j) = w_j u_{j-1} - u_j \tag{14}$$

becomes negative as j becomes sufficiently large. To prove this using (9) and (10), one shows that $G(j) < 0$ is equivalent to

$$B_j - B_{j-1} < 1 + u_{j-1}^{-1}, \tag{15}$$

where $B_j = u_j^{-1}S_j$. Since $\lim_{j\to\infty} B_j = u_\infty^{-1}M$ exists, the left-hand side of (15) approaches 0 as $j \to \infty$ while the right-hand side exceeds 1.

Properties (II) and (III) indicate the tendency for the STM primacy effect, if it exists, to become weaker as list length increases. This is because more and more of the normalized total STM activity S_i ($\leqslant M$) gets devoted to the recency effect as $i \to \infty$.

Special hypotheses on $u(t)$ are needed only to show that no more than one bow occurs in the STM patterns $(x_1, x_2, ..., x_L)$ no matter how long the list length L is chosen. These hypotheses constrain the differential amounts of attention that can be paid to items without creating more than one how. Without such constraints, a unimodal STM bow is not guaranteed, and this fact is experimentally important, as will be illustrated in Section 8. A sufficient condition that an STM bow is unimodal, if it occurs, is given by the hypotheses (13):

(IV) If the Invariance Principle and Normalization Rule both hold, and $u(t)$ satisfies (13), then all the STM patterns $(x_1, x_2, ..., x_L)$ are either increasing, decreasing, or possess a unimodal bow. To prove that the bow is unimodal, function $G(j)$ is extended to a function $G(t)$ of a continuous variable $t \geqslant 1$. It is then verified using (7), (8), and (13) that if $G(T) = 0$ then $(d/dt)\, G(T) \leqslant 0$; that is, once the primacy effect becomes a recency effect, it can never flip back to a primacy effect.

To assert that a unimodal bow occurs if (7), (8), and (13) hold, it therefore suffices to guarantee that a primacy effect exists. By (10), this is true if and only if

$$G(2) = w_2 u_1 - u_2 > 0. \tag{16}$$

If (9) and (12) hold, (16) is true if and only if

$$u_1\theta + M(1 - \theta) > 2[u_1\phi + u_\infty(1 - \phi)], \tag{17}$$

If (16) holds, the bow occurs at that index $j = J$ where $G(j) = w_j u_{j-1} - u_j$ changes sign from positive to negative.

7. Masking of STM Primacy by Normalization

The above sections note a behavioral rationale, a physiological mechanism, and some data in which an STM primacy effect is implicated. Why then is there so little evidence of STM primacy in interference experiments, wherein the primacy effect is little changed before or after interference with STM, yet the recency effect is almost entirely obliterated? One factor is that the STM primacy effect becomes smaller as the list length increases. I suggest, however, that even when a large STM primacy effect exists, it can be masked due to normalization.

To understand how this can happen, a brief review of how LTM is encoded in the present framework is needed. A psychophysiological theory of LTM encoding in response to a list of times is successively developed in Grossberg (1969), Grossberg and Pepe (1971), Grossberg (1974), and Grossberg (1978a). Many of these results were applied to the study of serial learning and paired associate learning, but they are readily adapted to the free recall paradigm. To fix ideas, suppose that two fields $\mathscr{F}^{(1)}$ and $\mathscr{F}^{(2)}$ of populations $v_i^{(1)}$ and $v_j^{(2)}$, respectively, are given. Suppose that the populations $\mathscr{F}^{(1)}$ can send signals to the populations $\mathscr{F}^{(2)}$ over directed pathways, or axons. Let an LTM trace z_{ij} be computed at the end of the pathway from $v_i^{(1)}$ to $v_j^{(2)}$. Assume that z_{ij} obeys the equation

$$(d/dt)z_{ij} = -C_{ij}z_{ij} + S_{ij}(x_i^{(1)})x_j^{(2)}, \tag{18}$$

where C_{ij} is the LTM decay rate, $S_{ij}(x_i^{(1)})$ is the sampling signal from $v_i^{(1)}$ to $v_j^{(2)}$, $x_i^{(1)}$ is the STM trace of $v_i^{(1)}$, and $x_j^{(2)}$ is the STM trace of $v_j^{(2)}$ (Fig. 7). If $x_i^{(1)}$ is sufficiently large to make $S_{ij}(x_i^{(1)}) > 0$, then z_{ij} can sample the STM trace $x_j^{(2)}$. Thus LTM in the model depends on a nonlinear mechanism that time-averages (via the term $-C_{ij}z_{ij}$) the products of sampling signals and STM traces (via the term $S_{ij}(x_i^{(1)})x_j^{(2)}$). Each $v_i^{(1)}$ controls all the LTM traces $z_{i1}, z_{i2}, z_{i3},...$ via its sampling signals $S_{i1}(x_i^{(1)})$, $S_{i2}(x_i^{(1)})$, $S_{i3}(x_i^{(1)})$,.... We therefore say that $v_i^{(1)}$ controls the LTM pattern $z_i = (z_{i1}, z_{i2}, z_{i3},...)$. All of the LTM patterns z_i can be different. These patterns are generated, via the products $S_{ij}(x_i^{(1)})x_j^{(2)}$, by the distributions of STM activity that evolve across $\mathscr{F}^{(1)}$ and $\mathscr{F}^{(2)}$ through time. The experimental paradigms of serial learning, paired associate learning, and free recall can all produce different STM patterns, and hence different LTM patterns.

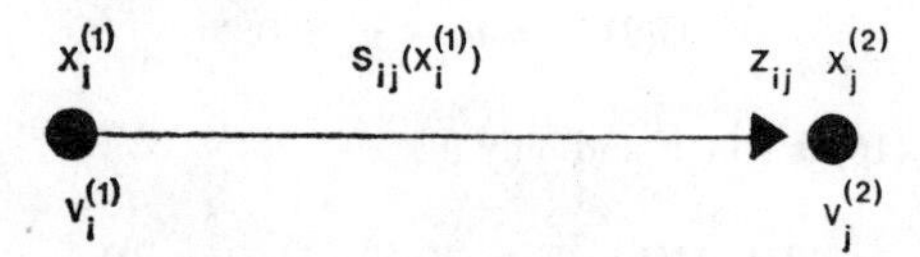

FIG. 7. The signal $S_{ij}(x_i^{(1)})$ from $v_i^{(1)}$ is gated by z_{ij} on its way to $v_j^{(2)}$.

How is the LTM pattern z_i read out to influence performance at some population $v_j^{(2)}$? This can only happen if $v_i^{(1)}$ is activated enough to elicit a sampling signal $S_{ij}(x_i^{(1)})$. This activation can, for example, be due to a probe stimulus that excites the STM trace $x_i^{(1)}$ of $v_i^{(1)}$, or to a lingering STM activity $x_i^{(1)}$ that is due to prior stimuli. Different experimental paradigms can generate different sequences of probes and hence different performance characteristics.

Readout from the LTM trace z_{ij} occurs when the sampling signal $S_{ij}(x_i^{(1)})$ from $v_i^{(1)}$ is gated by z_{ij} on its way to $v_j^{(2)}$. The net signal to $v_j^{(2)}$ is $S_{ij}(x_i^{(1)})z_{ij}$. This gating mechanism is also a nonlinear process. Every active population in $\mathscr{F}^{(1)}$ can produce such a gated signal to $v_j^{(2)}$. The total effect of $\mathscr{F}^{(1)}$ on $v_j^{(2)}$ is given by the sum of these gated signals; namely, by

$$T_j = \sum_i S_{ij}(x_i^{(1)})z_{ij}. \tag{19}$$

Thus LTM can affect $\mathscr{F}^{(2)}$ only if it is activated by STM-driven signals from $\mathscr{F}^{(1)}$.

The signal T_j in (19) influences $v_j^{(2)}$ by activating its STM trace $x_j^{(2)}$. Such a signal is received by every $v_j^{(2)}$, and hereby generates a pattern $(T_1, T_2, T_3,...)$ of inputs across $\mathscr{F}^{(2)}$. Competitive interactions within $\mathscr{F}^{(2)}$ transform this input pattern to produce the final STM pattern $(x_1^{(2)}, x_2^{(2)}, x_3^{(2)},...)$ across $\mathscr{F}^{(2)}$. Normalization of STM within $\mathscr{F}^{(2)}$ is a particular consequence of these competitive interactions. After the populations in $\mathscr{F}^{(2)}$ compete for STM activity, the effects of this competition feed back into each LTM pattern due to the STM term $x_j^{(2)}$ in (18). The STM competition hereby tends to produce ratio scales in LTM as well as in STM; cf., Eq. (4). This (approximate) LTM ratio scale helps to explain competitive retrieval rules in free recall experiments (Rundus, 1973). Note also that the sampling signal $S_{ij}(x_i^{(1)})$ both controls performance, via (19), and strengthens learning, via (18). This helps to physiologically explain how test trials can act as training trials (Lachman & Laughery, 1968; Tulving, 1967), since making $S_{ij}(x_i^{(1)})$ large enough to elicit performance also makes it large enough to strengthen the LTM trace z_{ij}. In more general physiological models, unbiased simultaneous sampling by many cues of the same event is impossible unless the performance signal is large only if the learning signal is also large. This constraint is called the *local flow* condition (Grossberg, 1972; 1974, Sect. VI).

What types of LTM patterns can evolve? Suppose for simplicity that each $v_i^{(1)}$ sends the same sampling signal $S_i(x_i^{(1)})$ to all cells in $\mathscr{F}^{(2)}$. Also suppose that the STM patterns across $\mathscr{F}^{(1)}$ and $\mathscr{F}^{(2)}$ exhibit either an STM recency gradient or an STM bow due to presentation of the list $r_1, r_2, r_3, \ldots, r_L$. Then the pattern z_1 learns an LTM primacy gradient; namely, $z_{11} > z_{12} > z_{13} > \cdots$ (The proof is in Grossberg (1977a, Sect. 32)). If a probe stimulus excites $v_1^{(1)}$ on performance trials, then the signals from $v_1^{(1)}$ to $\mathscr{F}^{(2)}$ satisfy $S_1(x_1^{(1)})\, z_{11} > S_1(x_1^{(1)})\, z_{12} > S_1(x_1^{(1)})\, z_{13} > \cdots$. These signals elicit an STM primacy effect across $\mathscr{F}^{(2)}$. Thus, even if $v_1^{(1)}$ samples an STM recency gradient across $\mathscr{F}^{(2)}$ on learning trials, it can perform an STM primacy gradient across $\mathscr{F}^{(2)}$ on performance trials. This is due to the nonlinear nature of Eqs. (18) and (19). The other populations $v_i^{(1)}$, $i \neq 1$, usually do not learn an LTM primacy gradient. For example, if $v_i^{(1)}$ and $v_i^{(2)}$ are simultaneously excited, $i = 1, 2, \ldots, L$, and both experience STM recency gradients, then the maximum LTM trace in pattern z_i is z_{ii}, and the other traces z_{ij} decrease as a function of $|i - j|$ to produce a generalization gradient that is centered at $v_i^{(2)}$. Moreover, if the STM recency gradients are inhibited by interfering activities right after the last list item is presented, then $z_{11} > z_{22} > z_{33} > \cdots$. Most LTM storage is therefore concentrated at the populations that are excited by the beginning of the list, especially in the LTM primacy gradient controlled by $v_1^{(2)}$. (This is not true in serial learning, where a bow in the LTM pattern of correct associations can occur.)

Given the above summary, we can now see how an STM primacy gradient can be masked by normalization. To fix ideas, suppose that there exist three fields $\mathscr{F}^{(1)}$, $\mathscr{F}^{(2)}$, and $\mathscr{F}^{(3)}$ of populations. Let $\mathscr{F}^{(1)}$ consist of acoustically coded populations, $\mathscr{F}^{(2)}$ consist of semantically coded populations, and $\mathscr{F}^{(3)}$ consist of motor control populations (Baddeley & Warrington, 1970; Bartlett & Tulving, 1974; Craik, 1970; Craik & Lockhart, 1972; Jacoby & Bartz, 1972; Maskorinec & Brown, 1974). Suppose that both $\mathscr{F}^{(1)}$ and $\mathscr{F}^{(2)}$ can send signals to $\mathscr{F}^{(3)}$, from which performance is controlled by a nonspecific source of motor arousal. Also let LTM traces z_{ij} occur in the pathways from $\mathscr{F}^{(2)}$ to $\mathscr{F}^{(3)}$, with z_1 coding a primacy gradient and the other z_i, $i \neq 1$, coding the type of gradients summarized above. Suppose moreover that prior presentation of a list $r_1, r_2, \ldots, r_L$ establishes either an STM recency gradient or an STM bow across $\mathscr{F}^{(1)}$, and either an STM primacy gradient or an STM bow across $\mathscr{F}^{(2)}$ (Fig. 8a). By (19), the STM primacy gradient across $\mathscr{F}^{(2)}$ magnifies the LTM primacy gradient that it reads out of LTM into STM at $\mathscr{F}^{(3)}$. This magnified STM primacy gradient at $\mathscr{F}^{(3)}$ has to compete, however, with the STM

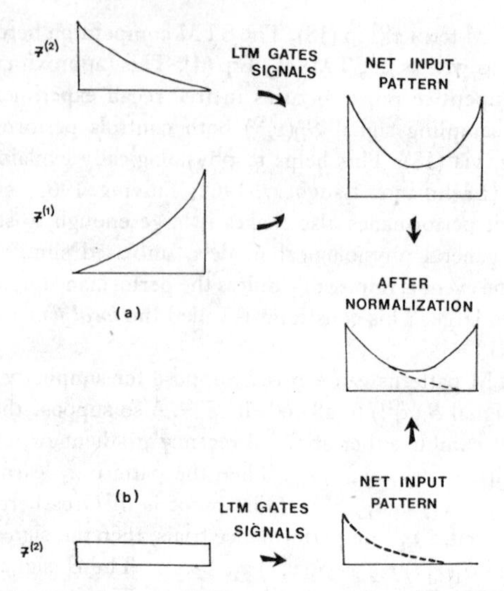

FIG. 8. Normalization can mask an STM primacy gradient by competition with an STM recency gradient.

recency gradient that $\mathscr{F}^{(1)}$ reads into $\mathscr{F}^{(3)}$. After normalization takes place across $\mathscr{F}^{(3)}$, the magnified STM primacy gradient is reduced due to competition by the STM recency gradient.

By contrast, consider what happens if an interfering event inhibits all STM in $\mathscr{F}^{(1)}$, $\mathscr{F}^{(2)}$, and $\mathscr{F}^{(3)}$. Correct performance is then impossible unless $\mathscr{F}^{(2)}$ is activated, since only the pathways from $\mathscr{F}^{(2)}$ to $\mathscr{F}^{(3)}$ contain LTM traces. Suppose that $\mathscr{F}^{(2)}$ is activated either randomly, or uniformly using a nonspecific sampling signal. In either case, the LTM primacy gradient will be read into an STM primacy gradient across $\mathscr{F}^{(3)}$ (Fig. 8b). This LTM primacy gradient is not reinforced by an STM primacy gradient across $\mathscr{F}^{(2)}$, so the net effect on $\mathscr{F}^{(3)}$ is smaller than before. However, the STM primacy gradient across $\mathscr{F}^{(3)}$ does not have to compete with an STM recency gradient. Hence normalization will amplify this STM primacy gradient, just as it amplifies the net gradient in Fig. 4. In all, normalization can mask an STM primacy gradient by differentially suppressing its effect when an STM recency gradient exists.

8. OTHER BINARY CODE DIFFICULTIES

The serial binary code assumption leads to several other beliefs that might be stated too strongly. For example, if STM storage is thought of as a binary event, then one can readily conclude that the number of rehearsals is the crucial parameter that determines whether an item remains or is reinstated in the STM buffer. For example, Bower writes "if the person is told during study of an item that its later recall will be worth a lot of money, he will concentrate harder (rehearse more, maintain that item in STM for a longer time) and remember it better" (Hilgard & Bower, 1975, p. 580). However "concentrating harder," "rehearsing more," and "maintaining that item in STM for a longer time" can all be achieved by distinct mechanisms. Concentrating harder can, for example, generate an unusually large arousal level that supplements the item's usual input to STM (Grossberg, 1975). The two input sources acting together can create an unusually large STM strength that, other things equal (which they usually are not), increases the probability of saying the item, and of saying it out of order; in particular, saying it at an earlier recall position, than would otherwise occur. These effects can be generated without rehearsing this item any more than any other item.

A boost in STM strength of one item can depress the STM strengths of related items that are simultaneously in STM; cf., Ellis *et al.*, 1971. This von Restorff-type of depression can sometimes be due to the Normalization Rule, rather than to less rehearsal.

The fact that reaction time increases with the number of items in STM does not imply that recognition memory is realized by a serial scanning process (Sternberg, 1966). In a normalized STM field, each item in the field—except possibly the last—has a smaller STM trace if a longer list perturbs the field. If reaction time increases as STM activity decreases, then reaction time will depend on how many items are stored in STM, even though the rehearsal operation is a parallel operation that simultaneously influences all populations in the STM field.

9. Concluding Remarks

During the last decade, experimental and theoretical studies of STM and LTM have been remarkably vigorous and productive. The use of theoretical analogies from other disciplines, such as the computer analogy, indicates a healthy desire to conceptually organize the vast array of experimental findings. However, the binary and serial nature of computer concepts leads to conceptual difficulties in the many situations where continuous and parallel brain processes are operative. Computer modelers often claim that details like whether a code is binary or continuous are unimportant, because the same global strategies, or wiring diagrams, will occur despite differences in the individual components. Similarly, Townsend's (1974) result showing the equivalence of serial and parallel reaction time models has sometimes been interpreted as saying that it does not matter which type of model is used when memory processes are studied. The above examples are a few of the growing number that show binary serial models and continuous parallel models to be fundamentally different in design and properties.

REFERENCES

Anderson, J. R. and Bower, G. H.: *Human associative memory*. Washington, D.C.: Hemisphere, 1974.
Atkinson, R. C. and Shiffrin, R. M.: Human memory: a proposed system and its control processes. In K. W. Spence and J. T. Spence (Eds.), *Advances in the psychology of learning and motivation, research and theory*, Vol. II. New York: Academic Press, 1968.

Atkinson, R. C. and Shiffrin, R. M.: The control of short term memory. *Scientific American,* August, 1971, 82.

Baddeley, A. D. and Warrington, E. K.: Amnesia and the distinction between long- and short-term memory. *Journal of Verbal Learning and Verbal Behavior*, 1970, **9**, 176–189.

Bartlett, J. C. and Tulving, E.: Effect of temporal and semantic encoding in immediate recall upon subsequent retrieval. *Journal of Verbal Learning and Verbal Behavior*, 1974, **13**, 297–309.

Bloomfield, T. M.: Behavioral contrast and the peak shift. In R. M. Gilbert and N. S. Sutherland (Eds.), *Animal discrimination learning.* New York: Academic Press, 1969. Pp. 215–241.

Cornsweet, T. N.: *Visual perception.* New York: Academic, 1970.

Craik, F. I. M.: The fate of primary memory items in free recall. *Journal of Verbal Learning and Verbal Behavior*, 1970, **9**, 143–148.

Craik, F. I. M. and Lockhart, R. S.: Levels of processing: A framework for memory research. *Journal of Verbal Learning and Verbal Behavior*, 1972, **11**, 671–684.

Craik, F. I. M. and Watkins, M. J.: The role of rehearsal in short-term memory. *Journal of Verbal Learning and Verbal Behavior*, 1973, **12**, 599–607.

Donchin, E. and Lindsley, D. B.: *Average evoked potentials: Methods, results and evaluations. Washington, D.C.: NASA, 1969.*

Ellis, N. R., Detterman, D. K., Runcie, D., McCarver, R. B., and Craig, E. M.: Amnesic effects in short-term memory. *Journal of Experimental Psychology*, 1971, **89**, 357–361.

Grossberg, S.: On the serial learning of lists. *Mathematical Biosciences*, 1969, **4**, 201–253.

Grossberg, S.: Pattern learning by functional-differential neural networks with arbitrary path weights. In K. Schmitt (Ed.), *Delay and functional differential equations and their applications.* New York: Academic Press, 1972. Pp. 121–160.

Grossberg, S.: Contour enhancement, short term memory, and constancies in reverberating neural networks. *Studies in Applied Mathematics*, 1973, **52**, 213–257.

Grossberg, S.: Classical and instrumental learning by neural networks. In R. Rosen and F. Snell (Eds.), *Progress in theoretical biology.* New York: Academic Press, 1974. Pp. 51–141.

Grossberg, S.: A neural model of attention, reinforcement, and discrimination learning. In C. Pfeiffer (Ed.), *International Review of Neurobiology*, 1975, **18**, 263–327.

Grossberg, S.: Adaptive pattern classification and universal recoding. I. Parallel development and coding of neural feature detectors. *Biological Cybernetics*, 1976a, **23**, 121–134.

Grossberg, S.: Adaptive pattern classification and universal recoding. II. Feedback, expectation, olfaction, illusions. *Biological Cybernetics*, 1976b, **23**, 187–202.

Grossberg, S.: Pattern formation by the global limits of a nonlinear competitive interaction in *n* dimensions. *Journal of Mathematical Biology*, 1977, **4**, 237–256.

Grossberg, S.: A theory of human memory: Self-organization and performance of sensory-motor codes, maps, and plans. In R. Rosen and F. Snell (Eds.), *Progress in theoretical biology.* New York: Academic Press, 1978a.

Grossberg, S.: A theory of visual coding, memory and development. In E. Leeuwenberg and H. Buffart (Eds.). *Formal theories of visual perception.* New York: Wiley, 1978b.

Grossberg, S.: Communication, memory, and development. In R. Rosen and F. Snell (Eds.), *Progress in theoretical biology*. New York: Academic Press, 1978c.

Grossberg, S. and Levine, D. S.: Some developmental and attentional biases in the contrast enhancement and short term memory of recurrent neural networks. *Journal of Theoretical Biology*, 1975, **53**, 263–327.

Grossberg, S. and Pepe, J.: Spiking threshold and over-arousal effects in serial learning. *Journal of Statistical Physics*, 1971, 3, 95–125.

Hilgard, E. R. and Bower, G. H.: *Theories of Learning*, 4th ed. Englewood Cliffs, N. J.: Prentice-Hall, 1975.

Hodgkin, A. L.: *The conducting of the nervous impulse.* Springfield, Ill.: C. C. Thomas, 1964.

Hogan, R. M. and Hogan, M. M.: Structural and transient components of memory. *Memory and Cognition*, 1975, **3**, 210–215.

Horowitz, L. W. and Izawa, C.: Comparison of serial and paired-associate learning. *Journal of Experimental Psychology*, 1963, **65**, 352–361.

Jacoby, L. L. and Bartz, W. H.: Rehearsal and transfer to long term memory. *Journal of Verbal Learning and Verbal Behavior*, 1972, **11**, 561–565.

Kahneman, D. and Beatty, J.: Pupil diameter and load on memory. *Science*, 1966, **154**, 1583–1585.

Katz, B.: *Nerve, muscle, and synapse.* New York: McGraw-Hill, 1966.

Lachman, R. and Laughery, K. R.: Is a test trial a training trial in free recall learning? *Journal of Experimental Psychology*, 1968, **76**, 40–50.

Lenneberg, E.: *Biological foundations of language.* New York: Wiley, 1967.

Levine, D. S. and Grossberg, S.: Visual illusions in neural networks: Line neutralization, tilt after effect and angle expansion. *Journal of Theoretical Biology*, 1976, **61**, 477–504.

Luce, R. D.: *Individual choice behavior.* New York: Wiley, 1959.

Maskorinec, A. S. and Brown, S. C.: Positive and negative recency effects in free recall learning. *Journal of Verbal Learning and Verbal Behavior*, 1974, **16**, 328–334.

Melton, A. W. and Martin, E. (Eds.): *Coding processes in human memory.* Washington, D.C.: Winston, 1972.

Miller, G. A.: The magic number seven, plus or minus two. *Psychological Review*, 1956, **63**, 81–97.

Milner, B.: Amnesia following operation on the temporal lobes. In C. W. M. Whitty and O. L. Zangwill (Eds.). *Amnesia.* London: Butterworths, 1956.

Restle, F., Shiffrin, R. M., Costellan, N. J., Lindman, H. R., and Pisoni, D. B. (Eds.): *Cognitive theory*, Vol. 1. Hillsdale, N. J.: Erlbaum, 1975.

Rundus, D. J.: Analysis of rehearsal processes in free recall. *Journal of Experimental Psychology*, 1971, **89**, 63–77.

Rundus, D.: Negative effects of using list items as recall cues. *Journal of Verbal Learning and Verbal Behavior*, 1973, **12**, 43–50.

Sternberg, S.: High-speed scanning in human memory. *Science*, 1966, **153**, 652–657.

Townsend, J. T.: Issues and models concerning the processing of a finite number of inputs. In Kantowitz, B. H. (Ed.): *Human information processing: Tutorials in performance and cognition.* Potomac, Md.: Erlbaum, 1974. P. 133.

Tulving, E.: The effects of presentation and recall of material in free-recall learning. *Journal of Verbal Learning and Verbal Behavior*, 1967, **6**, 175–184.

Tulving, E. and Donaldson, W. (Eds.): *Organization of memory.* New York: Academic Press, 1972.

Werblin, F. S.: Adaptation in a vertebrate retina: Intracellular recording in Necturus. *Journal of Neurophysiology*, 1971, **34**, 228–241.

Woodward, A. E., Bjork, R. A., and Jongeward, R. H., Jr.: Recall and recognition as a function of primary rehearsal. *Journal of Verbal Learning and Verbal Behavior*, 1973, **12**, 608–617.

Young, R. K.: A comparison of two methods of learning serial associations. *American Journal of Psychology*, 1959, **72**, 554–559.

Young, R. K.: The stimulus in serial learning. *American Journal of Psychology*, 1961, **74**, 517–528.

Young, R. K.: Tests of three hypotheses about the effective stimulus in serial learning. *Journal of Experimental Psychology*, 1963, **63**, 307–313.

Young, R. K.: Serial learning. In T. R. Dixon and D. L. Horton (Eds.), *Verbal Learning and General Behavior Theory.* Englewood Cliffs, N.J.: Prentice-Hall, 1968. Pp. 122–148.

Zeller, M. D.: The ratio theory of intermediate size discrimination. *Psychological Review*, 1963, **70**, 516–533.

RECEIVED: July 15, 1977

CHAPTER 12

ADAPTIVE PATTERN CLASSIFICATION AND UNIVERSAL RECODING

I: PARALLEL DEVELOPMENT AND CODING OF NEURAL FEATURE DETECTORS

PREFACE

This is the second of a three part series of articles on code development that appeared in 1976. The first article [36] responded to Malsburg's addition of a normalization rule to the equations of Chapter 7. Malsburg's rule directly constrains the total LTM strength of synaptic contacts to each cell. I realized that if this rule held in *all* learning cells, then classical conditioning would be impossible. I had realized a decade earlier (Chapter 2) that a direct LTM constraint can often be replaced by an STM constraint that influences LTM indirectly. Also my 1973 work on STM in shunting competitive networks was done, so I knew that the STM competition produces the normalization property for free, and does not contradict classical conditioning. My first article made these points, substituted shunting STM competition for additive competition and eliminated Malsburg's LTM normalization rule.

The first article then sketched how the new theory works. I later realized that the simplest coding process in the theory imitates a Bayesian decision rule for minimizing risk in the presence of ambiguous data. However, the theory of adaptive resonances goes way beyond Bayesian capabilities.

The first article also pointed out that the theory's formal computations are remarkably similar to certain reaction-diffusion models of morphogenesis, notably the model of Meinhardt and Gierer. This comparison represented another step towards the realization that neural design principles are often special cases of general principles of cellular design. This belief is developed more fully via a series of nonneural morphogenetic examples in [43], where I began to lay the groundwork for a universal developmental code, or evolutionary measurement theory. I have gradually come to realize, however, that many contemporary morphogenetic models omit important cellular design constraints, probably because they were derived from analogies with chemical reaction kinetics. This is especially true in their discussion of self-regulation [52].

The present paper takes a hard mathematical look at my theory of code development. The work of Malsburg and other scientists on coding was usually carried out by numerically simulating the reaction of small numbers of cells to small numbers of input patterns. These investigators hereby overlooked a fundamental fact. I was able to prove that the network can generate a stable code if there do not exist too many input patterns relative to the number of cells, as in the typical numerical set up, or even if many input patterns are distributed sparsely in pattern space. However, the code is unstable if too many input patterns are presented, which is the typical situation *in vivo*, or at best can be stabilized by a law which is insensitive to the behavioral meaning of the developing code. I realized that correcting this instability would require a context-sensitive mechanism for the termination of critical periods during development. I could see from my proofs that the instability was due to the feedforward flow of data patterns from the periphery inward. Feedback was necessary. What type of feedback?

Having been propelled this far by mathematics, the intuitive answer was now clear: The same feedback mechanisms that buffer adult attentional processing during overshadowing can buffer developing codes during the critical period. In other words, I did not have to add *any* further structure to stabilize the code. I just had to remember to include attentional mechanisms! This leap from adult attention to infant code development was most exhilarating, especially considering the fact that once I broke through the wall, I recognized my surroundings very well indeed.

The third article in the series [39] led by this line of reasoning to my theory of adaptive resonances. In this article, I related code development to adult attentional processing, and explained or predicted a variety of phenomena about development and perception.

One of my predictions was that a catecholaminergic cortical arousal system buffers the developing code by driving an antagonistic rebound from cortical on-cell activation to off-cell activation if erroneous filtering occurs. I later learned that about the same time, Pettigrew and Kasamatsu poisoned the cortical catecholaminergic arousal system with 6-hydroxydopamine (6-OHDA) in young cats and observed that plasticity was diminished during the critical period. In a 1978 paper, they added noradrenaline (NA) to the cortex of adult cats after their critical period had ended and found that plasticity was restored. These properties also formally occur in my networks, wherein the loss of plasticity can be traced to the occurrence of transmitter gates in the feedback loops that must resonate to drive the learning process.

The transmitter gates must occur in the feedback loops so that a rapid increment in nonspecific arousal can rebound the STM activities across on-cell off-cell dipoles and thereby reset STM to prevent adventitious recoding. The effect of 6–OHDA on these loops would be to block the resonant process by poisoning the transmitter (NA) terminals. Pettigrew's 1978 data is especially striking because the NA arousal system is still intact in adult cats. Why then does pouring some extra NA into the cortex restore plasticity? In my theory, the answer is: The extra NA disrupts the dynamic buffering that protects against recoding by making NA available for resonance in circuits that would normally be inhibited by the cortical (dipole) competitive process.

These phenomena are still not discussed in dynamic terms by most neurophysiologists. In fact, at the 1979 annual Society for Neuroscience meeting, the distinguished keynote speaker, W. M. Cowan, wondered aloud during his speech about what ingenious stroke had motivated Pettigrew's experiment and about what it could possibly mean. Since my theory also correlates a number of perceptual and psychophysiological phenomena with this neurophysiological mechanism, some further interdisciplinary experimental tests might greatly clarify the meaning of Pettigrew's challenging data.

Adaptive Pattern Classification and Universal Recoding:

I. Parallel Development and Coding of Neural Feature Detectors *

Abstract. This paper analyses a model for the parallel development and adult coding of neural feature detectors. The model was introduced in Grossberg (1976). We show how experience can retune feature detectors to respond to a prescribed convex set of spatial patterns. In particular, the detectors automatically respond to average features chosen from the set even if the average features have never been experienced. Using this procedure, any set of arbitrary spatial patterns can be recoded, or transformed, into any other spatial patterns (universal recoding), if there are sufficiently many cells in the network's cortex. The network is built from short term memory (STM) and long term memory (LTM) mechanisms, including mechanisms of adaptation, filtering, contrast enhancement, tuning, and nonspecific arousal. These mechanisms capture some experimental properties of plasticity in the kitten visual cortex. The model also suggests a classification of adult feature detector properties in terms of a small number of functional principles. In particular, experiments on retinal dynamics, including amacrine cell function, are suggested.

* Supported in part by the Advanced Research Projects Agency under ONR Contract No. N00014-76-C-0185

1. Introduction

This paper analyses a model for the development of neural feature detectors during an animal's early experience with its environment. The model also suggests mechanisms of adult pattern discrimination that remain after development has been completed. The model evolved from earlier experimental and theoretical work. Various data showed that there is a critical period during which experimental manipulations can alter the patterns to which feature detectors in the visual cortex are tuned (e.g., Barlow and Pettigrew, 1971; Blakemore and Cooper, 1970; Blakemore and Mitchell, 1973; Hirsch and Spinelli, 1970, 1971; Hubel and Wiesel, 1970; Wiesel and Hubel, 1963, 1965). This work led Von der Malsburg (1973) and Pérez et al. (1974) to construct models of the cortical tuning process, which they analysed using computer methods. Their models are strikingly similar. Both use a mechanism of long term memory (LTM) to encode changes in tuning. This mechanism learns by classical, or Pavlovian, conditioning (Kimble, 1967) within a neural network. Such a concept was qualitatively described by Hebb (1949) and was rigorously analysed in its present form by Grossberg (e.g., 1967, 1970a, 1971, 1974). The LTM mechanism in a given interneuronal pathway is a plastic synaptic strength which has two crucial properties: (a) it is computed from a time average of the product of presynaptic signals and postsynaptic potentials; (b) it multiplicatively gates, or shunts, a presynaptic signal before it can perturb the postsynaptic cell.

Given this LTM mechanism, both models invoke various devices to regulate the retinocortical signals that drive the tuning process. On-center off-surround networks undergoing additive interactions, attenuation of small retinocortical signals at the cortex, and conservation of the total synaptic strength impinging on each cortical cell are used in both models. Grossberg (1976) realized that all of these mechanisms for distributing signals could be replaced by a minimal model for parallel processing of patterns in noise, which is realized by an on-center off-surround recurrent network whose interactions are of shunting type (Grossberg, 1973). Three crucial properties of this model are: (a) normalization, or adaptation, of total network activity; (b) contrast enhancement of input patterns; and (c) short term memory (STM) storage of the contrast-enhanced pattern. Using these properties, Grossberg (1976) eliminates the conservation of total synaptic strength—which is incompatible with classical conditioning—and shows that the tuning process can be derived from *adult* STM and LTM principles. The model is schematized in Figure 1. It describes the interaction via plastic synaptic pathways of two network regions, V_1 and V_2, that are separately capable of normalizing patterns, but V_2 can also contrast enhance patterns and store them in STM. In the original models of Von der Malsburg and Pérez et al., V_1 was interpreted as a "retina" or "thalamus"

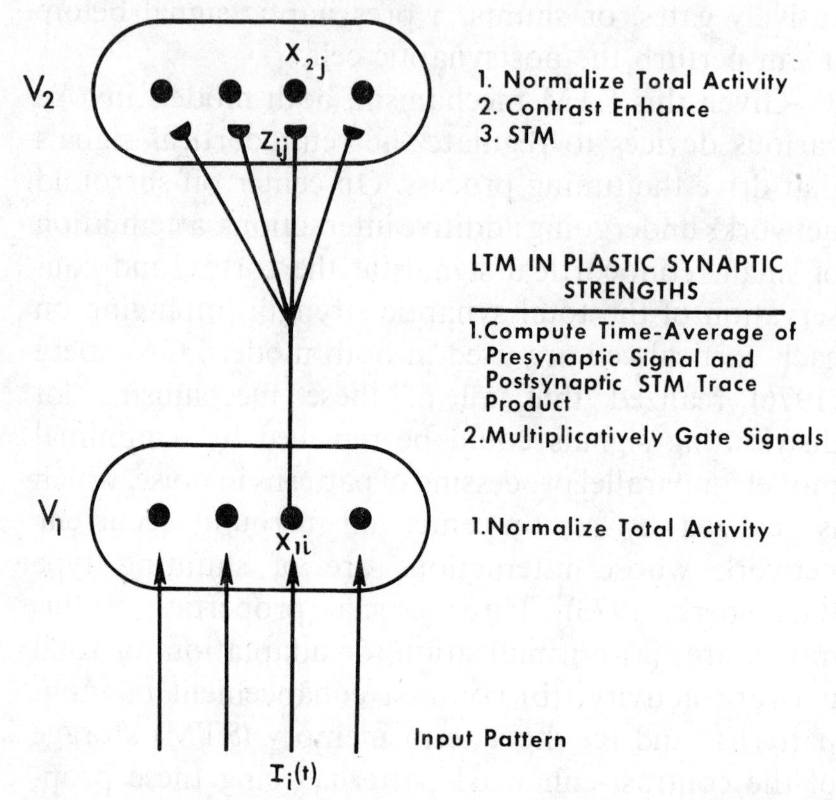

Fig. 1. Minimal model of developmental tuning using STM and LTM mechanisms

and V_2 as "visual cortex". In Part II, an analogous anatomy for V_1 as "olfactory bulb" and V_2 as "prepyriform cortex" will be noted. In Section 5, a more microscopic analysis of the model leads to a discussion of V_1 as a composite of retinal receptors, horizontal cells, and bipolar cells, and of V_2 as a composite of amacrine cells and ganglion cells. Such varied interpretations are possible because the same functional principles seem to operate in various anatomies.

Using this abstract structure, it was suggested in Grossberg (1976) how hierarchies of cells capable of discriminating arbitrary spatial patterns can be synthesized. Also a striking analogy was described between the structure and properties of certain reaction-diffusion systems that have been used to model development (Gierer and Meinhardt, 1972; Meinhardt and Gierer, 1974) and of reverberating shunting networks. This paper continues this program by rigorously analysing mathematical properties of the model, which thereupon suggest other developmental and adult STM and LTM mechanisms that are related to it. The following sections will describe these connections with a minimum of mathematical detail. Mathematical proofs are contained in the Appendix.

2. The Tuning Process

This section reviews properties of the model that will be needed below. Suppose that V_1 consists of n states (or cells, or cell populations) v_{1i}, $i=1, 2, \ldots, n$, which receive inputs $I_i(t)$ whose intensity depends on the presence of a prescribed feature, or features, in an external pattern. Let the population response (or activity, or average potential) of v_{i1} be $x_{1i}(t)$. The relative input intensity $\Theta_i = I_i I^{-1}$, where $I = \sum_{k=1}^{n} I_k$, measures the relative importance of the feature coded by v_i in any given input pattern. If the Θ_i's are constant during a given time interval, the inputs are said to form a *spatial pattern*. How can the laws governing the $x_{1i}(t)$

be determined so that $x_{1i}(t)$ is capable of accurately registering Θ_i? Grossberg (1973) showed that a bounded, linear law for x_{1i}, in which x_{1i} returns to equilibrium after inputs cease, and in which neither input pathways nor populations v_{1i} interact, does not suffice; cf., Grossberg and Levine (1975) for a review. The problem is that as the total input I increases, given *fixed* Θ_i values, each x_{1i} saturates at its maximal value. This does not happen if off-surround interactions also occur. For example, let the inputs I_i be distributed via a nonrecurrent, or feedforward, on-center off-surround anatomy undergoing shunting (or mass action, or passive membrane) interactions, as in Figure 2. Then

$$\dot{x}_{1i} = -Ax_{1i} + (B - x_{1i})I_i - x_{1i} \sum_{k \neq i} I_k \tag{1}$$

with $0 \leqq x_{1i}(0) \leqq B$. At equilibrium (namely, $\dot{x}_{1i} = 0$),

$$x_{1i} = \Theta_i \frac{BI}{A + I}, \tag{2}$$

which is proportional to Θ_i no matter how large I becomes. Since also $BI(A+I)^{-1} \leqq B$, the total activity $x_1 \equiv \sum_{k=1}^{n} x_{1k}$ never exceeds B; it is normalized, or adapts, due to automatic gain control by the inhibitory inputs. The normalization property in (2) shows that x_{1i} codes Θ_i rather than instantaneous fluctuations in I.

To store patterns in STM, recurrent or feedback

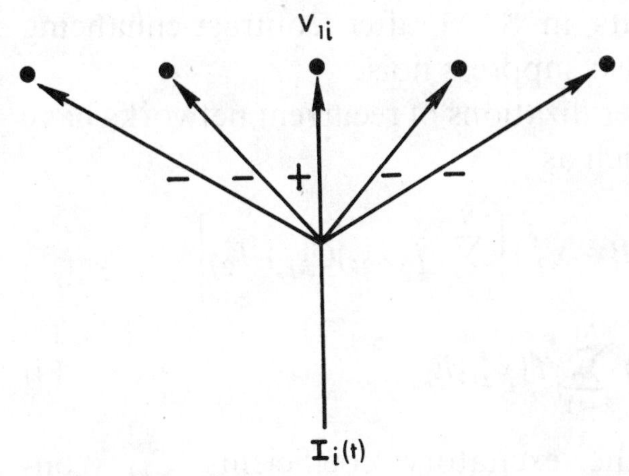

Fig. 2. Nonrecurrent, or feedforward, on-center off-surround network

pathways are needed to keep signals active after the inputs cease. Again the problem of saturation must be dealt with, so that some type of recurrent on-center off-surround anatomy is suggested. The minimal solution is to let V_2 be governed by a system of the form

$$\dot{x}_{2j} = -Ax_{2j} + (B - x_{2j})[f(x_{2j}) + I_{2j}] - x_{2j} \sum_{k \neq j} f(x_{2k}), \tag{3}$$

where $f(w)$ is the average feedback signal produced by an average activity level w, and I_{2j} is the total excitatory input to v_{2j} (Fig. 3a). In particular, v_{2j} excites itself via the term $(B - x_{2j})f(x_{2j})$, and v_{2k} inhibits v_{2j} via the term $-x_{2j}f(x_{2k})$, for every $k \neq j$. The choice of $f(w)$ dramatically influences how recurrent interactions within V_2 transform the input pattern $I^{(2)} = (I_{21}, I_{22}, \ldots, I_{2N})$ through time. Grossberg (1973) shows that a sigmoid, or S-shaped, $f(w)$ can reverberate important inputs in STM after contrast-enhancing them, yet can also suppress noise.

Various generalizations of recurrent networks have been studied, such as

$$\dot{x}_{2j} = -Ax_{2j} + (B - x_{2j})\left[\sum_{k=1}^{N} f(x_{2k})C_{kj} + I_{2j}\right] - (x_{2j} + D)\sum_{k=1}^{N} f(x_{2k})E_{kj}, \tag{4}$$

$D \geqq 0$, where the excitatory coefficients C_{kj} ("on-center") decrease with the distance between populations v_{2k} and v_{2j} more rapidly than do the inhibitory coefficients E_{kj} ("off-surround"). Levine and Grossberg (1976) show that, in such cases, the inhibitory off-surround signals $\sum_{k=1}^{N} f(x_{2k})E_{kj}$ to v_{2j} can be chosen strong enough to offset the saturating effects of inputs I_{2j} plus excitatory on-center signals $\sum_{k=1}^{N} f(x_{2k})C_{kj}$. Ellias and Grossberg (1975) study generalizations of (4) in which inhibitory interneurons interact with their excitatory counterparts.

Below we will consider networks in which the excitatory signals I_{2j} to V_2 are sums of signals from many populations in V_1. Moreover, the synaptic strengths of these signals can be trained. This fact suggests another reason for making V_2 recurrent. A recurrent anatomy is needed within V_2 to prevent saturation in response to trainable signals. To see this, note in the nonrecurrent network (1) that each excitatory input to v_{1i} is replicated as an inhibitory input to all v_{1k}, $k \neq i$. The size of a trainable signal to v_{2j}

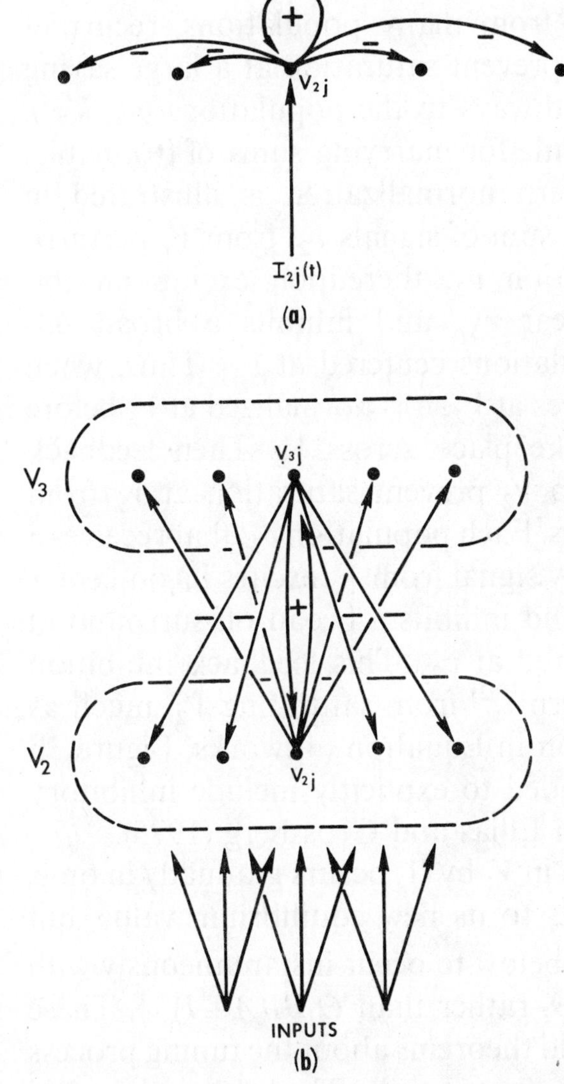

Fig. 3. Some recurrent, or feedback, on-center off-surround networks

depends on the activity at v_{2j}. This signal therefore cannot be replicated at populations v_{2k}, $k \neq j$, unless recurrent interactions within V_2 exist. Moreover, whether or not signals are trainable, whenever I_{2j} is a sum of signals from many populations, recurrent signals within V_2 prevent saturation at a large saving of extra signal pathways to the populations v_{2k}, $k \neq j$.

A related scheme for marrying sums of (trainable) signals with pattern normalization is illustrated in Figure 3b. Here a sum of signals I_{2j} from V_1 perturbs each v_{2j}. Population v_{2j} thereupon excites an on-center of cells near v_{3j}, and inhibits a broad off-surround of populations centered at v_{3j}. Thus, when a pattern $I^{(2)}$ arrives at V_2, it is normalized at V_3 before saturation can take place across V_2. Then feedback signals from V_3 to V_2 prevent saturation at V_2 from setting in as follows. Each population v_{3j} that receives a large net excitatory signal from V_2 excites its on-center of cells near v_{2j}, and inhibits a broad off-surround of populations centered at v_{2j}. This feedback inhibition prevents the pattern $I^{(2)}$ from saturating V_2, much as recurrent inhibition in Equation (4) works. Figure 3b can also be expanded to explicitly include inhibitory interneurons, as in Ellias and Grossberg (1975).

Normalization in V_1 by (1) occurs gradually in time, as each x_{1i} adjusts to its new equilibrium value, but it will be assumed below to occur instantaneously with x_{1i} approaching Θ_i rather than $\Theta_i BI(A+I)^{-1}$. These simplifications yield theorems about the tuning process that avoid unimportant details. The assumption that normalization occurs instantaneously is tenable because the normalized pattern at V_1 drives slow changes in the strength of connections from V_1 to V_2. Instantaneous normalization means that the pattern at V_1 normalizes itself before the connection strengths have a chance to substantially change.

Let the synaptic strength of the pathway from v_{1i} to the jth population v_{2j} in V_2 be denoted by $z_{ij}(t)$ (see Fig. 1). Let the total signal to v_{2j} due to the normalized pattern $\Theta = (\Theta_1, \Theta_2, \ldots, \Theta_n)$ at V_1 and the vector $z^{(j)}(t) = (z_{1j}(t), z_{2j}(t), \ldots, z_{nj}(t))$ of synaptic strengths be

$$S_j(t) \equiv \Theta \cdot z^{(j)}(t) \equiv \sum_{k=1}^{n} \Theta_k z_{kj}(t); \tag{5}$$

that is, each $z_{kj}(t)$ *gates* the signal Θ_k from v_{1k} on its way to v_{2j}, and these gated signals combine additively at v_{2j} (cf., Grossberg, 1967, 1970a, 1971, 1974). Since $z^{(j)}(t)$ determines the size of the input to v_{2j}, given any pattern Θ, it is called the *classifying vector* of v_{2j} at time t. *Every* v_{2j}, $j = 1, 2, \ldots, N$, in V_2 receives such a signal when Θ is active at V_1. In this way, Θ creates a pattern of activity across V_2.

Given any activity pattern across V_2, it can be transformed in several ways as time goes on. Two main questions about this process are: (a) will the *total* activity of V_2 be suppressed, or will some of its activities be stored in STM? and (b) which of the *relative* activities across V_2 will be preserved, suppressed, or enhanced? Several papers (Ellias and Grossberg, 1975; Grossberg, 1973; Grossberg and Levine, 1975) analyse how the parameters of a reverberating shunting on-center off-surround network determine the answers to these questions. Below some of these facts are cited as they are needed. In particular, if all the activities are sufficiently small, then they will not be stored in STM. If they are sufficiently large, then they will be contrast enhanced, normalized, and stored in STM. Figure 4 schematizes two storage possibilities. Figure 4a depicts a pattern of activity across V_2 before it is transformed by V_2. Given suitable parameters, if some of the initial activities exceed a quenching threshold (QT), then V_2 will *choose* the population having maximal initial activity for storage in STM,

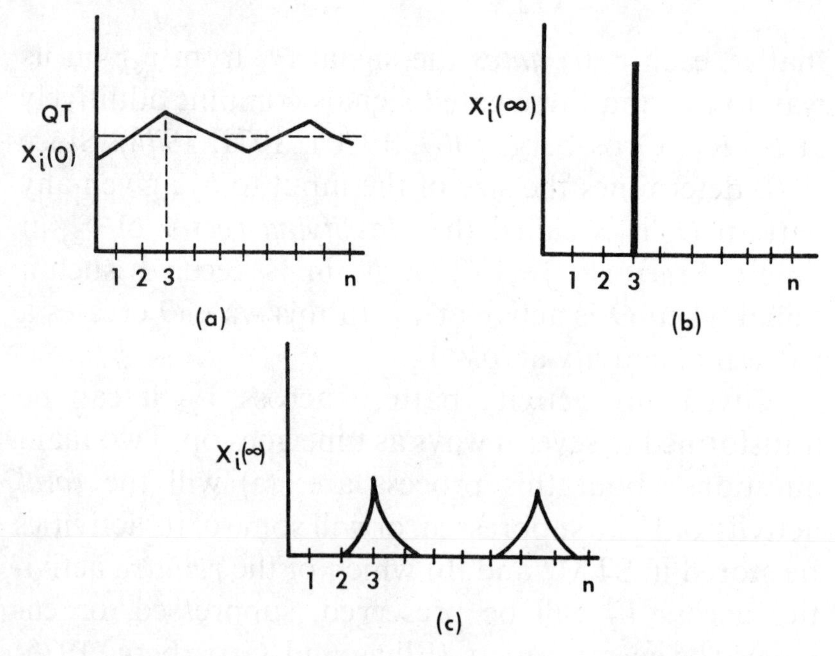

Fig. 4. Contrast enhancement and STM by recurrent network: (a) initial pattern; (b) choice; (c) partial contrast

as in Figure 4b. Under other circumstances, all initial activities below the QT are suppressed, whereas *all* initial activities above QT are contrast enhanced, normalized, and stored in STM (Fig. 4c); that is, *partial* contrast in STM is possible. Grossberg (1973) shows that partial contrast can occur if the signals between populations in a recurrent shunting on-center off-surround network are sigmoid (*S*-shaped) functions of their activity levels. Ellias and Grossberg (1975) show that partial contrast can occur if the self-excitatory signals of populations in V_2 are stronger than their self-inhibitory signals, and moreover if the excitatory signals between populations in V_2 decrease with inter-population distance faster than the inhibitory signals.

The enhancement and STM storage processes also occur much faster than the slow changes in connection strengths z_{ij}; hence, it is assumed below that these processes occur instantaneously in order to focus on the slow changes in z_{ij}.

The slow changes in z_{ij} are assumed to be determined by a time averaged product of the signal from v_{1i} to v_{2j} with the cortical response at v_{2j}; thus

$$\dot{z}_{ij} = -C_{ij}z_{ij} + D_{ij}x_{2j},$$

where C_{ij} is the decay rate (possibly variable) of z_{ij}, and D_{ij} is the signal from v_{1i} to v_{2j}. For example, if $C_{ij}=1$, the V_1 and V_2 patterns are normalized, and V_2 chooses only the population v_{2j} whose initial activity is maximal for storage in STM (Fig. 4b), then while

v_{2j} is active,

$$\dot{z}_{ij} = -z_{ij} + \Theta_i, \quad \text{for all } i = 1, 2, \ldots, n.$$

It remains to determine how these z_{ij} and all other z_{ik}, $k \neq j$, change under other circumstances. To eliminate conceptual and mathematical difficulties that arise if z_{ij} can decay even when V_1 and V_2 are inactive, we let *all* changes in each z_{ij} be determined by which populations in V_2 have their activities chosen for storage in STM. In other words, all changes in z_{ij} are driven by the *feedback* within the excitatory recurrent loops of V_2 that establish STM storage. Then

$$\dot{z}_{ij} = (-z_{ij} + \Theta_i)x_{2j} \tag{6}$$

where $\sum_{k=1}^{N} x_{2k}(t) = 1$ if STM in V_2 is active at time t, whereas $\sum_{k=1}^{N} x_{2k}(t) = 0$ if STM in V_2 is inactive at time t.

If V_2 *chooses* a population for storage in STM, as in Figure 4b, then

$$x_{2j} = \begin{cases} 1 & \text{if } S_j > max\,\{\varepsilon, S_k : k \neq j\} \\ 0 & \text{if } S_j < \max\,\{\varepsilon, S_k : k \neq j\}, \end{cases} \tag{7}$$

where as in (5), $S_j = \Theta \cdot z^{(j)}$ with $\Theta_i = I_i \left(\sum_{k=1}^{n} I_k \right)^{-1}$.

Equation (7) omits the cases where two or more signals S_j are equal, and are larger than all other signals and ε. In these cases, the x_{2j}'s of such S_j's are equal and add up to 1. Such a normalization rule for equal maximal signals will be tacitly assumed in all the cases below, but will otherwise be ignored to avoid tedious details. Equation (6) shows that z_{ij} can change only if $x_{2j}>0$. Equation (7) shows that V_2 chooses the maximal activity for storage in STM. This activity is normalized ($x_{2j}=0$ or 1), and it corresponds to the population with largest initial signal ($S_j > \max\{S_k : k \neq j\}$). No changes in z_{ij} occur if all signals S_j are too small to be stored in STM (all $S_j \leqq \varepsilon$).

If partial contrast in STM holds, as in Figure 4c, then the dynamics of a reverberating shunting network can be approximated by a rule of the form

$$x_{2j} = \begin{cases} f(S_j)\left[\sum\limits_{S_k>\varepsilon} f(S_k)\right]^{-1} & \text{if} \quad S_j > \varepsilon \\ 0 & \text{if} \quad S_j < \varepsilon \end{cases} \tag{8}$$

where $f(w)$ is an increasing nonnegative function of w such that $w=0$; e.g., $f(w)=w^2$. In (8), the positive constant ε represents the QT; the function $f(w)$ controls how suprathreshold signals S_j will be contrast enhanced; and the ratio of $f(S_j)$ to $\sum\{f(S_k) : S_k > \varepsilon\}$ expresses the normalization of STM.

3. Ritualistic Pattern Classification

After developmental tuning has taken place, the above mechanisms describe a model of pattern classification in the "adult" network. These mechanisms will be described first as interesting in themselves, and as a helpful prelude to understanding the tuning process.

They are capable of classifying arbitrarily complicated spatial patterns into mutually nonoverlapping, or partially overlapping, sets depending on whether (7) or (8) holds. These mechanisms realize basic principles of pattern discrimination using shunting interactions. An alternative scheme of pattern discrimination using a mixture of shunting and additive mechanisms has already been given (Grossberg, 1970b, 1972). Together these schemes suggest numerous anatomical and physiological variations that embody the same small class of functional principles. Since particular anatomies imply that particular physiological rules should be operative, intriguing questions about the dynamics of various neural structures, such as retina, neocortex, hippocampus, and cerebellum, are suggested.

First consider what happens if V_2 chooses a population for storage in STM. After learning ceases (that is, $\dot{z}_{ij} \equiv 0$), all classifying vectors $z^{(j)}$ are constant in time, and Equations (6) and (7) reduce to the statement that population v_{2j} is stored in STM if

$$S_j > \max \{\varepsilon, S_k : k \neq j\} . \tag{9}$$

In other words, v_{2j} *codes* all patterns Θ such that (9) holds; alternatively stated, v_{2j} is a *feature detector* in the sense that all patterns

$$P_j = \{\Theta : \Theta \cdot z^{(j)} > \max (\varepsilon, \Theta \cdot z^{(k)} : k \neq j)\} \tag{10}$$

are classified by v_{2j}. The set P_j defines a *convex cone* C_j in the space of nonnegative input vectors $J = (I_1, I_2, \ldots, I_n)$, since if two such vectors $J^{(1)}$ and $J^{(2)}$ are in C_j, then so are all the vectors $\alpha J^{(1)}$, $\beta J^{(2)}$, and

$\gamma J^{(1)}+(1-\gamma)J^{(2)}$, where $\alpha>0$, $\beta>0$, and $0<\gamma<1$. The convex cone C_j defines the *feature* coded by v_{2j}.

The classification rule in (10) has an informative geometrical interpretation in n-dimensional Euclidean space. The signal $S_j=\Theta \cdot z^{(j)}$ is the inner product of Θ and $z^{(j)}$ (Greenspan and Benney, 1973). Letting $\|\xi\|=\sqrt{\sum_{k=1}^{n} \xi_k^2}$ denote the Euclidean length of any real vector $\xi=(\xi_1, \xi_2, \ldots, \xi_n)$, and $\cos(\eta, \omega)$ denote the cosine between two vectors η and ω, it is elementary that

$$S_j=\|\Theta\| \, \|z^{(j)}\| \cos(\Theta, z^{(j)}) .$$

In other words, the signal S_j is the length of the projection of the normalized pattern Θ on the classifying vector $z^{(j)}$ times the length of $z^{(j)}$. Thus if all $z^{(j)}$, $j=1, 2, \ldots, N$, have equal length, then among all patterns with the same length, (10) classifies all patterns Θ in P_j whose angle with $z^{(j)}$ is smaller than the angles between Θ and any $z^{(k)}$, $k \neq j$, and is small enough to satisfy the ε-condition. In particular, patterns Θ that are *parallel* to $z^{(j)}$ are classified in P_j. The choice of classifying vectors $z^{(j)}$ hereby determines how the patterns Θ will be divided up. Section 8 will show that the tuning mechanism (6)–(7) makes the $z^{(j)}$ vectors more parallel to prescribed patterns Θ, and thereupon changes the classifying sets P_j. In summary

(i) the number of populations in V_2 determines the maximum number N of pattern classes P_j;

(ii) the choice of classifying vectors $z^{(j)}$ determines

how different these classes can be; for example, choosing all vectors $z^{(j)}$ equal will generate one class that is redundantly represented by all v_{2j}; and

(iii) the size of ε determines how similar patterns must be to be classified by the same v_{2j}.

If the choice rule (7) is replaced by the partial contrast rule (8), then an important new possibility occurs, which can be described either by studying STM responses to all Θ at fixed v_{2j}, or to a fixed Θ at all v_{2j}. In the former case, each v_{2j} has a *tuning curve*, or *generalization gradient*; namely, a maximal response to certain patterns, and submaximal responses to other patterns. In the latter case, each pattern Θ is *filtered* by V_2 in a way that shows how close Θ lies to *each* of the classifying vectors $z^{(j)}$. The pattern will only be classified by v_{2j}—that is, stored in STM—if it lies sufficiently close to $z^{(j)}$ for its signal S_j to exceed the quenching threshold of V_2.

For example, suppose that some of the classifying vectors $z^{(j)}$ are chosen to create large signals at V_2 when vertical lines perturb V_1, and that other $z^{(j)}$ create large signals at V_2 when horizontal lines perturb V_1. If a pattern containing both horizontal and vertical lines perturbs V_1, then the population activities in V_2 corresponding to both types of lines can be stored in STM, unless competition between their populations drives all activity below the QT. Now let V_3 be another "cortex" that receives signals from V_2, in the same fashion that V_2 receives signals from V_1. Given an appropriate choice of classifying vectors for V_3, there can exist cells in V_3 that fire in STM only if horizontal *and* vertical lines perturb a prescribed region of V_1; e.g., hypercomplex cells. The existence of tuning curves in a given cortex V_i hereby increases the discriminative capabilities of the next cortex V_{i+1} in a hierarchy; cf., Grossberg (1976).

The above mechanisms will now be discussed as cases of a general scheme of pattern classification. This is done with two goals in mind: firstly, to emphasize that these mechanisms might well exist in other than "retinocortical" analogs; and secondly, to generate explicit experimental directives in a variety of neural structures. One such directive will be described in Section 5.

4. Shunts vs. Additive Interactions as Mechanisms of Pattern Classification

The processing stages utilized in Section 3 are the following:

A) Normalization

Input patterns are normalized in V_1 by an on-center off-surround anatomy undergoing shunting interactions.

B) Partial Filtering by Signals

The signals S_j generated at V_2 by a normalized pattern on V_1 create the data base on which later computations are determined. The signal generating rule (5), for example, has the following important property. Suppose that an input $I_i(t) = \Theta_i I(t)$ is normalized to x_{1i}, as in (2), rather than to the approximate value Θ_i. The signal from V_1 to v_{2j} becomes

$$\tilde{S}_j = BI(A+I)^{-1} S_j$$

and (9) is replaced by the analogous rule

$$\tilde{S}_j > \max\{\varepsilon, \tilde{S}_k : k \neq j\}.$$

Then V_2 will classify a given pattern into the same class P_j no matter how large I is chosen. In other words, the signal generating rule is invariant under suprathreshold variations of the total activity at V_1. If I_i is the transduced receptor response to an external input J_i—that is, $I_i = g(J_i)$—then the signal-generating rule is invariant, given *any* $z^{(j)}$'s, if $g(w) = w^p$ for some $p > 0$.

C) Contrast Enhancement of Signals

The signals S_j are contrast enhanced by the recurrent on-center off-surround anatomy within V_2, and either a choice (Fig. 4b) or a tuning curve (Fig. 4c) results.

Two successive stages of lateral inhibition are needed in this model. The first stage normalizes input patterns. The second stage sharpens the filtering of signals.

Additive mechanisms can also achieve classification of arbitrarily complicated spatial patterns. These mechanisms also employ three successive stages A)–C) of pattern processing, with stage A) normalizing input patterns, stages A) and C) using inhibitory interactions, and stage C) completing the pattern classification that is begun by the signal generating rules of stage B). The additive model can differ in several respects from the shunting model:

(i) its anatomy can be feedforward; that is, there need not be a recurrent network in stage C);

(ii) threshold rules replace the inner product signal-generating rule (5) to determine partial filtering of signals; and

(iii) the responses in time of stages A)–C) to a sustained pattern at V_1 are not the same in the additive model. For example, sustained responses in the shunting model can be replaced by responses to the onset and offset of the pattern in the additive model (Grossberg, 1970b).

Mixtures of additive and shunting mechanisms are also possible. The additive mechanisms will now be summarized to illustrate the basic stages A)–C).

An additive nonspecific inhibitory interneuron normalizes patterns at V_1 (Fig. 5). Many variations on this theme exist (Grossberg, 1970b) in which such parameters as the lateral spread of inhibition, the number of cell layers, and the rates of excitatory and inhibitory decay can be varied. The idea in its simplest form is this. The excitatory input I_i excites a bifurcating pathway. One branch of the pathway is specific, and the other branch is nonspecific. The lateral inhibitory interneuron $v_{1,n+1}$ lies in the nonspecific branch. It

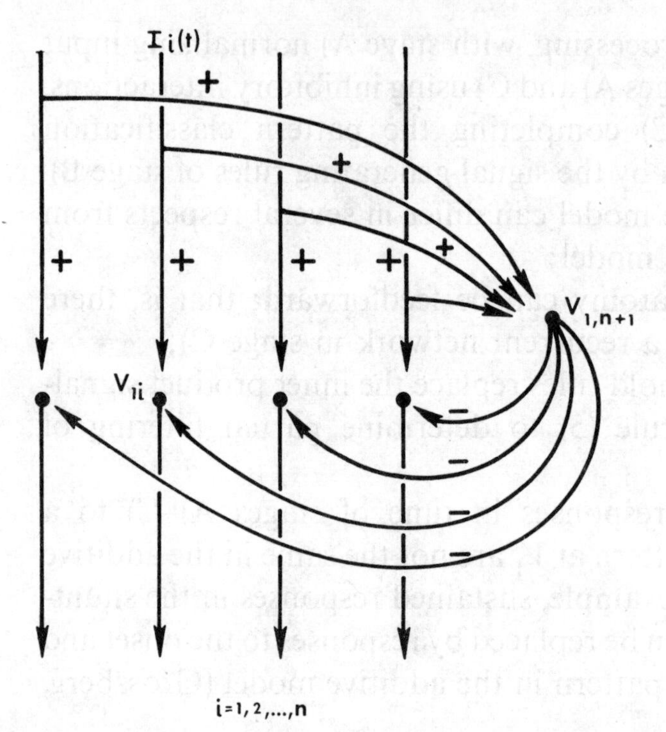

Fig. 5. Normalization and low-band filtering by subtractive nonspecific interneuron and signal threshold rules

sums the excitatory inputs I_i, and generates a nonspecific signal back to all the specific pathways if a signal threshold Γ is exceeded. Each input I_i also generates a specific signal from v_{1i} that is a linear function of I_i above a signal threshold. Each pathway from v_{1i} in V_1 to v_{2j} in V_2 has its own signal threshold Γ_{ij}. The net signal from v_{1i} to v_{2j} is

$$K_{ij} = [I_i - \Gamma_{ij}]^+ - \left[\sum_{k=1}^{n} I_k - \Gamma\right]^+ ,$$

where the notation $[u]^+ = \max(u, 0)$ defines the threshold rule. Define $\Theta_{ij}=\Gamma_{ij}\Gamma^{-1}$ and let the spatial pattern $I_i=\Theta_i I$ perturb V_1. Then

$$K_{ij}=[\Theta_i I-\Theta_{ij}\Gamma]^+-[I-\Gamma]^+ . \tag{11}$$

The net signal K_{ij} has the following properties:

(i) $K_{ij}\leqq 0$ for all values of $I>0$ if $\Theta_i\leqq\Theta_{ij}$;
(ii) $K_i>0$ for $I>\Theta_{ij}\Theta_i^{-1}$ if $\Theta_i>\Theta_{ij}$; and
(iii) $K_{ij}\leqq(\Theta_i-\Theta_{ij})\Gamma$ for all $I>0$.

In other words, by (i), no signal is emitted from v_{1i} to v_{2j} if $\Theta_i<\Theta_{ij}$; by (ii), if $\Theta_i>\Theta_{ij}$, a signal is emitted from v_{1i} if I exceeds a threshold depending on Θ_i and Θ_{ij}; and by (iii), the total activity in the cells v_{1i} is normalized. Partial filtering of signals is thus achieved by the choice of threshold pattern $\Theta^{(j)}=(\Theta_{1j},\Theta_{2j},\ldots,\Theta_{nj})$ rather than by the choice of classifying vector $z^{(j)}=(z_{1j},z_{2j},\ldots,z_{nj})$.

Stage C) is needed because the total signal to v_{2j} can be maximized by patterns Θ which are very different from the threshold pattern $\Theta^{(j)}$. This problem arises because the signals K_{ij} continue to grow linearly as a function of I after the threshold value $\Theta_{ij}\Theta_i^{-1}$ is exceeded. Grossberg (1970b) shows that the problem can be avoided by inhibiting each signal K_{ij} if it gets too large. For example, let the net signal from v_{1i} to v_{2j} be

$$S_{ij}^*=K_{ij}-\alpha[K_{ij}-\beta]^+ , \tag{12}$$

where $\alpha>1$ and $0<\beta\ll 1$. This mechanism inhibits the signal from v_{1i} to v_{2j} if it represents a Θ_i which is too much larger than Θ_{ij}. Equation (12) can be realized

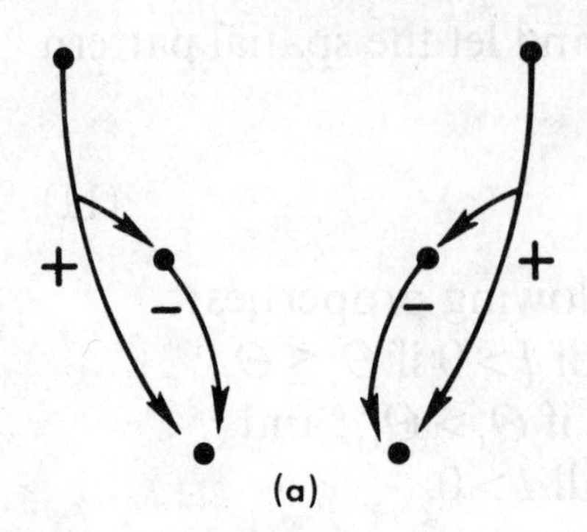

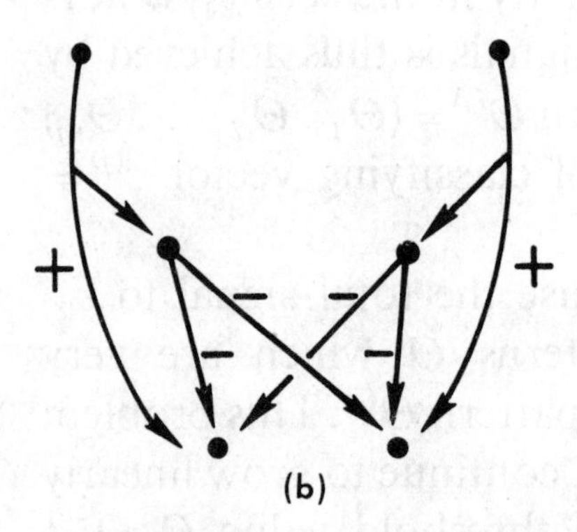

Fig. 6. (a) Specific subtractive inhibitory interneurons; (b) Non-specific inhibitory interneurons

by any of the several inhibitory mechanisms: a specific subtractive inhibitory interneuron (Fig. 6a), a switch-over from net excitation to net inhibition when the spiking frequency in the pathway from v_{1i} to v_{2j} becomes too large (Bennett, 1971; Blackenship et al., 1971; Wachtel and Kandel, 1971), or postsynaptic blockade of the v_{2j} cell membrane at sufficiently high spiking frequencies. Signal S_{ij}^* is positive only if Θ_i is sufficiently close to Θ_{ij} in value. Stage C) is completed by choosing the signal threshold of v_{2j} so high that v_{2j} only fires if *all* signals S_{ij}^*, $i=1, 2, \ldots, n$, are positive; that is, only if the input pattern Θ is close to the threshold pattern $\Theta^{(j)}$. The second stage of inhibition hereby completes the partial filtering process by choosing a population v_{2j} in V_2 to code $\Theta^{(j)}$, as in Figure 4b. If the specific inhibitory interneurons if Figure 6a are replaced by a lateral spread of inhibition, as in Figure 6b, then a tuning curve is generated, as in Figure 4c.

5. What Do Retinal Amacrine Cells Do?

This section illustrates how the principles A)–C) can generate interesting questions about particular neural processes. Grossberg (1970b, 1972) introduces a retinal model in which shunting and additive interactions both occur. In this model, retinal amacrine cells are examples of the inhibitory interaction in stage C). We will note that amacrine cells have *opposite* effects on signals if they realize a shunting rather than an additive model. In the retinal model of Grossberg (1972), normalization is accomplished by an on-center off-surround anatomy undergoing shunting interactions. Analogously, in vivo receptors excite bipolar cells (on-center) as well as horizontal cells, and the horizontal cells inhibit bipolar cells via their lateral interactions (off-surround). Partial filtering of the normalized inputs is accomplished by signal thresholds; for example, using the normalized x_{1i} activities in (2), the simplest signal function from v_{1i} to v_{2j} is $K_{ij}=[x_{1i}-\Gamma_{ij}]^{+}$. Stage C) is then accomplished by a mechanism such as (12), by which large signals are inhibited. Whether a choice (Fig. 4b) or a tuning curve (Fig. 4c) is generated depends, in part, on how broadly these lateral inhibitory signals that complete stage C) are distributed. This second stage of inhibition is identified with the inhibition that amacrine cells, fed by bipolar cell activity, generate at ganglion cells. Grossberg (1972) notes data that support the idea that stage C) is realized by an additive mechanism such as (12). In particular, amacrine cells often respond when an input pattern is turned on, or off, or both. Two questions about amacrine cells now suggest themselves.

(i) If this interpretation of amacrine cells is true, then they will shut off signals from the bipolar cells to the ganglion cells when these signals become too *large*; that is, they act as high-band filters. By contrast, inhibition in stage C) of the shunting model shuts off signals if they become too *small*. Opposite effects due to the second inhibitory stage can hereby create a similar functional transformation of the input pattern! If a shunting role for amacrine cells is sought, then the following types of anatomy would be anticipated: inhibitory bipolar-to-amacrine-to-bipolar cell feedback that contrast enhances the receptor-to-bipolar signals, or inhibitory ganglion-to-amacrine-to-ganglion cell feedback that contrast enhances the bipolar-to- ganglion cell signals, or some functionally similar feedback loop. To decide between these two possible roles for amacrine cells, one must test whether amacrine cells suppress large signals or small ones; in either case, if the model is applicable, contrast enhancement of the normalized and filtered retinal pattern is the result, so that this property cannot be used as a criterion.

(ii) Does the spatial extent of lateral amacrine interaction determine the amount of contrast, or the breadth of the tuning curves, in ganglion cell responses, as in Figures 4b and 4c? For example, there exist narrow field diffuse amacrine cells, wide field diffuse amacrine cells, stratified diffuse amacrine cells, and unstratified amacrine cells (Boycott and Dowling, 1969). Do these specializations guarantee particular tuning characteristics in the corresponding ganglion cells?

Grossberg (1972) also suggests a cerebellar analog based on the same principles. Thus at least formal aspects of various neural structures seem to be emerging as manifestations of common principles. These results suggest a program of classifying seemingly different anatomical and physiological data according to whether they realize similar functional transformations of patterned neural activity, such as total activity normalization, partial filtering by signals, and contrast enhancement of the signal pattern. Below are described certain properties of the shunting mechanism that will be needed when development is discussed.

6. Arousal as a Tuning Mechanism

The recurrent networks in V_2 all have a quenching threshold (QT); namely, a criterion activity level that must be exceeded before a population's activity can reverberate in STM. Changing the QT or, equivalently, changing the size of signals to V_2, can retune the responsiveness of populations in V_2 to prescribed patterns at V_1. For example, suppose that an unexpected, or novel, event triggers a nonspecific arousal input to V_2, which magnifies all the signals from V_1 to V_2 (see Part II). Then certain signals, which could not otherwise be stored in STM, will exceed the QT and be stored. For example, if V_2 is capable of partial contrast in STM and also receives a nonspecific arousal input, then (8) can be replaced by

$$x_{2j} = \begin{cases} f(\phi S_j)\left[\sum\limits_{\phi S_k > \varepsilon} f(\phi S_k)\right]^{-1} & \text{if } \phi S_j > \varepsilon \\ 0 & \text{if } \phi S_j < \varepsilon \end{cases} \tag{13}$$

where ϕ is an increasing function of the arousal level. Note that an increase in ϕ allows more V_2 populations to reverberate in STM; cf., Grossberg (1973) for mathematical proofs. In a similar fashion, if an unexpected event triggers nonspecific shunting inhibition of the inhibitory interneurons in the off-surrounds of V_2, then the QT will decrease (Grossberg, 1973; Ellias and Grossberg, 1975), yielding an equivalent effect. Equation (8) can then be changed to

$$x_{2j}=\begin{cases} f(S_j)\left[\sum_{S_k>\phi^*\varepsilon} f(S_k)\right]^{-1} & \text{if} \quad S_j>\phi^*\varepsilon \\ 0 & \text{if} \quad S_j<\phi^*\varepsilon \end{cases} \tag{14}$$

where ϕ^* is a decreasing function of the arousal level.

Reductions in arousal level have the opposite effect. For example, if (13) holds, and arousal is lowered until only one population in V_2 exceeds the QT, then a choice will be made in STM, as in Figure 4b. Thus a choice in STM can be due either to *structural* properties of the network, such as the rules for generating signals between populations in V_2 [cf., the faster-than-linear signal function in Grossberg (1973)], or to an arousal level that is not high enough to create a tuning curve. Similarly, if arousal is too small, then all functions x_{2j} in (13) will always equal zero, and no STM storage will occur.

Changes in arousal can have a profound influence on the time course of LTM, as in (6), because they change the STM patterns that drive the learning process. For example, if during development arousal level is chosen to produce a choice in STM, then the tuning of classifying vectors $z^{(j)}$ will be sharper than if the arousal level were chosen to generate partial contrast in STM.

The influence of arousal on tuning of STM patterns can also be expressed in another way, which suggests a mechanism that will be needed in Part II when universal recoding is discussed.

7. Arousal as a Search Mechanism

Suppose that arousal level is fixed during learning trials, and that a given pattern Θ at V_1 does not create any STM storage at V_2 because all the inner products $\Theta \cdot z^{(j)}$ are too small. If arousal level is then increased in (13) until some $x_{2j} > 0$, STM storage will occur. In other words, changing the arousal level can facilitate *search* for a suitable classifying population in V_2.

Why does arousal level increase if no STM storage occurs at V_2? This is a property of the expectation mechanism that is developed in Part II. Also in Part II a pattern Θ at V_1 that is not classified by V_2 will use this mechanism to release a subliminal search routine that terminates when an admissible classification occurs.

8. Development of an STM Code

System (6)–(7) will be analysed mathematically because it illustrates properties of the model in a particularly simple and lucid way. The first result describes how this system responds to a single pattern that is iteratively presented through time.

Theorem 1 (One Pattern)

Given a pattern Θ, suppose that there exists a unique j such that

$$S_j(0) > \max\{\varepsilon, S_k(0) : k \neq j\}. \tag{15}$$

Let Θ be practiced during a sequence of nonoverlapping intervals $[U_k, V_k]$, $k=1, 2, \ldots$. Then the angle between $z^{(j)}(t)$ and Θ monotonically decreases, the signal $S_j(t)$ is monotonically attracted towards $\|\Theta\|^2$ and $\|z^{(j)}\|^2$ oscillates at most once as it pursues $S_j(t)$. In particular, if $\|z^{(j)}(0)\| \leqq \|\Theta\|$, then $S_j(t)$ is monotone increasing. Except in the trivial case that $S_j(0) = \|\Theta\|^2$, the limiting relations

$$\lim_{t\to\infty} \|z^{(j)}(t)\|^2 = \lim_{t\to\infty} S_j(t) = \|\Theta\|^2 \tag{16}$$

hold if and only if

$$\sum_{k=1}^{\infty} (V_k - U_k) = \infty \,. \tag{17}$$

Remark. If $z^{(j)}(0)$ is small, in the sense that $\|z^{(j)}(0)\| \leqq \|\Theta\|$, then by Theorem 1, as time goes on, the learning process maximizes the inner product signal $S_j(t) = \Theta \cdot z^{(j)}(t)$ over all possible choices of $z^{(j)}$ such that $\|z^{(j)}\| \leqq \|\Theta\|$. This follows from the obvious fact that

$$\sup\{\Theta \cdot \psi : \|\psi\| \leqq \|\Theta\|\} = \|\Theta\|^2 \,.$$

Otherwise expressed, learning makes $z^{(j)}$ parallel to Θ, and normalizes the length of $z^{(j)}$.

What happens if several different spatial patterns $\Theta^{(k)} = (\Theta_1^{(k)}, \Theta_2^{(k)}, \ldots, \Theta_n^{(k)})$, $k=1, 2, \ldots, M$, all perturb V_1 at different times? How are changes in the z_{ij}'s due to one pattern prevented from contradicting changes in the z_{ij}'s due to a different pattern? The choice-making property of V_2 does this for us; it acts as a sampling device that prevents contradictions from occurring. A heuristic argument will now be given to suggest how sampling works. This argument will then be refined and made rigorous. For definiteness, suppose that M spatial patterns $\Theta^{(k)}$ are chosen, $M \leq N$, such that their signals at time $t=0$ satisfy

$$\Theta^{(k)} \cdot z^{(k)}(0) > \max \{\varepsilon, \Theta^{(k)} \cdot z^{(j)}(0) : j \neq k\} \tag{18}$$

for all $k=1, 2, \ldots, M$. In other words, at time $t=0$, $\Theta^{(k)}$ is coded by v_{2k}. Let $\Theta^{(1)}$ be the first pattern to perturb V_1. By (18), population v_{21} receives the largest signal from V_1. All other populations v_{2j}, $j \neq 1$, are thereupon inhibited by the off-surround of v_{21}, whereas v_{21} reverberates in STM. By (6), none of the synaptic strengths $z^{(j)}(t)$, $j \neq 1$, can learn while $\Theta^{(1)}$ is presented. As in Theorem 1, presenting $\Theta^{(1)}$ makes $z^{(1)}(t)$ more parallel to $\Theta^{(1)}$ as t increases. Consequently, if a different pattern, say $\Theta^{(2)}$, perturbs V_1 on the next learning trial, then it will excite v_{22} more than any other v_{2j}, $j \neq 2$: it cannot excite v_{21} because the coefficients $z^{(1)}(t)$ are more parallel to $\Theta^{(1)}$ than before; and it cannot excite any v_{2j}, $j \neq 1, 2$, because the v_{2j} coefficients $z^{(j)}(t)$ still equal $z^{(j)}(0)$. In response to $\Theta^{(2)}$, v_{22} inhibits all other v_{2j}, $j \neq 2$. Consequently none of the v_{2j} coefficients $z^{(j)}(t)$ can learn, $j \neq 2$; learning makes the coefficients $z^{(2)}(t)$ become more parallel to $\Theta^{(2)}$ as t increases. The same occurs on all learning trials. By inhibiting the postsynaptic part of the learning mechanism in all but the chosen V_2 population, the on-center off-surround network in V_2 samples one vector $z^{(j)}(t)$ of trainable coefficients at any time. In this way, V_2 can learn to

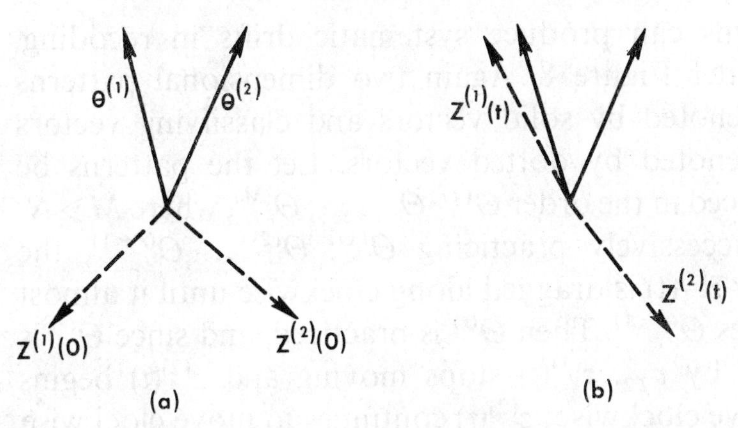

Fig. 7. Practicing $\Theta^{(1)}$ brings $z^{(1)}(t)$ closer to $\Theta^{(1)}$ and $\Theta^{(2)}$ than $z^{(2)}(0)$

distinguish as many as N patterns if it contains N populations.

This argument is almost correct. It fails, in general, because by making (say) $z^{(1)}(t)$ more parallel to $\Theta^{(1)}$, it is also possible to make $z^{(1)}(t)$ more parallel to $\Theta^{(2)}$ than $z^{(2)}(0)$ is. Thus when $\Theta^{(2)}$ is presented, it will be coded by v_{21} rather than v_{22}. In other words, practicing one pattern can recode other patterns. A typical example of this property is illustrated in Figure 7. Figure 7a depicts the two dimensional patterns $\Theta^{(1)}$ and $\Theta^{(2)}$ as solid vectors, and the two classifying vectors $z^{(1)}(0)$ and $z^{(2)}(0)$ as dotted vectors. Clearly (18) holds for $j=1, 2$. As a result of practicing $\Theta^{(1)}$ during a fixed interval, Figure 7b is produced. Note that $\Theta^{(2)} \cdot z^{(1)}(t) > \Theta^{(2)} \cdot z^{(2)}(t)$ after the practice interval terminates. Consequently, v_{21}, rather than v_{22}, codes $\Theta^{(2)}$ when $\Theta^{(2)}$ is practiced. This property can be iterated to show how systematic trends in the sequence of practiced patterns can produce systematic drifts in recoding. Consider Figure 8. Again two dimensional patterns are denoted by solid vectors and classifying vectors are denoted by dotted vectors. Let the patterns be practiced in the order $\Theta^{(1)}, \Theta^{(2)}, \ldots, \Theta^{(M)}$, where $M \gg N$. By successively practicing $\Theta^{(1)}, \Theta^{(2)}, \ldots, \Theta^{(r-1)}$, the vector $z^{(1)}(t)$ is dragged along clockwise until it almost reaches $\Theta^{(r-1)}$. Then $\Theta^{(r)}$ is practiced, and since $\Theta^{(r)}$ is coded by v_{22}, $z^{(1)}(t)$ stops moving and $z^{(2)}(t)$ begins to move clockwise; $z^{(2)}(t)$ continues to move clockwise while $\Theta^{(r+1)}, \Theta^{(r+2)}, \ldots, \Theta^{(2r-1)}$ are practiced. Then $z^{(3)}(t)$ begins to move clockwise, and so on. The clockwise drift in the practice schedule hereby shifts each $z^{(j)}(t), j = 1, 2, \ldots, M-1$, to a position that is close to the one $z^{(j+1)}(0)$ occupied. In other words, essentially *all* vectors in V_2 are reclassified. If the same practice schedule $\Theta^{(1)}, \Theta^{(2)}, \ldots, \Theta^{(M)}$ is repeated on a second learning trial, then essentially all v_{2i} are recoded by

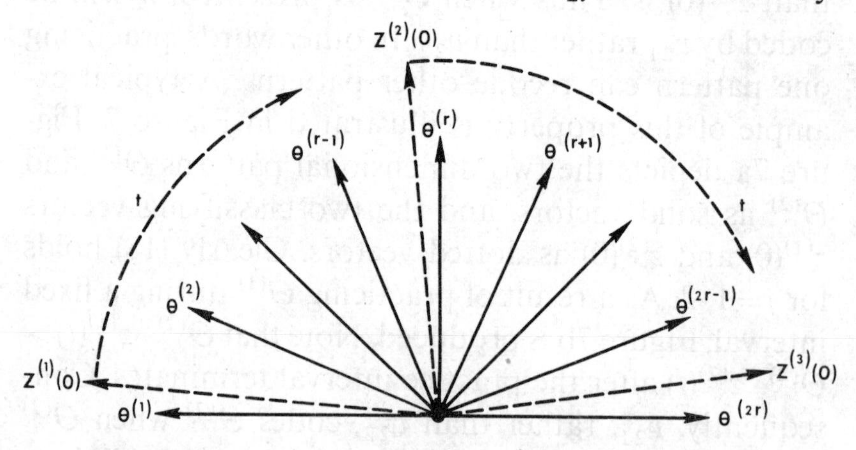

Fig. 8. Practicing a sequence of spatial patterns can recode all the populations

$v_{2,i+2}$, and so on. Each learning trial recodes V_2 until all the N populations in V_2 code one of the N most clockwise vectors $\Theta^{(k)}$. This asymptotic coding of V_2 is stable, except for a wild oscillation in the coding of v_{21} on each learning trial, if the same practice schedule is always repeated. If, however, a counter-clockwise drift in practiced patterns is then imposed, all of V_2 will be recoded until the N most counter-clockwise vectors $\Theta^{(k)}$ are coded. In general, if there are many patterns relative to the number of populations in V_2, and if the statistical structure of the practice sequences continually changes, then there need not exist a stable coding rule in V_2. This is quite unsatisfactory.

By contrast, if there are few, or sparse, patterns relative to the number of populations in V_2, then a stable coding rule does exist, and the STM choice rule in V_2 does provide an effective sampling technique. Such a situation is approximated, for example, when the network is exposed to a "visually deprived" environment, in imitation of experiments on young animals. A theorem concerning this case will now be stated, if only to suggest what auxiliary mechanisms will be needed to establish a stable coding rule in the general case. This theorem shows how populations learn to code convex regions of features. In particular, if v_{2j} learns to code a certain set of features, then it automatically codes *average* features derived from this set.

The following nomenclature will be needed to state

the theorem. A *partition* $\bigoplus_{k=1}^{K} \mathscr{P}_k$ of a finite set $\mathscr{P}$ is a subdivision of $\mathscr{P}$ into nonoverlapping and exhaustive subsets $\mathscr{P}_j$. The *convex hull* $\mathscr{H}(\mathscr{P})$ of a finite set $\mathscr{P}$ is the set of all convex combinations of elements in $\mathscr{P}$; for example, if $\mathscr{P}=\{\Theta^{(1)},\Theta^{(2)},\ldots,\Theta^{(M)}\}$, then

$$\mathscr{H}(\mathscr{P})=\left\{\sum_{k=1}^{M} \lambda_k\Theta^{(k)}: \text{each } \lambda_k \geqq 0 \text{ and } \sum_{k=1}^{M} \lambda_k=1\right\}.$$

Given a set $\mathscr{P}$ with subset $\mathscr{Q}$, let $\mathscr{R}=\mathscr{P}\backslash\mathscr{Q}$ denote the set of elements in $\mathscr{P}$ that are not in $\mathscr{Q}$. If the classifying vector $z^{(j)}(t)$ codes the set of patterns $\mathscr{P}_j(t)$, let $\mathscr{P}_j^*(t)=\mathscr{P}_j(t)\cup\{z^{(j)}(t)\}$. The *distance* between a vector P and a set of vectors $\mathscr{Q}$, denoted by $\|P-\mathscr{Q}\|$, is defined by

$$\|P-\mathscr{Q}\|=\inf\{\|P-Q\|:Q\in\mathscr{Q}\}.$$

Theorem 2 (Sparse Patterns)

Let the network practice any set $\mathscr{P}=\{\Theta^{(i)}: i=1,2,\ldots,M\}$ of patterns for which there exists a partition $\mathscr{P}=\bigoplus_{k=1}^{N}\mathscr{P}_k(0)$ such that

$$\min\{u\cdot v: u\in\mathscr{P}_j(0), v\in\mathscr{P}_j^*(0)\}>\max\{u\cdot v: u\in\mathscr{P}_j(0), v\in\mathscr{P}^*(0)\backslash\mathscr{P}_j^*(0)\} \tag{19}$$

for all $j=1,2,\ldots,N$. Then $\mathscr{P}_j(t)=\mathscr{P}_j(0)$ and the functions

$$D_j(t)=\|z^{(j)}(t)-\mathscr{H}(\mathscr{P}^{(j)}(t))\| \tag{20}$$

are monotone decreasing for $t\geqq 0$ and $j=1,2,\ldots,N$. If moreover the patterns in $\mathscr{P}^{(j)}(0)$ are practiced in intervals $[U_{jm}, V_{jm}]$, $m=1, 2, \ldots,$ such that

$$\sum_{m=1}^{\infty} (V_{jm} - U_{jm}) = \infty \tag{21}$$

then

$$\lim_{t \to \infty} D_j(t) = 0\,. \tag{22}$$

Remarks. In other words, if the classifying vectors initially code the patterns into sparse classes, in the sense of (19), then this code persists through time, and the classifying vectors approach a convex combination of their coded patterns. As (20) and (22) show, learning permits each v_{2j} to respond as vigorously as possible to its class of coded patterns.

The above results indicate that, given a fixed number of patterns, it becomes easier to establish a stable code for them as the number of populations in V_2 increases. Once V_2 is constructed, however, it is not possible to increase its number of populations at will. Moreover, *in vivo*, an enormous variety of patterns typically barrages the visual system. How can a stable code be guaranteed no matter how many patterns perturb V_1?

One way is to assume that a biochemically determined *critical period* exists during which the z_{ij}'s are capable of learning; once the critical period terminates, some chemical factor is removed and the z_{ij}'s remain fixed in the last code to be established. The existence of a critical period has been reported (Hubel and

Wiesel, 1970), but whether it is due to a chemical factor, or *merely* to a chemical factor, is as yet unknown. From a formal point of view, such a mechanism suffers from several significant related disadvantages. The most obvious one is that all the coded information that is learned throughout the critical period can be obliterated if its last phase exhibits an unlikely statistical trend. In addition, a repetitive statistical trend can prevent many patterns from being coded at all. For example, in Figure 8, once the classifying vectors code the N most clockwise patterns, many of the other $M-N$ patterns might be too far away from $z^{(1)}$ to satisfy the ε-condition in (7); they will then not be coded by any population. Yet each of these $M-N$ patterns has been presented as frequently as the N patterns that are coded. More generally, because populations which are already coded can be recoded so easily, it is hard to search for as yet uncommitted populations to code as yet uncoded patterns. This problem prevents a universal recoding from being achieved (see Part II).

These negative remarks can be supplemented by intriguing positive observations. Stabilizing the code seems to require the same formal machinery that is needed in models of adult attention and discrimination learning (Grossberg, 1975). This machinery, in turn, is highly evokative of data concerning attentional modulation of olfactory patterns by the prepyriform cortex of cats (Freeman, 1974). Auxiliary mechanisms for stabilizing the code will therefore be motivated below. It is understood that a biochemically triggered critical period can coexist with these mechanisms, or indeed can preempt them in sufficiently primitive organisms.

Various mechanisms can be contemplated which partially stabilize the code, but which are not sufficient. A satiation mechanism will be sketched below to clarify what is needed. Consider (6) with

$$x_{2j}(t)=\begin{cases} G_j(t) & \text{if} \quad S_j(t)G_j(t)>\max\{\varepsilon, S_k(t)G_k(t): k\neq j\} \\ 0 & \text{if} \quad S_j(t)G_j(t)<\max\{\varepsilon, S_k(t)G_k(t): k\neq j\} \end{cases} \tag{23}$$

where

$$G_j(t)=g\left(1-\int_0^t x_{2j}(v)K(t-v)dv\right). \tag{24}$$

In (24), $g(w)$ is a monotone increasing function such that $g(0)=0$ and $g(1)=1$. $K(w)$ is a monotone decreasing function such that $K(0)=1$ and $K(\infty)=0$; for example, $K(w)=e^{-w}$. Equation (23) says that persistent activation of v_{2j} causes its STM response to satiate, or adapt; if v_{2j} is active during a sufficiently long interval, its activity approaches zero. Correspondingly, $z^{(j)}$'s fluctuations are damped within a time interval of fixed length. Such a mechanism is inadequate if the training schedule allows v_{2j} to recover its maximal strength. Figure 9 shows, for example, an ordering of patterns that permits recoding of essentially all populations in V_2.

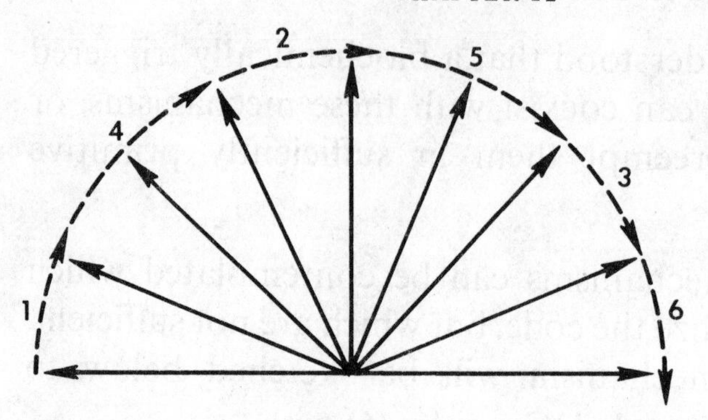

Fig. 9. Practicing in the order 1, 2, 3, 4, 5, 6 can recode all the populations even if satiation exists

This problem is only made worse by replacing the choice rule in (23) by a partial contrast rule such as

$$x_{2j} = \begin{cases} \dfrac{f(S_j G_j)}{\sum\limits_{S_k G_k > e} f(S_k G_k)} & \text{if} \quad S_j G_j > \varepsilon \\ 0 & \text{if} \quad S_j G_j < \varepsilon \, . \end{cases}$$

Here, if a prescribed pattern Θ causes a maximal STM response at v_{2j}, then the activity x_{2j} is suppressed by G_j more rapidly than the activities of other Θ-activated populations. There can consequently be a shift in the locus of maximal responsiveness even to a single pattern—that is, recoding—in addition to the difficulty cited in Figure 9.

Such examples clarify what is essential:

(A) *Before* $z^{(j)}(t)$ learns a pattern, or class of related patterns, it must be able to fluctuate freely in response to pattern inputs in search of a classification.

(B) *After* $z^{(j)}(t)$ learns a pattern, it must be prevented from coding very different patterns, no matter what the training schedule is. In particular, satiating $z^{(j)}$'s ability to change through time does not suffice, since a very different pattern can still be coded by $z^{(j)}$ if this pattern elicits a larger signal at v_{2j}, say due to the size of $\|z^{(j)}\|$ rather than the direction of vector $z^{(j)}$, than at any of the uncommited populations.

Requirements (A) and (B) constrain the interaction of STM and LTM mechanisms, given that (6) holds. For example, by (6), if a pattern Θ creates signals while v_{2j} is active in STM, then $z^{(j)}(t)$ will change. Suppose that a sequence $\Theta^{(1)}$, $\Theta^{(2)}$ of two very different patterns is successively presented to V_1, and that $z^{(1)}(t)$ codes $\Theta^{(1)}$. In response to $\Theta^{(1)}$, v_{21} is activated, but $z^{(1)}(t)$ does not substantially change because it already codes $\Theta^{(1)}$. Now let $\Theta^{(2)}$ perturb V_1. By requirement (B), $z^{(1)}(t)$ must not be allowed to change. By (6), $z^{(1)}(t)$ will change unless either no signal is emitted from V_1 when v_{21} is active, or a signal is emitted from V_1 only after v_{21} is inactivated. These two cases will be separately considered in the next two paragraphs.

In the former case, some type of feedback to V_1 must suppress the V_1-to-V_2 signals that would otherwise be generated by $\Theta^{(2)}$. This feedback somehow tells V_1 that $\Theta^{(2)}$ is very different from the pattern $\Theta^{(1)}$ that is presently coded in STM. By (A), however, $\Theta^{(2)}$ can generate V_1-to-V_2 signals at *some* time, either to search for a classifying vector, or to activate its already learned STM representation. Thus after V_1-to-V_2 signals are suppressed long enough for STM activity in v_{21} to also be suppressed, then V_1-to-V_2 signals are reactivated.

In the latter case, changing $\Theta^{(1)}$ to $\Theta^{(2)}$ somehow suppresses the STM activity that codes $\Theta^{(1)}$; in particular, somehow the network can tell when the spatial patterns that perturb V_1 are changed. In both cases, the same general issue is raised: how does the network process a temporal succession $\Theta^{(1)}, \Theta^{(2)}, \ldots, \Theta^{(k)}, \ldots$ of spatial patterns $\Theta^{(k)} = (\Theta_1^{(k)}, \Theta_2^{(k)}, \ldots, \Theta_n^{(k)})$; that is, a *space-time pattern.* Space-time patterns are the typical inputs to a receptive field *in vivo.* The problem of stabilizing the STM code forces us to consider their processing in some detail. Part II of this paper considers this problem.

Appendix

Proof of Theorem 1. Consider the case in which

$$|\Theta|^2 > S_j(0) > \max\{\varepsilon, S_k(0): k \neq j\}. \tag{A1}$$

The case in which $S_j(0) \geqq |\Theta|^2$ can be treated similarly. First it will be shown that if the inequalities

$$|\Theta|^2 > S_j(t) > \max\{\varepsilon, S_k(t): k \neq j\} \tag{A2}$$

hold at any time $t = T \in \bigcup_{m=1}^{\infty} [U_m, V_m]$, then they hold at all times $t \in [T, \infty) \cap \bigcup_{m=1}^{\infty} [U_m, V_m]$. By (A2), $x_{2j}(T) = 1$ and $x_{2k}(T) = 0$, $k \neq j$. Consequently, by (6),

$$\dot{z}_{ij}(T) = -z_{ij}(T) + \Theta_i \tag{A3}$$

and

$$\dot{z}_{ik}(T)=0 \tag{A4}$$

for $k \neq j$ and $i=1, 2, \ldots, n$. By (A2)–(A4),

$$\begin{aligned}\dot{S}_j(T) &= -S_j(T)+|\Theta|^2 \\ &> 0=\dot{S}_k(T),\end{aligned} \tag{A5}$$

$k \neq j$. Thus (A2) holds for all $t \in [T, \infty) \cap \bigcup_{m=1}^{\infty} [U_m, V_m]$. By (A2) and (A5), for all $t \in \bigcup_{m=1}^{\infty} [U_m, V_m]$, $S_j(t)$ increases monotonically towards $|\Theta|^2$ and (16) holds if and only if (17) holds. For $t \notin \bigcup_{m=1}^{\infty} [U_m, V_m]$, all $\dot{S}_k(t)=0$, $k=1, 2, \ldots, n$.

Letting $N_j=|z^{(j)}|^2$ and $C_j=\cos(z^{(j)}, \Theta) \equiv S_j N_j^{-1/2}|\Theta|^{-1}$, it readily follows from (A5) that for all $t \in \bigcup_{m=1}^{\infty} [U_m, V_m]$,

$$\dot{N}_j=2(-N_j+S_j) \tag{A6}$$

and

$$\dot{C}_j=|\Theta| N_j^{-1/2}(1-C_j^2). \tag{A7}$$

Equation (A7) shows that the angle between $z^{(j)}(t)$ and Θ closes monotonically as Θ is practiced. Since $S_j(t)$ is a monotonic function, (A6) shows that $N_j(t)$ oscillates at most once.

In particular, suppose $\|z^{(j)}(0)\| \leqq \|\Theta\|$. Then $S_j(0) \leqq \|\Theta\|^2$, since otherwise

$$\Theta \cdot z^{(j)}(0) > \Theta \cdot \Theta \geqq z^{(j)}(0) \cdot z^{(j)}(0)$$

which implies

$$1 \geqq C_j(0) > \|\Theta\| \, \|z^{(j)}(0)\|^{-1} \geqq \|z^{(j)}(0)\| \, \|\Theta\|^{-1},$$

and thus

$$\|z^{(j)}(0)\| > \|\Theta\| > \|z^{(j)}(0)\|,$$

which is a contradiction. By (A5), therefore $\|z^{(j)}(0)\| \leqq \|\Theta\|$ implies that $S_j(t)$ is monotone increasing

Proof of Theorem 2. Inequality (19) is based on the fact that, if a fixed set of patterns $\Theta^{(j_1)}, \Theta^{(j_2)}, \ldots, \Theta^{(j_k)}$ is classified by $z^{(j)}(t)$ for all $t \geqq 0$, then

$$z^{(j)}(t) \in \mathscr{H}(\Theta^{(j_1)}, \Theta^{(j_2)}, \ldots, \Theta^{(j_k)}, z^{(j)}(0)), \tag{A8}$$

for all $t \geqq 0$. For example, suppose that the patterns are practiced in the order $\Theta^{(j_1)}, \Theta^{(j_2)}, \ldots, \Theta^{(j_k)}$ during the nonoverlapping intervals $[U_1, V_1], [U_2, V_2], \ldots, [U_k, V_k]$. Except during these intervals, $\dot{z}^{(j)} = 0$. Thus for $t \in [U_1, V_1]$,

$$\dot{z}^{(j)} = -z^{(j)} + \Theta^{(j_1)},$$

or

$$z^{(j)}(t) = z^{(j)}(0)e^{-(t-U_1)} + \Theta^{(j_1)}(1 - e^{-(t-U_1)}),$$

so that

$$z^{(j)}(t) \in \mathscr{H}(\Theta^{(j_1)}, z^{(j)}(0)) \subset \mathscr{H}(\Theta^{(j_1)}, \ldots, \Theta^{(j_k)}, z^{(j)}(0)).$$

For $t \in [U_2, V_2]$,

$$z^{(j)}(t) = [z^{(j)}(0)e^{-(V_1-U_1)} + \Theta^{(j_1)}(1 - e^{-(V_1-U_1)})]e^{-(t-U_2)} + \Theta^{(j_2)}(1 - e^{-(t-U_2)}). \tag{A9}$$

Hence

$$z^{(j)}(t) \in \mathscr{H}(\Theta^{(j_1)}, \Theta^{(j_2)}, z^{(j)}(0)) \subset \mathscr{H}(\Theta^{(j_1)}, \ldots, \Theta^{(j_k)}, z^{(j)}(0)),$$

and so on.

Condition (19) is then applied using the fact that, for any $U \in P_j(0)$, $V \in \mathscr{H}(P_j^*(0))$, and $W \in \mathscr{H}(P^*(0) \backslash P_j^*(0))$,

$$U \cdot V > \max\{\varepsilon, U \cdot W\} \tag{A10}$$

because

$$U \cdot V \geqq \min\{u \cdot v : u \in P_j(0), v \in P_j^*(0)\}$$

and

$$\max\{u \cdot v : u \in P_j(0),\ v \in P^*(0) \backslash P_j^*(0)\} \geqq U \cdot W.$$

Until a pattern is reclassified, however, (A8) shows that $z^{(j)}(t) \in \mathscr{H}(P_j^*(0))$ and that $z^{(k)}(t) \in \mathscr{H}(P^*(0) \backslash P_j^*(0))$ for any $k \neq j$. But then, by (A10), reclassification is impossible.

That $D_j(t)$ in (20) is monotone decreasing follows from iterations of (A9). That (21) implies (22) follows just as in the proof of Theorem 1.

References

Barlow, H. B., Pettigrew, J. D.: Lack of specificity of neurones in the visual cortex of young kittens. J. Physiol. (Lond.) **218**, 98–100 (1971)

Bennett, M. V. L.: Analysis of parallel excitatory and inhibitory synaptic channels. J. Neurophysiol. **34**, 69–75 (1971)

Blackenship, J. E., Wachtel, H., Kandel, E. R.: Ionic mechanisms of excitatory, inhibitory, and dual synaptic actions mediated by an identified interneuron in abdominal ganglion of Aplysia. J. Neurophysiol. **34**, 76–92 (1971)

Blakemore, C., Cooper, G. F.: Development of the brain depends on the visual environment. Nature (Lond.) **228**, 477–478 (1970)

Blakemore, C., Mitchell, D. E.: Environmental modification of the visual cortex and the neural basis of learning and memory. Nature (Lond.) New Biol. **241**, 467–468 (1973)

Boycott, B. B., Dowling, J. E.: Organization of the primate retina: light microscopy. Phil. Trans. roy. Soc. B. **255**, 109–184 (1969)

Ellias, S. A., Grossberg, S.: Pattern formation, contrast control, and oscillations in the short term memory of shunting on-center off-surround networks. Biol. Cybernetics **20**, 69–98 (1975)

Freeman, W. J.: Neural coding through mass action in the olfactory system. Proceeding IEEE Conference on biologically motivated automata theory 1974

Gierer, A., Meinhardt, H.: A theory of biological pattern formation. Kybernetik **12**, 30–39 (1972)

Greenspan, H. P., Benney, D. J.: Calculus. New York: McGraw-Hill 1973

Grossberg, S.: Nonlinear difference-differential equations in prediction and learning theory. Proc. nat. Acad. Sci. (Wash.) **58**, 1329–1334 (1967)

Grossberg, S.: Some networks that can learn, remember, and reproduce any number of complicated space-time patterns, II. Stud. appl. Math. **49**, 135–166 (1970a)

Grossberg, S.: Neural pattern discrimination. J. theor. Biol. **27**, 291–337 (1970b)

Grossberg, S.: Pavlovian pattern learning by nonlinear neural networks. Proc. nat. Acad. Sci. (Wash.) **68**, 828–831 (1971)

Grossberg, S.: Neural expectation: cerebellar and retinal analogs of cells fired by learnable or unlearned pattern classes. Kybernetik **10**, 49–57 (1972)

Grossberg, S.: Contour enhancement, short term memory, and constancies in reverberating neural networks. Stud. appl. Math. **52**, 213–257 (1973)

Grossberg, S.: Classical and instrumental learning by neural networks. In: Rosen, R. and Snell, F. (Eds.): Progress in Theoretical Biology, pp. 51–141. New York: Academic Press 1974

Grossberg, S.: A neural model of attention, reinforcement, and discrimination learning. Int. Rev. Neurobiol. **18**, 263–327 (1975)

Grossberg, S.: On the development of feature detectors in the visual cortex with applications to learning and reaction-diffusion systems. Biol. Cybernetics **21**, 145–159 (1976)

Grossberg, S., Levine, D.S.: Some developmental and attentional biases in the contrast enhancement and short term memory of recurrent neural networks. J. theor. Biol. **53**, 341–380 (1975)

Hebb, D. O.: The organization of behavior. New York: Wiley 1949

Hirsch, H. V. B., Spinelli, D. N.: Visual experience modifies distribution of horizontally and vertically oriented receptive fields in cats. Science **168**, 869–871 (1970)

Hirsch, H. V. B., Spinelli, D. N.: Modification of the distribution of receptive field orientation in cats by selective visual exposure during development. Exp. Brain Res. **12**, 509–527 (1971)

Hubel, D. H., Wiesel, T. N.: The period of susceptibility to the physiological effects of unilateral eye closure in kittens. J. Physiol. (Lond.) **206**, 419–436 (1970)

Kimble, G. A.: Foundations of conditioning and learning. New York: Appleton-Century-Crofts 1967

Levine, D. S., Grossberg, S.: Visual illusions in neural networks: line neutralization, tilt aftereffect, and angle expansion. J. theor. Biol., in press (1976)

Meinhardt, H., Gierer, A.: Applications of a theory of biological pattern formation based on lateral inhibition. J. Cell. Sci. **15**, 321–346

Pérez, R., Glass, L., Shlaer, R.: Development of specificity in the cat visual cortex. J. math. Biol. (1974)

Von der Malsburg, C.: Self-organization of orientation sensitive cells in the striate cortex. Kybernetik **14**, 85–100 (1973)

Wachtel, H., Kandel, E. R.: Conversion of synaptic excitation to inhibition at a dual chemical synapse. J. Neurophysiol. **34**, 56–00 (1971)

Wiesel, T. N., Hubel, D. H.: Single-cell responses in striate cortex of kittens deprived of vision in one eye. J. Neurophysiol. **26**, 1003–1017 (1963)

Wiesel, T. N., Hubel, D. H.: Comparison of the effects of unilateral and bilateral eye closure on cortical unit responses in kittens. J. Neurophysiol. **28**, 1029–1040 (1965)

Received: October 6, 1975

In revised form: December 16, 1975

CHAPTER 13

A THEORY OF HUMAN MEMORY: SELF-ORGANIZATION AND PERFORMANCE OF SENSORY-MOTOR CODES, MAPS, AND PLANS

PREFACE

This article suggests a psychophysiological foundation for cognitive theory, and more generally for goal-oriented or purposive behavior. Of all my articles, this is the one which drives deepest into uncharted territory. I say this partly because new implications of my own concepts and constructions in the article are still crystallizing in my mind.

The theory's main contribution is to show how a temporal stream of data patterns can drive the formation of globally consistent cognitive representations and purposive behavioral plans despite the abysmal ignorance of individual cells. To build this theory, I needed all the conceptual and mathematical machinery that I had been accumulating over the past twenty years. The lesson of the article is that all the pieces fit together.

The paper indicates how principles and mechanisms which seem to be related to such general concerns as stability, adaptability, and sensitivity lead to a theory of word recognition and parsing; a theory wherein subtle relationships between chunking, recognition, masking, and rehearsal occur; a theory that distinguishes reset by rehearsal vs reset by interchunk competition and relates this distinction to the possibility of learning to predict future events and to G. A. Miller's Magic Number 7; a theory of item vs order information; a theory of motor synergies and of how performance velocity and motor coordination are regulated; a theory for coding by Gestalts vs coding by individual items wherein tradeoffs between input speed and information content help to choose the code; a theory wherein circular reactions emerge as particular types of adaptive resonances; and so on. In other words, the paper shows how intelligent properties can be the epiphenomena of general constraints on the possibility of doing accurate information processing at all.

The theory also contains unimplemented designs for self-organizing sensory pattern recognition machines, for language processors, and for intelligent robots. Some younger colleagues and I have recently begun to plan a program of implementations.

The paper's structure simulates the evolutionary method by studying the problem of serial order in behavior in a succession of ever-more-demanding environmental variations. Its last section summarizes some of my scientific hopes for the future by enumerating problem areas wherein the application of evolutionary concepts promises to influence our science right down to its philosophical boots.

A Theory of Human Memory: Self-Organization and Performance of Sensory-Motor Codes, Maps, and Plans*

1. Introduction

A psychophysiological theory of the self-organization and performance of sensory-motor codes, maps, and plans is derived herein. This general topic includes a variety of phenomena in many species, ranging from the imprinting of sensory-motor coordinates in an infant to complex goal-oriented serial behavior in an adult. The theory approaches the problem of biological diversity by seeking organizational principles that have evolved in response to environmental pressures to which a surviving species must adapt. These organizational principles are translated into explicit neural networks that realize the principles in a minimal way. Once the properties of the minimal mechanisms are understood, variations of them can be more readily recognized and analyzed in species-specific examples.

The result is a small number of robust principles and mechanisms that form a common substrate for coping with many tasks. These mechanisms are useful because their collective, or interactive, behavior causes most of their interesting properties. In effect, when simple principles are joined interactively, they can generate enormously complicated properties. Such collective properties often appear in psychophysiological data, and often generate erroneous interpretations because they are not obvious consequences of the dynamics of single cells. An analysis of single cells alone does not reveal which single-cell properties generate important collective properties, or whether prescribed collective properties are insensitive to wide variations in certain single-cell properties. The types of parallel, nonlinear, self-organizing, hierarchical, and feedback interactions that are commonplace in psychophysiological data must be explicitly modeled in order to derive accurate conclusions about them.

The present theory suggests a way of thinking no less than a series of

* This work was supported in part by the Advanced Research Projects Agency of the Office of Naval Research (N00014-70-C-0350).

mechanisms. In particular, it suggests how the probabilistic and computer models that have been used, in somewhat complementary fashion, to analyze memory data can be modified, unified, and strengthened. Psychophysiological models provide a natural framework for this synthesis because they must routinely deal with the evolution of patterned activities within hierarchically organized networks. Such models also synthesize serial and parallel processing properties into a unified framework, and weave together phenomena about development, perception, learning, and cognition into an interactive portrait. If nothing else, the method of deriving complex phenomena and predictions from simple environmental pressures confronts us with unexpected and nontrivial consequences of our present beliefs, and provides a rigorous and transparent conceptual superstructure with whose aid new concepts can be more effectively fashioned.

Another basic property of much psychophysiological data is their evolutionary character, whether due to the development of species, the development of individuals, or individual learning. The theory tries to respect the wisdom of evolution by imitating it. At each successive stage of theory construction, prescribed environmental pressures determine a definite class of network principles and mechanisms, and mathematical analysis shows what these mechanisms can and cannot do. As more sophisticated pressures are considered, the earlier principles and mechanisms provide a substrate on which newer principles and mechanisms are superimposed. Similarly, by imposing ever-more-demanding variations on the same problem, we find a sequence of related networks capable of ever-higher levels of behavioral sophistication. Such a sequence illustrates the evolution of a network principle in response to an environmental pressure. Of particular interest in the present work is the evolution of serial order in behavior.

The paper's structure imitates this evolutionary method, subject to space limitations. It is self-contained and written for an audience of nonspecialists. The remainder of this section motivates some central themes of the paper in a heuristic fashion.

A. Does Memory Preserve Order?

We shall consider a maze as illustrative of the many situations in which there exists a succession of choice points leading to a goal, such as in walking from one room of a house to another room at the other end of the house, or from home to school. (see Fig. 1). Suppose that one leaves the filled-in start box and is rewarded with food in the vertically hatched goal box. Every successful transit from the start box to the goal

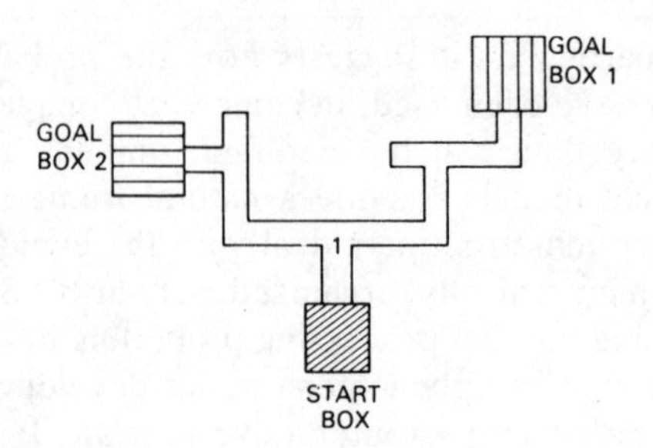

FIG. 1. Correct performance from the start box to a goal box is always order-preserving, with the goal box occurring last.

box requires the same sequence of turns at choice points in the maze. Correct performance is therefore *order-preserving,* and the goal box *always* occurs *last*. In some sense, therefore, correct performance requires that our memory traces remember the order in which events occur. The most naive possibility is that choice points are somehow organized in a chain, as in Fig. 2. Such an encoding is clearly insufficient, however, if the sequence of choices is triggered within the start box by the desire to attain the goal. For example, if I am sitting in my office and decide to go to the cafeteria for lunch, I can then elicit a characteristic series of sensory-motor coordinations that end by eating lunch. This could never happen using the mechanism of Fig. 2. In such a world, if a friend stopped me while I was walking down the hall and asked where I was going, I could only say, "I don't know. I'll tell you when I get there," because the goal in Fig. 2 always occurs last and is unaccessible to me until I reach the last link in the chain. In Figure 2, there is no behavioral *plan*. In goal-directed behavior, by contrast, an internal representation of the goal occurs *first*, and this representation somehow triggers the behavior that can lead to goal attainment.

This state of affairs can be rephrased as the *Goal Paradox:* How can the goal representation occur both last and first? More precisely, in *all* of our experiences with the goal, it is the last event to occur. This makes it plausible that our memory-traces order our choices so that behavior appropriate to the goal occurs last. However, if these memory traces are

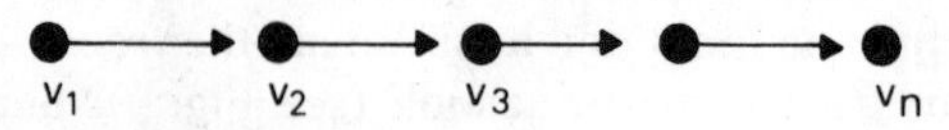

FIG. 2. A chain of associations can accurately code order, but it is insufficient to achieve goal-oriented serial behavior.

order-preserving, as is necessary to actually reach the goal, and if the goal always occurs last, how can an internal goal representation also be activated first, as is necessary for this representation of the goal to trigger a compatible behavioral plan?

We want an internal representation of the goal to trigger a plan that controls a sequence of acts leading to the goal. What we are demanding is schematically drawn in Fig. 3a, where we indicate by points and arrows, respectively, the minimal dimensions of the problem and directed influences between these dimensions. The events (for example, choices in a maze) have internal representations that are designated by states $v_1, v_2, v_3, \ldots, v_n$, where n is the index of the goal. The plan is a state that somehow organizes the order in which the events will occur—hence the arrows from plan to events. The state corresponding to the plan must be determined by the events themselves, since during a correct sequence of choices on a learning trial, only these events occur. This dependence is schematized by the upward-directed arrow in Fig. 3b. Thus the events determine the state that will represent the plan, and this state thereupon gains control of the event-representations themselves. Simultaneously, an internal trace of the goal gains control of the plan. Given such a picture, albeit vague at this stage, several definite design problems emerge:

(*a*) What mechanism maintains the activity of the plan throughout the

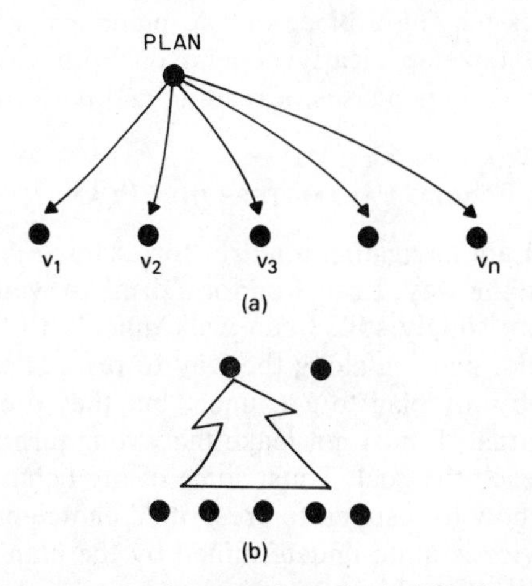

FIG. 3. (a) The goal representation organizes the individual commands v_i, $i = 1, 2, \ldots, n$; (b) the individual commands help to choose the plan.

presentation of all the events so that the plan knows which events to control?

(*b*) What mechanism tells the plan which of the events came first, so that it will be able to perform them in the proper order?

Such questions apply to a host of related situations. To illustrate the breadth of the problem, we consider a similar problem in language learning and performance. Suppose that I wish to say a long word, such as MAGNETOHYDRODYNAMIC. There is a clear intuitive sense in which I am ready to say the whole word at a given instant of time; that is, the "idea" or "plan" of saying the word is active in my mind at that instant, yet the actual elicitation of the plan occurs serially, one behavioral unit after another. A similar problem arises when we say the name of a familiar person or object that is visually perceived. How can I establish a command that can "see" the whole word at a given time, yet also organizes the serial performance of its parts? Clearly, a picture such as Fig. 3 is again called for, and the same formal problems must be solved in synthesizing a mechanism that justifies the picture.

In the hall-walking problem, we are considering how to control muscles in our arms, legs, eyes, etc., using visual and proprioceptive feedback, etc. In the word-elicitation problem, we are considering how to control muscles in our mouth, throat, larynx, diaphram, etc., using auditory and proprioceptive feedback, etc. Both situations address common problems about how sensory-motor loops between particular modalities are serially organized by command structures or plans. A related problem is playing a piece on the piano, in which the sensory-motor loops that develop clearly depend on both visual and auditory modalities during various phases of reading and performing the piece.

B. Free versus Forced Parameters in a Plan

Suppose that I am navigating a maze (for example, a hallway) on my way to lunch. On the way, I can stop for a drink of water or to chat with a friend. Or, more simply still, I can walk quickly to the cafeteria, or I can stroll leisurely, pausing along the way to rest. These events are not preprogrammed by my plan to get lunch, but they are compatible with the plan. By contrast, I must not make the wrong turn at a choice point, or I will never reach the goal. Thus, some of my behavior is under tight control, such as how to respond to prescribed choice-point cues, but the rest of my behavior is quite undetermined by the plan, in particular the velocity with which the plan is executed.

A similar temporal freedom occurs when I say a word or play a piece on the piano. Within rather broad limits, I can say certain syllables more slowly than others, or play certain passages more slowly than others. The *order* information in the plan is not tightly coupled to the *velocity* of performance.

Uncontrolled parameters in a plan are also of a more subtle type. For example, a common phrase in language might have a rigidly controlled order, yet the plan might allow a wide choice of nouns or verbs to be fit into the phrase between the rigidly controlled items—comparable to the rigid control of choice points in a hallway versus the freedom to do other things between the choice points. Or a certain number of equivalent phrases can be chosen to express an idea, just as a variety of techniques for executing the correct turns in the hallway will all lead to the goal. Similar remarks can be made about playing a piece on the piano: Certain passages can be played as tightly coupled units, so much so that it is difficult to start playing them in the middle.

C. Circular Reactions

To execute a sequence of sensory-motor coordinations, one must first be able to execute one member of the sequence. Even at the level of individual sensory-motor acts, there is a decoupling of order information (or positional information) from velocity information. For example, I can plan to move my hand to a fixed terminal position, and can move it there at a wide range of velocities. "Knowing" where I want to move my hand and "willing" it to move are not the same operation. Similarly, being "ready" to turn right in the hallway when I see a certain cue does not determine how fast I will turn right.

What is the cue that tells my hand where to move? Suppose that my eyes are focused on a certain object. The tilts of my neck, head, and eyes, along with the vergence of my eyes, etc., establish proprioceptive coordinates that determine the relative position of the object from my body. Somehow these coordinates must get translated into commands to the muscles in my hand and arm that correspond to the correct terminal coordinates of the hand on the object. In other words, the *proprioceptive map* of the head, neck, eyes, etc., excites a *terminal motor map* of the hand. "Willing" the hand to move releases the information in the motor map, and makes the hand move.

How does the transformation between maps get established? Because there exists so many individual differences in body parameters between individuals, it seems clear that much of the transformation must be learned. Piaget (1963) has carefully observed the development of the ability in young children. He notes that at first an infant's hand makes a series of unconditional motions, which the infant's eyes unconditionally follow. As the hand occupies a variety of positions that the eye fixates upon, a map is learned from the proprioceptive coordinates of the hand–arm system to the motor coordinates of the eye–head–neck system, and *conversely,* from the proprioceptive coordinates of the eye–head–neck system to the motor coordinates of the hand–arm system. Using the map from eye–head–neck proprioception to hand–arm motor coordinates, we can move our hands to a fixed object. During the learning trials, the eyes try to continuously follow the motions of the hand, or if they fall behind, they must try to catch up by leaping to the correct position via saccades. Since the eye always tries to fixate the present position of the hand, the two transformations between proprioceptive maps and motor maps code only the (approximately) present motor positions. During performance trials, the transformation from eye–head–neck proprioception (or where we are now looking) to hand–arm motor coordinates therefore determines only the *terminal* position of the hand (or where we *want* the hand to go). If the initial position of the hand is very different from the desired terminal position, then the directed motion of the hand can be viewed as a saccade of the hand. To say that positional and velocity information are decoupled translates into the statement that the saccadic velocity is not preprogrammed in this system.

The above observations can be reformulated to emphasize an important point. Since the terminal motor map suffices to guide the hand throughout its trajectory from initial to terminal position, all intermediate positions of the arm–hand system must be derivable from this information. What auxiliary feedback mechanisms within the hand–arm system translate the terminal motor map into a physically realizable trajectory of this system? More precisely, the transformation between maps codes only where the hand is destined to go, but not how it gets there. It ignores the properties of the arm–hand system as a mechanical system, and codes only the plan. In particular, on each performance trial aimed at extending the hand to a fixed position, any of the free parameters—such as hand velocity—can be chosen differently, and can thereby alter the forces on the system, even though the plan remains the same. Somehow these varying mechanical properties must be controlled by auxiliary mechanisms, which average them away, so that the invariant plan can be realized.

D. The Internal Structure of Maps and Their Transformations

If directed reaching for an object is controlled by a transformation between maps, then a tremendous reduction in the amount of information that must be stored has been achieved. Indeed, suppose that the transformation has been learned up to a given level of accuracy on a finite number of learning trials. Then, without *any* further learning, any of the infinitely many reachable positions of an object can be approached by the hand, under eye–head–neck guidance, up to this level of accuracy.

This assertion tacitly assumes that, if a set of proprioceptive coordinates $\mathcal{P}_1$ has been associated with a terminal motor map $\mathcal{M}_1$, and a different set of proprioceptive coordinates $\mathcal{P}_2$ has been associated with a terminal motor map $\mathcal{M}_2$, then a new set of proprioceptive coordinates $\mathcal{P}_3$ will have the following effects:

(*a*) It partly excites $\mathcal{P}_1$ and $\mathcal{P}_2$ both, with an intensity that depends on how similar $\mathcal{P}_3$ is to $\mathcal{P}_1$ and $\mathcal{P}_2$.

(*b*) $\mathcal{P}_1$ and $\mathcal{P}_2$ will excite $\mathcal{M}_1$ and $\mathcal{M}_2$, respectively, with an intensity that depends on how excited they are by $\mathcal{P}_3$.

(*c*) The mixture of motor excitation will form a hybrid terminal motor map $\mathcal{M}_3$ that is between $\mathcal{M}_1$ and $\mathcal{M}_2$, which moves the arm closer to the position that excited $\mathcal{P}_3$ than either $\mathcal{M}_1$ or $\mathcal{M}_2$ could have separately (see Fig. 4). In other words, each proprioceptive representation has a *generalization gradient*. Representations excite each other with an intensity that depends on how close they lie with respect to each other on their gradients. Each motor map also has a generalization gradient. The fact that $\mathcal{M}_3$ can be synthesized from $\mathcal{M}_1$ and $\mathcal{M}_2$ to yield a position close to the one determined by $\mathcal{P}_3$ means that the transformation from

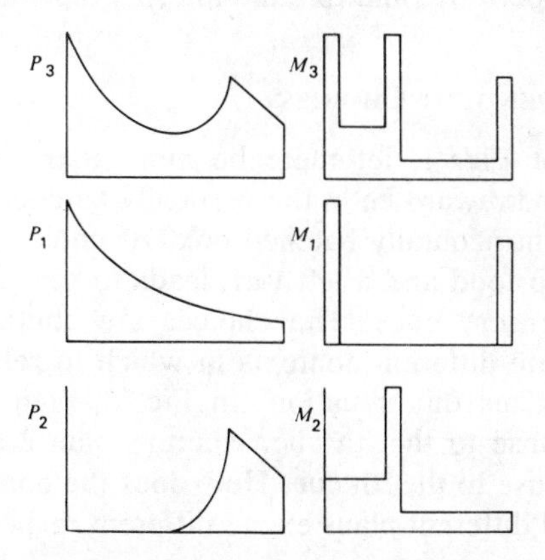

FIG. 4. Filtering of $\mathcal{P}_3$ as a weighted average of $\mathcal{P}_1$ and $\mathcal{P}_2$ followed by synthesis of $\mathcal{M}_3$ as a weighted average of $\mathcal{M}_1$ and $\mathcal{M}_2$.

proprioceptive maps to terminal motor maps preserves (at least approximately) the distances between representations that are defined by the generalization gradients.

An instructive example occurs in language learning. Again there exists a period during which unconditional behavior is emitted (Fry, 1966). Instead of unconditional motions of the eye–hand system, there exists a period when the infant babbles various simple sounds. Instead of the eye following the hand so that a transformation from proprioceptive feedback to terminal motor position can be learned, the internal trace of the auditory sensory feedback elicited by a babbled sound is conditioned to the motor coordinates that produced the sound. The babbling phase cannot go on forever; if it did, the unconditional urge to babble would forever interfere with the desire to say something interesting. When babbling stops, a certain number of connections exist between prescribed auditory representations and the motor controls that produced them. Exciting one of these representations can elicit the appropriate sound, so that simple imitation begins to be possible. How does an infant learn more complex sounds than the ones that occurred during babbling? One way is to suppose that more complex sounds are decomposed into weighted combinations of the simpler sounds that already are capable of eliciting speech sounds. If these simpler sensory representations map into a motor speech space that preserves their mutual distances, then the speech sound that is synthesized in this way will be closer to the heard sound than any of the simpler sounds. The system can hereby try to imitate more complex sounds than are originally in its repertoire, and to build internal representations of these.

E. Context-Dependent Choices

In the maze of Fig. 1, let there be more than one goal box. For example, let a food reward be in the vertically hatched box and a sexual reward be in the horizontally hatched box. At choice point number 1, a right turn leads to food and a left turn leads to sex, in response to the same external sensory cues. The choices are controlled by different plans, which create different contexts in which to respond to the cues. Figure 5 schematizes this situation. In Fig. 5, plan 1 excites a given pathway in response to the *i*th cue, whereas plan 2 excites a different pathway in response to the *i*th cue. How does the convergence between a given event and different plans excite different responses as a result of learning trials?

This problem also occurs in many situations. For example, how can

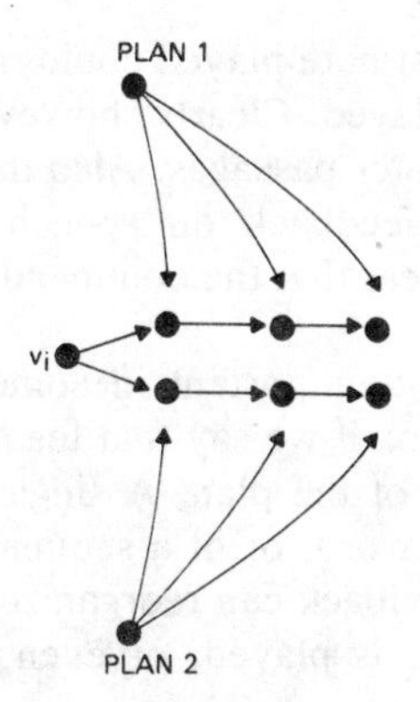

FIG. 5. Plans 1 and 2 excite different pathways in response to the ith cue representation v_i.

we serially learn both lists ABCDEFGH and ABAFALAK? In the two lists, the responses to B are different even though both lists have the same beginning AB, and in the second list the letter A is a precursor of four different letters. Clearly the stimulus for saying the next letter is not merely the previous letter, or even the previous two letters. Longer subsequences of letters somehow distinguish the two lists in our memory. Again Fig. 5 is called to mind.

F. CHUNKING AND FEEDBACK

The above remarks can be rephrased by saying that the units that control behavior are not necessarily representations of individual stimuli or responses, but can be built up from lists or other aggregates of these units. Such composites, or chunks (Miller, 1956), make possible a recoding of memory that enables ever-more-complex commands to form. For example, if an adult had to pay attention while walking to every step taken as a complex juxtaposition of motor events in different joints, guided by sensory feedback, then it would be difficult to pay attention to anything else while walking. Once these events are organized by higher-order commands or chunks, by contrast, the details of walking become simple, and attention can be devoted to other tasks. Similarly, once a long word is organized by a higher-order command, it can be treated as a single behavioral unit rather than as a complicated series of mouth, tongue, larynx, diaphragm, and related motor activities, with attendent sensory feedback, whose conscious control would interfere with thinking about other things.

Lashley (1951) noted that a pianist can play successive notes so fast that feedback from the last note played could not possibly influence how the next note would be played. Clearly, however, making a mistake can influence our playing of later passages. Also there are many data on the importance of auditory feedback on speech production (Lenneberg, 1967), even though it is clear that the commands controlling language are not individual letters.

How can feedback be unimportant in some cases and important in others? There is no paradox if we say that feedback is important when it reorganizes the structure of the plan. A single command can guide the performance of an entire word, or of a sequence of notes on the piano, without feedback. But feedback can reorganize which commands will be active after this sequence is played, or even before it is played in its entirety.

G. Performing the First or the Last Item

A picture such as Fig. 3 does not tell us how events are performed in a given order. Daily experience suggests some important constraints on this mechanism. Suppose that the letters ABARFD are said to me one at a time. I can be told to do any of at least four things:

(*a*) Repeat each letter aloud as soon as I hear it.

(*b*) Listen passively to the list and, after hearing it, repeat the list items in their proper order.

(*c*) Repeat each letter aloud as soon as I hear it, and repeat the list items in their proper order after the whole list is presented.

(*d*) After completing task (*a*), I can be asked to repeat the whole list in its proper order.

In task (*a*), I am being asked to repeat the *last* thing that I have heard. In some sense, the last item must have the greatest "weight," so that I can choose it above all others. By iterating this requirement, each item has greater weight than the preceding items. In task (*b*), I am being asked to repeat the *first* thing that I have heard. In some sense, the first item must also have the greatest "weight," so that I can choose it above all others (see Fig. 6a). Furthermore, after saying the first item, its "weight" must be decreased, so that I can say the second item, whose weight is then greater than all other weights. After saying the second item, its "weight" must decrease, etc. But this means that the last item has the *least* weight in task (*b*). How can it also have the *greatest* weight in task (*a*)? Moreover, tasks (*c*) and (*d*) mix tasks (*a*) and (*b*), so that the

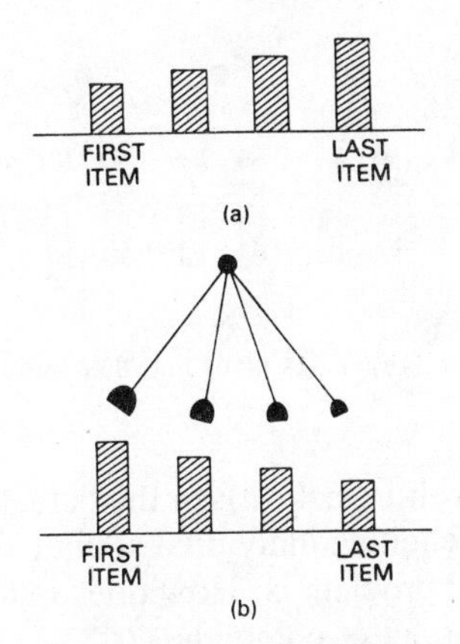

FIG. 6. (a) The most recent item has the greatest weight in task (*a*); (b) the least recent item has the greatest weight in task (*b*).

last item must be able to have the least and the greatest weight in the same situation! How can this be?

It is intuitively clear that the most recent item "should" have the greatest weight in the sense of task (*a*). Surely the most recent events are more "salient" than earlier events. It is also clear that something like a command, or plan, develops to perform tasks (*b*), (*c*), or (*d*), since the letter A leads to B or R, depending on the context. The dilemma of performing tasks (*a*) and (*b*) therefore can be rephrased as follows: How does activation of the plan *reverse* the weights of its individual events? (See Fig. 6b.)

In the above example, I could also be told to do the following:

(*e*) Repeat every pair of successive letters when it occurs. Then repeat the whole list in its proper order.

To repeat a given pair of items, the first item in the pair must have the greater weight. This is true for *every* pair, so that we can no longer talk about weights that increase monotonically or decrease monotonically with list position as in Fig. 6. It is tempting to instead draw a picture such as that shown in Fig. 7. Task (*a*) differs from task (*e*) only in their rehearsal strategies. How does rehearsal reorganize plans as in Figs. 6 and 7? Actually, familiar letter sounds, such as the sound of A, can be composed of more than one sound component, as slow pronunciation clearly indicates. Prior experience has organized these components into a single letter via a suitable command. In this sense, Fig. 7 is a

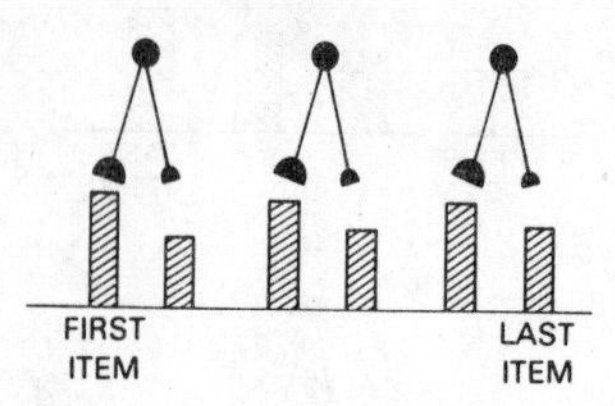

FIG. 7. Rehearsal organizes plans and their order information.

refinement of Fig. 6 even in task (*a*); in this situation, the paired arrows are the commands for eliciting individual sound components of the same letter. Thus the general problem is: How does rehearsal group individual motor acts into a hierarchy of commands?

2. Stimulus Sampling of Spatial Patterns

The neural networks of this paper are built out of network components that have been derived elsewhere from psychological postulates and mathematically analyzed (see, for example, Grossberg 1967, 1969a,b, 1970a, 1971a, 1972a). They are sketched herein for completeness. The material in Sections 2 through 5 is more completely reviewed in Grossberg (1974).

The first stage of the theory analyzes the simplest concepts of classical conditioning: How does pairing of a conditioned stimulus (CS) with an unconditioned stimulus (UCS) on learning trials enable the CS to elicit a conditioned response (CR), or UCS-like event, on performance trials? This analysis yields a psychophysiological theory operating in real time. Psychological inputs, or stimuli, representing particular experiments perturb a neural network that elicits definite outputs or responses. The network dynamics are described by interactions between the *short-term memory* (STM) *traces* $x_i(t)$ of cell body populations v_i, and the *long-term memory* (LTM) *traces* $z_{jk}(t)$ of the axonal pathways e_{jk} from v_j to v_k, as in Fig. 8. The simplest realization of these interactions among n

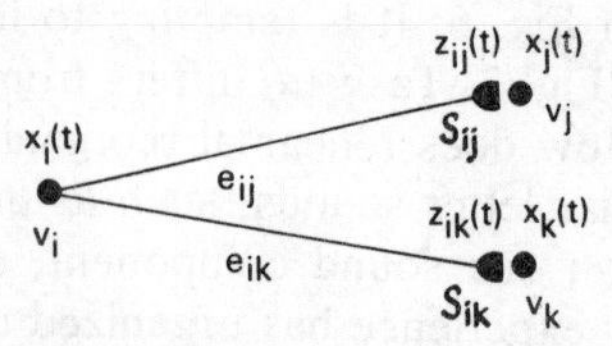

FIG. 8. STM traces $x_i(t)$ and LTM traces $z_{jk}(t)$ in cell body populations and axonal pathways, respectively.

populations $v_1, v_2, \ldots, v_n$ is given by the system

$$\dot{x}_i = -A_i x_i + \sum_{k=1}^{n} B_{ki} z_{ki} - \sum_{k=1}^{n} C_{ki} + I_i(t) \tag{1}$$

and

$$\dot{z}_{jk} = -D_{jk} z_{jk} + E_{jk} x_k \tag{2}$$

where $i, j, k = 1, 2, \ldots, n$. The terms in (1) and (2) have the following interpretation. Function A_i in (1) is the decay rate of the STM trace x_i. Function B_{ki} in (1) is a performance signal from v_k to the synaptic knobs S_{ki} of e_{ki}. Two typical choices of B_{ki} are

$$B_{ki}(t) = b_{ki}[x_k(t - \tau_{ki}) - \Gamma_{ki}]^+ \tag{3}$$

where $[\xi]^+ = \max(\xi, 0)$ for any number ξ, or

$$B_{ki}(t) = f(x_k(t - \tau_{ki}))b_{ki} \tag{4}$$

where $f(\xi)$ is a sigmoid (S-shaped) function of ξ with $f(0) = 0$. In (3), signals leave v_k only if x_k exceeds the signal threshold Γ_{ki} (Fig. 9a) and reach S_{ki} after τ_{ki} time units; in (4), the signal threshold Γ_{ki} is replaced by attenuation of the signal at small x_k values (Fig. 9b). Such a population signal from v_k is generated, for example, if the signal thresholds of cells in v_k are Gaussianly distributed around a mean

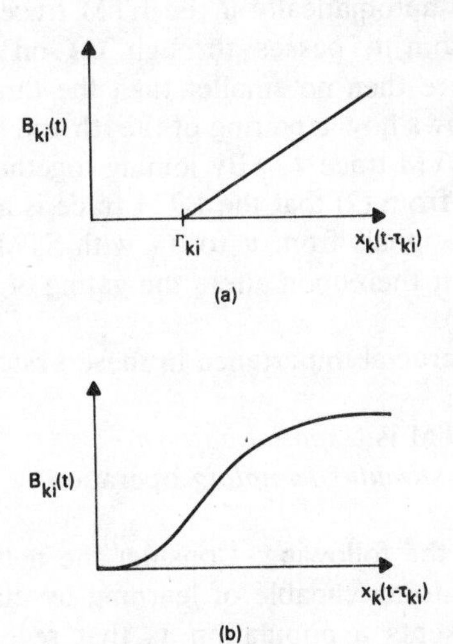

FIG. 9. (a) Signal $B_{ki}(t)$ is positive only if $x_k(t - \tau_{ki})$ exceeds the threshold Γ_{ki}; (b) signal $B_{ki}(t)$ is attenuated at small $x_k(t - \tau_{ki})$ values.

threshold value. Term $B_{ki}z_{ki}$ in (1) says that the signal B_{ki} from v_k to S_{ki} interacts with the LTM trace z_{ki} at S_{ki}. In particular, z_{ki} *gates* signal B_{ki}, so that the signal strength that perturbs x_i at v_i is $B_{ki}z_{ki}$ rather than B_{ki}. Thus, even if an input to v_k excited equal signals B_{ki} in all the pathways e_{ki}, only those v_i such that z_{ki} is large will be appreciably excited by v_k. All such gated signals from populations v_k combine additively at v_i in term $\Sigma_{k=1}^{n} B_{ki}z_{ki}$. The term $\Sigma_{k=1}^{n} C_{ki}$ in (1) describes the total effect of inhibition from cells v_k upon v_i. The choice

$$C_{ki}(t) = g(x_k(t - \sigma_{ki}))c_{ki}$$

with $g(\xi)$ a sigmoid signal function, is illustrative. The function $I_i(t)$ in (1) is the input corresponding to presentations of the ith event through time; $I_i(t)$ is large when the stimulus is presented, and otherwise equals zero. In all, the STM trace x_i can spontaneously decay, be excited by external stimuli, and interact with other populations via sums of gated excitatory signals and inhibitory signals. The net size of x_i after all these processes operate determines whether v_i will generate an output.

Function D_{jk} in (2) is the decay rate of the LTM trace z_{jk}. Function E_{jk} in (2) describes a learning signal from v_j to S_{jk} which drives the LTM changes in z_{jk} at S_{jk}. In other words, v_j *samples* v_k by turning on E_{jk}. In the simplest cases, E_{jk} is proportional to B_{jk}, but this is not necessary. It is only necessary to prevent B_{jk} from being large over a sustained time interval if E_{jk} is small over that interval (Grossberg, 1969b, 1971a, 1972a). This occurs automatically if the LTM trace is computed at S_{jk}; since the signal from v_j passes through S_{jk} on its way to v_k, the thresholds for B_{jk} are then no smaller than the thresholds for E_{jk}. The term $E_{jk}x_k$ in (2) shows how a pairing of the jth and kth events influences the growth of the LTM trace z_{jk}. By joining together terms $-D_{jk}z_{jk}$ and $E_{jk}x_k$, we conclude from (2) that the LTM trace is a time average of the product of learning signals from v_j to S_{jk} with STM traces at v_k. When z_{jk} changes in size, it thereupon alters the gating of signals from v_j to v_k via term $B_{jk}z_{jk}$ in (1).

Two facts are of crucial importance in these systems:

A. The unit of LTM is a *spatial pattern*.
B. There exists a *stimulus sampling* operation.

By (A) we mean the following: Consider the network in Fig. 10a. It has the minimal anatomy capable of learning by classical conditioning. The network represents a population v_0 that receives a CS-activated input. Population v_0 can send signals to its axon collaterals, which abut on the UCS-activated populations $v_1, v_2, \ldots, v_n$. The LTM traces z_{0i}

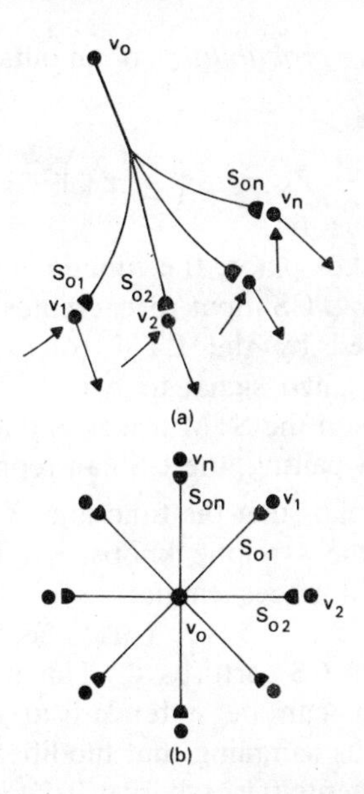

FIG. 10. (a) The CS-activated population v_0 samples populations $v_1, v_2, \ldots, v_n$; (b) the *outstar* is the minimal network capable of classical conditioning.

are computed at the synaptic knob terminals S_{0i} of the v_0 axon collaterals e_{0i}. Such a network is called an *outstar* because it can be redrawn as in Fig. 10b.

An outstar can learn an arbitrary spatial pattern. A *spatial pattern* is a UCS to the cells $v_1, v_2, \ldots, v_n$ whose intensities have a fixed relative size through time; that is, $I_i(t) = \theta_i I(t)$, for some $\theta_i \geq 0$ such that $\Sigma_{k=1}^n \theta_k = 1$. For example, suppose that the UCS is a picture playing across the "retina" of cells $v_1, v_2, \ldots, v_n$. The total intensity of white light that illuminates the picture can be varied through time without changing the picture itself. The *relative* intensities of light (or *reflectances*) reflected from various points in the picture characterize it, and these remain constant through time (Cornsweet, 1970). The function $I(t)$ is the total UCS input intensity, which can fluctuate wildly through time. The constant relative intensities $\theta = (\theta_1, \theta_2, \ldots, \theta_n)$ characterize the spatial pattern. In short, an outstar can learn an arbitrary spatial pattern of relative figure-to-ground. Thus, the unit of LTM cannot be determined by measurements from just one population v_i; *parallel* measurements are needed to test whether the *relative* intensities are changing through time.

The *stimulus sampling probabilities* of an outstar are the relative LTM traces

$$Z_{0i} = z_{0i}\left(\sum_{k=1}^{n} z_{0k}\right)^{-1} \tag{5}$$

As CS–UCS pairing takes place, the functions Z_{0i} approach θ_i. During later performance trials, a CS input to v_0 creates signals in the e_{0i} axons. These signals are gated by the LTM traces z_{0i}. Since the z_{0i} are proportional to θ_i, the gated signal to v_i is proportional to θ_i. The CS hereby elicits responses in the STM traces x_i that are proportional to θ_i. In short, after CS–UCS pairing, the CS can reproduce the pattern θ.

Stimulus sampling means that the functions Z_{0i} can change *only* when signals from v_0 reach the synaptic knobs S_{0i}. Unless the CS perturbs these knobs, their LTM traces cannot "see" what UCS patterns are received at the cells $v_1, v_2, \ldots, v_n$. This is because the learning signals E_{0i} in (2) vanish unless a CS perturbs v_0. This interpretation of stimulus sampling in an outstar can be extended to a more general neural interpretation of stimulus sampling that modifies and generalizes Estes' theory of amplifier elements (Grossberg, 1972b,c).

3. Sensory Codes and Motor Synergies

The outstar is a general-purpose device. It can learn a spatial pattern of activity playing across whatever cells its knobs S_{0i} sample. For example, suppose that the cells $v_1, v_2, \ldots, v_n$ are feature detectors in a sensory cortex of a network. By this we mean the following. When a picture is presented to the network's retina, the picture is analyzed in such a way that each v_i responds most vigorously to particular features (for example, color, orientation, disparity) in a prescribed retinal region. Each picture hereby generates a spatial pattern of activity across the feature detectors of $v_1, v_2, \ldots, v_n$. This pattern is a coded internal representation of the picture. An outstar can learn and reproduce *any* such representation with complete fidelity.

Alternatively, suppose that the cells $v_1, v_2, \ldots, v_n$ are motor control cells. In this case, each v_i can excite a particular group of muscles, and a larger signal from v_i causes a faster contraction of its target muscle group. A spatial pattern across $v_1, v_2, \ldots, v_n$ then codes fixed relative contraction rates across many muscle groups: for example, playing a chord on the piano with prescribed fingers; or withdrawing a hand with fixed relative speeds of wrist, elbow, and shoulder motion; or forming a particular configuration of lips and tongue when uttering a sound. In other words, the outstar can learn any motor synergy in which prescribed relative rates exist across a family of muscle groups. An increase of the CS input speeds up all the muscle contractions at their fixed relative rates; that is, the CS can perform the synergy at an arbitrary rate.

In summary, a single outstar can coordinate, through its parallel pathways e_{0i}, the learning and reproduction of a distributed sensory code or a synergistic pattern of motor commands. Not all sensory and motor acts have fixed relative figure-to-ground, but using the results on outstars shows how to approach the coding and performance of arbitrary sequences of events.

4. Ritualistic Learning of Arbitrary Acts

The properties of stimulus sampling and of encoding in spatial pattern units show how to learn an arbitrary act, such as a piano recital, a dance, or a sequence of sensory images, in a minimal way (Grossberg, 1969c, 1970a, 1974). The simplest example describes a ritualistic encoding, wherein performance is insensitive to environmental feedback. In this case, only one cell is needed to encode the memory of an arbitrary act. This fact shows that the encoding of complexity per se is relatively easy. In fact, nervous systems with few cells can activate complicated behaviors, as is well known in invertebrates (Dethier, 1968; Kennedy, 1968; Willows, 1968). The ritualistic construction is also universal; such a cell can encode *any* act. The genetic code for such a cell need not concern itself with which act will eventually be encoded. The ritualistic construction focuses our attention on deeper questions concerning the global organization of memory when environmental feedback is operative, and suggests mechanisms of encoding that are sensitive to environmental feedback.

Suppose that the act to be learned is controlled by the cells $v_1^{(1)}$, $v_2^{(1)}, \ldots, v_n^{(1)}$ in a field of cells $\mathcal{F}^{(1)}$. Each $v_i^{(1)}$ might be a feature detector, a motor control cell, a hormonal source, an interneuron—anything you like. The number n of cells being controlled can be chosen arbitrarily large. Let each cell $v_i^{(1)}$ receive a nonnegative and continuous input $I_i(t)$, $t \geq 0$, $i = 1, 2, \ldots, n$. Any such input is covered by our analysis. A particular choice

$$J(t) = (I_1(t), I_2(t), \ldots, I_n(t)), \qquad t \geq 0$$

of inputs controls a given act. In intuitive terms, $J(t)$ describes a moving picture playing on the cells $v_1^{(1)}, v_2^{(1)}, \ldots, v_n^{(1)}$ through time. The movie shall be learned and performed as a sequence of still pictures that are smoothly interpolated in time.

Because each $I_i(t)$ is continuous, the functions

$$\theta_i(t) = I_i(t) \left[\sum_{k=1}^{n} I_k(t) \right]^{-1} \tag{6}$$

are also continuous. As in the case of moving pictures, any continuous function $\theta_i(t)$ can be arbitrarily well approximated by a sequence of its values

$$\theta_i(\zeta),\ \theta_i(2\zeta),\ \theta_i(3\zeta),\ \theta_i(4\zeta),\ \ldots$$

sampled every ζ time units, if ζ is chosen so small that the functions $\theta_i(t)$ do not change too much in a time interval of length ζ. For every fixed k, the numbers

$$\theta^{(k)} = (\theta_i(k\zeta), \qquad i = 1, 2, \ldots, n)$$

sampled across all the cells $v_1^{(1)}, v_2^{(1)}, \ldots, v_n^{(1)}$ at time $t = k\zeta$ form a spatial pattern. To learn and perform the movie $J(t)$, $t \geq 0$, it therefore suffices to learn and perform the spatial patterns $\theta^{(1)}, \theta^{(2)}, \theta^{(3)}, \ldots$ in the correct order. This can be done if a sequence of outstars $\mathcal{O}_1, \mathcal{O}_2, \mathcal{O}_3, \ldots$ is arranged so that $\mathcal{O}_k$ samples just spatial pattern $\theta^{(k)}$ on successive learning trials, and is then briefly activated in the order $\mathcal{O}_1, \mathcal{O}_2, \mathcal{O}_3, \ldots$ on performance trials. An *avalanche*-type anatomy, such as that in Fig. 11, accomplishes this by using the minimum number of spatial dimensions. In Fig. 11, a brief CS-activated sampling pulse travels along the long axon of cell $v_1^{(2)}$ from left to right, and down its serially arranged

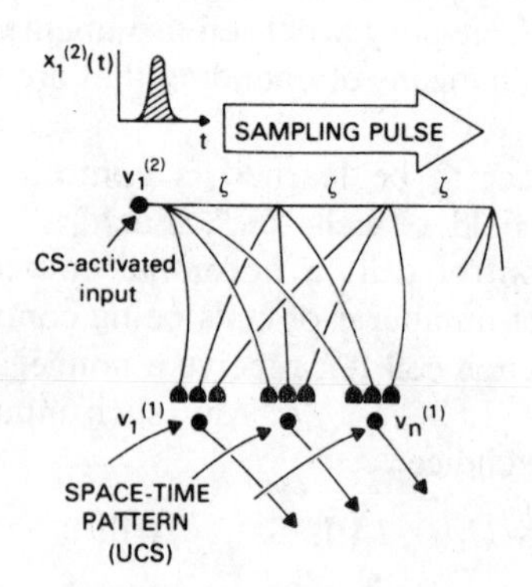

FIG. 11. An avalanche is the minimal network that can ritualistically learn any space-time pattern.

bouquets of axon collaterals. Each bouquet (really an outstar) can learn a spatial pattern, and successive bouquets are activated every ζ time units. No bouquet can see more than one pattern $\theta^{(k)}$, because of the property of stimulus sampling. On performance trials, the CS-activated pulse runs along the axon, serially exciting the bouquets and their learned spatial patterns; the space–time pattern is hereby synthesized from its ordered spatial components $\theta^{(k)}$. Thus one cell, albeit in possession of many axon collaterals, can code the memory of an arbitrary act.

5. Nonspecific Arousal as a Command

Once a pulse is emitted by $v_1^{(2)}$, there is no way to stop it in Fig. 11. If, for example, the avalanche controlled the performance of a long dance, and the stage on which the dance was being performed began to burn, there would be no way to stop the dance in mid-course to escape the flames. Sensitivity to environmental feedback is possible only if the pulse can be abruptly terminated as it travels along the avalanche axon (Grossberg, 1969c, 1970a, 1971b, 1974). By considering the minimal way to do this, we find an anatomy that is isomorphic with that discovered for command neurons in various invertebrates—for example, in the control of the rhythmic beating of crayfish swimmerets (Stein, 1971).

The avalanche must be modified so that performance can be terminated at loci all along the axon of the CS-activated cells $v_1^{(2)}$. Consider Fig. 12a. In Fig. 12a, cell bodies $v_1^{(2)}, v_2^{(2)}, v_3^{(2)}, \ldots$, forming a field of $\mathscr{F}^{(2)}$ of cells, are interpolated at every outstar source. Performance still cannot be terminated if a signal from $v_i^{(2)}$ suffices to fire a signal at $v_{i+1}^{(2)}$. Figure 12b remedies this situation in a minimal way. The new population $v_1^{(3)}$ can supply a signal that reaches all the populations $v_1^{(2)}, v_2^{(2)}, \ldots$ (approximately) simultaneously. Require that $v_{i+1}^{(2)}$ can fire a signal *only* if it receives an input from $v_i^{(2)}$ *and* $v_1^{(3)}$ simultaneously. Withdrawal of the signal from $v_1^{(3)}$ can therefore abruptly terminate output from the avalanche, since $v_{i+1}^{(2)}$ cannot fire even if it receives input from $v_i^{(2)}$. In this sense, $v_1^{(3)}$ supplies an *arousal* input to the avalanche. Because this input is delivered to all populations $v_1^{(2)}, v_2^{(2)}, \ldots$, it is a *nonspecific* arousal input. Population $v_1^{(3)}$ supplies a *command* signal that prepares the avalanche to fire in response to the CS input to $v_1^{(2)}$.

Such command neurons are familiar in the control of behavioral acts by invertebrates (Dethier, 1968; Kennedy, 1968; Willows, 1968). If changes in the LTM traces of the avalanche in Fig. 12 are prevented [set $D_{jk} = E_{jk} = 0$ in (2)], then this network is capable of performing

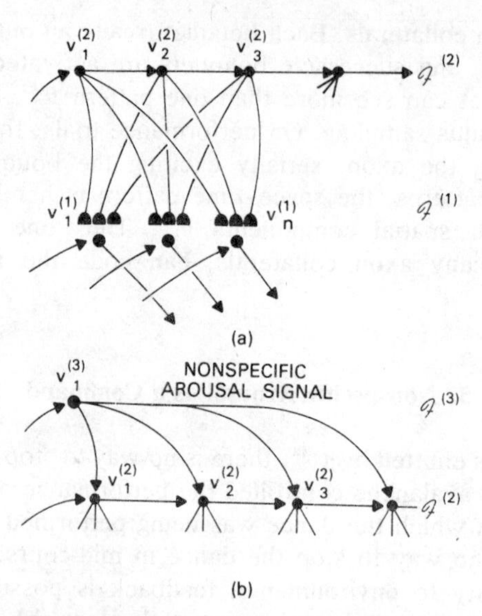

FIG. 12. (a) Interpolating cells $v_i^{(2)}$ along the avalanche cannot terminate performance unless (b) a nonspecific signal is also needed to fire the next cell.

arbitrary acts whose pattern weights are permanently encoded in its synaptic knobs.

Note that varying the size of the arousal signal through time can continuously modulate the speed of performance. Suppose that a given cell $v_i^{(2)}$ starts to receive a signal from $v_{i-1}^{(2)}$ when arousal from $v_1^{(3)}$ is large. Then the total input to $v_i^{(2)}$ is large, so its STM trace grows rapidly, and quickly exceeds its signal threshold. By contrast, if $v_i^{(2)}$ receives a signal from $v_{i-1}^{(2)}$ when arousal from $v_1^{(3)}$ is small, then the total input to $v_i^{(2)}$ is small, so its STM trace grows slowly, and takes longer to exceed its signal threshold. If the arousal signal is too small, the signal threshold is never reached, and the reaction time of $v_i^{(2)}$ is infinite.

6. Self-Organization of Codes and Order Information

Several important themes are made evident by the avalanche example. They are introduced below to motivate our later network constructions.

A. INSTRUMENTAL CONDITIONING

Not every competing event should be able to switch off nonspecific arousal. To cite a colorful example: It is one thing to stop dancing at your debut if the stage is consumed by flames, and quite another to risk your career because a mosquito is hovering above. Only more important events should be able to shut off the arousal that supports a given act. Knowing what is important to an organism requires, in particular, that we know what events are rewarding or punishing to the organism. What is the relationship between reinforcement and a cue's ability to trigger nonspecific arousal? Grossberg (1971b, 1972a,b, 1975) develops a reinforcement theory that suggests an answer to this question.

From the discussion of arousal as a command, we expect a process akin to that depicted in Fig. 13. In Fig. 13, cue CS_1 excites arousal source $\mathscr{A}_1$ plus its avalanche $v_1^{(1,2)}$, $v_2^{(1,2)}$, $v_3^{(1,2)}$, . . . to elicit sequential performance of its encoded act. When CS_2 occurs, it excites arousal source $\mathscr{A}_2$ plus its avalanche $v_1^{(2,2)}$, $v_2^{(2,2)}$, $v_3^{(2,2)}$, ... Arousal sources $\mathscr{A}_1$ and $\mathscr{A}_2$ mutually inhibit each other. If $\mathscr{A}_2$ is excited more than $\mathscr{A}_1$, performance of CS_1's act abruptly terminates and performance of CS_2's act commences. How do the cues CS_1 and CS_2 gain control over their arousal sources $\mathscr{A}_1$ and $\mathscr{A}_2$ in cases where such control is not genetically preprogrammed?

B. Serial Learning of Order Information

In the avalanche of Fig. 12, a chain of connections from $v_1^{(2)}$ to $v_2^{(2)}$ to $v_3^{(2)}$, and so on, exists in the network at all times. This chain determines the order in which spatial patterns will be performed. Such preprogrammed chains of cells do not generally exist before we learn a sequence of successive acts, such as a piano sonata. For example,

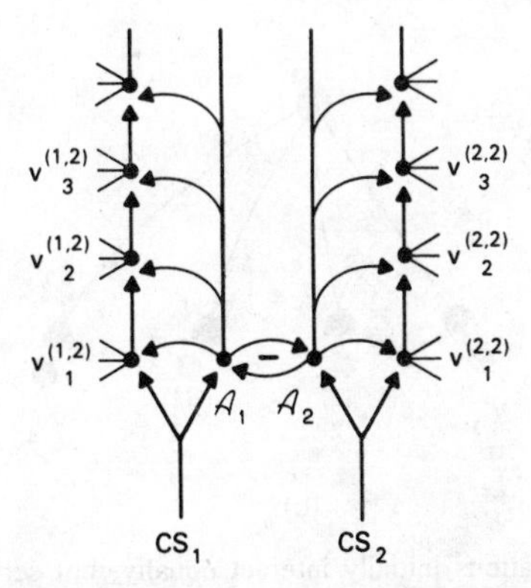

FIG. 13. Competition between arousal sources decides which avalanche will be triggered.

suppose that each $v_k^{(2)}$ controls a different chord in the sonata. Surely there does not exist in every mind a chain corresponding to every possible sequence of chords! This order information must be learned. There exist two conceptually distinct ways in which this can happen. In Fig. 14a, each $v_i^{(2)}$ is initially connected to all cells $v_k^{(2)}$, and eventually becomes differentially connected to $v_{i+1}^{(2)}$ as a result of practicing the sequence $v_1^{(2)}, v_2^{(2)}, v_3^{(2)}, \ldots$. This is a problem in serial learning. In Fig. 14b, somehow a higher-order cell population looks at the sequence $v_1^{(2)}, v_2^{(2)}, v_3^{(2)}, \ldots$ as it is practiced, and learns to reproduce a spatial pattern of activity across these cells such that the earliest cells have the largest activity. When these differential activity levels are translated into speed of performance, $v_1^{(2)}$ is performed before $v_2^{(2)}$, $v_2^{(2)}$ before $v_3^{(2)}$, and so on. This is again a problem of serial learning. Which strategy of serial encoding is used? The theory of serial learning in Grossberg (1969d) and Grossberg and Pepe (1971) provides a foundation for answering this question.

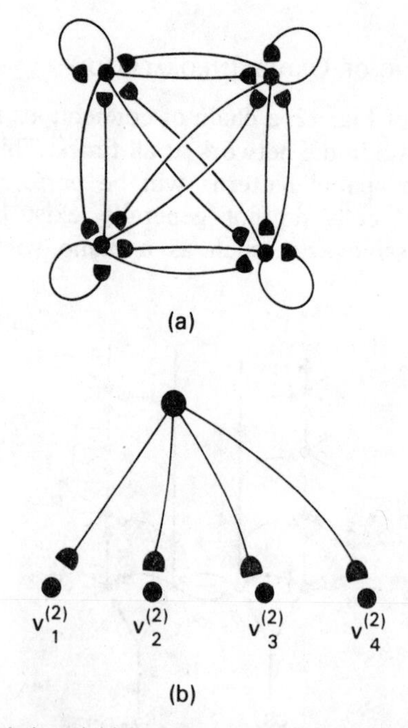

FIG. 14. (a) All populations initially interact equally, but serial learning differentially weights a chain of LTM traces; (b) a command population elicits a graded pattern of activity across a population field, which translates the activities into order of performance.

C. SELF-ORGANIZATION OF CONTEXTUAL COMMANDS OR CHUNKS

If the order information from $v_1^{(2)}$ to $v_2^{(2)}$ to $v_3^{(2)}$, and so on, is not genetically preprogrammed, then the command $v_1^{(3)}$ in Fig. 12 that nonspecifically arouses these cells is also not genetically preprogrammed. In other words, given that the particular sequence $v_1^{(2)}$, $v_2^{(2)}$, $v_3^{(2)}$, . . . depends on the act that it controls, the command $v_1^{(3)}$ that arouses this sequence must be chosen by the sequence itself. Otherwise, there would exist innately prewired commands for every possible sequence, which is absurd. How does a particular sequence code a particular command? How does this coding occur so that different orderings of the same set of cells $v_i^{(2)}$ do not all code the same command?

Given that the command $v_1^{(3)}$ is adaptively coded by its controlled sequence $v_1^{(2)}$, $v_1^{(2)}$, $v_3^{(2)}$, . . . , how does $v_1^{(3)}$ learn what sequence it codes? This question immediately calls Fig. 14b to mind. Is the higher-order cell in Fig. 14b a command population $v_1^{(3)}$? Does the learning process that teaches $v_1^{(3)}$ which sequence it controls *automatically* encode the order information needed to perform this sequence correctly? In more abstract terms, are adaptive coding (or chunking) and the learning of order information *dual* processes in a feedback system?

Given that sequence $v_1^{(2)}$, $v_2^{(2)}$, $v_3^{(2)}$, . . . adaptively codes its command, we must also realize that there is nothing special about this particular sequence. Every subsequence of $v_1^{(2)}$, $v_2^{(2)}$, $v_3^{(2)}$, . . . is also a sequence, and must be able to code a command. Not every subsequence will be able to code a command with equal ease. Nonetheless, at every time, there will exist a field $\mathscr{F}^{(3)}$ of command cells, each excited to a different degree by its generating sequence. Each command cell in $\mathscr{F}^{(3)}$ samples the activity patterns that prevail across $\mathscr{F}^{(2)}$ while it is active. Simultaneously, the pattern of activity across $\mathscr{F}^{(3)}$ continually shifts through time as new events are rehearsed, and thereby generates new subsequences to be coded by $\mathscr{F}^{(3)}$. At every time, the total signal from the command field $\mathscr{F}^{(3)}$ to $\mathscr{F}^{(2)}$ helps to determine the order information among the cells in $\mathscr{F}^{(2)}$. Grossberg (1976a,b,c) provides a conceptual foundation for synthesizing the adaptive feedback relationships between the generating sequences and the commands that organize them.

Speaking intuitively, the command cells provide the *context* in which a particular pattern is performed. For example, after playing the first few bars of a piano sonata, a pianist is ready to play the next several bars. While the second movement of the sonata is being played, the first few bars no longer control performance; more recently played notes provide the command context for determining the next notes to be played.

Similarly, if I emerge from my bedroom door, I can decide to turn left to brush my teeth, or to turn right to go directly to breakfast. The difference is decided not by the sensory cues that confront me whenever I am at the door, but by the context, command, or plan that modulates these cues.

The above remarks clarify an important distinction. An adaptively coded command in $\mathscr{F}^{(3)}$ provides a type of *cognitive* context, or arousal, for the populations in its sequence in $\mathscr{F}^{(2)}$. This type of arousal is different from the *motivational* arousal that determines whether performance of a given act continues to be in the best interests of the network. With this distinction, the elicitation of performance by an avalanche would employ an anatomy such as that in Fig. 15. In Fig. 15, the command cell is activated by a particular cue plus motivational arousal. Once activated, the command cell determines a cognitive context that arouses certain populations in preparation for their firing in the correct order.

D. Adaptive Coding of Sensory-Motor Relationships

Similar coding problems occur on a more microscopic level. In particular, how is it determined which $v_k^{(2)}$ will be chosen to learn a particular chord? For example, while a piano piece is being learned, a $v_k^{(2)}$ is presumably excited by some combination of visual cues, from reading the piano music and seeing the keyboard, and auditory feedback cues, from having played the chord. At this point, characterizing the exact cue combination is unimportant. What is important is that $v_k^{(2)}$ is adaptively coded by its cues in a manner that is strikingly similar to the adaptive coding of a command by a sequence of $v_k^{(2)}$'s. Thus, we are dealing with a problem concerning the *hierarchical* organization of adaptive codes, or chunks, and the feedback signals that order these codes, among emergent fields $\mathscr{F}^{(1)}, \mathscr{F}^{(2)}, \mathscr{F}^{(3)}, \ldots$.

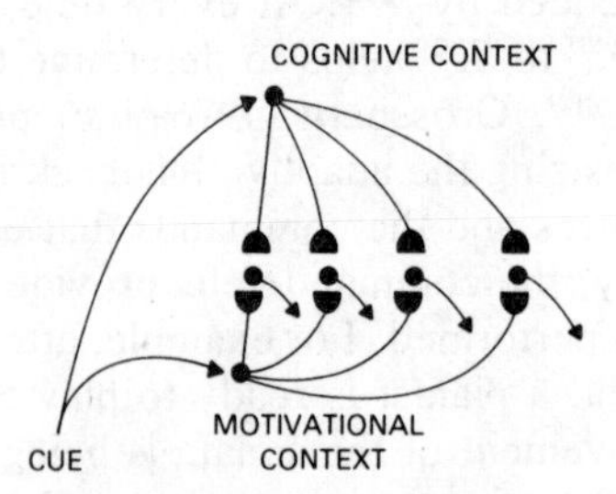

FIG. 15. A cue can excite a cognitive command as well as a nonspecific source of incentive-motivation.

E. Temporal Discrimination and Feedback Inhibition

The above remarks set the stage for analyzing how representations are turned on at appropriate times. How are they turned off?

For example, suppose that the CS that activates $v_1^{(2)}$ in Fig. 12 has a very long duration. If, consequently, each $v_i^{(2)}$ fires for a long time, then each synaptic knob will sample many spatial patterns, and will learn a weighted average of all the patterns (that is, "noise") rather than any particular pattern in the act. How is a prolonged input translated into a brief sampling signal? Grossberg (1970b) proves that a feedforward inhibitory interneuron can create a brief signal that is turned on by rapid changes in input level (Fig. 16b). A feedback inhibitory interneuron can turn off the signal only temporarily at best (Fig. 16c) and can allow a steady leakage of signal if inhibition is too weak (Fig. 16d). None of these mechanisms prevent a second input pulse from reactivating the avalanche while it is performing a later stage of the act. Then cells such as $v_1^{(2)}$ can sample and learn spatial patterns very much out of their correct order. Clearly a feedback inhibition mechanism is needed which prevents premature reactivation, or other perturbations of avalanche performance, unless more urgent environmental demands occur. In cases where the order information is not genetically preprogrammed,

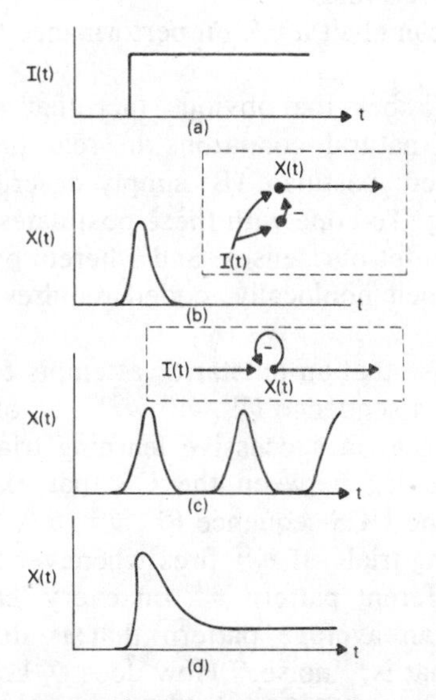

FIG. 16. The prolonged input $I(t)$ in (a) can be shut off by a feedforward inhibitory interneuron (b), but not by a feedback inhibitory interneuron, (c) or (d).

these inhibitory mechanisms must have an anatomy that is independent of any particular ordering that might be learned (cf. Grossberg, 1969c).

With the above heuristic remarks as motivation, we shall now construct a class of networks capable of adaptively synthesizing codes, maps, and plans.

7. Instrumental Conditioning

The mechanisms of drive, reward, and motivation that are needed herein can be derived from postulates about classical conditioning (Grossberg, 1971a, 1972a,b, 1975). This procedure demonstrates that classical and instrumental conditioning share certain mechanisms in common. In effect, these mechanisms embed, or buffer, the cells capable of learning in a network that prevents sampling except under appropriate circumstances. It is remarkable that explicit mechanisms for such nontrivial phenomena as self-stimulation, partial reinforcement acquisition effect, peak shift and behavioral contrast, and novelty as a reinforcer can be derived from such seemingly innocuous postulates as:

A. The time interval between CS and UCS presentation on successive learning trials can differ; and

B. The CS alone can elicit a CR on performance trials.

Postulate (A) describes the obvious fact that successive stimulus presentations under natural conditions in real time are not always perfectly synchronized; postulate (B) simply describes the outcome of classical conditioning. To cope with these postulates in a world wherein events continually buffet our senses, and wherein our long-term memories are spatially, albeit nonlocally, coded requires additional network structure.

To see this, suppose that an outstar $\mathcal{O}_1$ attempts to learn a prescribed spatial pattern $\theta^{(1)}$ in a sequence $\theta^{(1)}, \theta^{(2)}, \theta^{(3)}, \ldots$ of spatial patterns by practicing the sequence on successive learning trials. If postulate (A) holds, then the time lag between the CS that excites $\mathcal{O}_1$'s sampling population $v_1^{(2)}$ and the UCS sequence $\theta^{(1)}, \theta^{(2)}, \theta^{(3)}, \ldots$ can be different on successive learning trials. If $v_1^{(2)}$ fires whenever the CS occurs, then $\mathcal{O}_1$ can sample a different pattern $\theta^{(k)}$ on every learning trial. $\mathcal{O}_1$ will consequently learn an average pattern that is derived from all the sampled patterns—that is, "noise." How does $\mathcal{O}_1$ know when to sample the "important" pattern $\theta^{(1)}$? Somehow, the onset of sampling by $v_1^{(2)}$ and the arrival of the UCS at the field $\mathcal{F}^{(1)} = \{v_1^{(1)}, v_2^{(1)}, v_3^{(1)}, \ldots\}$ of sampled cells must be synchronized. This can happen only if the UCS lets $v_1^{(2)}$ know when it will arrive at $\mathscr{F}^{(1)}$ by sending a signal to $v_1^{(2)}$. Also $v_1^{(2)}$ must be prevented from eliciting a sampling signal unless a large CS *and* UCS signal converge at $v_1^{(2)}$. This UCS signal must arrive at $v_1^{(2)}$ before the UCS pattern activates $\mathscr{F}^{(1)}$, since $v_1^{(2)}$ must be able to send a signal to $\mathscr{F}^{(1)}$ in time to sample $\theta^{(1)}$. In other words, the UCS activates a bifurcating pathway; one branch *arouses* $v_1^{(2)}$, and the other branch delivers the UCS pattern a little while later. The same argument holds for every cell $v_i^{(2)}$ that is capable of being activated by a CS, since it is not known a priori which CS–UCS combination will be learned. Thus the UCS *nonspecifically* arouses the field $\mathscr{F}^{(2)} = \{v_1^{(2)}, v_2^{(2)}, v_3^{(2)}, \ldots\}$ of sampling cells. In summary, simultaneous convergence of the CS input and the UCS nonspecific arousal at a sampling cell are needed to fire this cell. This mechanism synchronizes the onset of CS-activated sampling from $\mathscr{F}^{(2)}$ and the arrival of UCS patterns at $\mathscr{F}^{(1)}$ on successive learning trials. Convergence of a specific input and a nonspecific input is also needed to fire sampling cells $v_i^{(2)}$ in the avalanche of Fig. 12b. It is the same mechanism derived from different considerations.

Postulate (B) shows that conditioning of the CS to the UCS arousal pathway occurs during learning trials. This is the basis for the emergence of "conditioned reinforcers" or "secondary reinforcement" in the networks. Conditioned arousal is necessary, since otherwise the CS alone could not elicit a CR on performance trials. This is because sampling cells can be fired only by the convergence of a CS input and an arousal input. Since the UCS is not present on performance trials to fire the arousal pathway, the CS must gain control over the arousal pathway by being paired with the UCS. An analysis of the minimal mechanism capable of conditioned arousal is shown in Fig. 17, wherein each

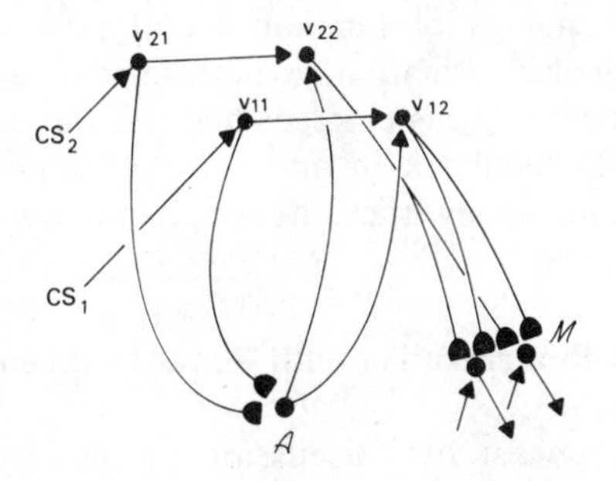

FIG. 17. Conditioning occurs in parallel at the arousal source $\mathscr{A}$ and at the motor command cells $\mathscr{M}$.

discriminated cue CS_i has a two-stage network representation $\{v_{i1}, v_{i2}\}$, $i = 1, 2, \ldots$. Consider the network response to a prescribed CS. Let the CS fire v_{11}. Then v_{11} sends signals to v_{12} and to the arousal population $\mathscr{A}$. Nothing else happens until the UCS arrives at v_{21}. This is because v_{12} can fire only if it receives an input from v_{11} *and* from $\mathscr{A}$, but the signal from v_{11} to $\mathscr{A}$ is initially too small to fire $\mathscr{A}$. When the UCS perturbs v_{21}, v_{21} sends a signal to v_{22} and to $\mathscr{A}$. The $v_{21} \rightarrow \mathscr{A}$ signals are large enough to fire $\mathscr{A}$, *because* the cue firing v_{21} is a UCS. When $\mathscr{A}$ fires, it releases nonspecific signals to all cells $v_{12}, v_{22}, v_{32}, \ldots$. Now three things happen. First, since v_{11} and $\mathscr{A}$ are both active, the LTM traces in the synaptic knobs of $v_{11} \rightarrow \mathscr{A}$ axons get stronger. When these traces get strong enough, the CS alone will be able to fire v_{12}. Second, the arousal signal from $\mathscr{A}$ combines with the UCS-derived signal from v_{21} at v_{22}, thereby firing signals from v_{22} to $\mathscr{M}$. These signals elicit the UCS pattern in the populations of $\mathscr{M}$. Third, because the arousal signal from $\mathscr{A}$ is nonspecific, it also combines with the CS-derived signal from v_{11} at v_{12}, thereby firing signals from v_{12} to $\mathscr{M}$. These signals sample the UCS-elicited pattern at $\mathscr{M}$. Consequently, the CS begins to acquire UCS properties, both by learning to control the arousal pathway $\mathscr{A}$, and by learning to elicit the UCS-induced pattern at $\mathscr{M}$.

The following psychological terms can be used to interpret the above interactions. The arousal pathway supplies "incentive motivation" to the cells v_{i2}. As a population v_{i1} gains control over the arousal pathway, it becomes a "conditioned reinforcer." As v_{i2} samples a pattern at $\mathscr{M}$, it is said to learn a "habit." Thus, a cue that excites v_{i1} can learn to control incentive motivation via the pathway $v_{i1} \rightarrow \mathscr{A} \rightarrow v_{i2}$ as it simultaneously learns to control a habit via $v_{i1} \rightarrow v_{i2} \rightarrow \mathscr{M}$.

Grossberg (1971b, 1972a,b, 1975) continues this derivation by imposing other simple psychological postulates that act as environmental pressures on an evolving network. These postulates lead to networks in which functional analogs of familiar neural regions appear, such as hippocampus, reticular formation, hypothalamus, septum, and cerebral cortex. Each of these regions emerges because the network tries to deal with environmental feedback in an ever-more-sophisticated way. The present paper cannot review these developments for lack of space.

8. STM Reverberation until Reward Influences LTM

For present purposes, our discussion of instrumental conditioning makes two essential points: (1) A cue can generate an STM response at certain cells v_{i1} without firing the cells v_{i2}; (2) the cells v_{i2} can be fired only if the specific signals from v_{i1} are supplemented by nonspecific arousal.

In many situations, an STM trace of a previous event must be kept active after the event itself terminates. For example, in an instrumental conditioning paradigm, the STM traces of previous events must be kept active long enough for later rewards to influence their storage in LTM. Yet these STM traces must also be capable of rapid decay if competing events occur. The two properties of sustained STM activity and rapid induced decay cannot both be achieved by a slow passive decay of STM (Grossberg, 1971b). An active reverberation from v_{i1} to excitatory interneurons v_{i3} and then back to v_{i1} has these two properties (Fig. 18). Excitation in the $v_{i1} \leftrightarrow v_{i3}$ loop can sustain itself indefinitely, even if the passive decay rates of the v_{i1} and v_{i3} populations separately are fast. If one of the links v_{i1} or v_{i3} in the loop is inhibited, then the potentials in v_{i1} and v_{i3} can rapidly decay. The reverberation in $v_{i1} \leftrightarrow v_{i3}$ can go on indefinitely without influencing any LTM changes. Only when arousal reaches v_{i2} can v_{i2} fire and induce sampling by its LTM traces of patterns at $\mathcal{M}$.

The very virtue of this mechanism introduces a difficulty. Unless the reverberation is inhibited, the loop will continue to reverberate even after arousal allows v_{i2} to fire. In short, reverberation can keep STM traces on, but cannot turn them off. When such a mechanism is used in a network trying to learn order information, chaos results unless there exist sources of inhibition to shut off the STM reverberations at appropriate times. To see how to do this, we embed the functional units of Fig. 18 into the simplest anatomy capable of learning order—namely, the avalanche.

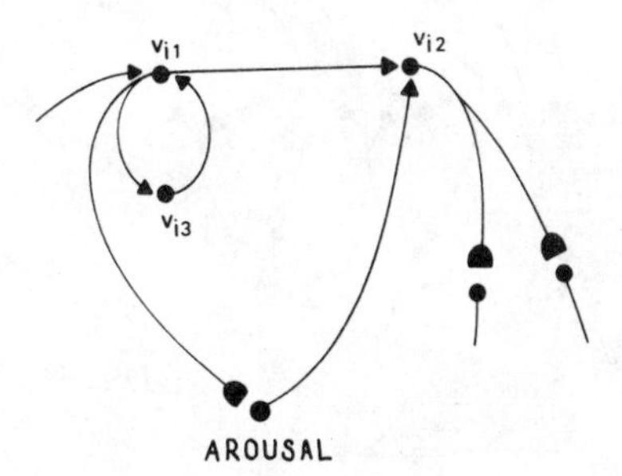

FIG. 18. Reverberation in the $v_{i1} \leftrightarrow v_{i3}$ loop keeps the STM trace active and permits it to be rapidly shut off by inhibition.

9. Rehearsal Resets STM Order Information Using Feedback Inhibition and Decouples Order and Velocity Information

The cells $v_i^{(2)}$ in Fig. 12 are analogous to the cells v_{i2} in Fig. 18. In effect, Sections 7 and 8 refine our understanding of how the cells $v_i^{(2)}$ are influenced by arousal. To construct the simplest example of how to combine the properties of sequential performance, STM reverberation, and modulation by arousal, we replace the populations $v_i^{(2)}$ of Fig. 12 by the units $\{v_{i1}, v_{i2}\}$, along with their interaction pathways, as in Fig. 19. In this figure, the following pathology occurs. When a signal from $v_{i-1,2}$ activates v_{i1}, the STM reverberation $v_{i1} \leftrightarrow v_{i3}$ is switched on, and is never shut off. The signal $v_{i1} \leftrightarrow v_{i3}$ plus arousal keeps the population v_{i2} firing at all future times. Consequently, v_{i2} will sample *every* pattern that reaches $\mathcal{M}$ after v_{i1} is switched on, thereby learning nothing. This is intolerable. The $v_{i1} \leftrightarrow v_{i3}$ loop can be allowed to reverberate until v_{i2} performs its LTM pattern, but the reverberation must then be inhibited, or else v_{i2}'s pattern will be washed away by the tide of future events. Thus, when v_{i2} fires, it not only sends excitatory signals to $\mathcal{M}$. It must also send feedback inhibitory signals to either v_{i1} or v_{i3} that terminate their STM reverberation (Fig. 20). As v_{i2} fires, it also excites $v_{i+1,1}$, which reverberates with $v_{i+1,3}$ and sends signals to $v_{i+1,2}$. Population $v_{i+1,2}$, in turn, fires when it is aroused. This system thus provides a simple example of how order and velocity information are decoupled;

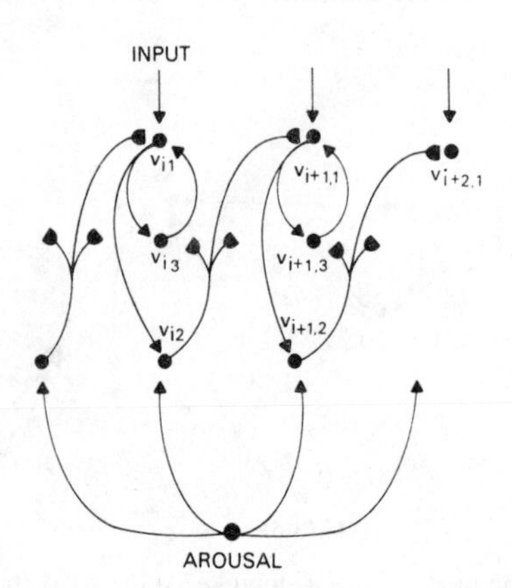

FIG. 19. Minimal synthesis of learned sequential performance, STM reverberation, and modulation by arousal.

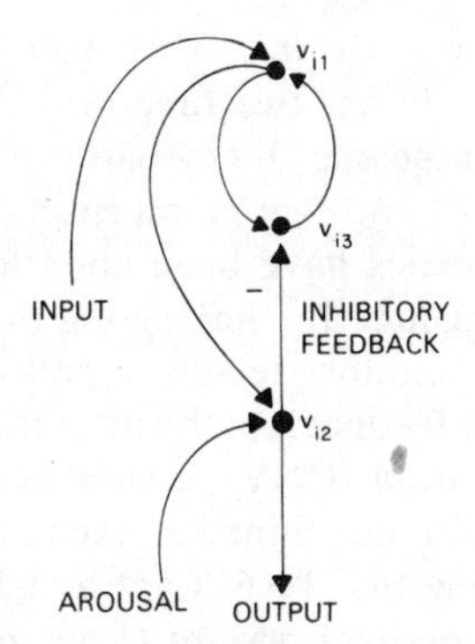

FIG. 20. Feedback inhibition elicited by performance signal prevents permanent sampling.

that is, the same sequence of acts can be performed with different rhythms by varying the size of the rehearsal wave on successive trials.

10. An Emergent Neocortical Analog

The units $\{v_{i1}, v_{i2}, v_{i3}\}$, $i = 1, 2, \ldots$, along with their auxiliary interneurons and pathways, form network "modules" or "chips" that carry out important "data processing" tasks. When many of these units are placed alongside each other, a natural laminar structure emerges, with the same type of processing going on in all the cells of a given layer. A column of layers is a functional unit in this laminar field.

There are many reasons to interpret this structure as an emergent neocortical analog. The ubiquitous laminar organization of neocortical tissue (Crosby *et al.*, 1962) and its organization into columnar functional units (Hubel and Wiesel, 1962, 1963) are two of the most casual similarities. Furthermore, the cells v_{i2} are often polyvalent cells, or cells that respond to more than one modality, such as an auditory CS (tone) and a visual UCS (visual presentation of food). These cells fire only in response to the sum of CS plus UCS inputs, and are importantly implicated in plastic network changes. John (1966, 1967) reports the existence of analogous polyvalent cells in neocortical tissue. Grossberg (1971b) summarizes related data.

In vivo, a slab of neocortex that has been isolated by cutting through the underlying white matter can maintain sustained reverberating activity (Burns, 1958). A similar persistence of reverberation occurs in the $v_{i1} \leftrightarrow v_{i3}$ excitatory loop if inhibitory $v_{i2} \rightarrow v_{i3}$ feedback is prohibited by cutting off arousal inputs to v_{i2}. This fact clarifies an interesting paradox. Arousal in Fig. 20 has two functions. First, it excites the cells v_{i2}, which thereupon rehearse their patterns. Second, it indirectly inhibits the loop $v_{i1} \leftrightarrow v_{i3}$, and thereupon resets STM.

Other functional properties have been cited for identifying this structure as a neocortical analog. In the theory of attention in Grossberg (1975), a conditionable excitatory feedback pathway modulates excitability in this structure. This feedback pathway is analogous to the source of *contingent negative variation* (CNV), a slow cortical potential shift that has been associated with an animal's expectancy, decision (Walter, 1964), motivation (Irwin *et al.*, 1966; Cant and Bickford, 1967), volition (McAdam *et al.*, 1966), preparatory set (Low *et al.*, 1966), and arousal (McAdam, 1969). Walter (1964) hypothesized that the CNV shifts the average base line of the cortex by depolarizing the apical dendritic potentials of its pyramidal cells and thereby priming the cortex for action. The arousal pathway of Fig. 19 is this CNV analog, although we have not, for lack of space, reviewed why this pathway should be conditionable. In brief, this conditionable pathway establishes a "psychological set" without which inappropriate acts can be elicited by any motivational source (Grossberg, 1969c, 1971b, 1975). Grossberg (1975) also notes that more than one type of arousal exists; for example, incentive–motivational arousal and the arousal triggered by novel events are conceptually and anatomically different mechanisms that, in fact, often compete with each other. The former system focuses attention on cues that have in the past yielded expected consequences; it blocks, or overshadows, irrelevant cues. The latter system frees irrelevant cues from overshadowing when unexpected consequences occur.

The constructions in later sections will further develop this analog. For the present, we draw Fig. 21 in a way that emphasizes known neocortical structure.

11. Control of Performance Duration by STM and Arousal

In Fig. 20, once a state v_{i2} receives a signal from v_{i1} plus arousal, it fires and thereupon accomplishes three things:

1. It inhibits $v_{i1} \leftrightarrow v_{i3}$ reverberation, and thus it fires only for a brief time.
2. It excites $v_{i+1,1}$, which can reverberate with $v_{i+1,3}$ for an indefinite interval of time, until arousal combines with the $v_{i+1,1} \rightarrow v_{i+1,2}$ signal, and thereupon inhibits the reverberation via the $v_{i+1,2} \rightarrow v_{i+1,3}$ inhibitory pathway.

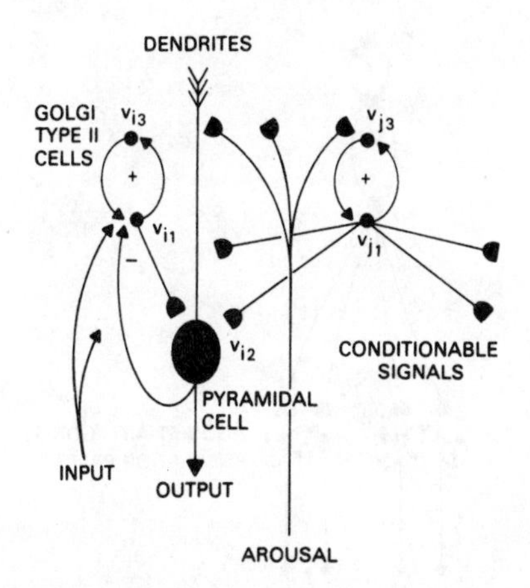

FIG. 21. A cortical analog of the minimal network module suggests interactions between phasic inputs, nonspecific arousal, pyramidal cells, and cortical interneurons—for example, Golgi type II cells.

3. It briefly excites an output pathway that leads, perhaps multisynaptically, to certain muscles.

Suppose that, after v_{i2} fires, arousal is terminated. Then v_{i2}'s brief signal moves down the output pathway, and the $v_{i+1,1} \leftrightarrow v_{i+1,3}$ reverberation labels the next pattern to be elicited without releasing this pattern. How does the brief v_{i2} output signal generate a new output configuration that can be maintained until arousal releases the next output pattern? For example, how can one *hold* a tone, or a note on a keyboard, or a phrase, until the next pattern is released? Somehow the brief output signal imposes a new pattern on the motor controls, and this new pattern also reverberates in STM until it is supplanted, or reset, by the next pattern. In other words, the motor controls maintain a *posture* until a new command changes the posture or terminates the reverberation. Figure 22 schematizes this relationship in terms of descending control by v_{i2} of pairs of agonist–antagonist muscle groups. The descending signal inhibits the excitatory reverberation due to previous patterns, and imposes a new pattern on all the pairs. This new pattern thereupon reverberates until a new disturbance occurs, such as a competing command or the removal of arousal.

The above discussions make plain the need to study two kinds of processes in greater detail. First, how is order information embedded in LTM when a sequence of events is presented to a network? Second,

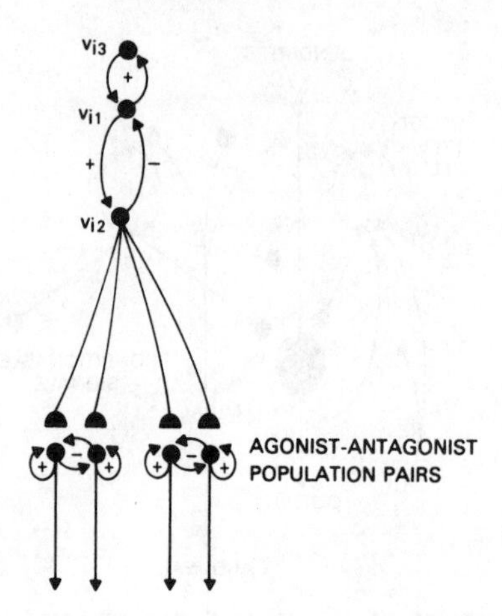

FIG. 22. The brief command from v_{i2} is stored in STM until it is reset by a new disturbance.

how is a graded pattern of STM activity maintained across many populations? We shall begin by summarizing some facts about the first problem, and then the second, before combining these facts to reconsider the first. The two problems are intertwined, and so therefore is our discussion of them.

12. Serial Learning and STM → LTM Order Reversal

This section summarizes the simplest results on how order information is transferred from STM to LTM. The basic references are Grossberg (1969d) and Grossberg and Pepe (1971). Grossberg (1974, Section VII) reviews some of these results. They show how temporal sequences of events are coded as spatial patterns of LTM activity.

Two types of serial learning are included in this analysis. In both types, order information is not innate. In type I (Fig. 14a) many $v_{i2} \to v_{j1}$ pathways of comparable strength exist before learning occurs. Somehow a sufficient amount of serial learning embeds a directed chain-like structure $v_{12} \to v_{21}$, $v_{22} \to v_{31}$, $v_{32} \to v_{41}$, . . . into this anatomy. Actually, we shall see that spatially distributed LTM patterns exist even in this case. Type II (Fig. 14b) exhibits another anatomical substrate of serial learning. Here a command state learns order information by sampling populations as they are sequentially activated. This latter anatomy has the advantage that serial order in $\mathscr{F}^{(2)}$ can be reorganized by changing which commands are active at any time. The chains in Fig. 14a, by contrast, rigidly constrain the possible performance order once they are entered (cf. Lenneberg, 1967, Chapter 3). Many of the same LTM patterns are learned in both types of anatomy. This fact is important, because, when both types of phenomena are operative, they yield self-consistent order information. The two cases are schematized more completely in Fig. 23. Figures 23a and 23b depict variants of the type of serial learning in Fig. 14a. In Fig. 23a, every state v_i is connected to all *other* states v_j by conditionable pathways. In Fig. 23b, every state v_i is connected to all states v_j by conditionable pathways. In both cases, we let the states $v_1, v_2, \ldots, v_L$ be sequentially excited by a list of

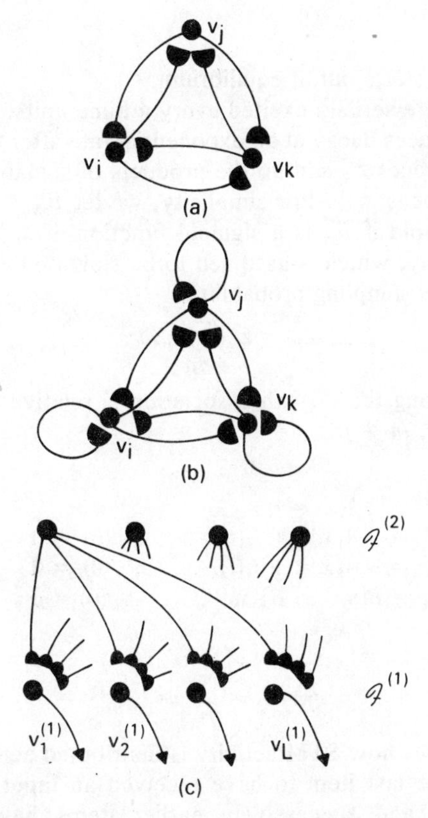

FIG. 23. Two anatomies in which serial learning builds up chains of LTM associations, (a) and (b). In (c), command states $v_j^{(2)}$ learn serial order of inputs presented to populations $v_k^{(1)}$.

inputs, with a time lag of w (intratrial interval) between successive list items. See Grossberg (1969d, 1974) and Grossberg and Pepe (1971) for a discussion of how w, L, and the intertrial interval between list presentations influence LTM. In Fig. 23c, the states $v_1^{(1)}, v_2^{(1)}, \ldots, v_L^{(1)}$ in $\mathscr{F}^{(1)}$ are sequentially excited by a list of inputs. The states in $\mathscr{F}^{(2)}$ are command states, or chunks, that are either directly excited by the list inputs, or indirectly excited by sets of active states in $\mathscr{F}^{(1)}$. For the remarks below to hold, it only matters that disjoint subsequences in $\mathscr{F}^{(2)}$ are sequentially excited at a uniform rate and with a uniform intensity through time. Generalizations of this situation will be clear once the basic mechanism is understood.

For definiteness, we shall consider Fig. 23c when an input to state $v_i^{(1)}$ also excites $v_i^{(2)}$. We shall discuss LTM under the following assumptions:

1. The system starts out at equilibrium.
2. The states are serially excited every w time units.
3. The STM traces decay at an exponential rate after they are excited.
4. The LTM traces z_{jk} add up the products of signals B_{jk} from $v_j^{(2)}$ to $v_k^{(1)}$ and STM traces $x_k^{(1)}$. For simplicity, we let $B_{jk} = [x_j^{(2)} - \Gamma]^+$, but the results also hold if B_{jk} is a sigmoid function of $x_j^{(2)}$. We ignore the rate of LTM decay, which is assumed to be slow on each learning trial.
5. The stimulus sampling probabilities

$$Z_{jk} = z_{jk}(\sum_m z_{jm})^{-1}$$

measure how strong the (j, k)th association is relative to competing (j, m)th associations, $m \neq j$.

In all,

$$\dot{x}_i^{(1)} = -A_1 x_i^{(1)} + I_i(t), \qquad x_i^{(1)}(0) = 0 \tag{7}$$

$$\dot{x}_i^{(2)} = -A_2 x_i^{(2)} + I_i(t), \qquad x_i^{(2)}(0) = 0 \tag{8}$$

$$\dot{z}_{jk} = B[x_j^{(2)} - \Gamma]^+ x_k^{(1)}, \qquad z_{jk}(0) = \alpha > 0 \tag{9}$$

and

$$Z_{jk} = z_{jk}(\sum_m z_{jm})^{-1} \tag{10}$$

Figure 24a shows how STM activity is distributed across the field $\mathscr{F}^{(1)}$ through time. The last item to have received an input is always most intensely active, and successively earlier items have progressively weaker STM traces (STM recency effect). A similar distribution of activity holds for the STM traces of the command states in $\mathscr{F}^{(2)}$; the last

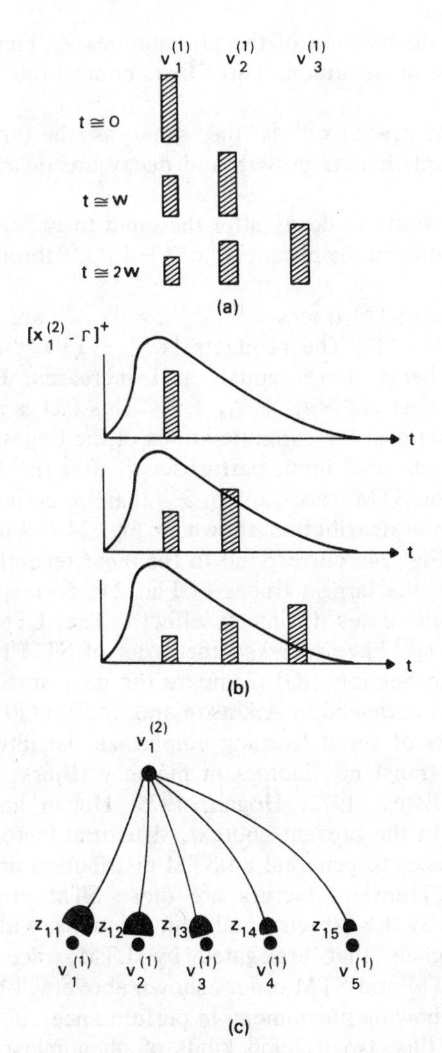

FIG. 24. Whereas STM shows a recency effect, LTM (at least as sampled by $v_1^{(2)}$) shows a primacy effect.

item to have received an input is always the most active, etc. What about the sizes of the LTM traces z_{jk}? By (9), z_{jk} grows faster when either $x_j^{(2)}$ or $x_k^{(1)}$ is increased in size, other things being equal, but does not grow if either $x_j^{(2)} \leq \Gamma$ or $x_k^{(1)} \cong 0$. Note, however, that the stimulus sampling probability Z_{jk} in (10) can decrease even when z_{jk} grows, if the competing terms $\Sigma_{m \neq j}\, z_{jm}$ grow more quickly.

Consider the distribution of the probabilities Z_{1i} emanating from $v_1^{(2)}$ after a single list presentation. Three facts control this distribution:

1. Each STM trace $x_i^{(1)}$ is the same as the previously excited trace $x_{i-1}^{(1)}$, except that its growth and decay are delayed in time by w because of (7).
2. Trace $x_1^{(2)}$ starts to decay after the input to $v_1^{(2)}$ terminates, by (8).
3. Trace z_{i1} adds up the products $[x_1^{(2)} - \Gamma]^+ x_i^{(1)}$ through time, by (9).

In Fig. 24b, the STM traces $x_1^{(1)}, x_2^{(1)}, x_3^{(1)}, \ldots$ are superimposed on the signal $[x_1^{(2)} - \Gamma]^+$. The products $[x_1^{(2)} - \Gamma]^+ x_i^{(1)}$ are clearly made smaller, other being things equal, as i increases. Hence, after the learning trial is over $z_{11} > z_{12} > z_{13} > \cdots$. This fact is illustrated in Fig. 24c by drawing largest the synaptic knobs of the largest LTM traces. In particular, if a rehearsal input perturbs $v_1^{(2)}$ after the STM traces have decayed, then the STM traces $x_i^{(1)}$ in $\mathscr{F}^{(1)}$ that are activated by the LTM traces z_{1i} have the distribution shown in Fig. 24c. Whereas the largest STM traces in Fig. 24a correspond to the *most* recently inputted states (recency effect), the largest traces in Fig. 24c correspond to the *least* recently inputted states (primacy effect). The LTM traces of the command state $v_1^{(2)}$ have *reversed* the order of STM trace strength, as was discussed in Section 1,G! Compare the data on STM recency and LTM primacy as reviewed in Atkinson and Shiffrin (1971).

Recent studies of serial learning emphasize the interplay of "structural" versus "transient" factors in memory (Bjork, 1975; Craik and Jacoby, 1975; Estes, 1972; Hogan, 1975; Hogan and Hogan, 1975; Shiffrin, 1975). In the present context, structural factors are those that use the LTM traces to generate an STM distribution on which performance is based. Transient factors are those STM properties that are *directly* induced by the experimental manipulations, without intervention of feedback signals that are gated by LTM traces. The opposite tendencies in LTM and STM orders shown above will be used below to explain various bowing phenomena in performance.

Before doing this, two related kinds of phenomena will be summarized: first, the relative LTM learning rates at different list positions; and second, the shape of the generalization gradients, or STM spatial patterns, controlled by the LTM traces at different list positions.

To discuss the learning rates at different list positions, we define the function

$$G(i, \Gamma, t) = Z_{ii}(tw) \tag{11}$$

and consider $G(i, \Gamma, i)$. This latter function measures the "correct association'' from $v_i^{(2)}$ to $v_i^{(1)}$ one time unit after sampling signals from $v_i^{(2)}$ reach $\mathscr{F}^{(1)}$, and thus before ''incorrect future associations'' such as $z_{i,i+k}$, $k \geq 1$, can develop. Figure 24a shows that, as i increases, there are more active ''past'' STM traces $x_{i-1}^{(1)}, x_{i-2}^{(1)}, \ldots$ to compete with the growth of Z_{ii} after the input to $v_i^{(2)}$ occurs. The function $G(i, \Gamma, i)$ correspondingly has the graph in Fig. 25a, at any threshold value Γ for which some learning occurs. By contrast, let t in $G(i, \Gamma, t)$ be allowed to increase to values that correspond to times long after the last list item is presented. Then $G(i, \Gamma, t)$ is no longer monotone decreasing. The nonoccurrence of any more list items after the Lth item is presented facilitates LTM growth at positions near the end of the list. If $\Gamma = 0$ and t is allowed to become arbitrarily large, this facilitation propagates backward through list items until the middle of the list is reached. That is, the minimum of $G(i, 0, \infty)$ occurs at $i = L/2$ or $(L - 1)/2$, whichever is an integer (Fig. 25b). The middle of the list is consequently hardest to learn in this case (bowing); the proof is in Grossberg (1969d). If t increases by a finite amount beyond $(L - 1)w$, then a curve between Fig. 25a and Fig. 25b is obtained by continuous interpolation (Fig. 25c). That is, *skewing* of the bowed curve occurs, if only because the intertrial interval is of finite duration. Skewing can also be caused by choosing Γ

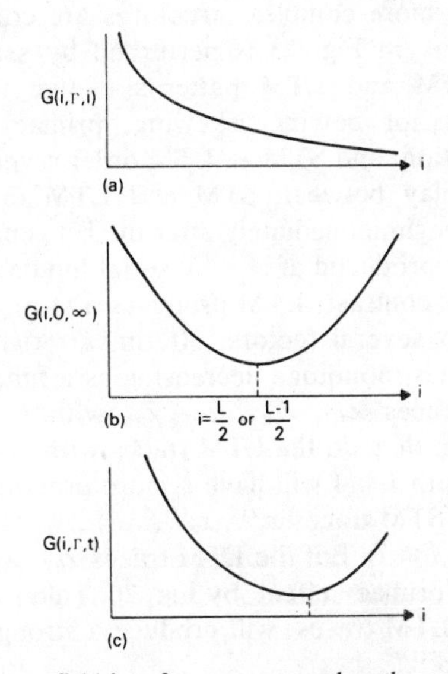

FIG. 25. (a) If no future field interference occurs, then the end of the list is hardest to learn; (b) if maximal future field interference occurs, then the middle of the list is hardest to learn (bowing); (c) if the signal threshold is finite, then skewing occurs.

> 0. This happens because, as Γ increases, or equivalently, nonspecific arousal decreases, population $v_i^{(2)}$ in $\mathscr{F}^{(2)}$ can sample fewer "future" populations $v_{i+2}^{(1)}, v_{i+3}^{(1)}, \ldots$ in $\mathscr{F}^{(1)}$, but can still sample all active populations $v_1^{(1)}, v_2^{(1)}, \ldots, v_{i-1}^{(1)}$ in the "past field" of $v_i^{(2)}$. Grossberg and Pepe (1971) prove, in addition, that, whereas the beginning of the list is easier to learn, given "normal" arousal or threshold levels (primacy dominates recency), eventually the end of the list is easier to learn at overaroused levels (recency dominates primacy). They compare these results to attentional problems that learning subjects have when they are overaroused. The result that the relative LTM trace strengths at the end versus the beginning of the list reverse as arousal level increases does not include many important STM and LTM interactions. It nonetheless emphasizes the importance of performance variables on what is encoded, and suggests a convenient learning measure by which the degree of overarousal might be quantified (cf. Section 27).

The LTM generalization gradients at various list positions have the familiar position-dependent form shown in Fig. 26. There is a forward distribution of associations $Z_{11}, Z_{12}, \ldots, Z_{1L}$, a backward distribution of associations $Z_{L1}, Z_{L2}, \ldots, Z_{LL}$, and a two-sided distribution of associations $Z_{L/2,1}, Z_{L/2,2}, \ldots, Z_{L/2,L}$ (L even).

These results illustrate some important facts that will reappear in a suitable form when more complex structures are considered. First, no matter what anatomy in Fig. 23 is perturbed by serial inputs, similar distributions of STM and LTM patterns evolve through time, and suggest mechanisms of bowing, skewing, primacy/recency balance, response generalization, and STM $\rightarrow$ LTM order reversal. Second, there is a delicate interplay between STM and LTM factors at $\mathscr{F}^{(1)}$. For example, let recall begin immediately after the list is presented. Then the STM pattern that is produced at $\mathscr{F}^{(1)}$ by serial inputs exhibits a recency effect (Fig. 24a). By contrast, LTM produces a strong primacy effect at $\mathscr{F}^{(1)}$. This is due to several factors. At times t right after the list is presented, $G(i, \Gamma, t)$ is monotone decreasing as a function of i. In other words, the LTM traces $Z_{i1}, Z_{i2}, \ldots, Z_{iL}$ with $i \cong 1$ code a more differentiated pattern than do the LTM traces with $i \cong L$. Thus, a signal to $\mathscr{F}^{(1)}$ from a $v_i^{(2)}$ with $i \cong 1$ will have a more pronounced effect on the relative sizes of the STM traces $x_1^{(1)}, x_2^{(1)}, \ldots, x_L^{(1)}$ than will a signal to $\mathscr{F}^{(1)}$ from a $v_i^{(2)}$ with $i \cong L$. But the LTM traces $Z_{i1}, Z_{i2}, \ldots, Z_{iL}$ with $i \cong 1$ exhibit a strong primacy effect, by Fig. 26. Thus, signals from $\mathscr{F}^{(2)}$ to $\mathscr{F}^{(1)}$, gated by their LTM traces, will produce a strong primacy effect in the STM pattern at $\mathscr{F}^{(1)}$.

On the other hand, if recall begins after the STM pattern at $\mathscr{F}^{(1)}$ has decayed, then a recency effect can still be obtained, but it is due to LTM

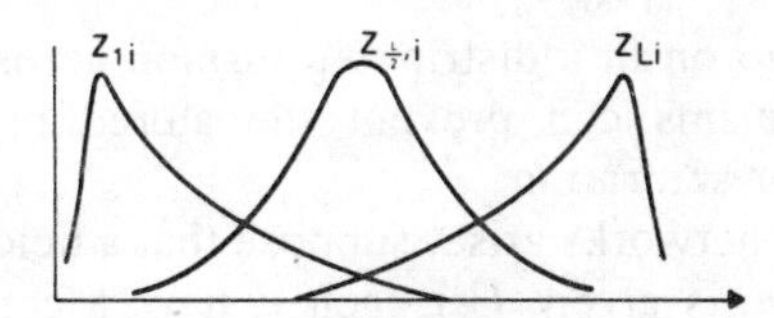

FIG. 26. Generalization gradients as a function of list position.

rather than STM. More precisely, if the STM traces at $\mathcal{F}^{(1)}$ are allowed to decay without experimental interference, then the LTM traces $Z_{ii}(t)$ will be bowed as a function of i. In other words, LTM traces near the end of the list can be large. Moreover, the distributions Z_{i1}, $Z_{i2}, \ldots, Z_{iL}$ with $i \cong L$ exhibit a recency effect by Fig. 26. Signals to $\mathcal{F}^{(1)}$ from states $v_i^{(2)}$, $i \cong L$, will be gated by these LTM traces and will, therefore, produce a recency effect at $\mathcal{F}^{(1)}$. This recency effect will be balanced by the primacy effect due to signals to $\mathcal{F}^{(1)}$ from states $v_i^{(2)}$, $i \cong 1$, whose LTM traces $Z_{i1}, Z_{i2}, \ldots, Z_{iL}$ code a primacy effect. Different experimental manipulations can differentially activate the states $v_i^{(2)}$ during recall trials. Such manipulations act as *probe* stimuli, in the sense that they determine which LTM patterns in the $\mathcal{F}^{(2)} \to \mathcal{F}^{(1)}$ pathways will be activated and thereby influence STM at $\mathcal{F}^{(1)}$. The net STM pattern elicited across $\mathcal{F}^{(1)}$ in a given time frame by all stimuli will determine the performance controlled by $\mathcal{F}^{(1)}$.

These remarks address some aspects of the complex interplay between activity at $\mathcal{F}^{(1)}$ that is due to serial inputs, and activity at $\mathcal{F}^{(1)}$ that is due to feedback signals from $\mathcal{F}^{(2)}$ that are gated by LTM traces. The remarks also raise some questions. *Do* the STM traces spontaneously decay? If not, what factors cause their ultimate termination? How are STM traces of command states in $\mathcal{F}^{(2)}$ activated in the first place? We now turn to the fundamental question of how STM activities are regulated.

13. Storing Spatial Patterns in STM

Several papers have appeared that analyze how spatial patterns of activity are stored in STM (Grossberg, 1973; Ellias and Grossberg, 1975; Grossberg and Levine, 1975; Levine and Grossberg, 1976). The general conclusion is that, in a field of populations capable of this task, each population excites nearby populations (recurrent on-center) and inhibits a broad expanse of populations (recurrent off-surround) by mechanisms that obey mass action laws. All these operations describe *parallel* computations that go on in a distributed fashion across the field at any time. Such mechanisms can prevent the stored pattern from being distorted by noise or saturation.

To see how these networks arise, suppose that a field $\mathscr{F}$ of populations v_i, $i = 1, 2, \ldots, n$, is given. Let each v_i have a certain number B of sites that can be in either an excited or an unexcited state. Let the STM activity $x_i(t)$ be the number of excited sites in v_i at time t. Suppose that each v_i is perturbed by a continuously changing input $I_i(t)$, which will excite a certain number of v_i's sites. How can such inputs $I_i(t)$ change through time? As Section 2 notes, two very different types of changes can be described in terms of the *total* input strength $I(t) = \Sigma_{k=1}^{n} I_k(t)$ and the *relative* input intensities $\theta_i(t) = I_i(t)/I(t)$ at each v_i. For example, let the v_i represent cells in a retina, and expose the retina to a picture in shades of white, gray, and black. Then $I(t)$ describes the intensity of background illumination of the picture. This intensity can vary wildly through time. The picture itself is characterized by the spatial pattern $\theta = (\theta_1, \theta_2, \ldots, \theta_n)$ of numbers (the *reflectances*), which do not change through time (Cornsweet, 1970). Thus, it is very important for a system to be able to tell what the pattern weights $\theta(t) = (\theta_1(t), \theta_2(t), \ldots, \theta_n(t))$ are, whether or not the total input $I(t)$ fluctuates through time. The weights $\theta(t)$ describe the "relative figure-to-ground" of the inputs at every time t.

Another reason for distinguishing $\theta(t)$ from $I(t)$ is that the unit of LTM is a spatial pattern. When an outstar performs a spatial pattern θ on a field $\mathscr{F}$, it can do so in response to a CS input that fluctuates wildly through time. Somehow certain network cells must be able to "read" θ independent of the fluctuations in $I(t)$, or else the pattern could never be decoded. Such considerations originally motivated the construction of network filters that can discriminate relative figure-to-ground (Grossberg, 1970b, 1972d, 1976b). Some of the minimal filters have anatomies that are strikingly "retinal," and they are capable of formal analogs of such perceptual constancies as hue, brightness, and lightness constancy (Grossberg, 1972d).

14. Gain Control and Adaptation in On-Center Off-Surround Networks

How can we design a system capable of distinguishing the pattern weights $\theta(t)$ from fluctuations in the background activity $I(t)$? First we consider a trial system to show what the difficulties are. If B is the total number of excitable sites in any population v_i, and $x_i(t)$ is the number of excited sites, then $[B - x_i(t)]$ is the number of unexcited sites at time t.

Suppose that there is an equilibrium point (say 0), such that, as excited sites spontaneously become unexcited at rate A, $x_i(t)$ approaches 0. Also suppose that unexcited sites $[B - x_i(t)]$ are excited at a rate jointly proportional to their number and the input intensity $I_i(t)$. This is a mass action law. Then

$$\dot{x}_i = -Ax_i + (B - x_i)I_i(t) \tag{12}$$

with $0 \leq x_i \leq B$. In other words, system (12) describes the switching-on and passive decay of excitation by mass action.

System (1) does not suffice for the following reason. Suppose that $x_i(t)$ approaches a steady state as t increases in response to inputs with fixed pattern weights θ_i and total activity I. A steady state, $\dot{x}_i = 0$, and (12) implies that

$$x_i = \frac{B\theta_i I}{A + \theta_i I} \tag{13}$$

Now keep the θ's fixed and vary I. In other words, study how (12) processes the *same* pattern θ given different background activity levels. By (13), as I increases, all x_i approach B, and all information about θ is lost because of saturation. By contrast, if the system also contains noise, then as I becomes small, the weights θ are lost in the noise. The system processes θ badly at both low and high I values. How can this be corrected?

System (12) fails because there are no interactions among the v_i. Each θ_i is defined by an interaction of *all* the inputs I_k, $k = 1, 2, \ldots, n$. Since neither the populations nor their inputs interact, they cannot possibly compute θ. What interactions are needed? Writing $\theta_i = I_i(I_i + \Sigma_{k \neq i} I_k)$, it is clear that increasing I_i increases θ_i and that increasing any I_k, $k \neq i$, decreases θ_i. In other words, I_i "excites" θ_i, whereas all I_k, $k \neq i$, "inhibit" θ_i; the inputs compete in order to prevent saturation and thereby to compute relative figure-to-ground. When this intuition is translated into mass action dynamics, we find the simplest example of a type of system that occurs throughout the nervous system—namely, a feedforward on-center off-surround network undergoing shunting, or passive membrane dynamics. Thus let each I_i excite population v_i and inhibit all populations v_k, $k \neq i$. The inputs then form a nonrecurrent (or feedforward) on-center off-surround interaction pattern (Fig. 27), and (12) is replaced by

$$\dot{x}_i = -Ax_i + (B - x_i)I_i - x_i \sum_{k \neq i} I_k \tag{14}$$

The new term $-x_i \Sigma_{k \neq i} I_k$ says that excited sites at v_i (which number x_i)

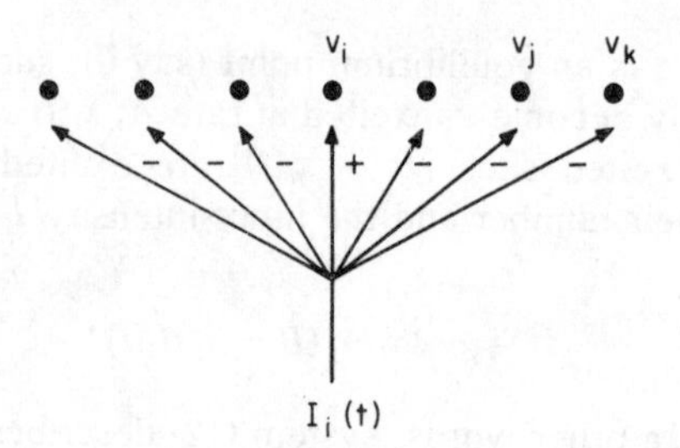

FIG. 27. Feedforward (nonrecurrent) on-center off-surround anatomy.

are inhibited (note the minus sign!) at a rate proportional to the total inhibitory input (which is a sum of inputs from the off-surround of v_i). Equation (14) is again a mass action law, or passive membrane equation, in which the off-surround automatically changes the *gain* of the system, because the inhibitory inputs multiply x_i.

How does inhibitory gain control change the system's steady state? At steady state, $\dot{x}_i = 0$, and Eq. (14) implies

$$x_i = \theta_i \frac{BI}{A + I} \tag{15}$$

In other words, no matter how large I becomes, each x_i is proportional to θ_i; there is no saturation. The system has an infinite dynamical range no matter how small B is! Furthermore, the total activity $x = \Sigma_{k=1}^{n} x_k$ satisfies $x = BI(A + I)^{-1} \leq B$; the maximal total activity B is independent of the number n of populations and of the total input intensity I. The off-surround hereby *normalizes*, or *adapts*, the total network response to fluctuations in total input. I have elsewhere suggested that this adaptation is akin to retinal light adaptation, say as studied by Werblin (1971) in the mudpuppy retina (Grossberg, 1972d), with the responses in (14) analogous to potential changes in retinal bipolar cells. Indeed, bipolar cell potential [cf. x_i in (15)] is sensitive to the ratio of on-center to off-surround excitation [cf. θ_i in (15)] and obeys a type of Weber–Fechner psychophysical law [cf. $BI(A + I)^{-1}$ in (15)]. Furthermore, bipolar potential is shifted to the right as a function of the logarithm of on-center input intensity when the off-surround input is parametrically increased. Similarly, rewrite (15) as $x_i(K, L) = Be^K(A + e^K + L)^{-1}$, where $K = \log I_i$ and $L = \Sigma_{k \neq i} I_k$. Check that, if L is changed from L_1 to L_2, then $x_i(K + S, L_1) = x_i(K, L_2)$ for all $K \geq 0$, where the shift $S = \ln[(A + L_1)(A + L_2)^{-1}]$.

The above example illustrates a point that repeatedly reappears in this paper. The teleology, or principles, of design that lead to network mechanisms are all based on a world in which continuous fluctuations exist. Binary codes are singular limits of the continuous models, and lose most of the properties, and the rationale, of the continuous models.

15. Contrast Enhancement in STM

System (14) cannot remember the pattern θ for long after the inputs are shut off, because each x_i then decays to 0. As was suggested in Section 8, recurrent (or feedback) signals among the populations v_i are needed to ensure STM. How should these feedback signals be distributed? In Section 8, we worried only about how to keep STM active until it is shut off. Now we have to worry also about what spatial pattern across $\mathscr{F}$ will be stored in STM. In particular, we have to prevent saturation of this pattern, so the signals should be distributed in an on-center off-surround anatomy, as in Fig. 28. The excitatory on-center signals $v_{i1} \leftrightarrow v_{i3}$ in Figs. 18 through 21 must be supplemented by inhibitory off-surround signals to prevent pattern saturation. Thus, given average activity $x_i(t)$, population v_i will generate a signal $f(x_i(t))$ to be distributed in an on-center off-surround anatomy among all the populations v_k, $k = 1, 2, \ldots, n$. Then (14) is replaced by the nonlinear system

$$\dot{x}_i = -Ax_i + (B - x_i)[f(x_i) + I_i] - x_i[\sum_{k \neq i} f(x_k) + J_i] \qquad (16)$$

$i = 1, 2, \ldots, n$. Term $(B - x_i)f(x_i)$ describes how a feedback signal $f(x_i)$ from v_i to itself excites the unexcited sites $B - x_i$ by mass action. The inhibitory term $-x_i \; \Sigma_{k \neq i} \; f(x_k)$ describes the switching-off of excitation at v_i by inhibitory signals $f(x_k)$ from all v_k, $k \neq i$. Term I_i is the total excitatory input and term J_i is the total inhibitory input at v_i. Often J_i is an off-surround input such as $J_i = \Sigma_{k \neq i} I_k$.

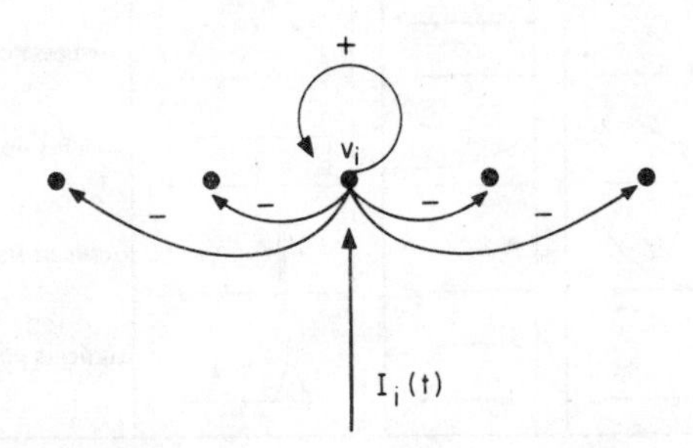

FIG. 28. Feedback (recurrent) on-center off-surround anatomy.

Much more complex versions than (16) of recurrent on-center off-surround networks have been studied (cf. Ellias and Grossberg, 1975; Levine and Grossberg, 1976; Grossberg, 1977a,b,c). But system (16), by being simple, focuses on a very important problem whose solution was a prerequisite to further progress. How does system (16) know the difference between behaviorally important patterns, which should be stored in STM by the feedback signals, and behaviorally unimportant data, such as noise, which should be suppressed? How should the average signal $f(w)$ be chosen as a function of the average activity w to distinguish between important and unimportant data? Grossberg (1973) solves this problem. The solution is summarized in Table I in terms of the total STM trace $x = \Sigma_{k=1}^{n} x_k$ and the relative STM traces $X_i = x_i x^{-1}$. After a brief pattern of inputs (I_i, J_i) is delivered to the network, does $x(t)$ converge to 0 (no STM) or to a positive limit that is bounded above by a value that is independent of n and I (normalization)? How do the relative activities X_i change through time? Do they remember the pattern θ? Do they enhance certain population activities and suppress others? Do they become more similar through time? All these cases can occur if $f(w)$ is suitably chosen.

A sigmoid, or S-shaped, $f(w)$ has all the desirable physical properties that we need. It is proved that $f(w)$ must grow faster-than-linearly at

TABLE I
INFLUENCE OF SIGNAL FUNCTION $f(w)$ ON PATTERN TRANSFORMATION AND STM STORAGE

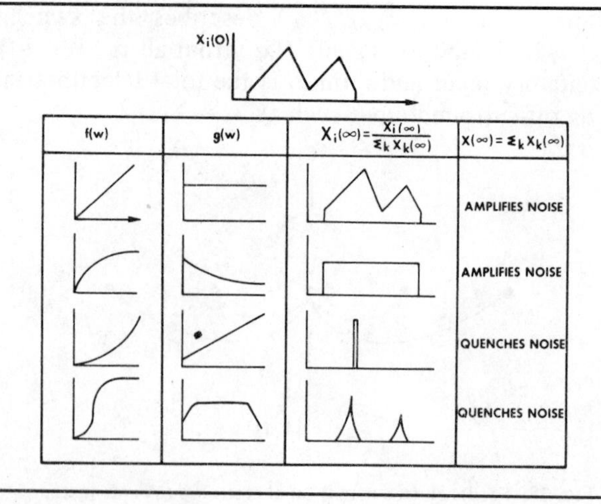

small w values in order not to amplify noise; that is, the function $g(w) = w^{-1}f(w)$ must be strictly increasing at small w values, as when $f(w) = w^2$. To prevent the system from choosing the population with maximal initial activity for STM storage and suppressing all other populations, $f(w)$ must be approximately linear—that is, $f(w) \approxeq Cw$—at intermediate w values. At large W values, all signal functions are necessarily bounded. Piecing the three regions together yields a sigmoid $f(w)$. Grossberg and Levine (1975) review in detail the reasons for using a sigmoid signal function. Other signal functions, such as faster-than-linear signal functions, can produce a network that behaves like a finite-state machine.

Given a sigmoid signal function, a *quenching threshold* (QT) exists. As Figs. 29a and 29b depict, if an initial activity $x_i(0)$ is smaller than the QT, then the activity of v_i will be quenched, or masked, by the STM reverberation. A population's activity will be stored in STM only if its initial activity exceeds the QT. The pattern of suprathreshold activities is contrast-enhanced, as in Fig. 29b. Thus, a sigmoid $f(w)$ is capable of *partially* contrast-enhancing a pattern. Its QT determines the cutoff between significant and insignificant data. If the QT is pathologically small, then the network can bootstrap into STM disturbances that do not represent behaviorally meaningful inputs; that is, a "seizure" occurs. Grossberg (1973) computes the QT in a special case and shows what parameters can lower it.

16. Tuning of STM and Releasing Subliminal Maps by Arousal

Given the existence of a QT, varying the arousal level can dramatically change what will be stored in STM, if arousal modulates the

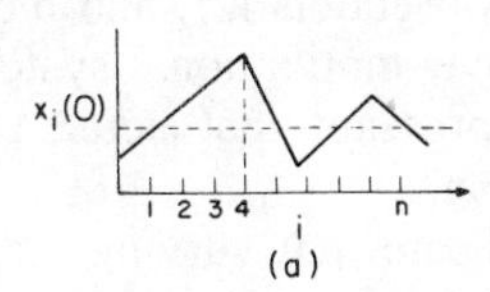

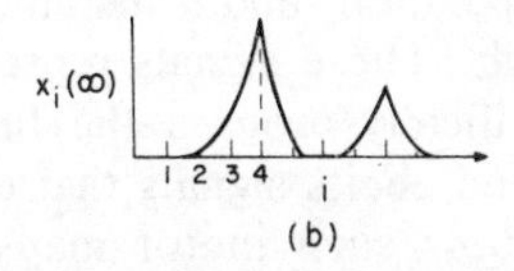

FIG. 29. Influence of the quenching threshold (QT) on STM storage of patterns.

excitability of cell responses to inputs. Arousal either can directly amplify the response of excitatory cells to inputs, or can shunt the excitability of recurrent off-surround interneurons to yield an equivalent effect. In the former case, arousal is excitatory; in the later case, it is inhibitory, and turning it on disinhibits excitatory cells.

One important example shows how a choice can be made in one parallel processing step among complex data. For example, suppose that inputs from several sensory sources stream into the network and excite a complicated pattern of activity across billions of populations, where each population can be thought of as a feature detector to fix ideas. How can the dominant feature be chosen in one step of parallel processing? If the arousal level is lowered until only one population's activity exceeds the *QT*, then the network chooses that population for STM storage and suppresses all other activities.

By contrast, if arousal suddenly increases in response to an unexpected event, then all recently presented cues can be amplified and stored in STM until the network can use all the available data to cope with the unexpected event. Grossberg (1975) applies this property to the analysis of various attentional processes, such as overshadowing. This analysis suggests that two feedback systems continually retune STM during behavior, as noted in Section 10. One system—an incentive motivational and CNV system—focuses attention on cues that are expected to generate prescribed consequences of behavior. This system can overshadow irrelevant cues. The competing system is triggered by unexpected events (novelty) and allows the network to redefine the set of relevant cues to avoid unexpected consequences. Overarousal of either system can yield attentional deficits, but the exact nature of the deficit and its proper treatment depends on the particular system that is overaroused. For example, a schizophrenic-like syndrome of punning, fuzzy response categories (Section 12), and blocking can be elicited by overarousal of the incentive–motivational system, but would not necessarily be cured by a depressant that acted primarily on the novelty (reticular formation) system.

The above remarks indicate that varying the arousal level can determine which sensory chunks will reverberate in STM and thereupon influence behavior. Similar properties hold in the motor system. For example, let an active population send a sustained pattern of signals to a field of motor control cells. These signals represent a subliminal motor map. If arousal nonspecifically arouses the field, then the pattern is bootstrapped into STM and elicits signals that determine motor output. Section 48 will discuss how such motor maps are learned. Here we merely note the following example to fix ideas. If I look at a given object that is within my reach, I can decide to touch it with either hand, either foot, my nose, etc. Somehow the proprioceptive coordinates of my eyes, head, and neck can be mapped into a terminal position of either hand, either foot, my nose, etc. "Willing" to move my right hand arouses my right hand--arm system, which thereupon moves to its terminal position.

A more sophisticated version of the tuning process is depicted in Fig. 30, which shows that there can be several (in fact, any number) of equilibrium points of total STM activity $x(t)$. As Fig. 30 shows, every other solution $E_2, E_4, \ldots$ of the equation $g(w) = A(B - w)^{-1}$ represents a stable equilibrium point of total STM activity. As the total initial input size increases, so can the asymptotic activity $x(\infty)$.

In each population v_i, successive pairs of roots of $g(x_i) = A(B - x_i)^{-1}$ are produced by bumps in the graph of the signal function $f(x_i)$. Each bump corresponds to a subpopulation of v_i; for example, the signal thresholds of cells in a population can be distributed around several preferred mean values. Then the population can fire preferentially not only to particular features, but also to particular energy levels to which these features are excited. Grossberg (1977b,c) proves that a definite STM pattern is achieved no matter how many random factors exist within each population, no matter how many populations interact, and no matter how the average level of interpopulation competition is chosen. However, the decision, or enhancement, steps whereby the asymptotic STM pattern is reached can be incredibly complicated, and in general cannot be computed. This lack of computability does not

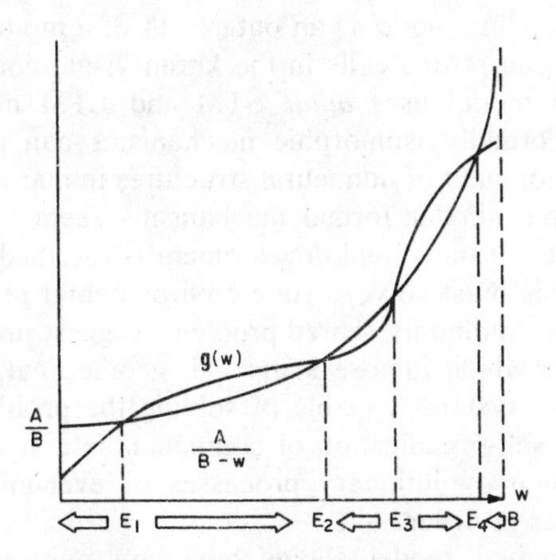

FIG. 30. The even solutions $E_2, E_4, \ldots$ of $f(w) = Aw(B - w)^{-1}$ are stable equilibrium points of $x(\infty)$. If $x(0) < E_1$, then $x(\infty) = 0$; thus E_1 defines the level below which $x(t)$ is treated as "noise."

prevent the asymptotic STM pattern from reliably being read into LTM. Here is thus an example wherein individual differences can yield subtle behavioral changes, yet the individual differences are expressions of a robust and stable principle of behavior.

The network (16) is a simple version of more complex networks that capture further properties of interacting neurons (cf. Ellias and Grossberg, 1975; Levine and Grossberg, 1976). In these networks, the properties of contrast enhancement and normalization go through in a modified form, say because the strength of recurrent excitatory and inhibitory interactions both decrease as a function of distance. When this is true, feature detectors are enhanced or suppressed only by prescribed subsets of other feature detectors. For present purposes, we note that the STM traces in the field $\mathscr{F}^{(1)}$ of Fig. 24a are not normalized. We must check how normalization of total STM activity, and tuning of it by shifting the arousal level, change our discussion of serial learning.

17. Adaptive Coding and the Emergence of Command Chunks

Before reanalyzing how normalization and tuning influence serial LTM, we shall discuss a mechanism whereby chunks are adaptively synthesized, or coded, by their defining sequences. This must be done first, because we can then ask what kinds of sequential operations on STM leave the code invariant. In particular, we can discuss whether, and how, normalization and tuning leave the code invariant.

The adaptive coding model is an outgrowth of a model that describes how experience can retune cells in the kitten visual cortex (Grossberg, 1976a,b,c). This model uses *adult* STM and LTM mechanisms in a minimal way. Formally isomorphic mechanisms can also be used to discuss the development of nonneural structures in many species (Grossberg, 1978a). That similar formal mechanisms seem to arise in many species and stages of individual development is ascribed to the fact that all living creatures must solve certain environmental problems in order to survive. These commonly shared problems suggest universal developmental principles which impose statistical, geometrical, and dynamical constraints on all systems capable of solving the problems. From this perspective, the self-organization of commands is a special case of the emergence through evolutionary processes of ever-more-complex, or "higher," degrees of order.

The developmental model shows how any number of arbitrarily complex spatial patterns can be adaptively coded by emergent command populations, and indeed recorded into any other spatial patterns using the command populations as sampling sources. In particular, the network does not have to preprogram *all* the codes that it will ever need; they can be adaptively synthesized by the environmental demands that confront each individual. Each command population responds to a prescribed convex set of spatial patterns. This convex set is determined, in part, by the affinities of all the other command populations in a field of populations; the populations in a field compete to determine which patterns will succeed in exciting them. The convex set to which a given command population can respond defines the *features* associated with the population. The command population is thus a *feature detector* in a well-defined sense. Since the population responds to a convex set of features, it automatically responds to *average* features chosen from the set, even if the average features have never been experienced. Experience can retune a command population to respond to different features. This is accomplished by changing the competitive balance of STM and LTM activity across the field of populations in response to input patterns. In effect, the model illustrates how experience can generate and globally organize a field of command states, or chunks.

The feedforward version of the model describes the interaction via trainable synaptic pathways (LTM traces) from a field of cell populations $\mathscr{F}^{(1)}$ (for example, lateral geniculate nucleus) to a field of cell populations $\mathscr{F}^{(2)}$ (for example, visual cortex). Fields $\mathscr{F}^{(1)}$ and $\mathscr{F}^{(2)}$ are separately capable of normalizing and contrast enhancing their activity, but $\mathscr{F}^{(2)}$ can also store the contrast enhanced pattern in STM (Fig. 31). In the simplest case, $\mathscr{F}^{(1)}$ consists of a nonrecurrent on-center off-surround anatomy undergoing mass action interactions, as in (14). Denote the STM trace of $v_i^{(1)}$ by $x_i^{(1)}$. Then

$$\dot{x}_i^{(1)} = -Ax_i^{(1)} + [B - x_i^{(1)}]I_i - x_i^{(1)} \sum_{k \neq i} I_k \tag{17}$$

with $0 \leq x_i^{(1)} \leq B$. As in (15), at equilibrium

$$x_i^{(1)} = \theta_i \frac{BI}{A + I}$$

Normalization in $\mathscr{F}^{(1)}$ by (17) occurs gradually in time, as each $x_i^{(1)}$ adjusts to its new equilibrium value, but it will be assumed below to occur instantaneously with $x_i^{(1)}$ approaching θ_i rather than $\theta_i BI(A + I)^{-1}$ to avoid unimportant details. This assumption of instantaneous normalization is tenable because the normalized pattern at $\mathscr{F}^{(1)}$ drives slow changes in the LTM traces that gate the signals from $\mathscr{F}^{(1)}$ to $\mathscr{F}^{(2)}$. Instantaneous normalization means that the pattern at $\mathscr{F}^{(1)}$ normalizes

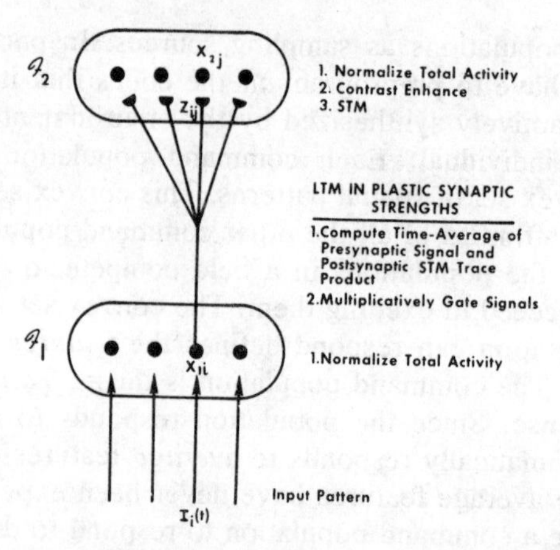

FIG. 31. The input pattern is normalized at $\mathscr{F}_1$. The normalized signals are gated and added on their way to each population in $\mathscr{F}_2$. The $\mathscr{F}_2$ inputs compete (adaptation and contrast enhancement) for storage in STM. The STM activities drive LTM traces via feedback, so that there appears to be a competition for synaptic sites.

itself before the LTM traces have a chance to change substantially; that is, it is "fast" relative to the "slow" LTM time scale.

Let the synaptic strength of the pathway from $v_i^{(1)}$ to $v_j^{(2)}$ be denoted by $z_{ij}(t)$. By (1), the total signal to $v_j^{(2)}$ due to the normalized pattern $\theta = (\theta_1, \theta_2, \ldots, \theta_n)$ at $\mathscr{F}^{(1)}$ and the vector $z_j(t) = (z_{1j}(t), z_{2j}(t), \ldots, z_{nj}(t))$ of synaptic strengths is [in the simplest case of $B_{ij}(w) = w$]

$$S_j(t) \equiv \theta \cdot z_j(t) \equiv \sum_{k=1}^{n} \theta_k z_{kj}(t) \tag{18}$$

Since $z_j(t)$ determines the size of the input to $v_j^{(2)}$, given any pattern θ, it is called the *classifying vector* of $v_j^{(2)}$ at time t. *Every* $v_j^{(2)}$, $j = 1, 2, \ldots, N$, in $\mathscr{F}^{(2)}$ receives such a signal when θ is active in $\mathscr{F}^{(1)}$. In this way, θ creates a pattern of activity across $\mathscr{F}^{(2)}$.

Suppose that $\mathscr{F}^{(2)}$ is endowed with a recurrent on-center off-surround network undergoing mass action interactions. In particular, let $\mathscr{F}^{(2)}$ normalize and contrast-enhance its signals before they are stored in STM. In effect, the populations in $\mathscr{F}^{(2)}$ *compete* for STM activity. These STM operations also occur much faster than the slow changes in connection strengths z_{ij}; hence, it is assumed below that these operations occur instantaneously in order to focus on the slow changes in z_{ij}. These slow changes in the z_{ij}'s will pick out populations in $\mathscr{F}^{(2)}$—the feature detectors—to code the spatial patterns perturbing $\mathscr{F}^{(1)}$.

We now summarize the coding process. Suppose for simplicity that all changes in z_{ij} are driven by the *feedback* within the excitatory recurrent loops in $\mathscr{F}^{(2)}$ that establish STM storage. In other words, the "fast" competition for STM activity feeds back as a "slow" competition for LTM activity, or for synaptic site efficacy. Then (2) becomes

$$\dot{z}_{ij} = (-z_{ij} + \theta_i)x_j^{(2)} \tag{19}$$

where $\Sigma_{k=1}^N x_j^{(2)}(t) = 1$ if STM in $\mathscr{F}^{(2)}$ is active at time t, whereas $\Sigma_{k=1}^N x_k^{(2)}(t) = 0$ if STM in $\mathscr{F}^{(2)}$ is inactive at time t. Two important cases can be distinguished.

If tuning is adjusted so that $\mathscr{F}^{(2)}$ *chooses* a population for storage in STM, then

$$x_j^{(2)} = \begin{cases} 1 & \text{if } S_j > \max\{\epsilon, S_k: k \neq j\} \\ 0 & \text{if } S_j < \max\}\epsilon, S_k: k \neq j\} \end{cases} \tag{20}$$

except if two or more populations have a maximum signal; in the latter case, share the total STM activity (= 1) among the maximally excited populations. The parameter ϵ in (20) is the *QT*; if no signal exceeds the *QT*, no STM storage occurs.

If *partial contrast* in STM holds at $\mathscr{F}^{(2)}$, then the dynamics of the network can be approximated by a rule of the form

$$x_j^{(2)} = \begin{cases} f(S_j)[\sum_{S_k > \epsilon} f(S_k)]^{-1} & \text{if } S_j > \epsilon \\ 0 & \text{if } S_j < \epsilon \end{cases} \tag{21}$$

In (21), ϵ represent the *QT*; $f(w)$ controls how suprathreshold signals will be contrast-enhanced [for example, $f(w) = w^2$]; and the ratio of $f(S_j)$ to $\Sigma\{f(S_k): S_k > \epsilon\}$ expresses the normalization of STM.

18. Feature Detectors

To see how these rules classify patterns, first hold all z_{ij} constant. Then Eqs. (18) and (20) reduce to the statement that population $v_j^{(2)}$ is stored in STM if

$$S_j > \max\{\epsilon, S_k; k \neq j\} \tag{22}$$

where $S_j = \theta \cdot z_j$ and $\theta_i = I_i(\Sigma_k I_k)^{-1}$. In other words, $v_j^{(2)}$ *codes* all patterns θ such that (22) holds. Stated alternatively, $v_j^{(2)}$ is a *feature detector* in the sense that all patterns

$$P_j = \{\theta: \theta \cdot z_j > \max\{\epsilon, \theta \cdot z_k: k \neq j\}\} \tag{23}$$

are classified by $v_j^{(2)}$. The set P_j defines a *convex cone* C_j in the space of nonnegative input vectors $J = (I_1, I_2, \ldots, I_n)$, since if two such vectors J_1 and J_2 are in C_j, then so are all the vectors αJ_1, βJ_2, and $\gamma J_1 + (1 - \gamma)J_2$, where $\alpha > 0$, $\beta > 0$, and $0 < \gamma < 1$. The convex cone C_j defines the *features* coded by $v_j^{(2)}$.

The classification rule in (22) has an informative geometrical interpretation in n-dimensional Euclidean space. The signal $S_j = \theta \cdot z_j$ is the inner, or dot, product of θ and z_j (Thomas, 1968). Letting $\|\xi\| = (\Sigma_{k=1}^n \xi_k{}^2)^{1/2}$ denote the Euclidean length of any real vector $\xi = (\xi_1, \xi_2, \ldots, \xi_n)$, and $\cos(\eta, \omega)$ denote the cosine of the angle between two vectors η and ω, it is elementary that

$$S_j = \|\theta\| \, \|z_j\| \cos(\theta, z_j)$$

The signal S_j is thus the length of the projection of the normalized pattern θ on the classifying vector z_j times the length of z_j. In effect, θ is projected on all the cells v_{2j} by their classifying vectors z_j. Thus if all z_j, $j = 1, 2, \ldots, N$, have equal length, then (22) classifies all patterns θ in P_j whose angle with z_j is smaller than the angles between θ and any z_k, $k \neq j$, and is small enough to satisfy the ϵ-condition. In particular, patterns θ that are *parallel,* or proportional to z_j, are classified in P_j. The choice of classifying vectors z_j hereby determines how the patterns θ will be divided up. Section 19 describes how the adaptive coding mechanism (18) through (20) makes the z_j vectors more parallel to prescribed patterns θ, and thereupon changes the classifying sets P_j. In summary, given (20):

(*i*) The number of populations in $\mathscr{F}^{(2)}$ determines the maximum number N of pattern classes P_j.

(*ii*) The choice of classifying vectors z_j determines how different the sets P_j can be; for example, choosing all vectors z_j equal will generate one class that is redundantly represented by all $v_j^{(2)}$.

(*iii*) The size of ϵ determines how similar patterns must be to be classified by the same $v_j^{(2)}$ if the z_k's are not the same.

If the choice rule (20) is replaced by the partial contrast rule (21), then an important new possibility occurs, which can be described by studying STM responses either to all θ at fixed $v_j^{(2)}$, or to a fixed θ at all $v_j^{(2)}$. In the former case, each $v_j^{(2)}$ has a *tuning curve*; namely, a maximal response to certain patterns, and submaximal responses to other patterns. In the latter case, each pattern θ is *filtered* by $\mathscr{F}^{(2)}$, or generates a *generalization gradient,* in a way that shows how close θ lies to *each* of the classifying vectors z_j. The pattern will be classified by $v_j^{(2)}$—that is, stored in STM—only if it lies sufficiently close to z_j for its signal to exceed the quenching threshold of $\mathscr{F}^{(2)}$. It can be shown that the existence of tuning curves in a given cortical field $\mathscr{F}^{(i)}$ increases the discriminative capabilities of the next cortex $\mathscr{F}^{(i+1)}$ in a hierarchy (cf. Grossberg, 1976a).

19. Development of an STM Code

System (18) through (20) will be discussed below because it illustrates important properties of the coding process in a lucid way. First, consider how the system responds to a single pattern θ that is iteratively presented through time. Suppose that there exists a unique j in $\mathscr{F}^{(2)}$ such that

$$S_j(0) > \max\{\epsilon, S_k(0): k \neq j\}$$

That is, $v_j^{(2)}$ is the population that initially receives the largest signal from $\mathscr{F}^{(2)}$. In particular, the signal $S_j(0)$ might be only slightly larger than the other signals $S_k(0)$, $k \neq j$. Then, as θ is presented to $\mathscr{F}^{(1)}$, the angle between $z_j(t)$ and θ, as n-vectors, monotonically decreases to zero. In other words, coding makes z_j parallel to θ. Simultaneously, the signal $S_j(t)$ monotonically approaches $\|\theta\|^2$; in other words, the coding process maximizes the inner product signal $S_j(t) = \theta \cdot z_j(t)$ over all possible choices of z_j such that $\|z_j\| \leq \|\theta\|$. Thus, whereas the initial signal $S_j(0)$ to $v_j^{(2)}$ might be only slightly larger than the signals $S_k(0)$, $k \neq j$, after coding takes place, θ generates a maximal signal to $v_j^{(2)}$. The other signals $S_k(t)$, $k \neq j$, remain constant.

By changing $z_j(t)$, the coding process changes the class of patterns θ that will be coded by $v_j^{(2)}$. For example, patterns θ that were originally more parallel to $z_j(0)$ than any $z_k(0)$, $k \neq j$, and which are therefore coded by $v_j^{(2)}$ at $t = 0$, can become more parallel to some $z_k(0)$ than $z_j(T)$, and are therefore coded by $v_k^{(2)}$ at $t = T$. Conversely, patterns that were originally more parallel to some $z_k(0)$ than $z_j(0)$, and which are coded by $v_k^{(2)}$ at $t = 0$, can become more parallel to $z_j(T)$ than $z_k(0)$, and are therefore coded by $v_j^{(2)}$ at $t = T$. In effect, presenting pattern θ at $\mathscr{F}^{(1)}$ has shifted the convex set of patterns—the features—that will be coded by $v_j^{(2)}$.

In the terminology of development biology, the z_j define *positional gradients* between $\mathscr{F}^{(1)}$ and $\mathscr{F}^{(2)}$. The initial positional gradients determine what features $\mathscr{F}^{(2)}$ will try to classify. Thus, if $\mathscr{F}^{(1)}$ feeds signals in parallel to several adaptive fields, each with different positional gradients, then each field will try to classify different features of the data base at $\mathscr{F}^{(1)}$, even though all fields use the same computational machinery. A serial hierarchy $\mathscr{F}^{(1)}, \mathscr{F}^{(2)}, \mathscr{F}^{(3)}, \ldots$ of adaptive fields can also be constructed, since each field $\mathscr{F}^{(i)}$ possesses the requisite condition of normalization that is used to prepare data for field $\mathscr{F}^{(i+1)}$. As i increases, the data that $\mathscr{F}^{(i)}$ feeds to $\mathscr{F}^{(i+1)}$ include increasingly abstract convex sets of features (cf. Rocha-Miranda *et al.*, 1975). For example, what will a field $\mathscr{F}^{(3)}$ code if it receives patterned signals from $\mathscr{F}^{(2)}$? A population in $\mathscr{F}^{(3)}$ will code convex sets of spatial patterns across $\mathscr{F}^{(2)}$. Each population in $\mathscr{F}^{(2)}$ codes a set of features, so a spatial pattern across $\mathscr{F}^{(2)}$ is a global construct that describes how much of each feature is in the input pattern to $\mathscr{F}^{(1)}$. A convex set of spatial patterns across $\mathscr{F}^{(2)}$ describes the tolerated changes, or "fuzziness," in each feature's activity that are compatible with unchanged coding at $\mathscr{F}^{(3)}$. Cells in $\mathscr{F}^{(3)}$ can hereby generate stable responses as the input patterns to $\mathscr{F}^{(1)}$ undergo significant global transformations. If the positional gradients from $\mathscr{F}^{(i)}$ to $\mathscr{F}^{(i+1)}$ only excite a localized cluster of populations, then the populations in $\mathscr{F}^{(i+1)}$ will respond to features only in prescribed regions of $\mathscr{F}^{(i)}$—for example, in the simplest case, prescribed retinal regions. If also the on-center and off-surround interactions within $\mathscr{F}^{(i+1)}$ decrease with distance, so that normalization only holds approximately and among prescribed subsets, or channels, of populations, then features in prescribed regions will be enhanced and/or suppressed only by particular classes of related features in nearby regions. Changing the distribution of positional gradients and on-center off-surround interactions can thus dramatically influence which features are computed and which features mutually influence one another.

What happens if several different spatial patterns $\theta^{(k)} = (\theta_1{}^{(k)}, \theta_2{}^{(k)}, \ldots, \theta_n{}^{(k)})$, $k = 1, 2, \ldots, M$, all perturb $\mathscr{F}^{(1)}$ at different times? How are changes in the z_{ij}'s due to one pattern prevented from contradicting changes in the z_{ij}'s due to a different pattern? The choice-making property of $\mathscr{F}^{(1)}$ helps to do this; it acts as a sampling device that often prevents contradictions from occurring. The following argument suggests how sampling works. This argument is, however, not entirely correct. For definiteness, suppose that M spatial patterns $\theta^{(k)}$ are chosen, $M \leq N$, such that their signals at time $t = 0$ satisfy

$$\theta^{(k)} \cdot z_k(0) > \max\{\epsilon, \theta^{(k)} \cdot z_j(0): j \neq k\} \tag{24}$$

for all $k = 1, 2, \ldots, M$. In other words, at time $t = 0$, $\theta^{(k)}$ is coded by $v_k{}^{(2)}$. Let $\theta^{(1)}$ be the first pattern to perturb $\mathscr{F}^{(1)}$. By (24), population $v_1{}^{(2)}$ receives the largest signal from $\mathscr{F}^{(1)}$. All other populations $v_j{}^{(2)}$, $j \neq 1$, are thereupon inhibited by the off-surround of $v_1{}^{(2)}$, whereas $v_1{}^{(2)}$ reverberates in STM. By (19), none of the synaptic strengths $z_j(t)$, $j \neq 1$, can learn while $\theta^{(1)}$ is presented. Presenting $\theta^{(1)}$ makes $z_1(t)$ more parallel to $\theta^{(1)}$ as t increases. Consequently, if a different pattern, say $\theta^{(2)}$, perturbs

$\mathscr{F}^{(1)}$ on the next learning trial, then it will excite $v_2^{(2)}$ more than any other $v_j^{(2)}$, $j \neq 2$: it cannot excite $v_1^{(2)}$ because the coefficients $z_1(t)$ are more parallel to $\theta^{(1)}$ than before; and it cannot excite any $v_j^{(2)}$, $j \neq 1, 2$, because the $v_j^{(2)}$ coefficients $z_j(t)$ still equal $z_j(0)$. In response to $\theta^{(2)}$, $v_2^{(2)}$ inhibits all other $v_j^{(2)}$, $j \neq 2$. Consequently none of the $v_j^{(2)}$ coefficients $z_j(t)$ can learn, $j \neq 2$; learning makes the coefficients $z_2(t)$ become more parallel to $\theta^{(2)}$ as t increases. The same occurs on all learning trials. By inhibiting the postsynaptic part of the learning mechanism in all but the chosen $\mathscr{F}^{(2)}$ population, the on-center off-surround network in $\mathscr{F}^{(2)}$ samples one vector $z_j(t)$ of LTM traces at any time. In this way $\mathscr{F}^{(2)}$ can learn to classify as many as N patterns if it contains N populations. (When tuning curves or resonant feedback exist, each population can share in the coding of several patterns.)

This argument is almost correct. It fails, in general, because by making (say) $z_1(t)$ more parallel to $\theta^{(1)}$, it is also possible to make $z_1(t)$ more parallel to $\theta^{(2)}$ than $z_2(0)$ is. Thus when $\theta^{(2)}$ is presented, it will be coded by $v_1^{(2)}$ rather than $v_2^{(2)}$. In other words, practicing one pattern can recode other patterns. This property can be iterated to show how systematic trends in the sequence of practiced patterns can produce systematic drifts in recoding (Grossberg, 1976b). Moreover, if the statistical structure of the practice sequences continually changes, then there need not exist a stable coding rule in $\mathscr{F}^{(2)}$. This is quite unsatisfactory.

By contrast, if there are few, or sparse, patterns relative to the number of populations in $\mathscr{F}^{(2)}$, then a stable coding rule does exist, and the STM choice rule in $\mathscr{F}^{(2)}$ does provide an effective sampling technique. In effect, given any *fixed* class of patterns at $\mathscr{F}^{(1)}$ and sufficiently many populations in $\mathscr{F}^{(2)}$; the $\mathscr{F}^{(1)}$ patterns can induce a stable STM code in $\mathscr{F}^{(2)}$. By contrast, the problem of stabilizing the STM code given a *fixed* number of cells in $\mathscr{F}^{(2)}$ and *arbitrarily* many patterns θ at $\mathscr{F}^{(1)}$ requires additional network mechanisms. This problem is studied in Grossberg (1976b,c). In this case the cells in $\mathscr{F}^{(2)}$ can continually be recoded by patterns at $\mathscr{F}^{(1)}$. No stable hierarchy of codes could develop using only this mechanism, since the coded meaning of the signals from one level to the next would be continually changing. Below are reviewed relevant aspects of how a developing code can be stabilized in an arbitrary environment. In passing, we note that rules such as (20) and (22) define discriminant functions of a type that is familiar in pattern classification studies, and is related to Bayesian decision rules that make choices to minimize risk (Duda and Hart, 1973, Chapter 2). If terms such as $\theta \cdot z_j$ are generalized to $f(\theta) \cdot z_j$, where $f(\theta) = (f_1(\theta_1), f_2(\theta_2), \ldots, f_n(\theta_n))$, then the decision boundaries of the discriminant functions are not necessarily convex. Our method departs from the classical development in several ways. One way is described by the rules whereby the $z_j(t)$ vectors shift, owing to learning. A more fundamental way is described in the next section, which shows how local discriminants, or features, are synthesized into a global code by adaptive resonance between two fields of cells. The local properties of the field, by themselves, neither define feature detectors nor when adaptation of feature detectors will occur. The functional unit of coding is a global feedback module that I call an adaptive resonance.

20. Stabilizing the STM Code: Expectation, Resonance, Rebound, and Search

To stabilize the code, it suffices to use attentional mechanisms. These mechanisms were introduced in Grossberg (1975). The reverse statement is also true: the minimal mechanisms for stabilizing the STM code can also generate various attentional phenomena. Why is there a relationship between code stability and attention? The next example motivates this relationship.

Suppose that a population $v_j^{(2)}$ in $\mathscr{F}^{(2)}$ already codes a given class of patterns $\mathscr{P}_j$ at $\mathscr{F}^{(1)}$, and that a pattern θ not in this class succeeds in activating $v_j^{(2)}$. If this activation is not rapidly terminated, then recoding of the LTM traces will occur, since by (19) z_j can learn θ while $v_j^{(2)}$ is active. Somehow, sustained activation of $v_j^{(2)}$ by an erroneous pattern θ must be prevented; activity in $v_j^{(2)}$ is somehow inhibited. This can happen only if the network can determine that $v_j^{(2)}$ codes a pattern class that is incompatible with θ. Furthermore, the operation that inhibits $v_j^{(2)}$ cannot inhibit all populations in $\mathscr{F}^{(2)}$; otherwise θ could not find any population in $\mathscr{F}^{(2)}$ to code it. Somehow the network selectively inhibits the erroneously activated populations $v_j^{(2)}$ before it searches for an uncommitted population with which to code θ. Given these remarks, it is not surprising that STM code stability is related to attention; stability requires the network to selectively activate populations whose codes are compatible with the sensory data of the moment.

The mechanisms in Grossberg (1976c) describe how a test pattern θ at $\mathscr{F}^{(1)}$ can tentatively activate feature detectors in $\mathscr{F}^{(2)}$, which thereupon generate feedback signals either to $\mathscr{F}^{(1)}$ or to a field $\mathscr{E}^{(1)}$ that acts in parallel with $\mathscr{F}^{(1)}$. These feedback signals represent an expectation, or template, with which the afferent test pattern at $\mathscr{F}^{(1)}$, or its parallel representation at $\mathscr{E}^{(1)}$, is compared (cf. Section 44). This expectation is a spatial pattern that can be learned by the $\mathscr{F}^{(2)} \rightarrow \mathscr{F}^{(1)}$ (or $\mathscr{F}^{(2)} \rightarrow \mathscr{E}^{(1)}$) LTM traces at the same time that chunks are being coded by the $\mathscr{F}^{(1)} \rightarrow \mathscr{F}^{(2)}$

LTM traces. If the test pattern matches the expectation, then the patterned STM activity in $\mathscr{F}^{(1)}$ and $\mathscr{F}^{(2)}$ is amplified and can resonate between $\mathscr{F}^{(1)}$ and $\mathscr{F}^{(2)}$. This resonant activity can activate the STM of other fields, say of higher-order feature detectors or of motor commands; it can also drive slow LTM changes in synapses that sample $\mathscr{F}^{(1)}$ or $\mathscr{F}^{(2)}$.

Suppose, however, that the expected and test patterns are very different. This mismatch means that an erroneous classification has occurred at $\mathscr{F}^{(2)}$. An alarm system is thereupon triggered that generates a nonspecific wave of input activity across $\mathscr{F}^{(2)}$. The alarm system acts nonspecifically because, at the place where the mismatch is computed, no data is available concerning *which* populations in $\mathscr{F}^{(2)}$ have erroneously been activated. The nonspecific signal must somehow selectively inhibit, or reset, the active populations of $\mathscr{F}^{(2)}$ without preventing inactive populations from being tentatively activated during the next time interval. In effect, the populations whose activity set off the alarm must have been erroneously classified, so they should be selectively suppressed. This idea realizes a kind of probabilistic logic operating in real time, with activity level replacing truth value.

The inhibition of active populations must be enduring as well as selective. Otherwise, the inhibited populations could immediately be reactivated by the pattern θ. Given such an inhibitory mechanism, the network automatically searches for a population which is not compatibly classified. When one is found, say $v_i^{(2)}$, STM at $v_i^{(2)}$ can stay on long enough to drive the "slow" coding process in the z_i LTM traces.

How is a selective and enduring inhibition at $\mathscr{F}^{(2)}$ effected? I suggest that it is due to the organization of $\mathscr{F}^{(2)}$ into antagonistic pairs, or dipoles, of populations. In effect, the antagonistic population to an active population $v_i^{(2)}$ is turned on when the nonspecific alarm goes off, and thereupon selectively inhibits $v_i^{(2)}$; that is, if "yes" at $v_i^{(2)}$ is wrong, turn on "no" at $v_i^{(2)}$, but in a graded fashion.

The idea of population dipoles was not originally introduced to stabilize the STM code, although this is a fundamental reason for their existence. Originally the idea arose in a neural theory of reinforcement, wherein cell dipoles regulate net incentive motivation through time (Grossberg, 1972b,c, 1975); also see Wise *et al.* (1973) for compatible data. Properties of the rebound from positive (negative) incentive to negative (positive) incentive through time are analogous to many paradoxical phenomena about reinforcement—for example, how an amphetamine can calm an agitated syndrome that is really a form of underaroused emotional depression, whereas overaroused depression can yield indifference to the emotional meaning of cues.

More generally, suppose that the on-cell of a dipole is activated persistently by the presence of its external cue, whereas the off-cell is activated transiently by the offset of the cue. Otherwise expressed, offset of the cue elicits a transient antagonistic rebound. This transient activity can be used to sample STM patterns at the synaptic knobs of the off-cell and encode these patterns in LTM. Hereby the offset of a cue can elicit learned behavior.

When the antagonistic rebound is explicitly modeled, one is led to postulate the existence of slowly varying transmitter substances that multiplicatively gate all signals before they can reach the on-cells and off-cells. Among these signals are a tonically active nonspecific arousal that is distributed uniformly across all the cells. The arousal signal regulates the size of the off-cell rebound when the cue to the on-cell terminates. This happens as follows. When the cue is on, the total signal in the on-cell channel exceeds the total signal in the off-cell channel. Both signals are gated by transmitter before they reach their targets. Because the on-cell signal is larger, transmitter is depleted more in the on-cell channel than in off-cell channel. The on-cell nonetheless receives the larger input because of the multiplicative effect of signal and transmitter on the cells: the equilibrium transmitter level has the form $A[B + CS]^{-1}$ in response to a steady signal S, and therefore decreases as S increases; but the equilibrium input has the form $ADS[B + CS]^{-1}$, which increases as S increases. When the cue is removed, equal arousal signals remain in both channels. Since transmitter level changes slowly, there is more transmitter in the off-cell channel. The multiplicative coupling of arousal signal to transmitter now gives the off-cell a larger input, thereby causing the rebound. Gradually, in response to the equal arousal signal in both channels, the transmitter levels also equalize, and both channels receive equal inputs, so that the rebound eventually terminates.

A *property* of this system is that a rapid increment in nonspecific arousal can, by itself, reverse or rebound the relative activities in a dipole. Thus, if an on-cell is active when arousal increases, then it can be inhibited by its off-cell, whereas if neither cell is active when arousal increases, then neither on-cell nor off-cell receives any relative advantage. The rebound therefore selectively inhibits active populations. If the on-cells are now hooked into a recurrent network capable of STM, and the off-cells are similarly organized, then it follows that a transient arousal increment can selectively, and in a graded fashion, inhibit active populations by shifting the STM pattern across both fields. When this mechanism acts on various fields of formal feature detectors, phenomena analogous to negative afterimages and spatial frequency adaptation are found (Grossberg, 1976c). The properties of antagonistic rebound in a dipole of populations are useful for understanding many psychophysiological processes in which rapid shifts of specific cues and/or nonspecific arousal signal important events.

21. Pattern Completion, Hysteresis, and Gestalt Switching

The concept of two fields $\mathscr{F}^{(1)}$ and $\mathscr{F}^{(2)}$ joined together by reciprocal trainable signal pathways is also relevant to many psychophysiological processes, if only because it describes the minimal network module that can stabilize its STM code in a rich input environment. The two LTM processes—code-learning in $\mathscr{F}^{(1)} \rightarrow \mathscr{F}^{(2)}$ LTM traces and template-learning in $\mathscr{F}^{(2)} \rightarrow \mathscr{F}^{(1)}$ LTM traces—are partners in establishing a stable state of resonant STM activity in $\mathscr{F}^{(1)}$ and $\mathscr{F}^{(2)}$ when the active LTM channels are compatible. I call this module an *adaptive resonance*. Grossberg (1976c) summarizes some examples of this concept in olfactory coding, in the regulation of attention by the matching of presently available cues (conditioned reinforcers) with feedback from compatible drive sources (expressed through the contingent negative variation), and in a search and lock mechanism for stabilizing eye position. The next two sections describe several other important examples of this concept. These examples illustrate how closely related ''perceptual'' and ''cognitive'' properties can be.

First, feedback from $\mathscr{F}^{(2)}$ to $\mathscr{F}^{(1)}$ can deform what ''is'' perceived into what ''is expected to be'' perceived. Otherwise expressed, the feedback is a prototype, or higher-order Gestalt, that can deform, and even complete, activity patterns across lower-order feature detectors. For example, suppose that a sensory event is coded by an activity pattern across the feature detectors of a field $\mathscr{F}^{(1)}$. The $\mathscr{F}^{(1)}$ pattern is then coded by certain populations in $\mathscr{F}^{(2)}$. If the sensory event has never before been experienced, then the $\mathscr{F}^{(2)}$ populations that are chosen are those whose codes most nearly match the sensory event, because the pattern at $\mathscr{F}^{(1)}$ is projected onto $\mathscr{F}^{(2)}$ by the positional gradients in the $\mathscr{F}^{(1)} \rightarrow \mathscr{F}^{(2)}$ pathways. If no approximate match is possible, then these $\mathscr{F}^{(2)}$ populations will be inhibited and a search procedure will be elicited. If an approximate match is possible, however, then feedback signals from $\mathscr{F}^{(2)}$ to $\mathscr{F}^{(1)}$ will elicit the template of the sensory events that are *optimally* coded by the $\mathscr{F}^{(2)}$ pattern. These feedback signals gradually deform the $\mathscr{F}^{(1)}$ pattern until this pattern is a mixture of feedforward codes and feedback templates. Otherwise expressed, $\mathscr{F}^{(2)}$ tries to *complete* the $\mathscr{F}^{(1)}$ pattern using the prototype, or template, that its active populations release. I suggest that many Gestalt-like pattern completion effects are special cases of this feedback mechanism, and therefore consequences of STM stability.

Two important manifestations of the completion property are hysteresis and Gestalt switching. Once an STM resonance is established, it resists changing its codes in response to small changes in the sensory event; this is hysteresis. Hysteresis occurs because the active $\mathscr{F}^{(2)} \to \mathscr{F}^{(1)}$ template keeps trying to deform the shifting $\mathscr{F}^{(1)}$ pattern back to one that will continue to code the $\mathscr{F}^{(2)}$ populations that elicit this template.

If, however, the sensory event changes so much that the mismatch of test and template patterns becomes too great, then the alarm-and-reset mechanism is triggered and a new code in $\mathscr{F}^{(2)}$ will be activated. This new code feeds its template back to $\mathscr{F}^{(1)}$ and deforms the $\mathscr{F}^{(1)}$ pattern toward its optimal pattern. A dramatic switch between global percepts can hereby be effected. The global nature of the switch is due to the fact that $\mathscr{F}^{(2)}$ contains codes that synthesize data from many feature detectors in $\mathscr{F}^{(1)}$, and the templates of these codes can reorganize large segments of the $\mathscr{F}^{(1)}$ field. I suggest that an analogous mixture of hysteresis and switching is operative in various visual illusions, such as Necker's cube (Graham, 1966). In fact, one can think of such illusions in the following light. Often when one slowly scans a visual scene, small shifts in perspective imply small shifts in higher-order codes. These illusions are *designed* so that small shifts in perspective imply large shifts in higher-order codes after the hysteresis range is exceeded and the reset mechanism is triggered. Also of interest are the "spontaneous" switches that can occur when ambiguous figures are persistently examined. By Section 20, a prolonged STM reverberation at a given population can partially deplete its transmitter, and thereby shift the relative balance of on-cell to off-cell excitation across the field in favor of other populations. If this shift is sufficiently large, it can induce a cyclic drift in activity across the field as populations cyclically deplete and accumulate their transmitter stores (cf. Section 40).

Hysteresis can also occur between two reciprocally connected fields that are not hierarchically organized. In particular, suppose that each eye activates a field of monocularly coded feature detectors. Suppose that each monocular field is endowed with a recurrent on-center off-surround anatomy. Also let corresponding detectors be capable of sending each other signals. It does not matter *what* features are coded by these detectors to draw the following conclusion. Once a resonance is established between the two fields, hysteresis will prevent small changes in input pattern from changing the coded activity. Julesz (1971) introduced a field of physical dipoles to model binocular hysteresis. Resonance between two recurrent on-center off-surround networks undergoing mass action dynamics provides a neural model of the phenomenon.

In passing, we note that "reverberation" and "resonance" are not interchangeable concepts. Reverberation between $\mathscr{F}^{(1)}$ and $\mathscr{F}^{(2)}$ occurs even as $\mathscr{F}^{(2)}$ is continually reset in search of an admissible population. Resonance occurs only when the test pattern at $\mathscr{F}^{(1)}$ and the active codes at $\mathscr{F}^{(2)}$ are compatible.

22. Context-Dependent Coding and Restricted Conditions for Recoding

Can a field of feature detectors uniquely classify a sensory event even if no single detector is uniquely activated by that event? The answer is "yes" if feedback exists within the system. In particular, a resonant state of activity need not be established between $\mathscr{F}^{(1)}$ and $\mathscr{F}^{(2)}$ unless the pattern of activity across feature detectors in $\mathscr{F}^{(1)}$ correctly classifies the sensory event. Otherwise expressed, the code is context-dependent; each $\mathscr{F}^{(1)}$ population shares in "multiple meanings" by participating in the coding of many events (Chung *et al.*, 1970).

Not all sensory patterns need recode the network's feature detectors. Only patterns that generate a resonant state of activity can generate such a change. The resonant state explicates the intuitive idea that the network is attending to the pattern and has stored it in STM, whereupon it can induce recoding via LTM changes. In particular, mere passive presentation of patterns need not recode any feature detectors. Adaptational differences between passive and active responses to cues have been experimentally described (Held and Hein, 1963) and might be one factor that explains why certain experiments seem to reliably recode feature detectors, whereas others do not (Stryker and Sherk, 1975). From this perspective, terminating the critical period for developmental plasticity seems to depend on the switching on of attentional factors in addition to the possible switching off of a chemical agent. The paper by Grossberg (1977a) discusses this possibility for the visual system. Some themes are differential amplification or attenuation of lateral geniculate nucleus (LGN) activity depending on whether afferent sensory data match or mismatch cortical feedback to LGN; lateral inhibition in LGN as a matching mechanism, and thus the growth of LGN inhibitory pathways as a precursor of critical period termination; temporal modulation of catecholaminergic cortical arousal as a rebound trigger mechanism; and cortical dipole organization, possibly realized by feature detector pairs that code complementary features, and embedded in recurrent on-center off-surround networks, as a mechanism to maintain the rebounded STM pattern.

23. Reset, Reaction Time, and P300

When a mismatch between test pattern and expected pattern occurs, STM is reset by a nonspecific mechanism that inhibits active populations until an admissible population is found. Several properties of the reset mechanism are of interest. First, STM does not spontaneously decay at a rapid rate, but can endure until it is actively reset by a variety of mechanisms: competing inputs, change of expectations, shift in arousal level, etc. The durability of STM in a recurrent network is in marked contrast to the rapid passive decay of activity in nonrecurrent networks. *In vivo,* one expects to find recurrent networks more centrally (for example, in neocortex) and nonrecurrent networks more peripherally (for example, in retina).

Second, unexpected events trigger STM reset by a nonspecific mechanism that inhibits ongoing activity. In average evoked potential experiments, one often finds an inhibitory wave, the P300, that accompanies unexpected events (Rohrbaugh *et al.*, 1974; Squires *et al.*, 1976). To discuss the model's relationship to P300, we consider an idealized example in Fig. 32. The active population $v_j^{(2)}$ in Fig. 32 generates a subliminal pattern at $\mathscr{F}^{(1)}$ that acts as a sensory expectation. Suppose that $v_j^{(2)}$ also generates a subliminal pattern at the motor control cells $\mathscr{M}$. This latter pattern acts as a motor expectation, or subliminal motor map.

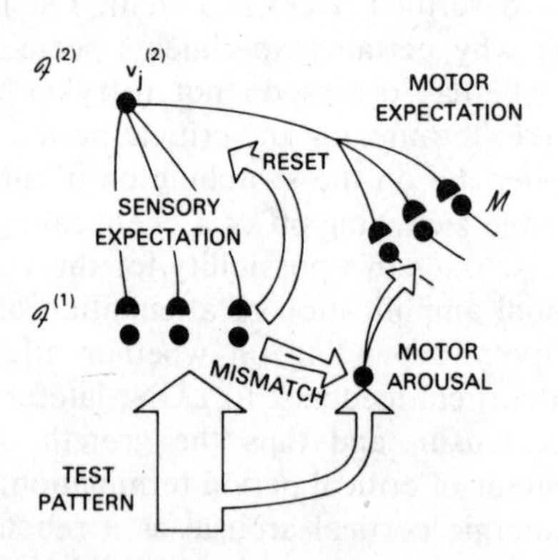

FIG. 32. Sensory and motor expectations are reset or released by mismatch or match, respectively, of sensory expectation with sensory test pattern.

Let a test pattern perturb $\mathscr{F}^{(1)}$. If the test pattern does not match the sensory expectation, then nonspecific arousal is triggered. How does this happen? We assume (Grossberg, 1975) that *every* test pattern can activate this arousal source. If the test pattern matches the sensory expectation, then a signal is released by $\mathscr{F}^{(1)}$ which inhibits the arousal on its way to $\mathscr{F}^{(2)}$. We also suppose that there is a complementary arousal system in Fig. 32. If the test pattern matches the sensory expectation, then the motor map in $\mathscr{M}$ is aroused, and therefore sends signals to motor effectors. If the test pattern does not match the sensory expectation, then the motor arousal is inhibited. In summary, a mismatch can excite the reset arousal system as it inhibits the goal-oriented motor arousal system (it says "no"), whereas a match can inhibit the reset arousal system as it excites the goal-oriented motor arousal system (it says "yes"). There are many variations on this theme; for example, the sensory expectation might merely predispose $\mathscr{F}^{(1)}$ to excite a particular $\mathscr{F}^{(2)}$ population which thereupon excites $\mathscr{M}$; or a match between patterns might activate a signal that inhibits a nonspecific inhibition, which in turn disinhibits pacemakers in $\mathscr{F}^{(2)}$ that drive dipole rebounds. The main points are independent of such details.

If the reset mechanism is indeed related to P300, then P300 is conceptually different from the model's CNV, which subserves a conditionable "psychological set." Donchin *et al*. (1975) report compatible data. Furthermore, if P300 resets STM, it can do so without eliciting motor activity. See Donchin *et al*. (1972) for compatible data.

If the test pattern matches the sensory expectation, then the system's reaction time is less than if the test pattern does not match the sensory expectation. This is due to two effects working together: In the former case, the motor pattern can be released without a prior reset of STM, and the subliminal expectation reduces the reaction time for suprathreshold signals to be emitted from $\mathscr{F}^{(1)}$. Moreover, the reaction time will be a monotone increasing function of the inhibitory rebound size, since the latter index measures how extensively STM must be reset before an appropriate response to the test pattern can be elicited. Analogously, reaction time is an increasing function of P300 size (Squires *et al*., 1976). Finally, Section 36 shows that a given population can code a sequence of events, that longer sequences can be coded by more cellular sites, at least up to some maximal length, and that these long-sequence codes are harder to inhibit than short-sequence codes. This leads to the prediction that resetting the population of a long sequence can take longer than resetting the population of a short sequence, and that P300 should again be a monotone increasing function of reaction time. Analogous data have been reported by Remington (1969) and Squires *et al*. (1976).

Various other properties of adaptive resonances can be cited, but go beyond the scope of this paper. One of the deepest and least understood concerns the periodic oscillations of STM activity that can arise in an adaptively resonating network. These oscillations are due to response lags of the inhibitory cells that generate off-surround signals relative to the excitatory cells that generate on-center signals. These periodic oscillations are formal analogs of brain rhythms in the model. They are discussed in Grossberg (1977a). Section 35 discussed a possible effect of these slow waves on interference due to delayed auditory feedback.

24. Hierarchical Critical Periods and Retrograde Amnesia

Several fundamental points should be emphasized about the role of feedback in stabilizing the STM code. The coded data are not locally defined; they is expressed by patterns across many cells. Hence no local signal is sufficient to determine *what* is being coded, or whether the code is a useful one behaviorally. Nonetheless, Grossberg (1974, Section VII) shows that each network cell that is capable of LTM is a chemical dipole. This dipole can, in principle, determine when LTM-like activity is turned on, and can therefore turn it off a fixed time later using an internal cellular clock. Once this happens, no reset of LTM can occur. Switching off LTM changes would end the cell's critical period of developmental plasticity.

Such a clock mechanism would seem to be satisfactory only if the coded data do not depend on individual experience. Within fields $\mathscr{F}^{(i)}$ that are at least partly determined by experience, I prefer the view that LTM stabilization by a clock is supplemented, or even supplanted, by signals from a higher-order field $\mathscr{F}^{(i+1)}$ which has learned templates of feedback signals. The present paper argues that this feedback mechanism is useful in generating a self-consistent code whether or not a chemical clock exists. Moreover, it explains several facts parsimoniously: why it takes a while for LTM in $\mathscr{F}^{(i)} \rightarrow \mathscr{F}^{(i+1)}$ synapses to become stabilized (critical period); why the highest-order code remains plastic; why lower-order codes can be reset by sufficient disconfirmation despite their stability; why the oldest memories are the most stable, since they are the best buffered by feedback; and thus (one reason!) why newer memories are often easier to erase than older ones (retrograde amnesia).

Given the above results on adaptive coding, the question of how sequences of events can be coded by prescribed populations can be broached.

25. Invariance of the Past Code under Future Sequential Inputs

We now make a series of crucial observations. Suppose as in Fig. 24 that a certain number of states $v_1^{(1)}, v_2^{(1)}, \ldots, v_i^{(1)}$ have already received serial inputs, and therefore have positive STM traces. At a given time, these traces define a spatial pattern. In other words, the temporal order of inputs is coded spatially. Imagine that $\mathscr{F}^{(1)}$ sends trainable signals to $\mathscr{F}^{(2)}$, as in the coding model of Sections 17 through 19. Then a certain population in $\mathscr{F}^{(2)}$ will be excited by this particular spatial pattern, and will reverberate in STM. Suppose that the same states had been perturbed by a different serial ordering. Then the activities of the populations $v_1^{(1)}, v_2^{(1)}, \ldots, v_i^{(1)}$ would be permuted, and a very different population in $\mathscr{F}^{(2)}$ could be excited by the new spatial pattern. In other words, the spatial pattern of activity across these cells carries temporal, or order, information, whereas the adaptive coding process whereby each $v_i^{(1)}$ is excited carries item information. The fact that different spatial patterns on the same set of cells in $\mathscr{F}^{(1)}$ can excite different cells in $\mathscr{F}^{(2)}$ is a crucial one in our analysis. Knowing only *that* the cells $v_1^{(1)}, v_2^{(1)}, \ldots, v_i^{(1)}$ are excited does *not* determine the order in which they were excited, nor which cells in $\mathscr{F}^{(2)}$ will fire. The enormous loss of information when a graded spatial pattern is replaced by a binary on–off code weakens algebraic theories of sequential coding by preventing activity per se from generating order information. Our parallel code for order information must then be replaced by a serial code that depends on an item's position in a serially organized buffer (cf. Atkinson and Shiffrin, 1968). As soon as item codes must move through a buffer, one is beleaguered by problems concerning how the code shifts along the buffer, how the code can elicit the same motor output at different buffer positions, and how a given buffer position can elicit the motor output of every item. None of these difficulties arises in the present theory. More generally, the entire design of the system must be changed when its code is changed.

Once a sequence $v_1^{(1)}, v_2^{(1)}, \ldots, v_i^{(1)}$ is already presented, its spatial pattern represents "past" order information. Presenting a new input to $v_{i+1}^{(1)}$ can reorganize the *total* pattern of coded STM activity at $\mathscr{F}^{(2)}$, but we shall assume that it does not recode that part of this coded activity which involves only past order information. In other words, new inputs can weaken the strength of past codes, but do not deny the fact that the past events did occur. If this were not the case, the code in $\mathscr{F}^{(2)}$ would become very unstable through time, since every little perturbation at $\mathscr{F}^{(1)}$ could destroy the entire past record of events. We therefore impose the following basic constraint on STM coding.

Invariance Principle. The spatial patterns at $\mathscr{F}^{(1)}$ are generated by sequential inputs in such a way as to leave the codes in $\mathscr{F}^{(2)}$ of past sequences invariant.

This principle implies rules for generating spatial patterns across $\mathscr{F}^{(1)}$ in response to input sequences. Suppose that at time t, the STM activities at $v_1^{(1)}, v_2^{(1)}, \ldots, v_i^{(1)}$ form a pattern $P_i(t) = (x_1^{(1)}(t), x_2^{(1)}(t), \ldots, x_i^{(1)}(t))$. By (18), the signal to any cell $v_j^{(2)}$ in $\mathscr{F}^{(2)}$ is $S_j(t) = P_i(t) \cdot z^{(j)}(t)$. In order not to recode *any* of these populations in response to this sequence, $P_i(t)$ must not change its direction, as a Euclidean i-vector, at later times t. In other words,

$$P_i(t) = P_i(T) f(t, T) \tag{25}$$

where $f(t, T)$ is a scalar function of $t \geq T$. The invariance principle thus implies that, after $x_1^{(1)}, x_2^{(1)}, \ldots, x_i^{(1)}$ are excited by sequential inputs, they thereafter undergo proportional changes through time. Table II describes rules for generating such changes. It displays STM activities at discrete times which are synchronized with the input presentation rate. In Table II, μ_i is the activity of the last item to be presented. Thus, when $v_1^{(1)}$ is first excited, it reaches activity μ_1. Then $v_2^{(1)}$ is excited with activity μ_2, and $v_1^{(1)}$'s activity is changed from μ_1 to $\mu_1\omega_2$ to satisfy the Invariance Principle. Then $v_3^{(1)}$ is excited with activity μ_3. The "past" information is scaled down by ω_3 to yield activity $\omega_3\omega_2\mu_1$ at $v_1^{(1)}$ and activity $\omega_3\mu_2$ at $v_2^{(1)}$. And so on, until an item's STM trace falls below the QT, whence it is quenched. Physical intuition suggests, at least in cases where all the list items are equivalent in all important respects, that μ_k and ω_k are functions of the total activity in the $(k-1)$st time frame, and of a parameter ν that represents input strength; that is,

$$\mu_k = U(\nu, S_{k-1}) \tag{26}$$

and

$$\omega_k = V(\nu, S_{k-1}) \tag{27}$$

TABLE II
INVARIANCE PRINCIPLE CONSTRAINS POSSIBLE SEQUENTIAL STM VALUES

	$x_1^{(1)}$	$x_2^{(1)}$	$x_3^{(1)}$	$x_4^{(1)}$
$t \cong 1$	μ_1	0	0	0
$t \cong 2$	$\mu_1\omega_2$	μ_2	0	0
$t \cong 3$	$\mu_1\omega_2\omega_3$	$\mu_2\omega_3$	μ_3	0

where the total activity of the STM field in the jth time frame is

$$S_j = \sum_{m=1}^{j} \mu_m \prod_{r=m+1}^{j} \omega_r \tag{28}$$

Since these relations are merely approximate guides to construct the networks that actually perform the computations, we content ourselves with discussing some special cases that provide important physical insights.

First we note what the rules do *not* say. They do not say whether or not STM activity at $\mathscr{F}^{(1)}$ spontaneously decays. Spontaneous decay is permitted, but not required. For example, if the $\mathscr{F}^{(1)}$ activities exponentially decay, then (25) is satisfied, with $f(t, T) = \exp[-A(t - T)]$. If no decay occurs, then all changes in $\mathscr{F}^{(1)}$ are induced by the presentation of new items. Individual cells in $\mathscr{F}^{(2)}$ then code order information, but not the time spacing between items. If spontaneous decay does occur, then different time spacings of the same items in the same order can generate different spatial patterns across $\mathscr{F}^{(1)}$. When this occurs, individual cells in $\mathscr{F}^{(2)}$ can code a mixture of order and spacing information.

If two different input sequences at $\mathscr{F}^{(1)}$ create similar (distinct) STM patterns at $\mathscr{F}^{(2)}$, then the "distance" between these sequences is small (large). The above remarks illustrate how different distance functions on the space of input sequences at $\mathscr{F}^{(2)}$ can be generated by different decay rules for $\mathscr{F}^{(1)}$.

26. Bowing of the STM Pattern

Two extreme cases of the sequence-generating rules will now be considered. More realistic cases will often be mixtures of the two extremes, as Section 27 illustrates.

Case I. Let items be presented at a fixed rate such that the most recent item has unit strength, and each item's strength decays at a fixed rate in time. Then all $\mu_i = \mu$ and $\omega_i = \omega$. Set $\mu = 1$ for convenience. The distribution in Table IIIA is found. This distribution describes the situation in Fig. 24a, in which new items can excite their states with a strength that is independent of the amount of prior STM activity in the field.

A very different result occurs if total field activity is normalized, say due to a recurrent on-center off-surround network. Let all $\omega_i = \omega$, and choose the total activity equal to 1. Then Table IIIB is generated. Parts A and B of Table III share certain similarities. In both, the last item in any list of length at least two has the same STM activity no matter how

TABLE III
(A) PASSIVE DECAY OF STM TRACES

	$x_1^{(1)}$	$x_2^{(1)}$	$x_3^{(1)}$	$x_4^{(1)}$
$t \cong 1$	1	0	0	0
$t \cong 2$	ω	1	0	0
$t \cong 3$	ω^2	ω	1	0

(B) NORMALIZATION OF AN STM PATTERN THAT OBEYS THE INVARIANCE PRINCIPLE

	$x_1^{(1)}$	$x_2^{(1)}$	$x_3^{(1)}$	$x_4^{(1)}$	$x_5^{(1)}$
$t \cong 1$	1	0	0	0	0
$t \cong 2$	ω	$1 - \omega$	0	0	0
$t \cong 3$	ω^2	$(1 - \omega)\omega$	$1 - \omega$	0	0
$t \cong 4$	ω^3	$(1 - \omega)\omega^2$	$(1 - \omega)\omega$	$1 - \omega$	0

many prior items are in the list, which makes good intuitive sense. Also, if $0 < \omega < \frac{1}{2}$, the earliest items have the weakest STM strengths.

An important new phenomenon occurs if $\frac{1}{2} < \omega < 1$. Then the STM pattern that is produced by serial inputs can develop a bow without any intervention of LTM. To see this, consider time step $t = 2$. Since $\omega > 1 - \omega$, the *earlier* item has a larger STM trace. If $t = 3$, then $\omega^2 > (1 - \omega)\omega < 1 - \omega$, so that there is a bow in the STM pattern at the second list item. An important parameter is the maximal k such that $\omega^k > 1 - \omega$. Denote it by K. The longest list length for which the first item has a larger STM activity than any other item is then $K + 1$. Every list length $k \leq K + 1$ is said to exhibit a *primacy* effect. Lists of length $k > K + 1$ exhibit a *recency* effect (Fig. 33).

27. Regulation of STM Primacy, Recency, and Bowing by Lateral Inhibition

The size of parameter ω in Table IIIB measures the relative balance between STM maintenance by recurrent intrafield interactions and STM

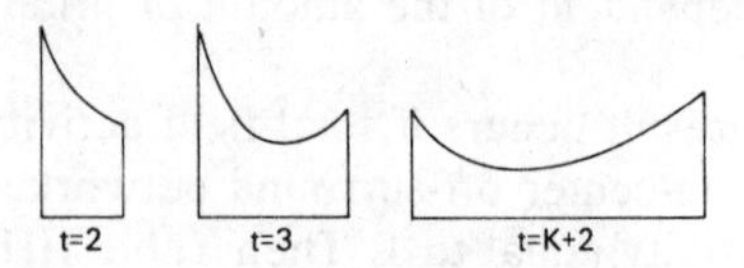

FIG. 33. Bowing of the STM pattern at list length 3, and recency effect at list length $K + 2$.

reset due to inhibition triggered by new inputs. A small value of ω represents powerful reset inhibition. Strengthening reset inhibition decreases ω, which decreases K, which enhances recency by making it easier for the STM traces of recent items to become stronger than the STM traces of early items. In the absence of STM reverberation, only a recency effect is possible, since the earliest items will always have decayed in strength more than recent items. Because the Invariance Principle constrains inhibition to change the past field activities by a multiplicative factor, it argues for *shunting* (mass action) rather than *subtractive* inhibition.

In general, an STM pattern that mixes properties of Tables IIIA and IIIB can be expected. This can happen, for example, if normalization holds only partially. In Eq. (16), normalization holds because each $v_i^{(1)}$ in $\mathscr{F}^{(1)}$ excites only itself and inhibits *all* other $v_k^{(1)}$, $k \neq i$, with equal strength. Often the strength of recurrent excitation and inhibition from $v_i^{(1)}$ to other populations $v_k^{(1)}$ decreases with the interpopulation distance (cf. Ellias and Grossberg, 1975; Levine and Grossberg, 1976). These distance-dependent connections form the anatomical substrate of the generalization gradients that join feature detectors together. If a given $v_i^{(1)}$ does not inhibit *all* other $v_k^{(1)}$, then certain $v_k^{(1)}$ can build excitation without interference from $v_i^{(1)}$, so that the normalization property is weakened. If partial normalization holds, then a bow in the STM pattern can coexist with a gradual increase in total STM activity as more items are presented. (Also recall the multiple STM equilibrium points in Fig. 30.) Furthermore, the bow does not, in general, have to occur at the second list item. By comparison, the statistical model of Hogan and Hogan (1975) for structural and transient components of memory describes data in which the transient memory distribution is bowed at list positions other than the second (see their Fig. 3).

We now solve a coding process that marries the Invariance Principle with the Partial Normalization Property to show that the bow can occur at any list position if system parameters are properly chosen. More precisely, consider codes for which:

1. The Invariance Principle holds.
2. The last item to be presented has an STM strength that is independent of list length.
3. The total STM activity grows in a negatively accelerated fashion from an initial value μ to a finite asymptotic value M that is independent of list length, but not necessarily of the stimulus materials from which a given class of lists is constructed.

Property 2 is a natural approximation because, no matter how many past items have occurred, it should still be possible to attend to the most recent item under normal stimulus conditions. Postulate 3 merely interpolates the simplest continuous curve between the initial and asymptotic values of total STM strength. Independent evidence that total STM strength has such a qualitative curve is found in studies of pupillary dilation in short-term memory tasks (Kahneman and Beatty, 1966). The properties of this code are described by the following theorem.

Theorem 1 (STM Buffer): Let all $\mu_i = \mu$. Let the total STM strength in a list of length i satisfy

Invariance Principle:

$$S_i = \mu \sum_{m=1}^{i} \prod_{r=m+1}^{i} \omega_r \tag{29}$$

as in Table II, as well as

Partial Normalization:

$$S_i = \lambda^{i-1}\mu + M(1 - \lambda^{i-1}) \tag{30}$$

for some λ such that $0 < \lambda < 1$, and $\mu < M$. Then letting $R = \mu^{-1}M$, for every $i > 1$,

$$\omega_i = \frac{\lambda^{i-1} + R(1 - \lambda^{i-1}) - 1}{\lambda^{i-2} + R(1 - \lambda^{i-2})} \tag{31}$$

and the STM strength in a list of length j of the kth item, $k < j$, is

$$x_{kj} = \mu \prod_{m=k+1}^{j} \omega_m \tag{32}$$

Every STM pattern $(x_{1j}, x_{2j}, x_{3j}, \ldots, x_{jj})$ is either monotone decreasing, monotone increasing, or bowed. The longest list length J for which the STM pattern is monotone decreasing is given by the maximal j such that

$$(R - 1)(1 - \lambda)\lambda^{j-2} > 1 \tag{33}$$

In every list of length greater than J, the bow occurs at list position J.

The proof is found by equating (29) and (30) and solving for ω_2, $\omega_3, \ldots$, etc., by iteration. The theorem shows that a bow can occur at *any* list position J if R and λ are properly chosen. Also, given this code and prescribed stimulus materials that are homogeneous with respect to one another, the bow always occurs at the same list position, independent of list length. Moreover, since the last item always has strength μ, and the total STM strength never exceeds M, a strong recency effect develops as the list length becomes large. This result is generalized in Grossberg (1978b).

Do experimental operations that change the relative strengths of arousal and inhibition determine whether and where an STM bow occurs *in vivo*? Grossberg and Pepe (1971) show that variations in arousal level can change the list position where the bow occurs in network LTM traces (cf. Section 12). Theorem 1 suggests that arousal can change the list position at which the network's STM pattern bows by shunting inhibitory interaction strengths. Indeed, if some of the weights ω_i exceed 1, then STM at a given population can *increase* through time. This is a form of behavioral contrast due to lateral inhibition; cf. Grossberg (1975, Section 12), where peak shift and behavioral contrast of a generalization gradient also are explained by shunting inhibition. These network results suggest that performance variables, such as motivational or attentional state, can influence information processing constraints, such as primacy and recency.

28. Feedback Inhibition by Rehearsal in an Opaque STM Field

We now broach the question of how order information is read out of an STM pattern. To motivate the discussion, consider the task of repeating a telephone number that you have just heard. At no time are all the digits simultaneously rehearsed, and there can exist times during which no digit is consciously in mind. Moreover, the telephone number can be rehearsed at various rates, which can be controlled at will, within limits. At times when no digit is consciously available, the sequence of digits is *opaque* to the individual (Estes, 1972). Somehow the STM buffer organizes order information so that, when a rehearsal act perturbs the buffer, the correct item is elicited. Because the STM code is opaque, rehearsal *nonspecifically* activates all the possible item representations. The internal organization of activity patterns in the buffer codes the order information, and the nonspecific activation translates this activity into output signals. This use of arousal is analogous to its use in the avalanche of Fig. 12. There also arousal controls the readout of ordered signals.

Once a given item is rehearsed, the buffer must reset its activity so that the next item can be rehearsed; otherwise, the nonspecific rehearsal wave would cause an endless repetition of the first item. In some way, rehearsal of an item deletes its STM trace from the buffer (cf. Section 9).

The new STM activity pattern is then stored (or decays, etc.) until the next rehearsal wave perturbs it.

Consider the case in which order information is stored by a spatial pattern. Let each population v_i in a field $\mathscr{F}$ code a certain command. The relative sizes of the STM activities x_i then determine in what order these commands will be elicited. Suppose that $x_1 > x_2 > x_3 > \cdots > x_n$. Whenever a nonspecific rehearsal wave perturbs $\mathscr{F}$, all the populations begin to emit signals to the next processing stage. The most active population v_1 reaches its firing threshold soonest, so that it begins to fire earliest. It is likely that emergent signals can inhibit each other via feedforward on-center off-surround interactions. This would prevent leakage of signals corresponding to later list items. Using this mechanism, v_1 wins the lateral inhibitory competition, since its STM trace x_1 is largest. Thus v_1 firest its command signal first. As v_1 fires, it also activates a feedback inhibitory signal to its STM source. This feedback inhibition continues to act until it self-destructs by quenching suprathreshold STM activity at v_1, but not necessarily subthreshold activity (Fig. 34). After v_1 is deactivated, the population v_2 has the largest STM activity. It can therefore fire signals through the feedforward on-center off-surround network, and it continues to do so until it self-destructs via feedback inhibition. The process continues until either all items are rehearsed, or arousal is terminated.

29. Transient Memory Span and Free Recall

Given the above rehearsal mechanisms, items that are presented earlier must have larger STM activities in order to be rehearsable in their

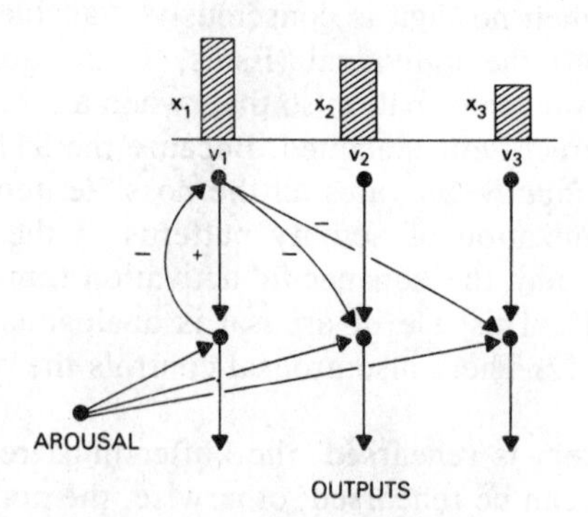

FIG. 34. STM traces compete before the maximal trace elicits a signal and self-inhibits via feedback inhibition. The process then repeats itself.

correct order. Below we first consider how this can happen if no LTM effects occur. That is, we consider only the primary effect of serial inputs on the STM pattern. By Sections 26 and 27, an earlier item can then have a larger STM activity only if inputs to the field are influenced by lateral inhibitory signals from previously stored items; without at least partial normalization, earlier items will always have smaller STM traces. In order to be reliably rehearsed in its correct order, after just one presentation, using only transient STM data, a list must be so short that it has not yet bowed. We define the *transient memory span* (TMS) to be the length of the longest list with a monotone decreasing transient STM pattern. This length is J in Theorem 1. The TMS can depend on the materials from which the list is constructed, since these will determine the strength of inputs and the distribution of lateral inhibition among items, as well as performance variables, such as the magnitude of shunting arousal.

If a list is so long that STM bowing takes place, then there will be a tendency to guess items near the beginning and the end of the list before guessing items in the middle, as is often observed in free recall experiments (Hogan and Hogan, 1975, Fig. 3). As ever longer lists are chosen, a strong recency effect develops, as Theorem 1 illustrates.

30. Parallel versus Serial Search in STM

The activation of order information using a nonspecific rehearsal wave is a parallel operation. In a normalized STM field, however, each item in the field—except perhaps the last—has a smaller STM trace if a longer list perturbs the field. Since an item's reaction time depends on its STM activity, as in the avalanche of Section 5, reaction time can vary with list length. This fact has often been used to support the idea that recognition memory is realized by a serial scanning process that exhaustively searches all stored items (Sternberg, 1966). Townsend (1974) has pointed out that the longer reaction times do not unambiguously implicate a serial process, by describing statistical parallel processing completion times that are indistinguishable from their serial processing counterparts. Whereas items are read out of the field in a prescribed order, the field operations that accomplish this are parallel rather than serial operations. Sections 60 and 61 discuss the questions of reaction time and (apparently) serial processing in greater detail.

31. The Influence of Rehearsal on Chunking

Given that rehearsal deletes an item from STM, it follows that rehearsal organizes which combinations of events will generate codes, or chunks. Naive experience also suggests that rehearsal should have such an effect, since rehearsing a particular combination of events defines that combination as a behaviorally meaningful unit.

Imagine that a series of events sequentially activates populations $v_{i_1}^{(1)}, v_{i_2}^{(1)}, \ldots, v_{i_k}^{(1)}$ before these items are rehearsed. These events establish a spatial pattern of STM activity in $\mathscr{F}^{(1)}$. This spatial pattern generates $\mathscr{F}^{(1)} \rightarrow \mathscr{F}^{(2)}$ signals which begin to code a population $v_{i_1 i_2 \ldots i_k}^{(2)}$ in $\mathscr{F}^{(2)}$. Simultaneously, $v_{i_1 i_2 \ldots i_k}^{(2)}$ can send signals to $\mathscr{F}^{(1)}$ which learn the spatial pattern of activity across $v_{i_1}^{(1)}, v_{i_2}^{(1)}, \ldots, v_{i_k}^{(1)}$. This pattern represents a sensory expectation, as shown in Section 20. What happens when a rehearsal wave perturbs $\mathscr{F}^{(1)}$? If $k <$ TMS, then $x_{i_1} > x_{i_2} > \cdots > x_{i_k}$, so that the individual items can be rehearsed in their proper order. Two remarks are pertinent here. First, by grouping items into sublists of length less than the TMS, it *is* possible to rehearse them in their correct order. Second, we must ask how chunk $v_{i_1 i_2 \ldots i_k}^{(2)}$ learns to reproduce this order. Does the chunk activate order information only via its learned signals to $\mathscr{F}^{(1)}$ (sensory expectation), or can the chunk directly sample motor representations? The next sections will show that the latter alternative must hold. After the chunk $v_{i_1 i_2 \ldots i_k}^{(2)}$ is adaptively coded and learns to reproduce the order of its coded sequence, it acts as a new functional unit of the network.

After all the items $v_{i_1}^{(1)}, v_{i_2}^{(1)}, \ldots, v_{i_k}^{(1)}$ are rehearsed, suppose that another series of events occurs which serially activates the populations $v_{i_{k+1}}^{(1)}, v_{i_{k+2}}^{(1)}, \ldots, v_{i_{k+m}}^{(1)}$. Since the first k items no longer reverberate in STM, signals from $\mathscr{F}^{(1)}$ to $\mathscr{F}^{(2)}$ will begin to adaptively code a chunk $v_{i_{k+1} i_{k+2} \ldots i_{k+m}}^{(2)}$ that depends only on the second series of events. Simultaneously $v_{i_{k+1} i_{k+2} \ldots i_{k+m}}^{(2)}$ begins to learn the sensory expectation that characterizes the second series of events. Suppose that $m <$ TMS, so that the second series of events can be rehearsed in its correct order. As this happens, each item's STM representation is deleted and the chunk learns order information by sampling the items' motor representations.

Two important phenomena occur together in this scheme. Grouping items into lists of length less than than the TMS allows them to be rehearsed in their correct order, and simultaneously defines a code for the list that is capable of learning to perform the list in its correct order. Time enters this mechanism in a subtle fashion. Rehearsal occurs *after* the time interval in which $v_{i_1 i_2 \ldots i_k}^{(2)}$ is coded by its defining sequence and prevents *future* events from being chunked with this sequence. However, the rehearsal act presents $v^{(2)}_{i_1 i_2 \ldots i_k}$ with the data for sampling that enables it to control overt behavior on future trials.

32. Immediate Memory Span, and Readout of LTM Order Information by Feedback Signals from Commands

Section 27 shows that a bowed STM pattern can be elicited by serial inputs across a field $\mathscr{F}^{(1)}$ whose populations are joined by a recurrent on-center off-surround network. Section 12 shows that an STM bow can also be elicited by a combination of serial inputs and feedback signals. These feedback signals are generated by a field $\mathscr{F}^{(2)}$ and are gated on their way to $\mathscr{F}^{(1)}$ by LTM traces. Section 17 shows that the $\mathscr{F}^{(2)}$ populations that generate the feedback signals to $\mathscr{F}^{(1)}$ can be activated by STM patterns at $\mathscr{F}^{(1)}$. The adaptive coding process that accomplishes this also uses LTM traces. The process whereby event sequences at $\mathscr{F}^{(1)}$ are coded at $\mathscr{F}^{(2)}$, and codes at $\mathscr{F}^{(2)}$ learn order information at $\mathscr{F}^{(1)}$, is clearly a special type of adaptive resonance. The order information is the expectation, or template, of the resonance.

Before further analyzing this adaptive coding process, we can generalize the serial learning model of Section 12 by using the Invariance Principle. We want to see what kinds of feedback patterns from $\mathscr{F}^{(2)}$ to $\mathscr{F}^{(1)}$ can arise. In particular, under what circumstances does a population $v_i^{(2)}$ correctly code the order with which populations in $\mathscr{F}^{(1)}$ were excited? This code is carried by the LTM pattern $(Z_{i1}, Z_{i2}, \ldots, Z_{in})$ of stimulus sampling probabilities Z_{ij} from $v_i^{(2)}$ to $v_j^{(1)}$.

To discuss this problem, we introduce some convenient nomenclature. Consider the LTM pattern $(Z_{i1}(t), Z_{i2}(t), \ldots, Z_{in}(t))$ at any time t. Let $j_i(t)$ be the smallest j such that $Z_{ij}(t) > 0$, and let $J_i(t)$ be the largest j such that $Z_{ij}(t) > 0$. The integers $j_i(t)$ and $J_i(t)$ define the range of positive $\mathscr{F}^{(1)}$ activities that $v_i^{(2)}$ has sampled by time t. Restrict attention to the LTM pattern formed by $(Z_{ij}(t)\colon j_i(t) \le j \le J_i(t))$. If the function $M(j;\, i,\, t) \equiv Z_{ij}(t)$, for fixed i and t, is monotone decreasing in j, for $j_i(t) \le j \le J_i(t)$, we say that $v_i^{(2)}$'s LTM pattern is *monotone decreasing* (Fig. 35a). If $M(j;\, i,\, t)$ has a single maximum, neither at $j_i(t)$ or at $J_i(t)$, we say that $v_i^{(2)}$'s LTM pattern is *unimodal* (Fig. 35b); if $M(j;\, i,\, t)$ has two local maxima, we say that $v_i^{(2)}$'s LTM pattern is *bimodal* (Fig. 35c). Population $v_i^{(2)}$ codes the order in which a list, or sublist, perturbs $\mathscr{F}^{(1)}$ only if its LTM pattern is monotone decreasing at all times after the list has been presented.

At any given time, these LTM patterns influence $\mathscr{F}^{(1)}$ by gating signals

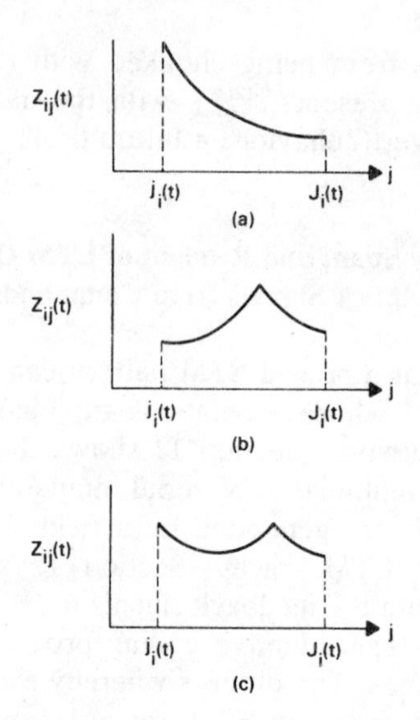

FIG. 35. (a) Monotone decreasing LTM pattern; (b) unimodal LTM pattern; (c) bimodal LTM pattern.

from $\mathscr{F}^{(2)}$. Let $F_j(t)$ be the *total* feedback signal from $\mathscr{F}^{(2)}$ to $v_j^{(1)}$ at time t. Signal $F_j(t)$ perturbs $v_j^{(1)}$ and thereupon influences the STM trace $x_j^{(1)}$ of $v_j^{(1)}$. It is possible for the pattern $(F_1(t), F_2(t), \ldots, F_n(t))$ to code the order in which the list was presented, even though the LTM patterns of individual $v_i^{(2)}$ do not. That is, the total feedback pattern can be monotone decreasing, with $F_1(t) > F_2(t) > \cdots > F_n(t)$, even though certain LTM patterns $(Z_{i1}, Z_{i2}, \ldots, Z_{in})$ are not monotone decreasing. Each $v_i^{(2)}$ has a *local view* of the serial experiment at $\mathscr{F}^{(1)}$. The *global synthesis* of all these views is expressed by the signal pattern $(F_1(t), F_2(t), \ldots, F_n(t))$, which controls behavior due to $\mathscr{F}^{(2)}$ at $\mathscr{F}^{(1)}$. Thus we shall ask what types of adaptive resonances code correct order information.

In Section 29, the concept of transient memory span (TMS) was introduced to discuss the effect of serial inputs on the STM pattern at $\mathscr{F}^{(1)}$. When feedback signals from $\mathscr{F}^{(2)}$ also perturb $\mathscr{F}^{(1)}$, they can change the list length at which the total STM pattern of $\mathscr{F}^{(1)}$ becomes bowed. We define the *immediate memory span* (IMS) to be the maximal list length at which the STM pattern at $\mathscr{F}^{(1)}$ is monotone decreasing when feedback signals from $\mathscr{F}^{(2)}$ are operative.

Our first results apply to the case wherein a list is presented once to the network, which thereupon tries to repeat it in its correct order. We study the LTM patterns that develop in $\mathscr{F}^{(2)} \to \mathscr{F}^{(1)}$ synapses, but do not yet let these patterns influence STM at $\mathscr{F}^{(1)}$. The first result shows that the IMS can only be longer than the TMS, under weak conditions. In effect, feedback signals can only make it easier for lists to be repeated in their correct order.

Theorem 2 (Primacy). Let $\mathscr{F}^{(1)}$ obey the Invariance Principle. Also let the STM activities of populations in $\mathscr{F}^{(2)}$ decrease when new inputs perturb $\mathscr{F}^{(1)}$. Then the IMS is at least as long as the TMS, because *every* active $v_i^{(2)}$ codes a monotone decreasing LTM pattern after a list no longer than the TMS is presented to $\mathscr{F}^{(1)}$. In fact, this conclusion holds if the Invariance Principle is replaced by the weaker conditions

$$x_i^{(1)}[(i+m)\tau] \geq x_{i+k}^{(1)}[(i+k+m)\tau], \qquad k > 0, \quad m \geq 0 \tag{34}$$

where τ is the intratrial interval.

If the list is longer than the TMS, then not all LTM patterns can be monotone decreasing. For example, in Fig. 35b a population $v_i^{(2)}$ that is excited after many list items have been presented will code more recent items more strongly than early items. The next result generalizes the statements of Section 12 about STM → LTM order-reversal to the case where $\mathscr{F}^{(1)}$ obeys the Invariance Principle. After that, we shall suppose that $\mathscr{F}^{(2)}$ obeys the Invariance Principle to show how feedback signals can lengthen the IMS. Theorem 3 is conveniently stated in terms of the following definition.

Definition. The *past (future) field* of $v_i^{(2)}$ is the set of populations $v_j^{(1)}$ that are excited by inputs before (after) $v_i^{(2)}$ is excited.

Theorem 3 (Generalization Gradients). Let $\mathscr{F}^{(1)}$ obey the Invariance Principle. Let its invariant parameters $\mu_j^{(1)}$ and $\omega_k^{(1)}$ satisfy

$$\mu_1^{(1)} \geq \mu_2^{(1)} \geq \mu_3^{(1)} \geq \cdots \tag{35}$$

and

$$\omega_2^{(1)} \geq \omega_3^{(1)} \geq \cdots \tag{36}$$

Also let the STM activities of populations in $\mathscr{F}^{(2)}$ decrease when new inputs perturb $\mathscr{F}^{(1)}$. Let a list of length L serially perturb $\mathscr{F}^{(1)}$. If $L \leq$ TMS, then every active $v_i^{(2)}$ population has a monotone decreasing LTM pattern. If $L >$ TMS, then the LTM pattern of each $v_i^{(2)}$ is either monotone decreasing, unimodal, or bimodal. In all cases, each $v_i^{(2)}$ codes in LTM the ordering in its past field, and a monotone decreasing LTM of its future field. In particular, the LTM pattern of $v_1^{(2)}$ can be monotone decreasing no matter how large L is, or how small the TMS is, if $v_1^{(2)}$ is excited before $v_2^{(1)}$ is excited (primacy). A unimodal pattern can be generated only if the past field of $v_i^{(2)}$ is monotone increasing, either because there is no lateral inhibition within $\mathscr{F}^{(1)}$, or because the STM traces of the TMS items have become subthreshold. The maximum Z_{ij} then occurs at the $v_j^{(1)}$ that is excited most simultaneously with $v_i^{(2)}$. A bimodal pattern occurs if the past field of $v_i^{(2)}$ has a bowed STM pattern. Then one local maximum occurs at Z_{i1}, and the other occurs as in the unimodal case.

The above theorem makes no assumption about $\mathscr{F}^{(2)}$, except that its STM activities decrease when new inputs perturb $\mathscr{F}^{(1)}$, owing either to competitive inhibition by newly activated $\mathscr{F}^{(2)}$ populations, or to passive decay. It is natural to assume, moreover, that $\mathscr{F}^{(2)}$ obeys the invariance principle, if only because $\mathscr{F}^{(2)}$ can be a source of adaptively coded inputs to another field in a hierarchy of codes, and a stable coding of the past field in each level of the hierarchy must be guaranteed. Section 41 will show, however, that $\mathscr{F}^{(2)}$ cannot in general be constructed from a single, homogeneous pool of populations; several distinct population types can exist in a single field. In the simple case that only one population type exists, $\mathscr{F}^{(2)}$ possesses a single set of invariant parameters $\mu_j^{(2)}$, $j = 1, 2, \ldots$, and $\omega_k^{(2)}$, $k = 2, 3, \ldots$. In the next, more general case, each population type, or subfield, of $\mathscr{F}^{(2)}$ possesses its own set of invariant parameters. By defining invariant parameters for $\mathscr{F}^{(2)}$, we temporarily sidestep the adaptive coding problem. Instead of studying how $\mathscr{F}^{(2)}$ populations are activated by the ith serial input to $\mathscr{F}^{(1)}$, we demand that the ith population $v_i^{(2)}$ in $\mathscr{F}^{(2)}$ is excited according to the rule of Table II. Given this rule, we can study the total feedback pattern $(F_1(t), F_2(t), \ldots, F_n(t))$ as it evolves through time. In particular, we can study how the feedback pattern tends to produce a primacy effect that balances the recency effect that is produced directly at $\mathscr{F}^{(1)}$ by a long list.

Whenever $\mathscr{F}^{(2)}$ has one population type that obeys the Invariance Principle, it also possesses a transient memory span (TMS_2). Denote the transient memory span of $\mathscr{F}^{(1)}$ by TMS_1 to avoid confusion. To understand how $\mathscr{F}^{(2)}$ can lengthen the IMS of $\mathscr{F}^{(1)}$, suppose that $TMS_2 \geq TMS_1$. This makes sense intuitively, because often the populations in $\mathscr{F}^{(2)}$ will represent commands that stay active for a long time in order to sample long sequences of events. Consequently the STM pattern in $\mathscr{F}^{(2)}$ can be monotone decreasing long after the STM pattern in $\mathscr{F}^{(1)}$ has bowed. In particular, the populations that are excited earlier in $\mathscr{F}^{(2)}$ will have greater STM activity than those excited later on. These early populations also tend to code a monotone decreasing LTM pattern, by Theorems 2 and 3. Thus, they sample a monotone decreasing STM pattern at $\mathscr{F}^{(1)}$ until the list exceeds TMS_1. After that time, populations such as $v_1^{(2)}$ continue to sample a monotone decreasing LTM pattern. The later $\mathscr{F}^{(2)}$ populations have smaller STM activities, so they sample the bowed STM activities at $\mathscr{F}^{(1)}$ less vigorously. Since $\mathscr{F}^{(2)}$ has a long TMS_2, the STM signals from the early populations in $\mathscr{F}^{(2)}$ will be stronger than those from $\mathscr{F}^{(2)}$ populations that are excited later on. Hence the *total* feedback signal from $\mathscr{F}^{(2)}$ will more heavily weight the LTM patterns coded by the early $\mathscr{F}^{(2)}$ populations. This tends to make the $(F_1(t), F_2(t), \ldots, F_n(t))$ pattern monotone decreasing (primacy effect). When these feedback signals act at $\mathscr{F}^{(1)}$, they tend to make the IMS longer than the TMS.

Another point of interest can be made here before it is developed in Section 61. The STM buffer at $\mathscr{F}^{(1)}$ can, in principle, store a list much longer than the TMS, albeit with STM activities that code incorrect order information. Given any search task that must be performed under time pressure before STM is reset, many of these items can be masked by items with larger STM activities.

33. A Minimal Model of Structural versus Transient Components of Memory

The above analysis suggests how feedback signals from $\mathscr{F}^{(2)}$ can induce a primacy effect at $\mathscr{F}^{(1)}$ even if presenting a long list to $\mathscr{F}^{(1)}$ tends to produce a recency effect. The total feedback pattern $(F_1(t), F_2(t), \ldots, F_n(t))$ induces an STM response at $\mathscr{F}^{(1)}$ that can be called the *structural* component of memory. It is that part of the total input to $\mathscr{F}^{(1)}$ that is controlled by LTM, whether via $\mathscr{F}^{(1)} \rightarrow \mathscr{F}^{(2)}$ adaptive coding, or via $\mathscr{F}^{(2)} \rightarrow \mathscr{F}^{(1)}$ readout of order information. By contrast, the *transient* component of memory is the STM response at $\mathscr{F}^{(1)}$ due to serial inputs. Many papers have tried to understand the interplay of structural memory with transient memory (for example, Atkinson and Shiffrin, 1968; Estes, 1972; Hogan and Hogan, 1975). All these theories are weak in at least one respect. None of them gives an explicit description of how STM and LTM patterns are generated, coded, and mutually transform one another in real time. The present theory suggests a class of minimal models that is capable of approaching this task.

The simplest model discretizes and generalizes Eqs. (7) and (8) of Section 12. This generalization assumes that both $\mathscr{F}^{(1)}$ and $\mathscr{F}^{(2)}$ obey the Invariance Principle. By Theorem 1, the STM activities $x_j^{(1)}(k)$ and $x_i^{(2)}(k)$ of $v_j^{(1)}$ and $v_i^{(2)}$ at time $t = k$ satisfy

$$x_j^{(1)}(k) = \mu_j^{(1)} \prod_{m=j+1}^{k} \omega_m^{(1)} \tag{37}$$

and

$$x_i^{(2)}(k) = \mu_i^{(2)} \prod_{m=i+1}^{k} \omega_m^{(2)} \tag{38}$$

where $\prod_{m=l}^{k} \omega_m^{(p)} = 0$ if $l > k$, $p = 1, 2, \ldots$. Consequently the LTM trace $z_{ij}(k)$ from $v_i^{(2)}$ to $v_j^{(1)}$ at time $t = k$ satisfies

$$z_{ij}(k) = z_0 + \sum_{m=1}^{k} f(x_i^{(2)}(m))x_j^{(1)}(m)D(k-m) \tag{39}$$

where the signal function $f(w)$ either is a sigmoid function of w, or describes a threshold cutoff at Γ; $D(k-m)$ describes any LTM decay that might occur between time $t = m$ and $t = k$ (cf. Grossberg, 1974, Section IV); and $z_{ij}(0) = z_0$. The total feedback signal from $\mathscr{F}^{(2)}$ to $\mathscr{F}^{(1)}$ at time $t = k$ is

$$F_j(k) = \sum_{i=1}^{N} f(x_i^{(2)}(k))z_{ij}(k) \tag{40}$$

The pattern $(F_1(k), F_2(k), \ldots, F_n(k))$, $k = 1, 2, \ldots$, describes the effect of structural memory on $\mathscr{F}^{(1)}$ at any time $t = k$. The patterns generated at $\mathscr{F}^{(1)}$ according to Table II describe the transient memory through time. A weighted average of the two patterns describes the total STM pattern at $\mathscr{F}^{(1)}$. This latter pattern determines the order in which items will be rehearsed from $\mathscr{F}^{(1)}$ in response to a nonspecific rehearsal wave.

This total STM pattern replaces the probabilities of performance that are used in statistical learning models. In Section 25, we noted that, in the Atkinson–Shiffrin model, a computer analogy suggested a binary code for an item to be either in the STM buffer or not. Then item representations had to move through the buffer to remember order information. The binary code did not, however, meet STM order data. Somehow recency and primacy effects had to be generated. Recency gradients were generated by supposing that there exists a probability for the item to fall out of the buffer. Then the probability that an item is in the buffer decreases as a function of how long ago it entered the buffer (recency). Thus a hybrid mixture of binary and probabilistic concepts was wed together to achieve order information and a recency gradient. However, in each individual this hybrid scheme predicts that an item is either in the buffer with unit strength, or not in the buffer. The recency gradient is an intersubject construct. In the present theory, each item can be in the buffer of a single individual with variable STM activity, and item motion is unnecessary to code order information. These two conceptions can be differentiated by experiments that test whether an item's ability to influence probe stimuli depends on its position in the buffer.

When a nonspecific rehearsal wave perturbs an STM buffer, the buffer's order information manifests itself. This order information is the net effect of all STM and LTM interactions among buffer item representations and higher-order commands. We now turn to the problem of globally synthesizing these interactions to achieve the order needed to perform prescribed tasks. Two general classes of task impose different requirements on the buffer. The task of prediction looks into the future. The task of imitation looks into the past. The task of naming stands somewhere between. By endeavoring to harmonize these demands, a deeper insight into the global structure of the field of command populations is achieved. The prediction task is particularly useful as a probe of this structure.

34. Prediction

Suppose that a pianist has learned to play a long series of chords. After having played several of the chords, how does the pianist know what chords come next? How does playing the previous chords generate commands capable of eliciting the future chords in their proper order? Consider Fig. 36. Suppose that the pianist learns the piece by playing one chord at a time. For the moment, let motor commands for playing each chord already be coded at $\mathscr{F}^{(4)}$. As each chord is played, it generates a sensory feedback pattern at the sensory field $\mathscr{F}^{(1)}$. This pattern is then adaptively coded and stored in STM by a population in $\mathscr{F}^{(2)}$, as in Sections 17 through 19. Intuitively, $\mathscr{F}^{(2)}$ codes spatial patterns of sensory data, or item information.

As a long sequence of chords is elicited by the motor commands $v_1{}^{(4)}$, $v_2{}^{(4)}, \ldots, v_i{}^{(4)}$, their sensory feedback creates a spatial pattern across $\mathscr{F}^{(2)}$. By the Invariance Principle, once this spatial pattern is established, it does not subsequently change. It therefore activates a well-defined command population $v_{12\ldots i}^{(3)}$ in $\mathscr{F}^{(3)}$ by adaptive coding. Intuitively, $\mathscr{F}^{(3)}$ codes temporal sequences of sensory data, or order information. Population $v_{12\ldots i}^{(3)}$ thereupon reverberates in STM and begins to emit sampling signals via trainable pathways. The sampling signals to $\mathscr{F}^{(2)}$ learn a sensory expectation that codes order information. How is motor order information learned? Suppose for definiteness that $\mathscr{F}^{(3)}$ can sample $\mathscr{F}^{(4)}$. This will not ultimately be tenable, and the following argument shows

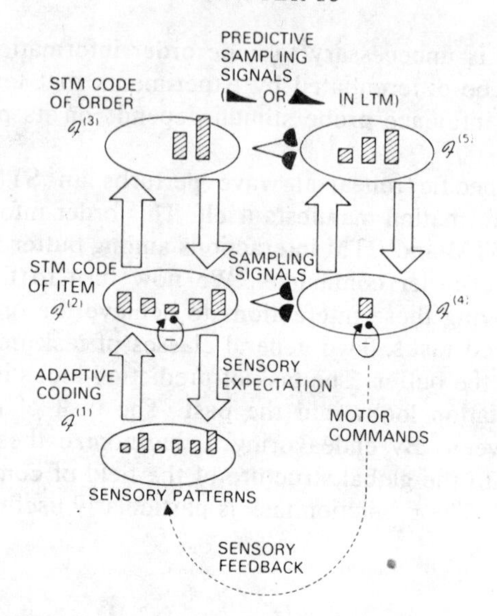

FIG. 36. Auditory feedback to $\mathcal{F}^{(1)}$ is coded at $\mathcal{F}^{(2)}$. Order information at $\mathcal{F}^{(2)}$ is coded at $\mathcal{F}^{(3)}$. Active populations in $\mathcal{F}^{(3)}$ can sample their future field of motor commands at $\mathcal{F}^{(5)}$.

why. Since rehearsal deletes motor representations after they are performed, population $v_{12\ldots i}^{(3)}$ can sample all the future motor commands $v_{i+1}^{(4)}, v_{i+2}^{(4)}, \ldots$ that are activated while it reverberates in STM. By contrast, $\mathcal{F}^{(3)}$ can also sample STM from past events in $\mathcal{F}^{(2)}$, since $\mathcal{F}^{(3)}$ is excited by sensory feedback owing to rehearsal at $\mathcal{F}^{(4)}$ (cf. Section 31). How does $v_{12\ldots i}^{(3)}$ encode the order information of the motor representations $v_{i+1}^{(4)}, v_{i+2}^{(4)}, \ldots$ in its LTM traces?

As in (2), the LTM trace from $v_{12\ldots i}^{(3)}$ to $v_j^{(4)}$ time-averages the product of the signal from $v_{12\ldots i}^{(3)}$ to $v_j^{(4)}$ with the STM trace of $v_j^{(4)}$. This signal is a monotone increasing function $f(w)$ of STM activity $w = x_{12\ldots i}^{(3)}$ at $v_{12\ldots i}^{(3)}$. As in Table II, the STM activity $x_{12\ldots i}^{(3)}$ decreases monotonically through time after it is excited. This is due, for example, to lateral inhibition from other chunks in $\mathcal{F}^{(3)}$ that are activated as new chords are played. Hence the signal $f(x_{12\ldots i}^{(3)})$ also decreases through time. Suppose, moreover, that $\mathcal{F}^{(4)}$ is normalized. Then each motor command at $\mathcal{F}^{(4)}$ has unit STM activity during its brief activation interval. The activation interval is brief because each motor command self-destructs its STM activity via feedback inhibition, as in Section 9. Consequently, the LTM trace from $v_{12\ldots i}^{(3)}$ to $v_{i+1}^{(4)}$ is larger than the trace to $v_{i+2}^{(4)}$, the LTM trace to $v_{i+2}^{(4)}$ is larger than the trace to $v_{i+3}^{(4)}$, and so on. The LTM pattern from $v_{12\ldots i}^{(3)}$ to the set $\{v_{i+1}^{(4)}, v_{i+2}^{(4)}, \ldots\}$ is thus monotone decreasing. On a later performance trial, after the chords $v_1^{(4)}, v_2^{(4)}, \ldots, v_i^{(4)}$ are played, $v_{12\ldots i}^{(3)}$ is activated by sensory feedback and thereupon elicits across $\mathscr{F}^{(4)}$ a monotone decreasing STM pattern that codes the correct order of future chord performance. The network is then ready to play a sequence of future chords. Turning on the motor rehearsal wave at $\mathscr{F}^{(4)}$ releases the chords in their correct order.

35. Sensory Feedback and Interference by Its Delay

The above mechanism works if sensory feedback due to motor performance does not excite other $\mathscr{F}^{(3)}$ populations and thereby change the total pattern of $\mathscr{F}^{(3)} \rightarrow \mathscr{F}^{(4)}$ signals. How is order information organized when sensory feedback continually excites new $\mathscr{F}^{(3)}$ populations? This question will lead to the conclusion that $\mathscr{F}^{(3)}$ samples $\mathscr{F}^{(5)}$ rather than $\mathscr{F}^{(4)}$, where $\mathscr{F}^{(5)}$ codes sequences of motor items just as $\mathscr{F}^{(3)}$ codes sequences of sensory items. To see what goes wrong if $\mathscr{F}^{(3)}$ samples $\mathscr{F}^{(4)}$, consider Fig. 37. Suppose that a given $\mathscr{F}^{(3)}$ population excites a monotone decreasing STM pattern at $\mathscr{F}^{(4)}$. Let motor arousal at $\mathscr{F}^{(4)}$ initiate performance of these items. The sequential motor perform-

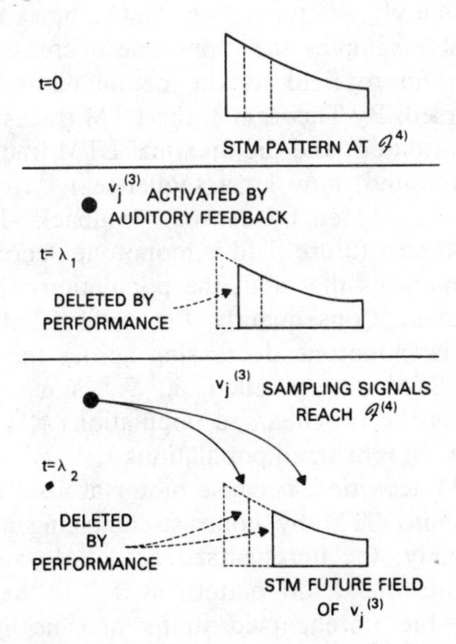

FIG. 37. $\mathscr{F}^{(3)}$ populations sample a monotone decreasing future field irrespective of how long it takes auditory feedback to activate them.

ance generates sensory feedback at $\mathscr{F}^{(2)}$, which is gradually coded by populations in $\mathscr{F}^{(3)}$. These $\mathscr{F}^{(3)}$ populations thereupon send sampling signals to $\mathscr{F}^{(4)}$. What patterns do these sampling signals see? Since motor performance at $\mathscr{F}^{(4)}$ deletes its motor command, they see only the activities of commands that have not yet been performed. The $\mathscr{F}^{(3)}$ populations therefore start to learn the monotone decreasing pattern that codes the correct order of the future items. If this trend continued through time, all would be well. It does not continue because, as each future item is rehearsed, its representation at $\mathscr{F}^{(4)}$ is deleted. As the $\mathscr{F}^{(3)}$ population continues to sample, it therefore tends to learn a monotone increasing pattern, rather than a monotone decreasing pattern, across its future field.

This difficulty arises only because the $\mathscr{F}^{(4)}$ representations are deleted when they are rehearsed. Somehow $\mathscr{F}^{(3)}$ must be able to sample representations that are linked to the motor commands in $\mathscr{F}^{(4)}$, but are not deleted when rehearsal occurs.

Let $\mathscr{F}^{(5)}$ code order information from $\mathscr{F}^{(4)}$. In particular, when an individual command $v_i^{(4)}$ is active at $\mathscr{F}^{(4)}$, it generates a code $v_i^{(5)}$ at $\mathscr{F}^{(5)}$, which in turn learns to activate $v_i^{(4)}$ via its feedback template (cf. Section 47). As a sequence of commands $v_1^{(4)}, v_2^{(4)}, \ldots, v_i^{(4)}$ is rehearsed at $\mathscr{F}^{(4)}$, it elicits a monotone decreasing STM pattern across $v_1^{(5)}$, $v_2^{(5)}, \ldots, v_i^{(5)}$. Then $v_{12\ldots i}^{(3)}$ is turned on, and samples this STM pattern. Population $v_{12\ldots i}^{(3)}$ also samples the monotone decreasing pattern that is elicited across its future field as the commands $v_{i+1}^{(5)}, v_{i+2}^{(5)}, \ldots$ are sequentially rehearsed. By Theorem 3, the LTM traces from $v_{12\ldots i}^{(3)}$ to $\mathscr{F}^{(5)}$ are unimodally distributed, and the maximal LTM trace abuts $v_{i+1}^{(5)}$.

Given this background, now let a sequence $v_1^{(3)}, v_{12}^{(3)}, \ldots, v_{12\ldots i}^{(3)}$ of $\mathscr{F}^{(3)}$ populations be activated by sensory feedback. The LTM pattern across each population's future field is monotone decreasing. In particular, *all* the LTM patterns that abut the populations $v_{i+1}^{(5)}, v_{i+2}^{(5)}, \ldots$ are monotone decreasing. Consequently the total STM pattern due to sensory feedback is monotone decreasing across the unrehearsed $\mathscr{F}^{(5)}$ populations. This STM pattern elicits at $\mathscr{F}^{(4)}$ a monotone decreasing STM pattern across the unrehearsed populations $v_{i+1}^{(4)}, v_{i+2}^{(4)}, \ldots$. The signals from $\mathscr{F}^{(5)}$ to the rehearsed populations $v_1^{(4)}, v_2^{(4)}, \ldots, v_i^{(4)}$ do not generate large STM activities, because motor arousal at $\mathscr{F}^{(5)}$ bootstraps subliminal signals into STM by contrast-enhancing activities that are already large—namely, the unrehearsed activities—and quenching the small activities. The total STM pattern at $\mathscr{F}^{(4)}$ is therefore monotone decreasing across the unrehearsed items and negligible across the rehearsed items. The correct order of performance is hereby predicted, no matter how many $\mathscr{F}^{(3)}$ populations are activated by sensory feedback.

The above mechanism is robust because it does not depend on any particular rate of generating sensory feedback in response to motor performance. It requires only that the feedback delay on performance trials be the same feedback delay that is experienced during learning trials. If feedback is artificially delayed by electronic means, then the motor command elicited by the sensory feedback can interfere with proper motor performance.

The amount of delay will influence how severe the interference is. Suppose, for example, that when the feedback is not artificially delayed, it excites $v_i^{(4)}$ as the next command to be performed. Then the feedback enhances performance of the $v_i^{(4)}$ command. If feedback is slightly delayed, then it tends to prolong performance of $v_i^{(4)}$. Figure 38 shows how this happens. Before $v_i^{(4)}$ is performed, the STM pattern in $\mathscr{F}^{(4)}$ is given by Fig. 38a. Suppose that performance of $v_i^{(4)}$ begins, and with it feedback inhibition of activity at $v_i^{(4)}$ (Fig. 38b). Once the activity at $v_i^{(4)}$ is less than that at $v_{i+1}^{(4)}$, performance at $v_{i+1}^{(4)}$ can begin. Suppose that the delay in sensory feedback to $v_i^{(4)}$ is so brief that feedback re-excites $v_i^{(4)}$ just before this can happen (Fig. 38c). Then $v_i^{(4)}$'s activity stays maximal for a longer time than usual and prolongs performance of the $v_i^{(4)}$ chunk.

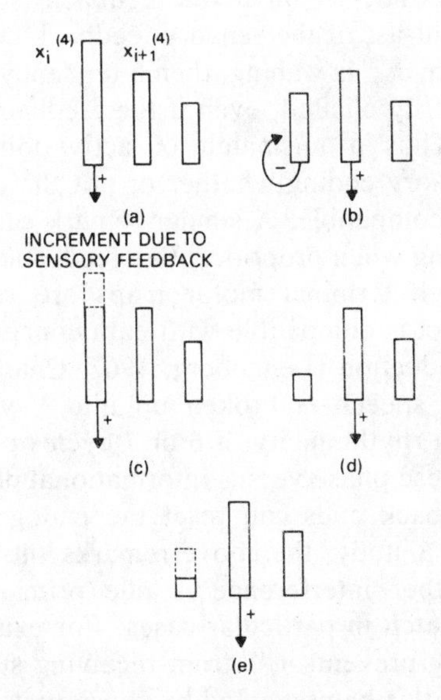

FIG. 38. Sensory feedback in (*c*) suffices to prolong performance at $v_i^{(4)}$, but in (*e*) $v_{i+1}^{(4)}$ continues to control performance.

If, however, self-inhibition of $v_i^{(4)}$ is almost complete when sensory feedback to $v_i^{(4)}$ arrives (Fig. 38d), then $v_{i+1}^{(4)}$ still has the maximal activity after the feedback signals take effect (Fig. 38e), so that the interference due to a longer delay is smaller. This phenomenon is analogous to the Lee effect, wherein auditory feedback causes maximal interference with speech production—primarily a drawing out of vowels—if it is delayed by about 180 msec, but less interference at smaller or later delays (Lenneberg, 1967, p. 109).

Other possible effects of delayed feedback go beyond the scope of the present discussion, because they all involve rhythmic activity in the network. As Section 23 notes, periodic oscillations in network activity can occur in recurrent networks due to the lagging behind of inhibitory activities relative to their excitatory counterparts (Ellias and Grossberg, 1975; Grossberg, 1977a). Whenever this happens, different effects can be elicited by a delay of sensory feedback that coincides with the waxing rather than the waning of STM activity. If we interpret the rhythmic STM activity as an analog of brain waves, then different effects can occur when feedback is in or out of phase with brain activity. The following effect is of this type. Let a sensory feedback cue reach $\mathscr{F}^{(1)}$ when the feedback expectation from $\mathscr{F}^{(2)}$ to $\mathscr{F}^{(1)}$ is waxing. Then the nonspecific alarm is not set off if the feedback is compatible with the expectation. By contrast, if the sensory feedback cue reaches $\mathscr{F}^{(1)}$ when the expectation from $\mathscr{F}^{(2)}$ is waning, then a tendency to reset STM at $\mathscr{F}^{(2)}$ is at least momentarily elicited, even if the feedback is compatible with the expectation. Thus a mismatch of activation *phases* can cause interference in sensory coding whether or not the expectations and the feedback cues are compatible. A similar remark can be made about the reset of motor coding when proprioceptive feedback is delayed, or about the rate with which terminal motor maps are reset by new motor commands. This fact is compatible with data concerning the rhythmnicity of language production (Lenneberg, 1967, Chapter V), in particular with the idea that speech is broken up into "syllables" or "breath pulses" that have a rhythmnicity of 6 or 7 cycles per second. A proper understanding of these phase versus informational phenomena requires a study of how feedback cues can reset the endogenous STM rhythm. Even without such a study, the above remarks show the importance of distinguishing whether interference is due primarily to phase or to informational mismatch in particular cases. For example, in Fig. 38e, a long feedback delay prevents $v_i^{(4)}$ from receiving sufficient feedback to dominate $v_{i+1}^{(4)}$. Can this be prevented by appropriate amplification of the feedback signal, or is the reduction in interference due primarily to a matching of phases? Testing this alternative unambiguously will be difficult, if only because amplification of the feedback signal might also reset the phase of the STM rhythm.

The above mechanisms work well as far as they go. They clarify some points, but also raise new questions. On the clarification side, they begin to show why sensory feedback is not needed to play the *next* chord, but is nonetheless important. *After* a command such as $v_{12...i}^{(3)}$ is activated, it can predict the correct performance order of a long sequence of chords $v_{i+1}^{(4)}, v_{i+2}^{(4)}, \ldots$. Thus, sensory feedback from $v_{i+1}^{(4)}$ is not needed to perform $v_{i+2}^{(4)}$. Nonetheless, sensory feedback does determine which commands, such as $v_{12...i}^{(3)}$, will be activated. A sequence from the past hereby determines a sequence in the future.

On the question side, we note the following:

1. Population $v_{12...i}^{(3)}$ is not the only active chunk in $\mathscr{F}^{(3)}$. Every active subsequence across $\mathscr{F}^{(2)}$ can, in principle, be coded at $\mathscr{F}^{(3)}$. A spatial pattern of activity therefore exists across $\mathscr{F}^{(3)}$ at every time. How should the relative activities of these chunks be determined so that the *total* signal from $\mathscr{F}^{(3)}$ to $\mathscr{F}^{(4)}$ correctly codes the order information of future chords?
2. What advantages are gained by using higher-order chunks like $v_{12...i}^{(3)}$ rather than $v_i^{(3)}$? The LTM traces of chunk $v_i^{(3)}$ can also encode the future commands $v_{i+1}^{(4)}, v_{i+2}^{(4)}, \ldots$ in their proper order. Why bother using higher-order commands at all?

36. Greater Weight and Longer Duration of Higher-Order Chunks

These questions suggest a general principle with far-reaching consequences. Consider question (2) to start off. Chunk $v_i^{(3)}$ can be activated by *every* chord sequence in which chord $v_i^{(4)}$ is played. Chunk $v_i^{(3)}$ might therefore already code a different chord sequence at $\mathscr{F}^{(4)}$ before the new piece is learned. By contrast, chunk $v_{12...i}^{(3)}$ is better defined by the piece that the pianist is playing, and can therefore more reliably predict the correct sequence of future chords. For example, if I hear the word "C" (= "see"), I can reply "See what?"; but if I hear "ABC," it is much more likely that I will reply "D." Higher-order chunks should therefore have greater STM activity than lower-order chunks in $\mathscr{F}^{(3)}$.

There is a related reason for giving higher-order chunks greater weight. It is desirable to keep the chunks that code longer sequences active for a longer time, so that they can predict far into the future. Such chunks can dramatically compress the encoding of data by each predicting a long sequence of chords. In particular, once a chunk in $\mathscr{F}^{(3)}$ enters STM, it can be actively quenched only by lateral inhibition due to later activation of other chunks in $\mathscr{F}^{(3)}$, assuming that arousal level and other performance variables stay fixed. Shutting off the $\mathscr{F}^{(2)}$ populations that originally activated an $\mathscr{F}^{(3)}$ chunk does not inhibit this chunk. This property is similar to the maintenance of STM activity by populations in $\mathscr{F}^{(2)}$ after their defining input patterns at $\mathscr{F}^{(1)}$ are deleted by rehearsal and replaced by new input patterns. These constraints are summarized by the following rule.

Self-Similar Coding Rule: Other things being equal, higher-order chunks have greater STM activity and longer duration than lower-order chunks.

37. Spatio-temporal Self-Similarity and the Resolution of Uncertainty

The above idea is a special case of the principle of *spatio-temporal self-similarity*, or STSS (Grossberg, 1969e). The network takes a risk by allowing any chunk to remain active for a long time. What if the chunk codes erroneous information? It can then cause errors until its activity can finally be quenched. The risk is minimized by letting the highest-order chunks remain active the longest. Since these chunks are better characterized by the temporal sequence in which they are embedded, it is more likely that this output will be the correct one. Thus, in the present situation, STSS means that chunks that are coded by long sequences can remain active for a long time to sample and control performance of long sequences.

How is STSS physically realized? The STSS concept constrains the *global* rules for building a field; what are the *local* rules whereby its individual populations are constructed? A mechanism is suggested by considering the concept of an *STSS cell type*. Suppose, statistically speaking, that a given cell type has a characteristic shape, including dendritic tree, cell body, and axons. If the cell type is STSS, than a small cell of this type is transformed into a large cell by blowing up all its spatial dimensions. What effects does this scale change have on a cell's functional capabilities? We now discuss certain neocortical cells using the teleology of STSS. Consider a cortical Betz cell that has a large dendritic tree (Crosby *et al.*, 1962). Such a dendritic tree gets a good view of input data in the vicinity of the cell. The firing of such a cell can thus be precisely controlled by prescribed patterns of sensory inputs. Because the cell fires only at (essentially) the correct times, it can be permitted to send a long axon all the way down to spinal motor centers (pyramidal tract), where it directly activates motor output. By contrast, if a cortical pyramidal cell has a small dendritic tree, then it can be fired by a much wider variety of input data. Its axon cannot be allowed to be a final command pathway to motor centers. By giving such cells a shorter axon, their effects on motor outputs will be manifested only over multisynaptic pathways (extrapyramidal tract), wherein further computations can be undertaken. In a similar fashion, if longer axons are also wider, then they can carry their signals more quickly, and thereby achieve the same transmission time lag as shorter and thinner axons. STSS is thus a structural constraint on individual cells that helps the aggregate network dynamics to resolve uncertain input data without taking untoward risks.

38. Order-Preservation in the Future Field of Motor Commands

The principle of SSTS suggests that higher-order chunks should be coded either by more cell sites—either more or bigger cells in a population—and/or by stronger signal pathways than lower-order chunks. This design will lead to fields $\mathscr{F}$ whose anatomy is in a dynamic equilibrium with the average spatial distribution of inputs that perturb $\mathscr{F}$ through time.

Before making this construction, we note that predicting the order of future motor commands could be achieved under weak constraints on $\mathscr{F}^{(4)}$, were it not for the fact that low-order chunks can sample many different chord sequences. This was demonstrated in Section 34, wherein it was proved that every chunk in $\mathscr{F}^{(3)}$ codes a monotone decreasing pattern across its future field in $\mathscr{F}^{(4)}$, independent of what items or sequences of items are coded by the chunk. Thus $\mathscr{F}^{(3)}$ can elicit the correct performance order at $\mathscr{F}^{(4)}$, no matter how many classes of chunks sample $\mathscr{F}^{(4)}$. We wish to suppress lower-order chunks only because they can sample and perform too many incompatible sequences.

39. Masking of STM by More Cell Sites or Amplified Signals

What STM patterns can exist in networks wherein some populations have more cell sites, stronger signals, or broader tuning curves than other populations? It is shown below that such networks have exactly the properties needed to satisfy STSS. Examples of such networks have been studied by Grossberg and Levine (1975) and Levine and Grossberg (1976). The former paper considers recurrent on-center off-surround networks of the form

$$\dot{x}_i = -Ax_i + (B_i - x_i)[f(x_i) + I_i] - x_i[\sum_{k \neq i} f(x_k) + J_i] \quad (41)$$

where the populations v_i can have different total numbers B_i of cell sites. System (41) is formally equivalent to

$$\dot{u}_i = -Au_i + (C - u_i)[f(D_i u_i) + I_i] - u_i[\sum_{k \neq i} f(D_k u_k) + J_i] \quad (42)$$

where both the excitatory and inhibitory signals $f(D_i u_i)$ from each population v_i are amplified by a scaling factor D_i. System (41) is transformed into (42) by the substitutions $x_i = D_i u_i$ and $B_i = CD_i$. A system

$$\dot{w}_i = -Aw_i + (C_i - w_i)[f(D_i w_i) + I_i] - w_i[\sum_{k \neq i} f(D_k w_k) + J_i] \quad (43)$$

in which both types of asymmetry exist is thus equivalent to system (41) with $x_i = D_i w_i$ and $B_i = C_i D_i$. In particular, if $C_1 \leq C_2 \leq \cdots \leq C_n$ and $D_1 \leq D_2 \leq \cdots \leq D_n$, then $B_1 \leq B_2 \leq \cdots \leq B_n$.

Such asymmetries introduce a new type of contrast enhancement into the system. For example, in (41), populations v_i with the largest B_i values tend to quench, or mask, the STM activity in populations v_j with smaller B_j values. If higher-order chunks are given larger B_i values in (41), then they will mask lower-order chunks, as we desire by STSS. This is the main idea.

More precisely, suppose that $B_1 \leq B_2 \leq \cdots \leq B_n$ in (41). If $f(w) = Ew$, then $x_i(\infty) = 0$ if $B_i < B_n$, and $x_i(\infty) = Kx_i(0)$ if $B_i = B_n$. In other words, given a linear signal function, all populations with nonmaximal B_i are masked, and the STM pattern of all populations with maximal B_i (that is, $B_i = B_n$) is stored faithfully in STM. No states are masked if all $B_i = B_n$. If $f(w)$ is, more realistically, chosen to be a sigmoid signal function, then an interesting phenomenon occurs. Once again there is the tendency for populations v_i with maximal B_i to mask other populations. In particular, if some populations v_i with $B_i = B_n$ get relatively large inputs, then all states v_j with $B_j < B_n$ will be masked. In general, however, there is a competition between the relative sizes of the B_i's and the relative sizes of the initial activities $x_i(0)$, the latter in turn being determined by the relative sizes of inputs to v_i before time $t = 0$. In all cases, only the STM traces corresponding to one B_i can be stored in STM. If certain v_i with $B_i < B_n$ have sufficiently large $x_i(0)$ compared with the $x_j(0)$ values of all v_j with $B_j = B_n$, then the subfield of populations with the nonmaximal weight B_i can mask all other populations. The STM pattern of this subfield is simultaneously contrast-enhanced and stored in STM. Grossberg and Levine (1975) interpret the competition between B_i and $x_i(0)$ in terms of developmental and attentional biases in the field. A developmental bias can, for example, give certain feature detectors larger B_i values than others. An attentional shunt can amplify the signals of one subfield more than others via larger D_i values. Either operation biases the field in favor of some subfield. In (43), the developmental biases C_i and the attentional biases D_i can create a complicated tug of war that favors the particular subfield having maximal $B_i = C_i D_i$ for STM storage. Nonetheless, a population with nonmaximal B_i can be stored if its features are present in the input display with relatively large saliency, or are coded by relatively strong pathways that amplify its inputs.

40. STM Drift toward a Norm: Primary Gradient Induces Secondary Gradient

The tendency of populations with maximal B_i to *totally* mask all other populations is due to the fact that each v_i can inhibit *all* v_k, $k \neq i$, with equal strength in (41). When the strength of recurrent excitatory and inhibitory signals decreases as a function of interpopulation distance, as in Section 27, then the masking effect can be partial, and can generate a slow drift by the spatial locus of maximal STM activity toward the populations having the largest B_i values. In the case that each v_i codes particular features, then the falloff with distance of recurrent signals defines generalization gradients between the feature detectors, and the detectors with the largest B_i act as "norms" toward which activity drifts across these generalization gradients.

For example, Levine and Grossberg (1976) study networks of the form

$$\dot{x}_i = -Ax_i + (B_i - x_i)\left[\sum_{k=1}^{n} f(x_k)C_{ki} + I_i\right] - (x_i + D)\left[\sum_{k=1}^{n} f(x_k)E_{ki} + J_i\right] \quad (44)$$

where the excitatory coefficients C_{ki} and inhibitory coefficients E_{ki} both decrease as a function of interpopulation distance $|i - k|$, with excitation ("on-center") decreasing faster than inhibition ("off-surround"). Suppose in addition that the B_i's are normally distributed around a given population v_I; that is, $B_i = Be^{-\lambda|i-I|^2}$. Then if an input perturbs a population v_i, $i \neq I$, the locus of maximal STM activity drifts toward v_I. The drift rate depends on how steep the slope of the function B_k is as a function of v_k for k values between i and I. If the slope is small, the drift rate is slow; if the slope is large, the drift rate is fast. Levine and Grossberg (1976) suggest that such a drift is responsible for the line neutralization effect that is perceived when a nearly vertical or horizontal line is inspected for a sufficiently long time (Gibson, 1933). In summary, if there exist generalization gradients among feature detectors, and if certain detectors are coded by more sites or broader tuning curves than other detectors, then STM activity drifts toward the nearest populations having the most sites or the most highly amplified signals.

The above mechanism can be described in terminology from developmental biology. The B_i's define a *primary* gradient; the input is an inducing stimulus; and the STM drift is a *secondary* gradient that is generated by field interactions in response to the inducing stimulus.

41. Masking of Lower-Order Codes

Which chunks in $\mathscr{F}^{(3)}$ will be masked? The answer depends on at least three factors. First, it depends on the spatial distribution of LTM traces across the $\mathscr{F}^{(2)} \to \mathscr{F}^{(3)}$ pathways. The LTM vectors z_i in these pathways define positional gradients that determine how close together in $\mathscr{F}^{(3)}$ are the populations that code two different sequences of events; for example, how close is the $\mathscr{F}^{(3)}$ code for sequence ABC to the code for ABCD? Second, it depends on how many sites code each population. If ABCD is coded by more sites than ABC, then ABCD's code will tend to mask ABC's code. Third, it depends on the breadth of recurrent excitatory and inhibitory signals within $\mathscr{F}^{(3)}$. Even if the codes for ABC and ABCD lie next to each other in $\mathscr{F}^{(3)}$, they are far away from each other, functionally speaking, if they do not fall within each other's generalization gradients. Section 39 shows that more sites and broader tuning curves have the same effect on STM masking. Hence we expect the chunks that have the most sites to have the broadest generalization gradients. This is immediately guaranteed, given STSS, if the largest cells are in the populations having the most sites.

It remains to determine how many sites will be given to $\mathscr{F}^{(3)}$ populations that code sequences of prescribed length. We seek a law that can plausibly be realized by simple rules of neuronal growth before the stage of adaptive coding takes place. The qualitative features of this law are already apparent. For example, given sequences A, AB, ABC, ABCD, . . . , it follows by STSS that the number of sites should increase monotonically with list length until a maximal length is reached. Thereafter, the number of sites should decrease with list length to prevent infinitely long (and infinitely unlikely) sequences from being coded (see Fig. 39). The simplest rule of this type is the Poisson distribution. Suppose that a population in $\mathscr{F}^{(2)}$ contacts a certain population in $\mathscr{F}^{(3)}$ with a prescribed small probability p. Let λ be the mean number of such contacts on all the cells of $\mathscr{F}^{(3)}$. Then the probability that exactly k contacts perturb a given population is

$$P_k = (\lambda^k/k!)e^{-\lambda} \tag{45}$$

(Parzen, 1960). If K is chosen so that $K < \lambda < K + 1$, then P_k increases for $1 \leq k \leq K$ and decreases for $k > K$. More sophisticated but related distributions, such as the hypergeometric distribution

$$P_{mk} = \binom{m}{k}\binom{N-m}{n-k}\Big/\binom{N}{n} \tag{46}$$

where ( ) is the binomial coefficient, are also discussed in Parzen (1960).

Consider the Poisson distribution in (45) for definiteness. Given this rule, sequences of length K will generate maximal STM activity at $\mathscr{F}^{(3)}$, other things being equal. For example, suppose that $K = 4$, and consider network response to the sequence of events ABCD . . . . As each item is processed, it excites a code (A); then codes (B), (AB), and (BA); then codes (C), (ABC), (BAC), (BC), . . . ; and so on. How close are the codes (A), (B), and (AB) to each other in $\mathscr{F}^{(3)}$? The code (AB) differs from the codes (A) and (B) separately only by one item. Since the items A and B are coded by being *projected* onto the LTM traces z_i in $\mathscr{F}^{(2)} \rightarrow \mathscr{F}^{(3)}$ pathways, (A) and (B) are closer to (AB) than, say, (F) is. This argument can be refined by taking into account the phonetic similarities that cause items to be similarly coded at the $\mathscr{F}^{(1)} \rightarrow \mathscr{F}^{(2)}$ stage of filtering; such extensions can be supplied once the main idea is clear. In summary, (A) and (B) lie close to (AB) in $\mathscr{F}^{(3)}$. Since (AB) has greater weight, it tends to mask (A) and (B) by lateral inhibition. Similarly, (ABC) masks (AB), and (ABCD) masks (ABC). Since $K = 4$, (ABCD) also masks (ABCDE), (ABCDEF) and so on. Thus, chunks of length four tend to dominate the STM activity in $\mathscr{F}^{(3)}$.

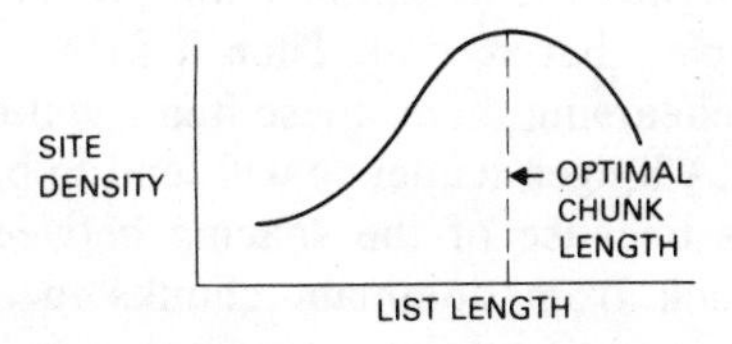

FIG. 39. STSS suggests that longer sequences are coded by more sites, up to some maximal sequence length.

42. Clustering, Compression, Spacing, and Completion

The above masking process has many interesting implications. Most obviously, it dramatically reduces, or compresses, the number of active chunks that are needed to control motor activity by suppressing the populations that code nondominant subsequences. It is important to recall here that the codes for A, AB, ABC, etc., are not wired into the network. Thus if a population for ABCD masks many of the populations that *would be* codes for BCD, CDB, CD, etc., before these codes can be learned, then the masked populations remain uncommitted and can be coded by other events; in particular, by events in which the items A, B, and C occur separately or in other dominant subsequences.

A second interesting consequence is a spacing effect in the chunks that get stored in $\mathscr{F}^{(3)}$. After the events A, B, C, and D are presented, (ABCD) actively masks the populations (BC), (CD), (BCD), etc. When event E occurs, there is a tendency for (BCDE) to be dominant, but this tendency is offset by two factors: First, all the populations related to (BCDE), such as (CD), which could supply it with recurrent excitation, have been suppressed; second, (ABCD) is sufficiently close to (BCDE) to suppress it, at least partially, by recurrent inhibition. This argument must be made with care, because it depends on the detailed choice of network parameters. It is, for example, possible for (BCDE) to be in the on-center of (ABCD), whence it is enhanced rather than suppressed, whereas (CDEF) is in the off-surround of (ABCD). The main robust point is clear, however. There is a tendency for the codes of dominant subsequences to be *spaced*, owing to mutual recurrent inhibition by their off-surrounds. For example, only the codes (ABCD), (DEFG), (GHIJ), . . . might have significant STM activity after recurrent inhibition acts. This spacing effect accomplishes a further compression of data encoding by $\mathscr{F}^{(3)}$.

The spacing effect generates a tendency to cluster responses into subsequences in order to maximize learning rate. As a sequence of events is presented to the network, it tends to generate spaced chunks of a dominant length. When these chunks control performance, they tend to group responses into the functional units coded by these chunks. Suppose, for example, that $K = 4$. Then if four consecutive items are listened to before rehearsing them, these items will be coded by a chunk of maximal weight. A longer sequence will tend to be broken into two or more subsequences because of the spacing between dominant chunks within $\mathscr{F}^{(3)}$. Feedback from dominant chunks in $\mathscr{F}^{(3)}$ to $\mathscr{F}^{(2)}$ and $\mathscr{F}^{(4)}$ thereupon tends to group items into rehearsal units of length four.

Does the length of the TMS at $\mathscr{F}^{(2)}$ influence the number K at $\mathscr{F}^{(3)}$ by determining the maximal length of sequences that can be rehearsed from the STM buffer with no LTM feedback? As Section 32 notes, the TMS is shorter than the IMS. Is the TMS $\cong$ 4 and the IMS $\cong$ 7 in most individuals?

In a similar fashion, suppose that free responding is controlled by dominant chunks in $\mathscr{F}^{(4)}$. Suppose that item A has just been elicited by chunk (ABCD). Since chunk (ABCD) is soon suppressed by nearby chunks via recurrent inhibition, there is a tendency not to say A again until after (ABCD) is released from inhibition. This creates an apparent refractoriness for emitting the same item again while its controlling chunk is suppressed by chunks that are activated by sensory feedback owing to performance of later items.

Estes (1972) suggests a coding theory in which inhibition is used to control the clustering effect. His theory does not, however, study the dynamics of coding, or the real-time mechanisms whereby the inhibition organizes itself across emergent codes. The above remarks also use inhibition, but provide a different and more complete theory of how inhibition works. Estes' paper also reviews various data that are related to the above mechanisms.

Finally, the masking mechanism provides a deeper insight into pattern completion. For example, let a spatial pattern at $\mathscr{F}^{(2)}$ be coded by a nondominant population in $\mathscr{F}^{(3)}$. The STM activity in $\mathscr{F}^{(3)}$ can then drift toward the nearest dominant population. The drift "completes" the pattern, or in the present case, the sequence, by activating a higher-order code. Then the dominant population can send its template of feedback signals back to $\mathscr{F}^{(2)}$ where they reorganize STM at $\mathscr{F}^{(2)}$ to code the "completed" pattern.

43. The Magic Number Seven and Self-Similar Coding

The coding of sequences by patterned activity across spaced dominant chunks accomplishes several tasks at once. One is to suppress predictive sampling by lower-order chunks that are excited by sensory feedback during motor performance. The discussion in Sections 36 through 41 shows that this mechanism is a consequence of STSS.

STSS also implies that the duration of higher-order chunks exceeds that of lower-order chunks. After recurrent inhibition causes spacing within the field of dominant chunks, it creates a TMS among the spaced dominant chunks. If a self-similar scaling of recurrent interaction strengths holds within every subfield of chunks, then the TMS of the spaced dominant chunks will be commensurate with that of the chunks

(A), (B), (C), ... Then the chunk for (ABCD) can remain active much longer than the chunk for (D), even if the recurrent interactions within each subfield of spaced chunks have the same effect on STM activity when a population in their subfield is excited. This is true because, as the length of the subsequences that are coded by a subfield increases, so does the time interval between successive activations of its spaced chunks.

By STSS, each subfield of chunks in a prescribed sensory field has the same TMS, *other things being equal*. Otherwise expressed, if sensory data are recoded by a different subfield, then they have the same TMS in the new code as they had in the old code. The existence of a commonly shared "magic number seven, plus or minus two," for the immediate memory span of various codes (Miller, 1956) thus supplies indirect support for STSS as a principle of code synthesis.

44. Suppression of Uniform Patterns and Edge Detection

The principle of STSS provides one mechanism whereby erroneous signals from lower-order chunks can be eliminated: Lower-order chunks, and their signals, are inhibited by higher-order chunks. Section 36 suggested one reason for doing this: If a given event occurs in different contexts, it should be able to elicit different responses. In particular, if a lower-order chunk, such as $v_i^{(3)}$, *were* allowed to vigorously sample every chord $v_j^{(5)}$ that succeeded it, then eventually $v_i^{(3)}$ could encode a nearly uniform pattern of activity across its LTM traces. Signals from $v_i^{(3)}$ to $\mathscr{F}^{(5)}$ would then be uninformative, or irrelevant, since they would not discriminate any population in $\mathscr{F}^{(5)}$ from any other. STSS helps to prevent this by restricting the circumstances under which sampling can occur.

Often cues should be allowed to sample even if their signals are uninformative. For example, before a chunk learns a pattern, its signals to $\mathscr{F}^{(5)}$ are uniformly distributed. If the chunk is not allowed to sample $\mathscr{F}^{(5)}$, then it can never learn a pattern. The problem is to allow uninformative cues to sample $\mathscr{F}^{(5)}$ under appropriate circumstances, and yet to prevent their uniform, or "noisy," signals from destroying the patterns that are driving $\mathscr{F}^{(5)}$. Whether a given chunk controls a uniform pattern or not in its LTM traces can be decided only after these traces elicit signals at $\mathscr{F}^{(5)}$. Moreover, even if each active chunk codes a nonuniform pattern, the total input pattern to $\mathscr{F}^{(5)}$ can be uniform, and therefore uninformative. Hence some mechanism within $\mathscr{F}^{(5)}$ must exist to deal with the noise in its total input pattern.

We now show how recurrent mass action networks suppress the "uniform part," or noise, in their total input patterns, and generate suprathreshold responses only to spatial differences, or "discriminations," in these patterns. This property implies that LTM sampling occurs among chunks when the sampling paths carry useful information, but not otherwise.

The simplest version of the uniform quenching property occurs in system

$$\dot{x}_i = -Ax_i + (B - x_i)I_i - (x_i + C) \sum_{k \neq i} I_k$$

with $-C \leq x_i \leq B$. If $C > 0$, by contrast with (14), the equilibrium response ($\dot{x}_i = 0$) to a pattern $I_i = \theta_i I$ is

$$x_i = \frac{(B + C)I}{A + I} \left(\theta_i - \frac{C}{B + C} \right) \tag{47}$$

If, for example, $B = (n - 1)C$, then $C/(B + C) = 1/n$. Now let the input pattern be uniform. Then all $\theta_i = 1/n$, so that no matter how intense I is, all $x_i = 0$. If not all $\theta_i = 1/n$, then the network quenches the "uniform part" of the pattern. More generally, whenever $B \leq (n - 1)C$, the x_i's are suppressed even more vigorously by inhibition than when $B = (n - 1)C$. Consequently only values of $\theta_i > C/(B + C) > 1/n$ can generate a supraequilibrium response. Increasing C hereby contrast-enhances the network's response to input patterns. It has been suggested that this contrast-enhancement property can influence the size of certain visual illusions, such as tilt aftereffect and angle expansion (Levine and Grossberg, 1976).

The quenching of uniform patterns is due to a competitive balance between a narrow on-center I_i that interacts with a relatively large excitatory saturation point B, and a broad off-surround that interacts with a relatively small inhibitory saturation point $-C$; such a relative size scaling between B and C often occurs in passive membranes (Hodgkin, 1964). This conclusion generalizes to systems

$$\dot{x}_i = -Ax_i + (B - x_i) \sum_{k=1}^{n} I_k C_{ki} - (x_i + D) \sum_{k=1}^{n} E_{ki} \tag{48}$$

wherein inputs I_i can excite populations v_j near to v_i via the coefficients C_{ij} ("on-center") and can inhibit populations v_j over a broad expanse of cells via the coefficients E_{ij} ("off-surround"). Since the equilibrium point of (48) is

$$x_i = \frac{I \sum_{k=1}^{n} \theta_k (BC_{ki} - DE_{ki})}{A + I \sum_{k=1}^{n} \theta_k (C_{ki} + E_{ki})} \tag{49}$$

a uniform pattern (all $\theta_i = 1/n$) is quenched (all $x_i \leq 0$) for any $I \geq 0$ whenever

$$B \sum_{k=1}^{n} C_{ki} \leq D \sum_{k=1}^{n} E_{ki}, \qquad i = 1, 2, \ldots, n \tag{50}$$

The breadth of excitatory and inhibitory interactions across the network determines the input patterns to which a population will respond. For example, let a vertical bar of light perturb the network. Suppose that the breadth of on-center and off-surround interactions is less than that of the bar. Then cells near the center of the bar will perceive a uniform field. Also, cells far away from the bar will perceive a uniform field. Both types of cells will be incapable of generating suprathreshold responses. Only cells near the transition regions of light and dark will respond. Such a network detects the edges of the bar.

The above mechanism can also be used as a matching mechanism, as in Section 20. To see this, consider the following question: Given a spatial pattern θ, how can a maximally mismatched pattern $\tilde{\theta}$ be generated? Clearly $\tilde{\theta}$ should be large where θ is small, and conversely. If both θ and $\tilde{\theta}$ are input patterns to the network, then their mismatched peaks and troughs will add to create an almost uniform net pattern. Network activity is consequently suppressed. By contrast, if $\tilde{\theta}$ is proportional (parallel) to θ, then the patterns add to amplify network activity.

45. The Growth of On-Center Off-Surround Connections

How can a balance between the ratio of excitatory and inhibitory saturation points, and between the distribution of on-center and off-surround coefficients, be effected? If it is not, and (say) the off-surround is too strong, then by (49) essentially all patterns will be suppressed. For example, suppose that lateral inhibition in the LGN not only contrast-enhances afferent sensory patterns, but also differentially amplifies LGN activity depending on how well the sensory patterns match cortical feedback. How is the excitatory–inhibitory balance that is needed for matching generated? Before considering structural substrates of this mechanism, we note that differential shunting of on-center or off-surround interactions can retune the network by shifting its criterion of how uniform a pattern must be to be suppressed (cf. Ellias and Grossberg, 1975, Section 3).

Simple growth rules are sufficient to formally explain some qualitative features of this balance. These are stated to illustrate how local data at each cell, such as the ratio BD^{-1}, can determine intercellular connections. Two related properties are desired: (1) the quenching of uniform patterns; (2) a narrow on-center and broad off-surround. By (50), the quenching of uniform patterns occurs if the ratio BD^{-1} equals all the ratios $E_i C_i^{-1}$, where C_i (E_i) is the total strength of excitatory (inhibitory) connections to v_i, $i=1,2, \cdots, n$. This can be achieved if "opposites attract." That is, suppose excitatory sites at v_i (whose number is proportional to B) support a process which attracts growing inhibitory connections, and inhibitory sites at v_i (whose number is proportional to D) support a process which attracts growing excitatory connections. Otherwise expressed, let the excitatory sites and inhibitory sites support processes which generate attractive *gradients* for inhibitory and excitatory connections, respectively. Then all the ratios $E_i C_i^{-1}$ will approach BD^{-1} if there exist enough intercellular connections to match the attracting sources. The hypothesis that cell growth follows some type of spatial gradient is a familiar one in developmental biology (cf. Gustafson and Wolpert, 1967; Grossberg, 1978a). Moreover, because $B>D$, the gradient attracting inhibitory connections will, other things being equal, be more uniform across space than the gradient attracting excitatory connections. A narrow on-center and broad off-surround will hereby tend to be produced. If this mechanism exists, then a change in BD^{-1} during the growth period, whether natural or experimentally controlled, should alter the relative spread of excitatory and inhibitory connections.

46. Goal Gradient and Plans

Using STSS and the quenching of uniform patterns, we can now study how associations develop among the chunks in a hierarchically coded field of populations. The need for such associations is clear from a variety of examples. In fact, we have already been using a simple version of this mechanism. In Section 21, the reciprocal trainable pathways between two fields $\mathscr{F}^{(1)}$ and $\mathscr{F}^{(2)}$ can be reinterpreted as a special case of reciprocal trainable pathways between two subfields of chunks in a recurrent network. The code-learning in $\mathscr{F}^{(1)} \to \mathscr{F}^{(2)}$ LTM traces is distinguished from template-learning in $\mathscr{F}^{(2)} \to \mathscr{F}^{(1)}$ LTM traces only by the fact that tuning in $\mathscr{F}^{(2)}$ might contrast-enhance its patterns more than tuning in $\mathscr{F}^{(1)}$ does, but even this distinction need not hold in general.

Sections 1,A and 6,C provide other examples of this concept. For example, how does the internal representation of a goal object, such as an apple, trigger a plan to get an apple from the refrigerator? Somehow the code for the apple, which is not as "abstract" as the code for the plan, can activate the code for the plan based on prior experience. This occurs, we assume, because the apple representation can sample the representation of the plan during learning, and can activate this representation during preformance.

Several remarks are important here. First, conditioning from representations of a goal to a plan should be possible given essentially any goal and any realizable plan. Thus, the possible sampling pathways should be distributed broadly throughout the network. This would create a devastating noise problem were it not for STSS and the quenching of uniform patterns. The quenching of uniform patterns eliminates all effects of sampling across irrelevant cues, and STSS amplifies the codes that are most informative in a given context of events. Second, if the goal representation is a lower-order chunk, its signals among other lower-order chunks can be suppressed by the quenching of uniform patterns, but its sampling of higher-order representations, such as plans, will not be quenched because of the built-in distinctiveness, or informativeness, of the activity patterns across higher-order chunks. Third, even if the plan is coded by a spatial pattern across higher-order populations, rather than by undifferentiated activity within a population, it can be accurately sampled by the goal representation. Fourth, when the goal occurs, it generates incentive–motivation that can amplify sampling by its representation of the plan (cf. Grossberg, 1975). At the time this occurs, the goal is the last event to have occurred. By Section 27, even if the goal is preceded by many prior events, it still has enough activity to elicit sampling when this activity is supplemented by incentive motivation. Fifth, because the incentive motivation is nonspecific, *all* active lower-order chunks can sample the plan, but with a strength that depends on their STM activities. For example, suppose that the STM pattern is monotone increasing across the most recent event representations, owing to the occurrence of many prior events. Then a classical goal gradient will be learned such that the most recent items will elicit the plan with the greatest efficacy. Sixth, this mechanism shows how *partial avalanche* structures can be embedded in the network, as in Fig. 40. The partial avalanches blend together, in a self-consistent fashion, descending order information from commands with ritualistic sequence information due to reliable, and therefore unquenched, occurrence of the same sequence through time. These partial avalanches are a type of primitive syntactic structure in the network. Figure 40 illustrates the importance of adaptation, or competition, of STM activity throughout the hierarchy of chunks. This competition prevents any population from receiving too large an input from any one source, and thereby preserves decision rules for signal generation that require convergence of specific and nonspecific inputs (Section 7). Finally, the hierarchy tunes itself. For example, suppose that chunks (A), (B), (C), . . . are activated during a particular stage of learning. Then plans and partial avalanches can be gradually built on these chunks during this learning stage. If, however, chunks (ABC), (DEF), . . . are activated during a later learning stage, then the earlier plans and partial avalanches will be masked by STSS. This observation can be expanded to explain various properties of transfer from serial learning to paired-associate learning, and vice versa. In effect, the rehearsal strategy helps to choose the base code on which higher codes and feedback relations will develop.

The above properties all depend on a rapid normalization, or adaptation, of STM activity throughout the hierarchy of chunks, followed by LTM sampling of informative STM patterns throughout the hierarchy. Competition within the STM hierarchy has been used to explain differences between simultaneous versus successive contrast, respectively, in the visual illusions of angle expansion and tilt aftereffect (Levine and Grossberg, 1976). These qualitative properties must eventually be supplemented by a mathematical classification of how parametric differences in the intrafield and interfield interactions of particular classes of feature detectors generate different STM and LTM patterns. This analysis must include systems of the form

$$\dot{x}_i = -A_i x_i + (B_i - C_i x_i)\left[\sum_{k=1}^{n} f_k(x_k) D_{ki} z_{ki} + I_i\right] - (E_i + F_i x_i)\left[\sum_{k=1}^{n} g_k(x_k) G_{ki} + J_i\right] \tag{51}$$

and (2), which generalize (1) and (2).

47. STM Order Reversal: Item Learning versus Order Learning

Having discussed aspects of the self-organization of codes, we must now focus more closely on what is coded. This problem can be

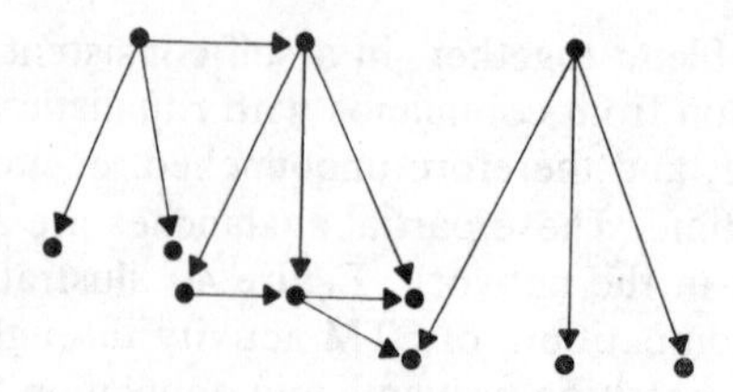

FIG. 40. Partial avalanches combine self-consistent order information from descending (contextual) commands with associational chains.

motivated by using the discussion in Section 31. Suppose that items are rehearsed one at a time as they occur. Let the $\mathscr{F}^{(2)}$ populations $v_{i_1}^{(2)}, v_{i_2}^{(2)}, \ldots, v_{i_k}^{(2)}, \ldots$ be sequentially activated by these items, and let rehearsal delete suprathreshold STM activity at $\mathscr{F}^{(2)}$ before the next item is presented. As this occurs, the populations $v_{i_1}^{(3)}, v_{i_2}^{(3)}, \ldots, v_{i_k}^{(3)}, \ldots$ in $\mathscr{F}^{(3)}$ are also sequentially activated (cf. Section 35). Although rehearsal deletes STM at $\mathscr{F}^{(2)}$, it does not delete STM at $\mathscr{F}^{(3)}$. If $k <$ TMS, then order information is accurately coded at $\mathscr{F}^{(3)}$ after rehearsal terminates. A nonspecific rehearsal wave to $\mathscr{F}^{(3)}$ can then trigger a readout from $\mathscr{F}^{(3)}$ of the items with their proper relative activities. This example illustrates how each item can be rehearsed during list presentation (repeat the last item), followed by a repetition of the whole list in its correct order (repeat the first item), even if no population codes order information in its LTM traces. This capability is important. Otherwise a telephone number could not be repeated unless it had already been encoded in LTM. This example again suggests the usefulness of studying how performance variables alter the arousal level and thus the form (decreasing, increasing, or unimodal) of the transient STM pattern.

Contrast the above experiment with one in which the items are presented in the same order, but they are rehearsed two at a time. Then a population $v_{i_1 i_2}^{(3)}$ in $\mathscr{F}^{(3)}$ will gradually code the first two items, and will learn order information as these items are rehearsed. As $v_{i_1 i_2}^{(3)}$ is coded, it gradually masks $v_{i_1}^{(3)}$ and $v_{i_2}^{(3)}$ if STSS holds. Eventually the chunks $v_{i_1 i_2}^{(3)}, v_{i_3 i_4}^{(3)}, \ldots$ will control readout from $\mathscr{F}^{(3)}$ of order information after rehearsal takes place. This argument shows how a combination of prior coding and the present rehearsal strategy determine which chunks will be active at $\mathscr{F}^{(3)}$; for example, how the network automatically groups items in a given rehearsal unit into familiar chunks. Thus, if the familiar letters Q and L are rehearsed together, then each letter can excite a previously coded $\mathscr{F}^{(3)}$ chunk even if the present rehearsal strategy starts to code a higher-order chunk. The dominant codes at a given time learn order information, but what codes are dominant can change as item learning continues.

The following sections consider the spatial analog of this temporal chunking process—namely, the self-organization of maps.

48. Circular Reaction and Map Formation

This section explicates the heuristic themes in Sections 1,C and 1,D. To fix ideas, imagine that an infant's hand–arm system is endogenously active and that its eye–head system tends to follow the motions of its hands. How does this unconditional process generate learned maps that can guide the hand–arm system to a terminal position, never before experienced, that is focused by the infant's eye–head system? Similarly, after an infant unconditionally babbles simple sounds, how does it imitate sounds that are more complicated than those it babbled? A complete answer to these questions would require a thorough analysis of the neural controls of motor behavior. Herein we note a minimal synthesis of resonance, sampling, and nonspecific arousal mechanisms acting on proprioceptive and terminal motor data that suffices to learn and perform sensory-motor maps and to maintain descending postural commands. Then we note that this model is really a variation of the adaptive coding model.

The following construction holds independently of what data are coded by a particular motor map. It focuses on the minimal operations that are needed to learn maps effectively. For example, we shall ignore the fact that different combinations of eye and head position can focus the eyes on the same physical position relative to the body. The construction will, however, be motivated by a familiar example: pursuit of the endogenously active hand by unconditional eye motions.

Let the image of the hand move across the retina. Suppose that its position on the retina (after compensatory computations of head position) determines a terminal eye position that will move the eye until the hand is viewed by the fovea. Clearly, no learned correlation between eye and hand position should be initiated until the eye foveally fixates the hand. Otherwise, an arbitrary eye position could continuously be correlated with all possible hand positions. How does the eye–head system know that it is fixating something? This happens when its terminal motor coordinates match its proprioceptive motor coordinates. The terminal coordinates code where the system *wants* to go; the proprioceptive coordinates code where the system *is*. When the eye actively fixates an object, these two sets of coordinates code the same position.

The following mechanisms explicate the idea that the eye is fixating something. First, the cells that code terminal coordinates send signals to the eye muscles that hold the eyes in position. As in Section 16, we assume that specific inputs to these cells determine their relative activities, and that a nonspecific arousal source shunts these activities into the suprathreshold range. In particular, let these cells be joined by on-center off-surround interactions so that the shunt can work. Second, the system must somehow know that the proprioceptive and terminal coordinates agree. Somehow data from the two coordinate sets must be brought together, and a characteristic dynamic state must be generated only when the two sets match. In the present theory, such a match triggers a resonant state. In all, sampling signals are emitted only if a resonant state exists between proprioceptive and terminal coordinates, and this resonant state is maintained as long as the nonspecific shunt is kept on.

The same conditions exist when a previously learned map is performed. Suppose that an eye fixates on an object that is to be touched. The eye–hand terminal coordinates then match their proprioceptive coordinates. If arousal is turned on, then the eye–hand system can send learned signals to the hand–arm system. These signals code the terminal hand position that was correlated with the eye position during learning trials.

Another constraint is needed before the minimal mechanism can be described. Often a motor system is directed to fixate on a different position from its present one. Then its terminal coordinates (where it wants to go) do not match its proprioceptive coordinates (where it is). Indeed, this is the typical situation when the hand is directed to touch an object. Obviously the motor system can do this, and it does it without interference from its present position. Thus, during performance trials, the terminal coordinates can release motor signals even if they do not match the proprioceptive coordinates. Arousal suffices to release signals from the terminal coordinates in this case.

In summary, sampling signals can be released only when terminal and proprioceptive coordinates match and are sustained by arousal; yet if terminal coordinates are activated by signals from another system, they can generate performance signals when they are activated by arousal, even if they do not match their proprioceptive coordinates. These constraints are summarized in Fig. 41. Figure 41 depicts two systems, (I) and (II), of proprioceptive and terminal motor coordinates; for example,

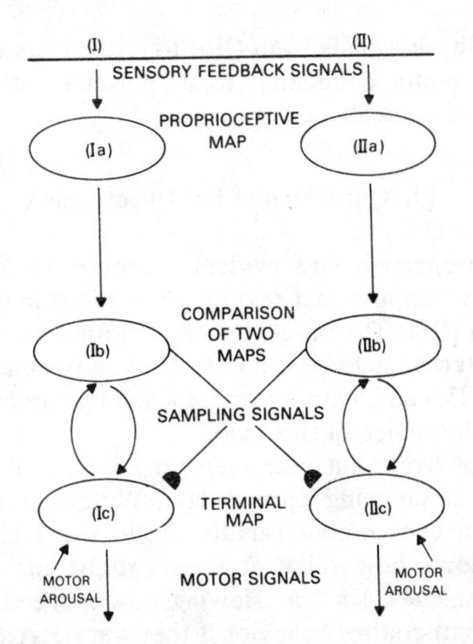

FIG. 41. Sampling signals are elicited only if proprioceptive and terminal maps agree. Motor signals are elicited if a subliminal terminal map is aroused.

let (I) be an idealization of the eye–head system, and let (II) idealize the hand–arm system. Consider (I) for definiteness. Proprioceptive coordinates are coded by the populations in field (Ia), and thereupon send signals to field (Ib). Terminal coordinates are coded in field (Ic), and can send signals in two directions if they are aroused. Signals can descend as motor commands to appropriate muscles. Signals can also go to (Ib). At (Ib), the proprioceptive and terminal maps are compared, as in Section 44, by an on-center off-surround network. If the maps match, then (Ib) amplifies their commonly shared pattern. If the maps do not match, then activity in (Ib) is suppressed. When the maps match, two things happen. Feedback signals go to (Ic), whereupon a resonance is established between (Ib) and (Ic). This resonance sustains sampling signals from (Ib) to the terminal coordinates at (IIc). In all, when proprioceptive coordinates at (Ia) match suprathreshold terminal coordinates at (Ic), then sampling signals are emitted from (Ib).

What happens to these sampling signals? Suppose that (IIc) receives signals from (Ib), which are thereupon amplified by arousal. Then (IIc) can emit motor commands to its muscles. Field (IIc) also sends signals to (IIb), but if the hand is not yet at the desired terminal position, the mismatch between proprioceptive and terminal signals at (IIb) quenches any possible feedback signals from (IIb) to (IIc). Thus the terminal map at (IIc) can emit motor commands to its muscles without interference due to proprioceptive feedback.

49. Quenching of Irrelevant Cues

An important property is now evident. Suppose that irrelevant signals reach (IIc); that is, suppose that several populations in (Ib) send uniform signal patterns to (IIc). The on-center off-surround network at (IIc) will quench these patterns, as in Section 44. Thus, any number of irrelevent cues can sample (IIc) and learn a map imposed by a relevant cue without distorting the performance of this map.

The property of irrelevant cue quenching is crucial wherever many cues are capable of sampling learned data. Without it, the omnipresent existence of such cues would rapidly erode the LTM pattern. This property also shows how relevant cues can be included in a plan, whereas irrelevant cues, such as slowing down, speeding up, taking a drink, etc., need not control behavior if they vary across trials.

50. Feedforward Reset of Sequential Terminal Maps

What happens if a sequence of motor commands iteratively perturbs (IIc)? How is the terminal map at (IIc) reset by the next command? Moreover, if no new commands occur, how can the terminal map at (IIc) be stored while its command is being executed? Sections 9 and 29 pointed out that a command is deleted after it is released to prevent continual iteration of the same command. Somehow the command data, which are now explicated as a terminal map, are then stored while performance proceeds.

Both the reset and STM storage properties follow from the uniform quenching property if the network at (IIc) is made recurrent (see Fig. 42). Suppose that no signals are being emitted by (IIb), because the terminal map at (IIc) has not yet been executed. Let a given terminal map be received at (IIci) via external signals. This map is then reproduced at (IIcii) via the (IIci) → (IIcii) pathways. If the external signals are shut off, then the terminal map can resonate at (IIc) as long as shunting arousal is maintained. Suppose, however, that a new terminal map is imposed at (IIci) by external signals. Feedback signals from (IIcii) to (IIci) still carry the old terminal map. If the two maps do not match, then activity across (IIc) will be momentarily quenched. All traces of the old terminal map signals are hereby eliminated, and the

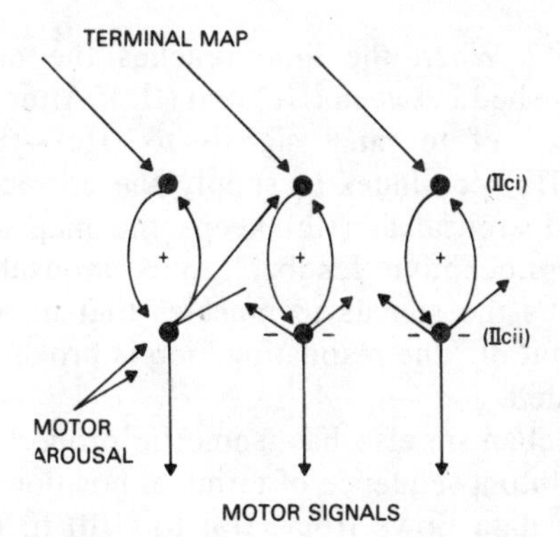

FIG. 42. Delayed feedback from (IIcii) to (IIci) briefly inhibits STM at (IIci) if the new terminal map does not match the old terminal map. The new terminal map then resonates in STM, and reset is completed.

new terminal map is instated. In short, delayed interneuronal feedback within a network that possesses the uniform quenching property implies the existence of an STM reset mechanism. Several variations on the theme of Fig. 42 exist. For example, *in vivo* do inhibitory interneurons also descend from (IIci) to (IIcii)? All these variations work better if the time needed to self-inhibit a descending command exceeds the time needed to reset a terminal map.

51. Posture, Isometrics, Saccades, and Feedforward Motor Control

The above construction suggests many insights and questions about motor control.

It suggests a mechanism of posture. There exist two main types of systems that must be distinguished by experiment in particular cases. The first type was described in Section 50. Stage (IIc) can maintain a terminal map using recurrent signals even if the descending motor command is silent. In this system, posture is the terminal position of a persistently aroused terminal map.

The second type cannot store a terminal map at stage (IIc). It requires a persistent source of terminal map signals during map performance. Is eye–head to hand–arm control of this type? For example, let a limb be controlled by system (II). Suppose that the limb moves to a desired position under the direction of sustained eye fixation at that position. How can this position be maintained after the source of its terminal signals is shut off? When the limb reaches the desired position, a resonance is established between (IIb) and (IIc). After resonance sets in, shut off the source of terminal signals to (IIc)—that is, the motor command. Field (IIb) continues to supply the correct terminal coordinates to (IIc), and arousal at (IIc) keeps the map suprathreshold. In other words, proprioceptive feedback plus arousal can sustain the resonance, and thus the signals to muscles that maintain the posture. When arousal is shut off, the resonating loop is broken, and the postural position is terminated.

The minimal mechanism also has isometric properties when a limb is passively dragged into a sequence of terminal positions. If arousal is off, the proprioceptive data flows from (IIa) to (IIb) to (IIc). There is no feedback to (IIb). When arousal is turned on, however, then the resonance between (IIb) and (IIc) resists further change due to hysteresis. Simultaneously, signals from (IIc) to the muscles tend to hold the present terminal position. Thus the size of the arousal signal determines the amount of isometric tension in this system.

Some light is also shed on the controversy about whether the motor commands of terminal position are feedforward, or are determined by proprioceptive feedback (Bizzi *et al.*, 1975). In fact, the answer even in the minimal network is subtle. After a signal is received at (IIc) from (Ib), then the motor command leaves (IIc) for the muscles. Proprioceptive feedback at (IIa) is blocked at (IIb) to prevent it from distorting the terminal map at (IIc). This system is therefore capable of reaching its terminal position by using feedforward control. However, as the limb approaches the desired terminal position, resonance is established between proprioceptive feedback and the terminal map. Then proprioceptive cues support the terminal map even though they are not needed to reach the terminal position. The subtlety arises because the existence of proprioceptive input at the terminal map cells does not imply the necessity of this input for computing terminal position. Rather it sets the stage for map learning and postural maintenance.

The systems also suggest interesting questions about saccadic versus continous map control. While a map is being *learned*, the arm and hand move together or are at rest. Only in this way can resonances be established between proprioceptive and terminal coordinates. Yet, while a map is being *performed,* its proprioceptive map almost never agrees with its terminal map. Indeed, when you fixate an object that your hand is already on, there is no need to turn on a map to move your hand onto the object. The situation during learning can be described as a *continuous* motion wherein proprioceptive and terminal coordinates are always very close. The situation during performance can be described as a *saccadic* motion wherein proprioceptive coordinates do not influence the motor act. This distinction is often made in the literature on motor control (Robinson, 1964; Yarbus, 1967). In the minimal network, the continuous and saccadic systems share some cells in common—for example, the (Ic) cells that code terminal maps. The present construction therefore suggests some interesting questions about the interpretation of oculomotor data.

The cells in (Ib) will fire only when the system moves in its continuous mode—that is, when proprioceptive and terminal coordinates are close to one another. Are these among the cells that are usually included in the continuous system? If they are, then cells such as those in (Ic) should exist that are active in both the continuous and the saccadic modes. A likely place to search for such cells in appropriate species is the superior colliculus (Goldberg and Wurtz, 1972a,b; Stryker and Schiller, 1975; Wurtz and Goldberg, 1972a,b), where maps of visual sensory data into motor eye-movement data have been experimentally described.

In the system as it stands, it is possible to learn spurious correlations between eye and hand positions. In principle, the eye can fixate one point and the hand can be held at another point while map learning takes place. However, since an infant's eye is drawn strongly toward moving objects (Piaget, 1963), it is very unlikely that this will ever happen under normal developmental conditions. Also, the tendency is not eliminated by building up maps using cells that are activated by dynamic limb motions rather than by spatial patterns. For example, the eye is often fixated on an object that it wants to touch; the sampling cells thus cannot be activated only by eye motion. Also the arm is often called upon to move to a prescribed terminal point from an arbitrary initial point; thus, the terminal map cannot be determined by directionally sensitive cells. Although the above system can in principle be fooled, it can also delete correlations that lead to erroneous, and therefore unexpected, consequences using attentional mechanisms, as in Section 20 and in Grossberg (1975).

The system can also be fooled, as it stands, because (Ib) can sample a suprathreshold pattern at (IIc) even if system (II) has not yet reached its terminal position. This property creates no difficulty if sampling usually occurs in the continuous mode.

The above shortcomings suggest possible limitations on the accuracy of map learning in infants, and emphasize the importance of an infant's sensitivity to moving objects. The last shortcoming can easily be formally overcome. Insert an interneuron in the path from (IIb) to (IIc), and let (Ib) sample the interneuron. Since the interneuron is inactive unless resonance holds between (IIb) and (IIc), map learning between (I) and (II) can then occur only if both systems are resonating. Such variations are not very useful in the absence of data that can discriminate between them.

There is a variation that is useful to consider, however. In principle, the proprioceptive and terminal coordinates of a given position need not generate the same pattern of neural activity. Where this occurs, stage (Ib) can be used to learn the match between proprioceptive and terminal maps that code the same position. In fact, stages (Ia), (Ib), and (Ic) then emerge as an adaptive coding scheme for bridging the gap between the proprioceptive and terminal coordinates within a given motor system. By way of illustration, suppose that the (Ia) → (Ib) pathways contain LTM traces as in Section 17. Then (Ib) can learn a code for the proprioceptive coordinates at (Ia). Feedback pathways from (Ib) to (Ia) would then exist to stabilize this code using learned templates. The coded proprioceptive map at (Ib) could then learn the corresponding terminal map at (Ic) using LTM traces in the (Ib) → (Ic) pathways, while the terminal map at (Ic) can learn the proprioceptive code at (Ib) using LTM traces in the (Ic) → (Ib) pathways.

Given any such adaptive rules within the systems (I) and (II), map learning *between* these systems then becomes a continuation of map learning *within* each system separately. The between-system stage cannot begin until resonances can be generated within each system. As in Section 24, each stage in this formal developmental sequence obeys similar principles, even though different stages code different behaviors.

Finally, we note that, if the networks learn only terminal positions, then auxiliary systems must exist that enable limbs to reach these positions. The α–γ system is a classical example of such an auxiliary system (Granit, 1966; Grillner, 1969; Thompson, 1967, Chapter 12) but it goes beyond the scope of this paper.

52. Feedforward versus Feedback Control of Sequential Map Performance

What is remembered when we learn a series of motor acts, such as a dance? Below we suggest that serially ordered terminal maps generate and are sampled by adaptive sensory-motor codes (cf. Sections 11, 50, and 51) under the guidance of internal and external feedback. A model of this process will now be described using mechanisms that have already been introduced.

First we suppose that a serial STM buffer for motor activity exists. In Fig. 36, fields $\mathscr{F}^{(1)}$, $\mathscr{F}^{(2)}$, and $\mathscr{F}^{(3)}$ model some stages of a sensory, in particular an auditory, STM buffer. Henceforth these fields are denoted by $\mathscr{F}_S^{(1)}$, $\mathscr{F}_S^{(2)}$, and $\mathscr{F}_S^{(3)}$. Now expand Fig. 36 as follows. Denote $\mathscr{F}^{(4)}$ in Fig. 36 by $\mathscr{F}_M^{(2)}$, since it is functionally analogous to $\mathscr{F}_S^{(2)}$, which also codes data from a given time frame. Let $\mathscr{F}_M^{(1)}$ denote the field of terminal motor maps, and denote $\mathscr{F}^{(5)}$ by $\mathscr{F}_M^{(3)}$, since it codes order information among these maps. Trainable pathways from $\mathscr{F}_S^{(3)}$ to $\mathscr{F}_M^{(3)}$ are introduced, as in Section 35, to enable sensory order information to activate corresponding motor order information. These sensory-motor signals will automatically be tuned at $\mathscr{F}_M^{(3)}$ by the properties of self-similarity and uniform quenching within the motor STM buffer (see Fig. 43). Indeed, all coding operations in the sensory buffer are assumed to have analogs in the motor buffer.

How are learned data read out of the motor buffer? To see this, suppose that a monotone decreasing pattern exists across the populations $v_1^{(2)}$, $v_2^{(2)}$, $v_3^{(2)}$, . . . of $\mathscr{F}_M^{(2)}$. When $\mathscr{F}_M^{(2)}$ is aroused by a nonspecific rehearsal wave, $v_1^{(2)}$ fires until its activity self-destructs via inhibitory feedback. The spatial pattern of $v_1^{(2)}$'s signals generates a terminal map at $\mathscr{F}_M^{(1)}$. As in Section 50, this pattern becomes suprathreshold and is stored in STM if $\mathscr{F}_M^{(1)}$ receives motor arousal. The terminal map thereupon elicits descending signals to the muscles that will execute it.

The above mechanism highlights a difficulty of traditional probabilistic models of behavior. How would a probability theorist interpret the control of motor commands? Suppose that STM activity at $v_1^{(2)}$ is interpreted as the probability of executing a terminal map. After $v_1^{(2)}$

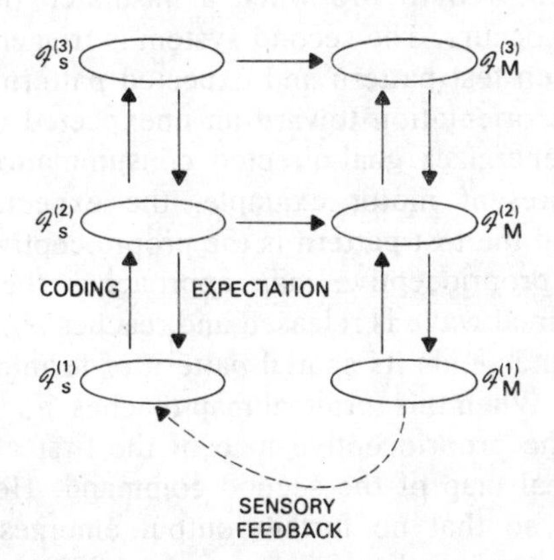

FIG. 43. Code learning and template learning in and between sensory and motor modalities.

self-inhibits, where does the probability go? The terminal map is still active, so the probabilistic modeler might say that $v_1^{(2)}$ stays active. This is, however, false and yields incorrect intuitions about the process by obscuring the distinction between coding of order and coding of terminal position.

How does field $\mathscr{F}_M^{(2)}$ know when the $v_1^{(2)}$ act is nearing completion? Otherwise expressed, how does $\mathscr{F}_M^{(2)}$ know when to turn on the next motor map? There are two answers in the model.

The first answer describes a feedforward control that is analogous to rapid arpeggio playing on a piano (cf. Lashley, 1951). If rehearsal arousal is kept on at $\mathscr{F}_M^{(2)}$, then each population $v_1^{(2)}$, $v_2^{(2)}$, $v_3^{(2)}$, . . . is excited in its proper order as the previous population self-inhibits. Each successive spatial pattern of signals descends to $\mathscr{F}_M^{(1)}$, where it controls motor action until it is reset by the next pattern. Feedback influences the system only indirectly via sensory feedback that is chunked before generating signals to $\mathscr{F}_M^{(2)}$.

The second answer describes a feedback control that is analogous to executing slow dance motions. Let $\mathscr{F}_M^{(2)}$ be briefly aroused. Then $v_1^{(2)}$ fires and self-inhibits. Motor arousal is maintained at $\mathscr{F}_M^{(1)}$, which stores the first terminal map and activates its muscles accordingly. As the act is executed, the proprioceptive coordinates approach the terminal coordinates. As the match between proprioceptive and terminal coordinates improves, a nonspecific rehearsal wave perturbs $\mathscr{F}_M^{(2)}$. This rehearsal wave was mentioned in Section 23, where the existence of a competition between two antagonistic motor arousal systems was postulated. The first system is allowed to fire when a mismatch of test pattern and expected pattern occurs. The second system is triggered by output from the match between test pattern and expected pattern. The first arousal system energizes orientation toward an unexpected event. The second arousal system energizes goal-directed consummation of an expected event. In the present motor example, the expected pattern is the terminal map, and the test pattern is the proprioceptive map.

Thus, as the proprioceptive map approaches the terminal map, a nonspecific rehearsal wave is released and reaches $\mathscr{F}_M^{(2)}$. Population $v_2^{(2)}$ thereupon fires and sends its spatial pattern of terminal signals to $\mathscr{F}_M^{(1)}$ as it self-inhibits. When this terminal map reaches $\mathscr{F}_M^{(2)}$, it resets $\mathscr{F}_M^{(2)}$ as in Section 50. The proprioceptive map of the first command does not match the terminal map of the second command. Hence the rehearsal wave terminates so that no further output emerges from $\mathscr{F}_M^{(2)}$. The second terminal map can therefore control performance until its goal is reached. Then the cycle of matching; resonance and rehearsal wave;

sampling and self-inhibition; and reset and mismatch begins again. After there are no further STM data in $\mathscr{F}_M^{(2)}$, the posture of the last gesture is held until motor arousal is withdrawn from $\mathscr{F}_M^{(1)}$.

The above mechanism can now be joined with previous results (for example, Section 35) concerning the reorganization of the future field due to sensory feedback. All the results go through because they hold for arbitrary spatial patterns, irrespective of what these patterns code.

53. Sequential Switching between Sensory and Motor Maps

The same general mechanisms seem to hold in many examples of goal-oriented behavior. Another example is briefly sketched below to illustrate how switching between sensory and motor map systems can occur when a plan is executed.

Suppose that the command "Touch the yellow ball" is given. I suggest that the verbal command is encoded and does (at least) two things. The verbally decoded message elicits signals that are gated by LTM traces. The gated signals subliminally activate sensory feature detectors (for example, in $\mathscr{F}_S^{(2)}$) which code a yellow, ball-like object because of prior learning. Since there is no match with this field, the message can also trigger a series of orienting reactions by activating the eye–head motor buffer. This series is perhaps subject to the spacing effect of Section 42, which tends to inhibit perseverative search in the same position. The search continues until the yellow ball is seen. The ball is then visually decoded and (approximately) matches the subliminal sensory pattern. The match induces a resonance that inhibits orienting motor arousal and, in a complementary fashion, triggers goal-oriented motor arousal.

What motor system is hereby aroused? When orienting arousal is inhibited, the eye–head system is fixated on the yellow ball. It can thereby generate a terminal map in the hand–arm systems. One of these is activated by motor arousal. The chosen hand then moves toward the ball until it touches it. In all, the plan has been executed by a sensory match, a motor match, and a switchover from orienting motor arousal to goal-oriented motor arousal.

54. Map Reversal by Antagonistic Rebound

Many details have been overlooked in the above description in order to emphasize its essential simplicity. Also, many new questions are raised. For example, after the hand surrounds the ball, how does it retrieve the ball? There exists an elementary answer that is probably most true in infants, but it also highlights how different adult control can be. After the hand surrounds the ball, suppose that an antagonistic rebound is triggered throughout the arm–hand system. For example, if touching the ball elicits nonspecific arousal to the arm–hand terminal map, then the agonist–antagonistic cells throughout the map will transform flexion into extension, and conversely, in a graded fashion across the affected muscle groups. In particular, the hand will close and the arm will be retrieved.

The main point is that retrieval need not, in principle, recode the act cell by cell, which would require a high-dimensional control. A one-dimensional increment in nonspecific arousal can be used instead.

55. Imprinting, Imitation, and Sensory–Motor Algebra

Consider the babbling behavior of an infant (Fry, 1966) as motivation for the following construction. Suppose that terminal maps at $\mathscr{F}_M^{(1)}$ are endogenously activated, analogous to the babbling of sounds. The execution of these maps elicits sensory feedback, such as sounds, which feed back to $\mathscr{F}_S^{(1)}$ and are coded at $\mathscr{F}_S^{(2)}$ on successive trials. The terminal motor map is adaptively coded at $\mathscr{F}_M^{(2)}$, and the motor code at $\mathscr{F}_M^{(2)}$ simultaneously learns the terminal motor map at $\mathscr{F}_M^{(1)}$ as a feedback template. As this happens, $\mathscr{F}_M^{(2)}$ is sampled by its sensory feedback code at $\mathscr{F}_S^{(2)}$. If more than one motor code is active at $\mathscr{F}_M^{(2)}$—say during predictive performance—then $\mathscr{F}_S^{(2)}$ samples its motor command in a context of temporally contiguous motor commands. Performance of the same command can thus be different in different contexts; compare the performance of a given speech sound in different words. As codes within $\mathscr{F}_S^{(2)}$ and $\mathscr{F}_M^{(2)}$ develop, repeated sequences of motor commands can be adaptively coded as sequential motor chunks at $\mathscr{F}_M^{(3)}$ and as sequential sensory feedback chunks at $\mathscr{F}_S^{(3)}$. Thereupon LTM traces from $\mathscr{F}_S^{(3)}$ to $\mathscr{F}_M^{(3)}$ can encode associations from the sensory to motor sequential chunks.

Endogenously active arousal sources supply the motivational support that drives these LTM processes. The combined effect of all these LTM processes is to *complete*, or *close*, the sensory-motor feedback loops that are endogenously activated during the babbling phase. Hereby the network can imitate a sound that is supplied by an external source, if that sound lies on the generalization gradient of some sound that was endogenously babbled. It does this by activating the sequence of fields

$$\mathscr{F}_S^{(1)} \to \mathscr{F}_S^{(2)} \to \mathscr{F}_M^{(2)} \to \mathscr{F}_M^{(1)}$$

if the sound is coded by a spatial pattern, or the sequence of fields

$$\mathscr{F}_S^{(1)} \to \mathscr{F}_S^{(2)} \to \mathscr{F}_S^{(3)} \to \mathscr{F}_M^{(3)} \to \mathscr{F}_M^{(2)} \to \mathscr{F}_M^{(1)}$$

if the sound is coded by a sequence of spatial patterns.

The endogenously active arousal sources *imprint* developing maps by using the same mechanisms that are driven by reactively activated motivational sources in the "adult" network. Imprinting in the young network and learning in the adult network are similar processes using different motivational triggers.

How does the network learn to imitate more complex sounds than those that are endogenously produced? This process follows automatically from properties of the adaptive coding model. A new sensory pattern at $\mathscr{F}_S^{(1)}$ is filtered, or decomposed, by $\mathscr{F}_S^{(1)} \to \mathscr{F}_S^{(2)}$ signals into a set of simpler patterns whose sensory-motor loops have been endogenously closed. Each simpler sensory pattern activates its corresponding motor pattern at $\mathscr{F}_M^{(1)}$. The total pattern that is hereby synthesized at $\mathscr{F}_M^{(1)}$ can produce a sound that is close to the new sensory pattern.

More precisely, suppose that m motor patterns O_i have been endogenously activated at $\mathscr{F}_M^{(1)}$ and thereby elicit the sensory feedback patterns I_i at $\mathscr{F}_S^{(1)}$, $i = 1, 2, \ldots, m$. Let I_i be adaptively coded by population $v_{S_i}^{(2)}$ in $\mathscr{F}_S^{(2)}$, and let O_i be adaptively coded by population $v_{M_i}^{(2)}$ in $\mathscr{F}_M^{(2)}$. Also let each population learn its template of feedback signals, and its $\mathscr{F}_S^{(2)} \to \mathscr{F}_M^{(2)}$ associations. Given this learned substrate, how does the network learn to imitate a sensory pattern I which has never been endogenously elicited at $\mathscr{F}_M^{(1)}$? It is shown below how input I at $\mathscr{F}_S^{(1)}$ is filtered by the $v_{S_i}^{(2)}$ and then resynthesized by the $v_{M_i}^{(2)}$ to produce at $\mathscr{F}_M^{(1)}$ a motor pattern O which elicits the sound I. In summary, during imprinting, endogenous motor commands at $\mathscr{F}_M^{(1)}$ elicit their sensory commands at $\mathscr{F}_S^{(1)}$ as external feedback; during imitation, sensory commands at $\mathscr{F}_S^{(1)}$ elicit their motor commands at $\mathscr{F}_M^{(1)}$ via network filtering.

The filtering mechanism uses elementary vector space properties (Thomas, 1968). As adaptive coding proceeds in response to a sensory pattern I_i, the classifying vector z_i becomes proportional to I_i, and the signal from $\mathscr{F}_S^{(1)}$ to $v_{S_i}^{(2)}$, namely $S_i = I_i \cdot z_i$, becomes proportional to $|I_i|^2$. Suppose that the set of I_i, $i = 1, 2, \ldots, n$, *spans* the vector space of input patterns I at $\mathscr{F}_S^{(1)}$; then any I can be written a linear combination $I = \sum_{k=1}^m \alpha_k I_k$ of the I_i's given suitable coefficients α_k. If, moreover, the I_i are mutually orthogonal (that is, $I_i \cdot I_j = 0$, $i \neq j$), then the signal to $v_S^{(2)}$ in response to I, namely $S_i = I \cdot z_i$, is proportional to α_i. Thus each $v_S^{(2)}$ is excited by a signal that is proportional to how close I_i is to I. If the signals from $\mathscr{F}_S^{(2)}$ to $\mathscr{F}_M^{(2)}$ are linear, then the signal to $v_{M_i}^{(2)}$ is also proportional to α_i. Since $v_{M_i}^{(2)}$ encodes pattern O_i in its LTM traces, $v_{M_i}^{(2)}$ generates a pattern at $\mathscr{F}_M{}^{(1)}$ that is proportional to $\alpha_i O_i$. The total pattern at $\mathscr{F}_M{}^{(1)}$ that is generated by $\mathscr{F}_M{}^{(2)}$ is therefore close to $O = \Sigma_{k=1}^m \alpha_k O_k$.

Does O elicit a sound close to I? This is true if the mapping from motor patterns at $\mathscr{F}_M{}^{(1)}$ to sensory feedback patterns (approximately) preserves the weights α_i. This important property should be tested experimentally.

Several aspects of this mechanism deserve comment. Most remarkably, the signal law $\Sigma_{k=1}^n B_{ki} z_{ki}$ in (1), which was originally derived from simple classical conditioning postulates, also implies a map formation property.

The crux of the argument is that, no matter what motor pattern is endogenously active at $\mathscr{F}_M{}^{(1)}$, and no matter what its sensory feedback pattern is at $\mathscr{F}_S{}^{(1)}$, the sensory feedback pattern gets associated via $\mathscr{F}_S{}^{(2)}$ and $\mathscr{F}_M{}^{(2)}$ with the motor pattern. In vector space terms, an arbitrary set of vectors in the vector space at $\mathscr{F}_M{}^{(1)}$ can be associated with an arbitrary set of vectors in the vector space at $\mathscr{F}_S{}^{(1)}$. This property lets each network adapt to individual differences in the structure of its sensory and motor modalities. Given these associations, imitation is achieved if the network can map the weights α_i as signal sizes from $\mathscr{F}_S{}^{(2)}$ to $\mathscr{F}_M{}^{(2)}$. This is relatively easy to do, since it reduces a global mapping problem between two multidimensional vector spaces to a local rule for signal transmission.

What kinds of coding difficulties can occur? First, the patterns I_i need not span the space of input patterns. Inputs I, which cannot be represented as a linear combination $\Sigma_{k=1}^m \alpha_k I_k$, will then be filtered, or projected, by $\mathscr{F}_S{}^{(1)} \rightarrow \mathscr{F}_S{}^{(2)}$ signals as the closest pattern I^* that can be represented in this way. Second, if the patterns I_i are not mutually orthogonal, then each signal $S_i = I \cdot z_i$ will include interference terms of the form $I_j \cdot I_i$. These terms will distort the relative activities of the $\mathscr{F}_S{}^{(2)}$ populations. How can these and other distortions be corrected? The next section suggests an answer.

56. Self-Tuning and Multidimensional Inference in a Parallel Processor

The size of the QT in $\mathscr{F}_S{}^{(2)}$ and $\mathscr{F}_M{}^{(2)}$ will determine which of the populations $v_{S_i}^{(2)}$ and $v_{M_i}^{(2)}$ will be supraliminally excited. In order to map the weights α_i accurately, the QT must be small. This can occur in two main ways: structurally or dynamically. Either lateral inhibition is weak, owing to a sparcity of inhibitory interneurons during the filtering stage, or shunting arousal is large. The latter mechanism is a special case of the self-tuning process in Section 20. It embodies a search procedure, or attentional mechanism, whereby the correct level of map tuning is stabilized. To illustrate the main idea, let input $I = \Sigma_{k=1}^{n} \alpha_k I_k$ be presented to $\mathscr{F}_S^{(1)}$, and suppose that $\alpha_1 > \alpha_2 > \cdots > \alpha_m$. If arousal starts out low, then only $v_{S_1}^{(2)}$ might initially be stored in STM. Consequently $v_{S_1}^{(2)}$ releases a template to $\mathscr{F}_S^{(1)}$ that is proportional to I_1. The mismatch between I_1 and I increases arousal at $\mathscr{F}_S^{(2)}$. The activity of $v_{S_1}^{(2)}$ starts to decrease owing to antagonistic rebound, and $v_{S_2}^{(2)}$ starts to become active. Then $v_{S_1}^{(2)}$ and $v_{S_2}^{(2)}$ both release templates to $\mathscr{F}_S^{(1)}$, and this hybrid template is approximately proportional to $\alpha_1 I_1 + \alpha_2 I_2$, which matches I better. Thus, if arousal increases again, the increment is smaller, and a hybrid template close to $\alpha_1 I_1 + \alpha_2 I_2 + \alpha_3 I_3$ is elicited. A few cycles of this reverberation can quickly retune $\mathscr{F}_S^{(2)}$ until the STM pattern across $\mathscr{F}_S^{(2)}$ generates a feedback template to $\mathscr{F}_S^{(1)}$ that almost matches I. The resonant STM pattern at $\mathscr{F}_S^{(2)}$ then automatically generates a motor pattern at $\mathscr{F}_M^{(2)}$ that approximately equals O. With O active at $\mathscr{F}_M^{(2)}$ while I is active at $\mathscr{F}_S^{(1)}$, coding and associative learning can gradually close the $I \to O \to I$ loop as codes for I and O are synthesized.

Many interesting developmental questions are posed by the concept of a self-tuning filtered map. For example, what keeps the reset mechanism from totally inhibiting $v_{S_1}^{(2)}$ before $v_{S_2}^{(2)}$ is excited? Is this due to the fact that the partial match of I_1 to I creates a small arousal increment, and of $\alpha_1 I_1 + \alpha_2 I_2$ to I an even smaller arousal increment, etc.? Or is $v_{S_2}^{(2)}$ quickly excited by arousal because of its prior subliminal excitation? Or is the rebound mechanism weak at this developmental stage?

The self-tuning process describes a type of multidimensional inference in a real-time parallel processor (Anderson, 1958). The successive switching-in, or reset, of channels $v_{S_i}^{(2)}$ is analogous to principal component analysis, or discriminant analysis, of a space–time pattern (Donchin and Herning, 1975). It is of particular interest that multidimensional techniques have successfully been used to analyze the P300, which is interpreted herein as an index of the amount of reset.

57. No Sensory Feedback Implies No Map Formation

Hein and Held (1967) and Held and Bauer (1967) have shown that, when young kittens and monkeys reach for objects without being able to see their hands, then no positional map develops. By contrast, if the eye can see where the hand goes, then a map does develop. This occurs in the model for a basic reason.

Let $\mathscr{F}_M^{(1)}$ be the field of terminal maps for the hand–arm system, and let $\mathscr{F}_S{}^{(1)}$ be stage (Ib) of the eye–head system (Section 48). If no sensory feedback reaches $\mathscr{F}_S{}^{(1)}$, then there is no way to build the associative bridges to $\mathscr{F}_M{}^{(2)}$ across which the coefficients α_i can be mapped by signals. Without these associative bridges, when the monkey looks at a new position, the position is not filtered into combinations of old positions that have the correct relative signal sizes. Hence no spatial map exists.

58. Does the Psychophysical Power Law Influence Imitation Errors?

Section 17 describes the simplest case of the adaptive coding model for purposes of exposition. Where a power law transformation (Mountcastle, 1967; Stevens, 1961) controls the filtering signals, the size of the power influences whether spatially localized or diffuse patterns will be preferentially weighted.

To illustrate this, suppose as in Section 17 that $\mathscr{F}^{(1)}$ normalizes its patterns. Let the input pattern I be transformed into a normalized pattern $\theta = (\theta_1, \theta_2, \ldots, \theta_n)$. Then $\Sigma_{k=1}^{n} \theta_k = 1$. In Section 17, the signal from $\mathscr{F}^{(1)}$ to $v_j{}^{(2)}$ was defined by $S_j = \Sigma_{k=1}^{n} \theta_k z_{kj}$. More generally, $S_j =$ $S_j = \Sigma_{k=1}^{n} f(\theta_k) z_{kj}$, and correspondingly (19) is replaced by

$$\dot{z}_{ij} = [-z_{ij} + f(\theta_i)]x_j{}^{(2)} \tag{52}$$

Introduce the notation $f(\theta) = (f(\theta_1), f(\theta_2), \ldots, f(\theta_n))$. Then

$$S_j = \|f(\theta)\| \cdot \|z_j\| \cos [f(\theta), z_j] \tag{53}$$

Often $f(w)$ is a sigmoid function of w, as in Section 15. The sigmoid can be approximated by a power law at small values of w, say $f(w) \cong f_p(w) \equiv w^p$. The size of p influences what patterns will be coded by each $v_j{}^{(2)}$ as follows.

Unbiased coding occurs if $p = \frac{1}{2}$. In this case, the normalization condition $\Sigma_{k=1}^{n} \theta_k = 1$ implies $\|f_{1/2}(\theta)\| = 1$. Furthermore (52) implies that $\|z_j\|$ approaches 1, owing to developmental tuning. Thus, after tuning takes place,

$$S_j \cong \cos [f_{1/2}(\theta), z_j]$$

so that S_j is maximized by the pattern $f_{1/2}(\theta) = z_j$. If $p > \frac{1}{2}$, there exists a tendency to code spatially localized patterns, because $\|f_p(\theta)\|$ is maximized by any point pattern: $\theta_i = 1$ and $\theta_j = 0, j \neq i$. If $p < \frac{1}{2}$, there exists a tendency to code spatially diffuse patterns, because $\|f_p(\theta)\|$ is maximized by the uniform pattern $\theta_i = 1/n$, $i = 1, 2, \ldots, n$. Thus if $p \neq \frac{1}{2}$, the signal law (53) mixes two maximizing tendencies: maximize cos $[f_p(\theta), z_j]$ by choosing $f_p(\theta)$ parallel to z_j; and maximize $\|f_p(\theta)\|$.

Given m normalized patterns ϕ_j, $j = 1, 2, \ldots, m$, suppose that z_j adapts to $f_p(\phi_j)$, and that pattern θ can be written as $\theta = \Sigma_{k=1}^n \alpha_k f_p(\phi_k)$. If the vectors $f_p(\phi_j)$ are mutually orthogonal, then the signal

$$S_j = \theta \cdot z_j = \alpha_j \|f_p(\phi_j)\| = \alpha_j \sum_{k=1}^{n} \phi_{jk}^{2p}$$

where $\phi_j = (\phi_{j1}, \phi_{j2}, \ldots, \phi_{jn})$. If $p = \frac{1}{2}$, then $S_j = \alpha_j$; hence the coefficients α_j are mapped without bias. However, if $p > \frac{1}{2}$, then populations $v_j^{(2)}$ are favored whose patterns ϕ_j are spatially localized; if $p < \frac{1}{2}$, then populations are favored whose patterns ϕ_j are spatially diffuse.

To test these effects experimentally will require a correlative analysis of coding at $\mathscr{F}_S^{(1)}$, of the (approximate) power laws of $\mathscr{F}_S^{(1)} \rightarrow \mathscr{F}_S^{(2)}$ and $\mathscr{F}_S^{(2)} \rightarrow \mathscr{F}_M^{(2)}$ signals, and of trends in imitation errors.

59. Rhythm and Phrasing

When a musician plays a piece, how are the relative velocities of each note controlled? More generally, many sequential sensory-motor skills are performed with a fixed order and a characteristic rhythm. Yet the rhythm can be modified without destroying the correct order. How is this flexible relationship between order and rhythm established?

Section 9 notes an example of this phenomenon wherein varying the size of rehearsal arousal through time can alter performance velocity in an avalanche. This mechanism is instructive, but is insufficient in general. Consider a sensory chunk whose template encodes a sequence of motor commands. Then the chunk must also encode performance velocity. How is this accomplished? The avalanche example, along with the discussion of instrumental conditioning in Section 7, suggests that the chunk sends a conditionable pathway to the arousal source. Then the amount of arousal, and hence performance velocity of the entire sequence, can be changed by conditioning. This mechanism allows order and velocity information to be decoupled; any other input to the arousal source can change the rhythm without changing the order. However, the mechanism only alters the overall performance velocity *within* a sequence. How is arousal calibrated so that fixed changes in arousal determine prescribed velocity changes in different sequences? In particular, how is the amount of arousal calibrated to compensate for differences in sequence length?

This problem can be restated as follows. To heighten intuition, call the items controlled by the given command a *phrase*. Is there a tendency to quantize time so that each phrase fills a unit time interval of performance? If this were so, then items coded in longer phrases would be performed faster, other things being equal, and variations in (conditioned) arousal level could shrink or expand this quantized time unit, and thereby increase or decrease performance velocity by fixed amounts. In the special case that phrases are words, time quantization is compatible with the existence of a breath pulse, or syllable (Lenneberg, 1967, p. 115).

The time quantization problem can be restated as a technical question. Given an STM buffer that contains k active items in a monotone decreasing pattern $x_1 > x_2 > \cdots > x_k$, how fast are the items performed as a function of k, other things being equal? Opposite answers can be derived if the design of the buffer is changed. Suppose for definiteness that total STM activity in the buffer is normalized. Then as k increases, each item has smaller activity, other things being equal. Smaller activities can imply slower or faster performance velocities, depending on other factors. For example, in the avalanche of Section 5, smaller activities imply slower velocities. This is because the time needed for activity to achieve suprathreshold values at $v_i^{(2)}$, given a fixed arousal base line, is a monotone decreasing function of signal size from $v_{i-1}^{(2)}$. In this situation, the rate at which excitation grows is the dominant effect. Whenever the rate of feedforward excitation growth is rate-limiting, longer phrases imply slower performance velocities.

Suppose by contrast that all STM activities are already actively reverberating at asymptotic levels before arousal is turned on. Suppose that these STM activities perturb the network's output cells, but that the output cells cannot fire until they are aroused. Let arousal act quickly when it is turned on, and when an output cell begins to fire, let it immediately begin to inhibit its STM source. In this situation, the rate at which feedback inhibition acts is rate-limiting, since as soon as one STM source is inhibited, the next fires, and so on. If feedback inhibition is rate-limiting, then smaller STM activities are more rapidly quenched. Smaller activities then imply faster velocities, and longer phrases imply faster performance. In summary, feedforward excitation and feedback inhibition have opposite effects on performance velocity.

A simple case of the feedback inhibition phenomenon is illustrated below. Let feedback inhibition grow at a rate proportional to suprathreshold STM activity, and let it inhibit STM activity at a rate proportional to its size. Denote by $x(t)$ the STM activity at time t, by $y(t)$ the amount of feedback inhibition, and by Γ the *QT*. Then $\dot{x} = -Ay$ and $\dot{y} = B[x - \Gamma]^+$ where $[w]^+ = \max(w, 0)$. Given initial STM activity I at time $t = 0$, it follows that STM reaches the *QT* at time $T(I) = \text{arc cos } [(I - \Gamma)^{-1}]$, which is a monotone increasing function of I.*

Many questions are raised by the above observations. If phrasing is an important factor in velocity control, then the number of phrases allowed in the performance buffer at any time must be carefully regulated. How is switching on of the next phrase accomplished in a way that prevents discontinuities in performance? Two different types of sensory feedback are probably important: sensory feedback that turns on new spaced sensory chunks which thereupon reset the motor buffer (Section 34), and reset of terminal maps when proprioceptive-terminal matching occurs (Section 52). A careful analysis of special cases is clearly needed, however.

For example, suppose that the buffer starts to renormalize its total activity after a population is quenched by feedback inhibition. If renormalization acts slowly relative to feedback inhibition, then items near the end of each phrase will have the smallest activities when they are performed, and performance rate will speed up as the phase is executed. If renormalization is fast relative to feedback inhibition, then items near the end of the phrase will have the largest activities when they are performed, and performance rate will slow down as the phrase is performed. If the two effects are balanced, then a uniform performance rate occurs. Is there a mechanism that automatically balances the two inhibitory effects to guarantee uniform performance rates in all cases? Such a mechanism would reduce the *QT*, or amplify STM, if the *total* buffer activity decreases. This will happen, for example, if nonspecific excitatory interneurons that are driven by total STM activity recurrently excite the off-surrounds of the STM buffer.

An entirely different kind of performance will occur if arousal is turned on only when matching between proprioceptive and terminal maps occurs. In this case, if the buffers can renormalize themselves faster than a terminal map is executed, then a uniform rate of performing "syllables" can be achieved (cf. Lenneberg, 1967, p. 115). By changing the relative balance between buffer reset by sensory feedback, arousal onset by feedback due to map matching, arousal onset by descending commands, feedforward excitation, feedback inhibition, and field normalization, one can change performance from item perseveration, to uniform rates, to rhythmical speeding up or slowing down, or to a wide

*Erratum: $T(I) = \pi / 2\sqrt{AB}$ is independent of I, which makes uniform phrase performance velocity easier to achieve.

range of phrase velocities. Are certain pathologies in speech production due to such changes of balance?

60. Reciprocal Intermodality Feedback, Internal Hearing, and Naming

This section lists some important implications of network mechanisms. One of them will be used to provide a unified explanation of recent data on serial versus parallel visual information processing in Section 61.

If two or more modalities are associatively related, then their effects on each other can be reciprocal. Figure 44 schematizes two examples. In Fig. 44a, $\mathscr{F}_S^{(1)}$ denotes an auditory field, and $\mathscr{F}_M^{(1)}$ denotes a field of motor commands for speech-related musculature. The other fields establish codes for their base fields in the usual way. Field $\mathscr{F}_S^{(1)}$ can be excited via at least two routes. Activating $\mathscr{F}_M^{(1)}$ can elicit sounds that excite $\mathscr{F}_S^{(1)}$ via sensory feedback. Learned template signals from $\mathscr{F}_S^{(2)}$ to $\mathscr{F}_S^{(1)}$ can also activate $\mathscr{F}_S^{(1)}$. This activation serves as a subliminal sensory expectation if $\mathscr{F}_S^{(1)}$ is not aroused. If $\mathscr{F}_S^{(1)}$ is aroused, then the subliminal pattern becomes supraliminal and creates the impression of

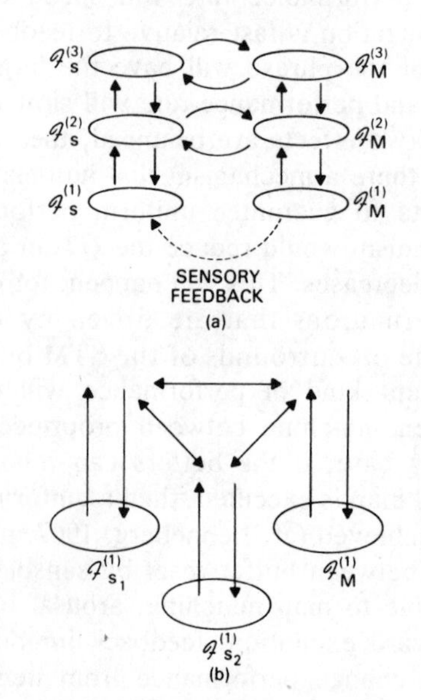

FIG. 44. Reciprocal intermodality associations.

hearing an internally generated sound. Thus, motor activity at $\mathscr{F}_M{}^{(1)}$ is not needed to excite internal sounds at $\mathscr{F}_S{}^{(1)}$. As noted in Section 52, signals from $\mathscr{F}_S{}^{(3)}$ to $\mathscr{F}_M{}^{(3)}$ can generate anticipatory motor expectations, which ultimately elicit motor acts. Reciprocal signals from $\mathscr{F}_M{}^{(3)}$ to $\mathscr{F}_S{}^{(3)}$ can, in principle, generate anticipatory sensory expectations. In general, the several levels of associationally joined sensory and motor systems can mutually support each other's performance.

Figure 44b schematizes an important special case of reciprocal associations between the visual, auditory, and motor systems. $\mathscr{F}_{S_1}^{(1)}$ represents an auditory field, $\mathscr{F}_{S_2}^{(1)}$ represents a visual field, and $\mathscr{F}_M{}^{(1)}$ represents a motor field for speech-related musculature. Suppose that an object is visually presented to the network as its name is spoken aloud. The sequence of sounds at $\mathscr{F}_{S_1}^{(1)}$ can be imitated by using filtering properties from system S_1 to M. The sounds can hereby gradually generate a sequential motor code at $\mathscr{F}_M{}^{(3)}$ on successive trials. Simultaneously, the visual image of the object is coded at $\mathscr{F}_{S_2}^{(2)}$. The visual code can then sample the sequential motor code. Later, the visual image can elicit motor performance of the name. Furthermore, as the name is practiced, it can generate a sequential auditory code at $\mathscr{F}_{S_1}^{(3)}$. This code can sample the object's visual code, which in turn has learned a visual template at $\mathscr{F}_{S_2}^{(1)}$. Later, hearing the name can create a visual expectation of seeing the object. Similarly, seeing the object can create at $\mathscr{F}_{S_1}^{(1)}$ an auditory expectation of hearing its name. This expectation can be fulfilled either indirectly by arousing the motor commands of the name, or directly by arousing the auditory field.

The network constructions also admit hierarchical variations, as Section 24 implies. Figure 45 illustrates a hypothetical case in which a sensory field $\mathscr{F}_S{}^{(1)}$ excites several parallel hierarchies of adaptive codes and feedback templates. Each hierarchy discriminates ever-more-refined features of its base code. Field $\mathscr{F}_S{}^{(n+1)}$ interacts with all the hierarchies via adaptive codes and feedback templates. In such a system, a dominant feature in a sensory pattern at $\mathscr{F}_S{}^{(1)}$ can bias the entire hierarchy to expect a global ensemble of features that has often contained the dominant feature. This happens as follows. When a sensory event perturbs $\mathscr{F}_S{}^{(1)}$, it is coded by a pattern across $\mathscr{F}_S{}^{(2)}$. The features in this pattern are processed in two ways. They are projected directly onto $\mathscr{F}_S{}^{(n+1)}$. Here they excite the code that is closest to their pattern. This code, in turn, reciprocally excites the entire hierarchy via its template. Since the features computed at $\mathscr{F}_S{}^{(2)}$ are "simple," the feedback template can bias the field to ignore higher-order features that occur in an unfamiliar configuration. Signals between the hierarchy and $\mathscr{F}_S{}^{(n+1)}$ continue to reset each other until a consensus is reached.

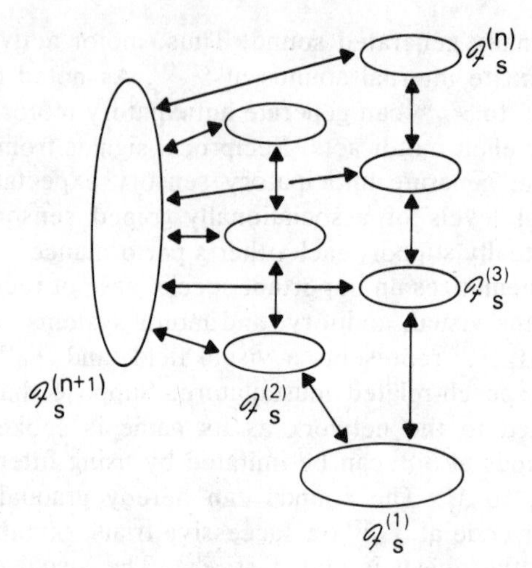

FIG. 45. The field $\mathcal{F}_S^{(n+1)}$ biases the entire hierarchy of fields to interpret patterns at $\mathcal{F}_S^{(1)}$ in terms of the "simple" features that are coded at $\mathcal{F}_S^{(2)}$.

These examples are set forth to emphasis a central problem for future theorizing: the classification of *which* features are computed by particular anatomical and physiological parameters.

61. Automatic versus Controlled Information Processing

Schneider and Shiffrin (1976) have described two complementary types of visual information processing in a series of interesting experiments. *Automatic* processing is said to be "a simultaneous, parallel, relatively independent detection process." *Controlled* processing is said to be a "serial terminating search" process. The authors argue that the two types of processing are associated with characteristic experimental paradigms, and that many earlier STM experiments about the serial or parallel nature of memory search can be classified into one or another of these paradigms. Below it is argued that both types of processing utilize common parallel operations, and that their apparent differences are due to shifts in the relative balance of these operations that are caused by experimental conditions. In particular, serial *properties* do not necessarily imply serial *operations*.

Both types of processing are studied by using a multiple frame search task. Four elements are presented simultaneously on a CRT screen. They are arranged in a square around a central fixation dot, and their joint presentation over a brief time interval is called a *frame*. A trial consists of 20 frames presented in immediate succession at a prescribed rate. The subject's task is to detect one of several items in a *memory set* that is presented before each trial. Items that are not in the memory set are called *distractors*. Items that are neither distractors nor in the memory set are called *masks*.

Automatic processing occurs when the subject has practiced at giving a consistent detection response to memory set items that are never distractors—for example, detecting digits among letter distractors. This is also called a *consistent mapping* (CM) condition. Controlled processing occurs when memory set items and distractors are mixed from trial to trial—for example, detecting digits among digit distractors. This is also called a *varied mapping* (VM) condition. CM performance is much better than VM performance. During CM performance, there is almost no effect of varying the number of distractors in a frame, or of the memory set size; VM performance is monotonically related to each of these variables. Also, during CM performance, false alarms (detections when no target was present) increase substantially at fast frame speeds, but this does not occur during VM performance.

These data are used to conclude that during VM search serial comparisons are made by comparing all display items with a given memory item before a switch to the next memory item occurs. Also each switch to a new memory item takes some time. Data are reported to suggest that VM attentional deficits and search time are intimately related.

During CM search, it is assumed that "a mapping of stimuli to an internal detection or attention response can be learned in long-term memory. . . . Thus in long-term memory an automatic attention response to each target will be learned: the subject can simply wait for the occurrence of one of the learned attention responses . . . the target is always matched or compared first, before any distractors. . . ." Schneider and Shiffrin support this view by showing that automatic search can be learned, and that substantial negative transfer is produced if the memory set and distractor set are then interchanged. This negative transfer is attributed to the learning of an automatic attention response that continues to operate even after the memory set items are used as distractors. They also suggest that CM subjects carry out a controlled search after reversal, and that categorization may facilitate controlled search by reducing the effective memory set size. Results of LaBerge (1973) are also cited to indicate that the automatic resonse is analogous to the process whereby unknown and unexpected characters that are originally matched feature by feature are eventually matched as unitary entities, or chunks, after they are repeatedly presented. More generally, their results support the dictum that "what is learned is what is attended," and that after automatic processing takes over at one level of behavioral organization, controlled processing can begin to organize the automatically processed behavioral units. This latter view is, in fact, the foundation on which the present theory of STM and LTM was originally constructed. In Grossberg (1969a), STM at previously coded behavioral units, which are represented by network vertices or nodes, drives associative LTM changes among these units until new units are thereby formed, whence the process repeats itself. Schneider and Shiffrin (1976) have a similar, albeit heuristic, model in mind when they write: "Suppose long-term store to consist of a collection of nodes that are associatively interrelated in a complex fashion. Each node may itself consist of a complex set of informational elements, including associative connections, programs for responses or actions, and directions for other types of information processing. The node is a distinguishable entity because it is unitized—when any of its elements are activated, all of them are activated."

The main effects found in the Shiffrin and Schneider data can be explained by the present theory. Consider VM search. Suppose that the memory set items are familiar. Then as they are read, they are recoded by their sequential auditory codes. This is an "automatic" process. If the memory set items are not familiar, they are filtered by those lower-order codes that are available. The explicit process of adaptive coding replaces the heuristic rubric of "unitization" in describing how these processes differ. Returning to the case of familiar items, an unfamiliar sequence of these items generates a spatial pattern of activity across the STM buffer of sequential auditory codes. This pattern stores order information in STM. As yet, there is no order information coded in LTM. A nonspecific rehearsal wave can read the items out of the STM buffer one at a time. This happens in a serial fashion (Section 28), and reset of the STM buffer takes some time, but the readout *operation* of nonspecific arousal is a *parallel* operation.

Suppose that a given item is read out of the auditory buffer into its visual code, where it forms a subliminal visual expectation. Then the items in a frame can be visually sampled until a match occurs. When this happens, a burst of activity from the visual code terminates search, as in Section 53. If no match occurs, a rehearsal wave can read the next item out of the auditory buffer, and so on.

Consider CM search. Repeated use of the same memory set gradually generates a higher-order auditory code that can sample the visual codes for all the items over successive trials. When the higher-order code is activated, the visual codes of *all* memory set items can be subliminally activated. Matching with any one of these codes generates a resonant burst. The process therefore seems to be more parallel than VM search. I claim, however, that this is primarily because the higher-order code must be established before the visual codes of all memory set items can be sampled by a single internal representation. In other words, the auditory-to-visual *codes* and *templates* that are activated in VM and CM conditions are different, but the two conditions otherwise share common mechanisms.

Attention enters the search process in several ways. The simplest attentional reaction is amplification of network response to expected items. When a memory set item matches a subliminal visual expectation, a resonant visual response is generated. This type of attentional reaction occurs even under tachistoscopic conditions (cf. Berlyne, 1970). The amplification process can also move the eyes toward prescribed regions of the visual field by generating amplified feedback to the appropriate terminal eye coordinates if there is enough time to execute these motions. The attention theory in Grossberg (1975) suggests how other processes, such as incentive–motivational, CNV, and novelty-triggered feedback, can regulate the amount and pattern of STM amplification. The present theory supports, indeed refines, the dictum that "what is learned is what is attended." As Section 20 notes, the very stability of the STM code requires feedback processes that explicate attentional mechanisms.

Several other aspects of the Schneider and Shiffrin data are interesting. The "time for automatic search is at least as long as that for a very easy controlled search." This is paradoxical if CM search is a more efficient processing scheme. Is partial normalization of the visual template one reason for this? If more cues are subliminally active during CM than during VM search, then each cue will then have less subliminal activity. The reaction time for supraliminal signals to be generated during a match will then be greater during CM than during VM.

Also of interest are the data concerning performance accuracy when a memory set item occurs 0, 1, 2, or 3 frames away from an identical, or different, memory set item. During CM search, no performance decrement occurs if two distinct memory items are separated by any frame interval. This follows if the memory set visual templates are all subliminally active at once. Matching one item does not require reset to match a different item. However, if two identical items occur simultaneously, then the first match can interfere with registration of the second match by either of two mechanisms: if the match elicits performance and feedback inhibition of the item's STM activity, then the higher-order code must be reactivated before the second match can occur; if the resonance due to the first match does not terminate before the second instance of the same item occurs, the resonant state cannot generate a distinguishable resonant burst. In summary, during CM search, two simultaneous distinct memory set items are recognized better than two simultaneous identical items.

The reverse is true during VM search. Here each display item is compared to the visual template of one memory set item at a time. Suppose that a display item matches the visual template. Then the item's visual code is amplified by resonance. This activates the automatic process of recoding the item auditorily. The visual resonance can excite the item's auditory code more strongly than would occur during casual reading of the item in a list of other items. Thus, a tendency exists to give the item maximal activity in the sequential auditory buffer. When the auditory buffer is then aroused, the item's visual template is reactivated. This reset effect makes it easier to detect two identical than two distinct memory set items in successive frames. Whether the items in successive frames are identical or distinct, the visual code is reset to detect the second item after a correct detection, and this takes time.

Similar reset effects can take place when an unexpected display item is processed. For example, suppose that an unexpected memory set item is visually scanned. Suppose that it is automatically recoded auditorily, and that its afferent auditory activity can summate with its activity in the sequential auditory buffer. Until this happens, the network has no way to tell that the display item is a memory set item. If summation does occur, then items with the largest prior STM activity should have the largest tendency to reset the visual code with their own template. This hypothesis should be tested experimentally. Analogously, during a VM search wherein CM items are used as distractors, a scanned CM item can activate its higher-order code, which then resets the visual template to expect CM items.

The above explanation of CM versus VM properties in terms of buffer reset can be tested by running the experiments again while also measuring P300. Also of interest is whether an expected memory set item can reset the auditory STM order more vigorously via visual resonance than an unexpected memory set item. This comparison might depend on subject strategies, since unexpected items can elicit a strong reset via the alarm system unless the alarm is already tuned by the search procedure to prevent this.

By explaining the Schneider and Shiffrin data in a unified way, we avoid several serious problems of their theory. They claim, and I agree, that automatic processing is used to rapidly code familiar behavioral units so that controlled processing can then build these units into new unitized elements. I disagree that the "automatic attention response" in the CM condition is a mechanism that is qualitatively different from mechanisms operating in the VM condition. If the two types of conditions used serial versus parallel *operations,* as Shiffrin and Schneider claim, then how does the brain tirelessly alternate between serial and parallel mechanisms as it practices any new list of unitized elements? How do the serial and parallel processes compete when a visual scene contains both unitized and unfamiliar but relevant objects? How does the switchover from serial to parallel processing take place as an item is unitized? These problems evaporate in the present theoretical framework.

62. Visual versus Auditory Processing and Cerebral Dominance

A pattern of activity across a field of populations at a given time is inherently ambiguous. Does the pattern code a single event in time, such as the features in a visual picture, or does it code the order information in a series of events? Because of this fundamental ambiguity, two distinct types of STM reset mechanisms have been posited in the present theory: (1) deletion of population activity by feedback inhibition; (2) deletion of patterned activity across populations by mismatch with a comparison pattern. Type 1 has been used to explain various facts about order information. To use type 1, the cells in each population are allowed to be broadly distributed across the network, but their activities must be distributed close to a single average value. If the average activities of localized cell clusters in a given population are not approximately the same, there does not exist a simple index of order information between populations. Mechanisms of type 2 are then needed to delete intrapopulation activity.

Given a mechanism of type 1 in a field $\mathscr{F}^{(i)}$, the cells in the next field $\mathscr{F}^{(i+1)}$ automatically code order information in a manner that is sensitive to rehearsal strategies, as in Section 31. Thus, merely changing the inhibitory mechanism within $\mathscr{F}^{(i)}$ causes the adaptive coding mechanism from $\mathscr{F}^{(i)}$ to $\mathscr{F}^{(i+1)}$ to code data about time rather than space. To make this temporal code usable, individual populations in $\mathscr{F}^{(i)}$ must code the controlling features of entire behavioral events.

This latter constraint suggests a reason why visual data are often recoded auditorily to achieve an IMS of significant length. The data that are derived from a typical visual scene are of very high dimension, including colors, myriad shapes, distance information, etc. Coding all the relevant dimensions of a typical visual object in a single population would require many stages of adaptive coding. Much simpler demands are placed on auditory coding, wherein many fewer dimensions covary in the sound spectrograms of simple sounds. Having coded such sounds in individual, albeit perhaps diffusely distributed, populations, order information among these codes is then readily learned by using STM reset by feedback inhibition. Thus the amount of data in a single perceptual frame (visual versus auditory) is traded off against the number of successive frames whose order can be coded.

A second distinction between visual and auditory coding emerges in the study of their circular reactions. The auditory-speech loop uses two modalities, each of which has low-dimensional codes—namely, sounds and speech motor acts. The visual system closes its loop with the motor systems of bodily position, but loop closure involves the motor system for moving the eyes as in Section 48, rather than scene analysis. The eye–neck–head system also has relatively simple codes. Order information among visual scenes can, however, be learned either if the codes for sequential eye movements sample visual representations, or if there does exist feedback inhibition of higher-order visual codes ("grandmother cells"), say in the inferotemporal cortex (Rocha-Miranda *et al.*, 1975).

The above remarks note that an activity pattern per se across a field of populations has an inherently ambiguous interpretation. Does it code data about time or data about space? In order to unambiguously decode temporal versus spatial data, somehow the populations that code the different types of data must be spatially segregated. The patterns themselves do not suffice to make this distinction; rather, the nature of their reset mechanisms accomplishes this. The ambiguity problem therefore suggests the need to spatially segregate the processing of sequential, including language-like, codes from codes that concern themselves primarily with spatial integration. Perhaps this dichotomy is one reason for the emergence of cerebral dominance (Gazzaniga, 1970, Chapter 8). Since visual and auditory representations are bilateral, the trend toward segregation of temporal versus spatial processing in separate hemispheres would be superimposed on relatively localized spatial and temporal processors in each hemisphere; that is, there might exist a subtle symmetry-breaking due to a drift of visual-like processing into the nondominant hemisphere and auditory-like processing into the dominant hemisphere.

The theory thus contains a tantalizing question as it stands. Does the ambiguity problem *necessarily* lead to distinct type 1 and type 2 mechanisms? If so, can anatomical traces of a type 1 mechanism (for example, negative feedback loops triggered by the output stage and feeding into the STM reverberation) be found in sequential STM buffers but not in spatial representations? If not, does there exist *in vivo* a unified mechanism—say a modified type 2 mechanism—that possesses both types of properties?

63. Concluding Remarks: Universal Adaptive Measurement

This section sketches a broader perspective within which to view the above results. Its central tenet is that the brain is a universal measurement device operating on the quantum level. By this is meant that data from all perceivable physical fields are translated into a common neural language, and that events on the quantum level, such as several photons, can be perceived. The universality of the neural language clarifies why results concerning the neural measurement process can have broad interdisciplinary implications.

A central result of the present theory is a description of an alternative to the probabilistic and computer memory models that have been used to explain cognitive data. In particular, probabilistic models are replaced by systems that undergo parallel interactions in real time. Why the formalism of probability theory works at all in describing physical processes is a nontrivial problem that is often overlooked because of the practical successes of probability models. In the present scheme, many probabilistic-like computations are described by competitive interactions among network codes. The universal problem of processing patterned data in noisy systems with finitely many sites requires the existence of such competitive interactions. Furthermore, the general problem of stabilizing adaptive codes in a fluctuating input environment requires that certain feedback relationships exist between the codes of individual events and the codes of various event combinations. Are such universal problems and their solutions by competitive systems one reason for the success of probability models? How generally can a more powerful alternative to probability theory be built up by using hierarchically organized competitive systems operating in real time? Especially in cases in which a system continually re-evaluates hypotheses based on disconfirming feedback does the present framework seem to be intrinsically richer than probability theory.

A related set of problems arises in the serial processing of lists. The spatial geometry of a list of events, represented as symbols on a serial tape, is not the same as the space–time geometry of the same list of events occurring in real time (Grossberg, 1969d). Indeed, suppose that a list $r_1, r_2, \ldots, r_n$ of events is presented to a subject with one item occurring every w time units. It is not until at least w time units after r_n occurs that the subject can know that r_n is the last list item. Only then can this past event be reclassified, via a "backward effect in time," as being the list's end. This fact implies that the types of properties and paradoxes that can occur in formal systems, such as classical logic or model theory, and in real-time parallel systems can be quite different. The results on serial learning in Section 12, and the real-time probabilistic logic across a field of populations in Section 20, provide two examples of how the approaches differ. Problems concerning the field representations of mathematical versus empirical data, of infinite operations in networks with finite numbers of coded populations, and of plans for which no digital algorithm exists are among the many that are worth investigating.

The evolutionary properties of the brain's measurement process suggest another class of problems. One of the triumphs of modern physics was to geometrize the dynamics of physical laws, as in Einstein's general relativity theory. One of the important tasks of brain theory is to reverse this procedure; namely, to explain the four-dimensional geometry of the world in terms of a dynamical system operating in a non-Euclidean network of very high dimension. As noted in Section 1,F, the high dimension of unfamiliar behavioral data seems to be successively reduced as new codes and commands for organizing this data evolve. The sections on hierarchical coding and map formation begin to show how these lower-dimensional representations emerge.

For almost a century, the measurement problems that concerned physics and biology diverged. Before that time, distinguished physicists, such as Helmholtz, Mach, and Maxwell, were also distinguished psychologists or physiologists. This then ceased to be true if only because profound insights concerning the measurement processes of physics could still be expressed by using the available linear mathematics, whereas it became clear that psychophysiological measurement processes involved nonlinear systems whose laws, and underlying principles, were at best dimly understood. Recently, both physics and biology have been driven toward processes in which nonlinear collective effects have been implicated. Indeed, analogs of such currently interesting physical phenomena as phase transitions (Grossberg, 1969f, 1974), globally irreversible but locally reversible interactions (Grossberg, 1969f), and backward effects in time (Grossberg, 1969d, 1974) are found even in simple neural networks. As both physical and biological theory incorporate measurement concepts that are explicated by parallel, nonlinear, self-organizing, hierarchical, and feedback interactions, we can anticipate a renewed flourishing of interdisciplinary studies and a deepening understanding of our interactions with the external world.

Appendix

Proof of Theorem 1. The proof is by induction. For $i = 2$, (29) and (30) imply

$$\mu(1 + \omega_2) = \mu\lambda + (1 - \lambda)M$$

or

$$\omega_2 = \lambda + (1 - \lambda)R - 1$$

as in (31). For $i > 2$, (29) and (30) imply

$$\mu[1 + \omega_i(1 + \omega_{i-1}(1 + \omega_{i-2}(\cdots)))] = \mu\lambda^{i-1} + (1 - \lambda^{i-1})M$$

or

$$\omega_i(1 + \omega_{i-1}(1 + \omega_{i-2}(\cdots))) = \lambda^{i-1} + (1 - \lambda^{i-1})R - 1 \tag{A1}$$

By the induction hypothesis, (31) can be used for all indices less than i, whence

$$1 + \omega_{i-1}(1 + \omega_{i-2}(\cdots))) = \lambda^{i-2} + (1 - \lambda^{i-2})R \tag{A2}$$

which along with (A1) proves (31) for all i.

Equation (32) shows that, for $2 \leq k \leq j$, $x_{k-1,j} > x_{kj}$ if and only if $\omega_k > 1$. By (31), $\omega_k > 1$ if and only if

$$(R - 1)(1 - \lambda)\lambda^{k-2} > 1$$

Since $0 < \lambda < 1$, $x_{1j} > x_{2j} > x_{3j} > \cdots > x_{jj}$ unless $j > J$, as (33) notes.

Proof of Theorem 2. By the Invariance Principle, the STM activities across $\mathcal{F}^{(1)}$ at successive times can be described by the rows

$$\begin{array}{lll} \mu & & \\ \omega_2\mu & \mu & \\ \omega_3\omega_2\mu & \omega_3\mu & \mu \end{array}$$

and so on. Let the sampling signals from a given population in $\mathcal{F}^{(3)}$ to the successive rows be $s_1, s_2, s_3, \ldots$. By hypothesis, $s_1 \geq s_2 \geq s_3 \geq \cdots$. The product of sampling signal and STM trace determines LTM growth in each time frame, as in (2). These products are

$$\begin{array}{lll} s_1\mu & & \\ s_2\omega_2\mu & s_2\mu & \\ s_3\omega_3\omega_2\mu & s_3\omega_3\mu & s_3\mu \end{array} \tag{A3}$$

Within each row, the entry in the ith column is at least as large as the entry in the $(i + 1)$st column, by the TMS hypothesis. Hence the $\mathscr{F}^{(2)}$ population samples a monotone decreasing pattern in every time frame, and then sums all the patterns to learn a net monotone decreasing pattern.

Inequalities (34) are discussed in the next proof.

Proof of Theorem 3. Deleting the superscripts (1) in the invariant parameters for simplicity, we find the chart

$$\begin{array}{ccc} s_1\mu_1 & & \\ s_2\omega_2\mu_1 & s_2\mu_2 & \\ s_3\omega_3\omega_2\mu_1 & s_3\omega_3\mu_2 & s_3\mu_3 \end{array} \tag{A4}$$

By (35), each row is either monotone decreasing (if all ω_i in the row exceed 1), monotone increasing (if all)ω_i in the row are less than 1), or unimodal (if some ω_i fall above and below 1). If a population in $\mathscr{F}^{(2)}$ starts sampling $\mathscr{F}^{(1)}$ when a given STM pattern is active, this pattern is encoded in its LTM traces. Because the STM pattern in the past field does not change, the same past STM pattern is encoded into LTM in every time frame. The total past field LTM pattern is the sum of these STM patterns, and hence has the same form as it had during its first sampling interval.

The future field LTM pattern is monotone decreasing because, by (35) and (36), $s_1\mu_1 > s_2\mu_2 > s_3\mu_3 \cdots$, $s_2\omega_2\mu_1 > s_3\omega_3\mu_2 > s_4\omega_4\mu_3 > \cdots$, etc. That is, the inequalities (34) hold. The sum of STM values in the ith column of (A4) thus exceeds the sum of STM values in the $(i + 1)$st column. Since the future field LTM pattern sums up column values for all columns that are first excited after it begins to sample, this LTM pattern is monotone decreasing.

REFERENCES

Anderson, T. W. (1958). "An Introduction to Multivariate Statistical Analysis." Wiley, New York.

Atkinson, R. C., and Shiffrin, R. M. (1968). *Adv. Psychol. Learning Motiv.* **2**, 89.

Atkinson, R. C., and Shiffrin, R. M. (1971). *Sci. Am.* p. 82, August.

Berlyne, D. E. (1970). *In* "Attention: Contemporary Theory and Analysis" (D. E. Mostofsky, ed.), p. 25. Appleton, New York.

Bizzi, E., Polit, A., and Morasso, P. (1975). *J. Neurophysiol.*

Bjork, R. A. (1975). *In* "Cognitive Theory (F. Restle *et al.*, eds.), Vol. 1, p. 151. L. Erlbaum Assoc., Hillsdale, New Jersey.

Burns, B. D. (1958). "The Mammalian Cerebral Cortex." Arnold, London.

Cant, B. R., and Bickford, R. G. (1967). *Electroencephalogr. Clin. Neurphysiol.* **23**, 594.

Chung, S.-H., Raymond, S. A., and Lettvin, J. Y. (1970). *Brain, Behav. Evol.* **3**, 72.

Cornsweet, T. N. (1970). "Visual Perception." Academic Press, New York.

Craik, F. I. M., and Jacoby, L. L. (1975). *In* "Cognitive Theory" (F. Restle *et al.*, eds.), Vol. 1, p. 173. L. Erlbaum Assoc., Hillsdale, New Jersey.

Crosby, E. C., Humphrey, T., and Lauer, E. W. (1962). "Correlative Anatomy of the Nervous System." Macmillan, New York.
Dethier, V. G. (1968). "Physiology of Insect Senses." Methuen, London.
Donchin, E., and Herning, R. I. (1975). *Electroencephalogr. Clin. Neurophysiol.* **38,** 51.
Donchin, E., Gerbrandt, L. A., Leifer, L., and Tucker, L. (1972). *Psychophysiology* **9,** 178.
Donchin, E., Tueting, P., Ritter, W., Kutas, M., and Heffley, E. (1975). *Electroencephalogr. Clin. Neurophysiol.* **38,** 1.
Duda, R. O., and Hart, P. E. (1973). "Pattern Classification and Scene Analysis." Wiley, New York.
Ellias, S. A., and Grossberg, S. (1975). *Biol. Cybernet.* **20,** 69.
Estes, W. K. (1972). *In* "Coding Processes in Human Memory" (A. W. Melton and E. Martin, eds.), p. 161. Holt, New York.
Fry, D. B. (1966). *In* "The Genesis of Language" (F. Smith and G. A. Miller, eds.). p. 187. MIT Press, Cambridge, Massachusetts.
Gazzaniga, M. S. (1970). "The Bisected Brain." Appleton, New York.
Gibson, J. J. (1933). *J. Exp. Psychol.* **16,** 1.
Goldberg, M. E., and Wurtz, R. H. (1972a). *J. Neurophysiol.* **35,** 542.
Goldberg, M. E., and Wurtz, R. H. (1972b). *J. Neurophysiol.* **35,** 560.
Graham, C. H. (1966). "Vision and Visual Perception" (C. H. Graham *et al.*, eds.), p. 548. Wiley, New York.
Granit, R., ed. (1966). "Muscular Afferents and Motor Control." Wiley, New York.
Grillner, S. (1969). *Acta Physiol. Scand., Suppl.* **327.**
Grossberg, S. (1967). *Proc. Natl. Acad. Sci. U.S.A.* **58,** 1329.
Grossberg, S. (1969a). *J. Math. Psychol.* **6,** 209.
Grossberg, S. (1969b). *J. Statist. Phys.* **1,** 319.
Grossberg, S. (1969c). *J. Math Mech.* **19,** 53.
Grossberg, S. (1969d). *Math. Biosci.* **4,** 201.
Grossberg, S. (1969e). *J. Theor. Biol.* **22,** 325.
Grossberg, S. (1969f). *J. Differ. Eq.* **5,** 531.
Grossberg, S. (1970a). *Stud. Appl. Math.* **49,** 135.
Grossberg, S. (1970b). *J. Theor. Biol.* **27,** 291.
Grossberg, S. (1971a). *Proc. Natl. Acad. Sci. U.S.A.* **68,** 828.
Grossberg, S. (1971b). *J. Theor. Biol.* **33,** 225.
Grossberg, S. (1972a). *In* "Delay and Functional Differential Equations and Their Applications" (K. Schmitt, ed.), p. 121. Academic Press, New York.
Grossberg, S. (1972b). *Math. Biosci.* **15,** 39.
Grossberg, S. (1972c). *Math. Biosci.* **15,** 253.
Grossberg, S. (1972d). *Kybernetik* **10,** 49.
Grossberg, S. (1973). *Stud. Appl. Math.* **52,** 217.
Grossberg, S. (1974). *Prog. Theor. Biol.* **3,** 51.
Grossberg, S. (1975). *Int. Rev. Neurobiol.* **18,** 263.
Grossberg, S. (1976a). *Biol. Cybernet.* **21,** 145.
Grossberg, S. (1976b). *Biol. Cybernet.* **23,** 121.
Grossberg, S. (1976c). *Biol. Cybernet.* **23,** 187.
Grossberg, S. (1977a). *In* "Formal Theories of Visual Perception" (E. L. J. Leeuwenberg and H. Buffart, eds.). Wiley, New York.
Grossberg, S. (1977b). *J. Math. Biol.* **4,** 237.
Grossberg, S. (1977c). *J. Math. Anal. Appl.* (in press).
Grossberg, S. (1978a). This volume.

Grossberg, S. (1978b). *J. Math. Psychol.* (in press).
Grossberg, S., and Levine, D. S. (1975). *J. Theor. Biol.* **53,** 341.
Grossberg, S., and Pepe, J. (1971). *J. Statist. Phys.* **3,** 95.
Gustafson, T., and Wolpert, L. (1967). *Biol. Rev. (Cambridge Pholos. Soc.)* **42,** 442.
Hein, A., and Held, R. (1967). *Science* **158,** 390.
Held, R., and Bauer, J. A., Jr. (1967). *Science* **155,** 718.
Held, R., and Hein, A. (1963). *J. Comp. Physiol. Psychol.* **56,** 872.
Hodgkin, A. L. (1964). "The Conduction of the Nervous Impulse." Thomas, Springfield, Illinois.
Hogan, R. M. (1975). *Mem. Cognit.* **3,** 197.
Hogan, R. M., and Hogan, M. M. (1975). *Mem. Cognit.* **3,** 210.
Hubel, D. H., and Wiesel, T. N. (1962). *J. Physiol.* **160,** 106.
Hubel, D. H., and Wiesel, T. N. (1963). *J. Neurophysiol.* **26,** 994.
Irwin, D. A., Rebert, C. S., McAdam, D. W., and Knott, J. R. (1966). *Electroencephalogr. Clin. Neurophysiol.* **21,** 412.
John, E. R. (1966). *In* "Frontiers in Physiological Psychology" (R. W. Russell, ed.), p. 149. Academic Press, New York.
John, E. R. (1967). *In* "The Neurosciences: A Study Program" (G. C. Quarton, T. Melnechuk and F. O. Schmitt, eds.), p. 690. Rockefeller Univ. Press, New York.
Julesz, B. (1971). "Foundations of Cyclopean Perception." Univ. of Chicago Press, Chicago, Illinois.
Kahneman, D., and Beatty, J. (1966). *Science* **154,** 1583.
Kennedy, D. (1968). *In* "Physiological and Biochemical Aspects of Nervous Integration" (F. O. Carlson, ed.), p. 285. Prentice-Hall, Englewood Cliffs, New Jersey.
LaBerge, D. (1973). *Mem. Cognit.* **1,** 268.
Lashley, K. S. (1951). *In* "Cerebral Mechanisms in Behavior: The Hixon Symposium" (L. P. Jeffress, ed.), p. 112. Wiley, New York.
Lenneberg, E. H. (1967). "Biological Foundations of Language." Wiley, New York.
Levine, D. S., and Grossberg, S. (1976). *J. Theor. Biol.* **61,** 477.
Low, M. D., Borda, R. R., Frost, J. D., and Kellaway, R. (1966). *Neurology* **16,** 771.
McAdam, D. W. (1969). *Electroencephalogr. Clin. Neurophysiol.* **26,** 216.
McAdam, D. W., Irwin, D. A., Rebert, C. S., and Knott, J. R. (1966). *Electroencephalogr. Clin. Neurophysiol.* **21,** 194.
Miller, G. A. (1956). *Psychol. Rev.* **63,** 81.
Mountcastle, V. B. (1967). *In* "The Neurosciences: A Study Program" (G. C. Quarton, T. Melnechuk, and F. O. Schmitt, eds.), p. 393. Rockefeller Univ. Press, New York.
Parzen, E. (1960). "Modern Probability Theory and Its Applications." Wiley, New York.
Piaget, J. (1963). "The Origins of Intelligence in Children." Norton, New York.
Remington, R. J. (1969). *J. Exp. Psychol.* **82,** 250.
Robinson, D. A. (1964). *J. Physiol. (London)* **174,** 245.
Rocha-Miranda, C. E., Bender, D. B., Gross, C. G., and Mishken, M. (1975). *J. Neurophysiol.* **38,** 475.
Rohrbaugh, J. W., Donchin, E., and Eriksen, C. W. (1974). *Percept. Psychophys.* **15,** 368.
Schneider, W., and Shiffrin, R. M. (1976). *In* "Basic Processes in Reading: Perception and Comprehension" (D. LaBerge and S. J. Samuels, eds.). L. Erlbaum Assoc., Hillsdale, New Jersey.
Shiffrin, R. M. (1975). *In* "Cognitive Theory" (F. Restle *et al.*, eds.), Vol. 1, p. 193. L. Erlbaum Assoc., Hillsdale, New Jersey.
Squires, K. C., Wickens, C., Squires, N. K., and Donchin, E. (1976). *Science* **193,** 1142.
Stein, P. S. G. (1971). *J. Neurophysiol.* **34,** 310.

Sternberg, S. (1966). *Science* **153,** 652.
Stevens, S. S. (1961). *In* "Sensory Communication" (W. A. Rosenblith, ed.), p. 1. MIT Press, Cambridge, Massachusetts.
Stryker, M. P., and Schiller, P. H. (1975). *Exp. Brain Res.* **23,** 103.
Stryker, M. P., and Sherk, H. (1975). *Science* **190,** 904.
Thomas, G. B., Jr. (1968). "Calculus and Analytic Geometry." 4th ed. Addison-Wesley, Reading Park, Massachusetts.
Thompson, R. F. (1967). "Foundations of Physiological Psychology." Harper, New York.
Townsend, J. T. (1974). *In* "Human Information Processing: Tutorials in Performance and Cognition" (B. H. Kantowitz, ed.), p. 133. Erlbaum, Potomac, Maryland.
Walter, W. G, (1964). *Arch. Psychiatr. Nervenkr.* **206,** 309.
Werblin, F. S. (1971). *J. Neurophysiol.* **34,** 228.
Willows, A. O. D. (1968). *In* "Physiological and Biochemical Aspects of Nervous Integration" (F. O. Carlson, ed.), p. 217. Prentice-Hall, Englewood Cliffs, New Jersey.
Wise, C. D., Berger, B. D., and Stein, L. (1973). *Biol. Psychiatry* **6,** 1.
Wurtz, R. H., and Goldberg, M. E. (1972a). *J. Neurophysiol.* **35,** 575.
Wurtz, R. H., and Goldberg, M. E. (1972b). *J. Neurophysiol.* **35,** 587.
Yarbus, A. L. (1967). "Eye Movements and Vision." Plenum, New York.

LIST OF PUBLICATIONS

of Stephen Grossberg

1. 'Nonlinear Difference-Differential Equations in Prediction and Learning Theory', *Proceedings of the National Academy of Sciences* **58** (1967), 1329–1334.
2. 'A Prediction Theory for Some Nonlinear Functional-Differential Equations, I. Learning of Lists', *Journal of Mathematical Analysis and Applications* **21** (1968), 643–694.
3. 'A Prediction Theory for Some Nonlinear Functional-Differential Equations. II: Learning of Patterns', *Journal of Mathematical Analysis and Applications* **22** (1968), 490–522.
4. 'Global Ratio Limit Theorems for Some Nonlinear Functional Differential Equations, I', *Bulletin of the American Mathematical Society* **74** (1968), 93–100.
5. 'Global Ratio Limit Theorems for Some Nonlinear Functional Differential Equations, II', *Bulletin of the American Mathematical Society* **74** (1968), 101–105.
6. 'Some Nonlinear Networks Capable of Learning a Spatial Pattern of Arbitrary Complexity', *Proceedings of the National Academy of Sciences* **59** (1968), 368–372.
7. 'Some Physiological and Biochemical Consequences of Psychological Postulates', *Proceedings of the National Academy of Sciences* **60** (1968), 758–765.
8. 'On the Global Limits and Oscillations of a System of Nonlinear Differential Equations Describing a Flow on a Probabilistic Network', *Journal of Differential Equations* **5** (1969), 531–563.
9. 'On Variational Systems of Some Nonlinear Difference-Differential Equations', *Journal of Differential Equations* **6** (1969), 544–577.
10. 'Embedding Fields: A Theory of Learning with Physiological Implications', *Journal of Mathematical Psychology* **6** (1969), 209–239.
11. 'On Learning, Information, Lateral Inhibition, and Transmitters', *Mathematical Biosciences* **4** (1969), 225–310.
12. 'On the Production and Release of Chemical Transmitters and Related Topics in Cellular Control', *Journal of Theoretical Biology* **22** (1969), 325–364.
13. 'On the Serial Learning of Lists', *Mathematical Biosciences* **4** (1969), 201–253.
14. 'Some Networks That Can Learn, Remember, and Reproduce Any Number of Complicated Space-Time Patterns, I', *Journal of Mathematics and Mechanics* **19** (1969), 53–91.
15. 'On Learning of Spatiotemporal Patterns by Networks with Ordered Sensory and Motor Components, I. Excitatory Components of the Cerebellum', *Studies in Applied Mathematics* **48** (1969), 105–132.
16. 'On Learning and Energy-Entropy Dependence in Recurrent and Nonrecurrent Signed Networks', *Journal of Statistical Physics* **1** (1969), 319–350.
17. 'A Global Prediction (or Learning) Theory for Some Nonlinear Functional-Differential Equations', in *Studies in Applied Mathematics, Advances in Differential and Integral Equations*, Vol. 5, (J.A. Nohel, Ed.), pp. 64–70. Phila.: SIAM, 1969.

18. 'Learning and Energy-Entropy Dependence in Some Nonlinear Functional-Differential Systems', *Bulletin of the American Mathematical Society* 75 (1969), 1238–1242.
19. 'Some Networks That Can Learn, Remember, and Reproduce Any Number of Complicated Space-Time Patterns, II', *Studies in Applied Mathematics* **49** (1970), 135–166.
20. 'Neural Pattern Discrimination', *Journal of Theoretical Biology* **27** (1970), 291–337.
21. 'Schizophrenia: Possible Dependence of Associational Span, Bowing, and Primacy Vs. Recency on Spiking Threshold, (with J. Pepe)', *Behavioral Science* **15** (1970), 359–362.
22. 'Embedding Fields: Underlying Philosophy, Mathematics, and Applications to Psychology, Physiology, and Anatomy', *Journal of Cybernetics* **1** (1971), 28–50.
23. 'Spiking Threshold and Overarousal Effects in Serial Learning, (with J. Pepe)', *Journal of Statistical Physics* 3 (1971), 95–125.
24. 'Functional-Differential Systems and Pattern Learning', in *Lecture Notes in Mathematics*, Vol. 206, (D. Chillingsworth, Ed.), pp. 147–150. Berlin: Springer–Verlag, 1971.
25. 'On the Dynamics of Operant Conditioning', *Journal of Theoretical Biology* **33** (1971), 225–255.
26. 'Pavlovian Pattern Learning by Nonlinear Neural Networks', *Proceedings of the National Academy of Sciences* **68** (1971), 828–831.
27. 'Neural Expectation: Cerebellar and Retinal Analogs of Cells Fired by Learnable or Unlearned Pattern Classes', *Kybernetik* **10** (1972), 49–57.
28. 'A Neural Theory of Punishment and Avoidance, I. Qualitative Theory', *Mathematical Biosciences* **15** (1972), 39–67.
29. 'A Neural Theory of Punishment and Avoidance, II. Quantitative Theory', *Mathematical Biosciences* **15** (1972), 253–285.
30. 'Pattern Learning by Functional-Differential Neural Networks with Arbitrary Path Weights', in *Delay and Functional-Differential Equations and their Applications*, (K. Schmitt, Ed.), pp. 121–160. N.Y.: Academic Press 1972.
31. 'Contour Enhancement, Short Term Memory, and Constancies in Reverberating Neural Networks', *Studies in Applied Math.* **52** (1973), 217–257.
32. 'Classical and Instrumental Learning by Neural Networks', in *Progress in Theoretical Biology*, (R. Rosen and F. Snell, Eds.), pp. 51–141, 1974.
33. 'A Neural Model of Attention, Reinforcement, and Discrimination Learning', in *International Review of Neurobiology*, (Carl Pfeiffer, Ed.), Vol. 18, pp. 263–327, 1975.
34. 'Short Term Memory of Recurrent Neural Networks (with D. S. Levine)', *J. Theoret. Biol.* **53** (1975), 341–380.
35. 'Pattern Formation, Contrast Control, and Oscillations in the Short Term Memory of Shunting On-Center Off-Surround Networks (with S. A. Ellias)', *Biol. Cybernetics* **20** (1975), 69–98.
36. 'On the Development of Feature Detectors in the Visual Cortex with Applications to Learning and Reaction-Diffusion Systems', *Biol. Cybernetics* **21** (1976), 145–159.
37. 'On Visual Illusions in Neural Networks: Line Neutralization, Tilt Aftereffect and Angle Expansion (with D. S. Levine)', *J. Theoret. Biol.* **61** (1976), 477–504.

38. 'Adaptive Pattern Classification and Universal Recoding. I: Parallel Development and Coding of Neural Feature Detectors', *Biol. Cybernetics* **23** (1976), 121–134.
39. 'Adaptive Pattern Classification and Universal Recoding. II: Feedback, Expectation, Olfaction, and Illusions', *Biol. Cybernetics* **23** (1976), 187–202.
40. 'Redundant Information in Auditory and Visual Modalities: Inferring Decision-Related Processes from the P300 Component', (with E. Donchin and K. and N. Squires), *J. of Experimental Psychol.* 3 (1977), 299–315.
41. 'Pattern Formation by the Global Limits of a Nonlinear Competitive Interaction in *n* Dimensions', *J. Math. Biol.* **4** (1977), 237–256.
42. 'A Theory of Human Memory: Self-Organization and Performance of Sensory-Motor Codes, Maps and Plans', in *Progress in Theoret. Biol.*, Vol. 5, (R. Rosen and F. Snell, Eds.), pp. 233–374. New York: Academic Press, 1978.
43. 'Communication, Memory, and Development', in *Progress in Theoret. Biol.*, Vol. 5, (R. Rosen and F. Snell, Eds.), pp. 183–232. New York: Academic Press, 1978.
44. 'A Theory of Visual Coding, Memory, and Development', in *Formal Theories of Visual Perception*, (E. Leeuwenberg and H. Buffart, Eds.), New York: Wiley, 1978.
45. 'Behavioral Contrast in Short Term Memory: Serial Binary Memory Models or Parallel Continuous Memory Models?', *J. Math. Psychol.* 3 (1978), 199–219.
46. 'Competition, Decision, and Concensus', *J. Math. Anal. and Applics* **66** (1978), 470–493.
47. 'Do All Neural Models Really Look Alike?', *Psychol. Review* **85** (1978), 592–596.
48. 'Decisions, Patterns, and Oscillations in Nonlinear Competitive Systems with Applications to Volterra-Lotka Systems', *J. Theoret. Biol.* **73** (1978), 101–130.
49. 'Adaptive Pattern Classification and Universal Recoding: Parallel Development and Coding of Neural Feature Detectors', in *Third European Conference on Cybernetics and Systems Research*, (R. Trappl, Ed.), pp. 375–383. Halstead, 1978.
50. 'How Does A Brain Build a Cognitive Code?', *Psychol. Review* **1** (1980), 1–51.
51. 'Biological Competition: Decision Rules, Pattern Formation, and Oscillations', *Proc. Nat'l. Acad. Sci.* 77 (1980), 2338–2342.
52. 'Intracellular Mechanisms of Adaptation and Self-Regulation in Self-Organizing Networks: The Role of Chemical Transducers', *Bull. of Math. Biol.* **42** (1980), 365–396.
53. 'Human and Computer Rules and Representations are not Equivalent', *Behavioral and Brain Sciences* **3** (1980), 136–138.
54. 'Direct Perception or Adaptive Resonance?', *Behavioral and Brain Sciences* **3** (1980), 385.
55. Editor, *Mathematical Psychology and Psychophysiology*. Providence, R. I.: Ameriican Mathematical Society, 1981.
56. 'Adaptive Resonance in Development, Perception and Cognition', in S. Grossberg (Ed.), *Mathematical Psychology and Psychophysiology*. Providence, R. I.: American Mathematical Society, 1981.
57. 'Psychophysiological Substrates of Schedule Interactions and Behavioral Contrast', in S. Grossberg (Ed.), *Mathematical Psychology and Psychophysiology*. Providence, R. I.: American Mathematical Society, 1981.
58. 'Adaptation and Transmitter Gating in Vertebrate Photoreceptors' (with G. A. Carpenter), *J. of Theoretical Neurobiology* **1** (1981), 1–42.
59. 'Psychophysiological and Pharmacological Substrates of a Developmental, Cognitive, and Motivational Theory', in J. Cohen, R. Karrer, and P. Tueting (Eds.), *Proceedings volume of the Sixth International Conference on Evoked Potentials of the Brain*, held at Lake Forest College, Ill., June 21–27, 1981. New York: New York Academy of Sciences, 1982.

60. 'Global Pattern Formation in Nonlinear Networks (with M. A. Cohen)', submitted for publication.
61. 'The Processing of Expected and Unexpected Events During Conditioning and Attention: A Psychophysiological Theory. *Psychol. Rev.*, in press.
62. 'Why Do Cells Compete: Some Examples from Visual Perception', UMAP Module, in press.
63. 'Transmitters, Expectancies, Extinction, and Avoidance', UMAP Module, in press.

INDEX